23rd International Laser Physics Workshop

(LPHYS' 14)

Journal of Physics: Conference Series
Volume 594

Sofia, Bulgaria
14 - 18 July 2014

ISBN: 978-1-5108-0251-3
ISSN: 1742-6588

Printed from e-media with permission by:

Curran Associates, Inc.
57 Morehouse Lane
Red Hook, NY 12571

Some format issues inherent in the e-media version may also appear in this print version.

Printed by Curran Associates, Inc. (2015)

For permission requests, please contact the Institute of Physics
at the address below.

Institute of Physics
Dirac House, Temple Back
Bristol BS1 6BE UK

Phone: 44 1 17 929 7481
Fax: 44 1 17 920 0979

techtracking@iop.org

Additional copies of this publication are available from:

Curran Associates, Inc.
57 Morehouse Lane
Red Hook, NY 12571 USA
Phone: 845-758-0400
Fax: 845-758-2634
Email: curran@proceedings.com
Web: www.proceedings.com

Table of contents

Volume 594

23rd International Laser Physics Workshop (LPHYS'14)
14–18 July 2014, Sofia, Bulgaria

Accepted papers received: 24 February 2015
Published online: 18 March 2015

Strong Field & Attosecond Physics

Biophotonics

Physics of Cold Trapped Atoms

Quantum Information Science

Fiber Optics

Extreme Light Technologies, Science, and Applications

Additional paper (Strong Field & Attosecond Physics)

Journal of Physics: Conference Series **594** (2015) 011001 doi:10.1088/1742-6596/594/1/011001

Dear Readers,

The 23rd annual International Laser Physics Workshop, LPHYS'14, took place in the City of Sofia, Bulgaria. 361 participants from 35 countries attended the conference. It was hosted by the Institute of Electronics at the Bulgarian Academy of Sciences.

This year's Workshop was dedicated to paying tribute to two major events:
- 50th anniversary of 1964 Nobel Prize in physics,
- 145th anniversary of the establishment of the Bulgarian Academy of Sciences.

LPHYS'14 has been taken under the High Patronage of Rosen Plevneliev, President of the Republic of Bulgaria.

The LPHYS'14 Steering Committee and the Advisory & Program Committee would like to extend their sincere gratitude to Professor Sanka Gateva (Co-Chair) and Professor Ekaterina Borisova (Head of the Local Organizing Committee) and to their team for the outstanding job performed in organizing, arranging, managing and putting in order the conference. Their combined efforts lead to a successful result.

In this volume of Journal of Physics: Conference Series you will find selected proceedings of the Workshop in Sofia.

Please make a note that the 24th annual International Laser Physics Workshop (LPHYS'15) will take place from August 21 to August 25, 2015 in the city of Shanghai, China hosted by Shanghai Institute of Optics and Fine Mechanics at the Chinese Academy of Sciences.

With kind regards,

Steering and Advisory & Program committees

LPHYS'14

Journal of Physics: Conference Series **594** (2015) 011002 doi:10.1088/1742-6596/594/1/011002

Peer review statement

All papers published in this volume of *Journal of Physics: Conference Series* have been peer reviewed through processes administered by the proceedings Editors. Reviews were conducted by expert referees to the professional and scientific standards expected of a proceedings journal published by IOP Publishing.

Journal of Physics: Conference Series **594** (2015) 012001 doi:10.1088/1742-6596/594/1/012001

Townes' contribution to nonlinear optics

Elsa Garmire

Dartmouth College, Thayer School of Engineering, Hanover, NH 03755

garmire@dartmouth.edu

Abstract. In honour of the Fiftieth Anniversary of the Nobel Prize in Physics, this talk introduced the contributions of Nicholas Basov and Alexei Prokhorov, who shared the prize with Charles Townes. The talk then detailed the quantum electronics research of Townes, particularly at MIT, which was related to nonlinear optics. The years from 1961 to 1968 were particularly exciting, as the ruby laser enabled a wide variety of new physics to be discovered and explored.

1. Introduction

The 1964 Nobel Prize in Physics was awarded "for the invention of the maser and the laser" to Charles Hard Townes, Nicolai Basov and Aleksandr Prokhorov (figure 1). At the ceremony it was stated, "Drs. Townes, Basov and Prokhorov: By your ingenious studies of fundamental aspects of the interaction between matter and radiation you have made atoms work for us in a new and most remarkable way." [1]

The Nobel Committee went on to say, "The first papers about the maser were published ten years ago as a result of investigations carried out simultaneously and independently by Townes and co-workers at Columbia University in New York and by Basov and Prokhorov at the Lebedev Institute in Moscow. ...Masers work as extremely sensitive receivers for short radiowaves. They are of great importance in radio astronomy and are being used in space research for recording the radio signals from satellites."

Figure 1. Nobel Prize winners in Physics from 1964: Townes, Prokhorov and Basov (left to right) Portraits from Nobel website http://www.nobelprize.org/nobel_prizes/physics/laureates/1964/

The independent work of Townes in the United States of America and of Basov and Prokhorov in the Union of Soviet Socialist Republics brought forward the field of *Quantum Electronics*. The term "quantum" ties the field to physics and "electronics" ties the field to electrical engineering. The USSR researchers were motivated by the development of synchrotron sources for spectroscopy. Townes was motivated by his work on radar sources during World War II. The state of their research at the end of the war is documented by Prokhorov and Basov's papers on frequency stabilization of synchrotron sources [2] while Townes published on cathode sputtering [3].

Research in both groups evolved into studies of microwave spectroscopy during the late 1940's and early 1950's. Prokhorov and student Basov were drawn to amplification of microwaves by the desire to achieve narrower spectral lines. Townes was drawn to the possibility of molecules as new microwave sources and fundamental molecular physics questions. The breakthrough idea, achieved independently and essentially simultaneously, was to create a molecular *oscillator* from a molecular *amplifier* by introducing microwave feedback. This concept developed from experience in electrical engineering building electron-tube amplifiers and oscillators at radio frequencies. The Russians called their new idea a "molecular generator" of microwaves and published it in Russian in a paper submitted January, 1954 [4]. Townes called his a MASER (microwave amplifier by stimulated emission). Townes independently developed the same ideas and demonstrated the first experimental microwave maser in a paper submitted in May, 1954 [5]. The Russians contributed their idea of a three-level maser soon thereafter [6].

These scientists were not the first to discuss molecular amplification. Stimulated emission had been introduced by Albert Einstein as the direct analogue to absorption in 1914 [7]. When a two-level population distribution is inverted (more in the excited state than in the ground state), incoming radiation is not absorbed, but stimulates emission from excited atoms, resulting in more light out than came in: electrical engineers call this *gain*. In 1951 Fabricant in the USSR predicted this result [8], while in the USA Purcell and Pound performed an experiment in a nuclear spin system and observed that immediately upon reversing the magnetic field the system had a negative temperature. The nuclear spin was reversed with respect to thermal equilibrium. They pointed out that the observed reversed deflection corresponds to induced radiation [9]. This radiation would be caused by stimulated emission. Later, in 1954, Joseph Weber pointed out that the inverted population in this system produces gain by stimulated emission and can become a source of microwaves [10].

"Quantum Electronics was born at the moment when an excited quantum system — a beam of specially sorted molecules — was placed into a cavity" [11]. This was the achievement that resulted in the Nobel Prize of 1964. Work on quantum electronics progressed rapidly in both the USA and USSR, as soon as Townes reported a working MASER; indeed, in Europe as well. The era after WWII was one of blossoming interest in science, both for its own sake and for its military applications. One of the lessons learned from WWII was the importance of forefront scientific effort to aid the military and political power of the advanced countries within Europe and the US. The focus of much of the maser research, besides studies of the device itself, was on microwave spectroscopy, using the high resolution that the maser made possible. Townes and his brother-in-law Arthur Schawlow published a book *Microwave Spectroscopy* in 1958, which presented a systematic, comprehensive account of the theory, techniques, experimental data, and interpretation involved in the study of microwave spectroscopy. The book is still available and continues to serve as an important reference [12].

2. Townes' Research after the Maser

From 1954 to 1959, Townes, a Professor at Columbia University, carried out a diversified research program that included both ongoing microwave spectroscopic studies and analysis of the performance of his maser [13]. In 1955 he and his student discovered oscillatory population transfer between two closely-spaced atomic/molecular levels that split the absorption spectra between the ground state and either of these levels [14]. This has been called Autler-Townes splitting (he called it an AC Stark effect), produced by a radio-frequency field tuned to spectral resonance.

After 1956 Townes was working on ways to push the maser to shorter and shorter wavelengths, first to mm waves. In 1958, he and Schawlow made the jump all the way to infrared and optical frequencies [15]. Their contribution was to suggest that a large, multimode "open" resonant optical cavity (in the Fabry-Perot geometry) could lead to extremely monochromatic and coherent light. The narrow atomic/molecular gain line serves to limit the output to only a few cavity modes. They suggested optical pumping in a three-level system as a way to invert the population and were the first to analyse theoretically the oscillation condition. Their numerical evaluation of a potassium vapour "optical maser" found the idea to be feasible (although it turned out to be experimentally very difficult).

The Schawlow-Townes analysis excited the quantum electronics community and the race was on to demonstrate a working laser. Their ground-breaking paper defined the laser as an inverted population of atoms/molecules within a large highly-multimode optical cavity that amplifies spontaneous emission noise to generate an output in the infrared or visible. For a directional output, they specified two parallel mirrors (the Fabry-Perot geometry). Townes called this an "optical maser" to affirm that the maser concept at optical frequencies is not different, in principle, from the microwave system.

This time period (mid-1950's) was during the Cold War, a mutual military build-up between the USA and the USSR. The US military asked Townes for technical advice about the future of the new field of quantum electronics. He agreed to chair two important committees: one was "to create interest in mm waves" and the other was "to urge continued support for research in the infrared." These commitments were unpaid and after-hours, but he took them very seriously. Thus he was solicited to take a leave of absence from Columbia as Vice President and Director of Research for the Institute for Defense Analysis in Washington D. C., a position he held from 1959-1961. As he said in an oral interview 25 years later, "I felt that there just were not enough good scientists in Washington, and we had a pressing problem with the Russian missiles and other things coming on, and it was just a part of my duty." [16]

In the fall of 1961, Townes accepted an offer to become Provost at Massachusetts Institute of Technology, a powerful position that reports directly to the President. Research in lasers became his after-hours unpaid activity. He took me on as his first PhD student; six months later he added Raymond Chiao. We were his only graduate students, but we had an active group that included post-docs and visiting scientists. During this time Townes also became Chairman of the Advisory Committee to NASA's Project Apollo (the man-to-the-moon mission). His responsibility was to secure support from the larger scientific community and to ensure that the moon flight would yield maximum benefits in scientific research. This was a highly political committee that put him in conflict with some at MIT and he was denied the open position of MIT president. So in 1967 he moved to the University of California in Berkeley to search for stimulated emission in astronomy.

Why did Townes leave laser research? He could see that the quantum electronics field was brimming over with eager and talented scientists/engineers. He told me that he wanted to work in unexplored research areas, where he thought his contribution would be greatest. Stimulated emission in astronomy was a wide-open field, eagerly awaiting his arrival. Indeed, he has since made many significant contributions to astronomy (outlined later).

The purpose of this paper is to focus on Townes' contributions to nonlinear optics (NLO), which mostly resulted from the years he was at MIT: 1961-1967.

3. Nonlinear Optics at MIT

It pays to be brilliant and active as a field is just beginning. In the space of 6 years Townes, along with his colleagues: 1) Proved that coherent molecular vibrations drive higher order stimulated Raman scattering (SRS) emission; 2) Identified SRS emission from individual filaments in Raman-active liquids; 3) Introduced stimulated Brillouin scattering (SBS) and observed it in crystals and liquids; 4) Introduced and theoretically analysed spatial solitons (SS), at that time called self-trapped optical beams; 5) Experimentally demonstrated self-focusing filament-formation in Kerr liquids; 6) Demonstrated small-scale filamentary beam break-up; 7) Introduced the concept of self-phase modulation (SPM) and 8) Introduced the concept of weak-wave retardation and gain in four-wave mixing (FWM).

It was my job to set up the Townes laser laboratory at MIT with the second commercially sold ruby laser (from a company called Trion – the first went to Peter Franken at University of Michigan). The first year was spent just getting the laser to work – the ruby had to be cooled to liquid nitrogen temperatures while excited by a flash lamp operating at kilovolts. Electrical shorts from condensation on the Dewar were a major problem! We evaporated a silver mirror on one end of the ruby rod and a partially silvered mirror on the other. Technological progress toward Q-switching was rapid: our Q-switch was a retro-reflecting prism mounted on a high-speed air-driven turbine (developed at Lincoln Laboratories); the ruby crystal was put into an external cavity; and dielectric mirrors replaced silver (which blew off with the high-power Q-switch). Now we were ready for high power experiments, which are described here.

3.1. Coherent Molecular Vibrations in Stimulated Raman Scattering

Townes' first contribution was to describe the stimulated Raman scattering process as introducing coherent molecular oscillations [17]. He had been inspired to delve into nonlinear optics at the Third Quantum Electronics Conference in Paris in Feb. 1963. He had heard Zeiger and Tannenwald's theoretical paper on optical phonons. He had heard that experiments on stimulated Raman scattering at Hughes Research Laboratories demonstrated Stokes frequencies shifted by up to three orders of vibrational frequency. He also heard about the intense anti-Stokes emission seen by both Terhune and Stoicheff. Because Stoicheff became a visitor at MIT, we had early access to his data. All this led Townes to realize that coherent laser light could drive coherent optical phonons (which we called molecular oscillations).

The Raman process was modelled as an ensemble of simple vibrating diatomic molecules with natural resonance frequency Ω_r [18]. Townes' innovation was to realize that coherent light interacting with matter transferred that phase coherence to the molecules. Periodic vibrations introduced into the medium by scattering into Stokes frequencies could be transferred back to the light wave through a coherent process, generating anti-Stokes light. This was a classic resonant parametric process.

The anti-Stokes could be explained by the requirement of phase coherence over the interacting length, a process called phase-matching, first described in second harmonic generation. Molecular vibration driven by Stokes generation from a laser photon has a wave vector **K**. A second laser photon scatters off **K** to produce anti-Stokes. Phase-matching means anti-Stokes generation is conical at an angle given by closing the wave-vector diagram (figure 2); i.e. conservation of momentum.

The theoretical paper was followed by an experiment with a Q-switched ruby laser transmitted through calcite that proved that the parametric process did, indeed, occur – that anti-Stokes light depleted a cone of stokes light at the correct phase-matching angle. Figure 3 shows the bright ring of anti-Stokes radiation along with a negative of the Stokes radiation pattern, with the arrow pointing to a weak ring showing depletion at the phase-matching angle [19].

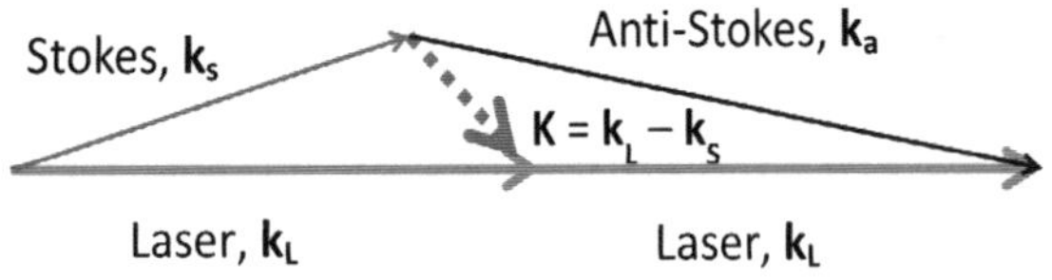

Figure 2. Phase-matching diagram for stimulated Raman creation of anti-Stokes radiation cone.

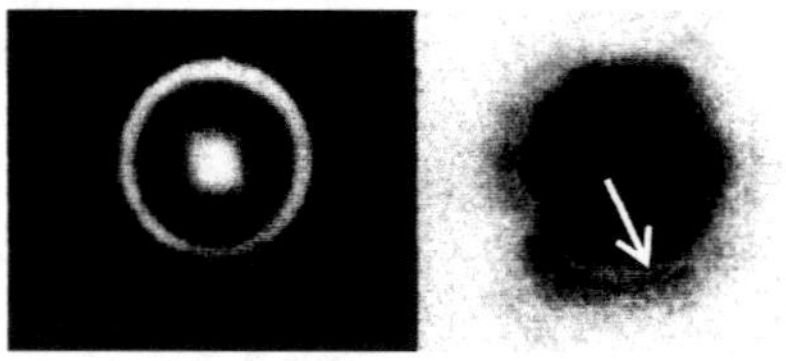

Figure 3. Depletion of Stokes (arrow) at phase-matching angle. From [19].

My research effort was to study SRS in Raman-active liquids using the Q-switched ruby laser [20]; now theory and experiment did not agree, as they did in calcite. This turned out to be due to filament-formation and self-phase modulation in liquids with nonlinear refractive indices.

3.2. Filament formation in SRS emission

Typical images of the radiating cones of anti-Stokes SRS generated in acetone are shown in figure 4a. The bright ring, always observed, violated the theory described above. It is the weak inner ring that corresponds to the above phase-matching angle. Further analysis showed that the bright ring at a larger cone angle corresponds to a theory describing phase-matching only along the direction of propagation. The reason was shown later to be due to the formation of self-focusing filaments. Because the Raman Stokes emission was strongly forward-directed, there was not enough Stokes emission at the angle required for anti-Stokes to be generated at the volume phase-matching angle. The measured anti-Stokes cone angle phase-matched in the direction of propagation only, and corresponded to a parametric interaction with the intense forward-going Stokes wave. The lack of lateral phase-matching was acceptable because this Raman interaction took place only in narrow filaments. This case can be called Cerenkov phase-matching, as opposed to volume phase-matching.

Several additional experimental results verified this discovery. When the 10-cm liquid cell was tilted off-axis, reflection from its glass faces provided feedback into an off-axis cavity. This produced a Raman Stokes laser with an intensity 10^4 times larger than that without feedback. The Stokes radiation at the required volume phase-matching angle (figure 4c) was now bright enough to parametrically generate anti-Stokes radiation at the expected angle (figure 4d, e). Proof of the parametric process is the decreased Stokes at the volume phase-matching angle is shown in figure 4f).

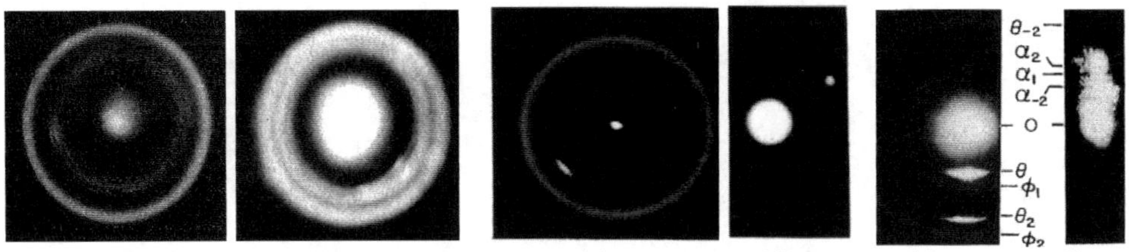

(a) (b) (c) (d) (e) (f)

Figure 4. Examples of stimulated Raman emission in liquids: (a) acetone and (b) cyclohexane both show two kinds of conical emission of anti-Stokes; the inner cones correspond to the volume phase-matching theory (above) and the outer cones correspond to Cerenkov phase-matching; (c) shows a full circle of Cerenkov phase-matched anti-Stokes as well as a weak beam of anti-Stokes at the volume phase-matching angle. The latter was made possible by an off axis resonator for Stokes radiation, that generated Stokes seen in (d) at the complementary angle from the anti-Stokes spot; (e) and (f) show how the anti-Stokes in (e) can sap energy from the Stokes in (f) at angles α corresponding to volume phase-matching theory [21, 22].

Independent proof that the anti-Stokes radiating into the large angle cones came from filaments (and thus thought of as Cerenkov radiation) is shown in figure 5a. This is the first direct evidence of self-trapped filaments emitting SRS. Further proof can be seen by comparing elliptical focussed volume phase-matching anti-Stokes in calcite (which has no self-trapped filaments) in figure 5b with anti-Stokes from benzene, where the same elliptical focusing geometry generated mostly circular emission from self-trapped filaments.

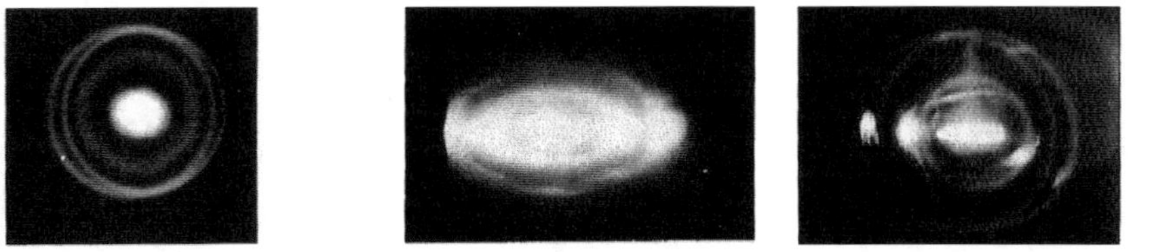

(a) (b) c)

Figure 5. Angular distribution of stimulated Raman anti-Stokes radiation; (a) Separate radiation from two different filaments caused side-by-side cones of anti-Stokes. A single anti-Stokes cone from volume phase-matching is also seen; (b) Elliptical volume phase-matching in calcite produced with a cylindrical lens; (c) Circular Cerenkov phase-matching from self-trapped filaments produced with the same conical lens [22, 23].

When the laser output was multimode, the anti-Stokes emission had a broad wavelength spectrum. Later it was realized that this was due to self-phase modulation in the nonlinear organic liquids. With a slit and an imaging spectrometer, the wavelength dependence of the output angles differs markedly between volume emission and filament emission (figure 6).

With a single-mode ruby laser, the anti-Stokes output was single-frequency, both with volume and surface emission. The spectrometer nicely separated out the different orders of anti-Stokes emission, although the lack of dynamic range of photographic film made it difficult to see several orders simultaneously.

3.3. Stimulated Brillouin Scattering

Townes was the first to predict stimulated Brillouin scattering (SBS) and his group was the first to demonstrate it. He understood that "Stimulated Brillouin scattering of an intense optical maser beam involves coherent amplification of a hypersonic lattice vibration and a scattered light wave. … It is analogous to Raman maser action, but with the molecular vibration replaced by an acoustic wave with frequency near 30 GHz…Both the acoustic and scattered light waves are emitted in specific directions" [24].

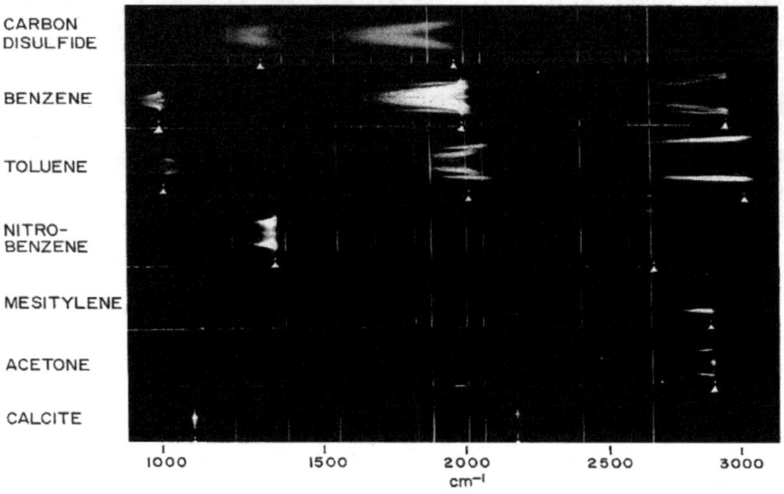

Figure 6. Angular-image spectra for stimulated anti-Stokes radiation excited by multi-mode ruby laser, with spectrum broadened by self-phase modulation. Sharp lines are from reference light source. Higher orders have successively more phase-modulation, with Cerenkov phase-matching changing much less in angle than volume phase-matching. Small white triangles indicate expected location of single-frequency Raman line [22, 23].

The Brillouin wave has a Stokes frequency shift Ω_s that depends on its emission angle θ as $\Omega_s = 2\omega_o(v_{ac}/v_{ph})\sin(\theta/2)$, where v_{ac} is the velocity of the acoustic wave and v_{ph} is the velocity of the photon. For retro-reflected waves, which are the most intense, the Stokes down-shift is $\Omega_s = 2\omega_o(v_{ac}/v_{ph})$. Stimulated Brillouin scattering can be considered parametric generation of an acoustic wave and a scattered light wave from an initial laser wave. With such a small frequency shift, SBS is most easily observed with a Fabry-Perot interferometer.

With solids quartz and sapphire, SBS occurred just as predicted [24]. This first observation of SBS in fused silica validated a process that was shown later to strongly limit the power that can be transmitted through fibers. Figure 7a shows typical Fabry-Perot spectra in these solids.

(a)
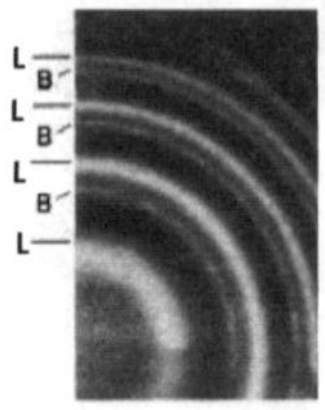

(b)
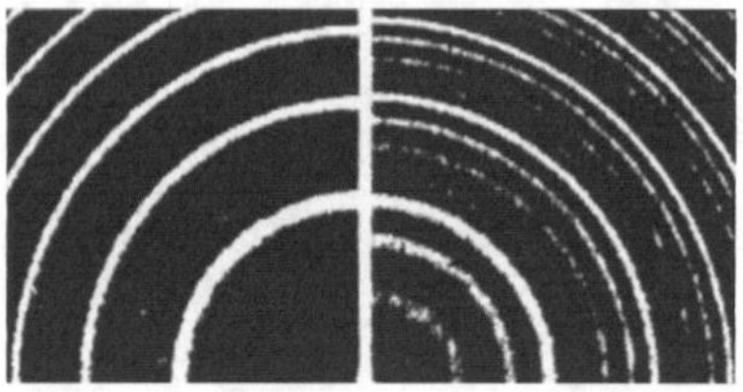
(c)

Figure 7. Fabry-Perot interferograms to demonstrate SBS: (a) SBS in quartz, observed with a composite illumination of 10% of incident laser (L) and the

Journal of Physics: Conference Series **594** (2015) 012001 doi:10.1088/1742-6596/594/1/012001

Brillouin-shifted retro-reflected beam (B); (b) Single-frequency laser light illumination; (c) SBS from a ruby laser focused in water [24, 25].

In liquids, a series of several orders of Stoke Brillouin shifts was observed, as shown in Fig. 7c compared to Fig. 7b, which could not be explained by the SBS theory. It turned out that the Stokes Brillouin wave was retro-reflecting into the inhomogeneously broadened ruby laser, where it found additional gain because it was at a new frequency [25]. It then came out of the laser as a new Brillouin-shifted frequency and entered the liquid cell where SBS caused another retro-reflected Brillouin shift. And so on. Many years later it was determined that the ability of the SBS Stokes light to go exactly back into the laser had occurred because of phase conjugation [26]. In 1964, while I was still a graduate student, I did not think sufficiently about why the retro-reflecting light should follow such an exact backward path through the lens, the mode selector and the imperfect ruby rod. If I had, I would have been the first to identify phase conjugation!

3.4. Spatial Solitons

In 1964, Dr. Townes became aware of the fact that when Mike Hercher at University of Rochester focused a powerful Q-switched ruby laser into a glass block, it left a trail of damage that did not appear to spread with distance by diffraction [27]. That led Townes to think about how optical nonlinearities could overcome diffraction and to our working out the theory, which we submitted in September and was published in October, 1964. We described the reported experimental data this way:

"Thin threads of damage in glass and other materials ...fairly easily demonstrated in glass by focusing a ruby-laser beam greater than a few megawatts inside good optical glass. Usually, though not always, there is extensive damage near the focal point and beyond the focal point a long straight filament of small bubbles and damage along the lens axis, accompanied by ionization. This filament may be as long as several centimeters and at the same time have a diameter of only a few wavelengths. This diameter is in some cases two orders of magnitude smaller than the focal diameter, assuming linear optics." [28]

The idea was as follows: "an electro-magnetic beam can produce its own dielectric waveguide and propagate without spreading. This may occur in materials whose dielectric constant increases with field intensity, but which are homogeneous in the absence of the electromagnetic wave." [28]

An estimate was obtained by considering diffraction of a circular optical beam of uniform intensity across diameter D in material for which the index of refraction may be quadratic in field with a coefficient n_2. A self-consistent waveguide will occur when the divergence angle of the finite beam, $\theta_D = 1.22\lambda/nD$, is set equal to the critical angle for total internal reflection, for a refractive index discontinuity of $\Delta n = n_2|E|^2$. This gives a threshold power in the laser beam of $P = (1.22\lambda)^2c/64n_2$, a total power independent of beam diameter. Inserting numbers, the beam should self-trap at approximately 1 Megawatt.

3.4.1. Exact solutions for confined beams

When lateral variation occurs only in one dimension (a slab-shaped beam), the nonlinear Helmholtz equation for the transverse field is given by $E_{yy} - \Gamma^2E_y + (\varepsilon_2/2)k_o^2E^2E = 0$, where the y subscript means the derivative in the y direction [28]. The analytic solution for the transverse field dependence is stable: $E(y) = E_o\text{sech}(\Gamma y)$, where $\Gamma = ½\, \varepsilon_2^{1/2}k_oE(0)$. Such self-trapped beams have, indeed, been generated in waveguides.

In two dimensions with a cylindrical beam, the Helmholtz equation (in dimensionless units) is

$$\frac{d^2E^*(r^*)}{dr^{*2}} + \frac{1}{r^*}\frac{dE^*(r^*)}{dr^*} - E^*(r^*) + E^{*3}(r^*) = 0,$$

There is no analytic solution, but Figure 8 shows the "Townes Profile," which was calculated by computer. When compared to the Gaussian beam, it can be seen to be close. Integration of this solution gives the critical power: $P = 5.763\ \lambda^2cn_{eff}/8\pi^3n_2n_0$, very close to the initial estimate.

We observed self-trapping experimentally by inserting microscope slide cover slips along the beam path and reflecting successive weak images as the beam traveled through carbon disulfide [29]. Figure 10 shows the near field of the beam at successive locations for (a) low power, with (b) and (c) representing successively higher powers. The beam is seen to not diffract but to

Journal of Physics: Conference Series **594** (2015) 012001 doi:10.1088/1742-6596/594/1/012001

maintain a constant beam diameter. At higher powers a higher order spatial soliton can be seen. It should be pointed out, however, that this early data was not at all definitive because other nonlinearities such as SRS and SBS were undoubtedly occurring at the same time. It would take considerable additional work before these effects began to be separately studied.

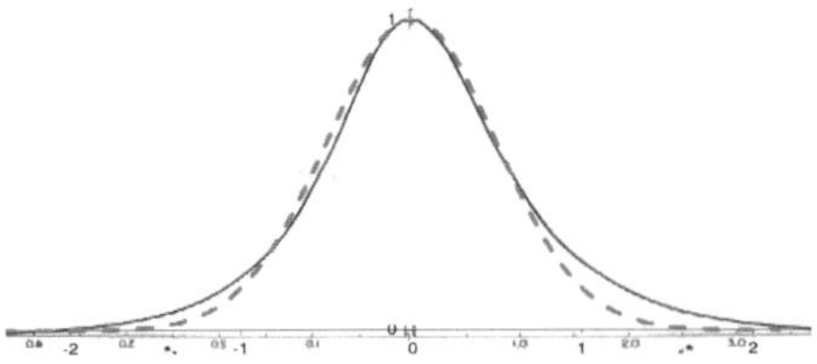

Figure 8. Numerical analysis for electric field in a spatial solitary wave, the "Townes Profile," compared to a Gaussian profile (dashed line).

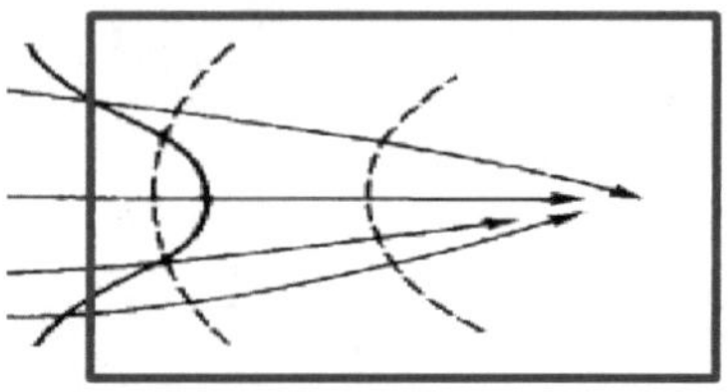

Figure 9. Self-focusing for beams above threshold power, in a manner similar to a graded index lens.

The self-trapped beam that was predicted is now known as a "spatial soliton" and has been seen in a number of situations. Today, an important application for self-trapped beams is in plasmas, particularly those created by ultra-short laser beams in air. Self-focusing in plasmas was, in fact, independently predicted before self-focusing in liquids or solids [30].

(a)
(b)
(c)

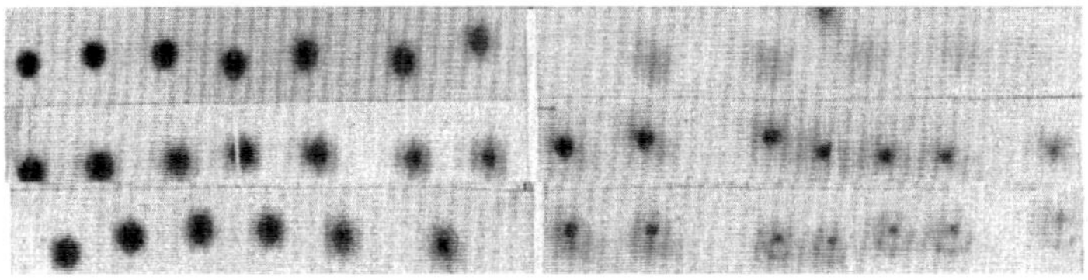

Figure 10. Experimentally observed images of an apparently self-trapped ruby laser beam, 60 μm in diameter, as it travelled through carbon disulphide [29]

3.4.2. Instabilities in self-trapped beams

Paul Kelley pointed out that the same nonlinearity that creates a self-trapped beam of just the right power will self-focus beams of powers higher than the critical threshold [31]. The nonlinearity overpowers diffraction, creating spherical surfaces of constant phase that focus the beam as shown in figure 9. The self-focusing process will cease when higher-order nonlinearities take over.

In order to achieve the results shown in figure 10, it was necessary to ensure that the ruby laser operated in a single spatial mode. With multi-spatial modes, large-scale trapping was not observed. In both cases, however, close observation showed that nonuniformities in the beam caused it to break up into many small-scale filaments on the order of only a few microns.

More careful analysis showed that superimposed on the large scale trapping were many small-scale filaments, only a few μm in size, apparently introduced by instabilities [32]. Such instabilities in χ_3 materials had been predicted by Talanov [33].

Later experiments working with Brewer from IBM demonstrated that these filaments last only $\sim 10^{-10}$ sec, after which they "blow up" [34], presumably because heating expands the liquid. With measured diameters of 4 μm in CS_2 and 12 μm in nitrobenzene, they were estimated to contain about 10kW peak power. This power level agreed with theory that assumed the Kerr effect is the principal mechanism for initiating filaments. For powers above the trapping threshold, filaments were seen to decrease in size and increase in intensity until higher-order terms in the electric field caused saturation of the nonlinear refractive index. Measured diameters were 10 × larger than theoretical limiting diameters based on Kerr effect saturation, which indicated that some other mechanism determined filament size. We measured lower than expected Raman gain in most of the filament length, which was explained by frequency broadening and dispersion.

Self-focusing and self-trapping have remained a very important factor in a number high-power laser experiments and many of the interrelated phenomena have been carefully disambiguated. A recent review offers an excellent and up-to-date review of the status today of these filaments [35].

While self-trapped filaments tend to be unstable in a material with a χ_3 nonlinearity, they are stable

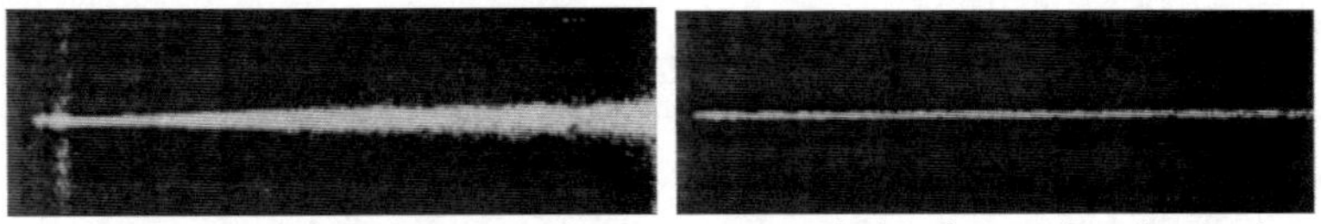

in photorefractive χ_2 materials. Figure 11 shows an example in photorefractive barium titanate.

Figure11. Photograph of green laser beam through photorefractive barium titanate: (a) normal diffraction at lower power; (b) self-trapped beam at higher power. [Courtesy Professor Z. Chen, https://www.physics.sfsu.edu/~laser/research.html]

3.5. Stimulated four-photon light scattering

Influenced by the realization of stimulated scattering from Raman and Brillouin effects, Townes' group also analyzed stimulated Rayleigh scattering [36]. They assumed a stationary, or near-stationary, periodic variation in refractive index change would occur as a result of interfering light beams and a nonlinear refractive index.

Stimulated Rayleigh-wing scattering discussions had not yet included the influence of stimulated light-by-light scattering and its associated weak-wave retardation. With both effects having large gains in the forward direction, the proper analysis required coupled waves. The paper analyzed the interaction between three waves: undepleted laser field E_o, weak wave E_1 at a small angle (or frequency difference) and amplified weak wave E_2. We were the first to show that weak wave E_2 experienced a gain per unit length of $g = 4\pi\Delta n/\lambda_o$. In nonlinear media, $g = \pi\varepsilon_2|\mathcal{E}_o|^2/n_o\lambda_o$ (cgs units). Since this is a parametric process, the weak wave E_1 experiences a loss per unit length equal to the gain of E_2 per unit length. When the waves experience 1/e loss in lengths L_1 and L_2, there will be a threshold for observation of this parametric amplification given by $4\pi\Delta n = \pi\varepsilon_2|\mathcal{E}_o|^2/n_o > \lambda_o/(L_1L_2)^{1/2}$. Because the weak waves are retarded relative to the strong wave (whose wave vector is extended by the nonlinearity), phase-matching gives rise to small angles for E_1 and E_2 with respect to E_o. The gain was shown to be strongly peaked at the phase-matching angle. Just as in SRS, transverse momentum conservation requires that if wave E_1 travels with angle θ with respect to the laser beam, wave E_2 travels at $-\theta$, both angles being very small. It is simple to show that this angle is $\theta = \pm (2\Delta n/n_o)^{1/2}$. This is easiest seen through picturing that interference causes stationary periodic layers in the medium, which can produce reflections in a manner similar to the Raman-Nath effect.

The assumption was local nonlinearity, so that degenerate gratings in the material are unshifted with respect to the optical interference fringes. Even so, this analysis showed that steady-state amplification was possible, via the parametric four-wave interaction. The phase-matching condition was met only for a strict angular window, and only there could net gain be achieved. An experiment was carried out with the Q-switched ruby laser in carbon disulfide to demonstrate that a weak wave did, indeed, experience parametric gain at the appropriate angle [37].

Considering only a weak and a strong wave, the nonlinear shift in the index of refraction induced by the pump wave on the probe wave was shown to be double that compared to the pump wave itself, which implies that the probe-wave vector is longer than the pump-wave vector. This additional lengthening of the weak-wave propagation vector had lasting impact on the thinking about two-wave, three-wave and four-wave mixing, particularly in photo-refractive media. It also provided early understanding of how instabilities grow and cause filamentation of high power beams traveling through nonlinear media.

3.6. Self-steepening of light pulses

Townes and his group predicted a change in temporal shape of light pulses traveling through a medium with an intensity-dependent refractive index [38]. They calculated the time required for a Gaussian pulse to steepen into a temporal optical shock through both analytic and numerical solutions for the pulse as it traveled through the medium. They investigated temporal development of pulses for both zero and nonzero relaxation times of the nonlinear χ_3 refractive index.

Journal of Physics: Conference Series **594** (2015) 012001 doi:10.1088/1742-6596/594/1/012001

The frequency spectrum calculated in the zero-relaxation time limit, found the largest peak intensities on the lower-frequency side (figure 12). The rate of steepening is modifed when pulse decay time is as short as the nonlinear refractive index; in a dispersionless medium the pulse decay time can become arbitrarily short. Estimates were made for both the thickness of the optical-shock region and the frequency spreading allowed by dispersion when relaxation was added. This phenomenon is now well-known and more familiarly called self-phase-modulation.

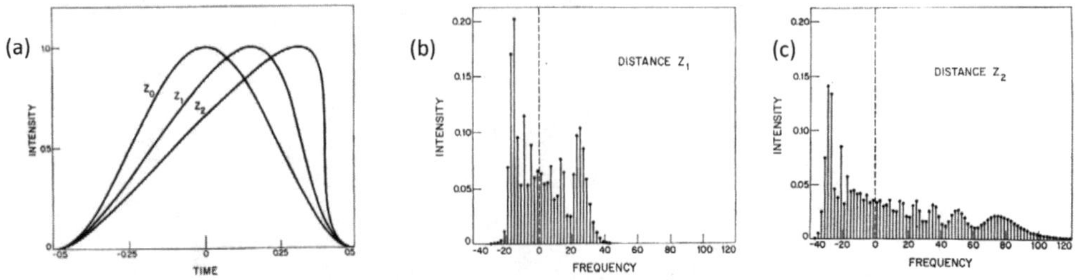

Figure 12. Self-steepening and frequency-broadening of a Gaussian laser pulse through χ_3 medium. (a) Time dependence of intensity of a sinusoidal input at z=0, z_1 and z_2; (b,c) Spectra of a sinusoidal pulse at z_1 and z_2 respectively. At z_2 the most intense peak is ~ 2000 cm^{-1} below the ruby laser wavelength. Adapted from [35].

4. Other MIT research in Nonlinear Optics

Professor Townes continued to interact with and inspire more work on self-trapped filaments of light. With Brewer at IBM he looked at standing waves in these filaments [39]; with Sacchi and Lifsitz he investigated anti-Stokes generated in trapped filaments [40]; phase modulation was investigated with Cheung et. al. [41]. As a result of the earlier work, spectral broadening of light in filaments was analysed [42], along with self-trapping in saturating media [43]. Similar studies broadened to new fields, such as thermal self-focusing [44] and coherent excitation of polaritons [45]. This list includes only work done in the field by those who had collaborated directly with Townes while he was at MIT and is not meant to be comprehensive (567 papers in nonlinear optics were published between 1961 through 1969, according to the Information Sciences Index).

5. Contributions of Townes to Astronomy (1967-2011)

Here is a list of some of the major topics of investigation in which Townes was involved at University of California, Berkeley:

- Water in interstellar space
- Ammonia in interstellar space
- Dust around stars
- Mechanisms in the galactic center
- Mechanisms of star formation in galaxies
- Far-infrared spectroscopy of galaxies
- A heterodyne stellar interferometer for the mid-infrared

Townes had long been interested in astronomy, ever since his initial work with radar during WWII and had a firm conviction that the nonlinear photon-atom processes that were observed terrestrially with lasers would be observable in space. He was rewarded by finding this so and continuing active research well into his 90's. The last research papers listing him as a co-author were in 2011 when he was 96 years old!

6. Contributions of Basov and Prokhorov (1967-2011)

This paper focused on Charles Townes, one of the Nobel Prizewinners. It did not include the important contributions that Nobelists Basov and Prokhorov have made to the field. Basov introduced the idea of semiconductor lasers as early as 1958 [46]. Prokhorov ran a laboratory with wide-ranging interests, with developments often proceeding in parallel with those in the US, including Q-switching, new pumping techniques, improving laser coherence and increasing power. Contributions were made in nonlinear optics, fiber and integrated optics, and in a variety of laser applications, especially in fiber-optic communication, laser technologies and the use of lasers in medicine and ecology. An excellent source for English translations of early papers describing quantum electronics research in the USSR can be found on-line [47].

References

[1] Edlén B 1964 Nobel Prize Award Ceremony Speech, retrieved at http://www.nobelprize.org/nobel_prizes/physics/laureates/1964/press.html
[2] Rytov S M, Prokhorov A M and Zhabotinsky M E 1945 On the theory of the frequency stabilization *Z. Eksperim. Teoretich. Fiziki* **15** 557
[3] Townes C H 1944 Theory of Cathode Sputtering in Low Voltage Gaseous Discharges *Phys. Rev.* **65** 319
[4] Basov N G and Prokhorov A M 1954 *Zh. Eksp. Teor. Fiz.* **27** 431, in Russian; available translated at http://www.quantum-electron.ru/laser50_eng.phtml.
Basov N G and Prokhorov A M 1955 The theory of a molecular oscillator and a molecular power amplifier *Discussions of the Faraday Society* **19** 96-99.
[5] Gordon J P, Zeiger H J and Townes C H 1954 Molecular microwave oscillator and new hyperfine structure in the microwave spectrum of NH_3 *Phys. Rev.* **95** 282
[6] Basov N G and Prokhorov A M 1955 Possible methods of obtaining active molecules for a molecular oscillator *Sov. Phys. JETP* **1** 184
[7] Einstein A 1916 Strahlungs-emission und -absorption nach der Quantentheorie *Verhandlungen der Deutschen Physikalischen Gesellschaft* **18** 318
[8] Fabrikant V A (1939) On experimental evidence for the existence of negative absorption *Doctor of Science (Habilitation) Thesis*, translation of pages 273 and 274. Found at http://www.quantum-electron.ru/pdfrus/laser50/11788_eng.pdf
[9] Purcell E M and Pound R V 1951 A nuclear spin Ssystem at negative temperature *Phys. Rev.* **81** 279
[10] Weber J 1953 Amplification of microwave radiation by substances not in thermal equilibrium *Trans. IRE Prof. Group on Electron Devices* pged-**3** 1
[11] Karlov N V, Krokhin O N and Lukishova S G 2010 History of quantum electronics at the Moscow Lebedev and General Physics Institutes: Nikolaj Basov and Alexander Prokhorov *Appl. Opt.* **49** F32-F46
[12] Townes C H and Schawlow A L 1958 *Microwave Spectroscopy* (NewYork, Dover, reprinted 1975)
[13] Gordon J P, Zeiger H J and Townes C H 1954 Molecular Microwave Oscillator and New Hyperfine Structure in the Microwave Spectrum of NH_3 *Phys. Rev.* **95** 282 (1954)
[14] Autler S H and Townes C H 1955 Stark Effect in Rapidly Varying Fields *Phys. Rev.* **100** 703
[15] Townes C H and Schawlow A L 1958 Infrared and Optical Masers *Phys. Rev.* **112** 1940
[16] Townes C H 1994 A Life in Physics: Bell Telephone Laboratories and World War II, Columbia University and the Laser, MIT and Government Service, California and Research in Astrophysics, *oral history* conducted by Riess S B, *Regional Oral History Office,* The Bancroft Library, *University of California*, *Berkeley* accessed at http://content.cdlib.org/view?docId=kt3199n627&brand=calisphere&doc.view=entire_text
[17] Chiao R Y, Garmire E and Townes C H 1963 Raman and Phonon Masers *Proceedings of the International School of Physics Enrico Fermi*, *Course XVII,* published as *Quantum Electronics and Coherent Light* ed P. Miles (New York: Academic Press). Preprint can be found at https://archive.org/details/nasa_techdoc_19640006830
[18] Garmire E, Pandarese F and Townes C H 1963 Coherently driven molecular vibrations and light modulation *Phys. Rev. Lett.* **11** 160
[19] Chiao R and Stoicheff B P 1964 Angular dependence of maser-stimulated Raman radiation in calcite *Phys. Rev. Lett.* **12** 290
[20] Garmire E 1964 *Bull. Am. Phys. Soc.* **9** 490

[21] Garmire E 1965 The angular distribution of stimulated Raman emission in liquids *Phys. Lett.* 1965 **17** 251
[22] "Stimulated Raman Scattering in Liquids" (E. Garmire) Physics of Quantum Electronics, Kelley, Tannenwald, eds. (McGraw-Hill, New York, 1966) pp. 167-179
[23] Garmire E 1965 *Nonlinear Optics in Liquids* PhD Thesis, Physics Department, MIT
[24] Chiao R Y, Townes C H and Stoicheff B P 1964 Stimulated Brillouin scattering and generation of intense hypersonic waves *Phys. Rev. Lett.* **12** 592
[25] Garmire E and Townes C H 1964 Stimulated Brillouin scattering in liquids *Appl. Phys. Lett.* **5**, 84
[26] Zeldovich B Ya, Popovichev V I, Ragulskii V V and Faizullov F S 1972 Connection between the wave fronts of the reflected and exciting light in stimulated Mandel'shtam-Brillouin scattering *Sov. Phys. JETP Lett.* **15** 109
[27] Hercher M 1964 *J. Opt. Soc. Am.* **54** 563. Townes also had a copy of a photograph by John Atwood at Perkin Elmer of a self-trapped track, which was later published as Figure 16 in the following: Minck R W, Terhune R W and Wang C C 1966 Nonlinear optics *Appl Opt.* **5** 1595
[28] Chiao R Y, Garmire E and Townes C H 1964 The self-trapping of optical beams *Phys. Rev. Lett.* **13**
[29] Garmire E, Chiao R Y and Townes C H 1966 Dynamics and characteristics of the self-trapping of intense light beams," *Phys. Rev. Lett.* **16**, 347
[30] Askarjan G A 1962 Effects of the Gradient of a Strong Electromagnetic Beam of Electrons and Atoms *Sov. Phys. JETP* **15**, No. 6
[31] Kelley P L 1965 Self-focusing of Optical Beams *Phys. Rev. Lett.* **15** 1006
[32] Chiao R Y, Johnson A, Krinsky S, Smith H A, Townes C H and Garmire E 1966 A new clss of trapped light filaments *IEEE J. Quantum Electr.* **QE-2**, 467
[33] Talanov V I 1965 Self-focusing of wave beams in nonlinear media *JETP Lett.* **2** 138. Also V. I. Talanov 1964 *Izv. Vysshikh Uchebn. Zavedenii, Radiofiz.* **7** 564 (1964).
[34] Brewer R G, Lifsitz J R, Garmire E, Chiao R Y and Townes C H 1968 Small-scale trapped filaments in intense laser beams *Phys. Rev.* **166** 326
[35] Boyd RW, Lukishova S G and Shen Y R, eds. (2009) Self-focusing: Past and Present Fundamentals and Prospects *Topics in Appl. Phys.* **114** (SpringerLink ebooks in Physics and Astronomy at http://link.springer.com/book/10.1007/978-0-387-34727-1)
[36] DeMartini F, Townes C H, Gustfson T K and Kelley P L 1967 Self-steepening of light pulses *Phys. Rev.* **164** 312
[37] Chiao R Y, Kelley P L and Garmire E 1966 Stimulated 4-photon interaction and its influence on stimulated Rayleigh-wing scattering *Phys. Rev. Lett.* **17** 1158
[38] Carman R L, Chiao R Y and Kelley P L 1966 Observation of degenerate stimulated 4-photon interaction and 4-wave parametric amplification *Phys. Rev. Lett.* **17** 1281
[39] Brewer R G and Townes C H 1967 Standing waves in self-trapped light filaments *Phys. Rev. Lett.* **18** 196
[40] Sacchi C H Townes C H and Lifsitz J R 1968 Anti-Stokes generation in trapped filaments of light *Phy. Rev.* **174** 439
[41] Cheung A C, Rank D M, Chiao R Y and Townes C H 1968 Phase modulation of Q-switched laser beams in small-scale filaments *Phys. Rev. Lett.* **20** 786
[42] Gustafson T K, Taran J P, Haus H A, Lifsitz J R and Kelley P L 1969 Self-modulation, self-steepening, and spectral development of light in small-scale trapped filaments *Phys. Rev.* **177** 306
[43] Gustafson T K, Kelley P L, Chiao R Y and Brewer R G 1968 Self-trapping in media with saturation of the nonlinear index *Appl. Phys. Lett.* **12** 165
[44] Carman R L, Mooradian A and Kelley P L 1969 Transient and steady state thermal self-focusing *Appl. Phys. Lett.* **14** 136
[45] Coffinet J P and F. DeMartini 1969). Coherent excitation of polaritons in gallium phosphide" *Phys. Rev. Lett.* **22** 60
[46] http://www.nobelprize.org/nobel_prizes/physics/laureates/1964/basov-lecture.html
[47] http://www.quantum-electron.ru/laser50_eng.phtml

23rd International Laser Physics Workshop (LPHYS'14) IOP Publishing
Journal of Physics: Conference Series **594** (2015) 012002 doi:10.1088/1742-6596/594/1/012002

Lasers: reminiscing and speculating

Michael Bass

Professor Emeritus, Creol, The College of Optics and Photonics, University of Central Florida, Orlando, Florida, USA

Perspective 

It is almost 50 years to the day when
Charles Townes,
Alexander Prokhorov and
Nikolay Basov
received the Nobel Prize for their work providing the scientific basis for the laser.

Journal of Physics: Conference Series **594** (2015) 012002 doi:10.1088/1742-6596/594/1/012002

Their comments in their Nobel Addresses

Townes – "solve a puzzle, understand something new, and its exhilarating"

Prokhorov – "many believed that we had gone crazy, that it was impossible"

Basov – "opened up vast new macroscopic and microscopic horizons"

Since then 20 Nobel Laureates have been honored for laser enabled research. Evidently Townes', Prokhorov's and Basov's outstanding contribution provided a means to amazing new fields of science.

A bit of history

You might think optics started when someone in Greece figured out the law of reflection (**Archimedes about 212 BC**).

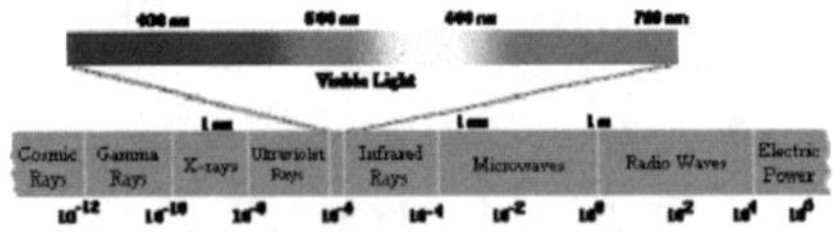

You might think **Newton** kick started it in the late 1600s when he separated white light into colors and couldn't make a color reform white light.

He also made the Newtonian telescope and Newton's rings and Devised the corpuscular theory of light.

But, in **1803 Thomas Young** performed the double slit experiment proving that light was wavelike. Developed gratings and a means to measure wavelength.

The experimental foundation for Maxwell's theory.

~**1814 Augustin-Jean Fresnel** required light be wavelike.

The last ~50 years or ~1/4 the history of modern optics

Some of us, myself included, have been contemporaneous with this piece of history and **so has the laser and nonlinear optics**.
- Either I have been very lucky or have lived very long.
- I would like to think I chose to work in a very exciting field.

It is worth noting that **James Clerk Maxwell** anticipated nonlinear optics when he wrote the polarization of a medium as a power series expansion in powers of the electric field. Then he dropped the terms of order higher than 1 since they would be too small.

He was right but he was also wrong.

Laser Pre-history

1916 – **Albert Einstein** realized the roles of spontaneous and stimulated emission.
- ~1900 **Henri Becquerel** had observed spectral line narrowing while studying the 694 nm emission from ruby. Made note of it in his notebook but did nothing about it.

1939 – **Valentin Fabrikant** proposed using stimulated emission to amplify light but did not pursue the idea.

1939 – I was born

1951 – **Charles Townes** proposed isolating excited ammonia in a resonant cavity so that stimulated emission could amplify at microwave frequencies.
- 1954 - Together with **James Gordon**, his student, Townes demonstrated the MASER.

1958 – **Charles Townes** and **Arthur Schawlow** (Townes' brother-in-law) published "Infrared and optical masers," in the December 15, 1958 *Physical Review.*
- *This paper was to serve as the blueprint for laser science.*

The patent fight

Gordon Gould was 37, an elderly graduate student of Polykarp Kusch at Columbia University.

On November 13, 1957 he took his notebook to a notary public and had his description of a Fabry-Perot resonator with a gain medium within notarized.

Gould discussed his idea with a patent attorney and developed an extensive list of possible gain media but neither he nor Townes and Schawlow were close to building one.

Gould left Columbia and went to work for TRG, Inc. and worked out a deal to share patent rights with the company.

Gould had to bring patent infringement suits against laser manufacturers and he won. The thirty year patent war **ended in 1985.** Gould was issued forty-eight patents including **optical pumping, collisional pumping, and applications patents** being the most important.

Laser history begins

In 1960 Theodore "Ted" Maiman

did what no one else expected - he used a pulsed flash lamp to excite lightly doped ruby.

- He silver coated both ends of a stubby ruby rod, scraped a small hole in the center of the silver at one end to serve as the output coupler and on May 16, 1960 demonstrated the first LASER.
- This took place at the Hughes Research Labs in Malibu, California.

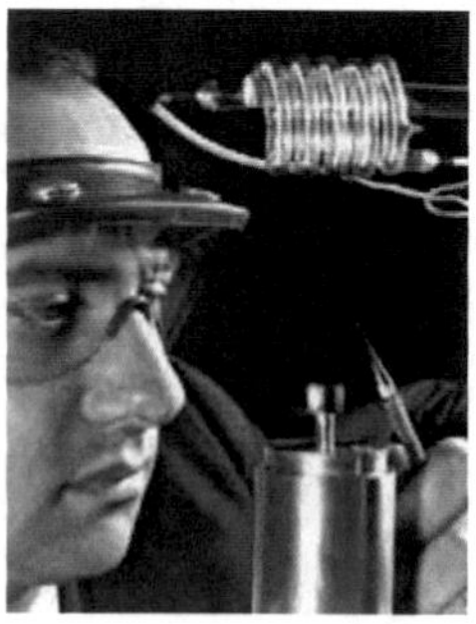

How could Maiman prove lasing?

Keep in mind that most of the diagnostics that we have today did not exist in 1960.

Maiman had to show:

1. Threshold
2. Spectral narrowing
3. Relaxation oscillations
4. Collimated beam

He did all of this and his paper was rejected by **Samuel Goudsmit**, editor of Physical Review Letters because there had been too many papers with the word Maser in the title.

Maiman had entitled his paper **"Optical Maser Action in Ruby"** – the word Laser had been coined by Gordon Gould in his not yet revealed notebook entry for his patent.

With that Hughes organized a press conference to announce the first laser and Maiman sent a note to Nature.

The Maiman laser

Because all lasers are essentially the same ever since:

In the 1960s I would give talks in which I would first define the acronym **LASER as Light Amplification by Stimulated Emission of Radiation** and then describe a laser as having three components – a means to excite the second component, a medium that when excited could amplify light and a third component, a resonator that could bounce the light back and forth many times to use the amplifier many times.

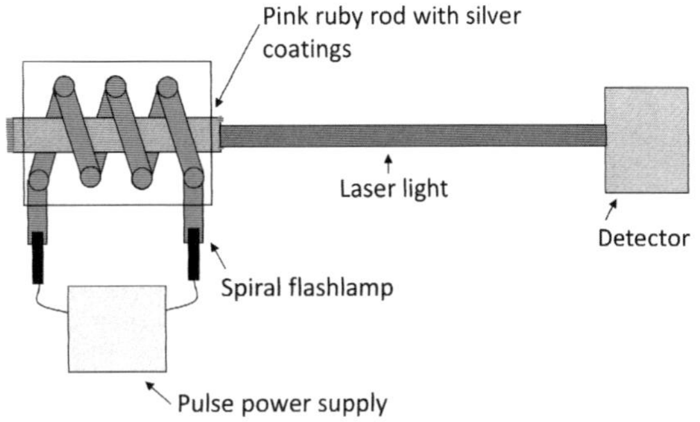

Then the dam burst

- **Theodore Maiman** at Hughes Res. Labs. - **Ruby** - 1960
- **Ali Javan** at Bell Laboratories - **HeNe** - 1960
- **Elias Snitzer** at American Optical – **Glass and fiber** – 1961
- **Robert Hall** at General Electric Research Laboratories – **Diode** – 1962
- **Kumar Patel** at Bell Laboratories - **Carbon Dioxide** - 1964
- **Joseph Geusic** at Bell Laboratories – **Nd:YAG** – 1964
- **George Pimentel** at U.C.-Berkeley – **Chemical** - 1965
- **Peter Sorokin** at IBM Watson Laboratories – **Dye** – 1966
- **Nikolai Basov** at the Lebedev Physical Institute – **Excimer** - 1970

An early application of the ruby laser

In the 1960s while people joked about the laser being a **SOLUTION IN SEARCH OF A PROBLEM** a problem was found.

(don't laugh – this is very major laser materials processing and without it there would be some very unhappy people)

drilling holes in baby bottle nipples

How else would you make the holes that the baby can suck on and the pressure relief holes that allow the formula to keep flowing.

Journal of Physics: Conference Series **594** (2015) 012002 doi:10.1088/1742-6596/594/1/012002

Something to notice

All but one of these new lasers were demonstrated by researchers in industry laboratories.

- Most likely because they could re-direct efforts more easily than could academics.
- Most likely because they had mixes of engineers and physicists.
- Most likely because they had materials scientists.
- Most likely because they had available equipment and supplies.

HOWEVER, something did happen at a university.

I started graduate school at the University of Michigan in the fall of 1960.

Peter Franken was the first year grad student advisor.

He had decided that there had to be a horse somewhere.

The horse was nonlinear optics!!!

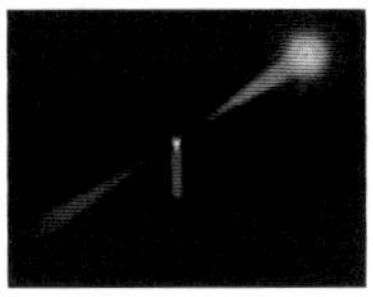

Second harmonic generation

Franken's idea was that lasers made the electric field large enough to make small terms detectable.

Weinreich pointed out that you had to have non centro-symmetric media.

Peters had the spectrograph and the dark room.

Allan Hill was a junior year physics student with unusual experimental skills.

- Besides, no graduate students were yet available and willing to take a risk on such a strange idea.

The experiment (Spring 1961)

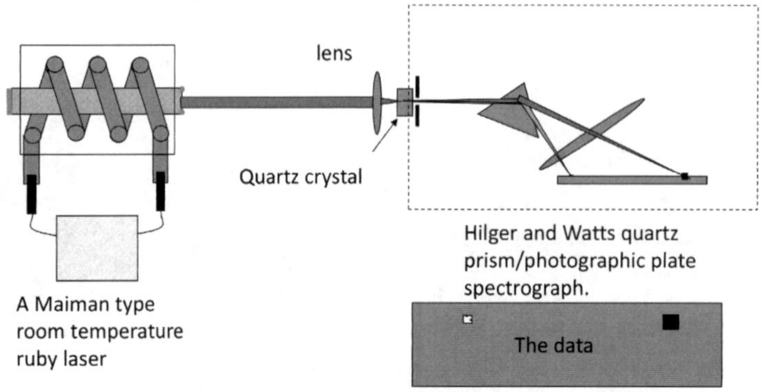

P. A. Franken, A. E. Hill, C. W. Peters and G. Weinreich, Phys. Rev. Letters, **7**, 118 (1961)

My involvement

I saw Alan Hill using the laser and decided anything that could make that much light had to be worthy of study so I made Peter Franken a deal.

- If I could work with the laser I would be his student.
- To my amazement he agreed – I think he planned this from when I first met him.

My first experiment was to demonstrate **optical sum frequency generation** - 1962

- The light from two different lasers was made to add together to produce the sum of their two frequencies. Now common but was not then.

My next was to show **optical rectification** -1962

- This had to await the **invention of the Q-switch by Robert Hellwarth** to allow clean temporal pulses.
- I will receive the 2014 R. W. Wood Prize for this discovery since people now recognize its importance as a means to generate teraherz waves.

Experimenting in the olden days

There were no:

Optical tables (used granite slabs from gravestone companies and vacuum hold downs)

Standard optical hardware (Ealing bench hardware only fit on Ealing benchs otherwise you made your own)

Optical components (most was military surplus – Edmunds Optical sold mostly surplus optics)

Alignment lasers (not yet invented) (you used an auto collimator)

High speed detectors (a few, very expensive photo diodes and photomultipliers with bandwidth up to about 1 GHz)

CCD cameras or detectors (you used photographic film, spectrographic plates or Polaroid film)

High speed oscilloscopes and all were analogue (a Tektronix 519 cost many dollars, had a few GHz bandwidth and required a signal of several volts to drive it. To record a scope trace you had to photograph it in real time usually with a scope camera and special Polaroid film.)

Calibrated energy meters (you were inventive – how many Gillettes could the laser burn through when focused with a certain lens?)

Computers, not even hand calculators (you used a slide rule)

The ABDP paper – my small role

In 1962 there was no such thing as e-mail so I mailed **Jack Armstrong** at Harvard a preprint of the sum frequency paper.

- He glanced at it and, I am told, put it in his trash basket.
- **Nicolas Bloembergen** came in and saw it, having read the first second harmonic generation paper by Franken et al, and realized the importance of nonlinear optics.

The next morning Bloembergen gathered Armstrong, Pershan and Ducuing in his office and assigned them their parts in **the now famous Armstrong, Bloembergen, Ducuing and Pershan paper (ABDP) that set down the theoretical underpinnings of nonlinear optics.**

Bloembergen says that he had to work on the theory because he didn't have a laser and couldn't afford to buy one.

- This theory may have been a large part of his Nobel Prize.

Franken should have shared in a Nobel prize for nonlinear optics but had taken a position at DARPA and that probably made him not politically correct enough for the Swedish committee.

Journal of Physics: Conference Series **594** (2015) 012002 doi:10.1088/1742-6596/594/1/012002

Lasers at Berkeley

From 1964-1966 **Y-R Shen and I** were the only laser people in physics at UC Berkeley.

We were constantly being quizzed about these strange new devices. At one lunch, **Owen Chamberlain**, Nobel Laureate, kept us for 2 hours with questions about fields in resonators, coherent properties of light and optical gain.

To try to have some answer about coherence I forced myself to read **Roy Glauber's** papers on coherence from a quantum point of view. It was a great struggle but when I finally figured out that he was solving the two slit diffraction problem it all became clear(?). He won a Nobel prize for this.

A curious side trip from an American Physical Society Meeting

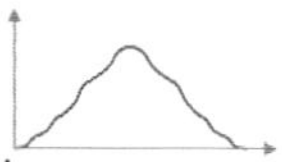

In 1965 I attended an APS meeting at Caltech in Pasadena, California.

I ran into **John "Jan" Hall**, who had been my undergraduate lab instructor at Carnegie and who would win a Nobel Prize in 2005.

During a break we decided to visit the laser group at Hughes in Culver City and were shown a laser pulse with steps as it rose and fell.

We said nothing because neither of us realized this was mode locking and their detector-oscilloscope combination was too slow to properly detect it.

Mode locking, though mentioned**, L.E. Hargrove** et al had reported it in a HeNe laser in 1964, had not yet been widely discussed in the literature.

- **Anthony deMaria's** work was not published until 1966. That is when mode locking became exciting.

Time in an industry lab - Raytheon

1966 -1973 at the Raytheon Research Division.

New solid state and dye lasers, and laser damage research.

We worked with our materials science people seeking an alternative to Nd:YAG.

- We discovered, studied and patented the host yttrium ortho aluminate, YAP or YALO.
- ZnSe for infrared windows.

Dye laser theory with **Marvin Weber** containing a criterion for cw lasing.

- In 1970 **Ben Snavely** and **Otis Peterson** may have used it to make the first cw dye laser.

Also at Raytheon

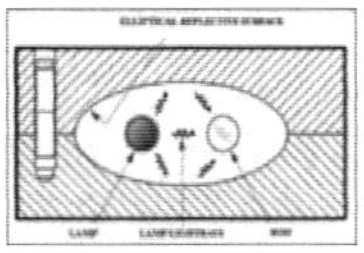

Colin Bowness patented the elliptical pump cavity 1965.

Dave Whitehouse built the first 1 kW CO_2.

Roy Paanenan built a 100 W argon ion laser.

Pim Andringa developed a passive method to prevent counter circulating beams in a laser gyroscope from locking.

I became involved in the optical properties of materials and introduced the concept of statistics to the laser damage community. We had to build our own lasers to do the experiments.

Years at USC and at CREOL

I then spent 1973-1987 at the **University of Southern California** where I developed the Center for Laser Studies

- and had the pleasure of working with outstanding colleagues, students and post docs some of whom have won very prestigious awards, are fellows of major societies and chaired professors at major universities.

Since the committee asked me to reminisce about the early days I leave out the amazing events of my time at **CREOL, the College of Optics and Photonics**, now 27 years old, which rose from just 7 hardy souls to over 250 faculty, students, post docs, research scientists and staff and became the first U.S. College of Optics and Photonics in less than 15 years.

Where are these lasers now?

The **ruby** laser is a historical footnote.

You can still buy a **HeNe** laser but visible diode lasers are cheaper, more rugged and more colorful.

The **diode** laser is the most widely used laser of all. It is so important that it gets its own slides.

Fiber lasers of many kilowatt outputs are now available for industrial applications. Fiber lasers also generate short pulses (sub pico-second). **Er doped fiber amplifiers** make fiber communication possible.

The **CO_2** laser can produce very high average power outputs (multi kW) and is a reliable tool in manufacturing.

The **Nd:YAG** laser with diode laser pump sources can be very small, light weight, and rugged and in other configurations very powerful for industrial and military applications.

Dye lasers were very important in short pulse studies were messy and sometimes dangerous. OPO's and tunable solid state lasers replaced dye lasers as tunable light sources.

Chemical lasers have dangerous exhausts but may be useful in certain military applications.

Excimer lasers are very important in surgery (LASIK) and some materials processing.

Diode lasers and a modern miracle

A miracle is something that happens that is unexpected and seemingly inexplicable.

The diode laser was first demonstrated in 1962 .
- It required cooling to LN2 temperature and lasted just a few minutes.

Meanwhile at the University of Michigan, **Wilbur Peters** had been experimenting with fiber optics.

At Corning, Inc. they were finishing work on Corningware and were casting about for a new project when at laser meetings AT&T people presented talks about optical communications.
- Optics had the bandwidth but they did not have a reliable low loss carrier.
- Corning took notice.

And so it came to pass that AT&T had a group working on making better diode lasers and Corning, a glass company, set to work on making low loss fibers.
- They did not plan this - it just happened that way.

In 1970 Corning was able to report a low loss fiber with only 17 db/km loss and so one part of the optical communications problem was in hand.

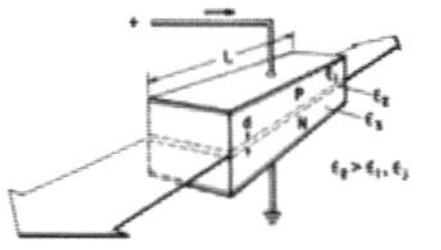

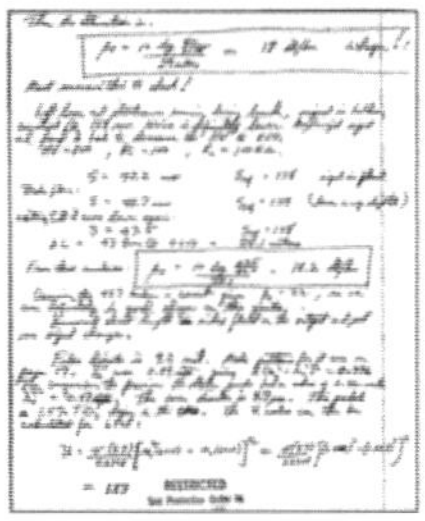

The million hour diode laser

In 1977, "**being very careful and very clean**" AT&T Bell labs reported diode lasers with expected lifetimes of 1 million hours.

With the improved Corning fibers, communications over optical fibers was feasible.

In 1980 the winter Olympics was broadcast over fiber optics links. Just 18 years from the first diode laser to the first optical communications system.

It was a MIRACLE!!!

Today there are nearly **2 x 10^{+9} km** of optical fiber in the telecommunications system with millions of diode lasers as sources and amplifier pumps.

Today diode lasers are made by the multi-millions per year and are –
- **in communications, supermarket scanners, laser printers, cd/dvd player/recorders, auto head lights, laser pointers, high power configurations for cutting and welding and solid state and fiber laser pumping, projection displays and ????.....**

B. C. DeLoach, R. W. Dixon and R. L. Hartman received the IEEE Gold Medal for Engineering Excellence For the Million Hour Laser

Speculation

Speculation is very risky business but I will try just don't hold me responsible:

Lasers will make possible:

1. Spectacular movie theaters but all too soon these will be adapted to home theaters.
2. Greatly improved surgical precision perhaps using pulsed UV laser light.
3. Exploration of extremely fast processes – the shortest pulsed laser is at CREOL and is ~65 atto-sec.
4. Studies of ultra-high field physics – the ELI will achieve 200 peta W and open up new physics.
5. The US National Ignition Facility (NIF) will achieve fusion threshold and find out how to optimize inertial confinement fusion.
   Plans for fusion power plants will be drawn up and then people will complain about tritium.
   There will be enough other energy sources for enough time at much lower cost that this will not be built.
6. Manufacturing technologies based on lasers that sustain Moore's law for another decade or two.

Far out speculation

Laser spacecraft propulsion.

Laser selective chemistry and biology.

X-ray lasers will find application in electronics.

- Will sustain Moore's law for several more decades.

Laser driven high energy physics – particle accelerators.

Space lasers for line of sight communications will become much more common.

- This will provide the bandwidth needed as the instrument packages become more capable.

Applications will be found for attosecond lasers.

Thank you

Any additional reminisces would be appreciated.

Any corrections too, but be nice about it.

Any additional speculations would be appreciated.

In 2016 OSA will publish a book commemorating its first 100 years. The Editor-in-Chief is **Paul Kelley** and the Associate Editors are **Carlos Stroud, Jeff Hecht, Michael Bass and Govind Agrawal**. I owe my colleagues in this effort thanks for involving me in this project as it made me more aware of the history of Optics and Lasers and the impacts they have made on how we live our lives.

23rd International Laser Physics Workshop (LPHYS'14) IOP Publishing
Journal of Physics: Conference Series **594** (2015) 012003 doi:10.1088/1742-6596/594/1/012003

Enrichment of nuclear spin isomers of molecules by bichromatic field

P L Chapovsky[1,2] and A D Wilson-Gordon[3]

[1]Insitute of Automation and Electrometry, Russian Academy of Sciences, Novosibirsk 630090, Russia
[2]Novosibirsk State University, 2 Pirogova St., Novosibirsk 630090, Russia
[3]Department of Chemistry, Bar-Ilan University, Ramat Gan 5290002, Israel

E-mail: chapovsky@iae.nsk.su; gordon@biu.ac.il

Abstract. A theoretical model for the enrichment of nuclear spin isomers of molecules by two electromagnetic fields is developed. The fields are resonant with the rotational transitions in the ground electronic and vibrational state of the molecule. Numerical calculations reveal a complicated dependence of the isomer enrichment on the radiation frequencies that consists of three peaks with different shapes, widths and amplitudes.

1. Introduction

Molecules with identical nuclei exist in Nature in the form of nuclear spin isomers [1]. This is a consequence of the quantum statistics that determine the symmetry of the molecular wave function under permutations of identical nuclei: bosons or fermions. The isomers have almost identical physical properties and appear to be extremely stable. The best known example is ortho and para H_2 which was comprehensively studied in the 1930s (see, for example, [2]) and found to have a lifetime of several months at ambient pressure and temperature [2]. Most of the studies of these isomers have been devoted to their enrichment and conversion under various conditions. The isomer enrichment appears to be easy to perform by the deep cooling of hydrogen gas. This unique property of hydrogen isomers results from the anomalously large rotational constant of H_2 and consequently large energy gap (170 K) between the ground states of the ortho and para H_2, which is significantly larger than the hydrogen boiling point, 20.4 K.

Polyatomic molecules with identical nuclei also have nuclear spin isomers. Since their structure is more complicated than that of diatomic molecules, polyatomic molecules can have two or more types of spin isomers. For example, CH_3F has two isomers (ortho and para) whereas methane has three and ethylene has four isomers. Enrichment of the isomers of polyatomic molecules appears to be much more difficult to perform than that of the hydrogen isomers. Nevertheless, a few methods for isomer enrichment have been developed. First, enrichment of the spin isomers of polyatomic molecules ($^{12}CH_3F$ and $^{13}CH_3F$) was performed using the light-induced drift effect [3]. Later, enrichment of the spin isomers of formaldehyde by UV photolysis was demonstrated [4]. Recently, the nuclear spin isomers of water and methane were separated using magnetic focusing of the cold molecular beams [5], and the spin isomer enrichment of ethylene was performed by catalyzed chemical synthesis [6].

Spin isomers have found important applications in NMR-based techniques due to their ability to significantly enhance the NMR signals. Thus, enhancement of more than three orders of magnitude was achieved with the help of enriched samples of hydrogen spin isomers [7, 8]. The NMR signal enhancement by the enriched isomers of ethylene was demonstrated recently [6]. There are also interesting applications of nuclear spin isomers in astrophysics where the spin isomer abundance in the cosmos is used to investigate the conditions of molecular formation and the properties of interstellar medium. The latest information on spin isomers in astrophysics can be found in the website of the Workshop: Nuclear Spin Effects in Astrochemistry, Goteborg, Sweden, June 9-11, 2014.

The applications of spin isomers rely on our knowledge of the isomers physics and on the available techniques for isomer enrichment. New access to isomer physics and enrichment can be based on the use of resonant electromagnetic fields. It was proposed in [9] that isomer enrichment can be performed by a resonant electromagnetic field that affects the population of the rotational states of the molecule. This idea was generalized in [10] where it was shown that isomer enrichment is affected not only by the level populations but also by the coherences induced by the radiation.

In laser spectroscopy, it is well-known that the use of two resonant electromagnetic fields instead of one significantly increases the variety of the field-induced modifications of the level populations and coherences [11]. For example, coherent population trapping (CPT) [12] has been demonstrated in a Λ system consisting of two lower levels that are close in energy (Zeeman sublevels or hyperfine levels) and an excited electronic level. Two laser fields interact with transitions from the lower levels to the excited level creating a coherence between the lower levels where all the population is trapped. This leads to a transparency window in the absorption spectrum at two-photon resonance. The existence of these dramatic effects motivated us to investigate isomer control by two fields instead of one. The paper is organized as follows. In section 2, we explain the essence of the specific type of isomer relaxation, *quantum relaxation*. This relaxation process is a key ingredient in the control of isomer conversion by radiation. In section 3, we formulate a theoretical model of isomer control by a bichromatic field in the Π-scheme. In sections 4 and 5, we present the 0th and 1st order perturbation theory of the problem. In section 6, we present the results of the numerical calculations of the isomer enrichment by a bichromatic field. In section 6 we draw conclusions.

2. Quantum relaxation of spin isomers

First, we note that direct electromagnetic transitions between the ortho and para states of the molecule are strongly forbidden [1]. Our approach to isomer control by electromagnetic radiation is not based on the direct ortho-para transitions and is applicable to the isomers whose relaxation is governed by the specific process first suggested in the theoretical paper by Curl *et al.* [13]. This process was later named *quantum relaxation* [14].

The essence of quantum relaxation can be explained as follows. Suppose that the Hamiltonian of the molecule under consideration can be divided into two parts,

$$\hat{H} = \hat{H}_0 + \hbar\hat{V}, \tag{1}$$

where $\hat{H}_0$ is the main part of the Hamiltonian. The eigenstates of $\hat{H}_0$ form two Hilbert subspaces of pure ortho and para molecular states, see Fig. 1. $\hat{V}$ is a small perturbation that creates coherent superpositions of the ortho and para states. The interaction of the molecule with the environment causes transitions of the molecule inside each ortho and para subspace but not between them.

The quantum relaxation of the isomers consists in the following. Suppose that initially the molecule was placed in the ortho subspace. Because of the interaction with the environment, the molecule starts to shuttle along the ladder of ortho states. It will continue to do so until, after a collision, the molecule arrives at the state $|m\rangle$. During the free flight after this collision, the intramolecular perturbation admixes the state $|\alpha\rangle$ from the para subspace with the state $|m\rangle$. Consequently, the next collision will have some probability to transfer the molecule to other para states thereby completing the relaxation process.

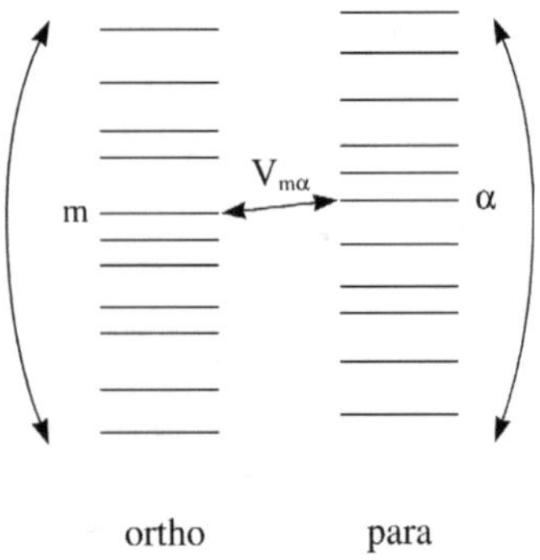

Figure 1. Quantum relaxation of spin isomers. Due to the interaction with the environment, the molecule undergoes relaxation inside the ortho and para rotational states in the ground electronic and vibrational state (indicated by the vertical bent arrows). Perturbation $\hat{V}$ mixes the $|m\rangle$ and $|\alpha\rangle$ states.

The quantum relaxation can be described alternatively using the notion of coherences between the molecular states. The intramolecular perturbation $\hat{V}$ creates the off-diagonal ortho-para density matrix element $R_{m\alpha}$. The ortho-para relaxation appears due to the interaction of the intramolecular perturbation $\hat{V}$ with the coherence $R_{m\alpha}$. Thus, ortho-para relaxation can be controlled by all the processes that affect $R_{m\alpha}$. These include molecular collisions that cause the decay of $R_{m\alpha}$, heating or cooling the gas thereby affecting the level populations, and Stark and Zeeman effects that change the molecular energy levels. It can also be done by resonant external electromagnetic fields that affect both the level populations and the mixing of ortho-para states by the perturbation $\hat{V}$.

The quantitative treatment of isomer relaxation in the framework of the density matrix formalism is given in [15]. For example, the nonequilibrium concentration $\delta n_o(0)$ of ortho isomers created at $t = 0$ decays exponentially,

$$\delta n_o(t) = \delta n_o(0)\exp(-\gamma_0 t), \tag{2}$$

with the rate

$$\gamma_0 = \frac{2\Gamma|V_{m\alpha}|^2}{\Gamma^2 + \omega_{m\alpha}^2}(W_m + W_\alpha), \tag{3}$$

in the case of just one ortho-para level pair mixed by the intramolecular perturbation $\hat{V}$, as in Fig. 1. In Eq. (3), Γ is the decay rate of the coherence between the states m and α, created by the perturbation $\hat{V}$, $\omega_{m\alpha}$ is the frequency gap between the states m and α, and W_m and W_α are the Boltzmann factors of the corresponding states. The equilibrium concentrations of ortho ($\overline{n}_o$) and para ($\overline{n}_p$) molecules are determined by the relations,

$$\overline{n}_o = N\frac{W_\alpha}{W_m + W_\alpha}, \quad \overline{n}_p = N\frac{W_m}{W_m + W_\alpha}, \tag{4}$$

where N is the total concentration of molecules. Note, that the equilibrium concentrations of ortho and para molecules (4) do not depend on $V_{m\alpha}$ and $\omega_{m\alpha}$.

In a series of studies (see, for example, [16, 17]) it was shown that the ortho-para mixing of states is performed by the intramolecular hyperfine interactions: nuclear spin-spin and spin-rotation. The quantum relaxation of spin isomers is a counter-intuitive process. Note, that each collision with the environment has the energy $k_BT \sim 10^{12}$ Hz at room temperature. Nevertheless, the isomer relaxation is governed by the hyperfine intramolecular coupling that is smaller than k_BT by seven orders of magnitude.

The goal of the present paper is to consider the isomer control by two electromagnetic fields. We have chosen the simplest cases: pure rotational excitation in the ground electronic and vibrational state of the molecule. We have performed the calculations for the three types of radiation schemes presented in Fig. 2. The most interesting results were found for the Π-scheme. This particular case will be considered in more detail below.

3. Isomer control in Π-scheme.

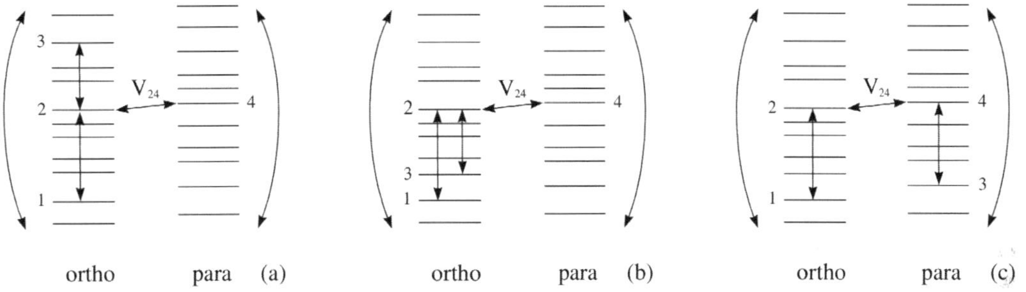

Figure 2. Control of spin isomers by two electromagnetic fields. (a) Ξ-scheme, (b) Λ-scheme, (c) Π-scheme.

The Π-scheme is presented in Fig. 2c. The molecular states are labeled by Roman or Greek letters for ortho and para states, respectively, or by numbers, as specified in Fig. 2c. The molecular Hamiltonian now consists of three parts,

$$\hat{H} = \hat{H}_0 + \hbar\hat{G} + \hbar\hat{V}, \tag{5}$$

where the terms $\hat{H}_0$ and $\hat{V}$ have the same meaning as in Eq. (1). The term $\hbar\hat{G}$ describes the molecular interaction with the bichromatic external field,

$$\hbar\hat{G} = -\mathbf{E}_1 \cdot \hat{\mathbf{d}} \cos\omega_1 t - \mathbf{E}_2 \cdot \hat{\mathbf{d}} \cos\omega_2 t, \tag{6}$$

where, $\mathbf{E}_1$ and $\mathbf{E}_2$ are the amplitudes of the two electromagnetic fields and $\hat{\mathbf{d}}$ is the molecular electric dipole-moment operator. Further, we will assume that the field $\mathbf{E_1}$ interacts only with the transition $1 \to 2$, and the field $\mathbf{E_2}$ interacts only with the transition $3 \to 4$. We will use the standard rotating-wave approximation. This leads us to the following nonzero matrix elements of $\hat{G}$,

$$G_{21} = -g_{21}\exp(-i\Omega_1 t), \quad g_{21} = \mathbf{E}_1 \cdot \overline{\mathbf{d}}_{21}/2\hbar, \quad \Omega_1 = \omega_1 - \omega_{21}, \tag{7}$$

$$G_{43} = -g_{43}\exp(-i\Omega_2 t), \quad g_{43} = \mathbf{E}_2 \cdot \overline{\mathbf{d}}_{43}/2\hbar, \quad \Omega_2 = \omega_2 - \omega_{43}. \tag{8}$$

Here, ω_{21} and ω_{43} are the frequencies of the corresponding transitions, g_{21} and g_{43} are the Rabi frequencies and the $\overline{\mathbf{d}}_{21}$ and $\overline{\mathbf{d}}_{43}$ are the time-independent dipole moments of the corresponding transitions.

The quantum kinetic equation for the density operator $\hat{R}$ reads [18]

$$\partial\hat{R}/\partial t = \hat{S} - i[\hat{G} + \hat{V}, \hat{R}], \tag{9}$$

where $\hat{S}$ is the collision integral and the rectangular brackets denote the commutator.

In our theory, we will use the strong collision model for the collision integral matrix, S. This model gives the diagonal terms of S in the form,

$$S_{ii} = -\nu R_{ii} + \nu n_o W_i, \quad n_o = \sum_i R_{ii}; \tag{10}$$

$$S_{\alpha\alpha} = -\nu R_{\alpha\alpha} + \nu n_p W_\alpha, \quad n_p = \sum_\alpha R_{\alpha\alpha}. \tag{11}$$

Here, ν is the rotational relaxation rate, and n_o and n_p are the total concentrations of ortho and para molecules, respectively. The collision integrals in the form of (10) and (11) give the Boltzmann distributions of the molecular level populations at equilibrium. Note, that our collision integral conserves the densities of ortho and para molecules,

$$\sum_i S_{ii} = 0, \quad \sum_\alpha S_{\alpha\alpha} = 0. \tag{12}$$

The off-diagonal elements of S are assumed to have only decay terms:

$$S_{ii'} = -\Gamma R_{ii'}, \quad S_{\alpha\alpha'} = -\Gamma R_{\alpha\alpha'}, \quad S_{i\alpha} = -\Gamma R_{i\alpha}, \tag{13}$$

where the decay rate Γ is independent of the states involved.

Using the properties of the collision integral and the kinetic equation (9) one can deduce the equations of change of the ortho and para densities. Thus, for example, for para molecules one has

$$\partial n_p/\partial t = 2Re(iR_{42}V_{24}). \tag{14}$$

This equation is derived on the assumption that the interaction $\hat{G}$ has no ortho-para matrix elements. Because of the conservation of the total molecular density, $\partial n_o/\partial t = -\partial n_p/\partial t$. We see from Eq. (14) that the ortho-para relaxation is determined by the ortho-para coherence R_{42} and the intramolecular mixing, $\hat{V}$.

We assume that the intramolecular perturbation $\hat{V}$ is weak in comparison with $\hat{G}$ and solve the kinetic equation (9) by perturbation theory. The density matrix is approximated by the sum,

$$\hat{R} \simeq \hat{\rho} + \hat{\epsilon}, \tag{15}$$

where $\hat{\rho}$ and $\hat{\epsilon}$ are the 0th and 1st order terms of $\hat{R}$, respectively. The kinetic equations for $\hat{\rho}$ and $\hat{\epsilon}$ read

$$\partial\hat{\rho}/\partial t = \hat{S}^{(0)} - i[\hat{G}, \hat{\rho}], \tag{16}$$
$$\partial\hat{\epsilon}/\partial t = \hat{S}^{(1)} - i[\hat{G}, \hat{\epsilon}] - i[\hat{V}, \hat{\rho}]. \tag{17}$$

4. The 0th order perturbation theory

In the 0th order perturbation theory, the ortho and para molecular subspaces are completely independent and one needs to solve the standard problem of the radiation interaction with the molecular rotational transition. Let us define the normalized density matrices, for ortho (ρ') and para (ρ'') molecules, as

$$\rho_{ii'} = n_o \rho'_{ii'}, \quad \rho_{\alpha\alpha'} = n_p \rho'_{\alpha\alpha'}. \tag{18}$$

The solution for ρ' was given in [10]. The diagonal elements of ρ' are given by

$$\rho'_{ii} = W_i + (W_1 - W_2)\frac{2\Gamma}{\nu}\frac{|g_{21}|^2}{\Gamma_o^2 + \Omega_1^2}(\delta_{i2} - \delta_{i1}), \quad \Gamma_o = \Gamma\sqrt{1 + 4|g_{21}|^2/\Gamma\nu}. \tag{19}$$

Here, we introduced the power-broadened line width of the $1 \to 2$ transition, Γ_o. Note that this broadening is caused by the finite decay rate Γ of the molecular coherence and the rotational level population relaxation rate ν.

The field $\mathbf{E}_1$ creates a nonzero off-diagonal matrix element on only the transition $1 \to 2$. This is given by

$$\rho'_{21} = \overline{\rho}'_{21} \exp(-i\Omega_1 t), \quad \overline{\rho}'_{21} = i g_{21}(W_1 - W_2)\frac{\Gamma + i\Omega_1}{\Gamma_o^2 + \Omega_1^2}. \tag{20}$$

The matrix elements of $\hat{\rho}''$ can be obtained from the expressions for the matrix elements of $\hat{\rho}'$ by the following index substitutions: $i \to \alpha, \quad 1 \to 3, \quad 2 \to 4$. Thus, one has the diagonal elements,

$$\rho''_{\alpha\alpha} = W_\alpha + (W_3 - W_4)\frac{2\Gamma}{\nu}\frac{|g_{43}|^2}{\Gamma_p^2 + \Omega_2^2}(\delta_{\alpha 4} - \delta_{\alpha 3}), \quad \Gamma_p = \Gamma\sqrt{1 + 4|g_{43}|^2/\Gamma\nu}. \tag{21}$$

Here, Γ_p is the power broadened line width of the $3 \to 4$ transition. The off-diagonal element is given by

$$\rho''_{43} = \overline{\rho}''_{43} \exp(-i\Omega_2 t), \quad \overline{\rho}''_{43} = i g_{43}(W_3 - W_4)\frac{\Gamma + i\Omega_2}{\Gamma_p^2 + \Omega_2^2}. \tag{22}$$

5. The 1st order perturbation theory

In first-order perturbation theory one has to solve the kinetic equation (17). One can check that in the first-order perturbation theory, the densities of the ortho and para molecules remain unchanged because of the relations

$$\partial\left(\sum_i \epsilon_{ii}\right)/\partial t = \partial\left(\sum_\alpha \epsilon_{\alpha\alpha}\right)/\partial t = 0. \tag{23}$$

Changes in the isomer densities n_p and n_o appear in the second-order approximation in $\hat{V}$ because they are now determined by the equations,

$$\partial n_p/\partial t = 2Re(i\epsilon_{42}V_{24}), \quad \partial n_o/\partial t = -\partial n_p/\partial t. \tag{24}$$

The density matrix element ϵ_{42} can be calculated from the system of four equations that follows from the 1st order kinetic equation (17),

$$\begin{aligned}
&\partial\epsilon_{42}/\partial t + \Gamma\epsilon_{42} + i(G_{43}\epsilon_{32} - \epsilon_{41}G_{12}) = -i n_o \rho'_{22} V_{42} + i n_p \rho''_{44} V_{42},\\
&\partial\epsilon_{41}/\partial t + \Gamma\epsilon_{41} + i(G_{43}\epsilon_{31} - \epsilon_{42}G_{21}) = -i n_o \rho'_{21} V_{42},\\
&\partial\epsilon_{32}/\partial t + \Gamma\epsilon_{32} + i(G_{34}\epsilon_{42} - \epsilon_{31}G_{12}) = i n_p \rho''_{34} V_{42},\\
&\partial\epsilon_{31}/\partial t + \Gamma\epsilon_{31} + i(G_{34}\epsilon_{41} - \epsilon_{32}G_{21}) = 0.
\end{aligned} \tag{25}$$

In isomer relaxation we have a clear separation in time scales: fast equilibration inside the ortho and para subspaces and slower, by many orders of magnitude, ortho-para relaxation. This allows us to eliminate the fast time-dependence in Eqs. (25) by the following substitutions:

$$\begin{aligned}
&V_{42} = \overline{V}_{42}e^{i\omega_{42}t}, \quad \epsilon_{42} = \overline{\epsilon}_{42}e^{i\omega_{42}t}, \quad \epsilon_{41} = \overline{\epsilon}_{41}e^{i(\omega_{42}-\Omega_1)t},\\
&\epsilon_{32} = \overline{\epsilon}_{32}e^{i(\omega_{42}+\Omega_2)t}, \quad \epsilon_{31} = \overline{\epsilon}_{31}e^{i(\omega_{42}-\Omega_1+\Omega_2)t}.
\end{aligned} \tag{26}$$

Here, the overlined terms are the time-independent factors. For the matrix elements $\overline{\epsilon}_{42}$, $\overline{\epsilon}_{41}$, $\overline{\epsilon}_{32}$, $\overline{\epsilon}_{31}$ one has a system of four algebraic equations that can be written in a compact matrix form as

$$A\hat{\overline{\epsilon}} = -i\overline{V}_{42}n_o B' + i\overline{V}_{42}n_p B'', \tag{27}$$

where we introduced the matrix A,

$$A = \begin{pmatrix} \Gamma + i\omega_{42} & ig_{12} & -ig_{43} & 0 \\ ig_{21} & \Gamma + i(\omega_{42} - \Omega_1) & 0 & -ig_{43} \\ -ig_{34} & 0 & \Gamma + i(\omega_{42} + \Omega_2) & ig_{12} \\ 0 & -ig_{34} & ig_{21} & \Gamma + i(\omega_{42} - \Omega_1 + \Omega_2) \end{pmatrix}, \tag{28}$$

and the vectors,

$$\hat{\bar{\epsilon}} = (\bar{\epsilon}_{42}, \bar{\epsilon}_{41}, \bar{\epsilon}_{32}, \bar{\epsilon}_{31})^T, \quad B' = (\rho'_{22}, \rho'_{21}, 0, 0)^T, \quad B'' = (\rho''_{44}, 0, \rho''_{34}, 0)^T. \tag{29}$$

The matrix element $\bar{\epsilon}_{42}$ that determines the isomer conversion can be calculated from Eq. (27),

$$\bar{\epsilon}_{42} = -i\overline{V}_{42}n_o(A^{-1}B')_1 + i\overline{V}_{42}n_p(A^{-1}B'')_1. \tag{30}$$

Substitution of $\bar{\epsilon}_{42}$ into Eq. (24) gives

$$\begin{aligned} \partial n_p/\partial t &= n_o|\overline{V}_{42}|^2F_o - n_p|\overline{V}_{42}|^2F_p, \\ F_o &= 2Re(A^{-1}B')_1, \quad F_p = 2Re(A^{-1}B'')_1. \end{aligned} \tag{31}$$

Because of the conservation of the total molecular density N, one has $n_o = N - n_p$ which allows us to transform Eq. (31) into its final form,

$$\partial n_p/\partial t = -|\overline{V}_{42}|^2(F_o + F_p)n_p + |\overline{V}_{42}|^2F_oN. \tag{32}$$

Here, we have assumed equal Boltzmann factors W_2 and W_4 and, consequently, equal equilibrium densities of ortho and para molecules, $n_o(0) = n_p(0) = N/2$, at $t = 0$, when the external field was switched on. The solution to Eq. (32) can be written as the sum of the time-independent part $\overline{n}_p$ and the time-dependent part $\delta n_p(t)$,

$$\overline{n}_p = \frac{F_o}{F_p + F_o}N; \quad \delta n_p(t) = \left(\frac{N}{2} - \overline{n}_p\right)e^{-\gamma t}, \tag{33}$$

where the isomer conversion rate,

$$\gamma = |\overline{V}_{42}|^2(F_o + F_p). \tag{34}$$

The steady state enrichment of the para molecules produced by the bichromatic field reads

$$\beta_p = 2\overline{n}_p/N - 1. \tag{35}$$

It should be noted that the isomer enrichment (35) does not depend on $\hat{V}$. This is a consequence of the assumption made in the model that there is only one channel of ortho-para tunneling.

6. The isomer enrichment

The resonant bichromatic field affects the densities of the ortho and para molecules and the conversion rate, γ. Below we will present the results of numerical calculations of the radiation-induced enrichment of the isomers. Where possible, the numerical parameters of the model will be chosen in such a way as to mimic the typical cases of isomer conversion in polyatomic molecules (see, for example, [16]). We will use the following set of parameters,

$$\begin{aligned} &\Gamma = 3\ \text{MHz}, \quad \nu = 0.1\Gamma, \quad \overline{V}_{24} = 0.01\ \text{MHz}, \quad g_{21} = g_{34} = 10\ \text{MHz}, \\ &\omega_{42} = 200\ \text{MHz}, \quad W_2 = W_4 = 1\cdot 10^{-2}, \quad W_1 = W_3 = 1.2\cdot 10^{-2}. \end{aligned} \tag{36}$$

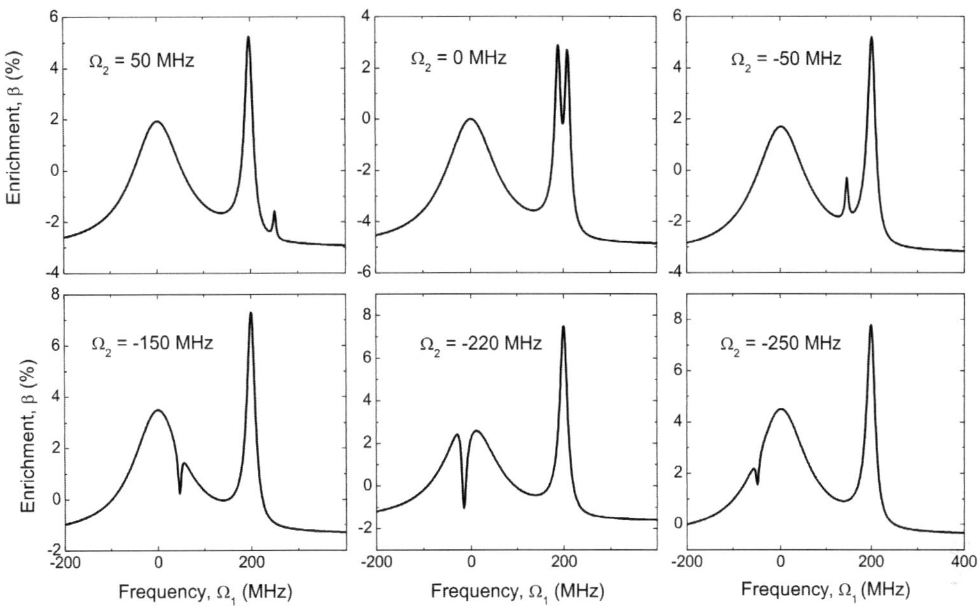

Figure 3. Dependence of the enrichment of para molecules β_p on the frequency of the first field Ω_1 at various frequencies of the second field Ω_2.

The values chosen for Γ and ν correspond to the particular case of CH_3F molecules at the gas pressure 0.1 Torr. The free conversion rate γ_0 (without an external field) calculated with the help of Eq. (3) using the parameters (36) is equal to $2 \cdot 10^{-3}s^{-1}$. This value is close to the conversion rate of $^{13}CH_3F$ isomers at the gas pressure 0.1 Torr.

The dependence of the enrichment of para molecules on the frequency Ω_1 at various frequencies Ω_2 is presented in Fig. 3. For all values of Ω_2 (except for the particular case $\Omega_2 = -200$ MHz, shown in Fig. 4), the dependence consists of three peaks. The wide peak at $\Omega_1 = 0$ can be associated with the population change of level $|2\rangle$ produced by the field $\mathbf{E}_1$. Population of this ortho state is transferred to the para state $|4\rangle$ by the intramolecular perturbation $\hat{V}$. The width of the wide peak is 55 MHz (HWHM). It is close (but not equal) to the power-broadened transition width Γ_o =63.3 MHz. On the other hand, we have checked that the absorption line width of the field $\mathbf{E}_1$ in our model is equal to Γ_o.

There are two narrow peaks in the curves of Fig. 3. Their physical nature can tentatively be explained as follows. The field $\mathbf{E}_1$ ($\mathbf{E}_2$) splits the level $|2\rangle$ ($|4\rangle$) into two sublevels having the energies [18],

$$E_{2,\pm} = E_2 + \frac{\hbar}{2}\left(\Omega_1 \pm \sqrt{\Omega_1^2 + 4|g_{21}|^2}\right), \quad (37)$$

and

$$E_{4,\pm} = E_4 + \frac{\hbar}{2}\left(\Omega_2 \pm \sqrt{\Omega_2^2 + 4|g_{43}|^2}\right). \quad (38)$$

Crossing of the level $E_{2,+}$ with the levels $E_{4,+}$ and $E_{4,-}$ produces the two narrow peaks in the isomer enrichment shown in Fig. 3. The width of the peak at $\Omega_1 \simeq \omega_{42}$ is 10 MHz. The width of the peak at $\Omega_1 \simeq \omega_{42} + \Omega_2$ is $\simeq$4.3 MHz. For the chosen parameters (36), the sign of this peak is positive (that is enrichment of para molecules) at $\Omega_2 > -\omega_{42}/2$, but becomes negative (depletion of the para molecules) at $\Omega_2 < -\omega_{42}/2$. For the case where $\Omega_2 = 0$ shown in Fig. 3, we can see from Eqs. (37) and (38) that the Stark split doublet appears at $\Omega_1 = \omega_{42} \pm |g_{43}|$.

A special case takes place at $\Omega_2 = -200$ MHz, Fig. 4. Here, there are only two peaks. The peak at $\Omega_1 = 0$ has a narrow triangular shape whereas the peak at $\Omega_1 = 200$ MHz has a width equal to $\simeq$4.3 MHz. A similar triangular shape has been reported but in a different context [19]. It was fitted to an arctan function.

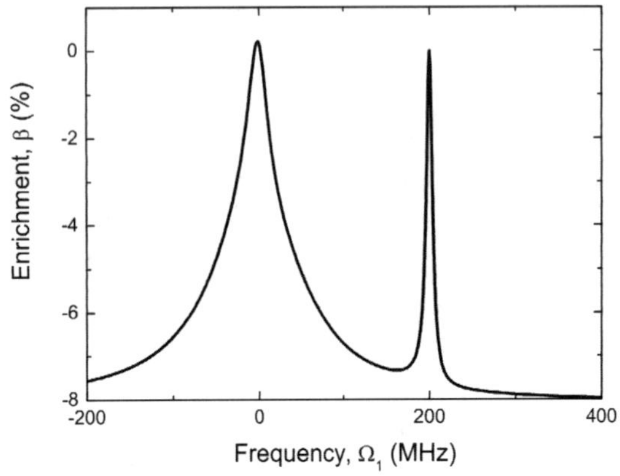

Figure 4. Dependence of the enrichment of para molecules β_p on the frequency of the first field Ω_1 at the frequency of the second field $\Omega_2 = -200$ MHz.

7. Conclusions

We have proposed a theoretical model for molecular spin isomer enrichment produced by two electromagnetic fields resonant with the molecular rotational transitions. Numerical calculations revealed richer enrichment spectra than in the case of one field [10]. In addition, our model predicts the existence of more narrow peaks in the enrichment spectra than in the case of one field excitation.

The maximum enrichment of the para molecules achieved for the parameters (36) is $\simeq$8%. This enrichment appears to be of the same order of magnitude as that created by one field. Thus there seems to be no advantage in using two fields rather than one for getting greater isomer enrichment. The main difference between the schemes explored here and those studied in atomic laser physics is that, in those systems, all the population can be trapped in the lower levels even although the electromagnetic fields are resonant with the one-photon transitions. In the molecular spin isomers considered here, the rotational levels that interact with the electromagnetic fields are part of a large manifold of levels and the rotational relaxation produces fast thermalization of the rotational level populations.

8. Acknowledgments

This work was partially supported by a Marie Curie International Research Staff Exchange Scheme Fellowship within the 7th European Community Framework Programme: Coherent optics sensors for medical applications-COSMA (grant agreement no PIRSES-GA- 2012-295264). One of the authors (P.L.C.) acknowledges the support from RFBR (grant No 12-02-01130), the Presidium of SB RAS (project No 61), and the RAS Physical Departments program Fundamental optical spectroscopy and applications.

9. References

[1] Landau L D and Lifshitz E M 1977 *Quantum Mechanics, 3rd ed.* (Oxford: Pergamon Press)
[2] Farkas A 1935 *Orthohydrogen, Parahydrogen and Heavy Hydrogen* (London: Cambridge University Press)
[3] Krasnoperov L N, Panfilov V N, Strunin V P and Chapovsky P L 1984 *JETP Lett.* **39** 122–124
[4] Kern J, Schwahn H and Schramm B 1989 *Chem. Phys. Lett.* **154** 292–298
[5] Kravchuk T, Reznikov M, Tichonov P, Avidor N, Meir Y, Bekkerman A and Alexandrowicz G 2011 *Science* **331** 319–321
[6] Zhivonitko V V, Kovtunov K V, Chapovsky P L and Koptyug I V 2013 *Ang. Chem. Int. Ed.* **52** 13251–13255
[7] Bowers C R and Weitekamp D P 1986 *Phys. Rev. Lett* **57** 2645–2648
[8] Bargon J, Kandels J and Woelk K 1993 *Z. Phys. Chem.* **180** 65–93
[9] Il'ichov L V, Hermans L J F, Shalagin A M and Chapovsky P L 1998 *Chem. Phys. Lett.* **297** 439–444
[10] Chapovsky P L 2001 *Phys. Rev. A* **63** 063402
[11] Wilson-Gordon A D 1993 *Phys. Rev. A* **48** 4639–4647
[12] Arimondo E 1996 *Progress in Optics* **35** 257–354
[13] Curl, Jr R F, Kasper J V V and Pitzer K S 1967 *J. Chem. Phys.* **46** 3220–3228
[14] Chapovsky P L 1996 *Physica A* **233** 441–448
[15] Chapovsky P L 1991 *Phys. Rev. A* **43** 3624–3630
[16] Chapovsky P L and Hermans L J F 1999 *Annu. Rev. Phys. Chem.* **50** 315–345
[17] Cacciani P, Cosleou J, Herlemont F, Khelkhal M and Lecointre J 2004 *Phys. Rev. A* **69** 032704(1–8)
[18] Rautian S G and Shalagin A M 1991 *Kinetic Problems of Nonlinear Spectroscopy* (Amsterdam: Elsevier Sci. Publ.) 439 pp.
[19] Taichenachev A V, Tumaikin A M, Yudin V I, Stähler M, Wynands R, Kitching J and Hollberg L 2004 *Phys. Rev. A* **69** 024501

Observations of complex frequency comb structure in a harmonically-pumped femtosecond optical parametric oscillator

Richard A McCracken[1], Karolis Balskus, Zhaowei Zhang and Derryck T Reid

Scottish Universities Physics Alliance (SUPA), Institute of Photonics and Quantum Sciences, Heriot Watt University, Riccarton, Edinburgh EH14 4AS, UK

E-mail: r.a.mccracken@hw.ac.uk

Abstract. Various schemes allow femtosecond optical parametric oscillators to produce pulses at harmonics of their pump laser repetition frequency, each apparently offering the possibility of generating widely-spaced, tunable frequency combs. Using a 100-MHz Ti:sapphire pump laser, we have compared two alternative optical parametric oscillator architectures, both leading to 300-MHz pulses but one configured in a cavity three times shorter than the pump laser and the other in a cavity one-third longer. Heterodyne measurements between the pump and each of these two systems show that they possess different carrier-envelope offset characteristics, with implications on the coherence and stabilization of the resulting combs.

1. Introduction

Synchronously-pumped optical parametric oscillators (OPOs) are used to convert femtosecond or picosecond pump pulses to signal and idler pulses of longer wavelengths in a resonant cavity whose optical length matches that of the pump laser [1]. The broadband wavelength coverage offered by OPOs has made them attractive as a source of laser frequency combs, with applications in precision spectroscopy [2,3] and coherent pulse synthesis [4]. Extending these OPO frequency combs to high repetitions rates would allow their use in applications where it is important to be able to resolve individual comb modes, such as in astronomical spectrograph calibration [5,6]. Strict synchronous pumping of an OPO limits its repetition frequency to exactly that of its pump laser, however operating the OPO at a harmonic of the pump repetition frequency is possible by employing a cavity whose length is an integer [7–9] or integer fraction [10–13] of the pump laser length. Here, we theoretically and experimentally compare two alternative routes to obtaining 300-MHz operation from an OPO pumped by a 100-MHz femtosecond pump laser, and show important practical differences in their frequency-comb characteristics, which have so far remained unexplored.

2. Harmonic operation of synchronously-pumped optical parametric oscillators

For a fundamentally modelocked pump laser of round-trip optical length L_{PUMP}, harmonic operation of a signal-resonant synchronously-pumped OPO is achieved when the OPO cavity length is

[1] To whom any correspondence should be addressed.

$$L_{OPO} = L_{PUMP} \times \frac{n}{Q} \tag{1}$$

where n and Q are positive integers and n/Q is an irreducible fraction [12]. Configured in this way, the OPO produces signal pulses at a repetition frequency Q times higher than that of the pump laser (f_{REP}), and fundamental synchronous pumping is obtained when $n = Q = 1$. As illustrated in Figure 1, two categories of harmonic operation ($Q > 1$) can be identified, depending on the choice of n.

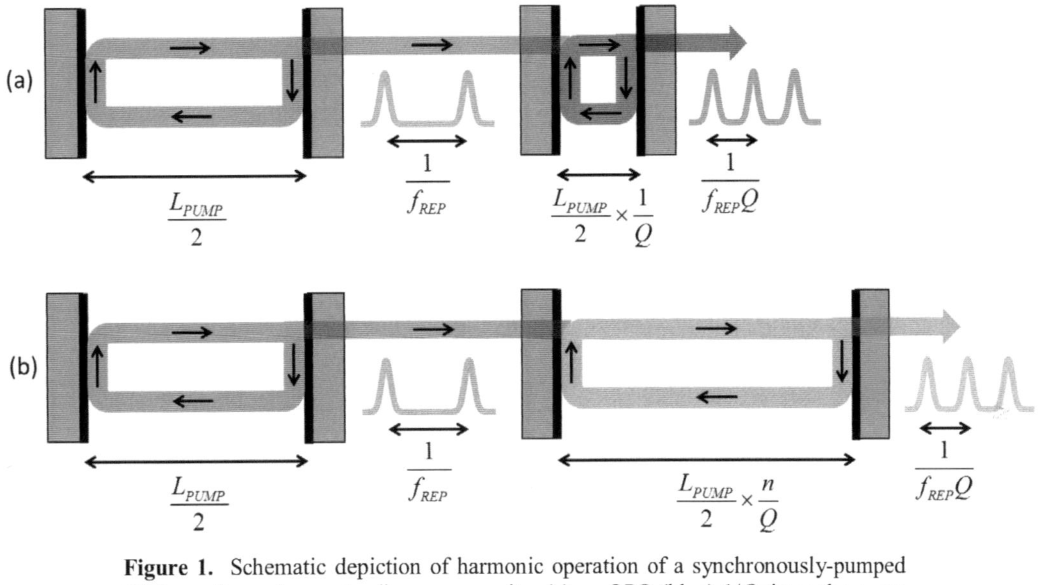

Figure 1. Schematic depiction of harmonic operation of a synchronously-pumped OPO, configured as a standing-wave cavity. (a) an OPO (blue) $1/Q$ times the pump cavity length (green); (b) an OPO (orange) n/Q times the pump cavity length (green). For clarity, the OPO gain medium is not shown.

2.1. Simple harmonic pumping

When $n = 1$, the OPO cavity is shorter than the pump cavity and a single pulse is resonant, experiencing gain when it coincides with a pump pulse every Q round-trips of the OPO cavity, as depicted in Figure 1a. Operation in this regime presents design challenges in obtaining good mode-matching at the small spot sizes needed to achieve the necessary intensities for parametric oscillation, in turn requiring the use of very small radius-of-curvature intracavity optics.

2.2. Vernier harmonic pumping

The case of $n > 1$ (Figure 1b) corresponds to n independent, equidistant pulses resonating in the OPO cavity, with each pulse interacting with a pump pulse every Q round trips, outputting an interleaved pulse train with a repetition frequency Qf_{REP}. If $Q > n$ the OPO is shorter than the pump cavity, but if $Q < n$ the OPO is longer. As demonstrated by Kokabee *et al* [14], designs where both Q and n are large allow high harmonic repetition frequencies to be obtained from a cavity which is sufficiently long to permit optimum pump-signal mode-matching and the insertion of dispersion-compensating optics.

3. Frequency-comb structure of harmonically-pumped optical parametric oscillators

It is well established that the Fourier transform of a regularly spaced train of pulses produced by a modelocked laser corresponds to a comb of equally spaced frequencies, separated by their repetition frequency [15]. The same is true for a strictly synchronously pumped OPO, which produces distinct pump, signal and idler frequency combs which all share the spacing of the pump repetition frequency

(Figure 2a). The frequency comb structure of both short and extended cavity OPOs producing pulse trains at harmonics of the pump repetition frequency Qf_{REP} will be examined theoretically.

In the case of simple harmonic pumping ($n = 1$ and $Q > 1$) there is a single circulating pulse in the OPO that experiences gain after every Q round trips (Figure 2b), with ring-down losses writing sidebands at $\pm f_{REP}$ onto comb lines spaced at Qf_{REP}. When $n > 1$ and $Q > 1$ there are n circulating pulses that each experience gain after every Q round trips; such a configuration can be considered as n OPOs operating in a single cavity (Figure 2c). Each distinct OPO comb has the same CEO frequency, which is determined by the OPO cavity dispersion. As with the simple harmonic case, sidebands are written onto the comb lines due to ring-down losses. Here, instability in the sideband structure is expected as the phases of the circulating pulses are unrelated; the initial phase of each pulse is random, determined as each signal pulse train builds up from quantum noise.

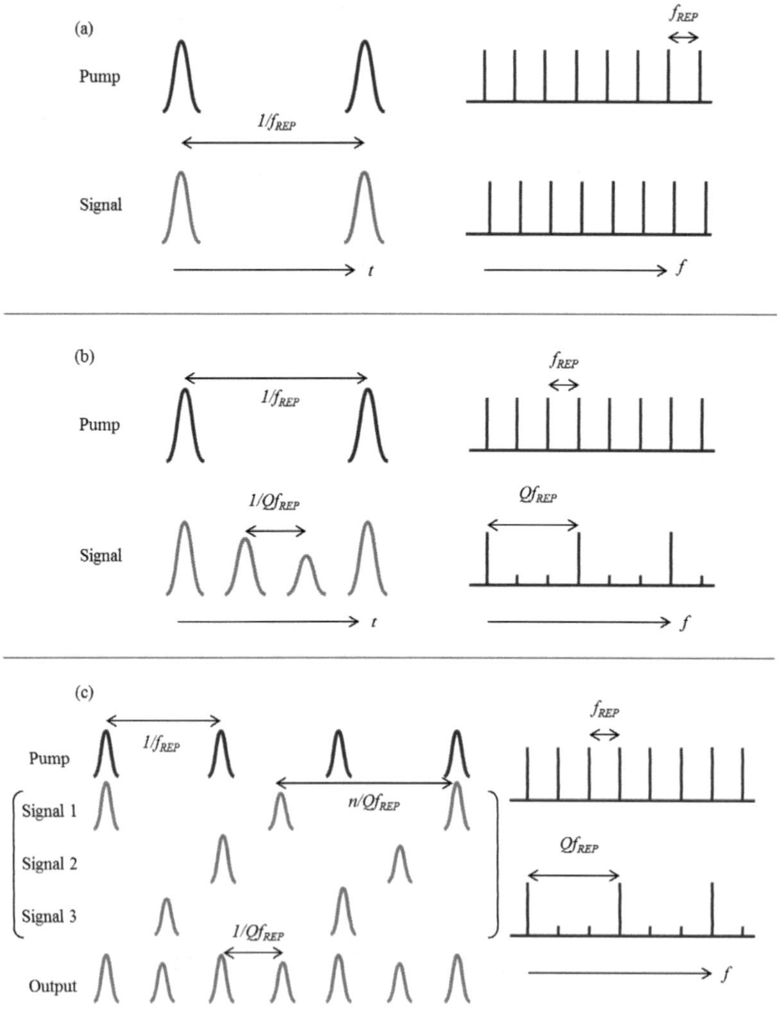

Figure 2. OPO frequency comb structure. (a) strictly synchronous pumping; (b) simple harmonic pumping; (c) Vernier harmonic pumping.

4. Experiment

We experimentally tested the description presented above by using the configuration illustrated in Figure 3. A 100-MHz Ti:sapphire laser (*Griffin-I, KM Labs*) was used to pump two PPKTP OPOs with both operated at 300 MHz, one in a cavity three times shorter than the pump laser ($n = 1$ and $Q = 3$) and the other in a cavity four-thirds longer ($n = 4$ and $Q = 3$). Practical measurement of the OPO comb CEO frequency normally proceeds by heterodyning a portion of the pump supercontinuum with common wavelengths in the OPO sum-frequency or second-harmonic outputs [16–18]. In our measurement, second-harmonic (SHG) light from the OPO was interfered with a pump supercontinuum generated in a 20-cm length of photonic crystal fiber (PCF; *NKT Photonics, NL-2.0-750*), producing a heterodyne beat observable on an avalanche photodiode (APD; *Hamamatsu, C5331-11*). Fast-photodiode measurements of the second-harmonic signal output were monitored to confirm the OPO repetition frequency was 300 MHz.

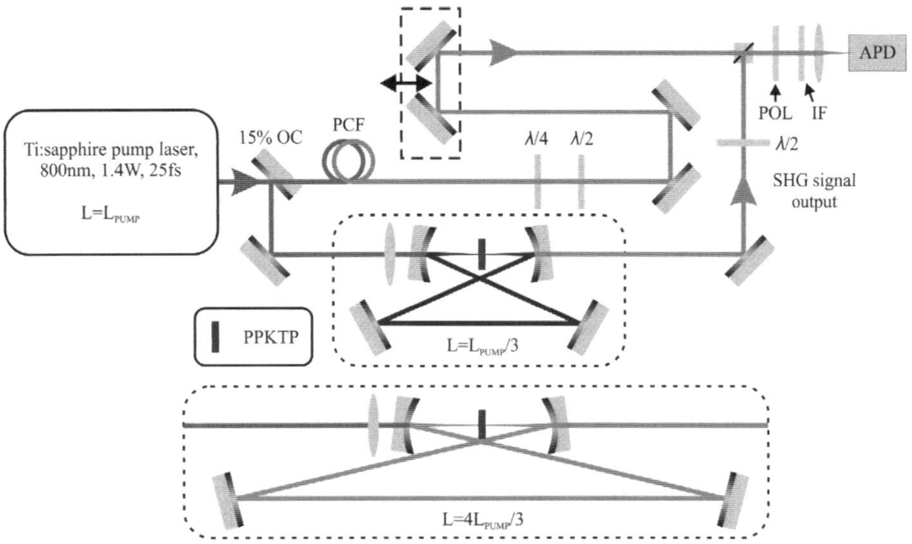

Figure 3 Experimental schematic. APD, avalanche photodiode; IF, interference filter; OC, output coupler; PBS, polarizing beam splitter; PCF, photonic crystal fiber; POL, polarizer.

Radio-frequency (RF) spectra detected by the APD were recorded from the two OPO cavities and are shown in Figure 4a,c. Pairs of sidebands encoding the difference between the pump and OPO CEO frequencies were observed centered around multiples of f_{REP}/n. For the short OPO cavity ($n = 1$ and $Q = 3$) sidebands were observed at multiples of 100 MHz ($f_{REP}/1$). In the extended OPO cavity ($n = 4$ and $Q = 3$) sidebands were observed at multiples of 25 MHz ($f_{REP}/4$). The dense RF spectrum produced when $n > 1$ presents greater practical challenges in isolating the beat frequency required for CEO locking than are encountered when $n = 1$. RF spectra of fast-photodiode measurements of the second-harmonic signal output appear in Figure 4b,d, showing that in the time domain the intensity of the fields produced by the two OPOs is equivalent.

The experimentally observed RF spectra can be explained as follows. The heterodyne signal between the pump super-continuum and the second-harmonic of the OPO comb contains frequencies described by the general equation

$$\left(Nf_{REP} + f_{CEO}^{p}\right) \pm \left(Mf_{CEP}^{2s} + f_{CEO}^{2s}\right) \tag{2}$$

where N and M are integers; $f_{REP} = 1/T$ is the repetition frequency of the pump pulses and f^{p}_{CEO} is their CEO frequency; and $f^{2s}_{REP} = Q/nT$ is the repetition frequency of the OPO signal second-harmonic pulses (equal to that of the fundamental signal pulses) and f^{2s}_{CEO} is their CEO frequency. Equation (2) can therefore be rewritten and its terms separated into those associated with the CEO and repetition frequencies:

$$\left(\frac{nN \pm Q}{n}\right) f_{REP} + \left(f^{p}_{CEO} + f^{2s}_{CEO}\right). \qquad (3)$$

For the experimental value of f_{REP} = 100 MHz, (3) predicts that the heterodyne signal when $n = 4$ and $Q = 3$ should contain pairs of frequencies $f_{CEO} \pm f^{2s}_{CEO}$ centered around integer multiples of 25 MHz, as observed experimentally. By comparison, when $n = 1$ and $Q = 3$, pairs of CEO frequencies are predicted at multiples of 100 MHz, again as observed experimentally.

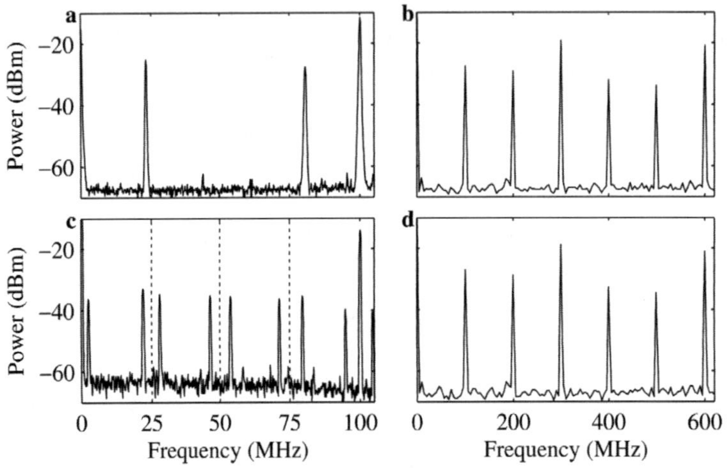

Figure 4 RF spectra from differently constructed 300-MHz harmonic OPOs. a, b, beat frequency and fast photodiode measurements for $n = 1$ and $Q = 3$. c, d, beat frequency and fast photodiode measurements for $n = 4$ and $Q = 3$. Dashed lines indicate multiples of 25 MHz.

5. Conclusions

We have investigated the frequency comb structure associated with two different approaches to obtaining harmonic OPO operation, presenting a general theoretical description of the comb structure produced under harmonic pumping and showing that it predicts the heterodyne frequencies produced experimentally when interfering the pump and OPO combs. The dense heterodyne beats produced when $n > 1$ can be expected to present practical problems when implementing CEO locking, therefore the method of choice for obtaining a harmonic OPO frequency comb should be simple harmonic pumping ($n = 1$) of an OPO cavity which is Q times shorter than its pump laser.

Acknowledgements
This research was carried out under the METROCOMB project, which has received funding from the European Community's Seventh Framework Programme (FP7/2007-2013) under Grant Agreement no. 605057.

References

1. D. C. Edelstein, E. S. Wachman, and C. L. Tang, "Broadly tunable high repetition rate femtosecond optical parametric oscillator," Appl. Phys. Lett. **54**, 1728–30 (1989).
2. N. Leindecker, A. Marandi, R. L. Byer, and K. L. Vodopyanov, "Broadband degenerate OPO for mid-infrared frequency comb generation," Opt. Express **19**, 6296–302 (2011).
3. Z. Zhang, T. Gardiner, and D. T. Reid, "Mid-infrared dual-comb spectroscopy with an optical parametric oscillator," Opt. Lett. **38**, 3148–3150 (2013).
4. J. Sun, B. J. S. Gale, and D. T. Reid, "Coherent synthesis using carrier-envelope phase-controlled pulses from a dual-color femtosecond optical parametric oscillator," Opt. Lett. **32**, 1396–8 (2007).
5. F. Quinlan, G. Ycas, S. Osterman, and S. A. Diddams, "A 12.5 GHz-spaced optical frequency comb spanning >400 nm for near-infrared astronomical spectrograph calibration," Rev. Sci. Instrum. **81**, 063105 (2010).
6. G. G. Ycas, F. Quinlan, S. A. Diddams, S. Osterman, S. Mahadevan, S. Redman, R. Terrien, L. Ramsey, C. F. Bender, B. Botzer, and S. Sigurdsson, "Demonstration of on-sky calibration of astronomical spectra using a 25 GHz near-IR laser frequency comb," Opt. Express **20**, 6631–43 (2012).
7. D. T. Reid, C. McGowan, W. Sleat, M. Ebrahimzadeh, and W. Sibbett, "Compact, efficient 344-MHz repetition-rate femtosecond optical parametric oscillator," Opt. Lett. **22**, 525–7 (1997).
8. B. Ruffing, A. Nebel, and R. Wallenstein, "All-solid-state cw mode-locked picosecond KTiOAsO4 (KTA) optical parametric oscillator," Appl. Phys. B Lasers Opt. **67**, 537–544 (1998).
9. K. A. Ingold, A. Marandi, C. W. Rudy, K. L. Vodopyanov, and R. L. Byer, "Fractional-length sync-pumped degenerate optical parametric oscillator for 500-MHz 3-μm mid-infrared frequency comb generation," Opt. Lett. **39**, 900–903 (2014).
10. J. Jiang and T. Hasama, "Harmonic repetition-rate femtosecond optical parametric oscillator," Appl. Phys. B Lasers Opt. **74**, 313–317 (2002).
11. J. Jiang and T. Hasama, "Synchronously pumped femtosecond optical parametric oscillator based on an improved pumping concept," Opt. Commun. **220**, 193–202 (2003).
12. A. Esteban-Martin, O. Kokabee, K. Moutzouris, and M. Ebrahim-Zadeh, "High-harmonic-repetition-rate 1 GHz femtosecond optical parametric oscillator pumped by a 76 MHz Ti:sapphire laser," Opt. Lett. **34**, 428–430 (2009).
13. O. Kimmelma, S. Chaitanya Kumar, A. Esteban-Martin, and M. Ebrahim-Zadeh, "Multi-gigahertz picosecond optical parametric oscillator pumped by 80-MHz Yb-fiber laser," Opt. Lett. **38**, 4550 (2013).
14. O. Kokabee, A. Esteban-Martin, and M. Ebrahim-Zadeh, "Extended-cavity, tunable, GHz-repetition-rate femtosecond optical parametric oscillator pumped at 76 MHz," Opt. Express **17**, 15635–15640 (2009).
15. S. T. Cundiff, "Phase stabilization of ultrashort optical pulses," J. Phys. D. Appl. Phys. **35**, R43–R59 (2002).
16. Y. Kobayashi and K. Torizuka, "Measurement of the optical phase relation among subharmonic pulses in a femtosecond optical parametric oscillator," Opt. Lett. **25**, 856–8 (2000).
17. J. Sun, B. J. S. Gale, and D. T. Reid, "Dual-color operation of a femtosecond optical parametric oscillator exhibiting stable relative carrier-envelope phase-slip frequencies," Opt. Lett. **31**, 2021–3 (2006).
18. T. I. Ferreiro, J. Sun, and D. T. Reid, "Locking the carrier-envelope-offset frequency of an optical parametric oscillator without f-2f self-referencing," Opt. Lett. **35**, 1668–70 (2010).

Nanocrystal fluorescence in photonic bandgap microcavities and plasmonic nanoantennas

Svetlana G. Lukishova[1], Justin M. Winkler[2], Dilyana Mihaylova[1], Andreas Liapis[3], Luke J. Bissell[4], , David Goldberg[5], Vinod M. Menon[5], Zhimin Shi[6], Robert W. Boyd[1,2,7], Guanuing Chen[8], Paras Prasad[8]

[1]The Institute of Optics, University of Rochester, Rochester, NY 14627, USA
[2]Department of Physics and Astronomy, University of Rochester, Rochester, NY 14627, USA
[3]Brookhaven National Laboratory, Upton, NY 11973, USA
[4]Air Force Research Laboratory, Wright-Patterson Air Force Base, OH 45433, USA
[5]Department of Physics, the City College of New York, New York, NY 10031, USA
[6]Department of Physics, University of South Florida, Tampa, FL 33620, USA
[7]Department of Physics and School of Electrical Engineering and Computer Science, University of Ottawa, Ottawa, Ontario K1N 6N5, Canada
[8]Institute for Lasers, Photonics and Biophotonics and Department of Chemistry, University at Buffalo, USA, Buffalo NY 14260, USA

E-mail: sluk@lle.rochester.edu

Abstract. Results are presented here towards robust room-temperature single-photon sources based on fluorescence in nanocrystals: colloidal quantum dots, color-center diamonds and doped with trivalent rare-earth ions (TR3+). We used cholesteric chiral photonic bandgap and Bragg-reflector microcavities for single emitter fluorescence enhancement. We also developed plasmonic bowtie nanoantennas and 2D-Si-photonic bandgap microcavities.

1. Introduction

Single-photon sources (SPSs) [1-12], producing nonclassical light, are pivotal components in quantum communication technology. They efficiently produce individual photons separated in time with no chance for overlap of photons (antibunching). For photon antibunching the second order coherence function $g^{(2)}(t) = \dfrac{<I(t)I(t+\tau)>}{<I(\tau)>^2}$ should have a minimum at interphoton time $t = 0$. If $g^{(2)}(0) <$ 0.5, it indicates the absence of photon pairs. Here $I(t)$ is intensity, t and τ are two moments of time, and averaging is time averaging. For common light sources (bulbs, sun light) and in lasers, multiple (nonseparated) photons cannot be avoided even in very faint sources. Using photons that are all temporally separated, absolutely secure quantum communication will prevent any potential eavesdropper (Eve) from intercepting a message without the receiver noticing. Bits of transferred information are coded in the photon polarization state. If the source emitted more than one photon at a time, it would be possible for Eve to extract one of these photons without disturbing the other photons that would reveal Eve's presence. There exist some schemes with classical sources, e.g., using decoy states to reduce leakage of information to an eavesdropper, but for long distance quantum communication with quantum repeaters single-photon sources are important. Some quantum computation protocols also require single photons. In addition, single photons are very important in metrology and fundamental physics experiments, such as in interference from uncoupled light sources and entanglement-state generation, in shedding light on wave-particle duality as well as in fundamental issues of quantum measurement and uncertainty.

Furthest in development are SPSs based on heterostructured quantum dots placed into 2D-photonic crystal microcavities or into micropillars. In spite of their many advantages in studying cavity quantum electrodynamics, these SPSs are very expensive in fabrication and require cryogenic temperatures. Therefore, many researchers aim for *room-temperature* SPSs. The most popular single emitters for those are dye molecules, colloidal nanocrystal quantum dots and rods, and various color-centers in nanodiamonds. To enhance the single-photon count rate (the bit rate of future quantum information systems) and improve SPS directionality, coupling efficiency into a fiber system and polarization selectivity, single emitters should be embedded in microcavities, photonic crystals and other micro- or nano-structures (e.g., plasmonic nanoantennas, metamaterials, etc.). In addition, an ideal SPS should afford ease of use in a Space environment for the future global quantum communication network.

It should be noted, that similar nano/microstructures can be also used for compact microlasers and on-chip integrated microlasers. The difference being the concentration of dopant: low dopant concentration for SPSs whereas microlasers require high concentration of the same emitters.

The enhancement of spontaneous emission in microcavities is characterized by the Purcell factor [13,14]. This factor is defined as $F_p = \gamma_c/\gamma_0$, where γ_c and γ_0 are the emitter decay rates with and without a cavity, respectively:

$$F_P = \frac{3Q(\lambda / n)^3}{4\pi^2 V_0}\left(\frac{|\boldsymbol{d}\cdot \boldsymbol{E}(r)|}{|d|\cdot|E(r)|}\right)^2, \qquad (1)$$

where Q is the cavity quality factor, λ is the free-space wavelength of light, n is the cavity refractive index, V_0 is the cavity mode volume, d is the emitter dipole moment, and $E(r)$ is the local electric field at the position r of the emitter.

Eq. (1) predicts enhancement or diminishing of F_p, depending on the ratio $Q\lambda^3/(V_0 n^3)$, and on how the incident field is aligned with respect to the dipole moment, d. Because high Q values and mode volumes that are small compared to $(\lambda/n)^3$ are required to obtain $F_p >>1$, it is challenging to increase γ_c at optical frequencies. In recent years, however, semiconductor processing technology and other innovative procedures have enabled enhanced radiative decay rates in microcavities at visible and IR wavelengths, using both cryogenic and room temperature implementations. In addition to increasing γ_c, another advantage of placing the emitter inside a microcavity is the increased coupling efficiency into a single-mode fiber and the polarization selectivity in certain geometries of microcavities or nanoantennas. Even though the Purcell factor is the standard figure of merit for spontaneous emission rate enhancement, care must be taken when using Eq. (1) to describe emission rate enhancements for plasmonic resonances. In the case of plasmonic nanoantennas, the Purcell factor is useful only as a qualitative design guideline.

The structure of this paper is as follows. Section 2 describes single-photon generation and characterization unit of the Institute of Optics, University of Rochester. In Section 3, our results towards room-temperature single-photon sources based on nanocrystal fluorescence (colloidal quantum dots and trivalent rare-earth ions) in 1-D chiral liquid crystal photonic bandgap structures are presented. Section 4 is devoted to a single-photon source based on a Bragg-reflector microcavity with a defect layer doped with nanocrystal quantum dots. In Section 5, gold bowtie plasmonic nanoantenna arrays and nanocrystal manipulation by an atomic force microscope tip are described. Section 6 contains the results of Si 2D-photonic bandgap structures for 1.5 μm wavelength.

Journal of Physics: Conference Series **594** (2015) 012005 doi:10.1088/1742-6596/594/1/012005

2. Single-Photon Generation and Characterization Unit

Figure 1 shows general schematic of experimental set up for single photon generation, detection and characterization. A single-photon generation unit consists of a confocal fluorescence microscope with a high numerical aperture (N.A.) objective that focuses a pulsed or cw-laser beam on a cover glass slip with a single-emitter in a microcavity. The sample is placed on a microscope table located on a piezo-translation stage for a raster scan of the sample around a focused laser beam. Fluorescence light is collected by the same objective. A dichroic mirror reflects laser light and transmits fluorescence light. Interference filters with 6-9 orders of magnitude attenuation further reject an excitation laser light. Single-photon detection and characterization unit represents a Hanbury Brown and Twiss intensity interferometer (correlator), which consists of a nonpolarizing beamsplitter, two single-photon counting, thermoelectrically cooled avalanche photodiode modules (APDs), and start-stop electronics to measure time intervals between two consecutive photons in pairs to prove photon antibunching. The characterization unit also contains electronics and software for confocal fluorescence imaging of single emitters. To increase confocal fluorescence imaging contrast, 150-μm detector areas serve as confocal microscope pinholes.

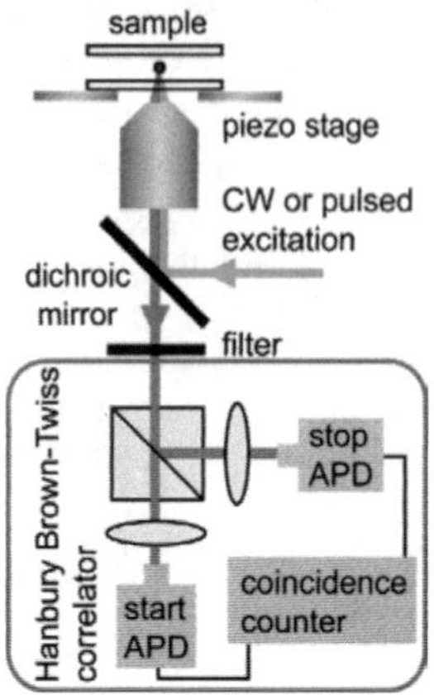

Figure 1. Schematic of a single-photon generation, detection and characterization unit (APD – single-photon counting avalanche photodiode module). A Hanbury Brown and Twiss correlator is used for photon antibunching measurements.

In our experiments we used a home-built confocal fluorescence microscope based on a Nikon TE2000-U inverted microscope with several output ports for diagnostics. We excite our samples with several laser sources: 76 MHz repetition rate, 6 ps pulse duration, 532-nm light from a Lynx mode-locked laser (Time-Bandwidth Products Inc.); 633-nm cw He-Ne laser; 976-nm diode laser; 514 nm cw argon ion laser. Two Si APDs SPCM AQR-14 (Perkin Elmer) were used as a single-photon counting detectors. The time intervals between two consecutively detected photons in separate arms of a Hanbury Brown and Twiss interferometer were measured by a TimeHarp 200 time correlated single-photon counting card with start and stop inputs. A low-light-level, electron-multiplying (EM), thermoelectrically cooled CCD-camera iXon DV 887 ECS-BV (Andor Technologies) was used both for spectral measurements with a Princeton Instruments (SpectraPro2150i) spectrometer, and for wide-field imaging of single-emitter fluorescence. Circular polarization measurements were accomplished using an achromatic quarter-wave plate and linear polarizer. Our polarization data were calibrated for the effect of the optical system on circularly-polarized light. We also performed a radiometric calibration of the optical system, including the spectrometer grating reflectivity dependence on polarization.

3. 1-D Photonic Bandgap Cholesteric Liquid Crystal Structures

Planar-aligned cholesteric liquid crystal (CLC) structures exhibit a chiral 1-D photonic bandgap for the handedness of circularly-polarized light where the electric field vector follows the rotation of the CLC molecular director. Circular polarization with the opposite handedness does not "see" a bandgap and propagates as in an isotropic medium. The stop band is centered at wavelength $\lambda_o = p(n_e + n_o)/2$, where p is the pitch of the CLC spiral structure, and n_e and n_o are extraordinary and ordinary refractive indexes. The bandwidth of the transmission stop band is given by $\Delta\lambda \sim p(n_e - n_o)$. Lasing in such CLC structures occurs at a band edge of this stop band [15-17]. In a similar way, to enhance single-emitter fluorescence in such a microcavity, the single-emitter fluorescence maximum should correspond to a band edge of a CLC structure stop band.

We prepared resonant planar-aligned cholesteric structures for single emitters with fluorescence in visible and near-IR. We used both (1) cholesteric oligomeric powders and (2) monomeric nematic fluids with chiral additives [18-29]. Prepared samples were analyzed using our homemade confocal microscope setup (Figure 1). By placing an achromatic quarter waveplate and linear polarizer in front of the spectrometer, we were able to filter for right handedness (RH) and left handedness (LH) of circularly polarized (CP) fluorescence.

3.1. Nanocrystal quantum dots in chiral photonic bandgap microcavities

Figure 2 shows our results on circularly polarized microcavity resonance in colloidal nanocrystal quantum dot (NQD) fluorescence doped in a CLC photonic bandgap microcavity [24]. We used left-handed cyclosiloxane oligomeric CLC powder from Wacker Chemie and produced a planar-aligned glassy CLC structure doped with CdSeTe NQD, Qdot 800 ITK organic, Invitrogen, with fluorescence maximum at 790 nm.

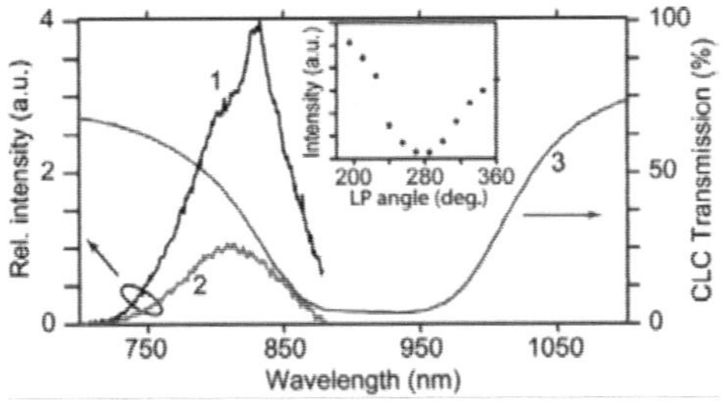

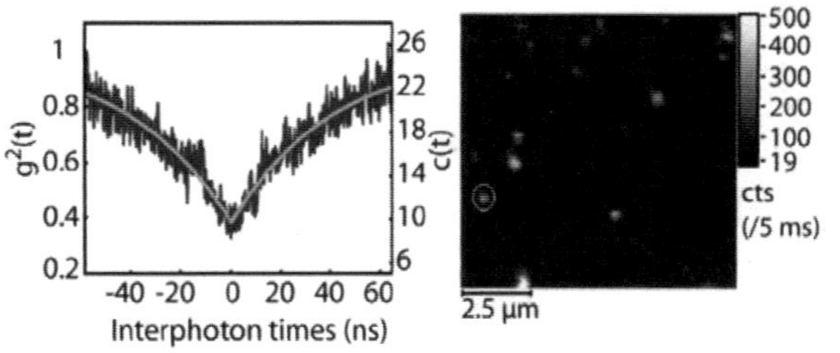

LEFT: **Figure 2**. Circularly-polarized fluorescence resonance from nanocrystal quantum dots (NQDs) doped in a glassy CLC microcavity (633-nm, cw laser excitation). Curve 1: LHCP fluorescence spectrum of the NQDs with a resonance at 833 nm. Curve 2: RHCP fluorescence spectrum for the same NQDs. Curve 3: selective transmission of LHCP light through CLC microcavity. Inset: Dependence of resonance peak intensity on rotation of a linear polarizer (LP) after a fixed quarter wave plate. CENTER and RIGHT: **Figure 3**. Center: Raw coincidence counts *c(t)* (right-hand scale) and $g^{(2)}(t)$ (left-hand scale), showing antibunching (dip at t = 0). Right: Confocal fluorescence microscopy image of single CdSeTe NQDs in a glassy CLC photonic bandgap microcavity.

The resulting fluorescence spectra can be seen in Figure 2. LHCP light experienced the photonic bandgap and therefore the black LHCP curve in Figure 2 (curve 1) shows the microcavity resonance, indicating that the LHCP light coupled to the cavity mode. The LHCP fluorescence had a center wavelength of 833 nm and a FWHM of 16 nm (Q ~ 50), as compared to a FWHM of 76 nm for RHCP fluorescence. The center wavelength of this resonance roughly matches the edge of the photonic stop band, centered at 910 nm, and shown in Figure 2 by the blue curve 3. The observed RHCP fluorescence is shown in red in Figure 2 (curve 2) and was less intense due to not experiencing the CLC microcavity, showing no sign of line narrowing. The maximum intensity ratio between LHCP and RHCP was a factor of 4.9 [24]. Figure 3, left shows antibunching with $g^{(2)}$ (0) = 0.382 of CdSeTe NQD fluorescence under 633 nm laser excitation in a similar glassy CLC photonic bandgap microcavity at lower concentration of NQDs than in the sample of Figure 2. A white circle in Figure 3, right marks a fluorescence image of a single NQD used for antibunching measurements. We also observed antibunching from NV color-center nanodiamonds in a monomeric (fluid) CLC chiral photonic bandgap microcavity [8, 25].

3.2. Rare-earth-doped nanocrystals in chiral photonic bandgap microcavity

A monomeric CLC microcavity made from a mixture of E7 and CB 15 was doped with rare-earth Er^{3+} and Yb^{3+} ions in 20 nm to 30 nm sized $NaYF_4$ nanocrystals with 20% Yb and 2% Er. When these ions were excited using a cw, 976 nm diode laser at incident powers of ~500 μW, we were able to observe upconverted fluorescence of Er^{3+} in visible, as shown in Figure 4, left (red solid curve). The blue dashed curve shows the spectral transmission of this CLC microcavity, measured with unpolarized light.

Energy levels of Er^{3+} and Yb^{3+} and the upconversion excitation scheme under 980 nm excitation are shown in Figure 4, right. The emission lines observed were attributed to the transitions $^2H_{11/2}$, $^4S_{3/2}$

→ $^4I_{15/2}$ (green) and $^4F_{9/2}$ → $^4I_{15/2}$ (red) of the Er^{3+} ions. The populations of upper levels in Er^{3+}occur due to an efficient energy transfer from the Yb^{3+} to the Er^{3+}. This fluorescence was measured to have a circular polarization at 680 nm. It should be noted that Er^{3+} ions also have a fluorescence line at ~1.5 μm, which is important for fiber optical communication.

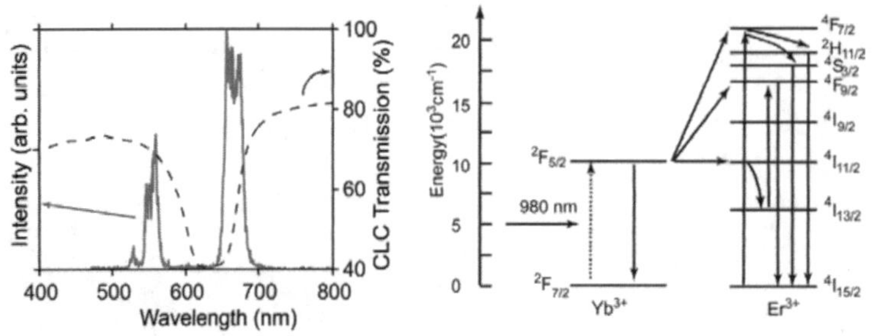

Figure 4. Left: Red solid lines: Fluorescence spectrum of Er^{3+} ions doped in $NaYF_4$ nanocrystals dispersed in a chiral CLC microcavity (E7 and CB15). The blue dashed curve shows the spectral transmission of a CLC microcavity measured with an unpolarized light. Right: Energy levels of Er^{3+} and Yb^{3+} and the upconversion excitation scheme under 980 nm excitation [see Z. Wang et al., Solid State Commun. 144, 255-258 (2007)].

4. Bragg Reflector Microcavity with a Defect Layer Doped with Nanocrystal Quantum Dots

We fabricated a 1-D photonic-bandgap microcavity consisting of a layer of colloidal CdSe/ZnS (core/shell) NQDs spin-cast between two distributed-Bragg reflector (DBR) mirrors (see Figure 5, left) [31]. The DBR mirrors, comprised of alternating layers of SiO_2 (n_{SiO2} ~ 1.54) and SiN_x (n_{SiNx} ~ 2.02).

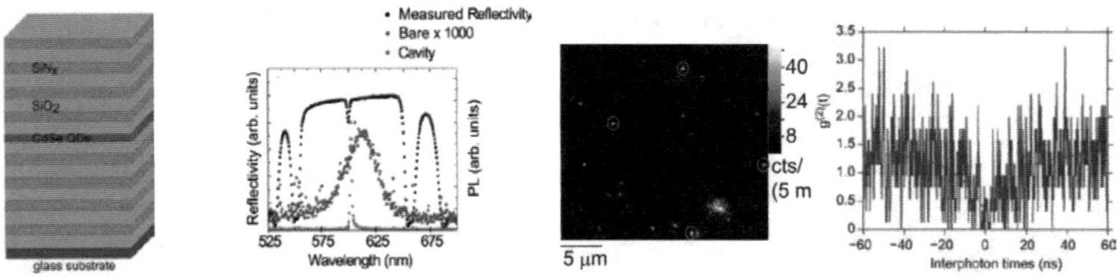

Figure 5. Left: Schematics of a Bragg-reflector microcavity with a defect layer with single NQDs. Center, left: Cavity reflectivity (black curve), fluorescence spectra from NQDs without a microcavity (blue curve) and between two Bragg mirrors (narrow red curve). Center, right: Confocal microscope fluorescence images of several single NQDs (white circles) in a Bragg-reflector microcavity. Right: Coincidence count histogram from single NQDs within the cavity, indicating photon antibunching.

We made two such DBR microcavities with NQD-doped defect layers, using high concentration (~ 1 μM) doping (sample 1), and low-concentration (~ 1 nM) doping (sample 2). Figure 5 (center, black curve) shows the measured reflectivity spectrum of sample 1, with a cavity mode at 600 nm. Figure 5 (center, blue curve) shows the NQD fluorescence spectrum without the second DBR mirror. Figure 5 (center, red curve) shows the spectrum of the NQDs in the DBR microcavity, indicating coupling to the cavity mode, with Q ~ 300. Figure 5 (center) shows confocal microscope fluorescence image of NQDs in a sample 2 under 532 nm laser excitation. For further details of the experimental setup, see Ref. [8]. Figure 5, right, shows the sum of the coincidence count histograms of six NQDs (some of them are marked by the white circles in Figure 5, center). The dip in the histogram at t = 0 indicates photon antibunching.

5. Plasmonic Nanoantennas for Single-Photon Source Applications

Using high-precision electron-beam lithography, we fabricated gold bowtie nanoantennas (Figure 6, left). Figure 6, center and right shows our results in antenna polarization selectivity. Figure 7, left shows confocal microscope fluorescence image of NQDs spin-coated on a cover glass slip with gold bowtie nanoantenna arrays. A photoluminescence from gold nano-antenna (periodic array in Figure 7, left) is weak in comparison with NQDs fluorescence. Some NQDs are located within the gaps of nanoantennas (e.g., marked with the stars). Figure 7, right shows fluorescence antibunching curve of one of these NQDs within a bowtie nanoantenna. The second dot fluorescence also reveals antibunching.

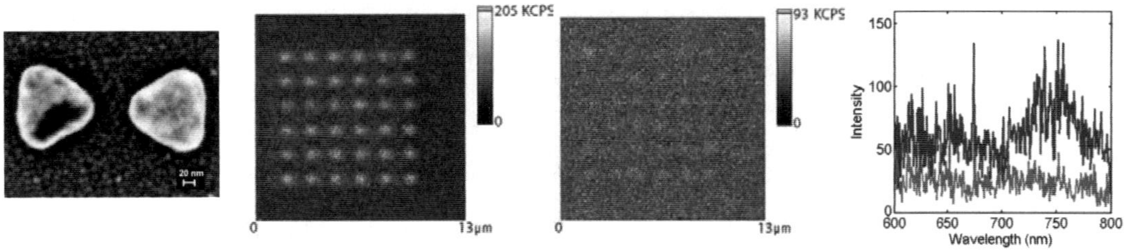

Figure 6. Left: Scanning electron micrograph of a lithographically defined gold bowtie nanoantenna structure with a 60-nm gap. Center: Confocal fluorescence microscope micrographs of gold bowtie nanoantenna array (30-nm-bowtie-gap) for polarization parallel to the bowtie axis (left) and perpendicular to it (right) under 633 nm laser excitation. Right: Photoluminescence spectra from a single gold bowtie nanoantenna under 532 nm laser excitation. The two curves show orthogonal photoluminescence polarizations.

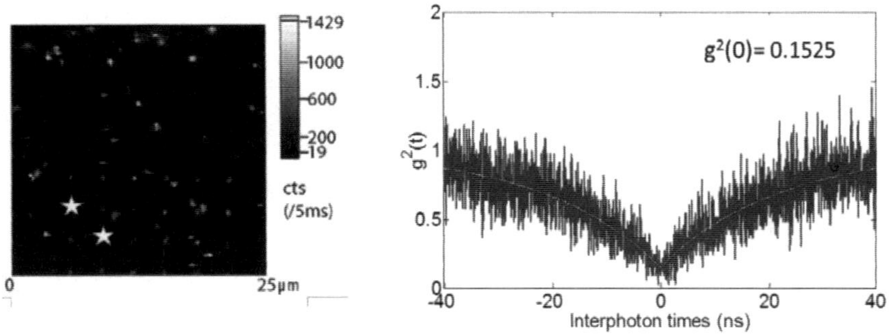

Figure 7. Left: Confocal microscope image of fluorescence from NQDs spin-coated on a cover slip with a bowtie nanoantenna array (633 nm, cw-laser excitation). Right: Fluorescence antibunching of NQD within a gap of a bowtie nanoantenna.

To place 40-50-nm-size nanodiamonds in the gap between the two triangular arms of this nanoantenna we employed a technique to manipulate nanodiamonds using an atomic force microscope (AFM) tip (lithography option of an AFM). Figure 8 shows obtained AFM-micrographs of the nanodiamond manipulation on a spin coated sample by an AFM tip ("pushing" the nanodiamonds). The positions of two nanodiamonds in the three top micrographs are indicated by 1 and 2. The two other micrographs of Figure 8 show nanodiamonds rearranged by an AFM tip to resemble a smiling face. Our next step will be to use this technique to place NV, SiV or Cr-center nanodiamonds or nanocrystals doped with TR3+ in a gap between the arms of the nanoantenna to enhance the photon count rate and directionality, diminish the fluorescence decay time, and provide polarization selectivity.

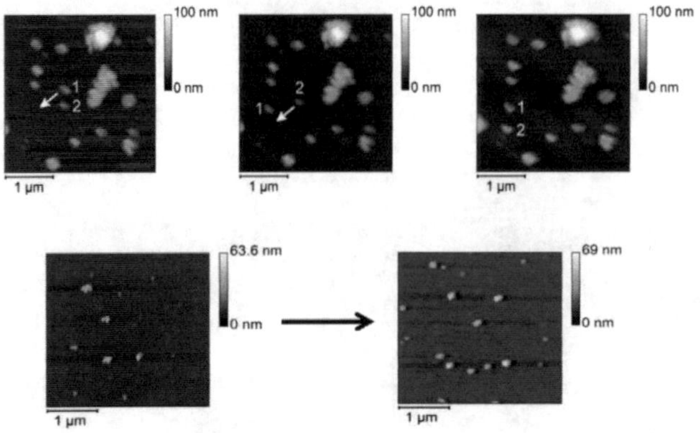

Figure 8. Top: Three AFM micrographs showing the manipulation of nanodiamond position using the AFM tip to "push" the nanodiamonds. Bottom: Two AFM micrographs showing rearrangement of the nanodiamonds by the AFM tip to resemble a smiling face.

6. 2D-Photonic Bandgap Microcavities

Using electron-beam lithography and reactive ion etching we prepared silicon 2D-photonic crystals (PhCs) with L3 defect nanocavities operating at optical communications wavelength around 1.55 μm (see Figure 9, left and center). This L3 nanocavity consists of a triangular lattice of air holes on a silicon membrane, with a line of three holes omitted.

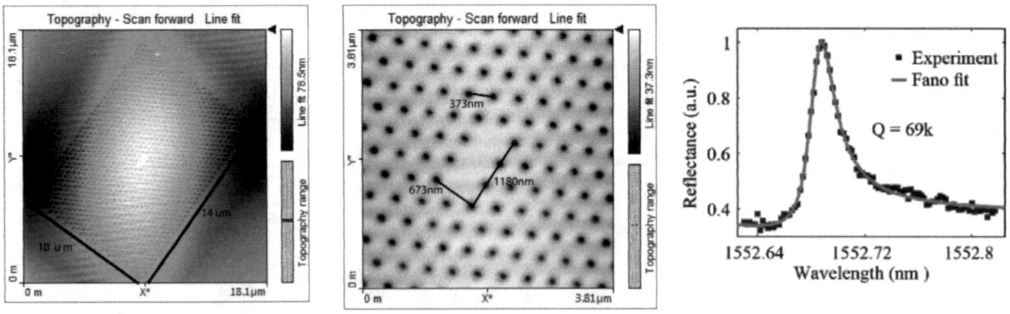

Figure 9. Left and Center: AFM micrographs with different scan area of the same 2D-PhC with a L3 nanocavity. Right: Nanocavity resonance at wavelength near 1.55 μm.

The holes with three slightly different diameters for the same PhC were fabricated on a commercial silicon-on-insulator wafer (SOITEC). The cavity's terminating holes were reduced in size and slightly displaced to increase the cavity Q-factor up to 70,000. Additionally, the radii of selected holes surrounding the cavity were slightly increased to modify the cavity's far-field emission pattern, and facilitate coupling in the out-of-plane direction. Figure 9, right, shows a measured resonance in a typical L3 PhC nanocavity at 1.55 μm wavelength. It should be mentioned that for a similar type of a nanocavity in GaAs with epitaxial quantum dots, a Purcell factor of 145 was reported in Ref. [32]. We are planning to use L3 PhCs with Er^{3+} doped nanocrystals to diminish fluorescence decay times of Er^{3+} fluorescence to a nanosecond range. The nanocavity's rectangular geometry will also provide polarization selectivity.

7. Conclusion

In summary, our results represent progress toward robust and efficient room-temperature SPSs based on fluorescence in nanocrystals. For different applications requiring different wavelengths, we considered colloidal nanocrystal quantum dots (NQDs), color-center nanodiamonds, and nanocrystals doped with trivalent rare-earth ions (TR^{3+}). For single-photon fluorescence control, we used cholesteric chiral photonic bandgap and Bragg-reflector microcavities, developed gold plasmonic bowtie nanoantennas, and Si 2D-photonic-crystal 3L nanocavity. Our next step will be placing a color-center nanodiamond and, separately, a single TR^{3+}-doped nanocrystal between nanoantenna lobes or within a 2D-photonic-crystal 3L-nanocavity to enhance the photon count rate and directionality, reduce the fluorescence decay time for TR^{3+} to tens of nanoseconds, and provide linear polarization selectivity with photon antibunching from nanocrystal within a nanoantenna or nanocavity.

Such SPSs will find applications both in free space and fiber-based quantum cryptography. Some quantum computation protocols also require single photons. In addition, single photons are very important in metrology and fundamental physics experiments, for instance, in interference from uncoupled light sources and entanglement-state generation, in shedding light on wave-particle duality as well as in fundamental issues of quantum measurement and uncertainty.

Acknowledgement

We thank Carlos Stroud and Ansgar Schmid for valuable discussions. Nanoantennas were prepared using the Cornell NanoScale Science & Technology Facility and the University of Rochester Integrated Nanosystems Center. S.G.L. was supported by the U.S. Army Research Office Award DAAD19-02-1-0285, National Science Foundation Awards ECS-0420888, DUE-0633621, DUE-0920500, EEC-1343673. J.M.W. is supported by a NASA Space Technology Research Fellowship NNX11AM82H. L.J.B. was supported by an Air Force SMART Fellowship. R.W.B was supported by the US Defense Threat Reduction Agency—Joint Science and Technology Office for Chemical and Biological Defense (Grant No. HDTRA1-10-1-0025) and by the Canada Excellence Research Chairs Program.

References

[1] Eisaman, M.D., Fan, J., Migdall, A. and Polyakov, S.V., "Invited Review Article: Single-photon sources and detectors", Rev. Scient. Instr. 82, 071101-1-25 (2011).

[2] Yamamoto, Y., Santori, C., Solomon, G., Vuckovic, J., Fattal, D., Waks, E., Diamanti, E., "Single photons for quantum information systems", Progress in Informatics, No 1, 5-37 (2005).

[3] Buckley, S., Kelley Rivoire, K., and Vuckovic, J., "Engineered quantum dot single-photon sources", Rep. Prog. Phys. 75, 126503-1-27 (2012).

[4] Lounis, B., and Orrit, M., "Single-photon sources", Rep. Prog. Phys. 68, 1129 (2005).

[5] Oxborrow, M. and Sinclair, A.G., "Single-photon sources", Contemp. Phys. 46, 173 (2005).

[6] Sheel, S., "Single-photon sources: an introduction", J. Mod. Opt. 56, 141 (2009).

[7] De Vittorio, M., Pisanello, F., Martiradonna, L., Qualtieri, A., Stomeo, T., Bramati, A., Cingolani, R., "Recent advances on single photon sources based on single colloidal nanocrystals", Opto-Electron. Rev. (Springer), 18(1), 1-9 (2010).

[8] Bissell, L.J., [Experimental Realization of Efficient, Room-Temperature Single-Photon Sources with Definite Circular and Linear Polarizations], Ph.D. Thesis, University of Rochester, Rochester, NY (2011).

[9] Lukishova, S.G., "Single-photon sources for secure quantum communication", Proceed. SPIE 9065, paper 90650C (2013).

[10] Santori, C., Fattal, D. and Yamamoto, Y., [Single-photon Devices and Applications], Wiley, Weinheim (2010).

[11] Michler, P., Ed., [Single Semiconductor Quantum Dots], Springer, Berlin (2010).

[12] S.G. Lukishova, J. Winkler, Luke J. Bissell, D. Mihaylova, A.C. Liapis, Z. Shi, D. Goldberg, V.M. Menon, R. Boyd, G. Chen. P.N. Prasad,"Room-temperature single photon sources based on nanocrystals in photonic/plasmonic nanostructures", Proceedings SPIE, Emerging Technologies in Security and Defence II; and Quantum-Physics-based Information Security III. Vol. 9254, paper 9254-05, doi: 10.1117/12.2066979 (October 2014).

[13] Novotny, L. and Hecht, B., [Principles of Nano-Optics], Cambridge University Press, Cambridge, New York (2012).

[14] Purcell, E.M., "Spontaneous emission at radio frequencies," Phys. Rev. 69, 681 (1946).

[15] Kopp, V.I., Fan, B., Vithana, H.K.M., Genack, A.Z., "Low-threshold lasing at the edge of a photonic stop band in cholesteric liquid crystals", Opt. Lett. 23, 1707–9 (1998).

[16] Wei, S.K.H., Chen, S.H., Dolgaleva, K., Lukishova, S.G., Boyd, R.W., "Robust organic lasers comprising glassy-cholesteric pentafluorene doped with a red-emitting oligofluorene", Appl. Phys. Lett. 94, 041111 (2009).

[17] Dolgaleva, K., Wei, S.K.H., Lukishova, S.G., Chen, S.H., Schwertz, K., and Boyd, R.W., "Enhanced laser performance of cholesteric liquid crystals doped with oligofluorene dye", J. Opt. Soc. Am. B 25, 1496-1504 (2008).

[18] Lukishova, S.G., Boyd, R.W., Stroud, C.R., "Efficient room-temperature source of polarized single photons", US Patent 7,253,871, Aug. 7, 2007.

[19] Lukishova, S.G., Schmid, A.W., McNamara, A.J., Boyd, R.W., and Stroud, C.R., "Room temperature single photon source: single dye molecule fluorescence in liquid crystal host", IEEE J. Sel. Top. Quant. Electron. 9, No 6, 1512-1518 (2003).

[20] Lukishova, S.G., Schmid, A.W., Supranowitz, C.M., Lippa, N., McNamara, A.J., Boyd, R.W., Stroud, C.R., "Dye-doped cholesteric-liquid-crystal room-temperature single photon source ", J. Mod. Opt. 51, No 9-10, 1535 -1547 (2004).

[21] Lukishova, S.G., Schmid, A.W., Knox, R.P., Freivald, P., McNamara, A., Boyd, R.W., Stroud, C.R., Marshall, K.L., "Single-photon source for quantum information based on single dye molecule fluorescence in liquid crystal host", Molec. Cryst. Liq. Cryst. 454, 403-416 (1-14) (2006).

[22] Lukishova, S.G., Bissell, L.J., Menon, V.M., Valappil, N., Hahn, M.A., Evans, C.M., Zimmerman, B., Krauss, T.D., Stroud, C.R., Boyd, R.W., "Organic photonic bandgap microcavities doped with semiconductor nanocrystals for room-temperature on demand single photon sources", J. Mod. Opt. 56, is 2 & 3, 167-174 (2009).

[23] Lukishova, S.G., Bissell, L.J., Stroud, C.R., Boyd, R.W., "Room-temperature single photon sources with definite circular and linear polarizations", Opt. and Spectr. 108, No 3, 417-424 (2010).

[24] Lukishova, S.G., Bissell, L.J., Winkler, J., and Stroud, C.R., "Resonance in quantum dot fluorescence in a photonic bandgap liquid crystal host", Opt. Lett. 37, Issue 7, 1259-1261 (2012).

[25] Lukishova, S.G., "Liquid crystals under two extremes: (1) high-power laser irradiation, and (2) single-photon level", Mol. Cryst. Liq. Cryst. 559, 127-157 (2012).

[26] Lukishova, S.G., Schmid, A.W., "Near-field optical microscopy of defects in cholesteric oligomeric liquid crystal films", Molec. Cryst. Liq. Cryst. 454, 15-21 (2006).

[27] Lukishova, S.G., Schmid, A.W., Knox, R., Freivald, P., Bissell, L., Boyd, R.W., Stroud, C.R., Marshall, K.L., "Room temperature source of single photons with definite polarization", J. Mod. Opt. 54, iss. 2 & 3, 417-429 (2007).

[28] Winkler, J.M., Lukishova, S.G., Bissell, L.J., Goldberg, D., Menon, V.M., Shi, Z., Boyd, R.W., "Room-temperature single photon sources: nanocrystals in photonic bandgap microcavities and plasmonic nanoantennae," Single Photon Workshop, Book of Abstracts, Oak Ridge National Laboratory, Oak Ridge, TN (October 2013).

[29] Lukishova, S.G., Liapis, A.C., Bissell, L.J., Gehring, G.M., and Boyd, R.W., "Single-photon experiments with liquid crystals for quantum science and quantum engineering applications", Liquid Crystal Reviews 2, No. 2, 111–129 (2014).

[30] Lukishova, S.," Efficient single-photon sources with definite polarization", 1 October 2014, SPIE Newsroom. DOI: 10.1117/2.1201409.005604.

[31] Goldberg, D. and Menon, V.M., "Enhanced amplified spontaneous emission from colloidal quantum dots in all-dielectric monolotic microcavities", Appl. Phys. Lett., 102, 081119 (2013).

[32] Hennessy, K, Badolato, A., Winger, M., Gerace, D., Atature, M., Gulde, S., Falt, S., Hu, E.L., Imamoglu, A., "Quantum nature of a strongly coupled single quantum dot-cavity system", Nature 445, 896-899 (2007).

Spin superradiance by magnetic nanomolecules and nanoclusters

V I Yukalov[1], **V K Henner**[2,3] **and E P Yukalova**[4]

[1]Bogolubov Laboratory of Theoretical Physics, Joint Institute for Nuclear Research, Dubna 141980, Russia
[2]Department of Physics, Perm State University, Perm 614190, Russia
[3]Department of Physics, University of Louisville, Louisville, Kentucky 40292, USA
[4]Laboratory of Information Technologies, Joint Institute for Nuclear Research, Dubna 141980, Russia

E-mail: yukalov@theor.jinr.ru

Abstract. Spin dynamics of assemblies of magnetic nanomolecules and nanoclusters can be made coherent by inserting the sample into a coil of a resonant electric circuit. Coherence is organized through the arising feedback magnetic field of the coil. The coupling of a magnetic sample with a resonant circuit induces fast spin relaxation and coherent spin radiation, that is, superradiance. We consider spin dynamics described by a realistic Hamiltonian, typical of magnetic nanomolecules and nanoclusters. The role of magnetic anisotropy is studied. A special attention is paid to geometric effects related to the mutual orientation of the magnetic sample and resonator coil.

1. Introduction

There exists a large class of magnetic nanomolecules and magnetic nanoclusters that can be considered as nanoparticles possessing high total spins (see review articles [1–9]). Below blocking temperature, the spin of such magnetic nanoparticles is frozen. For instance, the typical blocking temperature of magnetic nanomolecules is of order $1 - 10$ K. The blocking temperature for nanoclustres is $10 - 100$ K.

Magnetic properties of nanomolecules and nanoclusters are similar to each other. There are two main features distinguishing them. Magnetic molecules of the same chemical composition are identical and they can form crystals with almost ideal periodic lattice. While magnetic nanoclusters, even being made of the same element, say Fe, Ni, or Co, differ by their sizes, and they do not form periodic structures. Otherwise, the spin Hamiltonian for an ensemble of magnetic nanoparticles is of the same form for nanomolecules as well as for nanoclusters.

In the usual case, spin relaxation is due to spin-phonon interactions and, below the blocking temperature, is very slow. Thus for nanomolecules, the spin-phonon relaxation time is $T_1 \sim (10^5 - 10^7)$ s. But the spin relaxation time can be drastically shortened, if the magnetic sample is inserted into a coil of a resonant electric circuit. This is termed the Purcell effect [10]. In that case, the relaxation is caused by the resonator feedback field collectivizing moving spins and forcing them to move coherently. Coherent spin dynamics have been studied in several publications, e.g., [11–25].

Coherently moving spins produce coherent radiation, which, when it is self-organized, is called superradiance. It is worth stressing that spin superradiance is rather different from atomic superradiance. The latter is caused by the Dicke effect [26], while the cavity Purcell effect is secondary [27–29]. Contrary to this, spin superradiance is completely due to Purcell effect, with the Dicke effect playing no role [30]. The Purcell effect also enhances the signals of nuclear magnetic resonance [31,32] and of spin echo [33–35].

Here we consider the peculiarities of spin superradiance by magnetic nanomolecules and nanoclusters having strong magnetic anisotropy. We shall pay attention to the role of geometric effects related to the finiteness of the considered samples. Finite systems, as is known [36,37], can exhibit properties different from those of bulk systems. In the present case, we are interested in the geometric effects due to the mutual orientation of a finite magnetic sample and the resonator coil.

2. Spin Hamiltonian

An ensemble of magnetic nanomolecules or nanoclusters is described by the Hamiltonian

$$\hat{H} = \sum_i \hat{H}_i + \frac{1}{2} \sum_{i \neq j} \hat{H}_{ij} , \tag{1}$$

consisting of single-spin terms $\hat{H}_i$ and spin-interaction terms $\hat{H}_{ij}$, with the index $i = 1, 2, \ldots, N$ enumerating nanoparticles. The single-spin Hamiltonian

$$\hat{H}_i = -\mu_0 \mathbf{B} \cdot \mathbf{S} - D(S_i^z)^2 + D_2(S_i^x)^2 + D_4\left[(S_i^x)^2(S_i^y)^2 + (S_i^y)^2(S_i^z)^2 + (S_i^z)^2(S_i^x)^2\right] \tag{2}$$

is a sum of the Zeeman energy and single-site magnetic anisotropy terms. The total magnetic field, acting on each spin,

$$\mathbf{B} = B_0 \mathbf{e}_z + H \mathbf{e}_x , \tag{3}$$

includes an external field B_0 and a resonator feedback field H. Spins interact with each other through dipolar forces characterized by the Hamiltonian

$$\hat{H}_{ij} = \sum_{\alpha\beta} D_{ij}^{\alpha\beta} S_i^\alpha S_j^\beta , \tag{4}$$

with the dipolar tensor

$$D_{ij}^{\alpha\beta} = \frac{\mu_0^2}{r_{ij}^3} \left(\delta_{\alpha\beta} - 3 n_{ij}^\alpha n_{ij}^\beta \right) ,$$

where

$$r_{ij} \equiv |\mathbf{r}_{ij}| , \qquad \mathbf{n}_{ij} \equiv \frac{\mathbf{r}_{ij}}{r_{ij}} , \qquad \mathbf{r}_{ij} \equiv \mathbf{r}_i - \mathbf{r}_j .$$

The resonator feedback field is given by the Kirchhoff equation

$$\frac{dH}{dt} + 2\gamma H + \omega^2 \int_0^t H(t')\, dt' = -4\pi\eta \, \frac{dm_x}{dt} , \tag{5}$$

in which γ is resonator damping, ω is resonator natural frequency, η is filling factor, and

$$m_x \equiv \frac{\mu_0}{V} \sum_{j=1}^N \langle S_j^x \rangle \tag{6}$$

is the transverse magnetization density of the sample having volume V.

In addition to the resonator natural frequency ω, there are the following characteristic frequencies. The Zeeman frequency

$$\omega_0 \equiv -\frac{\mu_0}{\hbar} B_0 = \frac{2}{\hbar} \mu_B B_0 \tag{7}$$

and the anisotropy frequencies

$$\omega_D \equiv (2S-1) \frac{D}{\hbar} , \qquad \omega_2 \equiv (2S-1) \frac{D_2}{\hbar} , \qquad \omega_4 \equiv (2S-1) \frac{D_4}{\hbar} S^2 . \tag{8}$$

The resonator natural frequency has to be close to the Zeeman frequency, in order to satisfy the resonance condition

$$\left| \frac{\omega - \omega_0}{\omega} \right| \ll 1 . \tag{9}$$

And the Zeeman frequency has to be larger than the anisotropy frequencies that freeze spin motion,

$$\left| \frac{\omega_D}{\omega_0} \right| \ll 1 , \qquad \left| \frac{\omega_2}{\omega_0} \right| \ll 1 , \qquad \left| \frac{\omega_4}{\omega_0} \right| \ll 1 . \tag{10}$$

Among the anisotropy frequencies, the most important are ω_D and ω_2 that are close to each other. The frequency ω_4, up to spins $S \sim 10^3$, is much smaller than ω_D.

We analyze the Heisenberg equations of motion for spins in two ways, by employing the scale separation approach [4, 5] and by directly solving the spin evolution equations in semiclassical approximation. Both ways give close results. Finding the average spins as functions of time, we can calculate the radiation intensity.

3. Radiation intensity

The intensity of radiation, induced by moving spins, can be calculated in two ways. One possibility is the classical formula

$$I(t) = \frac{2\mu_0^2}{3c^3} \left| \sum_j \langle \ddot{\mathbf{S}}_j \rangle \right|^2 \tag{11}$$

that should provide good approximation for high spins $S \gg 1$. The other way is to use the quantum formula [5, 38, 39], according to which the radiation intensity

$$I(t) = I_{inc}(t) + I_{coh}(t) \tag{12}$$

is the sum of the incoherent radiation intensity

$$I_{inc}(t) = 2\omega_0 \gamma_0 SN[1 + s(t)] \tag{13}$$

and the coherent radiation intensity

$$I_{coh}(t) = 2\omega_0 \gamma_0 S^2 N^2 \varphi_0 w(t) , \tag{14}$$

where the natural width is

$$\gamma_0 \equiv \frac{2}{3} |\vec{\mu}|^2 k_0^3 = \frac{1}{3} \mu_0^2 k_0^3 \qquad \left(k_0 \equiv \frac{\omega_0}{c} \right) ,$$

φ_0 is a form-factor, and

$$s(t) \equiv \frac{1}{NS} \sum_{j=1}^{N} \langle S_j^z(t) \rangle \; , \qquad w(t) \equiv \frac{1}{N^2 S^2} \sum_{i \neq j}^{N} \langle S_i^+(t) S_j^-(t) \rangle \; . \tag{15}$$

If the wavelength is larger than the sample linear size, then $\varphi_0 \simeq 1$. But, when the wavelength is shorter than the system linear size, then the form-factor essentially depends on the sample shape [39, 40].

We have accomplished computer simulation for N magnetic nanomolecules possessing spin $S = 10$, such as Mn_{12} or Fe_8, employing the parameters typical of these nanomolecules, for which D_2 and D_4 are negligible. The spin system is prepared in a nonequilibrium initial state, with the external magnetic field directed along the initial spin polarization, so that the spins tend to reverse to the opposite direction. It is convenient to consider a dimensionless radiation intensity, expressed through the units of

$$I_0 \equiv \frac{2\mu_0^2}{3c^3}\, \gamma_2^4 \qquad \left(\gamma_2 \equiv \frac{1}{T_2} \right) \; ,$$

multiplied by the number of nanomolecules squared, N^2, where $\mu_0 = -2\mu_B = 1.855 \times 10^{-20}$ erg/G and $\gamma_2 = 10^{10}$ 1/s, which gives $I_0 = 0.852 \times 10^{-38}$ W. All frequencies are measured in units of γ_2. The resonance condition $\omega = \omega_0$ is assumed.

We have considered the influence of different factors on the radiation intensity. Thus, the role of the Zeeman frequency is exemplified in Fig. 1, showing that the larger the Zeeman frequency, the higher the radiation intensity. Figure 2 illustrates that the larger the initial spin polarization, the larger the radiation intensity. Figure 3 shows that increasing the magnetic anisotropy suppresses the radiation intensity. The role of dipole interactions is described in Fig. 4, demonstrating that they suppress the radiation intensity by a factor of 1.5. In Fig. 5, we study the role of the sample shape and its orientation, from where it follows that, under the same number of nanomolecules, the most favorable situation, with the highest radiation intensity, corresponds to the chain of nanomolecules along the resonator axis.

Calculations for the coherent radiation intensity (14) reduces to the solution of the evolution equation for the coherence function $w(t)$. Numerical solution yields the results close to the quasi-classical case, illustrated in Figs. 1 to 5. This is not surprising, since the coherent regime is known to be well represented by a quasi-classical approximation. The maximal number of coherently radiating spins can be estimated as $N_{coh} \sim \rho V_{coh}$, where ρ is the density of nanomolecules and V_{coh} is the coherence volume. The latter, for a cylindrical sample, is $V_{coh} \sim \pi R_{coh}^2 L$, where L is the cylinder length and R_{coh} is a coherence radius [39], which is of order $0.3\sqrt{\lambda L}$. This gives

$$N_{coh} \sim \rho \lambda L^2 \; .$$

The typical density of magnets, formed by nanomolecules, is $\rho \approx 0.4 \times 10^{21}$ cm^{-3}. For the Zeeman frequency $\omega \sim 2 \times 10^{13}$ 1/s, the wavelength is $\lambda \sim 10^{-2}$ cm. If $\lambda \sim L$, then $N_{coh} \sim 10^{14}$. The typical time of a superradiant pulse is 10^{-11} s.

In this way, magnetic nanomolecules and nanoclusters can be described by a similar macroscopic Hamiltonian. In the process of spin reversal from an initially prepared non-equilibrium state, there appears spin superradiance, due to the Purcell effect of the resonator feedback field. The radiation is mainly absorbed by the resonant coil surrounding the sample.

Acknowledgments
The authors are grateful for financial support to the Russian Foundation for Basic Research (grant 13-02-96018) and to the Perm Ministry of Education (grant C-26/628).

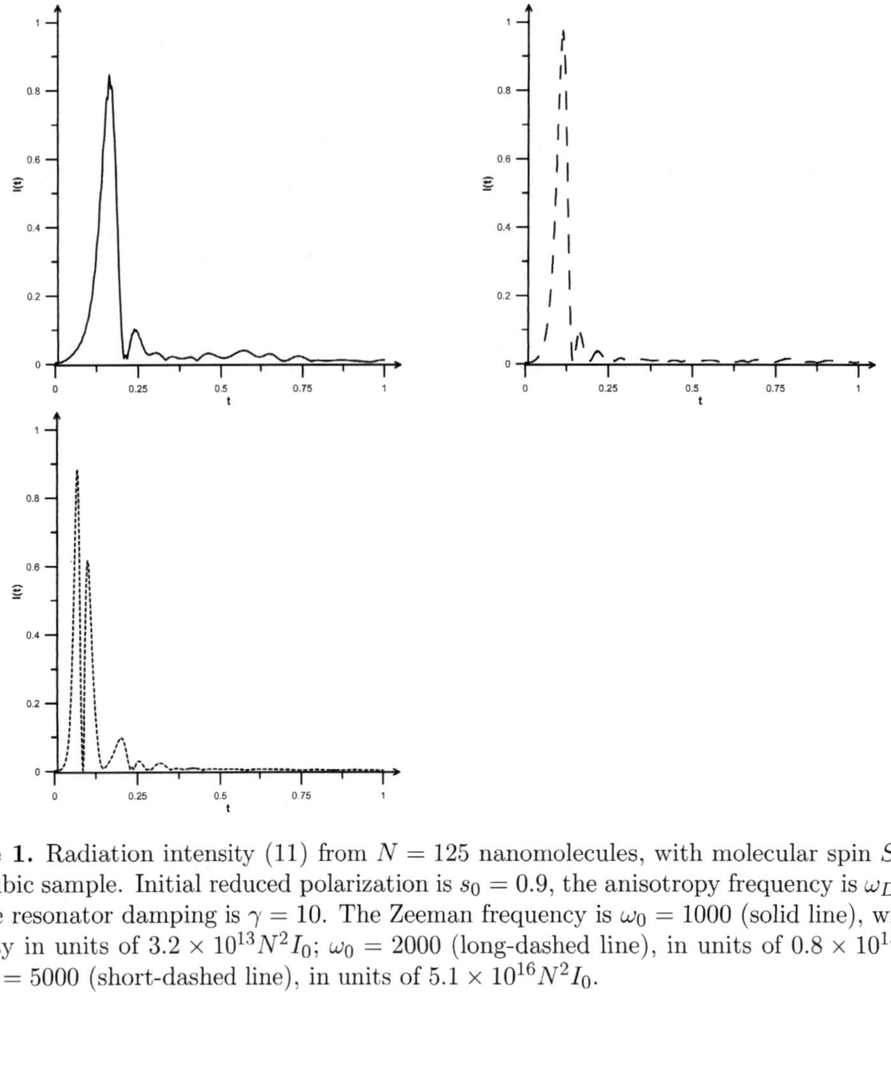

Figure 1. Radiation intensity (11) from $N = 125$ nanomolecules, with molecular spin $S = 10$, for a cubic sample. Initial reduced polarization is $s_0 = 0.9$, the anisotropy frequency is $\omega_D = 20$, and the resonator damping is $\gamma = 10$. The Zeeman frequency is $\omega_0 = 1000$ (solid line), with the intensity in units of $3.2 \times 10^{13} N^2 I_0$; $\omega_0 = 2000$ (long-dashed line), in units of $0.8 \times 10^{15} N^2 I_0$, and $\omega_0 = 5000$ (short-dashed line), in units of $5.1 \times 10^{16} N^2 I_0$.

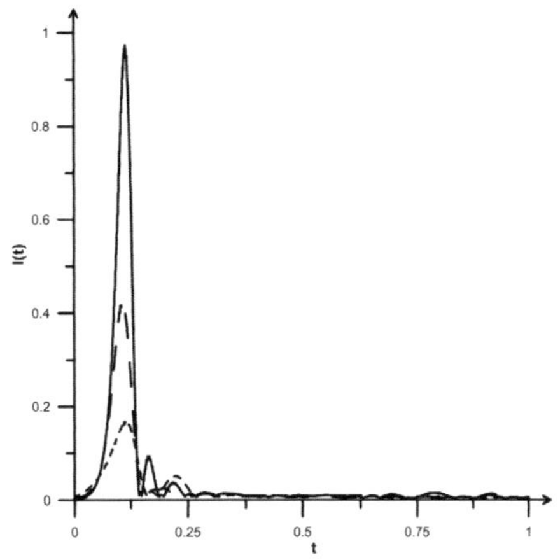

Figure 2. Radiation intensity (11) for a cubic sample of $N = 125$ nanomolecules, with spin $S = 10$, under the Zeeman frequency $\omega_0 = 2000$, anisotropy frequency $\omega_D = 20$, and the resonator damping $\gamma = 10$, for different initial polarizations: $s_0 = 0.9$ (solid line), $s_0 = 0.7$ (long-dashed line), and $s_0 = 0.5$ (short-dashed line). All intensities are in units of $0.8 \times 10^{15} N^2 I_0$.

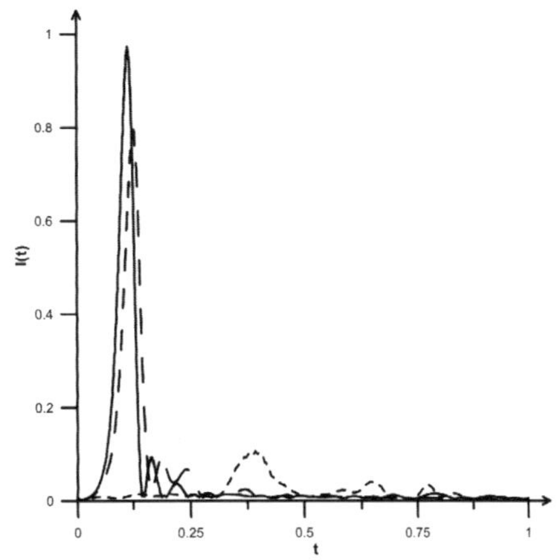

Figure 3. Radiation intensity (11) for a cubic sample of $N = 125$ nanomolecules, with spin $S = 10$, under the Zeeman frequency $\omega_0 = 2000$, resonator damping $\gamma = 10$, and the initial spin polarization $s_0 = 0.9$, for varying anisotropy frequency: $\omega_D = 20$ (solid line), $\omega_D = 50$ (long-dashed line), and $\omega_D = 100$ (short-dashed line). The values of the radiation intensity are in units of $0.8 \times 10^{15} N^2 I_0$.

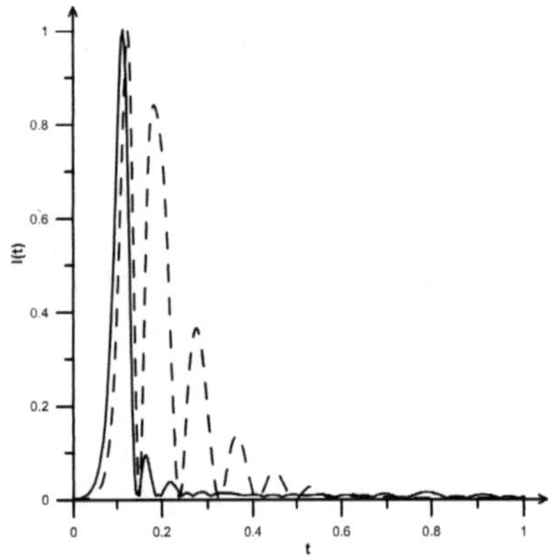

Figure 4. Radiation intensity (11) for a cubic sample of $N = 125$ nanomolecules, with spin $S = 10$, the Zeeman frequency $\omega_0 = 2000$, anisotropy frequency $\omega_D = 20$, resonator damping $\gamma = 10$, and the initial spin polarization $s_0 = 0.9$, for the cases with dipole interactions (solid line), in units of $0.8 \times 10^{15} N^2 I_0$, and without these interactions (dashed line), in units of $1.2 \times 10^{15} N^2 I_0$.

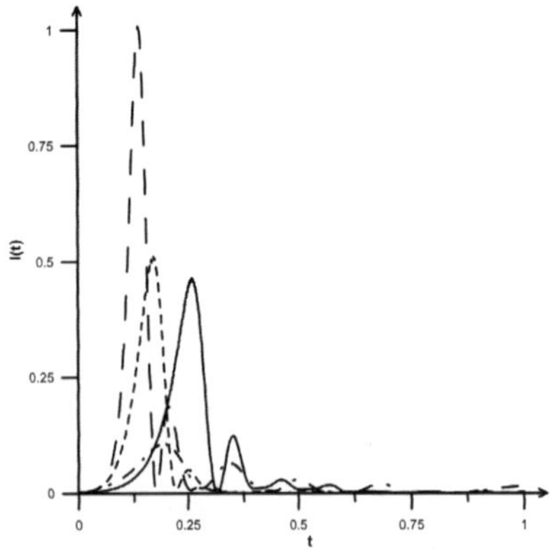

Figure 5. Radiation intensity (11) for a cubic sample of $N = 144$ nanomolecules, with spin $S = 10$, the Zeeman frequency $\omega_0 = 2000$, anisotropy frequency $\omega_D = 20$, resonator damping $\gamma = 30$, and the initial spin polarization $s_0 = 0.9$, for different sample shapes and orientations: the chain of molecules along the z - axis (solid line), the chain along the x - axis (long-dashed line), the $y - z$ plane of molecules (short-dashed line), and the $x - y$ plane of molecules (dotted-dashed line). The intensities are in units of $1.2 \times 10^{15} N^2 I_0$.

[1] Barbara B, Thomas L, Lionti F, Chioresku I, and Sulpice A 1999 *J. Magn. Magn. Mater.* **200** 167
[2] Wernsdorfer W 2001 *Adv. Chem. Phys.* **118** 99
[3] Ferré J 2002 *Top. Appl. Phys.* **83** 127
[4] Yukalov V I 2002 *Laser Phys.* **12** 1089
[5] Yukalov V I and Yukalova E P 2004 *Phys. Part. Nucl.* **35** 348
[6] Bedanta S and Kleemann W 2009 *J. Phys. D* **42** 013001
[7] Berry C C 2009 *J. Phys. D* **42** 224003
[8] Beveridge J S, Stephens J R, and Williams M E 2011 *Annu. Rev. Annal. Chem.* **4** 251
[9] Hoang V V and Canguli D 2012 *Phys. Rep.* **518** 81
[10] Purcell E M 1946 *Phys. Rev.* **69** 681
[11] Kiselev J F, Prudkoglyad A F, Shumovsky A S, and Yukalov V I 1988 *Mod. Phys. Lett. B* **1** 409
[12] Belozerova T S, Henner V K, and Yukalov V I 1992 *Phys. Rev. B* **46** 682
[13] Belozerova T S, Henner V K, and Yukalov V I 1992 *Laser Phys.* **2** 545
[14] Yukalov V I 1992 *Laser Phys.* **2** 559
[15] Yukalov V I 1993 *Laser Phys.* **3** 870
[16] Yukalov V I 1995 *Phys. Rev. Lett.* **75** 3000
[17] Yukalov V I 1996 *Phys. Rev. B* **53** 9232
[18] Yukalov V I 1997 *Laser Phys.* **7** 58
[19] Belozerova T S, Davis C L, and Henner V K 1998 *Phys. Rev. B* **58** 3111
[20] Yukalov V I 2005 *Phys. Rev. B* **71** 184432
[21] Davis C L, Henner V K, Tchernatinsky A V, and Kaganov. I V 2005 *Phys. Rev. B* **72** 054406
[22] Yukalov V I, Henner V K, Kharebov P V, and Yukalova E P 2008 *Laser Phys. Lett.* **5** 887
[23] Yukalov V I, Henner V K, and Kharebov P V 2008 *Phys. Rev. B* **77** 134427
[24] Yukalov V I and Yukalova E P 2011 *Laser Phys. Lett.* **8** 804
[25] Henner V, Raikher Y, and Kharebov P 2011 *Phys. Rev. B* **84** 144412
[26] Dicke R H 1954 *Phys. Rev.* **93** 99
[27] Walther H, Varcoe B T, Englert B G, and Becker T 2006 *Rep. Prog. Phys.* **69** 1625
[28] Manassah J T 2011 *Phys. Rev. A* **83** 025801
[29] Manassah J T 2012 *Laser Phys.* **22** 738
[30] Yukalov V I 2005 *Laser Phys. Lett.* **2** 356
[31] Chen H Y, Lee Y, Bowen S, and Hilty C 2011 *J. Magn. Res.* **208** 204
[32] Krishnan V V and Murali N 2013 *Prog. Nucl. Magn. Res. Spectrosc.* **68** 41
[33] Shavishvili T M, Khutsishvili K O, Fokina N P, and Lavrentiev G V 1989 *J. Tech. Phys. Lett.* **15** 33
[34] Fokina N P and Khutsishvili K O 1990 *Phys. Met. Metallogr.* **69** 65
[35] Nazarova O V, Fokina N P, and Khutsishvili K O 1991 *Phys. Met. Metallogr.* **70** 44
[36] Jin F, De Raedt H, Yuan S, Katsnelson M I, Miyashita S, and Michielsen K 2010 *J. Phys. Soc. Jap.* **79** 124005
[37] Birman J L, Nazmitdinov R G, and Yukalov V I 2013 *Phys. Rep* **526** 1
[38] Yukalov V I and Yukalova E P 2005 *Laser Phys. Lett.* **2** 302
[39] Yukalov V I 2014 *Laser Phys.* **24** 094015
[40] Allen L and Eberly J H 1975 *Optical Resonance and Two-Level Atoms* (New York: Wiley)

Towards Lasing Without Inversion in Neutral Mercury

Benjamin Rein, Martin R. Sturm, Reinhold Walser and Thomas Walther

Institute for Applied Physics, TU Darmstadt, Schlossgartenstr. 7, D-64289 Darmstadt

E-mail: thomas.walther@physik.tu-darmstadt.de

Abstract. Currently, we are implementing a lasing without inversion (LWI) scheme in mercury based on calculations by Fry *et al.* [1]. A recent detailed analysis predicting exact experimental parameters shows the feasibility of our LWI scheme [2]. In this paper we report on the progress of our experiment and compare its parameters with the prior analysis.

1. Introduction

The development of cw lasers in the UV and VUV wavelength regions is a very active field of research as there are many applications, e.g. in quantum information processing, biomedicine and laser lithography [3, 4]. However, this development remains a challenging task, particularly since the pump power required for a population inversion at the laser transition scales at least with the laser frequency ω^4 and therefore makes it hard to build conventional lasers in this regime [5]. Common methods to generate continuous-wave laser radiation in the UV are second and fourth harmonic generation [6]. However, second and fourth harmonic generation are limited by the availability of materials transparent in the VUV. An alternative is four-wave mixing which even makes it possible to generate light in the VUV region [7]. All of these techniques exploit nonlinear effects.

A different approach is lasing without inversion (LWI) based on the coherent excitation of atomic transitions. It was first proposed in [8, 9, 10]. The main idea is to cancel the absorption of radiation on the lasing transition such that a few incoherently excited atoms are sufficient to achieve gain by stimulated emission of radiation. Successful demonstrations of continuous-wave LWI were shown in rubidium [11] and sodium [12, 13], but until now there is no LWI scheme where the lasing transition has a significantly shorter wavelength than the lasers used to generate the coherent excitation. The overall objective of our project is to generate a UV source based on LWI in neutral mercury. In this contribution we report on the experimental progress towards this goal.

In our experimental implementation, we employ an LWI scheme in mercury which was first proposed by Fry *et al.* [1]. In this scheme, the laser transition is at 253.7 nm and the shortest wavelength to build up the coherence is at 435.8 nm. Because of the 4-level structure and its favorable energy structure, it is even possible to cancel the Doppler effect, preventing the gain spike from being washed out.

In this paper, we present a short summary of new detailed calculations of LWI in mercury which were carried out by Martin Sturm *et al.* [2] and give a status update of the experimental realization.

2. Theoretical model of LWI in mercury

The four-level scheme used for LWI in mercury is depicted in Fig. 1(a) and involves the 6^1S_0 ground state as well as the 6^3P_1 state and the metastable 6^3P_2 state which are connected through the 7^3S_1 state employing the strong driving laser at 435.8 nm as well as the weak coupling laser at 546.1 nm with the Rabi frequencies Ω_s and Ω_w, respectively. The laser transition at 253.7 nm is between the ground state 6^1S_0 and the excited state 6^3P_1 with Ω_p. Because there could be population trapping into the metastable state 6^3P_0 an incoherent repumping laser at 404.7 nm is added, connecting the 6^3P_0 state to the 7^3S_1 state.

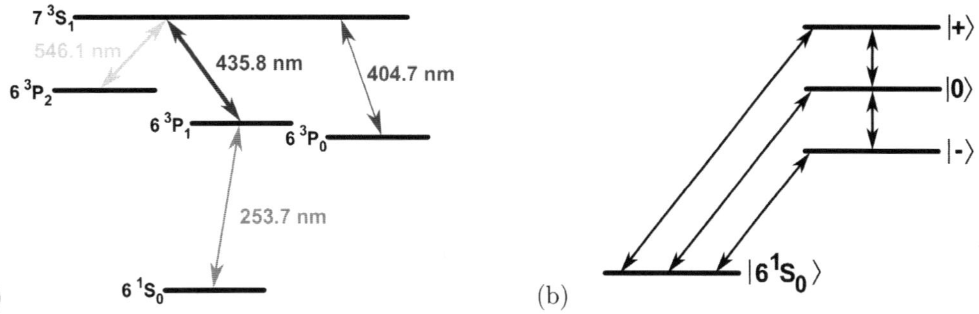

(a) (b)

Figure 1. (a) Mercury level scheme used for LWI at 253.7 nm. See text for details. (b) Dressed-state picture of the four-level scheme in mercury with the Rabi frequencies of the coupling lasers Ω_s, Ω_w. All detunings are assumed to be zero, i.e. $\Delta_{s,w} = 0$.

The Hamiltonian of this system in the dipole and rotating wave approximation is given by [2]

$$H = \begin{pmatrix} 0 & \Omega_p^* & \Omega_s & 0 \\ \Omega_p & -\Delta_p & 0 & 0 \\ \Omega_p^* & 0 & \Delta_s & \Omega_w^* \\ 0 & 0 & \Omega_w & \Delta_s - \Delta_w \end{pmatrix}, \tag{1}$$

in the basis of the atomic levels $\{|6^3P_1\rangle, |6^1S_0\rangle, |7^3S_1\rangle, |6^3P_2\rangle\}$ and $\Delta_{w,s,p}$ are the detunings of the lasers with respect to the corresponding atomic transitions. The origin of the LWI mechanism is easier to understand in the dressed-state picture. Therefore, we diagonalize the matrix and solve for the eigenvalues of the Hamiltonian:

$$|0\rangle = |6\ {}^3P_2\rangle - \frac{\Omega_w^*}{\Omega_s^*}|6\ {}^1S_0\rangle \tag{2}$$

$$|\pm\rangle = \frac{1}{\sqrt{2}}\left(|6\ {}^1S_0\rangle \mp \frac{|\Omega_s|}{\Omega_s}|7^3S_1\rangle + \frac{\Omega_w}{\Omega_s}|6\ {}^3P_2\rangle\right), \tag{3}$$

with all detunings $\Delta_{w,s,p} = 0$ and $\Omega_p = 0$. The corresponding dressed-state picture is shown in Fig. 1(b). For a vanishing coupling laser, $\Omega_w = 0$, the $|\pm\rangle$ states correspond to the usual Autler-Townes splitting in a three-state ladder system. With the coupling laser switched on, the dressed state $|0\rangle$ consists of a mixture of the bare states $|6^3P_2\rangle$ and $|6^3P_1\rangle$ and couples to the ground state $|6^1S_0\rangle$. In the bare-state picture this corresponds to a three-photon transition

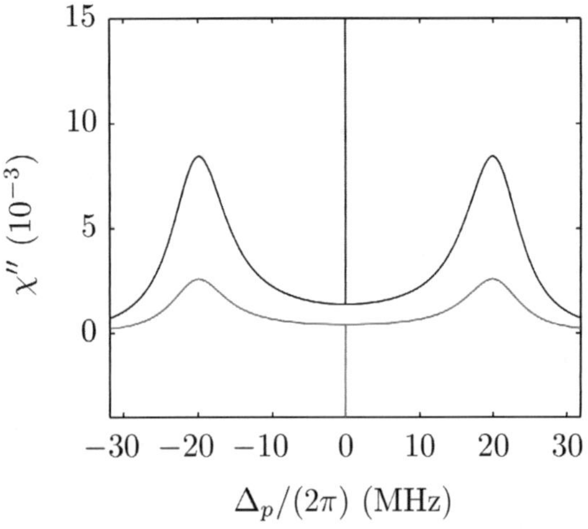

Figure 2. The imaginary part of the susceptibility χ, which corresponds to the absorption on the laser transition, versus the detuning Δ_p with enabled coupling lasers Ω_s and Ω_w. The black curve shows the absorption without an incoherent pump on the 253.7 nm transition. With the incoherent pump switched on (red curve), the sharp absorption spike at $\Delta_p = 0$ turns into a gain spike [2].

between the ground state $|6\,^1S_0\rangle$ and the upper state $|6^3P_2\rangle$ and leads to a sharp absorption feature (cf. Fig. 2).

Fig. 2 shows a simulation of the imaginary part of the susceptibility of the laser transition at 253.7 nm, solving the Bloch equations and employing the natural line widths of the corresponding transitions of mercury. By irradiating an incoherent pump on the lasing transition, the sharp absorption feature becomes a gain spike which could lead to LWI (c.f. red curve in Fig. 2).

The particular advantage of the 4-level scheme for LWI becomes evident when Doppler broadening is taken into account [1]. The photon momenta corresponding to the drive, coupling lasers and laser transition fulfill the triangle equation, i.e. one can find a geometrical orientation of the laser beams such that the momenta fulfill the equation $\vec{k_p} + \vec{k_s} - \vec{k_w} = 0$. Thus, for this alignment the Doppler effect is canceled in one direction for the laser field at 253.7 nm (cf. Fig. 3). This allows a much simpler experimental setup as it becomes feasible to use gas cells rather than much more complicated atomic beam setups.

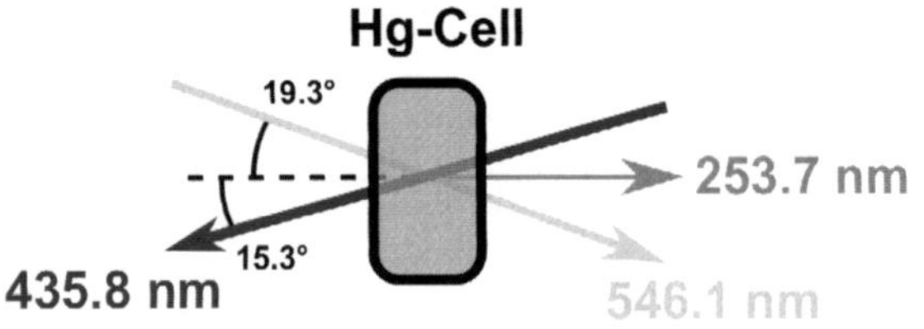

Figure 3. Geometrical orientation of the coupling lasers Ω_s and Ω_w such that the Doppler effect is canceled in one direction for the 253.7 nm laser field.

In order to arrive at a more realistic picture of the necessary experimental parameters, in the next step of the simulation we took into account the Doppler effect with respect to our particular alignment of the laser beams, the finite linewidths of the driving lasers and used the coupling scheme including all Zeeman sublevels and chose the polarization of the laser fields as proposed in [1]. The central part around the gain spike of the imaginary part of the susceptibility under these conditions is shown in Fig. 4. Clearly, the gain spike is dependent on atomic density, pump rate, the Rabi frequencies of the drive and coupling lasers as well as the linewidths of these two lasers. The latter is particularly striking as the dependence is very strong and for large linewidth the gain spike can even vanish. Thus, in an experimental implementation special attention must be paid to achieve the smallest possible linewidths for the corresponding laser radiation.

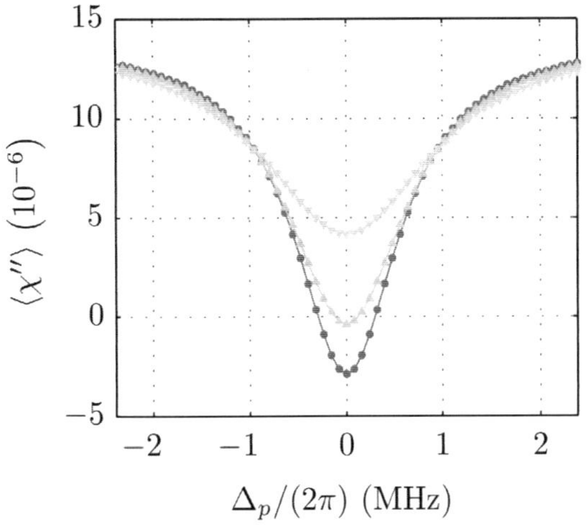

Figure 4. The imaginary part of the susceptibility χ of the laser transition versus the detuning Δ_p for different linewidths of the coupling lasers Ω_s and Ω_w. Red dots correspond to a linewidth of 0 kHz, green triangles to 8 kHz and cyan triangles to 32 kHz for both coupling lasers [2].

The simulations were extended to include the spatial dependence of the gain distribution for actual Gaussian beams (cf. Fig. 5). Based on these calculations with realistic laser parameters, it is possible to choose the optimal length of the mercury vapor cell in which the lasers are overlapped and form the LWI zone.

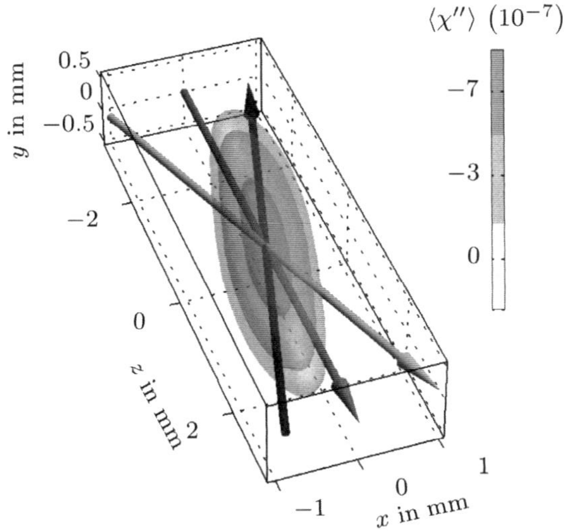

Figure 5. Three dimensional gain distribution of the LWI zone for Gaussian beams and realistic laser parameters [2].

3. Experimental setup

As the results of the previous section reveal, the most critical parts of the experimental setup of LWI in mercury are the driving lasers at 435.8 nm and 546.1 nm. As mentioned in section 2 special care has been taken to achieve small linewidths. Therefore, the two lasers are devised as frequency doubled external cavity diode lasers (ECDL) with a rugged design.

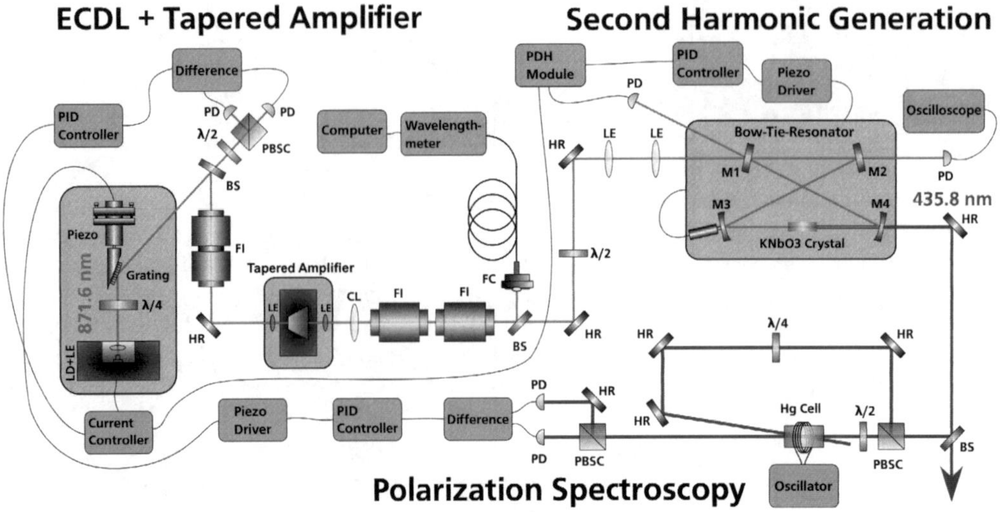

Figure 6. Setup of the 435.8 nm laser system. LD: Laser Diode, LE: Lens, BS: Beam sampler, PBSC: Polarizing Beamsplitter Cube, PD: Photo Diode, FI: Faraday Isolator, HR: High Reflecting Mirror, CL: Cylindrical Lens, FC: Fiber Coupler, M1: Input Coupler, M2-M3: High Reflecting Mirrors, M4: Output Coupler for second harmonic radiation. The $\lambda/4$ and $\lambda/2$ retarder plates, PBSC and photo diodes in the ECDL sector are used for the locking scheme introduced and discussed in [14].

A schematic view of the laser system producing the 435.8 nm radiation is shown in Fig. 6. The ECDL with a fundamental wavelength of 871.6 nm emits about 30 mW which are boosted up to 1.5 W by a tapered amplifier, while the laser diode is protected from backward radiation by a tandem faraday isolator. A novel locking scheme based on polarization spectroscopy [14] enables a stabilization of the ECDL and mode-hop free tuning up to 22 GHz.

After exiting the tapered amplifier the beam is shaped by a cylindrical lens and then coupled into a four-mirror home-built bow-tie ring resonator, consisting of two plane and two curved mirrors. A 10 mm long potassium niobate ($KNbO_3$) crystal is placed in the focal spot between the two curved mirrors. This crystal has a very high nonlinear coefficient of 13.8 pm V^{-1} and leads to an efficient second harmonic generation. To meet the non-critical phase matching condition the temperature of the $KNbO_3$ crystal must be kept very stable. Therefore, the whole resonator is embedded in a solid temperature stabilized aluminum block. Inside the block the $KNbO_3$ crystal has its separate temperature stabilization providing a stability of 9 mK over 2 hours at around 50 °C. The resonator itself is stabilized employing the Pound-Drever-Hall technique to the fundamental laser wavelength. The necessary frequency modulation of 20 MHz is directly modulated on the laser diode current through a home-built ultra low-noise current controller. This leads to 204 mW at 435.6 nm with a peak-to-peak stability of 2.2 % over 60 minutes.

The laser system is frequency stabilized on the 6^3P_1 - 7^3S_1 transition by polarization spectroscopy in a mercury vapor cell [15]. Since the aforementioned excited states are not populated, a constant gas discharge is induced in the cell to excite the mercury atoms. For this purpose the cell contains argon as a buffer gas at a very small pressure, which helps to ignite the discharge [16]. The circularly polarized pump beam induces an anisotropic saturation of the Zeemann sublevels of the transition. Therefore, the linearly polarized probe beam experiences a slight rotation of the polarization axis while scanning the laser frequency over the atomic transition. A polarizing beam splitter cube and two photodiodes detect this rotation and the difference of the two signals produces a dispersion like error signal, which is fed back to the ECDL grating piezo actuator through a PID controller. The frequency is stable to within 2.2 MHz peak-to-peak and a standard deviation of 275 kHz over 60 minutes for a 80 ms sampling interval.

The linewidth of the 435.6 nm radiation is measured with a delayed self-heterodyne beating technique [17] with a short delay fiber of about 410 m corresponding to a delay of 2.04 µs. Fig. 7 shows a beat-note averaged over 10 measurements; the corresponding fit leads to a short-time linewidth of 24.1 kHz.

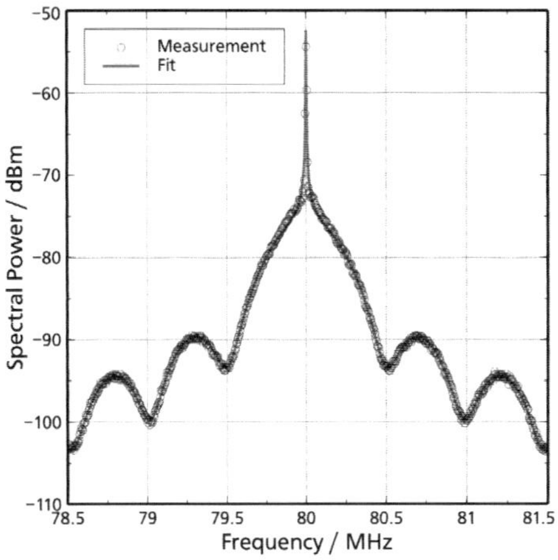

Figure 7. Beat-note of the 435.6 nm radiation averaged over 10 measurements, with fitted theory curve as described in [17].

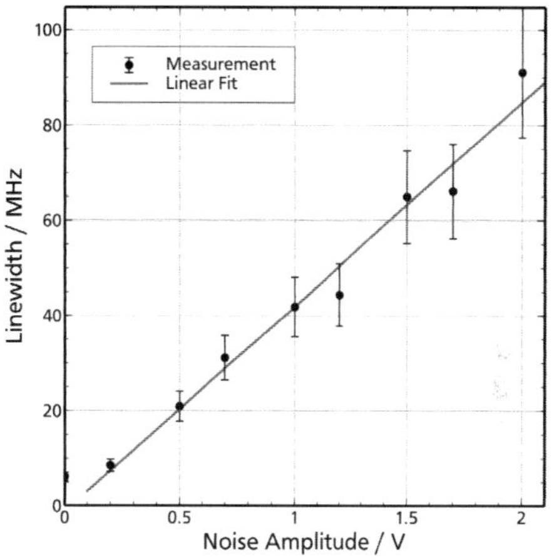

Figure 8. Dependence of the linewidth of the 404.7 nm laser on the white-noise modulation amplitude.

The setup of the 546.1 nm laser system is very similar. Because only 2 mW are required at this wavelength, a high-power laser diode in ECDL configuration, which emits 222 mW at the fundamental wavelength, is sufficient. The nonlinear medium for the second harmonic generation is a 20 mm long lithium niobate crystal with 5.5 % magnesium oxide doping (MgO:LiNbO3) and a theoretical nonlinear coefficient of 4.45 pm V^{-1}. With this configuration we achieved up to 6.3 mW and a conversion efficiency of 7.8 %. The short-time linewidth was measured to be 23.8 kHz.

The incoherent repump laser at 404.7 nm is also based on an ECDL configuration to provide enough power at the desired wavelength for saturating the atomic transition. To broaden the linewidth, the laser diode current is modulated with white-noise directly by a fast-modulation input of our ultra low-noise current controller. In Fig. 8 the dependence of the laser linewidth on the amplitude of the white-noise is shown. The laser frequency is also stabilized in a separate gas discharge by the polarization spectroscopy technique discussed earlier. The system provides up to 25 mW output power.

In order to verify the theoretical model of [1, 2], a laser at 253.7 nm based on fourth harmonic generation will be used to investigate the gain on the LWI lasing transition.

4. Conclusion and outlook

A short summary of our detailed theoretical calculations for lasing without inversion in neutral mercury were presented and the current progress of the experimental realization has been discussed.

The calculations clearly show the feasibility of LWI in mercury and yield the experimental parameters necessary for a successful implementation. The most crucial parameters are the linewidths of the driving lasers.

The presented laser systems meet these demands for the LWI scheme and are ready for first measurements to confirm the calculations of [1, 2]. Currently, we are setting up the gas cell for first measurements of amplification without inversion in neutral mercury.

The authors gratefully acknowledge support by the Deutsche Forschungsgemeinschaft under the research grant "Amplification and Lasing without Inversion in Mercury at 253.7 nm and 185 nm".

References

[1] Fry E S, Lukin M D, Walther T and Welch G R 2000 *Opt. Commun.* **179** 499–504
[2] Sturm M R, Rein B, Walther T and Walser R 2014 *J. Opt. Soc. Am. B* **31** 1964
[3] Kolbe, Daniel and Scheid, Martin and Walz J 2012 *Phys. Rev. Lett.* **109** 063901
[4] Watanabe Y Z, Sato Y, Watanabe N, Ananda R, Okada-Shudo Y, Watanabe M, Hyodo M, Wang X, Chen C, Kanai T and Shuntaro 2009 *Opt. Express* **17** 8119–8124
[5] Mompart J and Corbalán R 2000 *J. Opt. A Pure Appl. Opt.* **2** R7
[6] Südmeyer T, Imai Y, Masuda H, Eguchi N, Saito M and Kubota S 2008 *Opt. Express* **16** 1546–1551
[7] Scheid M, Kolbe D and Markert F 2009 *Opt. Express* **17** 11274–11280
[8] Kocharovskaia O and Khanin Y I 1988 *Pis'ma v Zhurnal Eksp. i Teor. Fiz.* **48** 581–584
[9] Harris S 1989 *Phys. Rev. Lett.* **62** 1033–1036
[10] Scully M, Zhu S and Gavrielides A 1989 *Phys. Rev. Lett.* **62** 2813–2816
[11] Zibrov A S, Lukin M D, Nikonov D E, Hollberg L, Scully M O, Velichansky V L and Robinson H G 1995 *Phys. Rev. Lett.* **75** 1499–1502
[12] Fry E S 1993 *Phys. Rev . Lett.* **70** 3235
[13] Padmabandu G, Welch G, Shubin I, Fry E, Nikonov D, Lukin M and Scully M 1996 *Phys. Rev. Lett.* **76** 2053–2056
[14] Führer T, Stang D and Walther T 2009 *Opt. Express* **17** 4991–4996
[15] Hughes C P P, Adams C S, Cox S G, Griffin P F, Smith D A and G I 2002 *J. Phys. B At. Mol. Opt. Phys.* **35** 5141–5151
[16] Veza D, Salit M L, Sansonetti C J and Travis J C 2005 *J. Phys. B At. Mol. Opt. Phys.* **38** 3739–3753
[17] Führer T and Walther T 2012 *Appl. Phys. B* **108** 249–253

Influence of diamond and graphite bonds on mechanical properties of DLC thin films

Petr Písařík [1,2], Miroslav Jelínek [1,2], Tomáš Kocourek [1,2], Jan Remsa [1,2], Josef Zemek [2], Jaroslav Lukeš [3] and Josef Šepitka [3]

[1] Czech Technical University in Prague, Faculty of Biomedical Engineering, nam. Sitna 3105, 272 01 Kladno, Czech Republic
[2] Institute of Physics ASCR, Na Slovance 2, 182 21 Prague 8, Czech Republic
[3] Czech Technical University in Prague, Faculty of Mechanical Engineering, Technicka 4, 166 07 Prague, Czech Republic

E-mail: petr.pisarik@fbmi.cvut.cz

Abstract. Mechanical properties of diamond-like carbon thin films with various ratios of sp^3/sp^2 bonds were studied. The films were prepared in argon atmosphere (0.25 Pa) by laser deposition method for laser energy densities from 4 J·cm^{-2} to 14 J·cm^{-2}. The sp^2 and sp^3 bonds were calculated by X-ray photoelectron spectroscopy. Films contained sp^3 bonds up to 70 %. Surface properties as roughness and atomic force microscopy topology were measured. Hardness (and reduced Young's modulus) were determined by nanoindentation and reached of 30 GPa (203 GPa). Films adhesion was studied using scratch test and was up to 12 N for biomedical alloy (titanium substrates - Ti-6Al-4V). Relations among deposition conditions and measured properties are presented.

1. Introduction

Diamond-like carbon (DLC) is a metastable form of amorphous carbon consists of hybridization in different percentage of sp^1 (acetylene-like), sp^2 (graphite-like), and sp^3 (diamond-like) bonding. DLC films are semiconductors with high mechanical hardness, chemical inertness, low coefficient of friction and optical transparency. In this work the characteristics of DLC films, possibilities of applications in orthopedics, cardiovascular surgery together with other medical field are presented [1, 2, 3, 4, 5, 6]. But adhesion of DLC films on biomedical alloy substrates (Ti-6Al-4V, Co-Cr-Mo and stainless steel) is poor [3, 7]. It is caused by stress in the film due to different hardness and different coefficient of thermal expansion of substrate and film. The content of sp^2, sp^3 bonds exhibits influence on optical, biological [8], electrical and mechanical properties [9]. Here is presented series of six samples with different ratio of sp^3 and sp^2 bonds and responsible mechanical properties.

2. Experimental details

2.1. Deposition

DLC films were prepared using KrF excimer PLD. Laser beam was focused onto a high purity graphite target with energy densities from 4 J·cm^{-2} to 14 J·cm^{-2}. Substrate (Si 111 and Ti-6Al-4V) was

Journal of Physics: Conference Series **594** (2015) 012008 doi:10.1088/1742-6596/594/1/012008

placed in a distance of 45 mm from the target. The high purity graphite target was rotated (0.5 Hz). The number of pulses was adjusted to reach approximately the same film thickness (400 nm for films on Ti-6Al-4V substrate - ISO 5832-3 and 100 nm on Si 111 substrate). The base vacuum of the coating system was 1×10^{-4} Pa. Before deposition the substrates were RF cleaned (13.56 MHz) in 5 Pa of argon for two minutes. The films were deposited in argon ambient (0.25 Pa), with gas (argon) flow of 10 sccm. The DLC films were created at room substrate temperature [4, 6, 8]. Deposition parameters are summarized in Table 1.

Table 1 Deposition conditions of DLC films created by PLD. Carbon atoms sp^3 hybridization content measured by XPS (C 1s).

Sample (Substrate)	Laser energy density ($J{\cdot}cm^{-2}$)	PFM thickness (nm)	PFM Ra (nm)	AFM Ra (nm)	AFM RMS (nm)	XPS (C 1s) sp^3	XPS (C 1s) sp^3/sp^2
DLC-01 (Si 111)	4	111	4.0	0.381	1.110	58	1.38
DLC-02 (Ti-6Al-4V)		426					
DLC-03 (Si 111)	6	113	6.6	0.356	0.771	59	1.44
DLC-04 (Ti-6Al-4V)		380					
DLC-05 (Si 111)	8	125	4.1	0.712	1.612	65	1.86
DLC-06 (Ti-6Al-4V)		253					
DLC-07 (Si 111)	10	94	4.3	0.471	1.633	64	1.78
DLC-08 (Ti-6Al-4V)		315					
DLC-09 (Si 111)	12	91	2.5	0.580	2.261	67	2.03
DLC-10 (Ti-6Al-4V)		400					
DLC-11 (Si 111)	14	113	1.5	0.635	0.975	70	2.33
DLC-12 (Ti-6Al-4V)		402					

2.2. *Ratio of sp^2 and sp^3 bonds - XPS*

The percentage of sp^2 and sp^3 hybridized carbon atoms was estimated by XPS. The C 1s spectra were measured by an ADES-400 photoelectron spectrometer (VG Scientific, UK) using Mg Kα excitation (1253.6 eV) and a hemispherical electron energy analyser. Photoelectron spectra were recorded for wide-survey and narrow scans in C 1s and O 1s regions with a pass energy 100 eV or 20 eV (C 1s line) [8]. Inelastic electron background was subtracted using the Shirley`s procedure [10].

2.3. *Thickness, film roughness and topology - PFM, AFM.*

The film thickness and the roughness parameter Ra (arithmetic average values) were determined using an Alphastep IQ mechanical profilometer (KLA Co.). Topology was characterized using atomic force microscopy (AFM) type NT-NDT Solver Next P7LS.

2.4. *Hardness and Reduced Young's modulus – Nano indenter*

A system for nanomechanical testing (hardness, reduced Young's modulus) Hysitron TI TriboIndenter 950™ (Berkovich tip) was used for analysis of films on titanium [11, 12]. To minimize the influence of the substrate on hardness, the indentation depth was kept to less than 10 % of coating thickness

(~40 nm). Trapezoidal shape of loading curve was employed – loading, dwell, unloading segment time 5-2-5 seconds. It was indented in matrix 4x4 indent with separation of 5 μm.

2.5. Adhesion – Scratch tester
For determination of adhesions DLC films on Ti-6Al-4V substrate we used macro scratch tester (REVETEST Scratch Tester - CSM co.). The normal load of the indenter (Rockwell, diamond type, Radius – 200 μm) was linearly ramped from the minimum to the maximum during scratching. The critical loads Lc2 (edge spallation) and Lc3 (spallation inside the groove) were determined by optical microscopy after testing. Figure 1 shows the CCD micrograph of the scratch tracks. Here, the minimum load and the maximum load were 1 N and 30 N, respectively. In the test, the scratch length was 8 mm and the scratch speed was 2.76 mm/min.

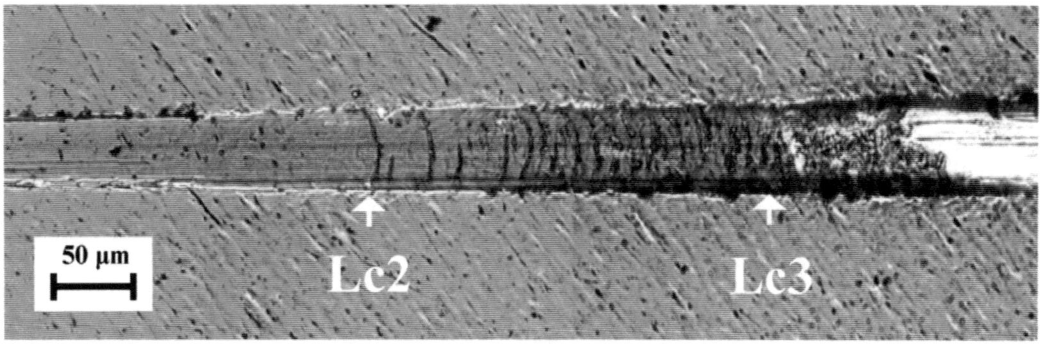

Figure 1 Optical micrograph of typical scratch tracks on DLC coated on Ti-6Al-4V substrate (The critical loads Lc2 - edge spallation and Lc3 - spallation inside the groove)

3. Results and discussion
DLC films were fabricated for laser energy densities of 4, 6, 8, 10, 12 and 14 J·cm^{-2}. Films prepared on Si substrates were used for thickness, AFM topology and XPS measurements. Films prepared on Ti-6Al-4V were used for determinated adhesion, hardness and Reduced Young's modulus.

3.1. Ratio of sp^2 and sp^3 bonds - XPS
The procedure are described in [6, 8] and the results are summarized in Table 1. The percentage of the sp^2 and sp^3 hybridized carbon atoms in the analyzed volume of the films was revealed from a peak-fit of the high-resolution C 1s spectra. As illustrated in Figure 2 three Gaussian lines correspond to the C sp^2 bonding peaked at 284.3 eV, C sp^3 at 285.2 eV, and the C-O at about 286 eV [8]. The percentage of the sp^3 bonds increased with increasing laser energy density. The maximum of sp^3 bonds (70 %) was measured by XPS for energy density of 14 J·cm^{-2}.

3.2. Thickness, film roughness and topology - PFM, AFM
Alpha-step profilometer with diamond tip (radius 5 μm, 60°) scanned the 2-mm film surface. With Gaussian filter (cutoff 250 μm) the arithmetic average of absolute values Ra parameter was calculated. Using AFM in noncontact regime the area of 10×10 μm was scanned and the roughness parameter Sq root mean squared (RMS) was determined. No distinct RMS changes were observed with changing deposition conditions, see Table 1. The films were generally smooth with rare droplets, usually tenths of nanometers in high, see Figure 3.

Journal of Physics: Conference Series **594** (2015) 012008 doi:10.1088/1742-6596/594/1/012008

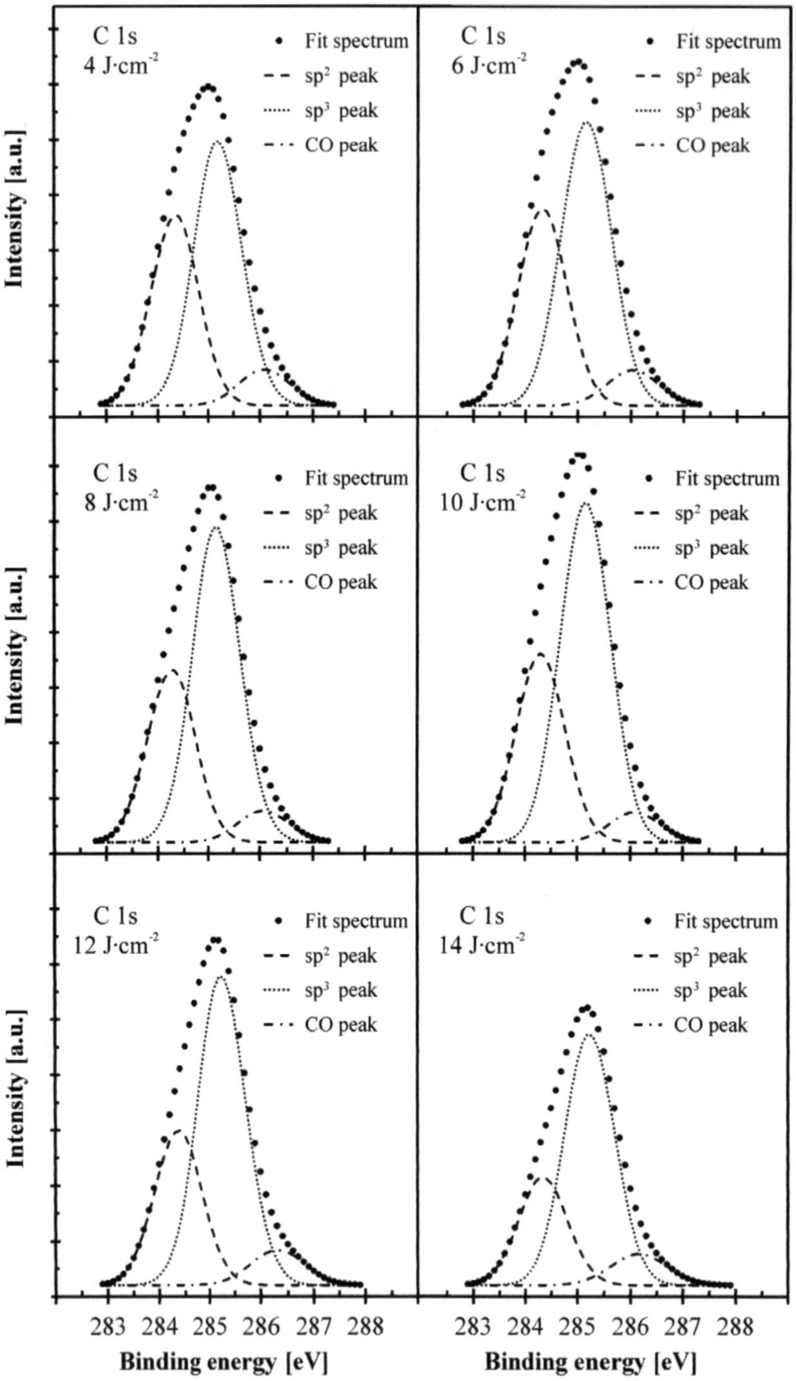

Figure 2 High-energy resolution C 1s peaks with illustration of peak fitting onto sp^2 at 284.2 eV, sp^3 at 285.2 eV, and C-O at 286.5 eV contributions for laser energy densities from 4 $J{\cdot}cm^{-2}$ to 14 $J{\cdot}cm^{-2}$

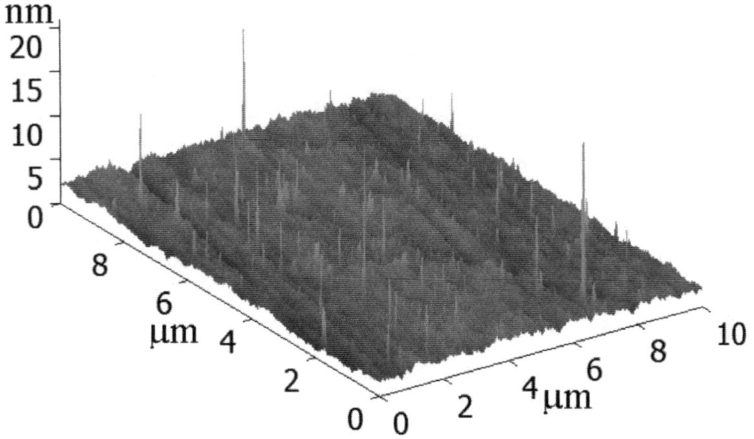

Figure 3 Typical AFM surface of DLC film (DLC-03 - energy density of 6 J·cm^{-2})

3.3. Hardness and Reduced Young's modulus – Nano indenter

Hardness and reduced Young's modulus of DLC films depends strongly on the deposition conditions and the type of deposition techniques. There are various values of hardness a and reduced Young's modulus reported in the literature [13, 14, 15, 16, 17, 18, 19], from units GPa to tens of GPa, from tens to hundreds of GPa respectively.

Hardness our layers grown from 0.8 GPa to 30.2 GPa, as a percentage of sp^3 bonds increased, as well as reduced Young's modulus of 17.8 GPa to 203.3 GPa - see Table 2. These results are similar to those of [13, 14].

Table 2 Hardness, reduced Young's modulus and adhesion (Lc2, Lc3) for laser energy densities from 4 J·cm^{-2} to 14 J·cm^{-2}

Sample (Ti-6Al-4V substrate)	DLC-02	DLC-04	DLC-06	DLC-08	DLC-10	DLC-12
Laser energy density (J·cm^{-2})	4	6	8	10	12	14
Lc2 (N)	8,3	5,5	9,1	5,5	7,6	6,8
Lc3 (N)	12,5	11,4	14,5	9,4	11,6	10,7
Hardness (GPa)	0,8	2,0	--	7,7	10,8	30,2
Reduced Young's modulus (GPa)	17,8	21,7	--	66,4	68,1	203,3

3.4. Adhesion – Scratch tester

The adhesion of DLC films on biomedical alloy substrates is poor [3, 7]. Sheeja found [20] that the critical load of the DLC films to silicon substrate was about 2.5 N. The higher value of the critical load in this study may be due to application of the Ti interfilm and control of the substrate temperature during deposition. Similar results were presented Baragetti [21]. DLC films on aluminium alloy exhibited critical load Lc3 about 3.5 N [21]. Jelínek [22] presented better results, the adhesion was about 14 N. This was probably caused by the use of RF cleaning and ion bombardment during growth of the layer.

Scratch test shows (Table 2) that the maximum critical load Lc3 of our DLC films was more than 14 N. This is a higher or same value published in the literature for similar DLC films (without buffer and doping). The same results as [22] is probably achieved by using RF surface cleaning before the deposition.

4. Conclusion

Diamond like carbon films with different content of sp^3 bonds were prepared by PLD using energy densities from 4 $J\cdot cm^{-2}$ to 14 $J\cdot cm^{-2}$. The percentage of the sp^3 bonds increased with increasing laser energy density. The maximum of sp^3 bonds (70 %) was measured by XPS for energy density of 14 $J\cdot cm^{-2}$. The films were generally smooth with rare droplets, usually tenths of nanometers in high.

Hardness our layers grown from 0.8 GPa to 30.2 GPa, as a percentage of sp^3 bonds increased, as well as reduced Young's modulus of 17.8 GPa to 203.3 GPa - see Table 2. These results are similar to those of [13, 14].

Scratch test shows (Table 2) that the maximum critical load Lc3 of our DLC films was more than 14 N. These results don't show any dependence with sp^3 content and adhesion.

Acknowledgements

This work was supported by the Grant Agency of the Czech Technical University in Prague, grant No. SGS14/096/OHK4/1T/17 and SGS14/168/OHK4/2T/17, the Grant Agency of the Czech Republic (grant P108/13-33056S).

References

[1] Robertson J 2002 *Mater. Sci. Eng.: R: Rep.* **37**(4-6) 129
[2] Grill A 1999 *Diam. Relat. Mat.* **8**(2-5) 428
[3] Narayan RJ 2002 *Functionally Gradient Hard Carbon Composites for Improved Adhesion and Wear* (PhD dissertation, North Carolina State University, Ralleigh, North Carolina, United State), *Mater. Sci. Eng.* p. 292.
[4] Kocourek T, Jelinek M, Vorlicek V, Zemek J, Janca T, Zizkova V, Podlaha J and Popov C 2008 *Appl. Phys. A* **93**(3) 627
[5] Jelinek M, Smetana K, Kocourek T, Dvorankova B, Zemek J, Remsa J, Luxbacher T 2010 *Mater. Sci. Eng. B* **169**(1-3) 89
[6] Jelinek M, Kocourek T, Remsa J, Miksovsky J, Zemek J, Smetana K Jr, Dvorankova B and Luxbacher T 2010 *Appl. Phys. A* **101**(4) 579
[7] Laube SJP, Voevodin AA, Keener KD, LeClair SR. 1998 *Eng. Appl. Artif. Intell.* **11**(5) 649
[8] Písařík P, Jelínek M, Smetana K Jr., Dvořánková B, Kocourek T, Zemek J, Chvostová D 2013 *Appl. Phys. A* **112**(1) 143
[9] Dwivedi N, Kumar S, Malik HK, Govind, Rauthan CMS, Panwar OS 2011 *Applied Surface Science* **257**(15) 6804
[10] Shirley DA 1972 *Phys. Rev. B* **5**(12) 4709
[11] Fischer-Cripps AC 2004 *Nanoindentation (Mechanical Engineering Series) 2nd edition* (New York: Springer)
[12] Oliver WC, Pharr GM 1992 *J. Mater. Res.* **7**(6) 1564
[13] Logothetidis S, Charitidis C, Patsalas R 2002 *Diam. Relat. Mat.* **11**(3-6) 1095
[14] Lu W, Komvopoulos K, Patsalas P, Charitidis C, Gioti M, Logothetidis S 2003 *Surf. Coat. Technol.* **168**(1) 12
[15] Bruno P, Cicala G, Losacco AM, Decuzzi P 2004 *Surf. Coat. Technol.* **180** 259
[16] Lemoine P, Zhao JF, Quinn JP, McLaughlin JA, Maguire P 2000 *Thin Solid Films* **379**(1-2) 166
[17] Ferrari AC, Robertson J 2000 *Phys. Rev. B* **61**(20) 14095
[18] Teo EHT, Chua DHC, Tay BK 2007 *Diam. Relat. Mat.* **16**(10) 1882
[19] Ferrari AC, Robertson J, Beghi MG, Bottani CE, Ferulano R, Pastorelli R 1999 *Appl. Phys. Lett.* **75**(13) 1893
[20] Sheeja D, Tay BK, Lau SP, Shi X, Shi J, Li Y, Ding X, Liu E, Sun Z 2000 *Surf. Coat. Technol.* **127**(2-3) 247
[21] Baragetti S, Lusvarghi L, Bolelli G, Tordini F 2009 *Surf. Coat. Technol.* **203**(20–21) 3078
[22] Jelínek M, Písarík P, Kocourek T, Zemek J, Lukeš J 2013 *Appl. Phys. A* **110**(4) 943

Two-photon resonance on the D_2 line of Cs enhanced by optical pumping

A. Krasteva[1], D. Slavov[1], A. Sargsyan[2], D. Sarkisyan[2], A. D. Wilson-Gordon[3] and S. Cartaleva[1*]

[1]Institute of Electronics, BAS, 72 Tzarigradsko Shosse bld, 1784 Sofia, Bulgaria
[2]Institute for Physical Research, Academy of Sciences of Armenia, Ashtarak-2, 0203, Armenia
[3]Department of Chemistry, Bar-Ilan University, Ramat Gan 5290002, Israel

e-mail: stefka-c@ie.bas.bg

Abstract. Sub-natural-width *N*-type resonance, observed in the D_2 line of Cs vapor has been reported recently. Three-photon, bi-chromatic excitation of Cs atomic vapor with 20 Torr of Neon was performed and a sub-natural-width resonance was observed, superimposed on the absorption profile of a set of hyperfine transitions starting from single hyperfine ground level. Here we report the study of a different three-photon, couple-probe- excitation scheme that allows the observation of a sub-natural-width resonance situated on a flat background, well outside the D_2 line absorption spectrum of Cs. The lack of a noisy background could be advantageous for many applications related to the narrow *N* resonance. In addition, a new broader feature of enhanced transparency is demonstrated that is centered at the ground state level that does not interacting with the coupling beam. This enhanced transparency resonance can be attributed to the atomic population loss in a narrow spectral interval occurring due to excitation of energy levels situated above the $6\,^2P$ levels.

1. Introduction

Coherent laser spectroscopy of alkali vapor contained in thermal optical cells is widely used for various applications, including frequency references, atomic clocks, precise optical magnetometers, slow and stored light, and photonic sensors based on coherent population trapping (CPT) and electromagnetically induced transparency (EIT) resonances, etc. For those applications, various research groups are still exploring suitable approaches aimed at a profitable combination of the advantages of CPT/ EIT with other processes. Optical systems based on Rb atoms, in which narrow band *N*-type resonances appear, have been extensively studied recently [1-4]. Such *N*-resonances can be considered as a type of three-photon resonance where a two-photon Raman excitation is combined with a resonant optical pumping field.

A sub-natural-width *N*-resonance has recently been observed in Cs vapor [5]. The two-photon Raman excitation in Cs atoms is less effective than in Rb vapor due to the higher ground-state hyperfine splitting in Cs. Nevertheless, a high-contrast *N*-resonance have been observed in Cs vapor that can be attributed to the higher rate of the optical pumping process in Cs than in Rb. Three-photon, bi-chromatic excitation of Cs atomic vapor, contained in a 10 mm long cell with 20 Torr of neon has been performed and successfully applied to measurement of the magnetic field over a large range [5]. A coupling laser was fixed at frequency 9.2 GHz lower than the position of the absorption profile of the $F_g = 4$ set of transitions, while the probe laser was tuned over the D_2 line ($\lambda = 852$ nm).

A narrow high-contrast enhanced absorption *N*-resonance has been observed in the probe light profile, superimposed on the absorption profile of the $F_g = 4$ set of transitions.

In this communication, we report a systematic study and analysis of a less explored scheme that has been applied to Rb vapor [1]. In the case of the D_2 line of Cs, such an approach allows the observation of a sub-natural-width resonance on a flat background, well outside of the D_2 line spectrum. Here, bi-chromatic excitation is also used, but the coupling laser frequency is fixed within the absorption profile of the $F_g = 3$ or $F_g = 4$ set of transitions, and the transmission of the coupling beam is measured while tuning the probe beam frequency over the D_2 line.

2. Experimental set up for sub-Doppler and sub-natural width resonance study

Figure 1 shows a schematic diagram of the experimental setup we used. The two narrow-band distributed feedback (DFBs) diode-lasers systems used in our experiment, are built as in Ref. [5] using two DFB diodes, both with an emission spectral width of about 2 MHz (FWHM). The two

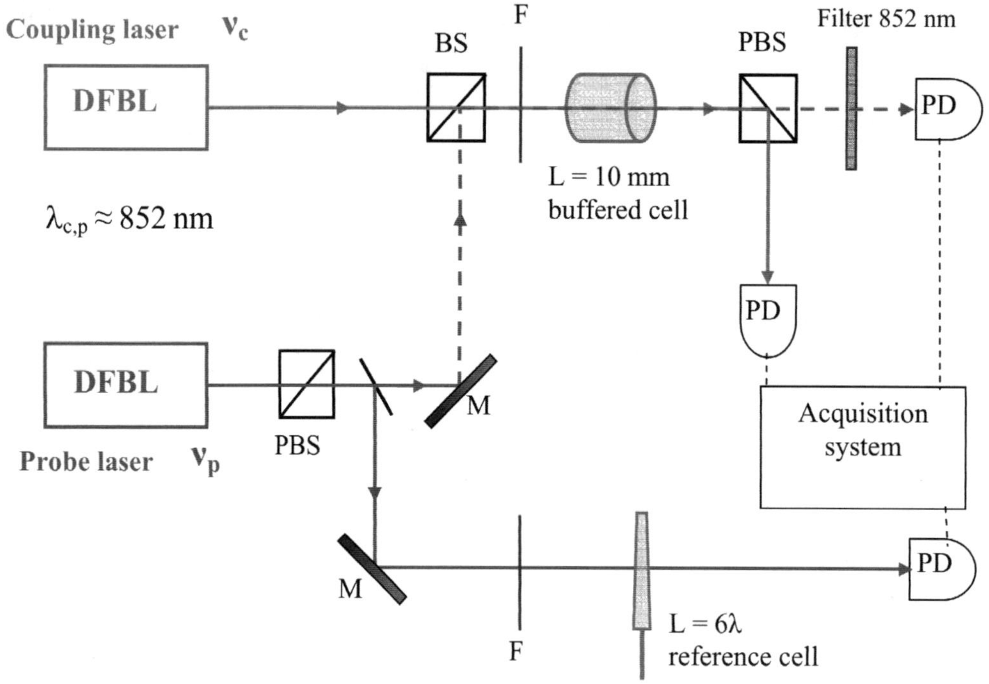

Fig.1 Schematic diagram of the experiment: DFBL - distributed feedback diode-laser: coupling and probe, PBS -polarizing beam splitters, PD - photo diodes, M-mirrors, F - neutral optical filters, interference filter of a wavelength of 852 nm, and acquisition system. Sub-natural-width resonance is registered in the transmission of the $L = 10$mm buffered cell. The $L = 6\lambda$ cell provides a reference spectrum.

laser systems show a mode-hop-free laser frequency detuning that largely exceeds the absorption spectrum of the Cs D_2 line. The coupling laser has a fixed frequency ν_c, while the probe laser with frequency ν_p is tuned over the D_2 line. Both laser beams have spot size 3x2mm and are carefully superimposed on a beam splitter BS and then directed at the cell containing atomic vapor. The intensities of both lasers falling on the $L = 10$ mm cell are equal, $I = 500$ mW/cm^2. The polarizations of the coupling and probe lasers are linear and mutually orthogonal. The $L = 10$ mm long optical cell is filled with atomic Cesium with 20 Torr of neon gas added that serves as a buffer. Part of the probe radiation is directed to an additional cell with thickness $L = 6\lambda$, which is used as a frequency reference [5]. In the experiments, the transmission of the coupling beam is registered with the probe beam scanned in a large frequency interval over the D_2 line absorption. The probe beam is separated from the coupling laser by a polarizing beam splitter.The broad transmission spectrum of the coupling laser is measured with a sub-natural-width two-photon resonance superimposed on it, for different Cs vapor densities and frequency positions of the coupling beam situated within the $F_g = 4$ transmission profile.

3. N_p resonance within absorption profile, observed in the probe beam transmission

For clarity, the formation of the *N*-type resonance reported in Ref. [5] is briefly discussed. A schematic diagram of the relevant atomic levels and optical transitions is shown in Fig.2. The lower levels used for formation of the required Λ system are the two ground $F_g = 3, 4$ levels and the upper level is $6P_{3/2}$ which consists of four hyperfine levels $F_e = 2, 3, 4, 5$. The probe laser frequency ν_p is tuned over the $F_g = 4 \rightarrow F_e = 3, 4, 5$ set of transitions and the coupling laser frequency ν_c is fixed. In the probe beam transmission spectrum, an *N*-type resonance involving two-photon process has been observed when the two-field frequency difference was $\nu_p - \nu_c = \Delta$, where Δ is the splitting ($\Delta = 9.2$ GHz) of the electronic ground state hyperfine levels. Due to the fact that the resonance is observed in the probe beam transmission, further on we denote it as N_p resonance.

An example of an experimentally observed N_p resonance is presented in Fig. 3. Here, the frequency of the coupling laser is fixed in the lower-frequency wing of the D_2 line absorption and the transmission of the probe laser is measured as a function of its frequency detuning. The absorption spectrum of $L = 6\lambda$ cell was also measured and used as a frequency reference. The probe beam frequency was scanned in a spectral region broader than 30 GHz, thus providing the transmission spectrum of the D_2 line and the frequency position of the coupling beam. A good contrast, sub-

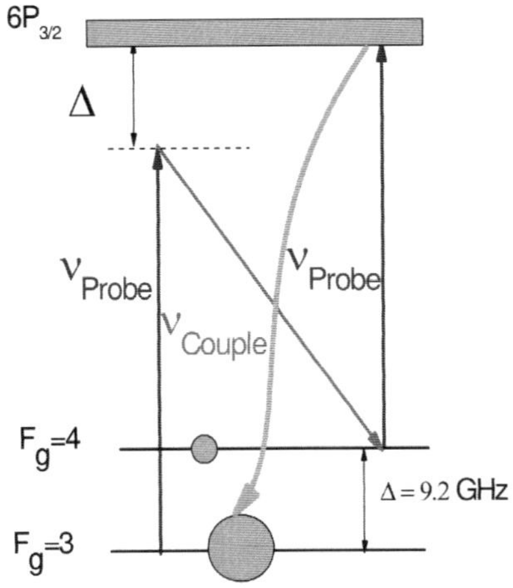

Fig.2 Relevant transitions of D_2 line of Cs. Probe frequency ν_P is scanned over the $F_g = 4 \rightarrow F_e = 3, 4, 5$ transitions, while the coupling frequency ν_c is fixed at a lower frequency [5].

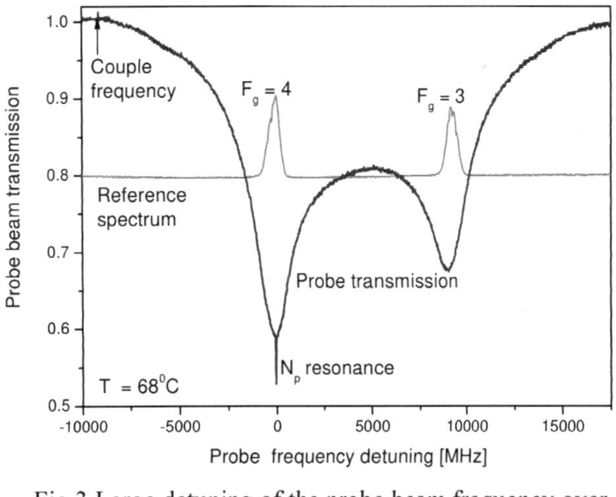

Fig.3 Large detuning of the probe beam frequency over the coupling-probe beating signal (showing the fixed-frequency position of the coupling beam) and both sets of hyperfine transitions starting from $F_g = 4$ and $F_g = 3$ levels (blue). Transmission spectrum of the D_2 line involving the N_p resonance situated within the $F_g = 4$ absorption set (magenta). The spectrum of the $L = 6\lambda$ cell is used as a frequency reference only.

natural (~ 1.5 MHz) line width N_p resonance is observed at the center of the absorption profile. The N_p-resonance frequency position is shifted from that of the coupling laser by the electronic ground-state hyperfine level splitting Δ. The atomic source temperature is kept at 68°C. Under this condition the density of cesium atoms was approximately $1.7\text{x}10^{12}$ atoms/cm^3.

4. Formation of narrow coherent resonance in the transmission of the fixed-frequency coupling beam

In the experimental study presented here, the frequency of the couple beam ν_c is fixed within the absorption profile of the $F_g = 4$ or $F_g = 3$ set of transitions. Now again the frequency of the probe beam is scanned over a large spectral interval. However, the transmission of the fixed-frequency coupling beam is measured.

Mainly, we consider the case where the coupling laser frequency is within the $F_g = 4$ absorption profile. The corresponding atomic levels, as well as the two-photon and the single-photon transitions, are shown in Fig. 4. The coupling beam is in resonance with the absorption profile of the $F_g = 4$ set of transitions. Under this condition, Cs atoms are accumulated by the coupling laser in the $F_g = 3$

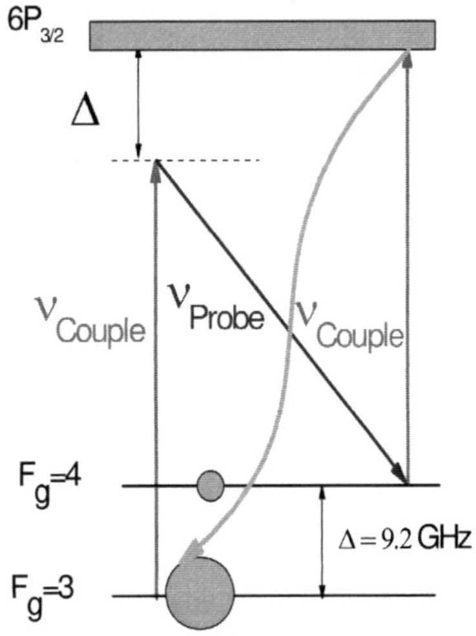

Fig. 4 Probe frequency ν_P is scanned over a large spectral interval, while ν_c is fixed within the $F_g = 3{,}4$ set of transitions.

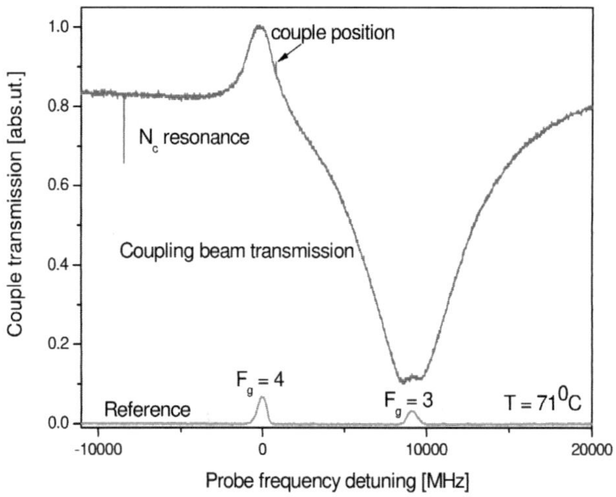

Fig. 5 Transmission spectrum of the fixed-frequency coupling beam, as probe is scanned. The N_c resonance is observed on a flat background, outside the absorption profile.

level due to the hyperfine optical pumping. During the probe beam detuning, when the Raman condition is not fulfilled, the $F_g = 4$ level is significantly depleted by the coupling beam. In case of two-photon Raman resonance, the atoms accumulated in the $F_g = 3$ level are transferred back to the $F_g = 4$ level by the coherent two-photon process, and are detected as enhanced absorption of the coupling beam. However now, the sub-natural-width enhanced absorption feature is observed for probe light frequency much lower than the frequencies of the $F_g = 4$ set of transitions, that is, outside the D_2 line absorption profile.

In Fig. 5, the experimentally observed sub-natural-width resonance is shown for cell temperature of T = 71^{0}C (i.e. Cs atom density is $2.1\text{x}10^{12}$ atoms/cm^3). Starting from the left side of the figure, first the very narrow resonance N_c can be seen, which is the enhanced absorption experienced by the coupling beam. The observed N_c resonance is well outside of the absorption profile of the $F_g = 4$ set of transitions. The frequency ν_p of the probe beam at which the N_c resonance is observed can be determined by the expression $\nu_p = \nu_c - \Delta$.

When the probe beam is tuned in resonance with the $F_g = 4$ set of transitions, the coupling beam experiences almost 100% transparency, due to the simultaneous depletion of the $F_g = 4$ level by the coupling and probe beams. Further increase in the probe frequency results in a reduction of the coupling beam transmission. The minimum transmission (i.e. maximum absorption) of the coupling beam is observed when the probe frequency is tuned to the $F_g = 3$ set of transitions (Fig. 5).

Note that the complete transmission resonance of the coupling beam, centered at $F_g = 4$, is narrower than the absorption resonance observed when the probe beam is detuned across the $F_g = 3$ set. The absorption resonance is attributed to the strong repumping of the $F_g = 4$ level caused by the probe light during its detuning over the $F_g = 3$ set of transitions. Thus, the $F_g = 4$ level depleting is within a narrower frequency interval than its repopulation by the probe light scanned around the $F_g = 3$ transitions. Such a difference can be related to velocity-changing collisions [6] between Cs and Ne atoms that take place during the fluorescence decay of Cs atoms from excited states to the $F_g = 4$ level after being excited from the $F_g = 3$ level by the probe beam. Our experimental study has shown that the spectral width of the enhanced absorption spectral region is strongly influenced by the Cs atom density (Fig. 6). The following measurement is performed for the coupling frequency fixed

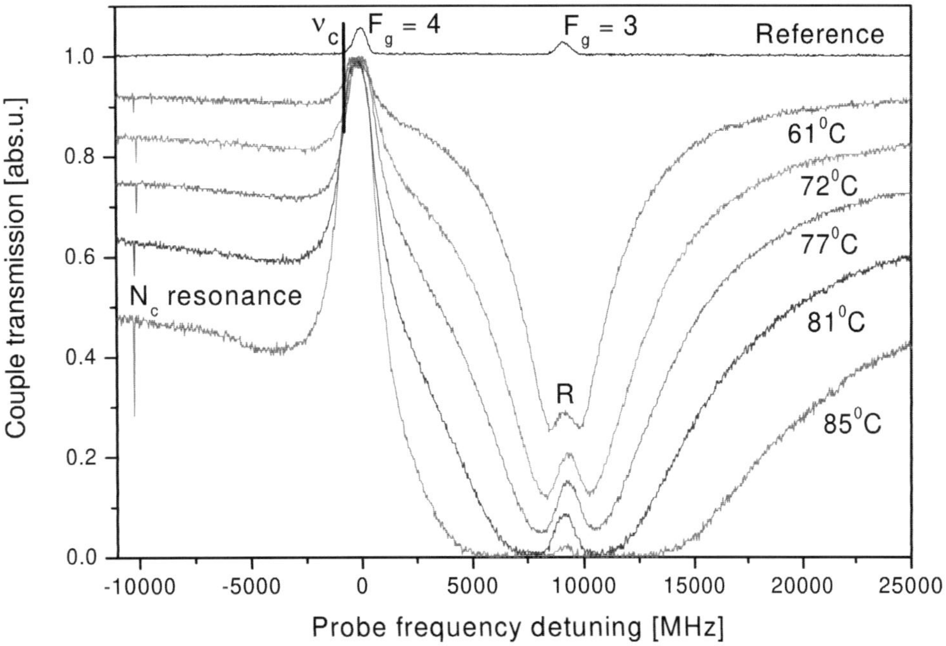

Fig. 6 Transmission spectra of the coupling beam when the probe is detuned over a spectral interval of more than 30 GHz. The coupling frequency ν_c is red detuned from the $F_g = 4$ absorption profile by 900 MHz. Here, in addition to the N_c resonance, a new much broader resonance (R) of reduced absorption appears superimposed on the coupling beam transmission, when the probe is resonant with the $F_g = 3$ level transitions. The spectra are shown for different atomic source temperatures; Cs atom densities vary from $(1 \div 5.6)\text{x}10^{12}$ atoms/cm^3.

close to the absorption maximum of the $F_g = 4$ set of transitions (ν_c is detuned by 900 MHz lower than the $F_g = 4$ absorption maximum). The narrow N_c resonance is observed for a large interval of atomic source temperatures, with strongly enhanced amplitude (up to 20% in transmission) for higher atomic density.

As the probe frequency approaches the absorption profile of the $F_g = 3$ set of transitions, and the atomic source temperature increases, the coupling beam experiences complete absorption over a very large spectral interval. The complete absorption of the coupling beam can be explained by the strong repumping to the $F_g = 4$ level caused by the probe beam.

In Fig. 6, we see a new broad resonance *R* of enhanced transparency centered at the $F_g = 3$ group. The behavior of the new resonance was investigated at different temperatures of the cell. Our experiments show that as the atomic density decreases, the amplitude of the new *R* resonance first increases significantly but then decreases at even lower temperatures, eventually disappearing. From Fig. 6, the following conclusion can be drawn: while the coupling absorption profile experiences strong broadening with increasing Cs atom density (for probe detuning around $F_g = 3$ set of transitions), the spectral width of the *R* resonance is practically unaffected by the atomic source temperature.

After studying the effect of the temperature, the next experiment was performed with a constant high atomic density. Here, the coupling frequency was changed, setting it at different frequencies lower than that of the $F_g = 4$ absorption maximum. Significant reduction of the N_c resonance amplitude takes place with the ν_c detuning from the $F_g = 4$ absorption profile maximum (Fig. 7). The amplitude of the N_c resonance is larger when the frequency ν_c is closer to the absorption profile

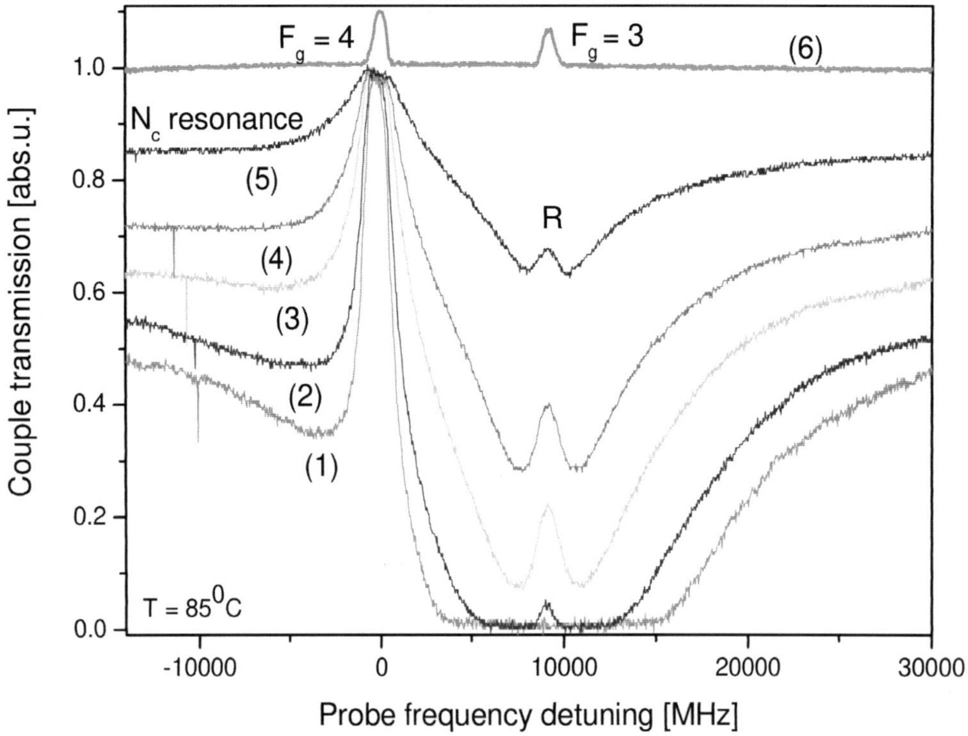

Fig. 7 Transmission spectrum of the coupling beam when the probe is scanned over a spectral interval of more than 40 GHz. The amplitude of the new *R* resonance depends on the coupling frequency ν_c detuning: (1) -869 MHz, (2) -1026 MHz, (3) -1484 MHz, (4) -2162 MHz, (5) -4214 MHz, (6) reference spectrum. The density of cesium atom is approximately $5.6\text{x}10^{12}$ atoms/cm^3.

maximum of the $F_g = 4$ group, but at the same time the slope of the coupling transmission background around the N_c resonance increases.

The position of the coupling frequency influences also the complete transparency resonance shape and amplitude. This resonance is observed when the probe beam frequency ν_p is tuned to the $F_g = 4$ set. In the case of ν_c detuning - 4214 MHz (i.e. large detuning of the coupling beam), the absorption of the couple beam is the lowest (hence the transmission is the highest, Fig.7, curve 5) and the depletion of the $F_g = 4$ level by the probe light results in low amplitude of the complete transparency resonance. The opposite is the case when the ν_c detuning is - 869 MHz, where the absorption rate of the coupling beam is much higher. This is the reason for the observation of the highest absorption background around the N_c resonance (Fig.7, curve 1). However, when the probe beam is in resonance with the $F_g = 4$ group, this ground level is strongly depleted and the coupling transmission approaches 100%. Thus, the highest amplitude, complete transparency resonance is observed.

In addition, a rather fast reduction of the coupling absorption around the $F_g = 3$ transitions occurs with the coupling frequency detuning. Let us first consider a small detuning of the coupling frequency (- 869 MHz). Here, complete absorption of the coupling beam is measured over a very large region, for probe frequency detuning around the $F_g = 3$ set of transitions. Under condition of high Cs atom density, the $F_g = 3$ level is effectively populated by the coupling beam, due to the hyperfine optical pumping. However, a strong repopulation of $F_g = 4$ level by the probe beam also takes place. In this way, Cs atoms are cycled between both ground hyperfine levels.

The situation is different for large detuning (- 4214 MHz) of the coupling frequency from the $F_g = 4$ group. In this case, the optical pumping of the $F_g = 3$ level by the coupling beam is much less effective, and hence $F_g = 4$ level repopulation by the probe light is lower. This results in significantly lower absorption of couple beam.

The new *R* resonance is also influenced by the coupling detuning. At small detuning (Fig.7, curve 1), the *R* resonance is completely washed out, probably due to complete absorption of the coupling beam over a large interval of probe detuning. Then as the detuning increases, the R resonance appears and significantly enhances its amplitude (Fig.7, curves 3, 4). At even greater detuning, the *R* resonance amplitude decreases (Fig.7, curve 5).

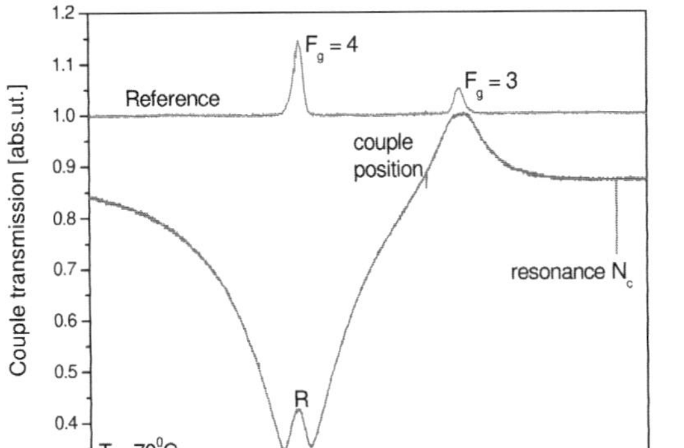

Fig. 8 Transmission of the coupling beam, for $\nu_c = 7360$ MHz. Here, the *R* resonance is formed at the $F_g = 4$ group. The sub-natural-width N_c resonance is situated at frequency higher than that of the $F_g = 3$ group, well outside the absorption profile and on a flat background. The density of cesium atom is ~ 1.9×10^{12} atoms/cm^3.

It is worth stressing that the *R* resonance appears at the position of the $F_g = 3$ set of transitions and do not follow the shift of the coupling frequency that is performed in a frequency interval larger than 3 GHz.

Analyzing the *R* resonance behavior, it can be assumed that it occurs when the cycling of Cs atom population between the $F_g = 3$ and $F_g = 4$ levels is most intense. Such cycling process can also provide very high population of the excited $F_e = 2, 3, 4, 5$ levels. In fact, under such conditions, a negligible amount of Cs atoms can stay optically pumped and accumulate in each of the two ground state levels.

A similar experimental study was performed for the observation of the N_c resonance in the high frequency wing of transmission of the $F_g = 3$ group of transitions (Fig. 8). For this transition set, the sub-natural-width resonance N_c also appears with very good contrast and superimposed on a flat background. The N_c resonance behavior with the atomic source temperature and coupling detuning is similar to that of the resonance observed in the lower frequency wing of the $F_g = 4$ group of transitions (see Fig.6 and Fig.7). The conditions for the formation the R resonance are also similar. Here it is observed when the frequency of the probe beam is tuned to the $F_g = 4$ set of transitions. Thus in this case also, the *R* resonance occurs when intensive cycling of Cs atoms between both hyperfine ground levels is provided due to the most efficient repumping of the $F_g = 3$ level, which takes place when the probe beam is in resonance with the $F_g = 4$ group of hyperfine transitions.

Special care should be taken to distinguish the *R* resonance from the EIT resonance which is based on a three-level Λ-system involving both hyperfine ground levels and an excited level. Hence the *R* resonance could be supposed to be an EIT one that can be strongly broadened under the conditions of our experiment.

In Fig. 9, the case of observation of all three (N_c, *EIT*, *R*) resonances is presented. Starting from low probe frequency, first the N_c resonance is observed at $\nu_p = \nu_c - \Delta$. After that, a very weak sub-

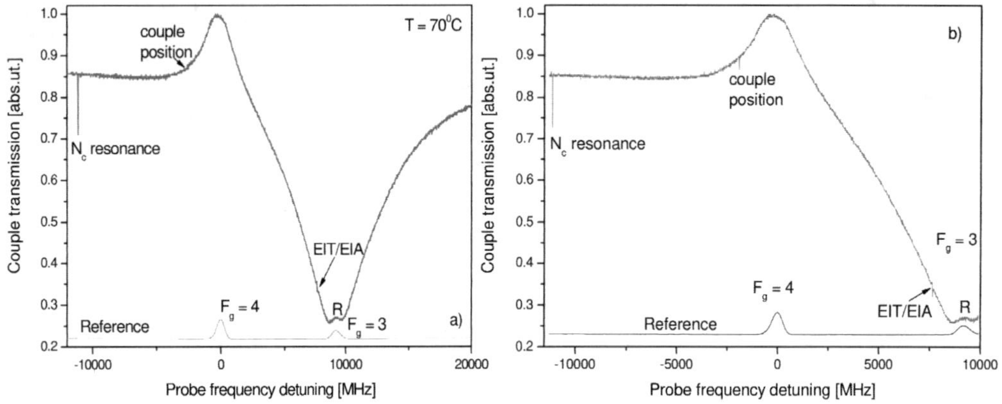

Fig.9. (a) Simultaneous observation of N_c, *R* resonance and EIT/EIA resonance in the couple beam transmission. (b) Zoom of (a) for a narrower interval of probe frequency detuning.

natural-width resonance can be seen at $\nu_p = \nu_c + \Delta$. Note that here, the EIT resonance is not characterized by the usual transparency window but instead has a dispersive shape. Such a dispersive shape can be observed due to the added buffer gas and the detuning of the probe and coupling frequencies from the center of the respective optical transitions [7]. From Fig. 9, it can be seen that the *R* resonance is situated at different values of ν_p from the EIT/EIA resonance. More specifically, for the EIT/EIA resonance formation the Raman condition should be satisfied, while the R resonance can be observed for different ν_c detunings provided the probe frequency ν_p is fixed at the maximum absorption of the $F_g = 3$ group of transitions. One can assume that under this condition, the $F_g = 4$ level repumping is the most effective.

Based on the experimental study of the behavior of the *R* resonance, it can be suggested that the physical process behind it could be related to optical transitions from the $F_e = 2, 3, 4, 5$ levels to higher Cs energy levels. Similar kinds of transitions have been realized in alkali atoms by the energy-pooling collision effect [8] which can be briefly summarized as follows. If hot Cs vapor is resonantly excited by high-intensity mono-mode laser light to the 6P levels, fluorescence from the 6D atomic levels will be observed. Note that the 6D atomic levels lie at nearly twice the frequency of the 6P levels. In Ref. [9], the population in the high-lying Cs levels was attributed to excited atom-excited atom collisions in which the two atoms pool their internal energy to produce one ground-state atom and one in a more highly excited state. Collisions of this type were mostly studied in Na, while there has been little work on Cs. However, the energy-pooling collision study in cesium is of current interest because such collisions can be an important loss mechanism in ultracold laser traps, in which the atomic density is very high (10^{11} to 10^{13} cm^{-3}).

The fluorescence produced by the energy-pooling collisions was systematically studied in Ref. [10]. There, by pumping the Cs D_2 line in a heated Cs atomic vapor cell, fluorescence containing about 30 spectral lines has been observed. The spectral lines were assigned to the transition lines of the cesium atom. In addition, the relative populations of the levels involved in the energy-pooling collision process have been determined. For instance, it has been shown that for Cs atomic density of $1.8x10^{14}/cm^3$, the relative populations of the $6D_{3/2}$ level compared to $6P_{3/2}$ is more than 1%. In our experiment the atomic density is less, but the light excitation is bi-chromatic so that it can provide more atomic population in the $6P_{3/2}$ levels. Thus, the *R* resonance can be considered as the beginning of population of energy levels higher than the $6P_{3/2}$ level. Further investigations will be needed in order to clarify if the bi-chromatic excitation of alkali hot atoms can result in more effective energy-pooling process.

5. Conclusion

Systematic experimental study of sub-natural-width resonance observed on the D_2 line of Cs vapor is presented. Three-photon, bi-chromatic excitation of Cs atomic vapor buffered by 20 Torr of neon is performed. The bi-chromatic excitation involves a fixed-frequency coupling beam and scanning the probe beam across the D_2 line. The transmission of the coupling beam is measured as a function of the probe frequency detuning. Such an approach allows the observation of very good contrast, sub-natural-width (less than 2 MHz) N_c resonance on a low signal flat background, which can be advantageous for many applications.

The probe light frequency is scanned in a large spectral interval (more than 30 GHz) that allows analyzing complex processes in the light-atom interaction such as velocity-changing collisions, hyperfine optical pumping and re-pumping, and their influence on the N_c resonance and the transmission of the constant frequency coupling laser.

The comparison between the N_c and EIT resonances shows that the N_c resonance contrast is much higher than that of the EIT resonance under the conditions of our experiment. A new spectral feature in the couple transmission, denoted as *R* resonance is observed and studied. Further theoretical analysis is needed to clarify the origin of the new resonance.

Acknowledgements: The authors greatly appreciate the fruitful discussions of experimental results with S. Gozzini, G. Grigoryan and Y. Malakyan. This work was supported by a Marie Curie International Research Staff Exchange Scheme Fellowship within the 7th European Community Framework Programme.

References

[1] Zibrov A S, Ye C Y, Rostovtsev Y V, Matsko A B, and Scully M O 2002 *Phys. Rev. A.* **65** 043817

[2] Zibrov S, Novikova I, Phillips D F, Taichenachev A V, Yudin V I, Walsworth R L and Zibrov A S, 2005 *Phys. Rev. A* **72** 011801(R)

[3] Novikova I, Phillips D F, Zibrov A S, Walsworth R L, Taichenachev A V and Yudin V I 2006 *Opt. Lett.* **31** 2353-2355.

[4] Novikova I, Phillips D F, Zibrov A S, Walsworth R L, Taichenachev A V and Yudin V I 2006 Opt. Lett. **31** 622-624.

[5] Slavov D, Sargsyan A, Sarkisyan D, Mirzoyan R, Krasteva A, Wilson-Gordon A D and Cartaleva S 2014 *J. Phys. B: At. Mol. Opt. Phys.* **47** 035001

[6] Tomasi F de, Allegrini M, Arimondo E, Agarwal G S and Ananthalakshmi 1993 *Phys. Rev. A* **48** 3820

[7] Alzetta G, Cartaleva S, Dancheva Y, Andreeva C, Gozzini S, Botti L, and Rossi A, 2001 *J. Opt. B: Quantum Semiclass. Opt.* **3** 181-188

[8] Allegrini M, Gozzini S, Longo I and Savino P 1982 *Nuovo Cimento* **1D** 49-69
[9] Jabbour Z J, Namiotka R K, Huennekens J, Allegrini M, Milosevic S and de Tomasi F 1996 *Phys. Rev. A* **54** 1372-1384
[10] Hoon-Soo Kang, Jae-Pil Kim, Cha-Hwan Oh and Pill-Soo Kim 2002 *J. Korean Phys. Soc.* **40** 220-223

Raman study of uniaxial deformation of single-crystal mats of ultrahigh molecular weight linear polyethylene

Yu V Zavgorodnev[1], S N Chvalun[2], G Yu Nikolaeva[3], E A Sagitova[3], P P Pashinin[3], S A Gordeyev[4] and K A Prokhorov[3]

[1]"MATI" - Tsiolkovsky Russian State Technological University, 3 Orshanskaya St., 121552 Moscow, Russia
[2]Karpov Physico-Chemical Institute, 10 Obukh St., 105064 Moscow, Russia
[3]Prokhorov General Physics Institute, Russian Academy of Sciences, 38 Vavilov St., 119991 Moscow, Russia
[4]UWI Technology Ltd, 3 Lochside Way, EH12 9DT Edinburgh, UK

E-mail: nikolaeva@kapella.gpi.ru

Abstract. We present for the first time a Raman spectroscopic study of the deformation process of solution-crystallized single-crystal mats of ultrahigh molecular weight linear polyethylene (UHMW PE). We study the deformed regions of the films, drawn only until the formation of the neck, and the films of much higher draw ratios, just before rupture starts. For comparison, we have also carried out Raman investigations of films produced by compression of UHMW PE powder. We have found that the uniaxial molecular orientation in the neck region of the single-crystal mat films develops more slowly as compared to the films, prepared by compression of the UHMW PE powder.

1. Introduction

Polyethylene (PE) with molecular weight $M_w < 10^6$ g/mol is one of the most important industrial polymers due to its excellent characteristics and reasonable cost. With regard to the strength, the chemical, temperature and wear resistance, the friction coefficient, and physiological inertness, ultrahigh molecular weight polyethylene (UHMW PE, $M_w \geq 10^6$ g/mol) possesses even better properties as compared to ordinary PE grades. As a matter of fact, UHMW PE has been successfully applied in the production of prostheses, light-weight and high-strength fibers and some other products. However, UHMW PE applications are limited because of difficulty of its processing, which is due to a steep increase in viscosity of UHMW PE melt at $M_w \geq 10^6$ g/mol. Ultrahigh molecular weight causes the large number of molecular entanglements, which prevent unfolding of macromolecules during drawing. To improve the UHMW PE processability, a number of special approaches are applied [1], including gel molding [2 - 4], solution drawing [5], solution growth of single-crystal mats [6, 7], synthesis of reactor blends [8 - 10], etc. Detailed investigations of the UHMW PE deformation mechanism could be very useful for overcoming this limitation and expanding the field of application of UHMW PE.

A number of techniques are traditionally used to monitor polymer structure as well as structural changes, which occur in polymers during deformation. The most common methods are wide angle (WAXS) and small angle (SAXS) X-ray diffraction technique, differential scanning calorimetry

(DSC), transmission (TEM) and scanning (SEM) electron microscopy, NMR, IR and Raman spectroscopy, and measurements of birefringence. Raman spectroscopy is one of the most suitable methods to evaluate polymer structure both in non-deformed and deformed states. This technique is non-destructive and highly informative about both crystalline and amorphous states of molecules. In particular, Raman spectra contain information on the conformational and phase compositions and the orientation of PE macromolecules [11 - 14]. Conventional Raman investigations do not require any sample preparation, which could affect the sample structure, and a spatial resolution of a few microns (both over a surface and in depth) can be achieved on solid samples.

Raman spectrum of PE is well investigated, and almost all PE Raman-active lines are reliably assigned [15, 16, Table 1]. Raman spectrum of a PE single macromolecule in the *all-trans* conformation consists of the lines with symmetry A_g, B_{1g}, B_{2g} and B_{3g}. Intermolecular interaction between two molecules in *trans*-conformation in the PE orthorhombic unit cell causes splitting of the Raman lines of a single PE molecule [17]. Therefore, Raman spectrum of the PE orthorhombic crystallites consists of doublets of the lines with symmetry A_g and B_{1g} or B_{2g} and B_{3g} [16, 17]. At room temperature this splitting can be observed only for the doublet of lines at 1415 cm^{-1} (symmetry A_g) and 1440 cm^{-1} (symmetry B_{1g}). This effect was successfully used to evaluate PE crystallinity in non-deformed [11, 12] and uniaxially drawn [14] samples. Besides, recording of polarized Raman spectra allows evaluation of the degree of uniaxial orientation of PE macromolecules in the *trans*- and *gauche*-conformations [13].

Table 1. Assignment of Raman-active lines of semicrystalline PE with orthorhombic crystalline phase.

Assignment	Frequency, cm^{-1}	Symmetry	Conformation	Phase
The asymmetric stretching vibration of C-C bonds	1062	$B_{2g}+B_{3g}$	*trans*	crystalline, amorphous
The stretching vibration of C-C bonds	1080	-	*gauche*	amorphous
The symmetric stretching vibration of C-C bonds	1130	A_g+B_{1g}	*trans*	crystalline, amorphous
The twisting vibration of CH_2 groups	1295	$B_{2g}+B_{3g}$	*trans*	crystalline, amorphous
The twisting vibration of CH_2 groups	1305	-	*gauche*	amorphous
The bending vibration of CH_2 groups	1415	A_g	*trans*	crystalline
The bending vibration of CH_2 groups	1440	B_{1g}	*trans*	crystalline
The symmetric stretching vibration of CH_2 groups	2848	A_g	*trans*	crystalline, amorphous
The asymmetric stretching vibration of CH_2 groups	2882	B_{1g}	*trans*	crystalline, amorphous

PE has complex semicrystalline structure [18], where macromolecules in the crystallites are in the *trans*-conformation, while the amorphous phase contains both *trans*- and *gauche*-conformers. It is well known, that uniaxial hot-drawing leads to preferred orientation of PE macromolecules along the direction of drawing. A number of structural models of the uniaxial deformation mechanism of PE were extensively discussed in literature [18 - 24]. During this process, the PE crystalline structure reorganizes from the lamellar to the fibrillar one [18], both having the orthorhombic unit cell. Simultaneously, the amount of *trans*-conformers in the amorphous phase grows, while the amount of *gauche*-conformers decreases. In particular, X-ray diffraction studies [19] have shown that the amorphous phase of uniaxially stretched PE contains significant amount of straightened tie molecules in *trans*-conformation, and most of these molecules are the direct continuations of the crystallite chains and connect adjacent crystallites along the axis of uniaxial orientation. With the growth of the degree of deformation, the number and lengths of such linear systems increase.

At uniaxial hot-drawing, transformation of the lamellar crystalline structure of original PE into the fibrillar one with preferred molecular orientation along the direction of drawing mainly takes place in the transition area of the neck region [18]. The neck region has spatially non-uniform supramolecular structure, altering from the non-deformed to the uniaxially oriented state. A length of the structure transformation region depends on both structure of non-deformed sample and the conditions of drawing (temperature and rate of deformation) and is usually of tens of micrometers. Further deformation leads to improving the chain orientation along the direction of drawing. All stages of the PE deformation process are of particular interest for fundamental and applied researches.

Most of published papers on UHMW PE study are related to investigation of the mechanisms of elastic and inelastic uniaxial deformation. Authors of paper [25] have studied structure of commercial and experimental UHMW PE powders, synthesized using different catalysts and under different conditions. Elastic uniaxial deformation of UHMW PE was studied by Raman spectroscopy [26], X-ray analysis [7, 27] and by EPR spectroscopy [28]. In particular, X-ray analysis was successfully applied to investigate features of uniaxial deformation of UHMW PE single-crystal mats [7]. Authors of the paper [7] have revealed that uniaxial deformation of the PE single-crystal mats leads to change in the dimensions of the PE unit cell and in the crystallite sizes. They also have observed anomalous relation between the micro- and macro-deformation characteristics of the samples, which was explained by non-uniform structure of the amorphous phase of the films, in particular, by wide distribution of lengths of the straightened tie molecules in the films, preliminarily hot-drawn up to low draw ratios. It was found that the deformation of UHMW PE crystalline lattice includes reversible and nonreversible components. The nonreversible deformation component was manifested in the reduction of the parameter a of the orthorhombic unit cell, and this effect was explained by conformational *gauche-trans* transitions under mechanical loading and nonreversible disappearance of these defects.

A number of investigations of inelastic deformation mechanism of UHMW PE have been reported [1, 29 - 41]. However, with the exception of two papers of our group [40, 41], there are no Raman spectroscopic studies of the PE neck region. We have found only two papers on micro-Raman study of neck region in polymers – for poly(vinylidene fluoride) [42, 43] and for polypropylene [42].

The mechanism of necking of thin UHMW PE films, prepared from dilute solution, was investigated by TEM [30]. DSC technique, IR spectroscopy, mechanical testing, measurements of density, birefringence and sonic velocity were applied to investigate uniaxial deformation of UHMW PE fiber, obtained by gelation/crystallization from solution [31]. Solid state drawing of solution-crystallized UHMW PE was studied by WAXS and SAXS techniques [32]. WAXS technique was also applied to analyze structure and alteration in crystallite orientation during melting of gel-drawn UHMW PE films [33, 34]. Polarized confocal Raman microscopy in combination with WAXS technique was used to characterize structure of UHMW PE fibers of various molecular weights, uniaxially drawn by solid-state extrusion [35]. WAXS technique, SEM, mechanical testing and numerical simulation were applied to study pre-swollen uniaxially drawn UHMW PE films [1]. WAXS and SAXS techniques, mechanical testing, and measurements of density were applied to uniaxially drawn UHMW PE films, produced by gelation/crystallization at various temperatures [36]. Series of papers [37-39] is connected with investigation of high-strength gel-spun UHMW PE fibers by WAXS technique, low frequency Raman and IR spectroscopy, and mechanical testing.

X-ray analysis investigation of PE single-crystal mat films has shown that samples, produced from PE with low molecular weight, are much more fragile than samples, produced from PE with medium and high molecular weights [29]. This effect was explained by very low amount of tie molecules, connecting lamellar crystallites in low molecular weight PE single-crystal mats.

Raman spectroscopy was applied to study alterations in the phase composition and molecular orientation in UHMW PE prosthesis components, subjected to various sterilization methods as well as after testing on movement-wear simulators [44 - 47]. Raman spectroscopy, WAXS, SAXS and DSC techniques were successfully applied to evaluate structural changes in UHMW PE of two different molecular weights [48] during γ irradiation in air, which is used as a sterilization method.

UHMW PE can be also applied to produce nanocomposites with multi-wall carbon nanotubes, having the improved mechanical properties [49]. Employment of Raman spectroscopy [15, 50] and X-ray analysis [15] to UHMW PE, subjected to elevated pressure and temperature, allowed to determine spectral features in the UHMW PE Raman spectrum, corresponding to the monoclinic, orthorhombic, and hexagonal crystalline modifications and melt.

In this work we present a Raman spectroscopic study of uniaxially hot-drawn films of solution-crystallized UHMW PE single-crystal mats. We study the deformed regions of the films, drawn only until the formation of the neck, and the films of much higher draw ratios just before rupture starts. The single-crystal mats are the stacks of folded-chain lamellae, which are sufficiently homogeneous in thickness and are interlinked by a small number of tie molecules [6, 19]. A very small amount of chain entanglements in the amorphous regions of UHMW PE single-crystal mats leads to the special mechanism of uniaxial deformation compared with PE films, produced by sintering of UHMW PE reactor powder. During uniaxial drawing, macromolecules in the single-crystal mats are able to relatively easy straighten and to be aligned along the direction of deformation. Uniaxially deformed UHMW PE single-crystal mat films contain large amount of straightened tie molecules and possess very high mechanical characteristics, in particular, draw ratio and the elastic modulus [6, 19].

For comparison, here we also discuss results of Raman study of UHMW PE films, produced by compression of UHMW PE powder [40, 41].

2. Experimental

We study two series of UHMW PE single-crystal mat films (molecular weight $M_w=10^6$ g/mol), which were subjected to uniaxial hot-drawing. The production method of these samples is described in detail in the paper [6]. Films, designated as series N 1 and N 2, were crystallized from solution of UHMW PE in xylene with UHMW PE concentration of 0.05 and 0.20 mass %, respectively, and were drawn at temperatures of 130 and 110°C, respectively. Length of the structure transition area in the neck for both series was about 600 μm.

As a result of the special production method and according to the X-ray diffraction data [6, 19], orientation of *c*-axes of orthorhombic crystallites in the non-deformed single-crystal mat films was predominantly perpendicular to the film surface. For the uniaxially drawn single-crystal mat films of both series N 1 and N 2, orientation of the *c*-axes was predominantly along the direction of drawing. In addition, for the drawn films from series N 2, the orientation of the *b*-axes of the orthorhombic crystallites was predominantly in the plane of the films. For the drawn films from series N 1, the orientation of the *a* and *b*-axes of the orthorhombic crystallites was random in the plane, perpendicular to the direction of drawing.

Series of the films N 3 was produced by compression of UHMW PE powder with $M_w=2.2\text{x}10^6$ g/mol at 200°C. These films were uniaxially drawn at 130°C, and length of the structure transition area in the neck was about 150 μm. According to the X-ray diffraction data, macromolecular orientation in the original film of series N 3 was isotropic. For the uniaxially drawn films from this series, molecular orientation of the *c*-axes was predominantly along the direction of drawing with random orientation of axes *a* and *b* in the plane, perpendicular to the direction of drawing.

The mechanical characteristics of the single-crystal mat films were much higher compared to that of the films, produced by compression of the UHMW PE powder. In particular, the ultimate draw ratio, that is the draw ratio λ, after which rupture of a sample occurs, was equal to 225 for the series N 1 and N 2 and only 16 for the series N 3.

Experimental setup for recording the Raman spectra included Ar^+-laser, double monochromator (U1000, Jobin Yvon), Olympus microscope, water-cooled photomultiplier, operating in the photon counting mode. Raman spectra were recorded in the back-scattering geometry with spectral resolution of 3 cm^{-1}. For excitation of the Raman spectra we used the laser line at 514.5 nm. Exciting and scattered radiation was collected using objective (100X, N.A. 0.95). Laser power at the sample surface was about 50 mW. Diameter of the laser spot on the sample surface was adjusted to be 10 μm in order to prevent local overheating of the sample and changing of its structure. To avoid influence of edge effects, Raman spectra were recorded in the central part in width of the samples. While considering results of the neck region investigation, l denotes distance from the non-deformed part (l=0, draw ratio λ=1) towards the oriented part along the direction of drawing. We analyzed two most informative regions in the PE Raman spectrum – from 1000 to 1550 cm^{-1} and from 2700 to 3050 cm^{-1}.

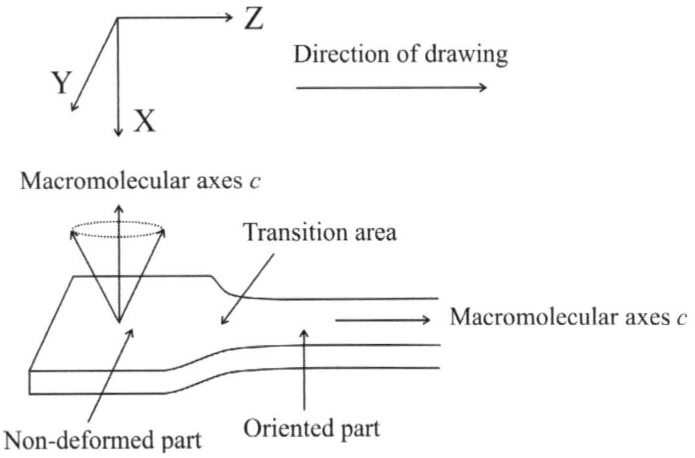

Figure 1. Experimental geometry for recording the Raman spectra of single-crystal mat films. For details, please, refer to the text.

Raman spectra were recorded in two scattering geometries - X(Z,Z)X and X(Z,Y+Z)X. In these notations axis X is normal to the sample surface, axis Z is in the plane of the sample surface and coincides with the direction of drawing (figure 1). In all experiments the direction of the laser polarization was along axis Z, and wave vectors of the exciting and scattered radiation were along axis X. For the X(Z,Z)X geometry the direction of polarization of scattered light was parallel to axis Z. In order to select required polarization of scattered light, polarizer was placed before the monochromator entrance slit. In the X(Z,Y+Z)X geometry we collected all scattered radiation and used $\lambda/4$ plate instead of a polarizer. This allowed avoidance of incorrect measurements of relative Raman intensities, which can be caused by influence of selectivity of Raman setup optical elements relative to different polarization directions of the scattered light.

3. Results and Discussion

Typical Raman spectra for the film N 1 of single-crystal mats, recorded in the neck region in the X(Z,Z)X and X(Z,Y+Z)X geometries, are presented in figures 2 and 3, respectively. The top spectra in these figures are the spectra of the same film, uniaxially deformed to draw ratio λ=200. Figures 4 and 5 show the corresponding spectra for the film N 2. Two top spectra in these figures are the spectra of the same film, uniaxially deformed to draw ratios λ=23 and 90. For illustrative purposes, the peak intensities of all lines in figures 2 – 5 are normalized to the peak intensity of the line at 1130 cm^{-1}. We chose this line as a reference line because its intensity changes relatively little with increase in l and in λ in both scattering geometries. Spectra of the film N 3 in the neck region can be found in [41].

For both scattering geometries, changes in the intensities of the PE Raman lines with increase in l are very similar for all three films under study. In accordance with selection rules for the X(Z,Z)X geometry, all PE Raman lines, that is of A_g, B_{1g}, B_{2g} and B_{3g} symmetry, can appear in the spectrum of PE film with isotropic molecular orientation. In case of complete uniaxial molecular orientation along the drawing direction (axis Z), only A_g symmetry lines can be observed in the X(Z,Z)X geometry. Therefore, intensity ratio of the lines with mixed symmetry $B_{2g}+B_{3g}$ and A_g+B_{1g} in the X(Z,Z)X geometry can serve as a measure of the degree of uniaxial molecular orientation.

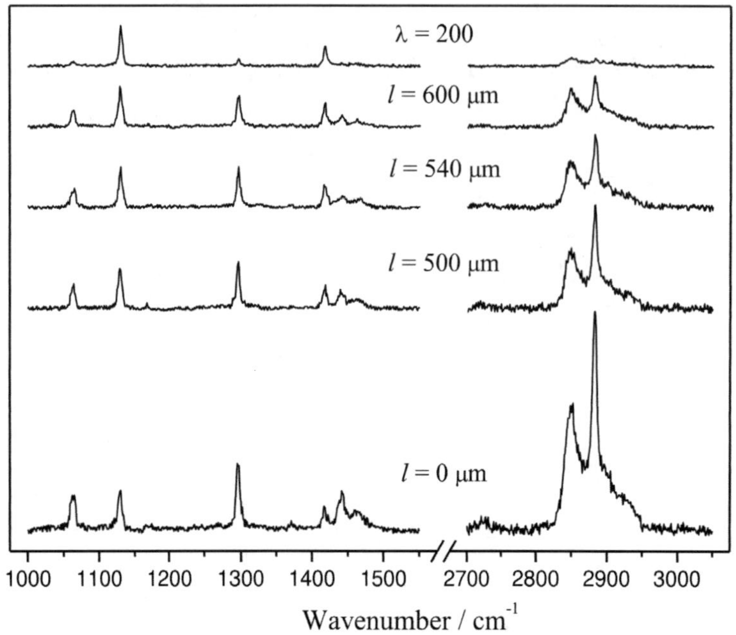

Figure 2. Raman spectra of the single-crystal mat film N 1 in the X(Z,Z)X geometry. The spectra were recorded in the neck region at l=0, 500, 540 and 600 μm. The top spectrum is the spectrum of the film N 1, deformed up to draw ratio λ=200.

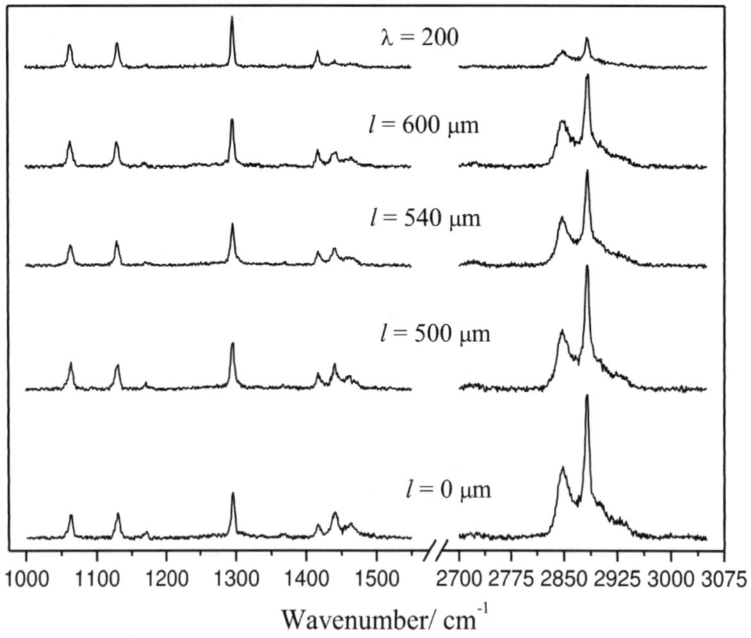

Figure 3. Raman spectra of the film N 1 in the X(Z,Y+Z)X geometry. For details, please, refer to the caption of figure 2.

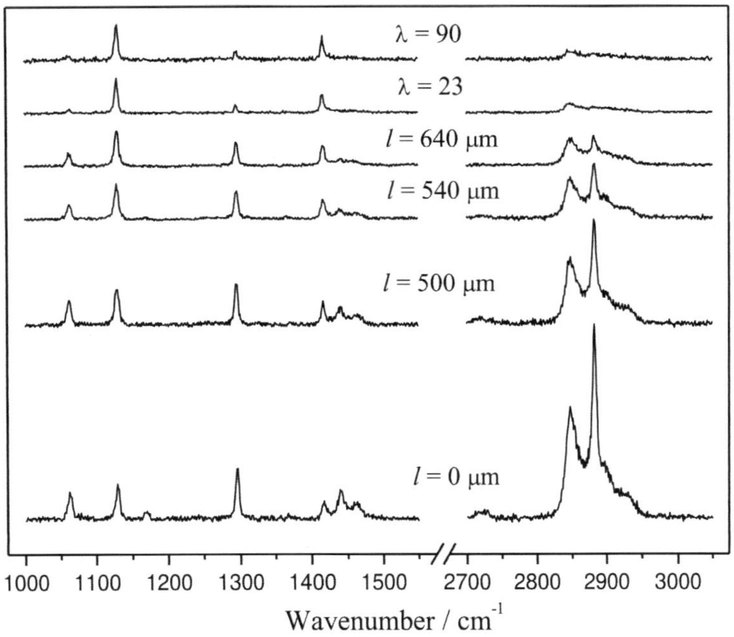

Figure 4. Raman spectra of the single-crystal mat film N 2 in the X(Z,Z)X geometry. The spectra were recorded in the neck region at *l*=0, 500, 540 and 640 μm. Two top spectra are the spectra of the film N 2, deformed up to draw ratios λ =23 and 90, respectively.

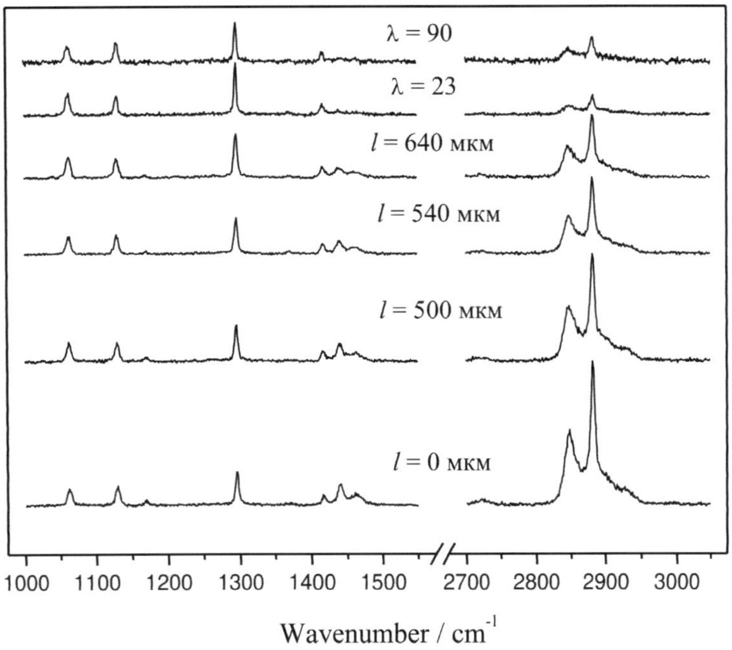

Figure 5. Raman spectra of the film N 2 in the X(Z,Y+Z)X geometry. For details, please, refer to the caption to figure 4.

As in the case of the X(Z,Z)X geometry, in the X(Z,Y+Z)X geometry Raman lines of A_g, B_{1g}, B_{2g} and B_{3g} symmetry can appear in the spectrum of a PE film with isotropic molecular orientation. In the case of complete uniaxial molecular orientation along the drawing direction in this geometry we should observe only A_g, B_{2g} and B_{3g} symmetry lines. Recording spectra in the X(Z,Y+Z)X geometry is informative, because we can prove that changes in the intensities of the lines in the X(Z,Z)X geometry are due to changes in the molecular orientation, rather than due to other changes in PE structure.

As can be seen from figures 2 and 4, in the X(Z,Z)X geometry increase in *l* leads to decrease in the intensities of the lines at 1062, 1295, 1440, 2848 and 2882 cm^{-1} compared with the intensities of the lines at 1130 and 1415 cm^{-1}. Simultaneously, reduction of the intensities of the lines at 1440, 2848 and 2882 cm^{-1} is observed in the X(Z,Y+Z)X geometry (figures 3 and 5).

The line at 1130 cm^{-1} has A_g+B_{1g} mixed symmetry and corresponds to the symmetric stretching vibration of C-C bonds of *trans*-conformers, localized in both crystalline and amorphous phases (table 1). The line at 1415 cm^{-1} with A_g symmetry relates to the bending vibration of CH_2 groups of macromolecules in the orthorhombic crystalline phase. Thus, these two lines should be observed in both used geometries irrespective of the degree of uniaxial molecular orientation.

The line at 1062 cm^{-1} corresponds to the asymmetric stretching vibration of C-C bonds, while the line at 1295 cm^{-1} - to the twisting vibration of CH_2 groups (table 1). Both the lines relate to the vibrations of *trans*-conformers, localized in the crystalline and amorphous phases, and possess B_{2g}+B_{3g} mixed symmetry. In the case of isotropic molecular orientation, these lines should be apparent in the both scattering geometries. In case of perfect chain orientation along the drawing direction, the lines at 1062 and 1295 cm^{-1} should be observed in the X(Z,Y+Z)X geometry and should disappear in the X(Z,Z)X geometry.

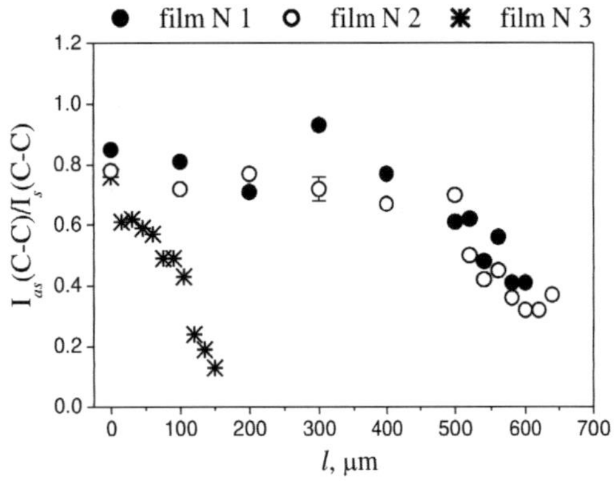

Figure 6. The ratio I_{as}(C-C)/I_s(C-C) of intensities of the asymmetric and symmetric stretching vibrations of C-C bonds in the X(Z,Z)X geometry for three UHMW PE films *vs. l.*

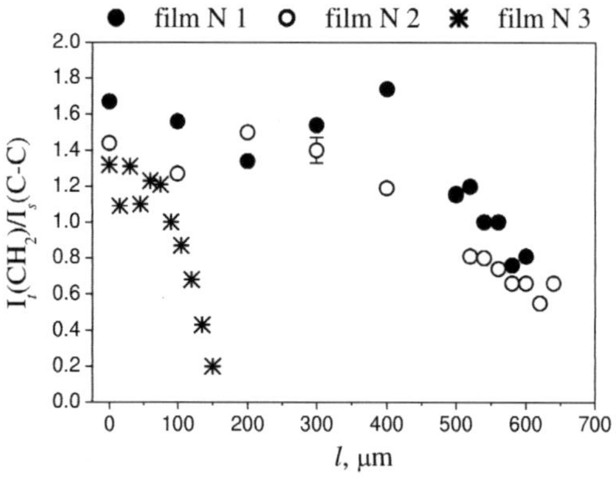

Figure 7. The ratio I_t(CH_2)/I_s(C-C) of intensities of the twisting vibration of CH_2 groups and the symmetric stretching vibration of C-C bonds in the X(Z,Z)X geometry for three UHMW PE films *vs. l.*

As can be seen from the spectra, presented in figures 2 and 4, intensities of the lines at 1062 and 1295 cm^{-1} significantly decrease in the neck region in the X(Z,Z)X geometry, and this trend continues for the films, drawn up to high draw ratios. Figures 6 and 7 compare the peak intensity ratios of the lines at 1062 and 1130 cm^{-1} (I_{as}(C-C)/I_s(C-C)) and the lines at 1295 and 1130 cm^{-1} (I_t(CH_2)/I_s(C-C)) in the X(Z,Z)X geometry at scanning the neck regions of three UHMW PE films under study. Scatter in the dependences of the intensity ratios can probably be explained by non-uniform spatial structure of the neck region.

The intensity ratios I_{as}(C-C)/I_s(C-C) and I_t(CH_2)/I_s(C-C) for the film N 3 are decreasing more rapidly than that for the films N 1 and N 2 (figures 6 and 7). This experimental fact is explained not only by different length of the structure transformation region for these samples, but also by different molecular orientation. The values of the intensity ratios I_{as}(C-C)/I_s(C-C) and I_t(CH_2)/I_s(C-C) at the end of the structure transformation region of the film N 3 is more than 2.5 times less than the same values for the films N 1 and N 2. On the contrary, in the X(Z,Y+Z)X geometry these intensity ratios are almost unchanged with increase in *l* (figures 3 and 5).

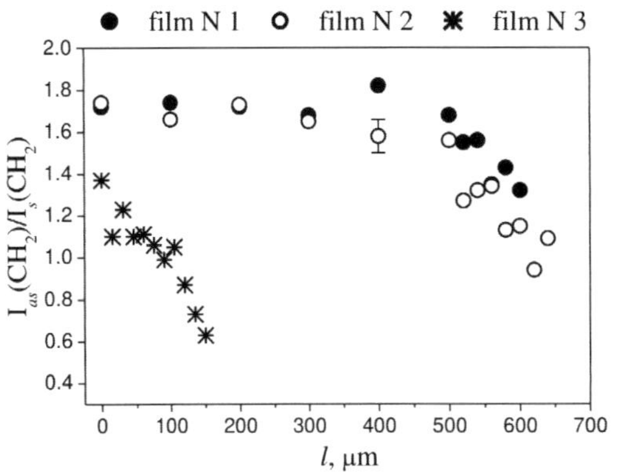

Figure 8. The ratio I_{as}(CH_2)/I_s(CH_2) of intensities of the asymmetric and symmetric stretching vibrations of CH_2 groups in the X(Z,Z)X geometry for three UHMW PE films *vs. l*.

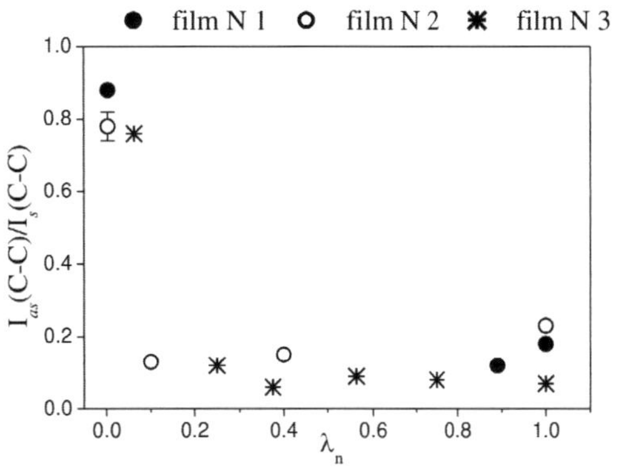

Figure 9. The ratio I_{as}(C-C)/I_s(C-C) of intensities of the asymmetric and symmetric stretching vibrations of C-C bonds in the X(Z,Z)X geometry for three series of UHMW PE films *vs.* λ_n.

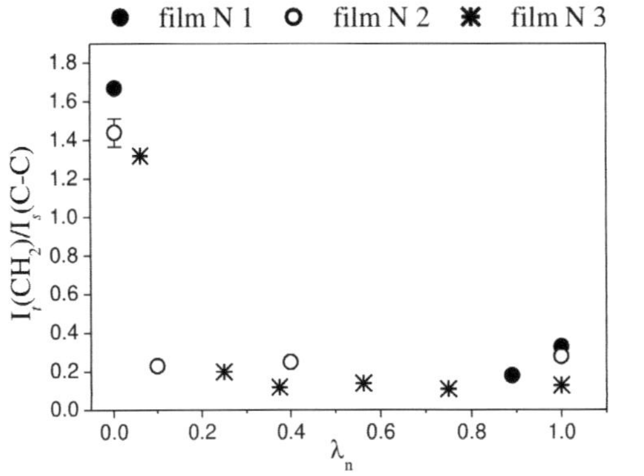

Figure 10. The ratio I_t(CH_2)/I_s(C-C) of intensities of the twisting vibration of CH_2 groups and the symmetric stretching vibration of C-C bonds in the X(Z,Z)X geometry for three series of UHMW PE films *vs.* λ_n.

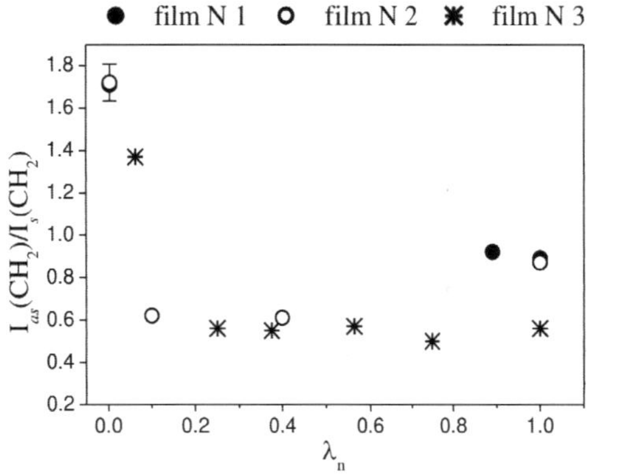

Figure 11. The ratio I_{as}(CH_2)/I_s(CH_2) of intensities of the asymmetric and symmetric stretching vibrations of CH_2 groups in the X(Z,Z)X geometry for three series of UHMW PE films *vs.* λ_n.

There are no significant changes in the ratio of peak intensities of the lines at 1415 and 1130 cm^{-1} with the growth of distance *l* and draw ratio λ in the both scattering geometries (figures 2 - 5). It means that the degree of crystallinity is not affected significantly by increase in *l* and λ. The line at 1440 cm^{-1} with B_{1g} symmetry is the second component of the doublet of the bending vibration of CH_2 groups of macromolecules in the PE orthorhombic crystalline phase (table 1). This line disappears in both scattering geometries with growth of distance *l* and draw ratio λ. As it is expected from the form of the derived polarizability tensor of B_{1g} symmetry vibration, with growth of *l* or λ intensity of this line is decreasing faster in the X(Z,Z)X geometry compared with the X(Z,Y+Z)X geometry (figures 2 - 5). The same effect is also observed for the line at 2882 cm^{-1}, also having B_{1g} symmetry.

The lines at 2848 and 2882 cm^{-1} are related to the symmetric and asymmetric stretching vibrations of CH_2 groups and have A_g and B_{1g} symmetry, respectively [51, table 1]. Figure 8 shows the ratio

$I_{as}(CH_2)/I_s(CH_2)$ of peak intensities of the lines at 2882 and 2848 cm^{-1} in the X(Z,Z)X geometry for three UHMW PE films as a function of *l*. Interesting experimental observation is the reduction of intensity of the line at 2848 cm^{-1} with increase in *l* in the both X(Z,Z)X and X(Z,Y+Z)X geometries (figures 2 - 5). We suppose that this effect is due to a small value of the component α_{cc} (here *c* is the chain axis) of the derived polarizability tensor compared with two other components for the vibration of A_g symmetry. This explanation is also confirmed by results, presented in [52, 53].

Figures 9, 10 and 11 present the ratios $I_{as}(C\text{-}C)/I_s(C\text{-}C)$, $I_t(CH_2)/I_s(C\text{-}C)$ and $I_{as}(CH_2)/I_s(CH_2)$, respectively, in the X(Z,Z)X geometry for three series of UHMW PE films as functions of normalized draw ratio λ_n, that is the draw ratio λ, divided by the ultimate draw ratio, after which rupture of a sample occurs. As it was mentioned early, the ultimate draw ratio was equal to 225 for the series N 1 and N 2 and only 16 for the series N 3. It is clear from figures 9 - 11, that when the relative degree of deformation is considered, the alterations in the molecular orientation for all three series are very similar. The most significant structural alterations take place in the neck region. Similar results were obtained for series of gelspun multifilament UHMW PE fibers with ultimate draw ratio, equal to 40 [40, 41].

Thus, in this work we used Raman spectroscopy for investigation of molecular orientation in the neck region of two single-crystal mat films, prepared from UHMW PE solution in xylene with UHMW PE concentration of 0.05 and 0.20 mass %, respectively. These data were compared with results of investigation of the film, produced by compression of the UHMW PE powder. We found out that uniaxial molecular orientation in the single-crystal mat films develops slower than that in the film, prepared by compression of the UHMW PE powder. Change in the molecular orientation along the transition area of the neck region for two single-crystal mat films was very similar. However, the degree of molecular orientation of *trans*-conformers in both the crystalline and amorphous phases was slightly higher at the end of the structure transformation area for the film, crystallized from more concentrated solution.

Different deformation mechanisms of the single-crystal mat films and the film, prepared by the UHMW PE powder compression, obviously could be explained by very different structure of the amorphous phase of these samples. In the films, produced by the powder compression, a large number of chain entanglements in the amorphous phase prevent the unfolding of chains during the uniaxial drawing and restrict the draw ratio. On the contrary, the amount of chain entanglements in the single-crystal mats is relatively small, and, hence, the chain unfolding of the lamellar crystallites can easy occur during deformation.

Acknowledgments

This work is supported by the Russian Foundation for Basic Research (project code 14-02-00832-a) and the Grant of the President of the Russian Federation for the Support of Leading Scientific Schools (project code 451.2014.2).

References

[1] Gao P, Mackley M R and Nicholson T M 1990 *Polymer* **31** 237
[2] Gordeev S A, Alekseev V G, Kaminskii V N, Tsaplin B A and Galitsyn V P 1993 *Fibre Chem.* **25** 101
[3] Smith P and Lemstra P J 1980 *J. Mater. Sci.* **15** 505
[4] Smith P, Lemstra P J and Booij H C 1981 *J. Polym. Sci.: Polym. Phys. Ed.* **19** 877
[5] Zwijnenburg A and Pennings A J 1976 *Colloid Polym. Sci.* **254** 868
[6] Konstantinopol'skaya M B, Chvalun S N, Selikhova V I, Ozerin A N, Zubov Yu A and Bakeev N F 1985 *Polym. Sci. U.S.S.R.* **27** 538
[7] Chvalun S N, Zubov Yu A and Bakeev N F 1989 *Polym. Sci. U.S.S.R.* **31** 2580
[8] Ushakova T M, Meshkova I N, Grinev V G, Ladygina T A, Arutyunov T A and Novokshonova L A 2002 *Polym. Sci. A* **44** 99
[9] Ushakova T M, Starchak E E, Krasheninnikov V G, Grinev V G, Ladygina T A and Novokshonova L A 2014 *J. Appl. Polym. Sci.* **131** 40151

[10] Meshkova I N, Ushakova T M, Gul'tseva N M, Grinev V G, Ladygina T A and Novokshonova L A 2008 *Polym. Sci. A* **50** 1161
[11] Strobl G R and Hagedorn W 1978 *J. Polym. Sci.: Polym. Phys. Ed.* **16** 1181
[12] Mutter R, Stille W and Strobl G 1993 *J. Polym. Sci. B* **31** 99
[13] Nikolaeva G Yu, Prokhorov K A, Pashinin P P and Gordeyev S A 1999 *Laser Phys.* **9** 955
[14] Sagitova E A 2008 *"Raman structural study of polyethylene – based nanocomposites"* PhD thesis, A M Prokhorov General Physics Institute, Moscow
[15] Kurelec L, Rastogi S, Meier R J and Lemstra P J 2000 *Macromolecules* **33** 5593
[16] Gall M J, Hendra P J, Peacock C J, Cudby M E A and Willis H A 1972 *Spectrochim. Acta A* **28** 1485
[17] Boerio F J and Koenig J L 1970 *J. Chem. Phys.* **52** 3425
[18] Oleinik E F 2003 *Polym. Sci. C* **45** 17
[19] Zubov Yu A, Chvalun S N, Selihova V I, Konstantinopol'skaya M B and Bakeev N F 1988 *J. Phys. Chem. USSR* **LXII** 2815
[20] Butler M F, Donald A M, Bras W, Mant G R, Derbyshire G E, Ryan A J 1995 *Macromolecules* **28** 6383
[21] Butler M F, Donald A M, Ryan A J 1998 *Polymer* **39** 39
[22] Butler M F, Donald A M, Ryan A J 1998 *Polymer* **39** 781
[23] Butler M F and Donald A M 1998 *Macromolecules* **31** 6234
[24] Peterlin A 1971 *J. Mat. Sci.* **6** 490
[25] Egorov V M, Ivan'kova E M, Kulik V B, Lebedev D V, Myasnikova L P, Marikhin V A, Radovanova E I, Yagovkina M A, Seydewitz V, Goerlitz S, Michler G H, Nochel U and Balta-Calleja F J 2011 *Polym. Sci. C* **53** 75
[26] Berger L, Kausch H H and Plummer C J G 2003 *Polymer* **44** 5877
[27] Collins M N, Dalton E, Schaller B, Leahy J J and Birkinshaw C 2012 *Polym. Test.* **31** 629
[28] Wang D, Klaassen A A K, Janssen G E, de Boer E and Meier R J 1995 *Polymer* **36** 4193
[29] Chvalun S N, Poshastenkova A B and Bakeev N F 1992 *Polym. Sci. USSR* **34** 161
[30] Kestenbach H-J and Petermann J 1994 *Polymer* **35** 5217
[31] Anandakumaran K, Roy S K and Manley R St J 1988 *Macromolecules* **21** 1746
[32] Van Aerle N A J M and Braam A W M 1988 *J. Mater. Sci.* **23** 4429
[33] Wasiak A and Sajkiewicz P 1993 *J. Mater. Sci.* **28** 6409
[34] Sajkiewicz P and Wasiak A 1999 *Colloid Polym. Sci.* **277** 646
[35] Fagnano C, Rossi M, Porter R S and Ottani S 2001 *Polymer* **42** 5871
[36] Ogita T, Suzuki N, Kawahara Y and Matsuo M 1992 *Polymer* **33** 698
[37] Pakhomov P M, Khizhnyak S D, Golikova A Yu and Galitsyn V P 2005 *Phys. Solid State* **47** 1028
[38] Pakhomov P M, Khizhnyak S, Reuter H, Galitsyn V and Tshmel A 2003 *Polymer* **44** 4651
[39] Pakhomov P, Khizhnyak S, Galitsyn V, Rogova E, Hartmann B and Tshmel A 2011 *Macromol. Symp.* **305** 63
[40] Gordeev S A, Nikolaeva G Yu and Prokhorov K A 1996 *Polym. Sci. A* **38** 517
[41] Gordeyev S A, Nikolaeva G Yu and Prokhorov K A 1996 *Laser Phys.* **6** 121
[42] Pastor J M, Jawhari T, Merino J C and Fraile J 1993 *Makromol. Chem. Macromol. Symp.* **72** 131
[43] Jawhari T, Merino J C, Rodriguez-Cabello J C and Pastor J M 1992 *Polymer* **33** 4199
[44] Bertoluzza A, Fagnano C, Rossi M, Tinti A and Cacciari G L 2000 *J. Mol. Struct.* **521** 89
[45] Puppulin L, Kumakura T, Yamamoto K and Pezzotti G 2011 *J. Orthop. Res.* **29** 893
[46] Taddei P, Affatato S, Fagnano C, Bordini B, Tinti A and Toni A 2002 *J. Mol. Struct.* **613** 121
[47] Takahashi Y, Puppulin L, Zhu W and Pezzotti G 2010 *Acta Biomater.* **6** 3583
[48] Barron D and Birkinshaw C 2008 *Polymer* **49** 3111
[49] Ciselli P, Zhang R, Wang Z, Reynolds C T, Baxendale M and Peijs T 2009 Eur. Polym. J. **45** 2741

Journal of Physics: Conference Series **594** (2015) 012010 doi:10.1088/1742-6596/594/1/012010

[50] Zhao Y, Wang J, Cui Q, Liu Zh, Yang M and Shen J 1990 *Polymer* **31** 1425
[51] Abbate S, Zerbi G and Wunder S L 1982 *J. Phys. Chem.* **86** 3140
[52] Prokhorov K A, Nikolaeva G Yu, Gordeyev S A, Pashinin P P and Nikolaev D P 2001 *Laser Phys.* **11** 86
[53] Nikolaeva G Yu, Semenova L E, Prokhorov K A and Gordeyev S A 1997 *Laser Phys.* **7** 403

Light-wave mixing and scattering with quantum gases

L. Deng and E.W. Hagley

National Institute of Standards and Technology, Gaithersburg, Maryland 20899 USA

C.J. Zhu
School of Physical Science and Engineering, Tongji University, Shanghai 200092, China

E-mail: lu.deng@nist.gov

Abstract: We show that optical processes originating from elementary excitations with dominant collective atomic recoil motion in a quantum gas can profoundly change many nonlinear optical processes routinely observed in a normal gas. Not only multi-photon wave mixing processes all become stimulated Raman or hyper-Raman in nature but the usual forward wave-mixing process, which is the most efficient process in normal gases, is strongly reduced by the condensate structure factor. On the other hand, in the backward direction the Bogoliubov dispersion automatically compensates the optical-wave phase mismatch, resulting in efficient backward light field generation that usually is not supported in normal gases.

1. Introduction

Since the invention of the laser in 1960, the field of nonlinear optics has exploded. In a typical nonlinear optical process, often referred to as coherent wave-mixing, several photons are absorbed simultaneously by an atomic or molecular medium, and a photon with a different frequency is produced. Under the right conditions, the generated radiation may be coherently amplified by the quantum mechanical mechanism of stimulated emission identified by Albert Einstein. Today, 95 years after Einstein's prediction, we are surrounded by practical applications of nonlinear optics, from multi-channel telecom lasers to the common laser pointer. Within the past decade, a new type of coherent wave mixing has emerged, which involves the interplay between light and matter waves.

In the field of nonlinear optics there is a class of novel effects that is based on quantum interference between different excitation pathways in an atomic or molecular medium resonantly driven by external fields [1]. The key concept is that a particular excited electronic state can be coupled simultaneously by external fields and an internally generated field produced by nonlinear optical-wave generation processes. The internally generated field, which is usually much closer to the single photon electronic transition, can thus drive the system in a more efficient manner, creating a different and yet very effective excitation pathway which competes with the resonant excitation by the external fields. The interference of these different pathways can lead to profound quantum interference effects that dominate the dynamics of the system response. One of the most well-known interference effects is the odd-photon destructive interference phenomenon discovered in the early 1980s and extensively studied throughout the 80s and 90s. In a typical three-photon-assisted multi-photon ionization (MPI) process in an inert gas, distinctive and sharp ionization signals were clearly observable at low medium concentrations. The researchers found [1], however, that the resonantly enhanced MPI signal disappeared completely at elevated medium concentrations. This observation, which is contrary to the naive expectation of a higher ion yield with elevated concentrations, was explained later by the quantum destructive interference between the ionization pathway resulting from the external fields and the pathway from the internally generated field. Payne *et al.* [2–4] showed analytically that excitations to the MPI enhancement state by different pathways are 180 degree out of phase, resulting in total cancellation of the excitation of the state that is actively and dynamically accessed by several strong light fields.

Recently light-wave mixing, the most widely studied nonlinear optics process in normal gases, has been extended to Bose-condensed media [5-7]. In this study, multidirectional nonlinear optical processes were examined for the first time, and both coherent wave propagation effects and the contribution from elementary excitations of condensed matter physics were investigated. The most startling findings were that there was pronounced enhancement or suppression of multidirectional high-gain hyper-Raman emission processes that are not possible in normal gases.

This study represents the first achievement that joins nonlinear optics with condensed matter physics of quantum gases, opening a new research field of *nonlinear optics with quantum gases*.

2. Theory

We consider the most studied nonlinear optical wave mixing process where the medium is a quantum gas such as a Bose-Einstein condensate or ultra-cold degenerate Fermion gas. Specifically, we focus on the third harmonic generation (THG) process well studied in the case of normal gas for the past 50 years [8]. We show that the THG can be generated in a quantum gas under the circumstances where it would have been impossible in the case of a normal gas.

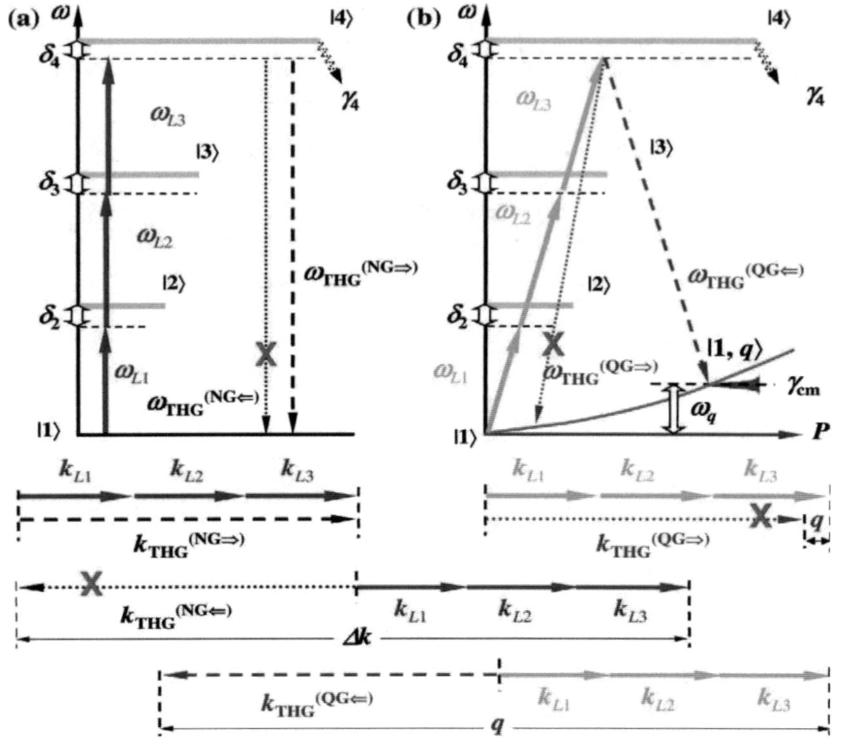

Figure 1: (a) Energy diagram and laser couplings for the THG in a normal gas where atomic center of mass motion is totally negligible. Phase-matched THG can be efficiently generated only in the presence of suitable buffer gas and only in the forward direction. (b) Energy diagram and laser couplings for the THG in a quantum gas where the atomic center of mass motion plays dominant role. Phase-matched THG can be efficiently generated only in the backward direction without the assistance of the buffer gas.

For a normal gas (see Fig. 1a), the THG field can be obtained by solving the Maxwell equation for the THG field amplitude which oscillates at 3ω (here, ω is the angular frequency of the fundamental field, i.e., the pump laser field). By solving the Maxwell equation and the material equation simultaneously one obtains, in the time Fourier transform space, the THG field [1]

$$\Lambda_{41}(z;\omega) = -\Lambda_{41}^{(3)}(0;\omega)\left[\frac{e^{i\Delta Kz} - e^{-i\kappa_{14}z/(\Delta_4+\omega)}}{1+\Delta K(\Delta_4+\omega)/\kappa_{14}}\right], \tag{1}$$

where ΔK is the general phase mismatch including contributions from both atomic medium and the buffer gas which provide an overall index compensation in order to achieve phase matched coherent constructive interference. Δ_4 is the third-photon detuning and κ_{14} depends on the atom number density [1]. Without the compensatory contribution from the buffer gas, it is not possible to make $\Delta K \sim 0$ and consequently the efficiency for THG is small. The key feature of this wave mixing process in a normal gas is that the THG field is sinusoidally dependent on the propagation distance in the forward direction (i.e., the direction of the pump laser) when $\Delta K = 0$ is satisfied. Only when the propagation distance z = L is much large than the decay characterized by $|\kappa_{14}L/(\Delta_4+\omega)| >> 1$, can one obtain $I_{41}(L;\omega)$ □ $I_{41}^{(3)}(0;\omega)$.

For a quantum gas (Fig. 1b) the atomic center of mass motion plays a dominant role and therefore overall system momentum transfer must be tracked to single photon recoil. Furthermore, density-density correlation plays a critical role in the overall THG efficiency and the most optimum direction for the phase matched wave to be produced. In the most case of quantum gas such as Bose condensate and degenerate Fermi gas, no buffer gas is allowed and the corresponding index compensation mechanism does not exist. All these novel circumstances associated with a quantum gas indicate that new physics of THG in a quantum gas is anticipated. Starting with the macroscopic atomic mean field wave function the polarization source term in the Maxwell equation for the THG frequency can be separated into the usual electronic contribution and a new term that arises from the elementary excitation rooted in quantum gas density fluctuation. The dominant contribution to the generation of the TH field is this elementary excitation since in the absence of the buffer gas the electronic contribution cannot achieve the phase matched coherent growth condition. Solving the Maxwell equation and the Gross-Pitaevskii equation simultaneously, in the Fourier space the THG field is given by [9,10]

$$\Lambda_{41}(z;\omega) = \Lambda_{41}^{(0)} \exp\left[i \frac{\kappa |\Psi_0|^2 S(q) z}{\omega_B - \Delta\omega + \omega + i\gamma_0} \right]. \tag{2}$$

where $S(q)$ is the quantum gas structure factor and ω_B is the Bogoliubov excitation spectrum ($\kappa|\Psi_0|^2 = \kappa_{14}$). Equation (2) is fundamentally different from the corresponding THG in normal gas given in Eq. (1). The first distinctive difference is that the THG process in a quantum gas is Raman-like (or hyper-Raman-like for multi-photon processes) with the typical exponential gain characteristics. It is this exponential gain characteristics of the THG (and indeed in any wave mixing processes) that prohibits any forward odd-photon destructive interference effect to occur in a quantum gas. These multi-photon interference effects in the forward direction have been widely observed and extensively studied in the last 30 years in the field of nonlinear optics. The second important difference is that in the case of a quantum gas, the dominant scattering and THG is in the backward direction (i.e., opposite the pump laser direction). This is a critically important condensed matter feature that does not have a correspondence in a normal gas and its impact to all nonlinear optical processes in a quantum gas can be seen from the condensate structure factor $S(q)$ contained in the THG gain parameter. It is well-known from condensed matter physics that when the quasi-momentum q □ 0, therefore the forward direction of THG, $S(q)$ □ 0, implying a strong suppression of the forward THG production. On the other hand and for an elongated quantum gas, $S(q)$ □ 1 in the backward direction, indicating that the most efficient direction of THG production, i.e., the overall phase matching direction is the backward direction. It is the quasi-momentum q from the Bogoliubov elementary excitation that provides the necessary momentum conservation that must be met in the wave generation process. In the case of normal gas, however, the backward direction can never be phase matched and efficient generation of THG is strongly prohibited. The third key difference shown in Eq. (2) in the case of a quantum gas is the resonant denominator that contains the Bogoliubov energy spectrum. It is this part of the elementary excitation that ensures the energy conservation, leading to a generalized matter-optical wave mixing phase matching condition. In the latter case, the condensed matter elementary excitations provide a novel phase matching role and all wave mixing processes become Raman/hyper Raman-like.

For a normal gas, the general phase matching conditions are purely optical and are given by

$$\Delta \mathbf{K} = \boldsymbol{k}_{3\omega} - 3\boldsymbol{k}_{\omega} - \boldsymbol{k}_{bg} = 0 \quad \text{and} \quad \omega_{THG} - 3\omega = 0.$$

For a quantum gas the generalized phase matching conditions include fundamental excitation and are given by

$$\Delta \mathbf{K} = \boldsymbol{k}_{3\omega} - 3\boldsymbol{k}_{\omega} - \boldsymbol{q} = 0 \quad \text{and} \quad \omega_B(\boldsymbol{q}) - [\omega_{THG} - 3\omega] = 0.$$

In Fig. 2 we show the relative THG field intensity in a normal gas in the forward direction (upper-left) and in the backward direction (upper-right) as the function of THG propagation distance for a typical rubidium vapor at room-temperature. These results show that forward THG is very efficient and later reaches a regime where the further growth of the THG is limited by the well-known three-photon destructive interference effect [1]. In addition, the THG in the backward direction is strongly prohibited.

Also shown in Fig. 2, as the comparison with the normal gas, the relative THG field intensity produced in a quantum gas. Here the rubidium atoms are cooled by laser cooling and evaporative cooling methods to below the momentum state condensation point, forming an exotic coherent matter material referred to as Bose-Einstein condensate. In the forward direction, i.e., the direction of the pump laser the THG is strongly suppressed by the condensate structure factor with small quasi-momentum transfer. In the backward direction, however, the generalized wave mixing and scattering phase matching conditions for the quantum gas are satisfied and the production of THG is most efficient.

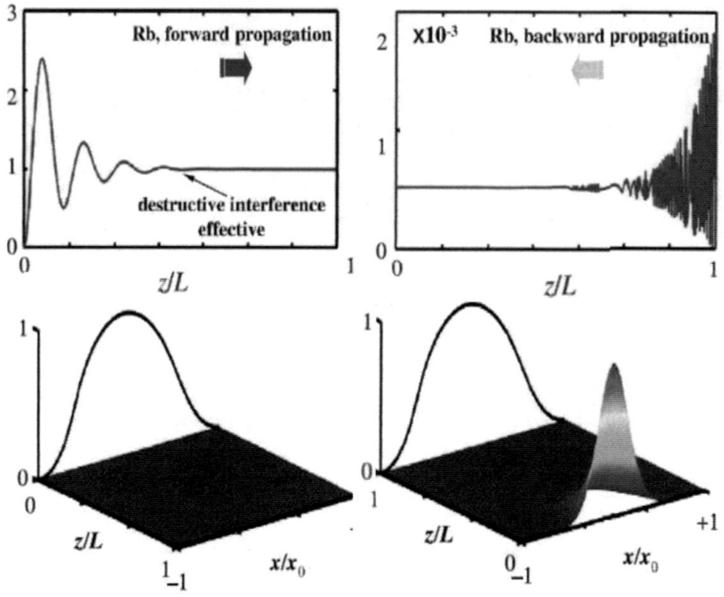

Figure 2. Right Panel: Upper plots: normal gas THG (vertical axis) as a function of normalized propagation distance. Lower plots: quantum gas THG (vertical axis) as a function of normalized propagation distance and transverse radius.

3. Conclusion

In conclusion, unlike the case of a normal gas where the medium passively participates in the nonlinear optical process, a quantum gas the medium actively participates the light generation and propagation process by exerting and enforcing effects arising from condensed matter properties of the medium. This opens a new chapter of light-matter interactions that can only be explained by applying nonlinear optics and condensed matter physics, and hence the new field of nonlinear optics with quantum gases. Indeed, all nonlinear optical processes studied in normal gas and solid-state materials must be revisited in the context of coherent matter wave and quantum gas.

References

[1] L. Deng, M.G. Payne, and W.R. Garrett, Phys. Rep. 429, 123 (2006).

[2]. M.G. Payne, W.R. Garrett, and H.C. Baker, Chem. Phys. Lett. 75, 468 (1980)

[3].M.G. Payne and W.R. Garrett, Phys. Rev. A 26, 356 (1982).

[4]. M.G. Payne and W.R. Garrett, Phys. Rev. A 28, 3049 (1983).

[5] M.H. Anderson et al., Science 269, 198 (1995).

[6]. K.B. Davis et al., Phys. rev. Lett. 75, 3969 (1995).

[7]. C.C. Bradley et al., Phys. Rev. Lett. 78, 985 (1997).

[8] Y.R. Shen, "The Principles of Nonlinear Optics", (John Wiley & Sons, New York, 1984).

[9] L. Deng, C.J. Zhu, and E.W. Hagley, *Phys. Rev. Lett.* **110**, 210401 (2013)

[10]. L. Deng, C.J. Zhu, E.W. Hagley, and W.R. Garrett, *Phys. Rev. A* 88, 043642 (2013)

Journal of Physics: Conference Series **594** (2015) 012012 doi:10.1088/1742-6596/594/1/012012

Restoration of distorted images, terahertz generation and terahertz interference based on multiple-coupled optical parametric oscillators

Yujie J. Ding*
Department of Electrical and Computer Engineering
Center for Photonics and Nanoelectronics
Lehigh University, Bethlehem, PA 18015, USA

**Corresponding author: yding300@gmail.com*

Abstract

We review our progress made on applications of coupled optical parametric oscillators based on a composite consisting of adhesive-free-bonded $KTiOPO_4$ stacks. By using the phase-conjugate output generated by the composite, we have demonstrated that the spatial profile after the beam propagates through the phase-distorted medium can be restored to the profile before the distortion. In addition, we have restored the images being blurred by the phase distortion. We have efficiently generated terahertz outputs by mixing the idler twins from the coupled optical parametric oscillators. By using the alternatively-rotated GaP plates as an output coupler for the coupled optical parametric oscillators, we have efficiently generated terahertz waves based on an intracavity configuration. By placing both the composite and a bulk $KTiOPO_4$ crystal in the same cavity, we have demonstrated the interference effect of the THz waves generated by using different pairs of the optical beams.

Number of our report according to workshop program: 1.13.1
Topic: Interaction of Laser Radiation with Matter
PACS: 42.65.Ky, 42.65.-k, 42.65.Re
Keywords: Coupled optical parametric oscillators, restoration of blurred images, phase conjugation, coupled parametric processes, difference-frequency generation, nonlinear-optical materials, terahertz sources, terahertz interference.

I. Introduction

A coherent infrared source being capable of generating dual wavelengths have a variety of applications such as differential absorption lidar [1], Doppler lidar measurement [2], coherent anti-Stokes Raman scattering microscopy [3], and generation of mid-infrared [4] and terahertz (THz) [5] waves. Although a conventional optical parametric oscillator (OPO) is capable of generating signal and idler wavelengths, spatial walk-off in a nonlinear crystal not only increases the threshold for the oscillation but also sacrifices the beam quality. Previously, it was demonstrated that periodically-orientated $KTiOPO_4$ (KTP) stacks based on adhesive-free-bonding (AFB) can be used to compensate spatial walk-off for the generation of about 2 μm due to quasi-phase-matching (QPM) [6]. In comparison, conventional QPM crystals have very small aperture sizes. Besides, compared with two separate nonlinear optical crystals [4,5], such a crystal composite has a significantly higher gain, i.e. a factor of $16/\pi^2$ higher [6]. In addition, because the dual-signal or dual-idler wavelengths (i.e. four output wavelengths) are generated simultaneously inside the same crystal, each pair has almost the same intensity and the temporal profiles which are completely synchronized.

Using a pump beam at the wavelength of 532 nm, singly resonant OPO's were implemented based on the AFB KTP stack composite [7]. Such coupled OPO's have a threshold pump power which is more than four times lower than that based on a single KTP crystal. Using the idler twins from such OPO's, we efficiently generated THz waves [7]. Besides THz generation, such a configuration allows us to restore the images being blurred by phase distortion based on phase conjugation [8]. In the past, phase-conjugation devices based on photorefractive effect had response times in the range of 100 μs – 1 s [9], i.e. they were extremely slow. Obviously, such response times make the corresponding scheme impracticable for compensating the blurred images due to atmospheric turbulence. It is well known that four-wave mixing in a third-order nonlinear medium can be used to achieve optical phase conjugation [10]. However, this scheme requires two pump beams. What is more, since third-order nonlinearity is weak, the pump powers must be extremely high in order to achieve a reasonably high nonlinear power reflection coefficient, i.e. nonlinear reflectivity (e.g. $R_{pc} \geq 1\%$). In comparison, our novel approach realized in Ref. [8] has the temporal response time being limited only by the propagation time of the interacting beams. Moreover, it is based on a second-order nonlinear process, which is much stronger than the third-order nonlinear process. Therefore, our approach overcomes the disadvantages of the slow response and the requirement of high pump intensities.

It is worth noting that such coupled OPO's have unique advantages. Indeed, frequency separation is insensitive to the fluctuations of pump frequency and temperature. In the past, all of the schemes based on nonlinear processes in the past suffer from a major drawback, i.e. optical phase conjugation works only for a specific polarization and wavelength of the input beam [11-13].

In this article, we summarize our results on the applications of coupled OPO's in corrections of distorted images, THz generation, and observation of the interference of THz waves.

II. Correction of spatial beam profile after phase distortion

In this Section, we summarize our result on the generation of a phase-conjugate beam based on AFB periodically-inverted KTP plates [14]. We have generated a phase-conjugated beam which is insensitive to the polarization of the input beam. We have demonstrated that using the phase-conjugated beam, we can recover the spatial beam profile after phase distortion.

A composite KTP stack has 20 pieces of x-cut KTP plates with the cutting angles of $\theta = 90^o$ and $\varphi = 0^o$, indicating that each plate is noncritically phasematched. As the 20 plates were cut from the same bulk KTP crystal, the crystal axes of these plates are aligned to the same directions initially. Before forming the stack, the crystal axes of every other plate are inverted such that the corresponding elements of the second-order nonlinear susceptibility tensor of KTP are all inverted. Such periodic modulations result in a QPM condition [6], i.e. $\Delta k l = \pm\pi$, where Δk is the wave-vector mismatch for each nonlinear parametric process and l is the thickness of each KTP plate. As a result, for each pump wavelength, there is a pair of the signals as well as a pair of the idlers that satisfy the condition above. Each KTP plate has a thickness of 1.19 mm ± 0.01 mm. When the pump wavelength is set to 532 nm, the wavelengths for the pair of the signals are 1034.8 nm and 1044.2 nm (ordinary wave - *o*) whereas those for the pair of the idlers are 1084.1 nm and 1093.9 nm (extraordinary wave - *e*), respectively, as shown in Fig. 1 [7]. As the pump wavelength is increased to 539 nm, the two nonlinear parametric processes reach the degenerate point. However, the wavelengths of the signal and idler are different, i.e. the wavelengths for the first pair of signal and idler are 1073 nm (e) and 1083nm (*o*) whereas the wavelengths for the second pair are 1083 nm (*e*) and 1073 nm (*o*), respectively. In contrast, at the degenerate point in a bulk KTP crystal, the signal and idler have exactly the same wavelength of 1078 nm.

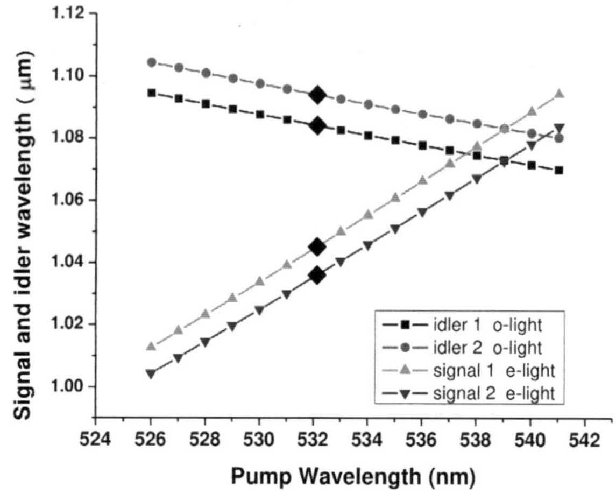

Fig. 1. Signal and idler wavelengths generated from AFB-KTP plates.

The generation of two pairs of the orthogonally-polarized signals and idlers can be exploited for generating the polarization-insensitive phase-conjugated beams. Indeed, assuming that the input beam at 1073 nm has both horizontal and vertical polarization components, the phase-conjugated beams for these two corresponding polarizations can be simultaneously generated by mixing them with the pump beam at 539 nm. Indeed, we mixed the pump beam at 539 nm with the input beam at 1073 nm, polarized in a horizontal or a vertical direction. Fig. 2(a) shows the spectrum for the horizontally-polarized input beam at 1073 nm after mixing it with the pump beam. A phase-conjugated beam in the vertical direction was observed at 1083 nm. Similarly, when we mixed the input beam at 1073 nm, polarized in the vertical direction, with the pump beam,

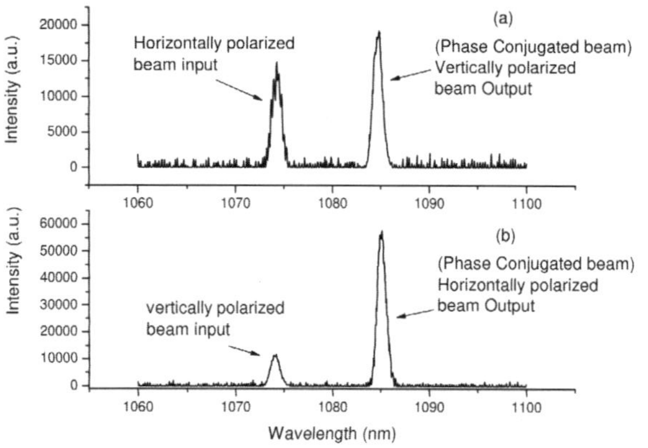

Fig. 2. Polarization-insensitive generation of phase conjugated beam.

the phase-conjugated beam at 1083 nm was observed, which was horizontally polarized, see Fig. 2(b).

In order to demonstrate the restoration of an image being blurred by a turbulent medium based on phase conjugation, we assembled a setup based on the nonlinear parametric processes in the AFB KTP plates, see Fig. 3. The pump beam at 539 nm is an output generated by a MOPO system (10 Hz, 5 ns). In order to generate the input beam for the phase conjugation, we set up an optical parametric oscillator (OPO) based on a bulk KTP crystal. The signal of such an OPO was then used as an input beam and mixed with the pump beam at 539 nm inside the AFB-KTP stack. To simulate a blurred image, a phase-distorting plate was inserted into the optical path. To recover the image from such distortion, the phase-conjugated beam was then reflected back along the previous path by a mirror and it subsequently propagated through the distorting plate at the same spatial location. A CCD camera was used to capture the image being generated by the phase-conjugated and residual input beams, respectively, by switching between the two positions #1 and #2. Correspondingly, the image being captured from position #1 has a significantly-improved image quality compared with that being captured at position #2, see Figs. 4(a) and 4(b).

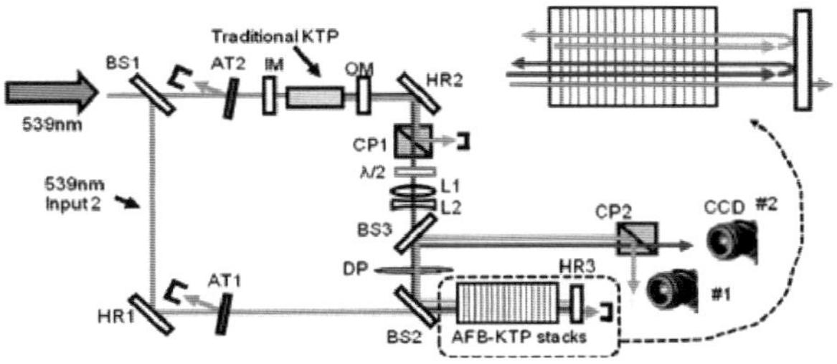

Fig. 3. Experimental setup for the image recovery.

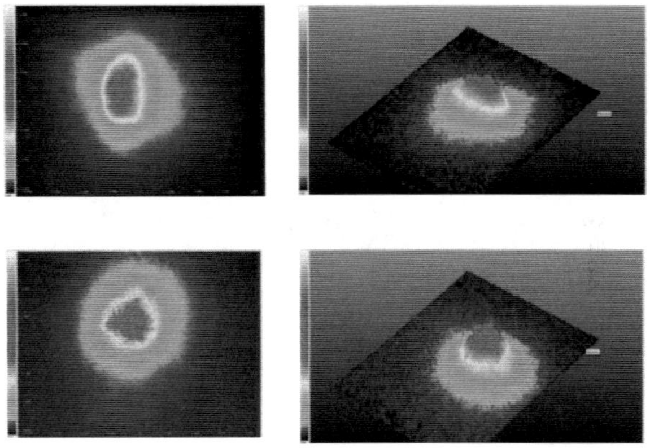

Fig. 4. (a) Spatial profile of phase-conjugated beam; (b) spatial profile of distorted input beam.

III. Restoration of blurred images

In this section, we review our previous experimental result on the restoration of the blurred images caused by the phase distortion [8].

The nonlinear composite used in the parametric interaction consists of 20 pieces of x-cut KTP plates with the cutting angles of $\theta = 90^\circ$ and $\varphi = 0^\circ$. Each plate has a thickness of 1.19 ± 0.01 mm and a cross-section of 5.6 mm×10.5 mm. All 20 pieces are bonded together by AFB technique. Since crystal axes of every other plate are inverted, the signs of nonlinear coupling coefficient d_{eff} are periodically inverted, this periodic alternation the sign of d_{eff} can compensate for the nonzero wave vector mismatch Δk, resulting in the quasi-phase matching (QPM) condition [6,7]. Such a QPM condition causes each pair of the signal and idler in a bulk KTP crystal splitting into two pairs of the orthogonally-polarized signal and idler beams. Therefore, QPM can be used to compensate for the spatial walk-off and allows us to use the largest nonlinear coefficient in the crystal for a nonlinear interaction [6].

Our experimental setup is shown in Fig. 5 for demonstration of image restoration based on broadband and polarization-insensitive phase conjugation. The laser beam (10 Hz repetition, 4.8 ns pulse width) is from an output of Master Oscillator/Power Oscillator (MOPO), with wide range wavelength tenability. The pump beam is split into two parts for each fixed pump wavelength in 535-542 nm. One part is used to pump the optical-parametric oscillator based on a bulk KTP crystal, whereas the other part is directed to the AFB-KTP plates. The idler beam generated from the bulk KTP passes through a cubic polarizer (CP1) and collimated by a pair of convex (L1) and concave (L2) lenses. It then transmits through a phase distortion plate (DP) and used as an input beam for DFG with the green pump beam in the AFB-KTP plates, where temporal and spatial overlaps of the input and pump pulses need to be maintained. Since the DP imposed phase distortion $\varphi(x, y, z)$ on the input signal, the generated phase-conjugate beam from the AFB-KTP plates also has the same phase distortion but with an opposite sign $-\varphi(x, y, z)$. Both the input and generated phase-conjugate beams are reflected back by the high-reflectivity mirror (HR3) and pass through the DP. The input beam suffers the phase distortion of $2\varphi(x, y, z)$ after the second pass of the DP, but for the phase-conjugated beam with the phase distortion of $-\varphi(x, y, z)$, passing through the DP just eliminates the phase distortion and recovers the wave front. Then, the signal and phase conjugate beams are directed to an imaging system by a beam splitter (BS3). The input beam is illuminated on the object and forms a blurring image on the CCD camera because of the phase distortion. While the image formed by the generated phase-conjugate beam will be restored since the phase distortion is removed. The phase conjugate beam has an orthogonal polarization with the input beam. If the input beam is *p*-polarized, it passes through a cubic polarizer (CP2) without reflection, and the generated phase conjugate beam is s-polarized reflecting from the CP2 with an angel ~90°, as shown in Fig. 5. The polarization of the input beam can be tuned by a λ/2 wave plate. The filter in the imaging arm is used to block the remaining green pump beam. By measuring this phase-conjugated beam, we have corrected the blurred images caused by the distortion plate.

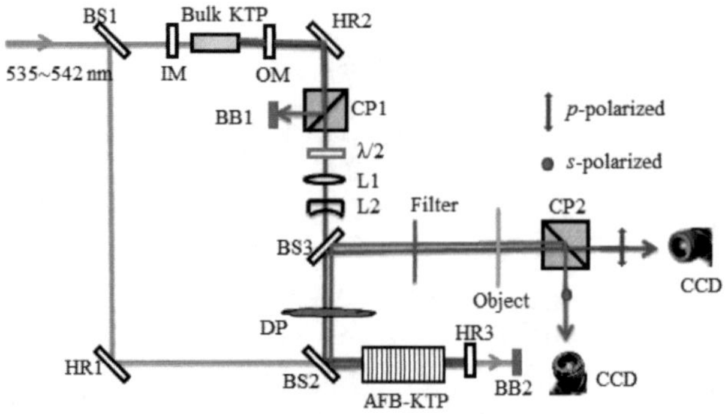

Fig. 5. Experimental setup for imaging based on phase conjugation. BS1-3: Beam splitter; HR1-3: high reflection mirror; IM: input mirror; OM: output mirror; BB1-2: beam block; L1-2: convex and concave lens; CP1-2: cubic polarizer; λ/2: half-wave plate; DP: distortion plate; CCD: CCD camera.

Fig. 6 shows our results of polarization-insensitive restoration of images. The input beam is *p*-polarized for Fig. 6(a) and *s*-polarized for Fig. 6(d) without the phase-distortion plate. After putting in the phase-distortion plate, these images are obviously blurring [Fig. 6(b) and Fig. 6(e)], since the wave front of the input beam is distorted. The phase-conjugated beam can correct the blurred images, and therefore,

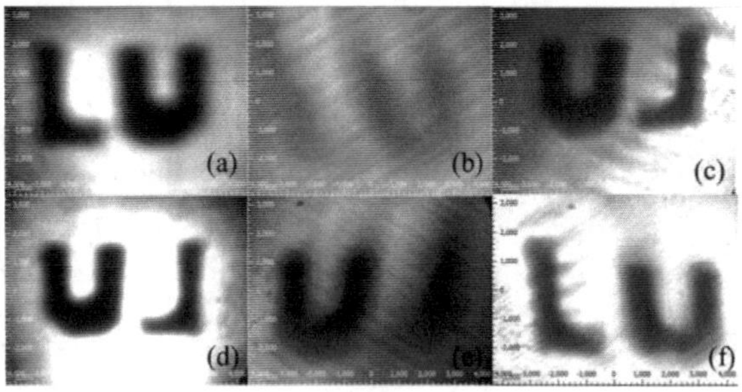

Fig. 6. *P*-polarized (a) and *s*-polarized (d) input beam without distortion images, after distortion images (b) and (e); restored images (c) and (f) by phase-conjugated beams at pump wavelength of 539 nm.

restores the images to the original quality [Fig. 6(c) and Fig. 6(f)]. Based on Fig. 6, we have estimated the degree of the image restoration to be 80±20% with 0% and 100% corresponding to no and complete restorations, respectively.

At around 539 nm pump, two pairs of the signal and idler become degenerate, as shown in Fig. 7(a). Therefore, instead of four different wavelengths, there are only two wavelengths of ~1073 nm and ~1083 nm. Each of these two wavelengths contains both *p*- and *s*-polarized components. Therefore, for the input beam wavelength at (or near) these two wavelengths, phase conjugated beams can be always generated for any polarization. Here, using the AFB-TTP crystal, we demonstrate that the imaging restoration based on polarization-insensitive phase conjugation is not only effective at the region of the degenerate pump point, but also in a broad wavelength range (535 -542 nm).

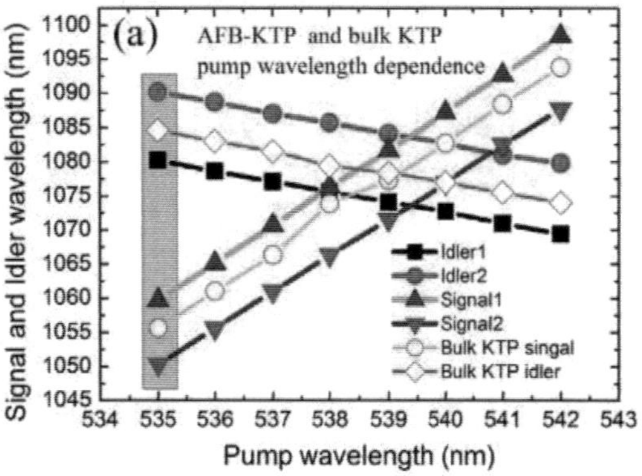

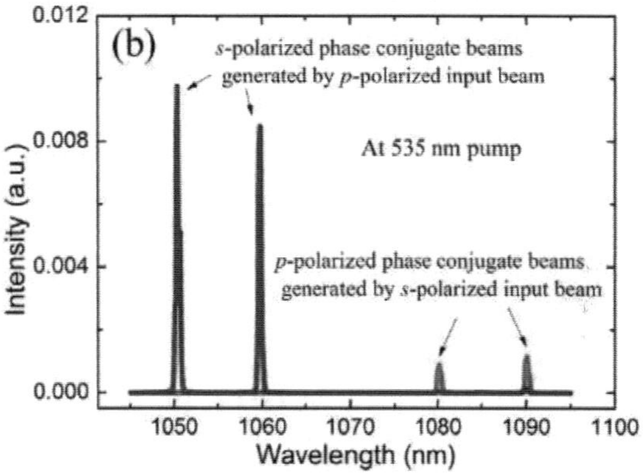

Fig. 7. (a) Measured signal and idler wavelengths of AFB-KTP and bulk KTP as a function of pump wavelength; (b) Phase conjugated beam spectra generated by *p*-polarized input beam (blue peaks) and *s*-polarized input beam (red peaks).

Fig. 8 shows the imaging restoration at 535 nm, i.e. away from the degenerate pump point. In such a case, the blurred images are also restored by the phase conjugated beams for both *p*-polarized and *s*-polarized input beams. Fig. 8(a) and Fig. 8(d) are the original images without the phase-distortion plate for the *p*-polarized and *s*-polarized input beams respectively. These images become blurring after the distortion plate is placed [Fig. 8(b) and Fig. 8(e)], and restored by the phase-conjugate beams [Fig. 8(c) and Fig. 8(f)]. Two pairs of signal and idler will be generated under the 535 nm pump, as seen the gray region in Fig. 7(a). Two idler beams are *p*-polarized and have wavelengths of ~1080 nm and 1090 nm, while the two signal beams are *s*-polarized and have wavelengths of ~1060 nm and 1050 nm. The polarization of the input beam (~1085 nm) can be tuned by the half-wave plate. If the input beam is *p*-polarized, the output phase conjugate beams are *s*-polarized and have exactly the same wavelength as the signal beams generated in the AFB-KTP crystal, as shown in Fig. 7(b) the blue peaks. Rotate the half-wave plate to make the input beam *s*-polarized, and then the phase conjugated beams become *p*-polarized and have exactly the same wavelengths of idler beams as

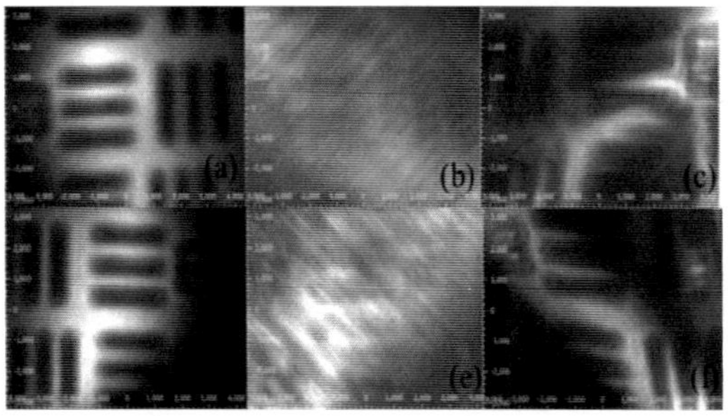

Fig. 8. Image restoration at pump wavelength of 535 nm. *p*-polarized (a) and *s*-polarized (d) input beam without distortion image, after distortion (b) and (e) and restored image (c) and (f) by phase conjugated beams.

shown by the red peaks in Fig. 7(b). It is worth noting that our experiment is different from the previous work on phase conjugation by DFG [6,8]. In their reports, strict restrains are imposed on polarizations and wavelengths of the input beams - only certain polarization direction and wavelength of the input beams can be used for the generation of phase conjugation beams. In contrast, in our case, phase conjugation beams can be generated by either *p*-polarized or *s*-polarized input beams, i.e. it can be generated by the input beam at any polarization direction. Also, this polarization-insensitive phase conjugation can be achieved at broad wavelength range (535 - 542 nm), besides the wavelength at the degenerate pump point (~539 nm). One limitation of our method is that two phase-conjugate wavelengths generated propagate with a very small angle. Therefore, the propagating distance of the phase-conjugate beams needs to be kept as short as possible, before they are divided into two separate beams and illuminate the object and camera.

IV. THz generation using output of coupled OPO's

In this Section, we review our new results on the oscillations at the dual-signal and dual-idler wavelengths of around 1 μm [7]. By mixing the dual-idler output at 1084.1 nm and 1093.9 nm from the coupled OPO's, we generated a THz wave at the wavelength of 118 μm based on difference-frequency generation (DFG).

Each KTP plate was cut according to $\theta = 90^o$ and $\varphi = 0^o$ (i.e. x-cut). As a result, at these angles the OPO can be noncritical-phase-matched at the pump wavelength of around 532 nm. All the crystal axes of the adjacent plates are opposite to each other in order to periodically switch all elements of second-order nonlinear susceptibility tensor of KTP to realize QPM based on d_{24}. The structure of the KTP plates studied here is quite different from that in Ref. [6]. Indeed, in Ref. [6], z axes of the adjacent KTP plates were bonded in a head-to-tail configuration (i.e. $\theta = 50.2^o$) to achieve not only the walk-off compensation but also QPM. It is worth noting that a singly resonant OPO is inherently more stable than a doubly resonant OPO implemented in Ref. [6]. Moreover, the thickness of each KTP plate, dictated by the frequency separation of the dual signals, is different from that in Ref. [6] (i.e. 2 mm).

The experimental setup for the singly-resonant OPO is shown by Fig. 9. The AFB KTP stacks, fabricated at Onyx Optics, consist of 20 x-cut KTP plates, each of which has a thickness of 1.19±0.01 mm, with their z axes being periodically switched as shown by Fig. 9; they were adhesive-free-bonded together as a composite for realizing QPM simultaneously at the dual-signal and dual-idler wavelengths. As a result, the QPM condition is given by $\Delta kl = \pm\pi$, where l is the thickness of each KTP plate. Consequently, a pair of the signals having slightly different wavelengths satisfies the QPM condition above. The two corresponding idler wavelengths satisfy the QPM condition.

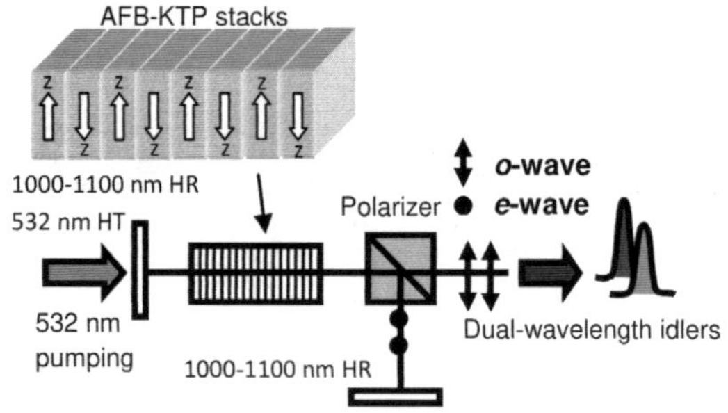

Fig. 9. Experimental setup for singly resonant OPO.

In comparison, only a single signal wavelength satisfies the birefringence-phase matching condition (i.e. $\Delta kl = 0$) if a single KTP crystal is used instead. Our composite has a clear aperture of 5.6 mm×10.5 mm and a total interaction length of 23.8 mm. As one can see, the aperture size is much larger than the QPM KTP crystals commercially available. As a result, such an AFB crystal composite holds great promise for the generation of high powers, which requires large laser beam sizes. For our composite, both the input and output facets were high-transmission (HT)-coated at the wavelengths of around 532 nm and 1000-1100 nm. To demonstrate the tunability of our OPO based on the KTP composite, we tuned the pump wavelength around 532 nm. Based on our experimental results, low-cost pulsed pump source at 532 nm, commercially available, was sufficient to achieve stable oscillations in the composite. The configuration for the OPO cavity is shown by Fig. 9. Two flat mirrors with high reflection (HR) coating in the range of 1000-1100 nm were designed as the cavity mirrors. To ensure the oscillations only at the signal wavelengths, a cubic polarizer was placed inside the cavity. It allows the o-polarized idler beams to pass through (i.e. $T \approx 97\%$) and reflects almost all of the e-polarized signal beams ($R > 99\%$). Since the polarization of each signal was perpendicular to that of the corresponding idler, we were able to oscillate just the dual signals inside the cavity. Almost all the powers of the dual-idler beams were transmitted through the polarizer. The inset to Fig. 10 illustrates the spectra of the dual signals and dual idlers generated by the OPO at the pump wavelength of 532 nm. The dual signals appearing in the spectra were caused by their weak transmissions through the polarizer. The wavelengths of the dual signals oscillating in the OPO cavity were measured to be 1034.8 nm and 1044.2 nm, whereas the wavelengths of the dual idlers were 1084.1 nm and 1093.9 nm, respectively. These measured wavelengths coincide with the calculated values.

When a pump beam at 532 nm with the repetition rate of 10 Hz and the pulse width of 4.8 ns was slightly focused onto the AFB KTP composite (a beam diameter of about 2 mm, measured by us), we measured the total output power from the OPO vs. the pump power, see Fig. 10. By linearly fitting the data points in Fig. 10, we deduced the threshold power for the OPO to be 12 mW (the corresponding energy of 1.2 mJ per pulse). Such a threshold power corresponds to an intensity of 8 MW/cm^2. At the pump power of 111.8 mW, the OPO output power reached 39 mW, corresponding to a conversion efficiency and slope efficiency of 34.9% and 37.1%, respectively. To demonstrate the advantage of a higher gain in the AFB KTP composite, we measured the power dependence from an OPO based on two traditional KTP crystals which were placed in the setup similar to Ref. [5], see Fig. 10.

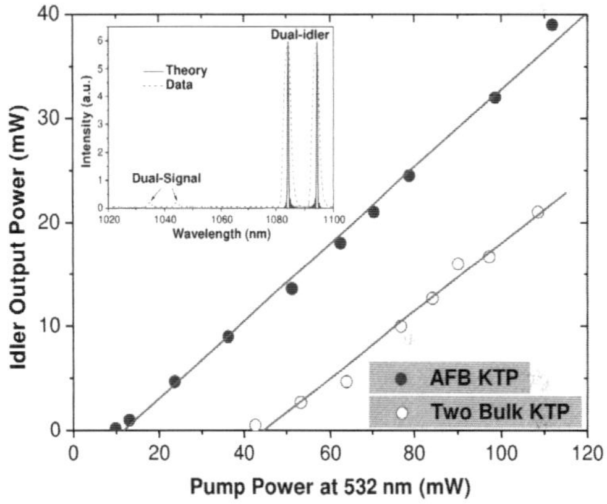

Fig. 10. Average output power of dual-idler beams vs. pump power at 532 nm based on both in AFB KTP and two bulk KTP crystals. Solid lines correspond to linear fits. Inset: spectra of signals and idlers generated by AFB KTP OPO.

Each of the two KTP crystals generated a pair of signal and idler at the slightly different wavelengths of around 1 μm. The length of each crystal is 10 mm such that the total length is 20 mm, which is close to the length of the AFB KTP. At the pump power of 108.5 mW, the dual-idler output power was 21 mW which was 53.8% of the output power generated from AFB KTP. This value is close to an estimate of 42.8%, considering the slightly length difference between the AFB KTP composite and two traditional KTP crystals. Since the implemented OPO is singly resonant, the spectrum and power of the dual-idler beams are rather stable based on our observation.

By mixing the dual-idler beams from the OPO based on DFG [15-17] in four-piece alternatively-rotated GaP plates, we generated the THz wave based on the setup shown by Fig. 11. Each GaP plate cut along the (110) plane had a thickness of 663 μm. The polarizations of the dual-idlers were along the same direction, which is different from the condition in Refs. [15,16]. As we rotated the azimuthal angle of a single GaP plate, the THz intensity exhibited a unique dependence, see Fig. 12. This is caused by the dependence of the second-order nonlinear coefficient on the polarization directions of the mixing beams relative to the azimuthal angle of the plate. Considering the above conditions and cut angle of GaP plate, the THz intensity can be shown to be $\propto \cos^4(\varphi)+\sin^2(2\varphi)$, where φ is the azimuthal angle, see Fig. 11. From Fig. 12, the measured data agree quite well with the theoretical values. We measured the THz power generated by the four-piece alternatively-rotated GaP stacks vs. the average power of the dual idlers from the OPO, see Fig. 13. At 37.2 mW, the THz peak power was measured to be 40.8 W. We fitted the data points using a quadratic dependence, which corresponds to the solid curve in Fig. 13. Obviously, the measured data clearly exhibit a quadratic dependence which is the typical characteristics of DFG. Based on the data presented in Fig. 13 the THz output power is stable.

By scanning a Si-based etalon, THz wavelength was measured to be 121 μm, which agrees well with the designed value of 118 μm (see inset to Fig. 12).

The THz frequency is determined by not only the phase-matching condition of the pump, signal and idler from the OPO's, but also the QPM condition due to the presence of the alternatively-inverted AFB KTP plates. The THz output frequency is given by $\nu_{THz}=\nu_{i1}-\nu_{i2}$, where ν_{i1} and ν_{i2} are the frequencies of the idler beams generated by the OPO. Since the pump wavelength of the OPO is 532 nm, the THz frequency is calculated to be 2.55 THz, which is close to 2.54 THz (118 μm). In addition, to explore the tunability of the THz generation, we varied the pump wavelength

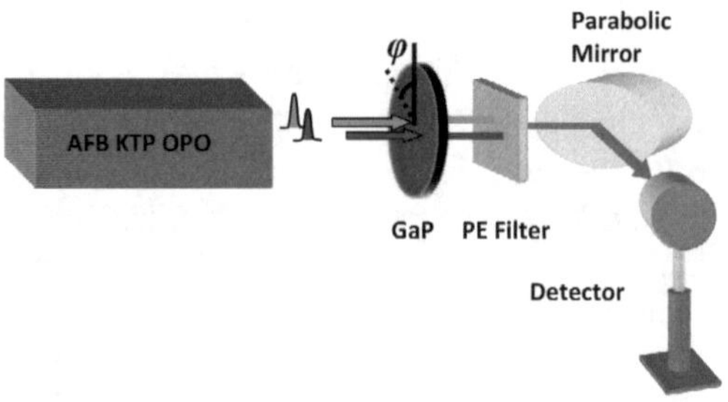

Fig. 11. An experimental setup for THz generation.

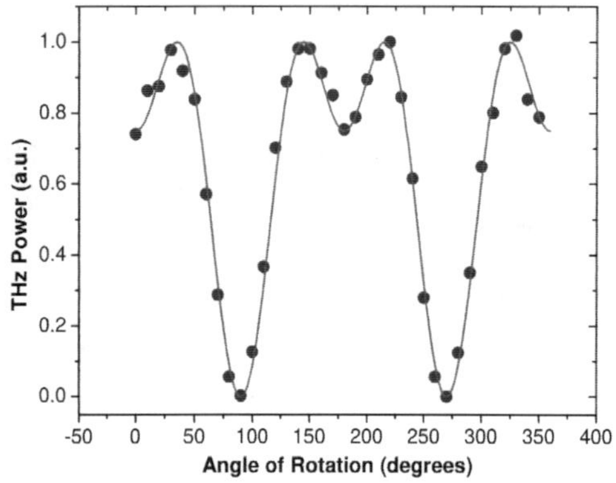

Fig. 12. THz power as a function of azimuthal angle in GaP plate. Dots correspond to experimental results. Solid curve corresponds to theoretical result.

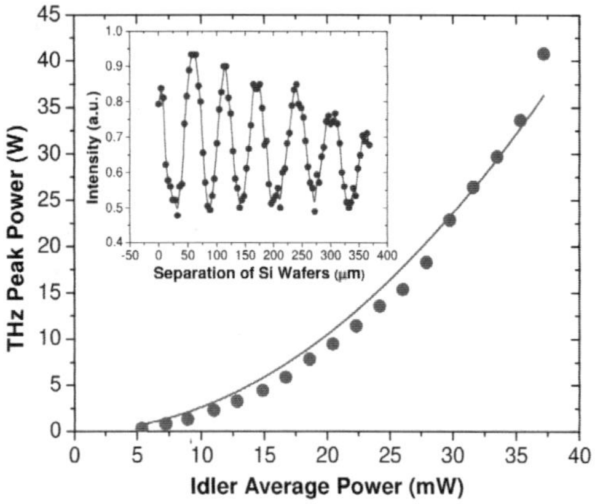

Fig. 13. Dependences of THz peak power on the dual-idler average power emitted from the OPO. Solid curve corresponds to the quadratic fit. Inset: the power being transmitted through a Si-based etalon versus the distance.

within the range of 512-552 nm. Fig. 14 illustrates the tuning characteristics. According to Fig. 14, the THz output wavelength can be tuned in the range of 105.0-133.7 μm, corresponding to the frequency range of 2.244-2.857 THz.

V. THz intracavity generation

In Section, we summarize our result on THz generation based on intracavity difference-frequency generation in stacked GaP plates as an output coupler of an optical parametric oscillator [18]. The total THz output peak power approaches 200 W.

In order to reach the QPM condition for the OPO's, let us consider the THz generation outside t he OPO cavity, shown in Fig. 15. This is similar to that in Ref. [7] except that the dual-signal beams are used for the THz generation. In such a case, the GaP plates were placed in the arm of the dual-wavelength signals which have the polarizations being perpendicular to that of the pump and idler beams. A cubic polarizer was placed in the cavity such that only the dual signal waves were reflected. Therefore, the damage of the KTP plates by the pump beam can be avoided. Before achieving the intracavity frequency conversion, we optimized the pump wavelength by using the external-cavity scheme shown by see Fig. 15. Different from the setup in Ref [7], the dual-idler beams were confined inside the OPO cavity while the dual signals were the output beams. Another key component inside the cavity is the adhesive-free bonded KTP plates, consisting of 20 pieces (each has its thickness of 1.19 mm), which were used as a gain medium for the OPO. It has a clear aperture of 5.6 mm × 10.5 mm and a total interaction length of 23.8 mm. Due to quasi-phase-matching (QPM), a single pair of the signal and idler in a bulk KTP crystal was split into two pairs of the signals and idlers (i.e. totally four output wavelength wavelengths) [6].

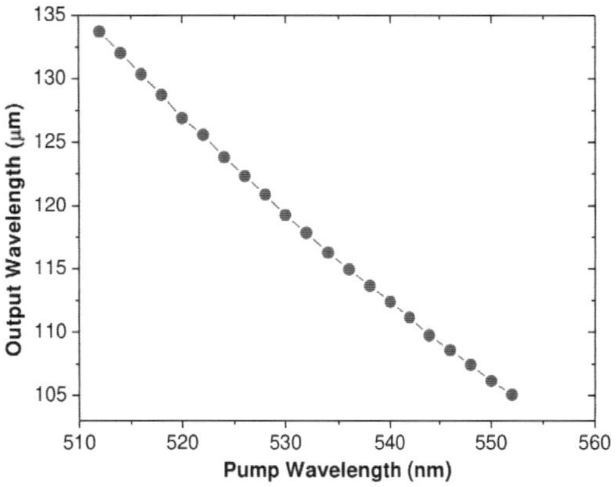

Fig. 14. An illustration of the tuning characteristics.

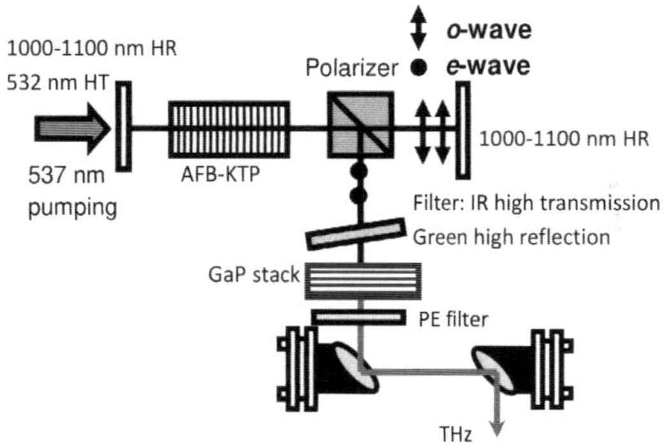

Fig. 15. Experimental setup for external-cavity THz generation.

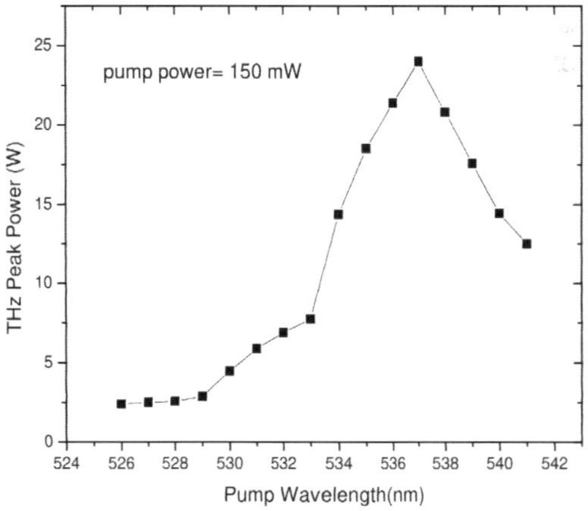

Fig. 16. THz peak power as a function of pump wavelength of OPO's, demonstrating QPM.

Each GaP plate used in our experiment has a thickness of 663 μm and a diameter of 48.5 mm. It was cut and polished on two (110) facets. The reflectivity from each GaP plate is about 20%, which is insufficient for being used as an output coupler. In order to reach a sufficiently high reflectivity, we stacked up to eight plates. While stacking, we alternatively rotated every the other plate to achieve QPM for the THz generation. We set the pump wavelength to one value within the range of 526-541 nm and measured the THz output power generated by the 8 pieces. We then set the pump wavelength to a different value and measured the output power again. According to Fig. 16, the THz output power reaches an optimal value at 537 nm (10 Hz, 5 ns), corresponding to the peak power of 24 W at the pump power of 150 mW. This is due to the reason that the QPM condition for the THz generation can be achieved for the set of the dual-signal wavelengths dictated by the pump wavelength of 537 nm and QPM for the OPO. When we set the pump to a value being different from 537 nm, the dual-signal wavelengths are shifted due to the QPM for the OPO, resulting in the change in the THz output wavelength.

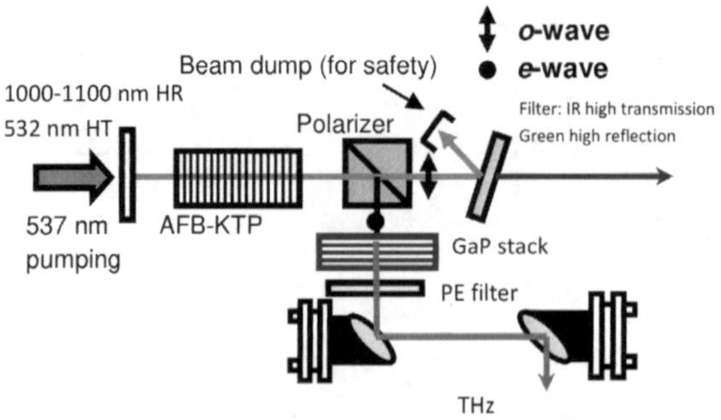

Fig. 17. Experimental setup for intracavity THz generation.

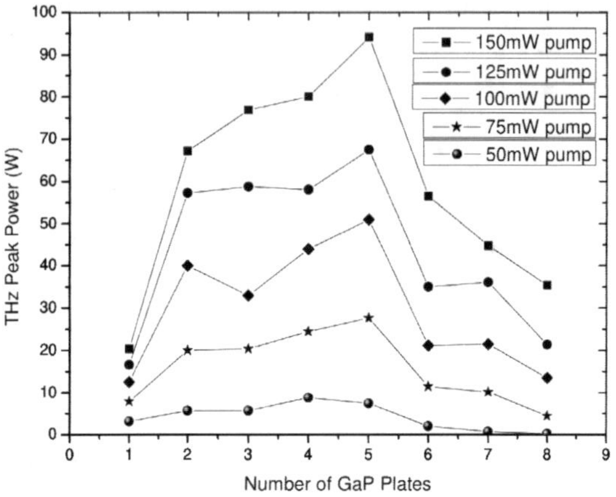

Fig. 18. THz power dependence on number of GaP plates.

For the intracavity THz generation, we set the pump wavelength to 537 nm such that the QPM condition for the THz generation is satisfied. We removed the HR output coupler for the dual-idler beams, see Fig. 17. Meanwhile, the stacked GaP plates were moved closer to the polarizer. It was well aligned for being used as an output coupler for the dual signals. As a result, the signals were resonated in the cavity consisting of the back reflector on the left side and the stacked GaP plates. The THz radiation was generated by the stacked GaP plates. This frequency-mixing process is another scheme for the intracavity THz generation, similar to Ref. [19]. We measured the THz output power as we set the pump power to five different values and set the number of the plates to the different values within 1-8, see Fig. 18. At the pump power of 50 mW, the highest output power was reached for four plates. However, for the pump powers of > 50 mW, the THz output power reached optimal values for five plates. At the pump power of 150 mW, the total THz power was measured to be about 188 W. Compared with the external-cavity [7], this power is a factor of 5 higher.

VI. Demonstration of THz interference

In this Section, we summarize our result following the observation of the THz interference. By mixing three nearly-even-spaced optical frequencies generated by coupled OPO's based in

KTP stacks and bulk KTP in a nonlinear medium, we demonstrate the coherent interference of the THz waves.

The experimental setup is shown in Fig. 19. This setup is different from the previous one. Indeed, we have inserted a half-wave plate between the polarizer and the GaSe crystal. To generate multiple THz wavelengths from the output beams of the coupled OPO's, the idler beam consisting three infrared wavelengths was first sent through a half-wave plate by the rotation by 45° to satisfy the polarization requirement, and then delivered onto a GaSe crystal. The adhesive-free-bonded (AFB) stacks [6] consist of 20 x-cut KTP plates, each of which has a thickness of 1.19 mm. The KTP stack composite has a clear aperture of 5.6 mm×10.5 mm. When the z axes of the KTP plates were periodically switched, they were bonded together via the AFB technique to form a strong composite such that the quasi-phase matching condition ($\Delta kl = \pm\pi$) is achieved for the OPO's. As a result, a single pair of the signal and idler frequencies is split into two signal wavelengths and two idler wavelengths at each pump wavelength. Hence, we introduced the terminology of coupled OPO's to distinguish them from the conventional OPO in where only a single pair of the signal and idler frequencies is generated. Beside the composite, an x-cut bulk KTP crystal (5 mm×5 mm×20 mm) was placed inside the OPO cavity. Consequently, the third set of signal and idler wavelengths was generated. In addition, the bulk KTP crystal was mounted inside an oven such that this set of the wavelengths can be precisely tuned by adjusting its temperature. The two cavity mirrors both have high-reflection coating in the range of 1000-1100 nm. The input mirror is coated for reaching the high transmittance at 532 nm (i.e. the pump wavelength of the OPO). To ensure the oscillations taking place only at three signal wavelengths, a cubic polarizer was placed inside the cavity. It reflects almost all of the *e*-polarized signal beams (R > 99%) such that most of the signal powers stayed inside the cavity whereas the *o*-polarized idler beam was efficiently coupled out of the cavity.

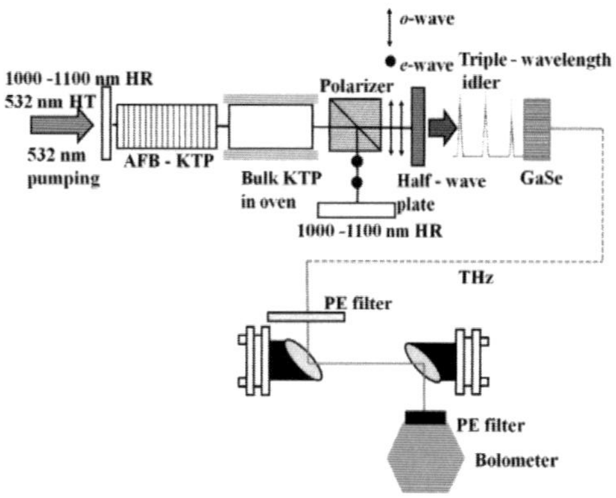

Fig. 19. Experimental Setup to generate triple idler beam and THz output.

Fig. 20 illustrates the spectrum of the idler output measured at room temperature. According to Fig. 20, there are three peaks at the wavelengths of 1084.3nm, 1089.3nm, and 1094.3nm, respectively. This means that the wavelength generated by the bulk KTP crystal was located exactly in the middle between the two idler wavelengths generated by the composite. The two difference

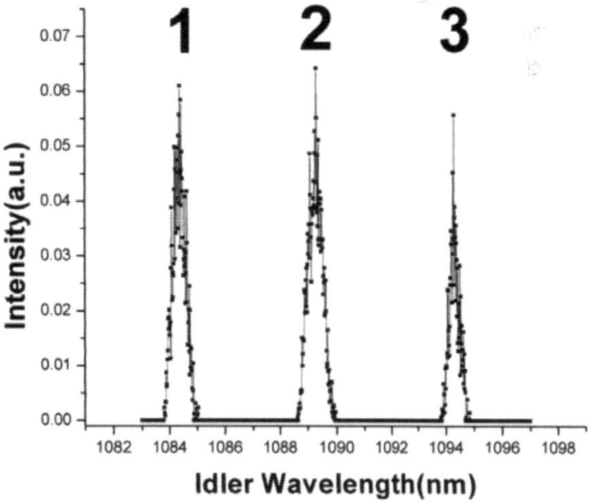

Fig. 20. Spectrum of the triple-wavelength idler beam measured at room temperature.

frequencies from such a set of the idler wavelengths can be calculated to be 1.2700 THz and 1.2584 THz, respectively. Due to the linewidths of the three idler peaks, the corresponding difference frequencies overlap in the THz spectrum. We measured the THz output power as a function of the phase-matching angle, see Fig. 21.

In Fig. 21, the peak at 119.88 μm (2.503 THz) originates from the mixing of the idler wavelengths labeled by 1 and 3 in Fig. 20. The stronger peak at 240.03μm (1.250THz) originates from the coherent interference of the two THz beams by mixing of those labeled by 1 and 2 and mixing of 2 and 3 in Fig. 20. As we increase the temperature of the bulk KTP crystal from 30 °C to 45 °C, the corresponding spectra of the THz output were measured by us, see Figs. 22(a)-(d). Fig. 22 illustrates a clear trend that when the temperature is increased, not only the strength of the interference peak decreases, but also the apparent peak is broadened. At the same time, the single peak at 119.88 μm remains unchanged. Such a trend indicates that we have indeed observed the coherent interference between the two THz beams having slightly different frequencies, which are generated by the GaSe crystal.

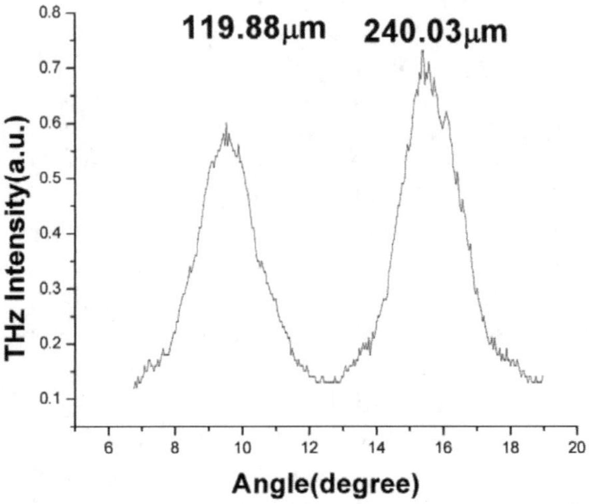

Fig. 21. Spectrum of the THz output beam as a function of phase-matching angle, generated by a GaSe crystal.

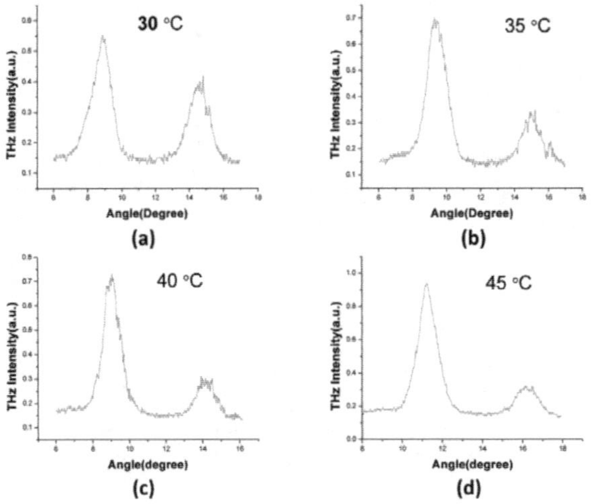

Fig. 22. Spectra of the THz output beam at different temperatures: a - 30 °C, b - 35 °C, c - 40°C, d - 45 °C.

VII. Conclusion

In conclusion, we have reviewed our progress on the applications of the coupled OPO's based on a composite consisting of the alternatively-rotated AFB KTP stacks. By using the phase-conjugate output generated by the composite, we have demonstrated that the spatial profile of the beam after propagating through the phase-distorted medium can be cleared up. In addition, we have restored the images after being blurred by the phase distortion. We have efficiently generated THz outputs by mixing the idler twins from the coupled OPO's. By using the alternatively-rotated GaP plates as an output coupler for the coupled OPO's, we have efficiently generated THz waves based on an intracavity configuration. By placing both the AFB KTP composite and a bulk KTP crystal in the same cavity, we have demonstrated the interference effects between the two THz beams generated by different pairs of the optical beams.

Our coupled OPO's possess unique properties compared with the conventional OPO. Indeed, the frequency separation within the idler twins or signal twins is insensitive to the fluctuations of the pump frequency and temperature.

Acknowledgements

We are indebted to X. Zou, P. Zhao, P. Hong, X. Lin, I. B. Zotova, X. Mu, H.-C. Lee, S. K. Meissner, and H. Meissner for making valuable contributions to some of the projects summarized in this article. This work has been partially supported by U.S. AFOSR.

References

[1]. U. Sharma, C. S. Kim, and J. U. Kang, "Highly stable tunable dual-wavelength Q-switched fiber laser for DIAL applications," IEEE Photon. Tech. Lett. **16**, 1277-1279 (2004).
[2]. R. Diaz, S. C. Chan, and J. M. Liu, "Lidar detection using a dual-frequency source," Opt. Lett. **31**, 3600-3602 (2006).
[3]. F. Ganikhanov, S. Carrasco, X. S. Xie, M. Katz, W. Seitz, and D. Kopf, "Broadly tunable dual-wavelength light source for coherent anti-Stokes Raman scattering microscopy," Opt. Lett. **31**, 1292-1294 (2006).
[4]. K. Miyamoto, and H. Ito, "Wavelength-agile mid-infrared (5-10 μm) generation using a galvano-controlled $KTiOPO_4$ optical parametric oscillator," Opt. Lett. **32**, 274-276 (2007).
[5]. H. Ito, K. Suizu, T. Yamashita, A. Nawahara, and T. Sato, "Random frequency accessible broad tunable terahertz-wave source using phase-matched 4-dimethylamino-N-methyl-4-stilbazolium tosylate crystal," Jap. J. Appl. Phys. **46**, 7321-7324 (2007).
[6]. X. Mu, H. Meissner, and H. C. Lee, "Optical parametric oscillations of 2 μm in multiple-layer bonded walk-off compensated KTP stacks," Opt. Lett. **35**, 387-389 (2010).
[7]. P. Zhao, S. Ragam, Y. J. Ding, I. B. Zotova, X. Mu, H.-C. Lee, S. K. Meissner, and H. Meissner, "Singly resonant optical parametric oscillator based on adhesive-free-bonded periodically inverted $KTiOPO_4$ plates: terahertz generation by mixing a pair of idler waves," Opt. Lett. **37**, 1283-1285 (2012).
[8]. X. Q. Zou, P. Zhao, P. Hong, X. Lin, Y. J. Ding, X. Mu, H.-C. Lee, S. K. Meissner, and H. Meissner, "Restoration of blurred images due to phase distortion based on polarization-insensitive phase conjugation in second-order nonlinear medium," Opt. Lett. **38**, 3054-3056 (2013).
[9]. P. Yeh, "Fundamental limit of the speed of photorefractive effect and its impact on device applications and material research," Appl. Opt. **26**, 602-604 (1987).
[10]. For a review, see various chapters in R. A. Fisher, ed., *Optical phase conjugation*, Academic, NY, 1983.
[11]. F. Devaux, E. Guiot, and E. Lantz, "Image restoration through aberrant media by optical phase conjugation in a type II three-wave mixing interaction," Opt. Lett. **23**, 1597-1599 (1998).
[12]. F. Devaux, G. le Tolguenec, and E. Lantz, "Phase conjugate imaging by type II parametric amplifcation," Opt. Commun. **147**, 309-312 (1998).
[13]. L. Lefort and A. Barthelemy, "Revisiting optical phase conjugation by difference-frequency generation," Opt. Lett. **21**, 848-850 (1996).
[14]. P. Zhao, Z. Liu, X. Lin, Y. J. Ding, X. Mu, H.-C. Lee, S. K. Meissner, and H. Meissner, "Polarization-Insensitive Optical Phase Conjugation Based on Adhesive-Free-Bonded Periodically-Inverted $KTiOPO_4$ Plates," CLEO 2013, CTu3E.7, June 9-14, 2013, San Jose, CA.
[15]. Y. Jiang, Y. J. Ding, and I. B. Zotova, "Power scaling of widely-tunable monochromatic terahertz radiation by stacking high-resistivity GaP plates," Appl. Phys. Lett. **96**, 031101 (2010).
[16]. Y. Jiang, D. Li, Y. J. Ding, and I. B. Zotova, "Terahertz generation based on parametric conversion: from saturation of conversion efficiency to back conversion," Opt. Lett. **36**, 1608-1610 (2011).
[17]. E. B. Petersen, W. Shi, A. Chavez-Pirson, N. Peyghambarian, and A. T. Cooney, "Efficient parametric terahertz generation in quasi-phase-matched GaP through cavity enhanced difference-frequency generation," Appl. Phys. Lett. **98**, 121119 (2011).

[18]. P. Zhao, X. Lin, Y. J. Ding, X. Mu, H.-C. Lee, S. Meissner, and H. Meissner, "Intracavity THz generation inside stacked GaP plates as output coupler of optical parametric oscillator," CLEO 2013, JTh2A.48.

[19]. P. Zhao, S. Ragam, Y. J. Ding, and I. B. Zotova, "Terahertz intracavity generation from output coupler consisting of stacked GaP Plates," Appl. Phys. Lett. **101**, 021107/1-3 (2012).M. V. Vasil′ev, V. Yu. Venediktov, and A. A. Leshchev, "Telescopic systems with dynamic nonlinear optical correction for distoritions," Quantum Electron. **31**, 1-15 (2001).

Journal of Physics: Conference Series **594** (2015) 012013 doi:10.1088/1742-6596/594/1/012013

Photon-phonon laser on crystalline silicon: a feasibility study

A A Zadernovsky

Moscow State Technical University of Radioengineering, Electronics and Automation, Department of Physics, 78 Vernadsky Ave., 119454 Moscow, RUSSIA

E-mail: zadernovsky@mirea.ru

Abstract. We discuss a feasibility of photon-phonon laser action in bulk silicon with electron population inversion. It is well known, that only direct gap semiconductors are used as an active medium in optical lasers. In indirect gap semiconductors, such as crystalline silicon, the near-to-gap radiative electron transitions must be assisted by emission or absorption of phonons to conserve the momentum. The rate of such two-quantum transitions is much less than in direct gap semiconductors, where the similar radiative transitions are single-quantum. As a result, the quantum efficiency of luminescence in silicon is too small to get it as a laser material. Numerous proposals to overcome this problem are aimed at increasing the rate of radiative recombination. We suggest enhancing the quantum efficiency of luminescence in silicon by stimulating the photon part of the two-quantum transitions by light from an appropriate external laser source. This allows us to obtain initially an external-source-assisted lasing in silicon and then a true photon-phonon lasing without any external source of radiation. Performed analysis revealed a number of requirements to the silicon laser medium (temperature, purity and perfection of crystals) and to the intensity of stimulating radiation. We discuss different mechanisms that may hinder the implementation of photon-phonon lasing in silicon.

1. Introduction

The possibility to exploit indirect band gap semiconductors as an active medium for lasers is discussed for many years. Already in the first publication on this subject [1], which appeared shortly after demonstration of the first semiconductor laser, it has been concluded that indirect band gap semiconductors are not suitable for light emitting devices and realizing a laser. In such semiconductors the bottom of the conduction band is not located directly above the top of the valence band in the momentum space of a crystal. This means that an electron near the conduction band edge can not to have the same momentum as a hole near the valence band edge. As a result, the near-to-gap radiative electron-hole recombination can only occur through two-quantum transitions with emission of a photon and simultaneous emission or absorption of a phonon to satisfy the laws of energy and momentum conservation. The rate of such two-quantum transitions is very low and therefore, indirect band gap semiconductors are highly inefficient as a host material for light sources.

Despite this disappointing conclusion researchers do not leave the efforts to find a solution to this problem [2–5]. The reason for such persistence is silicon – one of the most famous and attractive indirect band gap semiconductors. Silicon is the fundamental material used ubiquitously in the electronics industry. It is exploited as a substrate for most integrated circuits. Therefore, creation of silicon-based light sources (either light emitting diodes or lasers) would pave the way for closer integration of photonics and electronics.

The advent of nanotechnology has opened up entirely new possibilities for researchers. Optical properties of a tiny crystal with the size that are comparable to the electron de Broglie wavelength (for instance, less than 5 nm at room temperature) is fundamentally differ from its bulk counterpart. When the dimensions are so small, quantum effects start to play an important role and can fully alter the properties of the original material.

The quantum confinement effects may offer a solution to the problem of silicon. According to the Heisenberg uncertainty principle, when a particle is spatially localized, its momentum becomes uncertain. This means that in low-dimensional indirect band gap semiconductors (nanocrystals, thin layered structures, porous materials), the tails of the wavefunctions of an electron and a hole can now partially overlap, allowing the quasi-direct transitions to occur and thus increasing the probability of radiative recombination.

A number of groups have presented the experimental data on the observation of an enhanced luminescence and even an optical gain in nanostructured silicon, namely, in nanocrystals [6, 7], in silicon on insulator superlattices [8] and in nano-porous silicon [9–11]. Soon afterwards, the efficient silicon-based light emitting diodes [12, 13] have been demonstrated.

Silicon laser is the most challenging goal for researchers and a number of important breakthroughs have been made in the past decade [14]. Among the impressive achievements are the first demonstration of a pulsed silicon Raman laser [15] and the first successful demonstration of a continuous-wave silicon Raman laser shortly thereafter [16]. It should be noted, however, that Raman laser is a specific type of laser where the light-amplification mechanism is stimulated Raman scattering. The optical pumping in Raman lasers does not produce a population inversion. The pumping radiation is rather converted into stimulated radiation in nonlinear laser medium. In contrast, most conventional lasers rely on stimulated electronic transitions between inversely populated quantum states. Silicon laser with electron population inversion remains still elusive despite the efforts of many researchers [17–22].

In this paper we discuss a new approach to the implementation of laser action in crystalline silicon – two-quantum photon-phonon laser. Dynamics of the photon-phonon laser generation is investigated with aid of the rate equations. We formulate a set of requirements to the silicon laser medium (temperature, purity and perfection of crystals) and discuss different mechanisms that may hinder the development of photon-phonon lasing in silicon.

2. Dynamics of photon-phonon laser generation

Dynamics of the photon-phonon laser generation in a silicon crystal can be described by the rate equations for the electron concentration N_e in the conduction band, the photon concentration N_k in an optical mode, and for the phonon concentration N_q in an acoustic mode [23]. We restrict ourselves to the steady state solutions of the rate equations. One can see that there are two types of these solutions: below the phonon lasing threshold with $n_q = 0$ and above the phonon lasing threshold with $n_q \neq 0$. These solutions are shown in the figure 1 (curves of type I and type II, respectively) in the form of dependence of the normalized photon concentration n_k in the optical mode versus the normalized pumping rate W_p/W_0 with the normalized photon concentration n_0 of external stimulating radiation taken as a parameter of a family of curves.

It can be shown with the help of Routh-Hurwitz criterion that only increasing parts of the curves II in figure 1 correspond to the stable steady state solutions. Intersection in the point A between the stable branch of a curve II and a curve I with the same n_0 determines the phonon lasing pumping threshold $\left(W_p/W_0\right)_{\text{th}}$ and the threshold value for the normalized photon concentration $(n_k)_{\text{th}}$ inside the crystal. If the pumping is further increased the system goes on the stable branch of the curve II and the phonon laser action occurs. This leads to a growth of the phonon and the photon concentrations in the modes. If the pumping rate W_p/W_0 exceeds some threshold value W_s/W_0, the external source of stimulating radiation can be switched off and the system goes from the point C_1 to the working point C on the stable branch of the curve II with $n_0 = 0$. Thus, the system continues to operate with simultaneous lasing of both photons and phonons.

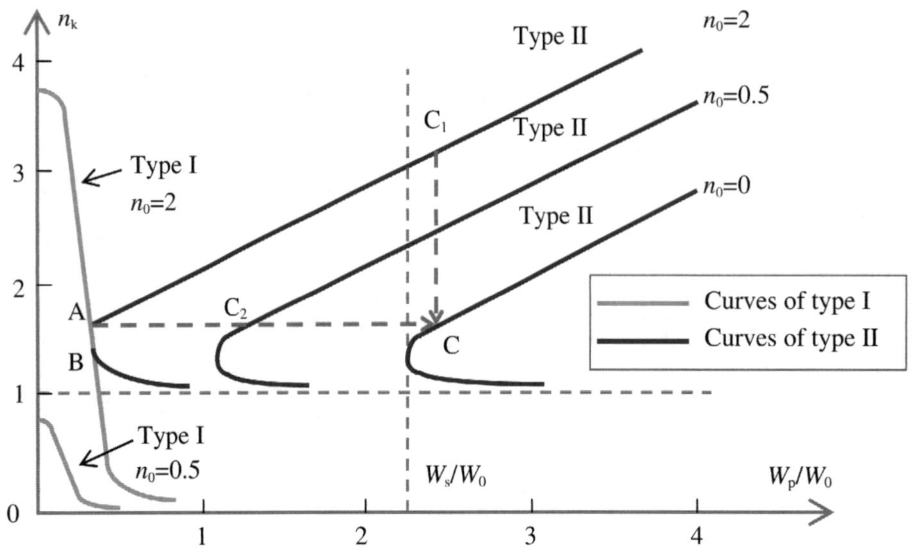

Figure 1. The steady state solutions of the rate equations.

The path $A \to C_1 \to C$ is not the only way to come to the working point C. Alternatively, we can use the path $A \to C_2 \to C$ shown in the figure 1. For this we should gradually reduce the intensity of external radiation according as the pumping rate increases so that the photon concentration $(n_k)_{th}$ in the optical mode remains constant.

Numerical estimates are collected in the table 1.

Table 1. Numerical estimates

Semiconductor	N, 10^{14} cm^{-3}	W_0, 10^{25} cm^{-3}c^{-1}	$(W_p/W_0)_{th}$, 10^{-3}	$(n_k)_{th}$	W_s/W_0	n_{ks}	n_{qs}	n_0
Si	1.6×10^2	5.6×10^2	4.2	1.6	2.4	3.0	2.4	2.0

N	the normalizing particle concentration
W_0	the normalizing pumping rate
$(W_p/W_0)_{th}$	the normalized threshold pumping rate for starting the phonon laser action
$(n_k)_{th}$	the normalized threshold photon concentration inside the crystal for starting the phonon laser action
W_s/W_0	the normalized threshold pumping rate for starting the simultaneous laser action of both photons and phonons
n_{ks}	the normalized photon concentration inside the crystal for starting the simultaneous laser action of both photons and phonons
n_{qs}	the normalized phonon concentration for starting the simultaneous laser action of both photons and phonons
n_0	the normalized photon concentration of external stimulating radiation

Returning from normalized to the physical variables, we obtain at the optimum photon energy of stimulating radiation, $\hbar\omega_{\mathbf{k}_0}$ =1.165 eV, the following estimate for the threshold light intensity (at the point A in figure 1) $I = c\hbar\omega_{\mathbf{k}_0} N(n_k)_{\text{th}}$ =2.7×10^7 W/cm^2, where c = 0.88×10^{10} cm s^{-1} is the speed of light in silicon.

3. Feasibility discussion of the silicon photon-phonon laser

In this section we discuss different mechanisms that may hinder the implementation of photon-phonon lasing in silicon. Performed analysis revealed a number of requirements to the active laser medium (temperature, purity and perfection of crystals) and to the intensity of stimulating radiation. First of all, we need cryogenically cooled silicon crystals in order to reduce the anharmonic phonon-phonon scattering of TA phonons down to an acceptable level. The TA phonons are long-lived at low temperatures, since the selection rules forbid the phonon decay into two other phonons with lower energies [24]. Under these conditions, the principal mechanism of the TA phonon losses from an acoustic mode becomes Rayleigh scattering by isotope impurities. For silicon with natural isotopic abundances the inverse lifetime τ_{i}^{-1} of the TA phonon is estimated by [25] τ_{i}^{-1} =7.2×10^9 s^{-1} . Further reduction of the phonon losses requires isotopically pure and perfect silicon crystals. The above obtained estimates presented in table 1 are based on the assumption that the natural concentration of isotopic impurities in silicon can be reduced by 4 orders of magnitude.

Even under such special conditions, the intensity of stimulating radiation, required to attain the photon-phonon lasing in bulk silicon, remains a rather high and may approach to the optical damage threshold. A study of optical resistance of silicon was made in connection with search for an upper limit for the light intensity restricting the performance of silicon Raman laser. It was found [26] that the optical damage threshold of crystalline silicon ranges from 1 GW/cm^2 to 4 GW/cm^2 depending on the radiation wavelength. In particular, at the operating wavelength of the silicon photon-phonon laser, λ= 1.06 μm, it is estimated to be 1 GW/cm^2. One can see that the above obtained threshold intensity of stimulating radiation, I=2.7×10^7 W/cm^2, is approximately two orders of magnitude lower and therefore cannot damage the silicon crystal. Furthermore, in an isotopically pure and perfect silicon crystal, where the phonon losses from an acoustic mode is mainly due to lattice absorption and due to phonon absorption in the anti-Stokes vibronic sideband, the intensity of stimulating radiation may be additionally reduced by an order of magnitude.

Another potentially detrimental process, that should be considered, is the two-photon absorption of stimulating radiation. This process which was not taken into account in the rate equations may have a significant impact on the dynamics of lasing, especially at a high optical intensity. Since the total energy of two photons of stimulated radiation is greater than the band gap energy of silicon, they can be simultaneously absorbed in indirect transitions with excitation of electrons from the valence band to the conduction band. The rate of the two-photon generation of free carriers in silicon was experimentally and theoretically investigated in numerous papers. To evaluate this rate, the coefficient β was introduced through the relation

$$\frac{\mathrm{d}N_{\text{e}}}{\mathrm{d}t} = \frac{\beta I^2}{2\hbar\omega_{\mathbf{k}_0}}. \tag{1}$$

The corresponding values of β were experimentally determined for silicon at different wavelengths and can be found in the review [27]. In particular, at the operating wavelength of the silicon photon-phonon laser, λ= 1.06 μm, the coefficient β is estimated to be 1.5 cm/GW . This leads to the following estimate $\mathrm{d}N_{\text{e}}/\mathrm{d}t = 2.9\times10^{24}$ cm^{-3}c^{-1} for the rate (1) at the threshold intensity of stimulating radiation $I = 2.7\times10^7$ W/cm^2 . One can see that this value is an order of magnitude lower than the threshold rate of pumping W_{p} = 2.3×10^{25} cm^{-3}c^{-1} for starting the phonon lasing (point A in figure 1) and almost four orders of magnitude lower than the threshold rate of pumping W_{s} = 1.3×10^{28} cm^{-3}c^{-1}

for starting the simultaneous lasing of photons and phonons (point C in figure 1). Therefore, the contribution of (1) to the rate equation for the electron concentration N_e is negligible.

As far as the rate equation for the photon concentration N_k is concerned, the photon losses due to the two-photon absorption should be compared with other channels of photon losses. It is well known that bulk photon losses in semiconductors are dominated by the free carrier absorption with the rate

$$W_e = \sigma c N_k N_e , \quad (2)$$

where σ is the photon capture cross section ($\sigma = 6\times10^{-18}$ cm^2 for silicon). The rate of photon losses due to radiation through the end crystal surfaces of reflectivity R (R = 0.30 for silicon) is equal to N_k/τ, where

$$\tau^{-1} = \frac{c}{L}\ln(1/R) \quad (3)$$

and L is the crystal length. Assuming L = 250 μm, we obtain $\tau^{-1} = 3.5\times10^{11}$ s^{-1}.

Since the generation of one electron-hole pair is accompanied by simultaneous absorption of two photons, the rate of the two-photon absorption, dN_k/dt, is two times greater than (1) and can be estimated by the value of 5.8×10^{24} cm^{-3}c^{-1}. The rate of photon absorption by free carriers W_e (2) at the electron concentration $N_e = 8.9\times10^{18}$ cm^{-3} and the rate of photon losses through the crystal facets N_k/τ (3) are 1.2×10^{28} cm^{-3}c^{-1} and 9.1×10^{27} cm^{-3}c^{-1}, correspondingly. One can see that the contribution of the two-photon absorption in the rate equation can be neglected.

Summarizing the above estimates, we come to a conclusion that the process of two-photon absorption does not affect the dynamics of the photon-phonon lasing in silicon.

Among the different mechanisms hindering the laser action in silicon, the parasitic non-radiative Auger recombination is of particular concern. This is a three-particle process where an excited electron recombines with a hole and the excess energy is transferred to another free carrier rather than emitted in the form of a photon. It occurs even in pure and perfect semiconductor crystals. The rate of competitive Auger recombination in bulk silicon is generally much higher than the rate of radiative recombination. The internal quantum efficiency of luminescence, which is defined as the ratio of the probability of radiative electron-hole recombination to the total probability of electron-hole recombination, is limited in silicon by the value of 10^{-5}–10^{-6} [14]. Although the efficiency can be improved to up to 10^{-3} by employing ultrapure silicon [28], it is still too small to consider silicon as a good light-emitting material.

The above obtained estimates for the rate of photon-phonon transitions in silicon confirm this conclusion. Indeed, the rate of Auger recombination is proportional to the third degree of concentration of excited carriers, that is $W_{Auger} = C_n N_e^3$ with $C_n \sim 10^{-31}$cm^{6}c^{-1} [29] for intrinsic silicon. Taken into account the electron concentration at the gain maximum, $N_e = 8.9\times10^{18}$ cm^{-3}, we obtain for the rate of Auger recombination the value of $W_{Auger} = 7.1\times10^{25}$ cm^{-3}c^{-1}, whereas the rate of spontaneous-spontaneous photon-phonon transitions W_{00} is estimated to be only 1.6×10^{23} cm^{-3}c^{-1}. Hence, most of the excited electron-hole pairs recombine non-radiatively and the corresponding quantum efficiency is

$$\eta = \frac{W_{00}}{W_{00} + W_{Auger}} = 2.2\times10^{-3}. \quad (4)$$

All proposals to circumvent this problem and to get silicon as a laser material are aimed at increasing the rate of radiative recombination and thus at raising the quantum efficiency. In the approach considered in this paper we suggest to stimulate the photon part of the two-quantum transitions in silicon by the light from an appropriate external laser source. This leads to enhancement the rate of radiative recombination. The rate of stimulated-spontaneous transitions W_{k0} at the threshold intensity of stimulating radiation, $I = 2.7\times10^7$ W/cm^2, is estimated to be 5.5×10^{25} cm^{-3}c^{-1}. Accordingly, the associated quantum efficiency,

$$\eta = \frac{W_{k0}}{W_{k0} + W_{\text{Auger}}}, \tag{5}$$

increases by more than two orders of magnitude and reaches the value of 0.44. This allows us to obtain initially an external-source-assisted lasing in silicon (point A in figure 1) and then a true photon-phonon lasing without any external source of radiation (point C in figure 1).

4. Conclusions

Most proposals for creation a silicon laser are based on overcoming or circumventing the problem of low rate of radiative recombination in indirect band gap semiconductors. Among them are [30]: 1) overcoming the indirect band structure by using spatial confinement of electrons, 2) introducing rare-earth impurities as optically active dopants, 3) applying epitaxy and hybrid integration with III-V-semiconductor lasers and 4) using Raman scattering to achieve optical gain. Our paper supplements this list with one more proposal. We suggest to make use of stimulated photon-phonon interband electron transitions in crystalline silicon to obtain phonon lasing and simultaneous lasing of both photons and phonons.

To drive the process, we start with stimulating the photon part of the two-quantum transitions with simultaneous emission of phonons and photons by the light from an appropriate laser source. This leads to a growth of the rate of phonon emission. At a high enough intensity of the external stimulating radiation and at a high enough level of pumping, the phonon gain turns out to be equal to the phonon losses in a particular acoustic mode and the phonon laser action appears. It is then possible to sustain the phonon laser action by gradual increasing the pumping rate with simultaneous decreasing the intensity of the external stimulating radiation. Finally, the source of stimulating radiation can be switch off and the system continues to operate with simultaneous lasing of both photons and phonons.

Dynamics of the photon-phonon laser generation has been investigated with aid of the rate equations. Numerical estimates of the threshold intensity of stimulating radiation, the threshold rate of pumping for starting the phonon lasing and the threshold rate of pumping for starting the simultaneous lasing of both photons and phonons have been made for pure (including isotopic purity) and perfect crystals of silicon. The feasibility discussion of the silicon photon-phonon laser allows us to make a conclusion about the self-consistency of our proposal.

Acknowledgements

This research was supported by the Ministry of Education and Science of the Russian Federation.

References

[1] Dumke W P 1962 Interband transitions and maser action *Phys. Rev.* **127** 1559–63
[2] Ossicini S, Pavesi L and Priolo F Light 2003 *Emitting Silicon for Microphotonics* (Berlin: Springer)
[3] Pavesi L and Lockwood D J 2004 *Silicon Photonics* (Berlin: Springer)
[4] Reed G T and Knights A P 2004 *Silicon Photonics: An Introduction* (Chichester UK: John Wiley)
[5] Reed G T and Knights A P 2008 *Silicon Photonics: The State of the Art* (Chichester UK: John Wiley)
[6] Pavesi L, Dal Negro L, Mazzoleni G, Franzo G and Priolo F 2000 Optical gain in silicon nano-crystals *Nature* **408** 440–4
[7] Pelant I 2011 Optical gain in silicon nanocrystals: current status and perspectives *Phys. Status Solidi* A **208** 625–30
[8] Lu Z H, Lockwood D J and Baribeau J M 1995 Quantum confinement and light emission in SiO_2/Si superlattices *Nature* **378** 258–60
[9] Cullis A G and Canham L T 1991 Visible light emission due to quantum size effects in highly porous crystalline silicon *Nature* **353** 335–8

[10] Gösele U and Lehmann V 1995 Light-emitting porous silicon *Mater. Chem. Phys.* **40** 253–9
[11] Hirschman K D, Tsybeskov L, Duttagupta S P and Fauchet P M 1996 Silicon-based visible light–emitting devices integrated into microelectronic circuits *Nature* **384** 338–41
[12] Wai Lek Ng, Lourenço M A, Gwilliam R M, Ledain S, Shao G and Homewood K P 2001 An efficient room-temperature silicon-based light-emitting diode *Nature* **410** 192–4
[13] Homewood K P and Lourenço M A 2005 Light from Si via dislocation loops *Materials Today* **8** No 1 34–9
[14] Di Liang and Bowers J E 2010 Recent progress in lasers on silicon *Nature Photonics* **4** 511–7
[15] Boyraz O and Jalali B 2004 Demonstration of a silicon Raman laser *Optics Express* **12** 5269–73
[16] Rong H, Jones R, Liu A, Cohen O, Hak D, Fang A and Paniccia M 2005 A continuous-wave Raman silicon laser *Nature* **433** 725–7
[17] Iyer S S and Xie Y H 1993 Light emission from silicon *Science* **260** 40–6
[18] Kimerling L C, Kolenbrander K D, Michel J and Palm J 1997 Light emission from silicon *Solid State Physics* **50** 333–6
[19] Trupke T, Green M A and Würfel P 2003 Optical gain in materials with indirect transitions *J. Appl. Phys.* **93** 9058–61
[20] Chen M J, Tsai C S and Wu M K 2006 Optical gain and co-stimulated emissions of photons and phonons in indirect bandgap semiconductors *Jpn. J. Appl. Phys.* **45** No 8B 6576–88
[21] Pavesi L 2008 Silicon-based light sources for silicon integrated circuits *Advances in Optical Technologies* **2008** Article ID 416926
[22] Pavesi L 2005 Routes towards a silicon-based laser *Materials Today* **8** 18–25
[23] Zadernovsky A A 2014 Two-quantum photon-phonon laser *Journal of Physics: Conference Series* **497** 012005
[24] Orbach R 1966 *Phys. Rev. Lett.* **16** 15
[25] Holland M C 1963 *Phys. Rev.* **132** 2461
[26] Jalali B and Fathpour S 2006 Silicon Photonics *Journal of lightwave technology* **24** No 12 4600–15
[27] Xinzhu Sang, En-kuang Tien and Ozdal Boyraz 2008 Applications of two-photon absorption in silicon *Journal of optoelectronics and advanced materials* **11** No 1 15-25
[28] Jalali B 2007 Making silicon lase *Scientific American* **296** No 2 58–65
[29] Svantesson K G and Nilsson N G 1979 The temperature dependence of the Auger recombination coefficient of undoped silicon *J. Phys.* C: *Solid State Phys.* **12** 5111–21
[30] Zhou Fang and Ce Zhou Zhao 2012 Recent progress in silicon photonics: a review *ISRN Optics* **2012** Article ID 428690

Photon polarization in Compton scattering: pulse shape effects

M. Boca, C. Stoica, A. Dumitriu, V. Florescu
Department of Physics, University of Bucharest, PoBox MG11, 077125, Magurele, Romania
E-mail: `madalina.boca@g.unibuc.ro`

Abstract.
We study in the framework of quantum electrodynamics the scattering of a plane wave electromagnetic field on free electrons in the low intensity limit. We derive analytic formulas describing the polarization properties of the emitted photons. We discuss and illustrate with a numerical example the effects of the electromagnetic pulse duration on their polarization.

1. Introduction

The study of inelastic scattering of radiation on electrons is deeply connected with the progress of quantum physics, starting with A. Compton major contribution (1923). The basis for the theoretical understanding of the process is the QED calculation of the scattering of low intensity monochromatic radiation on free electrons [1]. The availability of the first intense sources of radiation made necessary the distinction between the scattering in the low intensity regime (linear Compton scattering, LC) and the process at high intensity (nonlinear Compton scattering, NLC). The first studies of NLC are already almost fifty years old (see reviews [2], [3], [4]). In this paper, for more clarity, we shall also use the term "monochromatic linear Compton" (MLC) scattering.

The numerous recent studies of NLC have overcome the monochromatic case. Calculations are possible in the case of plane-wave pulses with fixed propagation direction but arbitrary duration and shape [5, 6] using in the Furry picture a Volkov solution for describing the state of the electron dressed by the laser field and treating the interaction between the electron and the quantized field of the emitted photon in first order perturbation theory. The papers published up to now present differential cross sections corresponding either to well defined states of polarization of the emitted photon and initial/final electrons, or summed/averaged over the polarization. Polarization properties are studied very little: in the monochromatic case a detailed analysis of polarization properties of the electrons and photons was presented by Ivanov *et al* [7], the calculation following the scheme used in the MLC case as developed in the textbook by Berestetskii, Lifshitz and Pitaevskii [8]. Polarization of the radiation emitted in nonlinear Thomson scattering (the classical limit of NLC) was also studied [9, 10].

The purpose of this paper is twofold: on the one hand we present an analytic expression of the transition probability valid in the low intensity limit in the case of a plane wave laser pulse with finite duration, on the other hand we adapt to this case the general formalism for the study of polarization properties of the scattered radiation. Our result is formally similar to

Journal of Physics: Conference Series **594** (2015) 012014 doi:10.1088/1742-6596/594/1/012014

the corresponding MLC expression, and we shall discuss the differences between them. Then, in the same conditions we write the equations for determining the Stokes parameters of the emitted photon and also compare with the case of MLC. The formalism developed here for the low intensity limit can be used also in the case of arbitrary intensity; the corresponding results in this case will published elsewhere.

In Sect. 2 we briefly review general textbook results for MLC in the case of unpolarized electrons in both initial and final state, following closely the formalism of [8], albeit in a slightly simplified version. Next, in Sect. 3, starting from the general formulas of the non-linear Compton scattering [5], valid for the incoming radiation described by an intense plane wave laser pulse of finite duration, we derive their low intensity limit; we shall denote this case by "plane-wave linear Compton" (PWLC), in order to distinguish it from MLC. The results are written such that the effect of the finite duration of the pulse are easily visible by comparison with the monochromatic case. Finally, we derive the Stokes parameters of the emitted photon in the same approximation. The Stokes parameters are defined with respect to a particular system of axes, chosen as in the case of MLC (Sect. 87 of [8]), and in order to simplify the results we shall describe the laser field in a particular convenient gauge. Our conclusions are summarized in Sect. 5. In Appendix is discussed the relation between MLC and PWLC.

2. The linear monochromatic Compton scattering

The analysis of the polarization state of the final particles in the linear Compton scattering in term of Stokes parameters is a textbook problem [8]. For further use, we introduce the notations and review the main equations leading to Eq. (12), a convenient way of presenting the LC transition probability for initially polarized photons as a function of the Stokes parameters of the emitted photon.

We denote the initial and final electron momenta by p_1 and respectively p_2 and, by $\hbar k_1$ and $\hbar k_2$ the momenta of the absorbed respectively emitted photon. Using the energy momentum conservation law it follows that the results can be expressed only in terms of two relativistic invariants which shall be chosen here as

$$X = \frac{p_1 \cdot \hbar k_1}{(mc)^2} = \frac{p_2 \cdot \hbar k_2}{(mc)^2}, \quad Y = \frac{p_2 \cdot \hbar k_1}{(mc)^2} = \frac{p_1 \cdot \hbar k_2}{(mc)^2}; \tag{1}$$

the remaining two invariants are

$$\hbar k_1 \cdot \hbar k_2 = (mc)^2(X - Y), \quad p_1 \cdot p_2 = (mc)^2(1 + X - Y). \tag{2}$$

In the case of unpolarized initial/final electrons, the transition probability between states with fixed initial and final photon polarization s_1 and respectively s_2 is, up to a factor dependent on the photon energy, but not relevant for the calculation of the Stokes parameters,

$$\begin{aligned} |\mathcal{Q}|^2_{s_1 s_2} = & \frac{(X-Y)^2}{2XY} - \frac{(X-Y)^2}{2XY}\left| s_1 \cdot s_2 - \frac{(s_1 \cdot p_1)(s_2 \cdot p_2)}{(mc)^2 X} + \frac{(s_1 \cdot p_2)(s_2 \cdot p_1)}{(mc)^2 Y}\right|^2 + \\ & \frac{(X+Y)^2}{2XY}\left| s_1^* \cdot s_2 - \frac{(s_1^* \cdot p_1)(s_2 \cdot p_2)}{(mc)^2 X} + \frac{(s_1^* \cdot p_2)(s_2 \cdot p_1)}{(mc)^2 Y}\right|^2 \\ & \equiv \mathcal{C}_{\alpha\beta} s_2^\alpha s_2^{*\beta}. \end{aligned} \tag{3}$$

We mention that the above equation is more general that formula (11.13) in the book by Jauch and Rohrlich [1] because it allows the initial and final polarization 4-vectors s_1 and s_2 to be complex. By gauge transformations, one can always make the two polarization vectors to obey the conditions

$$s_1 \cdot k_2 = s_2 \cdot k_1 = 0, \tag{4}$$

with the consequences

$$p_2 \cdot s_1 = p_1 \cdot s_1, \quad p_2 \cdot s_2 = p_1 \cdot s_2\,, \tag{5}$$

and then Eq. (3) reduces to the simpler form

$$\begin{aligned} |\mathcal{Q}|^2_{s_1 s_2} = & \frac{(X-Y)^2}{2XY} - \frac{(X-Y)^2}{2XY}\left| s_1 \cdot s_2 - \frac{(s_1\cdot p_1)(s_2\cdot p_1)(Y-X)}{(mc)^2 XY}\right|^2 \\ & + \frac{(X+Y)^2}{2XY}\left| s_1^* \cdot s_2 - \frac{(s_1^*\cdot p_1)(s_2\cdot p_1)(Y-X)}{(mc)^2 XY}\right|^2 \\ \equiv & \tilde{\mathcal{C}}_{\alpha\beta} s_2^\alpha s_2^{*\beta}\,. \end{aligned} \tag{6}$$

Next we want to express the transition probability to a final state in which the emitted photon is in a mixed polarization state. We choose a pair of reference four-vectors defined by (see [8], Sect. 87)

$$e^{(1)} = N/\sqrt{-N^2}\,, \quad e^{(2)} = -P/\sqrt{-P^2}, \tag{7}$$

$$N^\mu = \epsilon^{\mu\nu\lambda\rho} k_{1,\nu} k_{2,\lambda} p_{1,\rho} \quad P^\mu = \epsilon^{\mu\nu\lambda\rho} k_{1,\nu} k_{2,\lambda} e_{1,\rho} \tag{8}$$

with $\epsilon^{\mu\nu\lambda\rho}$ the four-dimensional Levi-Civita tensor, with respect to which the Stokes parameters of the final state will be expressed. The two basis vectors obey the condition (4), so the simpler form of the transition probability (6) will be used. We replace $s_2^\mu s_2^{*\nu}$ in the polarization matrix by the density matrix written in term of the Stokes vector $\boldsymbol{\xi}^{(2)}$ of the emitted radiation

$$\begin{aligned} \rho^{(2)\mu\nu} = & (e_1^\mu e_1^\nu + e_2^\mu e_2^\nu) + \xi_1^{(2)}(e_1^\mu e_2^\nu + e_2^\mu e_1^\nu) - i\xi_2^{(2)}(e_1^\mu e_2^\nu - e_2^\mu e_1^\nu) + \xi_3^{(2)}(e_1^\mu e_1^\nu - e_2^\mu e_2^\nu) \\ = & \rho_0^{(2)\mu\nu} + \sum_{i=1,3} \rho_i^{(2)\mu\nu} \xi_i^{(2)}, \end{aligned} \tag{9}$$

$$|\mathcal{Q}|^2_{s_1 s_2} \to |\mathcal{Q}|^2_{s_1 \boldsymbol{\xi}^{(2)}} = \tilde{\mathcal{C}}_{\alpha\beta}\, \rho^{(2)\alpha\beta}. \tag{10}$$

At the same time we decompose the initial polarization vector in the basis formed by e_1 and e_2,

$$s_1 = c_1\, e_1 + c_2\, e_2, \quad |c_1|^2 + |c_2|^2 = 1\,; \tag{11}$$

such an expansion is always possible in the particular gauge (4). Then, after a straight calculation, (6) becomes the desired expression of the transition probability between the initial state of fixed polarization described by the coefficients c_1 and c_2 and the final state characterized by the Stokes vector $\boldsymbol{\xi}^{(2)}$:

$$\begin{aligned} |\mathcal{Q}|^2_{s_1\boldsymbol{\xi}^{(2)}} = & \frac{X}{Y} + \frac{Y}{X} - 2|c_2|^2 \frac{(X-Y)(2XY - X + Y)}{X^2 Y^2} \\ & - \xi_1^{(2)}(c_1^* c_2 + c_2^* c_1)\frac{2(XY - X + Y)}{XY} + i\xi_2^{(2)}(c_1^* c_2 - c_2^* c_1)\frac{(X^2+Y^2)(XY - X + Y)}{X^2 Y^2} \\ & + 2\xi_3^{(2)}\left[1 - |c_2|^2 \frac{X^2Y^2 + (XY - X + Y)^2}{X^2 Y^2}\right] \\ \equiv & \mathcal{R}_0 + \sum_{i=1,3} \mathcal{R}_i \xi_i^{(2)}. \end{aligned} \tag{12}$$

This expression, containing c_1 and c_2 will be useful for the comparison of PWLC and MLC results in next section. If written in terms of the Stokes parameters of the incident photon, the expression reduces to a formula equivalent to Eq. (87.11) of [8].

If, instead of the emission probability for a Stokes vector fixed by the detector, one wants to calculate the Stokes vector *of the photon, before the detection*, the expressions of its components are (see also Eq. (87.17) of [8])

$$\xi_i^{(f)} = \frac{\mathcal{R}_i}{\mathcal{R}_0}. \tag{13}$$

3. The linear plane wave Compton scattering

3.1. The transition amplitude of NLC

In the case of nonlinear Compton scattering, we consider the scattering of a *laser pulse* with fixed propagation direction $\mathbf{n}_1$, defined by a vector potential which depends on coordinates and time only through the combination $ct - \mathbf{n}_1 \cdot \mathbf{r} \equiv n_1 \cdot x$, with $n_1 = (1, \mathbf{n}_1)$,

$$A^\mu(\phi) = A_0 f(\phi) \left[\zeta_1 a_1^\mu \cos(k_1\phi) + \zeta_2 a_2^\mu \sin(k_1\phi)\right] \equiv A_1(\phi) a_1^\mu + A_2(\phi) a_2^\mu \,. \tag{14}$$

In the above equation $f(\phi)$ is the laser pulse envelope, chosen such that

$$\lim_{\phi \pm \infty} f(\phi) = 0 \tag{15}$$

and the two four vectors a_1 and a_2 are mutually orthogonal, orthogonal on the four-vector n_1, and normalized

$$a_1 \cdot a_2 = a_1 \cdot n_1 = a_2 \cdot n_1 = 0, \quad (a_1)^2 = (a_2)^2 = -1 \,. \tag{16}$$

The parameter k_1 in (14) is related to the central frequency ω_1 of the pulse by $k_1 = \omega_1/c$. In the monochromatic limit the associated photons carry the momentum $\mathbf{n}_1 \hbar k_1$. The two parameters ζ_1 and ζ_2 obey the condition $\zeta_1^2 + \zeta_2^2 = 1$ and characterizes the laser polarization. The degrees of linear, respectively circular polarization of the laser are $P_L^{(L)} = (\zeta_1^2 - \zeta_2^2)^2$, $P_C^{(L)} = 4\zeta_1^2\zeta_2^2$, $P_C^{(L)} + P_L^{(L)} = 1$.

In [5] we have calculated the transition amplitude for the spontaneous emission of a photon by an electron dressed by the laser field of arbitrary intensity. In the initial state, at $t \to -\infty$ the electron has the momentum p_1 and spin index i_1, and the momentum and spin of the electron in the final state are denoted by p_2 and respectively i_2. The transition amplitude for the emission of a photon of momentum k_2 and polarization vector s_2, is given by Eqs. (27) and (28) of [5]. We write it here as

$$\begin{aligned} \mathcal{M}_{1\to 2} \;=\; & \frac{e}{i\hbar}\sqrt{\frac{\hbar}{2\epsilon_0\omega_2 V}}\frac{1}{V}\sqrt{\frac{(mc)^2}{p_1^0 p_2^0}\frac{\hbar}{mc}}(2\pi\hbar)^3 \\ & \times\delta(\mathbf{p}_{1,\perp} - \mathbf{p}_{2,\perp} - \hbar\mathbf{k}_{2,\perp})\,\delta[n_1\cdot(p_1 - p_2 - \hbar k_2)]\,\mathcal{Q}(2,1)\,, \end{aligned} \tag{17}$$

displaying the product of a two-dimensional δ function, expressing the conservation of the component of the total momentum orthogonal on the laser propagation direction, $\mathbf{P}_{tot\perp}$, with a one-dimensional δ function with the argument $n_1 \cdot P_{tot}$. The quantity $\mathcal{Q}(2,1)$ is [1]

$$\begin{aligned} \mathcal{Q}(2,1) \;=\; & \langle\overline{v}_{i_2}(p_2)|\left\{\mathcal{B}_0(2,1)\hat{s}_2^* + (mc)^2\frac{(n_1\cdot s_2^*)\mathcal{B}_2(2,1)}{(p_1\cdot n_1)(p_2\cdot n_1)}\hat{n}_1 + \right. \\ & \left. \frac{(mc)}{\sqrt{2}}\mathcal{A}_1(2,1)\left[\frac{\hat{n}_1\,\hat{a}_1\,\hat{s}_2^*}{(n_1\cdot p_2)} + \frac{\hat{s}_2^*\,\hat{s}_1\,\hat{n}_1}{(n_1\cdot p_1)}\right] + \frac{(mc)}{\sqrt{2}}\mathcal{A}_2(2,1)\left[\frac{\hat{n}_1\,\hat{a}_2\,\hat{s}_2^*}{(n_1\cdot p_2)} + \frac{\hat{s}_2^*\,\hat{a}_2\,\hat{n}_1}{(n_1\cdot p_1)}\right]\right\}|v_{i_1}(p_1)\rangle\,, \end{aligned} \tag{18}$$

where the Feynman "slash" was denoted using the symbol "hat" ($\hat{a} = a_\mu\gamma^\mu$). The previous result is expressed in terms of four integrals $\mathcal{B}_0$, $\mathcal{B}_2$, $\mathcal{A}_1$ and $\mathcal{A}_2$ defined as

$$\mathcal{B}_0(2,1) = \frac{mc}{\hbar}\int_{-\infty}^{\infty} d\phi\, e^{-\frac{i}{\hbar}G(\phi)} \qquad \mathcal{B}_2(2,1) = \frac{mc}{\hbar}\int_{-\infty}^{\infty} d\phi\,\frac{-e^2A^2(\phi)}{2(mc)^2}e^{-\frac{i}{\hbar}G(\phi)} \tag{19}$$

$$\mathcal{A}_1(2,1) = \frac{mc}{\hbar}\int_{-\infty}^{\infty} d\phi\,\frac{-eA_1(\phi)}{\sqrt{2}(mc)}e^{-\frac{i}{\hbar}G(\phi)} \qquad \mathcal{A}_2(2,1) = \frac{mc}{\hbar}\int_{-\infty}^{\infty} d\phi\,\frac{-eA_2(\phi)}{\sqrt{2}(mc)}e^{-\frac{i}{\hbar}G(\phi)} \tag{20}$$

[1] $\mathcal{Q}(2,1)$, differs from that in (Eq. (28) of [5] by the use of complex final polarization 4-vector s_2 and by the normalization convention of the free Dirac spinors $|v_i(p)\rangle$

with

$$G(\phi) = \frac{1}{2}\tilde{n}_1 \cdot (p_1 - \hbar k_2 - p_2)\phi + \Lambda(p_1;\phi) - \Lambda(p_2;\phi), \quad \tilde{n}_1 \equiv (1; -\mathbf{n}_1), \tag{21}$$

$$\Lambda(p;\phi) = -\frac{1}{2n_1 \cdot p}\int_{-\infty}^{\phi} d\chi \left[e^2 A^2(\phi) - 2eA(\phi)\cdot p\right]. \tag{22}$$

The four integrals are connected by the relation

$$\begin{aligned} -\frac{Q}{mc}\mathcal{B}_0(2,1) &= mc\left(\frac{1}{n_1\cdot p_1} - \frac{1}{n_1 \cdot p_2}\right)\mathcal{B}_2(2,1) + \\ &+\sqrt{2}\left[a_1\mathcal{A}_1(2,1) + a_2\mathcal{A}_2(2,1)\right]\cdot\left(\frac{p_1}{n_1\cdot p_1} - \frac{p_2}{n_1\cdot p_2}\right), \quad Q = \frac{\tilde{n}_1\cdot(\hbar k_2 + p_2 - p_1)}{2} \end{aligned} \tag{23}$$

which allows expressing the formally divergent integral $\mathcal{B}_0$ as linear combination of finite integrals $\mathcal{B}_2$, $\mathcal{A}_1$ and $\mathcal{A}_2$.

The major difference with respect to the case of linear Compton scattering is the presence of only three one-dimensional δ functions instead of four in the transition amplitude. As a consequence, there are three independent invariant in terms of which all the others can be expressed. We choose these parameters as

$$x = \frac{p_1\cdot n_1}{mc}, \quad y = \frac{p_2 \cdot n_1}{mc}, \quad z = \frac{p_1\cdot p_2}{(mc)^2} - 1, \tag{24}$$

mention the expression of the other invariants,

$$\hbar k_2 \cdot n_1 = (x-y), \quad \hbar k_2 \cdot p_1 = (mc)^2\frac{zx}{x-y}, \quad \hbar k_2 \cdot p_2 = (mc)^2 \frac{zy}{x-y}, \tag{25}$$

the identity $Q = (mc)z/(x-y)$. The correspondence of x, y, z with the quantities X and Y used in the monochromatic limit in Eq. (1) is

$$x \to X\frac{mc^2}{\hbar\omega_1}, \quad y \to Y\frac{mc^2}{\hbar\omega_1}, \quad z \to X - Y. \tag{26}$$

3.2. The low intensity limit

The dimensionless parameter

$$\eta \equiv \frac{-eA_0}{\sqrt{2}mc} \tag{27}$$

is a measure of the external field intensity. In the low intensity limit the terms of the order $\mathcal{O}(\eta^2)$ are neglected, so we neglect the integral $\mathcal{B}_2(2,1)$ and the dependence of the field in the exponent $G(\phi)$ defined in (21); then the integrals $\mathcal{A}_1(2,1)$ and $\mathcal{A}_2(2,1)$ are expressed only in terms of the Fourier transform Φ of the envelope $f(\phi)$:

$$\mathcal{A}_1(2,1) \approx \eta\zeta_1\int_{-\infty}^{\infty} d\phi f(\phi)\cos(k_1\phi)e^{-\frac{i}{2\hbar}\tilde{n}_1\cdot(p_1-\hbar k_2 - p_2)\phi} = \eta\zeta_1\pi\left[\Phi(\hbar k_1 - Q) + \Phi(-\hbar k_1 - Q)\right],$$

$$\mathcal{A}_2(2,1) \approx \eta\zeta_2\int_{-\infty}^{\infty} d\phi f(\phi)\sin(k_1\phi)e^{-\frac{i}{2\hbar}\tilde{n}_1\cdot(p_1-\hbar k_2 - p_2)\phi} = -i\eta\zeta_2\pi\left[\Phi(\hbar k_1 - Q) + \Phi(-\hbar k_1 - Q)\right],$$

where we defined the dimensionless Fourier transform of $f(\phi)$ by

$$\Phi(q) \equiv \frac{(mc)}{2\pi\hbar}\int_{-\infty}^{\infty} d\phi e^{-\frac{i}{\hbar}q\phi} f(\phi). \tag{28}$$

We observe here that usually the pulse envelope is real and symmetric, which makes the Fourier transform $\Phi(q)$ real. We shall write the transition amplitude in the low intensity limit as

$$\mathcal{Q}(2,1) = \sqrt{2}(mc)\pi\eta\left[\Phi(\hbar k_1 - Q))\mathcal{Q}^{(+)}(2,1) + \Phi(\hbar k_1 + Q)\mathcal{Q}^{(-)}(2,1)\right] \tag{29}$$

$$\begin{aligned}\mathcal{Q}^{(\pm)}(2,1) &= (mc)\langle\overline{v}_{i_2}(p_2)|\left\{\frac{1}{Q}\left(\frac{s_1^{(\pm)}\cdot p_1}{n_1\cdot p_1} - \frac{s_1^{(\pm)}\cdot p_2}{n_1\cdot p_2}\right)\hat{s}_2^* + \right. \\ &\left. \frac{1}{2}\left[\frac{\hat{n}_1\,\hat{s}_1^{(\pm)}\,\hat{s}_2^*}{(n_1\cdot p_2)} + \frac{\hat{s}_2^*\,\hat{s}_1^{(\pm)}\,\hat{n}_1}{(n_1\cdot p_1)}\right]\right\}|v_{i_1}(p_1)\rangle\,,\end{aligned} \tag{30}$$

using two complex vectors defined by

$$s_1^{(\pm)} = a_1\,\zeta_1 \pm i\,a_2\,\zeta_2\,, \quad s_1^{(-)} = s_1^{(+)*}\,. \tag{31}$$

The vectors $s_1^{(+)}$ and $s_1^{(-)}$ are normalized but they are not orthogonal, except for the case of circular polarization of the laser field

$$s_1^{(+)*}\cdot s_1^{(+)} = s_1^{(-)*}\cdot s_1^{(-)} = -1, \quad s_1^{(+)*}\cdot s_1^{(-)} = \zeta_2^2 - \zeta_1^2\,. \tag{32}$$

We mention that Eq. (30) is identical with the transition matrix element of MLC [2], written for the case when the incoming photon has direction $\mathbf{n}_1$, momentum Q and polarization vector $s_1^{(\pm)}$. More details are given in the Appendix, where is discussed the monochromatic limit.

Starting from the transition amplitude, we write now the probability of emission of a photon with the fixed polarization s_2, assuming, as before, that the polarization of the emitted photon is orthogonal on n_1 and also that the two vectors a_1 and a_2 are orthogonal on k_2, a situation that can be always reached by a gauge transformation:

$$\begin{aligned}|\mathcal{Q}|^2_{i,s_2} &= \frac{(x-y)^2}{2xy}\left[\Phi_+^2 + \Phi_-^2 - 2\Phi_+\Phi_- Re\{s_1^{(+)}\cdot s_1^{(+)}\}\right] \\ &- \frac{(x-y)^2}{xy}\left[\Phi_+^2|T(s_1^{(+)})|^2 + \Phi_-^2|T(s_1^{(-)})|^2 + 2\Phi_+\Phi_- Re\{T(s_1^{(+)})T(s_1^{(+)})\}\right] \\ &+ \frac{(x+y)^2}{xy}\left[\Phi_+^2|T(s_1^{(-)})|^2 + \Phi_-^2|T(s_1^{(+)})|^2 + 2\Phi_+\Phi_- Re\{T(s_1^{(-)})T(s_1^{(-)})\}\right]\end{aligned} \tag{33}$$

where we used

$$T(s_1^{(\pm)}) = s_1^{(\pm)}\cdot s_2 + \frac{(s_1^{(\pm)}\cdot p_1)(s_2\cdot p_2)(x-y)^2}{(mc)^2xyz}, \quad \Phi_\pm = \Phi(Q \mp \hbar k_1). \tag{34}$$

Equation (33) is the correspondent of Eq. (3) for the case of a plane wave laser field of very low intensity.

[2] The identity between the form usually written in the literature (see, e.g. R. Greiner, *Quantum electrodynamics*, third edition, Springer, 2003, Sect. 3.7) and the expression given in Eq. (30) can be immediately proven.

The next step in our calculation was to derive the equivalent of (12), i.e., the expression of the transition probability as a function of the Stokes parameters of the final photon. We have followed exactly the same procedure as in MLC using the same two basis vectors e_1 and e_2, defined by Eq. (7). We express the vectors $s_1^{(\pm)}$ in terms if e_1 and e_2 as

$$s_1^{(+)} = d_1 e_1 + d_2 e_2, \quad s_1^{(-)} = d_1^* e_1 + d_1^* e_2, \quad |d_1|^2 + |d_2|^2 = 1. \tag{35}$$

After a straight calculation the final result is

$$|\mathcal{Q}|^2_{i\,\boldsymbol{\xi}^{(2)}} = \mathcal{R}_0 + \sum_{k=1,3} \mathcal{R}_k \xi_k^{(2)} \tag{36}$$

with

$$\begin{aligned}
\mathcal{R}_0 &= (\Phi_+^2 + \Phi_-^2)\left[\frac{x}{y} + \frac{y}{x} + 2|d_2|^2 \frac{(x-y)^2(x^2 - 2xy(z+1) + y^2)}{x^2y^2z^2}\right] + \\
&+2\Phi_+\Phi_- \left[Re\{d_1^2\}\left(4 - \frac{x}{y} - \frac{y}{x}\right) + Re\{d_2^2\}\left(\frac{x}{y} + \frac{y}{x} - 4\frac{(x-y)^2}{xyz} + 2\frac{(x-y)^4}{x^2y^2z^2}\right)\right]; \qquad (37)\\
\mathcal{R}_1 &= \left[(\Phi_+^2 + \Phi_-^2)(d_1 d_2^* + d_2^* d_1) + 4\Phi_+\Phi_- Re\{d_1 d_2\}\right] \frac{2(x^2 - xy(z+2) + y^2)}{xyz}; \qquad (38)\\
\mathcal{R}_2 &= i(d_1^* d_2 - d_2^* d_1)(\Phi_+^2 - \Phi_-^2)\frac{(x^2+y^2)(xy(z+2) - x^2 - y^2)}{x^2y^2z}; \qquad (39)\\
\mathcal{R}_3 &= 2(\Phi_+^2 + \Phi_-^2)\left[1 - |d_2|^2 \frac{x^2y^2z^2 + (x^2 - xy(z+2) + y^2)^2}{x^2y^2z^2}\right] + \\
&4\Phi_+\Phi_- \left[Re\{d_1^2\} - Re\{d_2^2\}\frac{(x^2 - xy(z+2) + y^2)^2}{x^2y^2z^2}\right]. \qquad (40)
\end{aligned}$$

The Stokes parameters of the emitted photon are calculated according to the same formula (13), with the components $\mathcal{R}$ given by the previous equations (37-40). As discussed in the Appendix in the monochromatic limit, when $\Phi_- \to 0$ and the relations (26 are valid), the previous expressions reduce to (12); however, in the case of a finite pulse it is not possible to write the formula in terms of the Stokes parameters corresponding to the laser field, because of the terms $Re\{d_1^2\}$, $Re\{d_2^2\}$, $Re\{d_1 d_2\}$.

4. Numerical examples

We illustrate the previous results with an numerical example. For simplicity we choose to work in the reference frame of the incoming electron, with the Oz axis taken along the laser propagation direction $\mathbf{n}_1$ and assume that the laser field is linearly polarized, along the Ox axis of the reference frame; we take the central frequency of the laser field $\hbar\omega_1 = 0.8(mc^2)$ (or $\hbar k_1 = 0.8(mc)$). We shall only consider the case when the photon is emitted in the plane Oyz, at an angle θ_2 with respect to Oz. It is interesting to review the monochromatic predictions for this case: in the monochromatic limit, the frequency of the emitted photon is fixed by its direction; according to the well known Compton formula in the electron rest frame

$$\hbar\omega_2 = \hbar\omega_1 \frac{mc^2}{mc^2 + \hbar\omega_1(1 - \cos\theta_2)}, \tag{41}$$

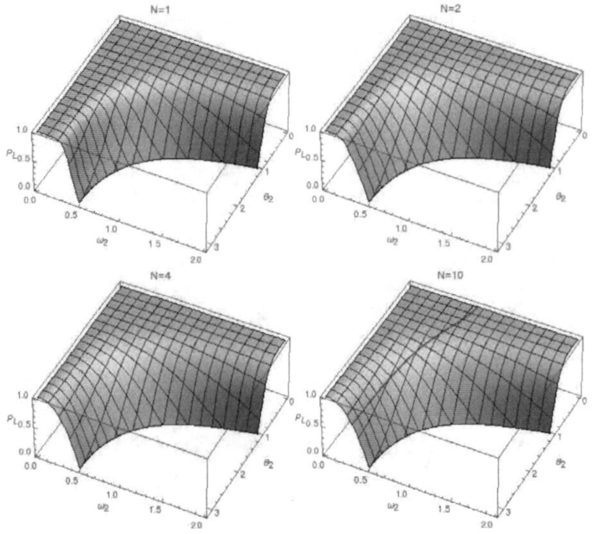

Figure 1. (Color online) The degree of linear polarization of the emitted photon as a function of ω_2, θ_2 for four pulse lengths. The red line on the last figure is the monochromatic prediction.

the polarization state is defined by Eqs. (12) and (13) from which it follows that the degree of circular polarization is 0, and the degree of linear polarization is

$$\xi_L^{(f)} = \frac{2}{\frac{\omega_2}{\omega_1} + \frac{\omega_1}{\omega_2}} . \tag{42}$$

In the plane wave case, the main difference is that for a fixed observation direction the frequency ω_2 of the emitted photon can take any value in the interval

$$\omega_2 \in \left[0, \frac{(mc^2)x}{1 - \mathbf{n}_1 \cdot \mathbf{n_2}}\right] . \tag{43}$$

We mention that the upper limit of the allowed interval can be obtained by imposing the condition $p_1 \cdot p_2 > (mc)^2$, i.e. $z > 0$ (see also Eq. (49) from [10] and the discussion there).

We have considered a Gaussian laser pulse with the envelope

$$f(\phi) = e^{-\frac{\omega_1^2 \phi^2}{N^2 c^2}}, \tag{44}$$

for four values of N (a parameter proportional to the FWHM of the pulse expressed in cycles): $N = 1, 2, 4, 10$.

The degree of linear polarization of the emitted photon, calculated according to (37-40), as a function of ω_2, θ_2 is represented in Fig. 1, for the four pulse lengths. One can see that in fact for the longer pulses ($N > 2$) the graphs are practically identical. In the monochromatic limit, due to the one-to-one correspondence between energy and direction, the figure reduces to a curve, represented in red on the last graph.

The representation used in Fig. 1 does not take into account the fact that, although the emission of a photon with any frequency in the allowed interval (43) is in principle possible, in fact the emission probability varies strongly with ω_2 and it also depends on the pulse shape.

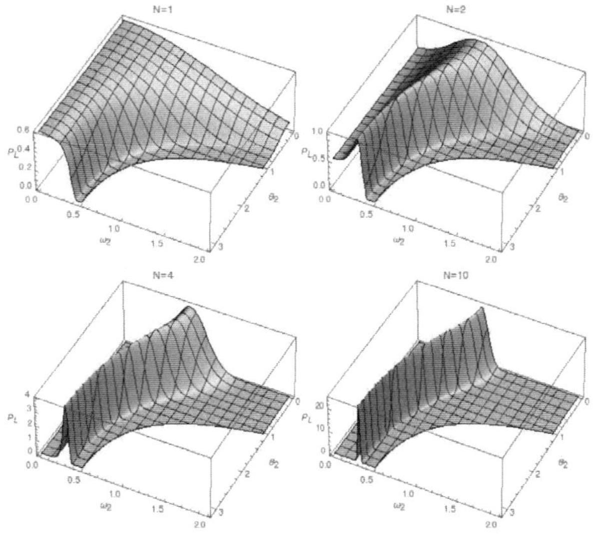

Figure 2. (Color online) The scaled degree of linear polarization of the emitted photon as a function of ω_2, θ_2 for four pulse lengths.

Then it could be useful to define a "scaled" polarization vector $\xi_s^{(f)}$, calculated as the product of the true Stokes vector $\xi^{(f)}$ and the density of probability for emission of a photon with the frequency ω_2 and in the direction $\mathbf{n}_2$. This quantity is represented in Fig. 2 for the same cases as before. One can see that, this time, the modifications due to the laser pulse duration are important. For longer pulses ($N = 4$, 10) the results are already close to the monochromatic limit, i.e. the photon detection probability has non-zero values only in a relatively small interval around the position of the monochromatic curve. On the other hand, in the case of the shorter pulse, since the emission probability is non-negligible for a large part of the allowed domain, the scaled Stokes parameters are non zero and almost constant in a relatively large domain.

5. Conclusions

We have reviewed the formalism used for the calculation of the Stokes parameters of the photon emitted in linear Compton scattering of monochromatic radiation and then we have considered the case of scattering of a plane wave pulse in the low intensity limit. We illustrated our results with a a numerical example. In the Appendix we have reobtained the MLC transition amplitude as a limiting case of the formulas valid for plane wave pulses.

Appendix A. The monochromatic limit

The monochromatic limit is obtained as the particular case of a constant envelope $f(\phi) \to 1$, whose Fourier transform is a δ function $\Phi(q) \to (mc)\delta(q)$. The photons associated to the monochromatic plane wave have the momentum $\hbar k_1 \mathbf{n_1}$, with k_1 defined in Eq. (14). Then with Eq. (29), the total transition amplitude (17) becomes

$$\mathcal{M}_{i\to f}^{\rm mono} = \frac{1}{\sqrt{2}}\pi\eta(mc)\hbar\frac{e}{i\hbar}\sqrt{\frac{\hbar}{2\epsilon_0\omega_2 V}}\frac{1}{V}\sqrt{\frac{(Mc)^2}{p_1^0 p_2^0}}(2\pi\hbar)^3\delta(\mathbf{p}_{1,\perp} - \mathbf{p}_{2,\perp} - \hbar\mathbf{k}_{2,\perp}) \times$$
$$\times\delta[n_1\cdot(p_1 - p_2 - \hbar k_2)]\left[\delta(Q - \hbar k_1)\mathcal{Q}^{(+)}(2,1) + \delta(Q + \hbar k_1)\mathcal{Q}^{(-)}(2,1)\right] \quad \text{(A.1)}$$

One can easily prove the identity

$$\delta(\mathbf{p}_{1,\perp} - \mathbf{p}_{2,\perp} - \hbar\mathbf{k}_{2,\perp})\,\delta[n_1 \cdot (p_1 - p_2 - \hbar k_2)]\delta(Q \mp \hbar k_1) = 2\delta^{(4)}(p_1 \pm \hbar k_1 - p_2 - \hbar k_2), \quad \text{(A.2)}$$

and if one notices that the δ function of argument $p_1 - \hbar k_1 - p_2 - \hbar k_2$ is always zero, the transition amplitude reduces to

$$\mathcal{M}_{i\to f}^{\mathrm{mono}} = \frac{1}{\sqrt{2}}\eta(mc)\frac{e}{i\hbar}\sqrt{\frac{\hbar}{2\epsilon_0\omega_2 V}}\frac{1}{V}\sqrt{\frac{(Mc)^2}{p_1^0 p_2^0}}(2\pi\hbar)^4\delta^{(4)}(p_1 \pm \hbar k_1 - p_2 - \hbar k_2)\mathcal{Q}^{(+)}(2,1) \quad \text{(A.3)}$$

More than that, in the expression of the matrix element $\mathcal{Q}^{(+)}(2,1)$, evaluated with the four conservation laws, we recognize easily the matrix element of the linear monochromatic Compton effect, although written in slightly different form.

So, we have obtained the standard expression of the transition amplitude of the linear Compton effect, except for the factor $(mc)\eta/\sqrt{2} \equiv |e|A_0/2$ which replaces another factor $|e|\sqrt{\frac{\hbar}{2\epsilon_0\omega_1 V}}$ present in MLC, the difference being a direct consequence of the classical description adopted for the laser field.

Acknowledgments

This work was suported by the contract No. 1 RNP/2012 from ANCS through CNCS-UEFISCDI

References

[1] Jauch J M, Rohlich F 1955 *The theory of photons and electrons:the relativistic quantum field theory of charged particles with spin one-half*, (Addison Wesley)
[2] Ritus V I 1979 *Trudy FIAN* **111** 5
[3] Ehlotzky F, Krajewska K and Kaminski J Z (2009) *Rep. Prog. Phys.* **72** 046401
[4] Di Piazza A, Muller C, Hatsagortsyan K Z and Keitel C H 2012 *Rev. Mod. Phys.* **84** 1177
[5] Boca M and Florescu V 2009 *Phys. Rev. A* **80** 053403.
[6] Seipt D and Kämpfer B 2011 *Phys. Rev. A* **83** 022101
[7] Ivanov D Yu, Kotkin G L and Serbo V G 2004 *Eur. Phys. J. C* **36** 127
[8] Berestetskii V B, Lifshitz E M and Pitaevskii L P 1965 *Quantum Electrodynamics*, (Interscience Publishers)
[9] Kraft A G A, Doyuran A and Rosenzweig J B 2005 *Phys. Rev. E* **72** 056502
[10] Boca M and Florescu V 2011 *Eur. Phys. J. D* **61** 449

Influence of laser pulse shape and spectral composition on strong-field Kapitza-Dirac scattering

Matthias M Dellweg and Carsten Müller

Institut für Theoretische Physik I, Heinrich-Heine-Universität, 40225 Düsseldorf, Germany

E-mail: `dellweg@tp1.uni-duesseldorf.de`

Abstract. Diffractive scattering of an electron beam from an intense standing wave of light (Kapitza-Dirac effect) is analyzed theoretically. The corresponding Schrödinger equation is solved in the Bragg regime by combined analytical and numerical means. Our goals are twofold: On the one hand, we examine the influence of the nature of the field's turning-on phase and demonstrate that a smooth, adiabatic envelope restricts the transitions to those satisfying the Bragg condition. On the other hand, we study signatures of quantum mechanical interference when the standing light wave comprises two commensurate frequencies. Characteristic modifications of the Rabi oscillation dynamics and a dependence on the relative phase between the frequency modes are revealed.

1. Introduction

Based on the newly established rules of quantum mechanics, Kapitza and Dirac predicted in 1933 that an electron beam can diffractively scatter from a standing wave of light, formed by two counterpropagating waves of equal frequency and intensity [1, 2]. The effect, which henceforth has been named after them, relies on the quantum wave nature of the electron. It represents an analogue of the classical diffraction of light on a periodic grating, but with the roles of light and matter interchanged. In its original version, the Kapitza-Dirac effect can be understood as a combined absorption and emission process involving two photons: the electron absorbs a photon of momentum $\vec{k}$ from one of the laser beams and emits a photon of momentum $-\vec{k}$ into the counterpropagating laser beam (stimulated Compton scattering). The incoming electron momentum thus changes by $2\vec{k}$, provided that it is incident under the Bragg angle which follows from the requirements of energy and momentum conservation.

The scattering probability predicted by Kapitza and Dirac was very small, though, because of the low intensities of light sources available at that time. Consequently, the interest in the Kapitza-Dirac effect was newly inspired in the 1960s when much higher light intensities were delivered by the first laser devices. This has led to several experimental attempts to observe the effect [3, 4, 5] and to seminal theoretical work on Kapitza-Dirac scattering in strong laser fields [6].

Nevertheless, a convincing experimental confirmation of the Kapitza-Dirac effect as originally proposed has been achieved only recently [7]. This successful observation has once again revived the interest of theoreticians in the effect. In recent years, various novel aspects of Kapitza-Dirac scattering have been examined. For example, a theory based on electronic wave packets was developed [8] and diffractive scattering in the case of counterpropagating laser beams of non-equal frequencies was studied [9]. Besides, a relativistic formulation of the Kapitza-Dirac effect based on the Dirac equation was presented [10, 11], which generalizes earlier treatments within Klein-Gordon theory [12, 13] and provides the proper framework to analyze the electron's spin dynamics in the process.

In this contribution, we study Kapitza-Dirac scattering in the nonrelativistic domain, focussing on the impact of the applied laser pulse shape and its frequency composition. After having established the theoretical framework in Section 2, we analyze the role played by the field's turning-on phase in Section 3. Using the leading order of perturbation theory we show that, in general, characteristic temporal oscillations are imprinted on the scattering amplitude. However, in the case when the switching duration is very long, the scattering amplitude for resonant electron momenta shows a linear growth in time, whereas off-resonant transitions are strongly suppressed. A tight interplay between the switching duration and the detuning from the resonance is revealed.

Afterwards, in Section 4 we consider Kapitza-Dirac scattering from a bichromatic standing wave which comprises, apart from the fundamental frequency ω, also its third harmonic 3ω. Due to quantum mechanical interference effects, the Rabi oscillations are modified in a characteristic manner which is revealed via comparison with the corresponding monochromatic scattering dynamics. Also the influence of the relative phase between the frequency modes is illustrated. We finish with some concluding remarks in Section 5.

2. Theoretical framework

The nonrelativistic evolution of the quantum mechanical wave-function ψ of an electron in the presence of an external electromagnetic field is described by the time-dependent Schrödinger equation

$$i\hbar\frac{\partial}{\partial t}\psi = \frac{1}{2m}\left(\frac{\hbar}{i}\vec{\nabla} + \frac{e}{c}\vec{A}\right)^2 \psi. \tag{1}$$

The mass of the electron beeing m, its charge $-e$. Besides, c is the speed of light and $\hbar$ the reduced Planck constant. A standing light wave of frequency $\omega = ck$ is described by the vector potential $\vec{A}$ in Coulomb gauge as

$$\vec{A} = f(t)A_0\vec{\varepsilon}\cos(\omega t)\cos(kz) \tag{2}$$

where the linearly polarized laser beams (counter-) propagate along the z-axis and $\vec{\varepsilon}$ is a polarisation vector of unit length perpendicular to $\hat{z}$. The constant amplitude A_0 is multiplied by an envelope function $f(t)$ to model entering and leaving of the electron in the interaction region, thereby giving it compact support in time. The intensity of the standing wave amounts to $I = \frac{ck^2A_0^2}{16\pi}$. From the spatial $\frac{2\pi}{k}$-periodicity of the potential in z-direction and its transverse invariance it follows, that only the subset of momentum eigenstates which differ by an integral multiple of the photon momentum $\hbar k$ can interact. Assuming that the incident electron momentum is perpendicular to the polarization vector $\vec{\varepsilon}$, the transverse dynamics can be separated, leaving an effectively one-dimensional situation. Therefore we use an ansatz in the form

$$\psi(t,z) = \sum_n b_n(t)\,\mathrm{e}^{inkz+ip_z z/\hbar} =: \sum_n b_n(t)\,|n\rangle \tag{3}$$

with time-dependent expansion coefficients $b_n(t)$. Note, that p_z is not necessarily the z-component of the initial electron momentum, but merely an offset from the integral multiples of $\hbar k$. We arrive at the equation of motion for b_n as the coupled system of differential equations:

$$i\hbar\dot{b}_n(t) = -\frac{(p_z + n\hbar k)^2}{2m}b_n(t) + \frac{e^2A_0^2}{8mc^2}f(t)^2\cos^2(\omega t)\left(b_{n-2}(t) + 2b_n(t) + b_{n+2}(t)\right) \tag{4}$$

This will be the foundation of the following considerations.

3. Adiabatic switching and frequency detuning

In this section we study to which extent the scattering dynamics of the electron is influenced by the switching-on phase of the interaction [14]. For algebraic simplicity, to simulate the adiabatic switching-on procedure, we use f as an evelope function such that

$$f(t)^2 = \begin{cases} \frac{t}{\tau} & \text{if} \quad 0 < t \le \tau \\ 1 & \text{if} \quad \tau < t \le T \\ \frac{T+\tau-t}{\tau} & \text{if} \quad T < t \le T+\tau \\ 0 & \text{elsewhere} \end{cases} \quad . \tag{5}$$

This way we have established, that $f(t)^2$ is continuous with compact support and gives an effective interaction duration of

$$\int dt f(t)^2 = T \quad . \tag{6}$$

Its derivative beeing bound by the inverse switching time $1/\tau$.

The time evolution operator $\mathcal{U}$, which solves (4) as a Green's function, can be expressed by the recurrent integral equation

$$\mathcal{U}(t) = \mathcal{U}_0(t) + \frac{1}{i\hbar}\int^t dt' \, \mathcal{U}_0(t-t')\frac{e^2}{2mc^2}\vec{A}(t')^2\mathcal{U}(t') \tag{7}$$

where the unperturbed version is given by

$$\mathcal{U}_0(t) = \sum_n \exp\left(\frac{1}{i\hbar}\frac{(\hbar n k + p_z)^2}{2m}t\right)|n\rangle\langle n| \quad . \tag{8}$$

The square of the vector potential can be written, with the help of the electron momentum eigenstates $|n\rangle$, as

$$\begin{aligned} \vec{A}(t)^2 &= \left[\frac{1}{2}f(t)A_0\vec{\varepsilon}\cos(\omega t)\sum_n\Big(|n+1\rangle\langle n| + |n-1\rangle\langle n|\Big)\right]^2 \\ &= \frac{1}{4}f(t)^2A_0^2\cos^2(\omega t)\sum_n\Big(|n+1\rangle\langle n-1| + 2|n\rangle\langle n| + |n-1\rangle\langle n+1|\Big) \quad . \end{aligned} \tag{9}$$

Equation (7) can be expanded in the time-dependent perturbation series for small values of A_0 as

$$\mathcal{U}(t) = \mathcal{U}_0(t) + \mathcal{U}_1(t) + \mathcal{U}_2(t) + \ldots \tag{10}$$

where the first-order term reads

$$\mathcal{U}_1(t) = \frac{1}{i\hbar}\int^t dt' \, \mathcal{U}_0(t-t')\frac{e^2}{2mc^2}\vec{A}(t')^2\mathcal{U}_0(t') \quad . \tag{11}$$

The transition amplitude from the state $|-1\rangle$ with momentum $p_z - \hbar k$ to the state $|1\rangle$ with momentum $p_z + \hbar k$ to the lowest order in A_0^2 is

$$\begin{aligned} &\langle 1|\mathcal{U}_1(T+\tau)|-1\rangle \\ &= \frac{e^2A_0^2}{2i\hbar mc^2}\int dt \exp\left(-i\frac{(\hbar k + p_z)^2}{2\hbar m}(T+\tau-t)\right)f(t)^2\cos^2(\omega t)\exp\left(-i\frac{(-\hbar k + p_z)^2}{2\hbar m}t\right) \\ &= \frac{e^2A_0^2}{2i\hbar mc^2}\exp\left(-i\frac{(\hbar k + p_z)^2}{2\hbar m}(T+\tau)\right)\int dt \exp\left(i\frac{2kp_z}{m}t\right)f(t)^2\cos^2(\omega t) \quad . \end{aligned} \tag{12}$$

The last integral in equation (12) can be evaluated for the resonant case ($p_z = 0$), where $|-1\rangle$ and $|1\rangle$ have the same energy, as

$$\int dt\, f(t)^2 \cos^2(\omega t) = \frac{T}{2} + \frac{\cos(2\omega\tau) - 1 - \cos\left[2\omega(T+\tau)\right] + \cos(2\omega T)}{8\omega^2\tau} \quad . \tag{13}$$

One can see, that the amplitude grows linearly in time with an oscillating part that is suppressed by $1/\tau$. In Fig. 1 the dependence on the duration of the switching is illustrated.

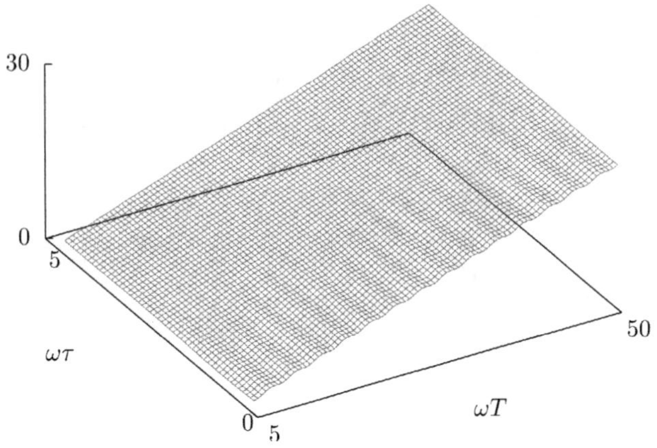

Figure 1. Values of the integral (13) for a range of different switching times τ and interaction times T.

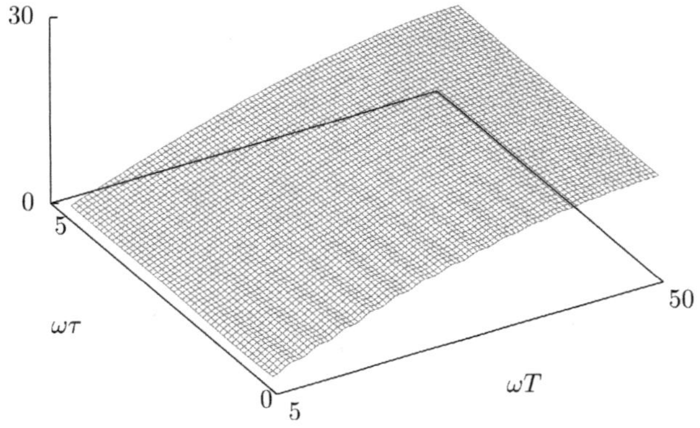

Figure 2. Absolute magnitude of the integral (14) for a range of different switching times τ and interaction times T, evaluated for $\eta = 0.06\,\omega$.

If we allow for the electron momentum to be slightly detuned from the resonant energy-conserving case, described by a non-zero detuning parameter $\eta := \frac{2kp_z}{m} \neq 0$, the integral (13) must be replaced by

$$\begin{aligned}\int dt\, f(t)^2 \cos(\omega t)^2 e^{i\eta t} = \frac{1}{2}\frac{\left(1 - e^{i\eta T}\right)\left(e^{i\eta\tau} - 1\right)}{\eta^2\tau} + \frac{1}{2\left(4\omega^2 - \eta^2\right)^2\tau}\times \\ \left\{\left[\left(4\omega^2 + \eta^2\right)\cos(2\omega t) - 4i\omega\eta\sin(2\omega t)\right] e^{i\eta t}\Big|_{t=0}^{\tau}\right. \\ \left. - \left[\left(4\omega^2 + \eta^2\right)\cos(2\omega t) - 4i\omega\eta\sin(2\omega t)\right] e^{i\eta t}\Big|_{t=T}^{T+\tau}\right\} \quad . \end{aligned} \tag{14}$$

The second term behaves qualitatively like the corresponding oscillations in (13). The first term shows that, for small interaction time T, the scattering amplitude of the detuned electron has the same growth as before. But for long interaction time, it is suppressed by a factor of the inverse switching time $1/\tau$ and a factor $1/\eta^2$. This dependence is shown in Fig. 2.

From these observations we infer, that for sufficiently smooth evelope the scattering process complies with the Bragg condition and can therefore be understood in a photon picture, where the electron absorbs and reemits a certain number of photons gaininig twice their momentum and no kinetic energy.

4. Interference effects in a bichromatic standing wave

In this section, we study Kapitza-Dirac scattering from a standing light wave which contains two different but commensurate frequency modes. The latter are assumed to be a fundamental frequency $\omega = ck$ and its third harmonic 3ω. The vector potential of the wave thus reads

$$\vec{A} = f(t)A_0 \left(\alpha\vec{\varepsilon}_1 \cos(\omega t)\cos(kz) + \frac{\sqrt{1-\alpha^2}}{3}\vec{\varepsilon}_2 \cos\left(3\omega t + \frac{\phi}{2}\right)\cos\left(3kz + \frac{\delta}{2}\right)\right) \tag{15}$$

where $\alpha \in [0,1]$ modulates the relative amplitudes of the waves in such a way that the overall laser intensity $I = \frac{ck^2A_0^2}{16\pi}$ does not depend on α. The relative phases of the frequency modes in time and space are respectively encoded in ϕ and δ. In such a field configuration, quantum mechanical interference effects can occur because it is indistinguishable whether the electron absorbs one big photon of wave vector $3\vec{k}$ or three small photons of wave vector $\vec{k}$ each. Similar two-pathway quantum interferences in strong bichromatic laser fields have also been studied with respect to photoionisation [15, 16], high-harmonic generation [17], electron-atom scattering [18, 19] and electron-positron pair production [20, 21, 22].

Instead of directly using the vector potential (15) in a minimal coupling scheme [see Eq. (1)], we shall apply here the approximation by a ponderomotive potential. The latter results from a suitable time average over the fast oscillations of the interaction Hamiltonian, leading to a space-periodic potential [23]. Then the Schrödinger equation becomes

$$i\hbar\frac{\partial}{\partial t}\psi = -\frac{\hbar^2}{2m}\vec{\nabla}^2\psi + \frac{e^2A_0^2}{4mc^2}f(t)^2\left[\alpha^2\cos^2(kz) + \frac{1-\alpha^2}{9}\cos^2\left(3kz + \frac{\delta}{2}\right)\right]\psi \tag{16}$$

which, similarly as before, can be transformed into a coupled system of ordinary differential equations for the momentum amplitudes $b_n(t)$. Compared to the last section, the envelope $f(t)^2$ here is changed to have $\sin^2$-edges. This system of equations has been solved by numerical integration, starting from the initial condition $b_{-3}(0) = 1$ and all other $b_n(0) = 0$. It describes an incident electron with longitudinal momentum $-3\hbar k$ which simultaneously satisfies the Bragg condition for both field frequencies. Consequently, it may be scattered into the mirrored momentum state with $3\hbar k$ (and momentum amplitude $b_3(t)$) by absorbing either one big photon or three small photons with a (total) wave vector of $3\vec{k}$ and, correspondingly, reemitting either one big photon or three small photons with a (total) wave vector of $-3\vec{k}$. A characteristic Rabi oscillation dynamics will occur between the electron momentum states $|\pm 3\rangle$ with momenta $\pm 3\hbar k$ due to the resonant coupling of these two states.

As an example, we consider Kapitza-Dirac scattering of an electron with incident transverse momentum $12\,\mathrm{eV}/c$ from a bichromatic standing wave with $k_1 = 4\,\mathrm{eV}/\hbar c$, $k_2 = 12\,\mathrm{eV}/\hbar c$, $I_1 = 8\times10^9\,\mathrm{W/cm^2}$ and $I_2 = 5.3\times10^9\,\mathrm{W/cm^2}$. Two different relative phase values have been considered ($\delta = 0$ and $\delta = \pi$) to demonstrate the dependence of the scattering dynamics on this quantity. Our results are shown by the lower two rows in Fig. 3. Note that our numerical data represent occupation probabilities *outside* the laser field; that is, after each value of T the interaction between the electron and the field has been faded out in the numerical simulation. For comparison, we have also calculated the corresponding scattering processes when only one of the two frequency modes is present, which are illustrated by the upper two rows in Fig. 3. We see that the scattering probability in the bichromatic field differs from the sum of the scattering probabilities in the respective monochromatic fields. Besides, the relative phase between the frequency modes can strongly affect the scattering dynamics. Both features indicate the occurence of quantum interference in the process.

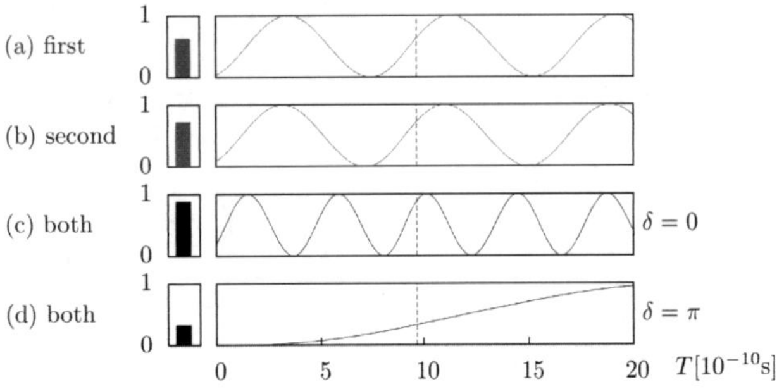

Figure 3. Scattering probabilities $|b_3(T)|^2$ after $T = 9.7 \times 10^{-10}\,\text{s}$ interaction time (left) and Rabi oscillation dynamics (right) for Kapitza-Dirac scattering from (a) a monochromatic standing wave with $k_1 = 4\,\text{eV}/\hbar c$ and $I_1 = 8 \times 10^9\,\text{W/cm}^2$, (b) a monochromatic standing wave with $k_2 = 12\,\text{eV}/\hbar c$ and $I_2 = 5.3 \times 10^9\,\text{W/cm}^2$, and (c) [(d)] a bichromatic standing wave with the combined parameters and $\delta = 0$ [$\delta = \pi$].

To emphasize the influence of the parameters α^2 and δ on the quantum interference, Figs. 4 and 5 show the Rabi frequency Ω for this resonant transition as a function of them. It can be seen in Fig. 4, that moving between the two phase independent monochromatic cases, while keeping the combined intensity fixed, leads to different minima in Ω for different phases. Correspondingly, Fig. 5 shows quite different phase dependencies for different colour mixtures.

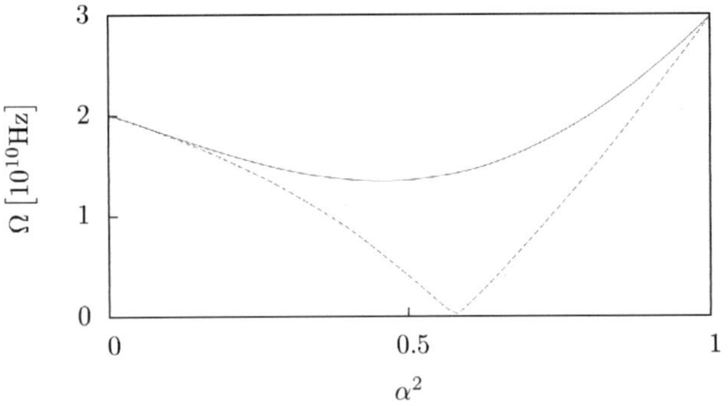

Figure 4. Rabi frequency Ω of the Kapitza-Dirac scattering from a bichromatic standing wave, as a function of the mixing parameter α^2. The wave numbers being $k_1 = 4\,\text{eV}/\hbar c$ and $k_2 = 12\,\text{eV}/\hbar c$, the combined intensity $I = 1.3 \times 10^{10}\,\text{W/cm}^2$, and the relative phase is $\delta = 0$ for the blue solid line and $\delta = \pi$ for the red dashed line.

5. Conclusion

Kapitza-Dirac scattering of electrons from a standing light wave was studied in the nonrelativistic regime governed by the Schrödinger equation. We have shown that a finite switching-on duration of the electron-light interaction influences the scattering amplitude in a characteristic way by introducing temporal oscillations. The latter are damped for long switching-on times and lead asymptotically to a linear growth of the scattering amplitude for incident electrons which satisfy the resonant Bragg condition, whereas scattering of electrons with off-resonant momenta is prohobited in this limit. The same conclusion was reached in [14]. From our Eq. (14) one may also deduce a tight interrelation between switching duration and detuning: For short switching

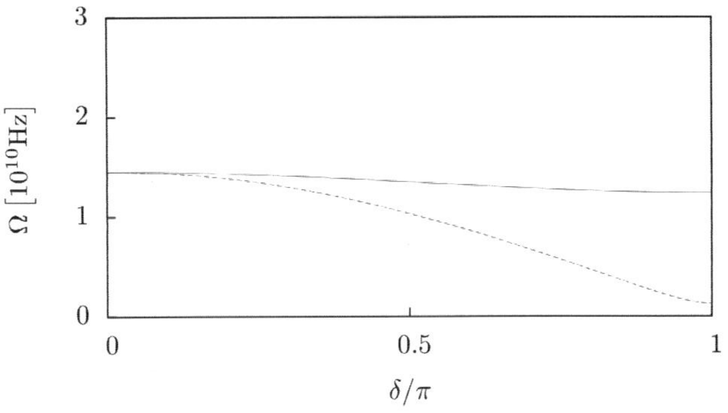

Figure 5. Rabi frequency Ω of the Kapitza-Dirac scattering from a bichromatic standing wave, as a function of the relative phase δ. The wave numbers and combined intensity are as in Fig. 4. The mixing parameter is $\alpha^2 = 0.3$ for the blue solid line and $\alpha^2 = 0.6$ for the red dashed line.

times, also electrons with (slightly) off-resonant momenta can acquire a sizeable scattering probability.

Besides, we have shown that quantum mechanical interference effects may arise in Kapitza-Dirac scattering from a bichromatic standing light wave. Considering the case where the frequency ratio is 1:3, we have demonstrated that the resonant scattering probability in the bichromatic case differs from the sum of the scattering probabilities in the respective monochromatic fields. Accordingly, the dependence of the scattering probability on the interaction time, which exhibits a typical Rabi oscillation dynamics, is modified by the presence of the second mode. The impact of the relative strength of the second mode on the Rabi frequency was examined and a very pronounced sensitivity on the relative phase between the frequency modes found.

Acknowledgments

This study was supported by SFB TR18 of the German Research Foundation (DFG) under project No. B11.

References

[1] Kapitza P L and Dirac P A M 1933 *Math. Proc. Camb. Phil. Soc.* **29** 297
[2] Batelaan H 2007 *Rev. Mod. Phys.* **79** 929
[3] Schwarz H 1967 *Z. Phys.* **204** 276–289
[4] Pfeiffer H 1968 *Phys. Lett.* A **26** 362–363
[5] Bartell L S 1968 *Phys. Lett.* A **27** 236–237
[6] Fedorov M 1967 *Sov. Phys. JETP* **25** 952
[7] Freimund D L, Aflatooni K and Batelaan H 2001 *Nature* **413** 142–143
[8] Efremov M and Fedorov M 2000 *J. Phys.* B **33** 4535
[9] Smirnova O, Freimund D L, Batelaan H and Ivanov M 2004 *Phys. Rev. Lett.* **92** 223601
[10] Ahrens S, Bauke H, Keitel C H and Müller C 2012 *Phys. Rev. Lett.* **109** 043601
[11] Ahrens S, Bauke H, Keitel C H and Müller C 2013 *Phys. Rev.* A **88** 012115
[12] Haroutunian V and Avetissian H 1975 *Phys. Lett.* A **51** 320–322
[13] Fedorov M and McIver J 1980 *Opt. Commun.* **32** 179–182
[14] Fedorov M 1974 *Opt. Commun.* **12** 205–209
[15] Schafer K J and Kulander K C 1992 *Phys. Rev.* A **45** 8026
[16] Véniard V, Taïeb R and Maquet A 1995 *Phys. Rev. Lett.* **74** 4161
[17] Milošević D B, Becker W and Kopold R 2000 *Phys. Rev.* A **61** 063403
[18] Kamiński J, Jaroń A and Ehlotzky F 1995 *J. Phys.* B **28** 4895
[19] Milošević D B, Ehlotzky F and Piraux B 1997 *J. Phys.* B **30** 4347
[20] Narozhny N and Fofanov M 2000 *J. Exp. Theor. Phys.* **90** 415–427
[21] Krajewska K and Kamiński J 2012 *Phys. Rev.* A **85** 043404
[22] Augustin S and Müller C 2013 *Phys. Rev.* A **88** 022109
[23] Batelaan H 2000 *Contemp. Phys.* **41** 369–381

Analytical formula for residual current density excited in the process of gas ionization by a few-cycle laser pulse in the low-intensity limit

A. A. Silaev and N. V. Vvedenskii

Institute of Applied Physics, Russian Academy of Sciences, Nizhny Novgorod 603950, Russia

University of Nizhny Novgorod, Nizhny Novgorod 603950, Russia

E-mail: silaev@appl.sci-nnov.ru, vved@appl.sci-nnov.ru

Abstract. This work is devoted to analytical study of excitation of the residual current density (RCD) in the process of gas ionization by a few-cycle laser pulse. The RCD remains in the laser-produced plasma after the passage of the laser pulse and is as an initial push leading to excitation of the plasma oscillations which can radiate terahertz waves. We derive simple closed-form analytical formula for RCD for relatively small peak intensity of few-cycle laser pulse, which corresponds to small final degree of ionization. The dependences of the RCD on laser pulse parameters are discussed.

1. Introduction

Recently, great attention has been given to the phenomenon of excitation of the residual current density (RCD) during gas ionization by few-cycle laser pulses [1–4]. The essence of this phenomenon can be explained as follows. In the process of ionization of atoms and molecules by an intense laser field, the newly born free electrons acquire, along with oscillatory velocity, the constant drift velocity. The value and direction of this velocity are determined by the phase of the electric field at the moment of ionization. As a result, after the passage of the laser pulse, quasi-constant RCD can arise in the produced plasma. The value of this RCD depends on the carrier-envelope phase (CEP) of few-cycle laser pulse and can reach high values at sufficiently short durations of the laser pulse [1–5]. The interest to this phenomenon is connected with the possibility of its application for generation of radiation in the terahertz (THz) range, which was demonstrated experimentally in several works [6–9]. The residual current serves as the initial push for plasma polarization and excitation of radiating plasma oscillations which have THz frequencies in a wide range of gas pressures and laser pulse intensities [1, 10].

Earlier studies of RCD excitation during gas ionization by few-cycle laser pulse were carried out by using numerical calculations. These calculations were based on semiclassical [1] and quantum-mechanical [3] approaches. The semiclassical approach is based on solution the hydrodynamic equation for the electron current density and the equation for free-electron density, which uses static-field probability of tunneling ionization per time unit. The quantum-mechanical approach is based on the *ab initio* numerical solution of the three-dimensional (3D) time-dependent Schrödinger equation. These approaches were used to calculate RCD dependences on the carrier-envelope phase (CEP), intensity, and duration of the laser pulse [2–5, 11–13]. In particular, comparison of the results obtained on the basis of semiclassical and quantum-mechanical approaches showed that the range of applicability of the semiclassical approach is limited by the parameters of the laser pulse, at which the conditions for tunneling ionization of the gas are fulfilled at the maximum of the pulse envelope [3, ?]. These conditions correspond to a low value of the Keldysh parameter γ [14].

In this work we derive analytical formula for RCD excited in the process of gas ionization by few-cycle laser pulse. To develop the analytical model, we use the semiclassical approach. We assume for simplicity that the intensity of the laser pulse is sufficiently small so that the final gas ionization dergee is much smaller than unity. Basing on the derived analytical formula, the dependences of RCD on CEP, intensity, duration and wavelength of linearly-polarized pulses are determined.

This paper is structured as follows. Section 2 contains the statement of the problem. In Section 3 the analytical formulas for RCD are derived. Then, in Section 4 we discuss the found analytical dependences of RCD on the main parameters of the laser pulses. Section 5 contains the summary and conclusions of the research.

2. Statement of the problem

We assume that the electric field $\mathbf{E}(t)$ of the laser pulse is polarized linearly along the z axis ($\mathbf{E}(t) = \hat{\mathbf{z}}E(t)$) and is given by the formula

$$\begin{cases} E(t) = \dfrac{E_0}{\omega_0}\dfrac{da}{dt}, & (1) \\ a(t) = f(t)\sin(\omega_0 t + \varphi_{\mathrm{CEP}}). & (2) \end{cases}$$

Here, E_0 is the maximum of the laser pulse envelope, ω_0 is the carrier frequency, φ_{CEP} is CEP, $f(t)$ is the slowly varying envelope of the pulse.

To derive analytical formula for RCD, we use the so-called semiclassical approach, which is based on solving the equation for the free-electron density $N(t)$ and the classical equation for the electron density $\mathbf{j}(t) = \hat{\mathbf{z}}j(t)$ in plasma with a variable number of particles

$$\begin{cases} \dfrac{\partial N}{\partial t} = (N_g - N)w(|E|), & (3) \\ j_{\mathrm{RCD}} = \dfrac{e^2}{m}\displaystyle\int_{-\infty}^{\infty} N(t)E(t)dt. & (4) \end{cases}$$

Here, $w(|E|)$ is the static-field probability of tunneling ionization per time unit, N_g is the unperturbed gas density, and e and m are the charge and mass of the electron, respectively. The probability of atom ionization per unit of time is assumed to be determined by the function of the electric-field strength. This function is taken from the solution of the quantum-mechanical problem of the ionization of a hydrogen atom from the ground state in a static electric field. We assume further that final ionization degree is sufficiently small and neglect the term responsible for atom depletion in Eq. 3.

We will normalize RCD to the maximum possible oscillatory current density in the pulse $j_{\rm osc} = e^2 E_0 N_g / m\omega_0$. The normalized current density obtained by this method,

$$j_{\rm norm} = j_{\rm RCD}/j_{\rm osc}, \tag{5}$$

does not depend on N_g and its squared value $j^2_{\rm norm}$ characterizes the efficiency of conversion of the laser pulse energy to the energy of low-frequency THz waves [1, 3]. Using integrating by parts in Eq. 4, we obtain equation for $j_{\rm norm}$:

$$j_{\rm norm} = -\int_{-\infty}^{\infty} a(t) w(|E(t)|) dt. \tag{6}$$

3. Analytical model

An important fact for the further study is that in the range of the values of E being of interest, the probability of ionization per time unit is a sharp function of the electric-field strength. The parameter that characterizes the sharpness of $w(E)$ is [15]

$$\eta(E) = \frac{w(E)}{E w'(E)}. \tag{7}$$

The lower is the quantity of η, the sharper is the behavior of $w(E)$ at the point E. At sufficiently low strength $|E|$ of the electric field, the value η is a small parameter, $\eta \ll 1$. Therefore the function $w(|E(t)|)$ consists of narrow peaks localized near the moments of maximums $|E(t)|$. In the case of sufficiently low peak intensity of the laser pulse these peaks can be approximated by the Gaussian functions

$$w(|E(t)|) \approx \sum_{k=-\infty}^{\infty} w(|F(t_{e,k})|) \exp\left(-\frac{2(t-t_{e,k})^2}{\tau^2_{e,k}}\right). \tag{8}$$

Here, $F(t) = E_0 f(t)$, k is an integer number, and $t_{e,k}$ are the moments of maximums $|E(t)|$, which are approximately equal to

$$t_{e,k} \approx \left(1 - 4(\omega_0\tau)^{-2}\right) t_{a,k}, \tag{9}$$

where

$$t_{a,k} = (\pi k - \varphi_{\rm CEP})\omega_0^{-1}. \tag{10}$$

The value $\tau_{e,k} = 2\sqrt{\eta(F(t_{e,k}))}\omega_0^{-1}$ is the characteristic width of the kth peak, $\tau = \sqrt{2/|\ddot{f}(0)|}$ (the dots means the second derivative with respect to time) is characteristic width of the pulse envelope. In what follows τ will be called the pulse duration. Since $\eta \ll 1$, $\tau_{e,k}$ is much shorter than the laser pulse period $T_0 = 2\pi/\omega_0$. Therefore, the spectrum of the function $w(|E(t)|)$ is concentrated near even harmonics of the laser pulse frequency. The function $w(|F(t_{e,k})|)$ from Eq. 8 is approximately equal to

$$w(F(t_{e,k})) \approx w(E_0) e^{-2t^2_{e,k}/\tau^2_i}. \tag{11}$$

$$\tau_i \approx \sqrt{2\eta_0}\tau, \quad \eta_0 = \eta(E_0) \tag{12}$$

is the characteristic time of the concentration rise, which will be called the *ionization time* in what follows [15]. Since $\eta_0 \ll 1$, τ_i is much shorter than the pulse duration.

Let us now obtain approximate analytical formula for the normalized RCD. The integral in formula 6 can be subdivided into a sum of integrals in the small vicinity of maximums $|E(t)|$. Thus, the normalized RCD is subdivided to the contributions Δj_k made by the electrons born on separate humps of $|E(t)|$,

$$j_{\mathrm{norm}} = \sum_{k=-\infty}^{\infty} \Delta j_k, \quad \Delta j_k = -\frac{\pi}{\omega_0} a(t_{e,k}) w(F(t_{e,k})). \tag{13}$$

The value of $a(t)$ at the time moment $t_{e,k}$ is

$$a(t_{e,k}) \approx -4(-1)^k f(t_{e,k})(\omega_0\tau)^{-2}\omega_0 t_{e,k}. \tag{14}$$

Substituting 14 to Eq. 13 and taking into account that $\tau \gg \tau_i$, we obtain

$$j_{\mathrm{norm}} \approx 4\pi\frac{w(E_0)}{(\omega_0\tau)^2}\sum_{k=-\infty}^{\infty}(-1)^k t_{e,k} e^{-2t_{e,k}^2/\tau_i^2}. \tag{15}$$

To calculate the sum in this formula, we will use approximate relationship

$$\sum_{k=-\infty}^{\infty}(-1)^k e^{-b(k-k_0)^2} \approx 2\sqrt{\pi/b}e^{-\pi^2/4b}\cos(\pi k_0), \tag{16}$$

which is valid at $\exp(-3\pi^2/b) \ll 1$. In order to apply Eq. 16 for calculation of the sum in Eq. 15, it is necessary to ensure fulfillment of inequality $\exp(-3\omega_0^2\tau_i^2/2) \ll 1$, which is knowingly fulfilled at the values of pulse intensity and duration being of interest. Using Eq.16 in formula 15, we obtain

$$j_{\mathrm{norm}} = -j_{\max}\sin(\varphi_{\mathrm{CEP}}). \tag{17}$$

Thus, j_{norm} is subdivided into the product of two factors, one of which depends only on CEP, and the other, on the other parameters of the laser pulse. The maximum normalized RCD, which corresponds to the optimal CEP, is equal to

$$j_{\max} = 2\sqrt{\eta_0}w(E_0)\frac{\tau_i^3}{\tau^2}\exp\left(-\frac{\omega_0^2\tau_i^2}{8}\right) = 4\sqrt{2}\eta_0^2 w(E_0)\tau\exp\left(-\eta_0\frac{\omega_0^2\tau^2}{4}\right). \tag{18}$$

Let us assume that $w(E)$ is approximated by the empirical formula proposed by Tong and Lin [16], which yields well approximation of the precise value w in both the regimes of tunneling and above-barrier ionization. In this regime, the tunneling formula $w_{TI}(E)$ [17, 18] yields strong overrating of the ionization probability. The adjusted formula proposed in [16] differs from $w_{TI}(E)$ in an additional adjustment factor:

$$w(E) = w_{TI}(E)e^{-\nu E/E_a}, \tag{19}$$

Here, $E_a = 5.14 \times 10^9$ V/cm is the atomic field strength, ν is the coefficient introduced to make formula 19 coincide with the dependence $w(|E|)$ found by means of direct numerical solution of the Schrödinger equation. The values of ν calculated for the hydrogen atoms and different noble gases can be found in [16]. For example, for hydrogen atoms $\nu = 12$. Since ν is sufficiently great, the adjustment factor in Eq. 19 is significant even at moderate electric field strength being a tenth of E_a. The tunneling formula $w_{TI}(E)$ entering into Eq. 19 has the following form [17, 18]:

$$w_{\rm TI}(E) = \alpha\Omega_a \left(\frac{E_a}{E}\right)^s e^{-\beta E_a/E}, \tag{20}$$

where $\Omega_a = 4.13 \times 10^{16}$ s^{-1} is the atomic unit of frequency. For hydrogen atoms, $\alpha = 4$, $\beta = 2/3$, and $s = 1$. In the case when w is specified by adjusted tunneling formula 19 the formula for $j_{\max}$ is

$$j_{\max} = \frac{4\sqrt{2}\alpha}{\beta^2}\Omega_a\tau \left(\frac{E_0}{E_a}\right)^{2-s} \exp\left(-\frac{\beta E_a}{E_0} - \frac{\nu E_0}{E_a} - \frac{E_0}{\beta E_a}\frac{\omega_0^2\tau^2}{4}\right). \tag{21}$$

4. Discussion

It follow from general formula 17 that a the case of low laser pulse intensities the dependence of RCD on CEP is sinusoidal,

$$j_{\rm RCD} \propto \sin(\varphi_{\rm CEP}), \tag{22}$$

independently of the shape of the function $w(E)$. Thus, the analytical formula confirm the results of numerical calculations of the RCD dependence on CEP in the tunneling regime of ionization [1–5, 11]. The optimal CEP is equal to $\varphi_{\rm opt} = \pi/2$, which corresponds to the asymmetric (sine-shaped) electric field of the laser pulse. This can be explained as follows. In the case of a symmetric pulse, the drift velocities of the electrons born at certain time moments $t = t_0$ and $t = -t_0$ are equal in their absolute values, but opposite in direction. As a result, since the ionization probability per time unit is the even function t, the currents of the electrons born at $t < 0$ and $t > 0$ compensate each other totally, and RCD is equal to zero. If $\varphi_{\rm CEP} \neq 0$, then the currents of the electrons moving in different directions do not coincide in the absolute values, which results in appearance of non-zero RCD. Since the dependence of RCD on CEP is sinusoidal, the optimal CEP is shifted from the nonoptimal one to $\pi/2$, i.e., it is equal to $\varphi_{\rm opt} = \pi/2$, which corresponds to the antisymmetric (sine-shaped) electric field of the laser pulse.

It is important to focuse on the following issue. It was noted in the introduction that the region of applicability of the semiclassical approach to RCD calculation is limited by small values of the Keldysh parameter, $\gamma = \sqrt{I_p/2U_p} \ll 1$ [14]. Here, $U_p = e^2E_0^2/4m\omega_0^2$ is the averaged ponderomotive energy of the electron in the pulse. Since $\gamma \propto 1/E_0$, γ increases as the intensity decreases. At low values of intensity, when final ionization degree is small, the parameter γ can be comparable with or exceed unity. In this case, the semiclassical approach goes beyond the region of applicability. However, when the laser pulse wavelength increases, the Keldysh parameter decreases in proportion to λ^{-1}, and the region of applicability of the semiclassical approach becomes wider. As a result, the semiclassical approach (and, in particular, analytical formula 17) is valid in the mid-infrared band including the case of low intensities corresponding to low final dergee of ionization.

Let us now consider the dependence of the maximum normalized RCD $j_{\max} = |j_{\text{norm}}(\varphi_{\text{opt}})|$, which corresponds to the optimal CEP, on the duration of the laser pulse. At low intensities $j_{\max} \propto \tau \exp(-\omega_0^2\tau_i^2/8)$. However, the pre-exponential factor affects the dependence $j_{\max}(\tau)$ only at very short durations of $\tau << T_0$ and is not important for the values of τ being of interest. Since the ionization time τ_i is proportional to τ, $j_{\max}$ decreases with an increasing duration obeying the Gaussian law:

$$j_{\max}(\tau) \propto \exp\left(-\eta_0 \frac{\omega_0^2\tau^2}{4}\right). \tag{23}$$

Since η_0 is a small parameter, the characteristic scale of the decrease in the function $j_{\max}(\tau)$ is $\sim 1/\sqrt{\eta_0}\omega_0$, which usually equals to several field periods. Another situation is realized at high intensities which corresponds to high final dergee of ionization. In this case, $j_{\max}(\tau)$ decreases obeying the exponential law [1, 3]. This is connected with the fact that at high intensities, the ionization time τ_i increases with an increase in τ approximately as $\tau_i(\tau) \propto \sqrt{\tau}$, rather than obeying the linear law.

Let us now pass over to analysis of the dependence of the value $j_{\max}$ on the central wavelength λ of the laser pulse. For the convenience of the analysis, we will fix the number $N_{cyc} = \tau/T_0$ of field periods in the pulse. In this case, $j_{\max}$ increases linearly with growing λ,

$$j_{\max}(\lambda) \propto \lambda. \tag{24}$$

This increase in $j_{\max}$ is connected with the increase in electron concentration, which is proportional to λ at a fixed ratio τ/T_0. The efficiency of excitation of the low-frequency current (estimated in [1] in the framework of the quasistatic approach) is proportional to the square of normalized RCD. Thus, in the regime of low intensities, this efficiency is proportional to λ^2. Hence, one can conclude that the use of few-cycle laser pulses in the near and middle IR wavelength range can result in a significant increase of terahertz wave generation efficiency.

Finally, consider the dependence of $j_{\max}$ on the peak intensity $I = cE_0^2/8\pi$ (where c is speed of light in vacuum) of the laser pulse. For the case, when $w(E)$ is specified by formula 19,

$$j_{\max}(I) \propto \exp(-\beta\sqrt{I_a/I} - \nu\sqrt{I/I_a}), \tag{25}$$

where $I_a = 3.51 \times 10^{16}\text{W/cm}^2$ is the atomic unit of intensity. The sharp increase in $j_{\max}$ at low intensities is connected with a sharp (exponential) increase in the free-electron density when the peak intensity grows.

5. Conclusions

To conclude, we have derived simple closed-form analytical expression for the residual current density (RCD) excited in the process of gas ionization by few-cycle laser pulse in the case of low peak intensity. The range of applicability of this formula is limited by the low values of the Keldysh parameter γ at the maximum of laser pulse envelope. This formula confirm the fact that at low dergee of ionization the dependence of RCD on the carrier-envelope phase (CEP) is a sinusoidal function and the optimal CEP is equal to $\pi/2$ and corresponds to the antisymmetric sine-like laser pulse.

It is shown that in the tunneling regime of ionization (at $\gamma \ll 1$), the value of maximal RCD $j_{\rm max}$ (normalized to the oscillatory current density), corresponding to the optimal phase, increases sharply (exponentially) with the growth of the peak pulse intensity. The character of the RCD decrease during the increase in the pulse duration τ depends on the field intensity. At low intensities, RCD decreases as τ increases obeying the Gaussian law. Also it is shown that the value of $j_{\rm max}$ increases linearly with an increasing wavelength (at a fixed number of field periods in the pulse). The efficiency of RCD excitation (determined by the squared value of $j_{\rm max}$) is proportional to λ^2. A fast increase in the efficiency of excitation of the low-frequency current with the increase in the pulse wavelength can be used to create sources of high-power THz radiation with the use of few-cycle laser pulses in the near and mid-infrared wavelength ranges.

This work was supported by the Government of the Russian Federation (Agreement No. 14.B25.31.0008), and the Russian Foundation for Basic Research (Grants No. 13-02-00964, No. 14-02-31044, and No. 14-02-00847).

References

[1] Gildenburg V B and Vvedenskii N V 2007 *Phys. Rev. Lett.* **98** 245002
[2] Wu H C, Meyer-ter Vehn J and Sheng Z M 2008 *New J. Phys.* **10** 043001
[3] Silaev A A and Vvedenskii N V 2009 *Phys. Rev. Lett.* **102** 115005
[4] Kim K Y 2009 *Phys. Plasmas* **16** 056706
[5] Silaev A A and Vvedenskii N V 2009 *Phys. Scr.* **T135** 014024
[6] Kress M, Loeffler T, Thomson M D, Doerner R, Gimpel H, Zrost K, Ergler T, Moshammer R, Morgner U, Ullrich J and Roskos H G 2006 *Nat. Phys.* **2** 327
[7] Li C, Wang D, Song L, Liu J, Liu P, Xu C, Leng Y, Li R and Xu Z 2011 *Opt. Express* **19** 6783
[8] Bai Y, Song L, Xu R, Li C, Liu P, Zeng Z, Zhang Z, Lu H, Li R and Xu Z 2012 *Phys. Rev. Lett.* **108** 255004
[9] Liu P, Li R and Xu Z 2015 Phase evolution and thz emission from a femtosecond laser filament in air *Progress in Ultrafast Intense Laser Science XI* Springer Series in Chemical Physics ed Yamanouchi K, Nam C H and Martin P (Springer International Publishing) pp 175–193
[10] Kostin V A and Vvedenskii N V 2010 *Opt. Lett.* **35** 247–249
[11] Chen M, Pukhov A, Peng X Y and Willi O 2008 *Phys. Rev. E* **78** 046406
[12] Silaev A A, Ryabikin M Y and Vvedenskii N V 2010 *Phys. Rev. A* **82** 033416
[13] Alexandrov L N, Emelin M Y and Ryabikin M Y 2013 *Phys. Rev. A* **87** 013414
[14] Keldysh L V 1965 *Sov. Phys. JETP* **20** 1307–1314
[15] Vvedenskii N V, Korytin A I, Kostin V A, Murzanev A A, Silaev A A and Stepanov A N 2014 *Phys. Rev. Lett.* **112** 055004
[16] Tong X M and Lin C D 2005 *J. Phys. B: At. Mol. Phys.* **38** 2593
[17] Perelomov A M, Popov V S and Terentev M V 1966 *Sov. Phys. JETP* **23** 924–934
[18] Ammosov M V, Delone N B and Krainov V P 1986 *Sov. Phys. JETP* **64** 1191–1194

Propagation and amplification of short radio-frequency pulses in a plasma channel created in gaseous media by the intense laser radiation

A V Bogatskaya[1,2,3], A M Popov[1,2,3], I V Smetanin[3] and E A Volkova[1]
[1] D.V. Skobeltsyn Institute of Nuclear Physics, Moscow State University, Moscow, Russia
[2] Department of Physics, Moscow State University, Moscow, Russia
[3] P.N. Lebedev Physical Institute, RAS, Moscow, Russia

E-mail: annabogatskaya@gmail.com

Abstract. The evolution of the electron energy distribution function (EEDF) in non-equilibrium plasma channel created in xenon by powerful KrF - femtosecond laser pulse is studied. It is demonstrated that such a plasma channel can be used as a waveguide for both transportation and amplification of the microwave radiation. The specific features of such a plasma waveguide are studied on the basis of the self-consistent solution of the kinetic Boltzmann equation for the EEDF in different spatial points of the gas media and the wave equation in slow-varying amplitude approximation for the microwave radiation guided and amplified in the channel.

1. Introduction

Impressive progress in high-power ultrashort laser techniques makes it possible to create long plasma channels in different gaseous media. The formation of a plasma channel by a high-intensity femtosecond laser pulses arises mainly from the process of multiphoton ionization of gas atoms and molecules, therefore the electron energy spectrum of such an plasma channel consists of a set of peaks, the positions of which are determined by absorption of certain quanta of energy. The specific electrodynamic features of such plasma structures cause particular interest in them for a number of applications, including generation of XUV attosecond pulses [1,2], triggering of high-voltage discharges [3-6], remote sensing of the atmosphere [7] and long distance transportation of microwave radiation [8,9]. In [10] it was proposed that such a plasma channel with strong nonequilibrium of the electron subsystem could be used to amplify electromagnetic radiation in a microwave frequency band. The possibility of using electron energy distribution functions (EEDF) containing an inverse population to amplify electromagnetic radiation was first discussed in [11,12]. The detailed analysis of EEDF relaxation in a plasma channel created by a femtosecond KrF laser pulse and calculation of the gain factor in dependence on initial plasma parameters and on the intensity and frequency of the RF radiation in a number of atomic and molecular gases is given in [13]. In particular, it was found that among both atomic and molecular gases that can be used for amplification of the RF radiation xenon plasma has some advantages. The time interval of the positive gain factor existence is the largest for xenon plasma and can reach hundreds of nanoseconds.

In this paper we discuss the fact that a strong nonequilibrium of photoelectron spectrum can also lead to unusual refractive properties of the plasma channel forming in the process of multiphoton ionization of gas media by femtosecond laser pulse. In particular, the real part of plasma permittivity can become larger than unity, i.e. plasma turns out to be optically more dense medium in comparison with the surrounding unionized gas. The aforesaid provides the possibility to use such a channel as a waveguide which enable to maintain both transportation and amplification of the microwave radiation. The process of guiding and amplification of the microwave radiation in the plasma channel formed by powerful KrF laser femtosecond pulse in xenon is studied numerically on the basis of the self-consistent solution of the Boltzmann equation for the electron energy distribution function (EEDF) evolution in the nonequlibrium plasma and the wave equation for the transported through the channel RF pulse. The conditions of effective propagation and amplification of the RF pulse in a plasma waveguide are discussed.

2. Photoionization by the ultrashort UV laser pulse. Multiphoton ionization

At first we consider the formation of a plasma channel in the process of photoionization of gas atoms by the femtosecond UV laser pulse. At figure 1 numerical calculations of the Schrodinger equation for probabilities of ionization (per pulse) of xenon and argon atoms are shown. For intensities of laser pulse 10^{12}-10^{14} W/cm^2 in accordance with the perturbation theory the ionization probability of xenon atoms is described by a cubic dependence on the radiation intensity. As regards to argon atoms, we have four-photon ionization therefore the dependence of ionization probability is four-power. In argon we can also observe saturation of the ionization probability, which is caused by the population trapping of the Rydberg states of atoms [14]. Figure 2 demonstrates the typical photoelectron spectra formed by the multihoton ionization of xenon and argon atoms. For intensities less than 10^{13} W/cm^2 the Stark shift of the continuum can be neglected, therefore the position of the first peak in the xenon energy spectrum corresponding to the absorption of three photons is determined by the energy $\varepsilon_0 = 3\hbar\omega - I_i$, I_i is the ionization potential for xenon atoms. In the case of argon atoms we have $\varepsilon_0 = 4\hbar\omega - I_i$. Above-threshold ionization peaks are negligible for such a range of radiation intensities. Therefore in fact we obtain one pronounced peak which evolves due to elastic and inelastic processes in plasma.

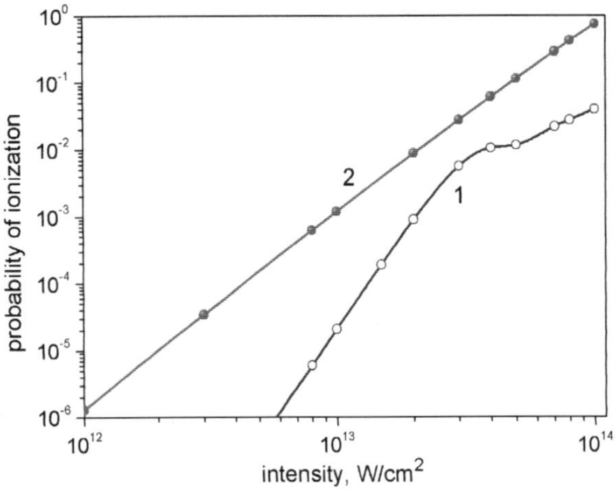

Figure 1. Ionization probabilities ν_i of argon (*1*) and xenon (*2*) atoms per pulse as functions of intensity I of laser radiation with $\lambda = 248$ nm.

It is worth to note that for such pulse durations (about 100 fs) the characteristic time between electron-atomic collisions which can be estimated as $T_c \approx 1/N\sigma v \sim 3\times10^{-13}$ s (here $N \cong 3\times10^{19}$ cm^{-3}

is the density of atoms (molecules) at the atmospheric pressure, $\sigma \cong 10^{-15}$ cm^2 is the collision cross section and $v \sim 10^8$ cm/s is the velocity of electrons resulting from photoionization) is more than the duration of the laser pulse. This signifies that the photoelectron energy spectrum created by the laser pulse no longer than 300 fs is determined only by the photoionization dynamics of gas atoms (molecules) and the evolution of the energy spectrum which is described by the Boltzmann equation takes place in the postpulse regime.

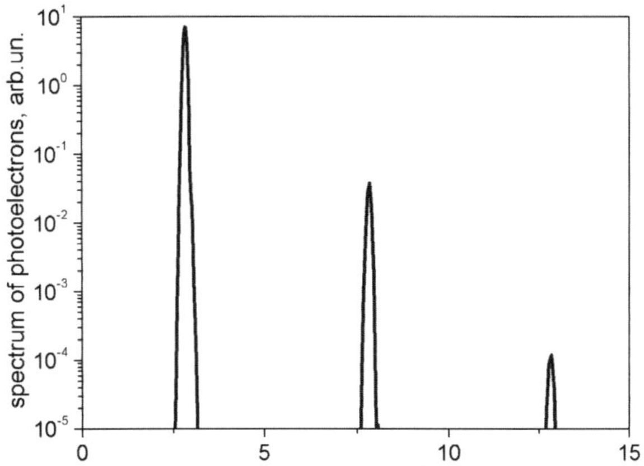

Figure 2. Photoelectron spectra in xenon for a radiation intensity $I = 10^{13}$ W/cm^2.

3. Boltzmann equation for the evolution of the EEDF

Evolution of the plasma channel with nonequilibrium EEDF can be studied using the kinetic Boltzmann equation:

$$\frac{\partial n(\varepsilon,t)}{\partial t}\sqrt{\varepsilon} = \frac{\partial}{\partial \varepsilon}\left(\frac{e^2 E_0^2 \nu_{tr}(\varepsilon)}{3m(\omega^2+\nu_{tr}^2)}\varepsilon^{3/2}\frac{\partial n}{\partial \varepsilon}\right) +$$

$$\frac{2m}{M}\frac{\partial}{\partial \varepsilon}\left(\nu_{tr}(\varepsilon)\varepsilon^{3/2}\left(n(\varepsilon,t)+T_g\frac{\partial n(\varepsilon,t)}{\partial \varepsilon}\right)\right)+Q_{ee}(n)+Q^*(n). \quad (1)$$

As for physical meaning of equation (1) it describes the diffusion process in the energy space. The first term stands for the influence of the transported RF field with frequency ω and the amplitude of electric field strength E_0, the second represents elastic losses of energy, the third one describes the effect of the gas temperature. Here T_g is the gas temperature (we assume $T_g \approx 0.03$ eV), m is the mass of electron, M is the mass of gas atom, $\nu_{tr}(v) = N\sigma_{tr}v$, σ_{tr} are the transport frequency and scattering cross section of electron-atomic collisions correspondingly. Last two terms of equation (1) are the integrals of electron-electron collisions $Q_{ee}(n)$ and inelastic collisions $Q^*(n)$. It is important to mention that in rare gases the energy lowest excitation thresholds for electronic states are equal to 8.31 eV for xenon atoms and 11.5 eV for argon atoms. This is significantly higher than the energy peak corresponding to the photoelectrons produced by the femtosecond laser pulse. Under the conditions involved, evolution of the energy spectrum in rare gases is determined by elastic and electron-electron collisions and by the influence of transported RF field. For the numerical analysis of Boltzmann equation we approximate the initial photoelectron peak by the Gaussian shape:

$$n(\varepsilon, t=0) = \frac{1}{\Delta\varepsilon\sqrt{\pi\varepsilon}}\exp\left(-\frac{(\varepsilon-\varepsilon_0)^2}{(\Delta\varepsilon)^2}\right) \quad (2)$$

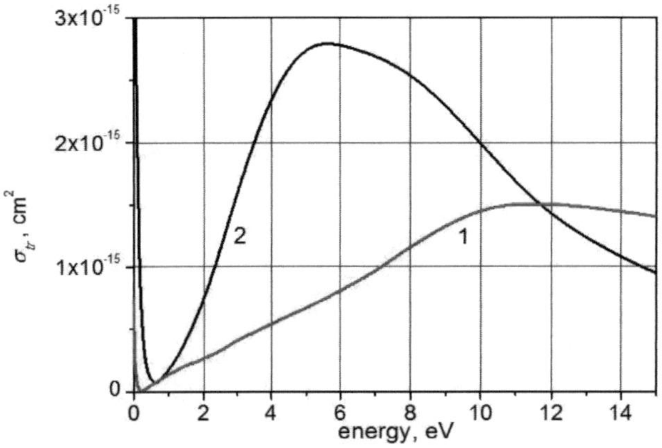

Figure 3. Transport scattering cross sections for argon (1) and xenon (2) atoms.

Figure 3 shows dependences of transport cross sections for xenon and argon atoms. Such rare gases are characterized by a Ramsauer minimum and as a consequence by the ranges of rapidly growing transport cross sections. It has to be emphasized that initial peak is located namely in the range of rapidly growing values of cross section. The numerical solutions of Boltzmann equation in xenon and argon obtained under the assumption that electron-electron collisions and RF field do not contribute to the evolution of the EEDF are presented at figure 4. It is clearly seen that in this case the initial peak gradually shifts towards lower energies keeping approximately the Gaussian shape and also one can notice the retardation of the velocity of peak displacement. Here the diffusion spreading is negligible, furthermore, the lowering of elastic collision frequency ν_{tr} with the deceleration of electrons leads to the opposite effect – the narrowing of the peak.

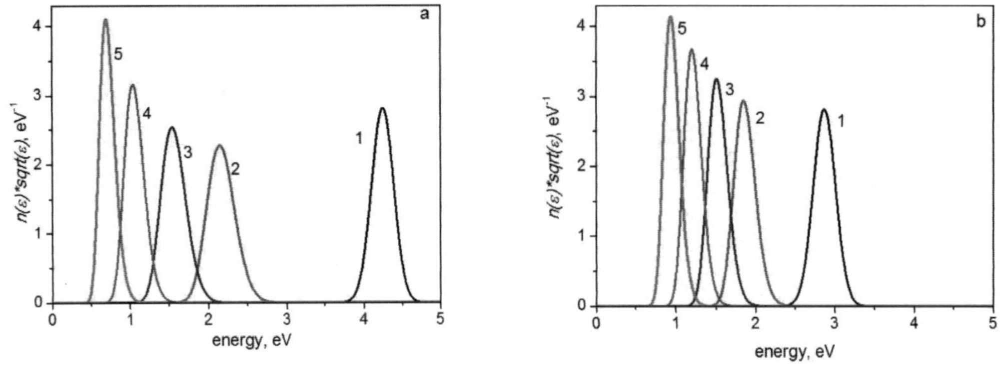

Figure 4. EEDF for argon (a) and xenon (b) at the instants of time (*1*) 0, (*2*) 25, (*3*) 50, (*4*) 100, and (*5*) 200 ns after plasma channel production by λ = 248 nm laser radiation.

4. Optical properties of the plasma channel

The knowledge of the EEDF makes it possible to obtain some optical properties of a plasma channel. The complex conductivity of plasma can be written in the following form [4]:

$$\sigma(\omega) = \sigma' + i\sigma'' = \frac{2}{3}\frac{e^2 N_e}{m}\int_0^\infty \frac{\varepsilon^{3/2}(\nu_{tr}(\varepsilon) + i\omega)}{\omega^2 + \nu_{tr}^2(\varepsilon)}\left(-\frac{\partial n(\varepsilon,t)}{\partial\varepsilon}\right)d\varepsilon. \qquad (3)$$

where N_e is the concentration of electrons in plasma. The real part of plasma conductivity stands for the energy dissipation in plasma; therefore we have the expression for the absorption coefficient:

$$\mu_\omega = \frac{4\pi\sigma'}{c} = \frac{8\pi}{3}\frac{e^2 N_e}{mc}\int_0^\infty \frac{\varepsilon^{3/2}\nu_{tr}(\varepsilon)}{\omega^2+\nu_{tr}^2(\varepsilon)}\left(-\frac{\partial n(\varepsilon,t)}{\partial\varepsilon}\right)d\varepsilon. \quad (4)$$

The EEDF usually decreases with the increase of energy, i.e. $\partial n/\partial\varepsilon < 0$ and consequently the value of integral (4) is positive: $\sigma' > 0$ and $\mu_\omega > 0$. However, the electron energy distribution which arises from the photoionization process contains energy ranges with the positive value of derivative $\partial n/\partial\varepsilon$, making a negative contribution to the integral (4). In [13] authors drew attention to the fact that in low-frequency range the value of absorption coefficient may become even negative in rare gases with a pronounced Ramsauer effect. In that case the medium is capable to amplify RF radiation and the amplification condition can be represented in the following form:

$$\frac{d}{d\varepsilon}\left(\frac{\varepsilon^{3/2}\nu_{tr}}{\omega^2+\nu_{tr}^2}\right) < 0 \,. \quad (5)$$

To obtain amplification in plasma the above condition should be fulfilled in the range of photoionization peak existence. When $\omega \ll \nu_{tr}$ one can obtain from (5):

$$\frac{d}{d\varepsilon}\varepsilon/\sigma_{tr} < 0, \quad (6)$$

which means that the transport cross section should increase with the energy faster than line dependence. Figure 5 shows the dependences $\sigma_{tr}(\varepsilon)/\varepsilon$ for xenon and argon atoms. One can figure out than xenon is more preferable as an amplifying media in comparison with argon. Actually, in xenon the amplification condition (6) is fulfilled in energy range of 1-4 eV, whereas for argon atoms the amplification effect may appear only for slow electrons in a narrow energy range $\varepsilon = 0.6 - 0.8$, which is hard to realize experimentally.

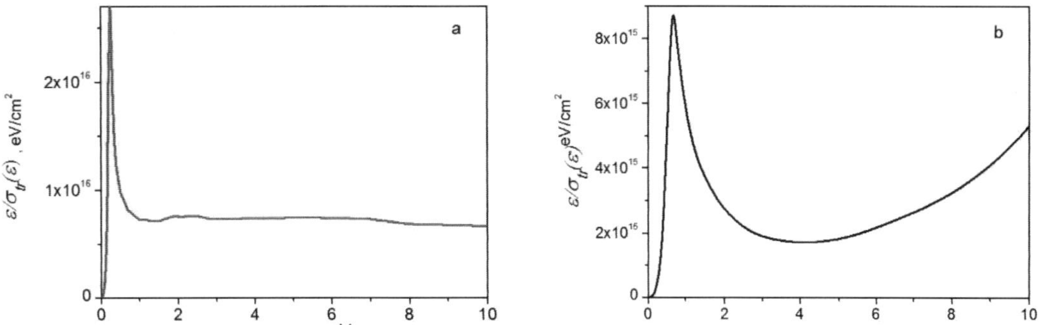

Figure 5. $\sigma_{tr}(\varepsilon)/\varepsilon$ as a function of electron energy for argon (a) and xenon (b) atoms.

In this work we would also like to pay attention to the fact that the inverse population of the EEDF also influences on the sign of the imaginary part of plasma conductivity (see equation (3)), which determines refraction features of a plasma channel. The common knowledge is that permittivity of weakly ionized plasma is less than unity, but in our case plasma channel with the nonequilibrium distribution function can be optically denser medium in comparison with unionized gas. We need the following inequality to be satisfied:

$$\frac{d}{d\varepsilon}\left(\frac{\varepsilon^{3/2}}{\omega^2+\nu_{tr}^2}\right) < 0\,, \quad (7)$$

which can be simplified in the low-frequency range: $\frac{d}{d\varepsilon}\varepsilon^{1/4}/\sigma_{tr} < 0$. The obtained result leads to the possibility to use such a plasma channel as a dielectric waveguide maintaining both effective transportation and amplification of RF radiation. The propagation process in a plasma channel is schematically shown at figure 6 and is based on the effect of total internal reflection at the interface with a unionized gas as an optically less dense medium [10].

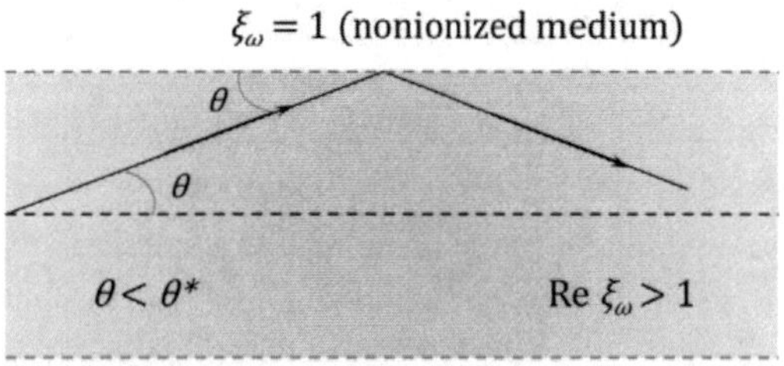

Figure 6. The sliding-mode waveguide scheme, θ is the angle of total internal reflection.

5. Analytics for optical properties of a plasma channel

In this section we discuss some analytics describing optical properties of a plasma channel. Dependences of transport cross sections in considered energy ranges can be approximated by power functions (see figure 7). We have the squared dependence of energy for xenon and a liner dependence for argon atoms, i.e. $\sigma_{tr}(\varepsilon) = \sigma_0(\varepsilon/\varepsilon_0)^\alpha$, where $\sigma_0 = 1.24\times10^{-15}$ см$^{-2}$, $\varepsilon_0 = 2.6$ eV, $\alpha = 2$ for Xe and $\sigma_0 = 2.3\times10^{-16}$ см$^{-2}$, $\varepsilon_0 = 1.7$ eV, $\alpha = 1$ for Ar.

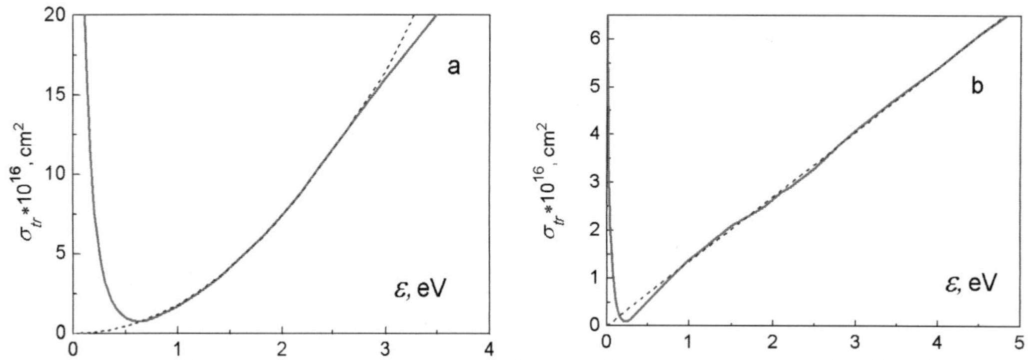

Figure 7. Transport scattering cross section (solid line) and its approximation (dashed line) in xenon (a) and argon (b).

In view of previous calculations of the EEDF evolution for qualitative analysis it is possible to assume the EEDF as a gradually shifted δ - peak, the temporal evolution of which is determined by the averaged over the spectrum energy of photoelectrons $\langle\varepsilon\rangle$:

$$\frac{d\langle\varepsilon\rangle}{dt} = -\frac{2m}{M}\nu_{tr}(\langle\varepsilon\rangle)\langle\varepsilon\rangle. \tag{8}$$

We can find the averaged energy $\langle\varepsilon\rangle$ also analytically:

$$\langle\varepsilon(t)\rangle = \frac{\langle\varepsilon(t=0)\rangle}{\left(1+\frac{2m}{M}(\alpha+1/2)\nu(t=0)t\right)^{1/(\alpha+1/2)}}. \tag{9}$$

As a result, using the δ - function properties we can easily obtain such expressions for the real and imaginary parts of plasma permittivity $\xi_\omega = \mathrm{Re}\,\xi_\omega + i\,\mathrm{Im}\,\xi_\omega$:

$$\operatorname{Re}\xi_\omega = 1 - \frac{\omega_p^2}{\omega^2 + \nu^2(t)} \frac{\omega^2 - \frac{2}{3}(2\alpha - 1/2)\nu^2(t)}{\omega^2 + \nu^2(t)}, \tag{10}$$

$$\operatorname{Im}\xi_\omega = \frac{2}{3}\frac{\omega_p^2\nu(t)}{(\omega^2 + \nu^2(t))\omega} \frac{(\alpha + 2)\omega^2 + (1-\alpha)\nu^2(t)}{\omega^2 + \nu^2(t)}. \tag{11}$$

Here ω_p^2 is the plasma frequency squared and $\nu(t) = \nu_{tr}(\langle\varepsilon(t)\rangle)$. In the low-frequency range ($\omega << \nu_{tr}$) one can simplify the expressions:

$$\operatorname{Re}\xi_\omega \approx 1 + \frac{2}{3}\frac{\omega_p^2}{\nu^2(t)}(2\alpha - 1/2) > 1 \quad \text{if } \alpha > 1/4, \tag{12}$$

$$\operatorname{Im}\xi_\omega \approx \frac{2}{3}\frac{\omega_p^2}{\nu(t)\omega}(1-\alpha) < 0 \quad \text{if } \alpha > 1. \tag{13}$$

It can be seen that the condition of effective propagation regime (12) is less strict that the amplification condition (13).

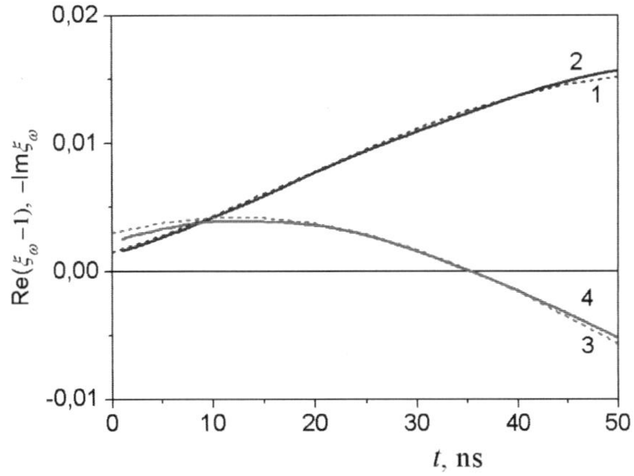

Figure 8. Time dependences of $\operatorname{Re}(\xi_\omega - 1)$ (1,2) and $-\operatorname{Im}\xi_\omega$ (3,4) in xenon obtained by analytical expressions (10), (11) (dashed line) and by numerical calculations (solid line), $n_e = 3\times10^{12}\text{cm}^{-3}$.

We provide the comparison between the analytics and numerical calculation for real and imaginary parts of permittivity for RF frequency $\omega = 5\times10^{11}$ s^{-1} at figure 8. One can see rather good agreement. As the imaginary part of permittivity determines the absorption (amplification) coefficient, one can observe the amplification zone and much more prolonged guiding zone in the plasma channel.

6. Electron – electron collisions and RF field influence on the evolution of the EEDF

The influence of the electron – electron collisions was not taken into account in the previous discussion. We can estimate, that the characteristic value of energy exchange in ee-collisions is about $\Delta\varepsilon_{ee} \sim \varepsilon$. The characteristic energy losses in electron – atomic collisions are about four orders of value less: $\Delta\varepsilon_{ea} \sim \frac{2m}{M}\varepsilon$. Thus, during the ee – collisions the effective energy exchange takes place even for rather low degrees of ionization.

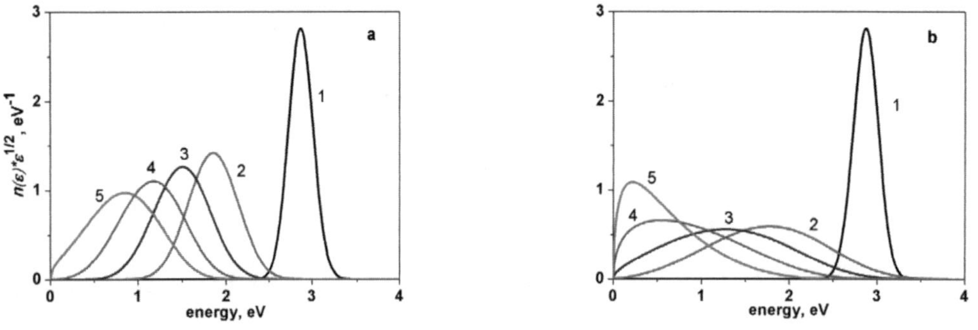

Figure 8. EEDF in xenon at the instants of time t=0 (1), t=25 (2), t=50 (3), t=100 (4), t=200 (5) ns after creation of the plasma channel by λ=248 nm radiation for electron densities 10^{12} (a) and 10^{13} cm^{-3} (b).

This process results in the fast maxwellization of the energy spectrum. The calculations confirm our predictions (see figure 8). As a result, that at the rather small instants of time the gain factor increases proportionally to the electron concentration, but the duration of the amplification in plasma decreases with the increase of N_e because of the speed up in relaxation of the EEDF.

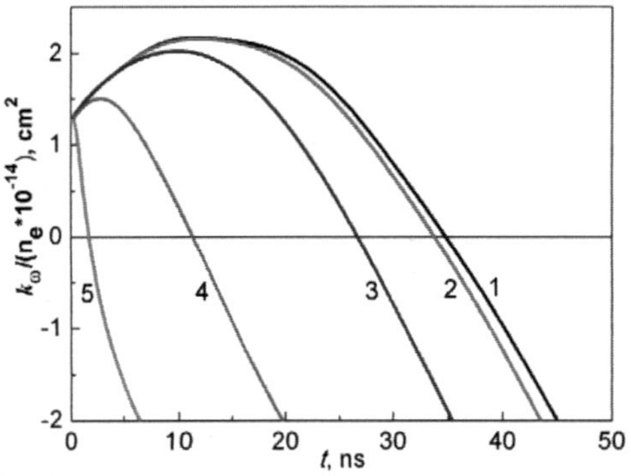

Figure 9. The gain factor per one electron in xenon plasma for different electronic densities (cm^{-3}): 1 - 10^{10}, 2 - 10^{11}, 3 - 10^{12}, 4 - 10^{13}, 5 - 10^{14}. Negative values correspond to the absorption on the RF radiation in plasma.

All the data discussed above were in the case when transported RF field doesn't influence on the evolution of the EEDF. Simple estimations reveal that one can neglect this influence if the relation $\frac{e^2 E_0^2}{3m\nu_{tr}^2} < \max\left\{\frac{2m}{M}T_g, \langle\varepsilon\rangle\frac{\nu_{ee}}{\nu_{tr}}\right\}$ is fulfilled. For $\langle\varepsilon\rangle \approx 2$ eV and $n_e = 3\times10^{12}$ cm^{-3} above inequality is valid for intensities less than 100 W/cm^2. The figure 10 show that the more intense RF field results in faster spreading of the initial peak due to the diffusion of electrons in energy space.

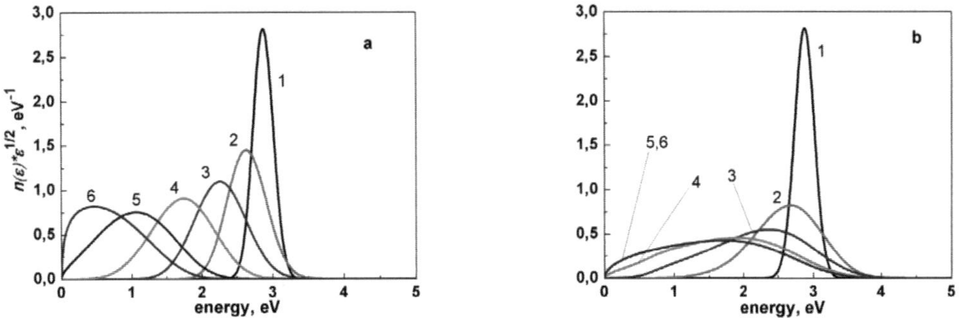

Figure 10. EEDF in xenon plasma at the instants of time t=0 (1), t=25 (2), t=50 (3), t=100 (4), t=200 (5) ns after the channel formation for RF field intensities (a) 0 and (b) 100 W/cm^2. Electronic density $N_e = 3\times10^{12}$ cm^{-3}.

The temporal evolution of averaged electron energy and the gain factor $k_\omega = -\mu_\omega$ for different values of the RF intensity is presented at figure 11. One can find out that the RF field presence leads to the additional cooling of the electron component of plasma which means the amplification if the RF signal. But with the increase of the RF field intensity the shortening of EEDF relaxation time takes place. This reduces the duration of the amplification (see figure 11a).

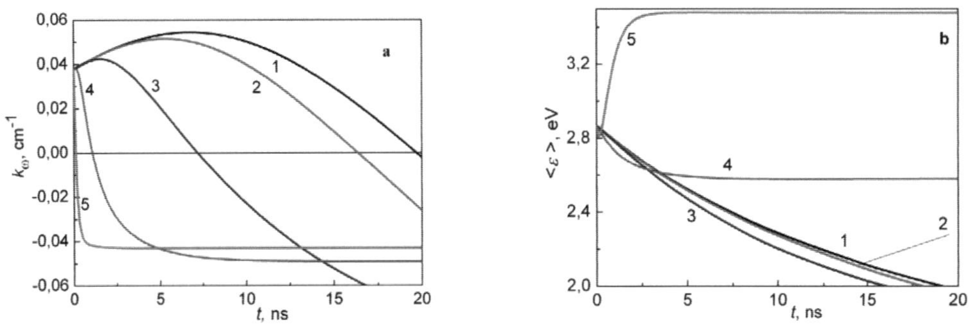

Figure 11. Time dependence of the gain factor of the electromagnetic radiation (a) and the average electron energy $\langle\varepsilon\rangle$ (b) for different intensities of the RF radiation: I=0 W/cm^2 (1), 10 W/cm^2 (2), 100 W/cm^2 (3), 10^3 W/cm^2 (4), 10^4 W/cm^2 (5). The data obtained for electronic density $n_e = 3\times10^{12}$ cm^{-3} and $\omega = 5\times10^{11}$ s^{-1}.

7. Propagation of the radio-frequency pulses in the plasma channel

In this chapter we will discuss the possibility to use the plasma channel as a waveguide for transportation and amplification of the RD radiation pulse. Our analysis is based on the self-consistent solution of the wave equation for the RF pulse and the Boltzmann equation for the EEDF in the plasma channel in different spatial points.

As it is known, propagation of the electromagnetic radiation in a medium is described by the wave equation:

$$\nabla^2\vec{E}(\vec{r},t) - \frac{1}{c^2}\frac{\partial^2\vec{E}}{\partial t^2} = \frac{4\pi}{c^2}\frac{\partial\vec{j}(\vec{r},t)}{\partial t}. \tag{14}$$

Here $\vec{E}$ is the electric field strength and $\vec{j}$ is the density of the electric current in a plasma. Further we will suppose the field is linearly polarized. To analyze the process of RF pulse propagation qualitatively we use the slow-varying amplitude and phase (SVAP) approximation for the solution of equation (14). According to this approximation for the pulse propagation along z-direction electric field E should be represented as

$$E(\vec{r},t) = E_0(\rho,z,t)\cdot\exp(i(kz-\omega t)). \quad (15)$$

Here E_0 is the envelope of the RF pulse, $k=\omega/c$ is the wave number and ρ is perpendicular spatial coordinate.

Neglecting the temporal dispersion it is possible to write down the following expression for the density of the electric current in plasma $j(\vec{r},t)=\sigma E(\vec{r},t)$ is the plasma conductivity. We also assume that the σ is slow varying function in space and time. Then using the SVAP approximation one can obtain the following equation for the electric pulse envelope

$$ik\left(\frac{\partial E_0}{\partial z}+\frac{1}{c}\left(1+i\frac{2\pi\sigma}{\omega}\right)\frac{\partial E_0}{\partial t}\right)=-\frac{1}{2}\nabla_\perp^2 E_0-\Delta n_\omega(t-z/c)k^2E_0+\frac{i}{2}k_\omega(t-z/c)kE_0. \quad (16)$$

Here $\Delta n_\omega=\mathrm{Re}(n_\omega-1)=-\dfrac{2\pi\sigma''}{\omega}>0$. The first term in the right part of the equation stands for the diffraction divergence of the pulse, the second one describes the refractive properties of plasma. In our case we have refractive index greater than unity, so plasma channel reveals the focusing features. The third term represents amplification or absorption of the RF radiation.

The propagation process is schematically shown at figure 12. The femtosecond laser pulse creates a plasma channel which is characterized by the amplifying zone and the longer guiding zone. If we launch the laser pulse and the RF pulse just one after another, the last one will continually locate in the amplifying zone of the laser pulse. The guiding regime can be realized if the focusing term in equation suppresses the diffraction divergence term, i.e. the relation

$$(n_\omega-1)k^2R^2>1 \quad (17)$$

(R is the plasma channel radius) is satisfied. For $\omega=5\times10^{11}$ s^{-1} and $n_e=3\times10^{12}$ cm^{-3} the guiding regime can be realized for channel radius $R\geq1\div2$ cm.

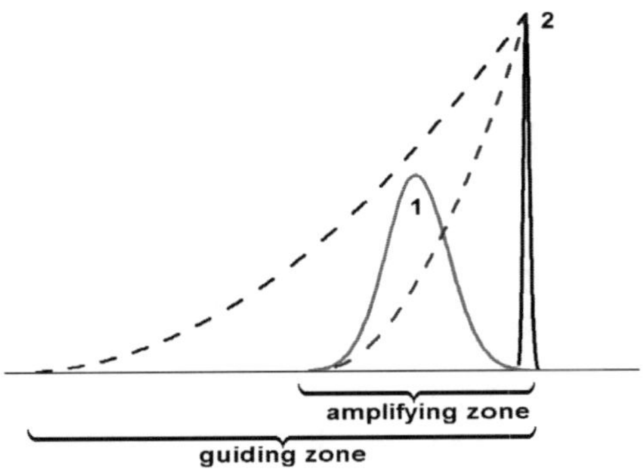

Figure 12. Spatial structure of radio (1) and laser (2) pulses for a given instant of time. Dash curves are the spatial profiles of the gain factor and the refractive index.

Numerical solution of equation (16) was realized the initial temporal profile of the RF pulse

$$E_0(\rho,z=0,\tau=t)=A\times R(\rho)\times\sin^2(\pi t/\tau_p) \quad (23)$$

where A is the amplitude of the RF pulse at the axis $\rho=0$, τ_p is its duration, and the function $R(\rho)$ gives the initial radial electric field distribution. Further we assume that $\tau_p=50T$ ($T=2\pi/\omega$), which corresponds to the RF pulse duration of $\tau_p\approx0.628$ ns for $\omega=5\times10^{11}$ s^{-1}. The initial radial distribution $R(\rho)$ was also chosen in the Gaussian form

$$R(\rho)=\exp(-\rho^2/2\rho_f^2), \quad (24)$$

where ρ_f is the RF beam radius. The radial profile of the electron density created by KrF laser pulse and determining the value of the gain factor (or absorption coefficient) and the plasma refractive index was also assumed to be Gaussian:

$$n_e(\rho) = n_e(\rho = 0) \times \exp\left(-\rho^2 / 2\rho_e^2\right), \quad (25)$$

At figures 13 and 14 we demonstrate the numerical results of the temporal profile of the intensity of RF pulse for different propagation distances for two values of the initial peak intensity.

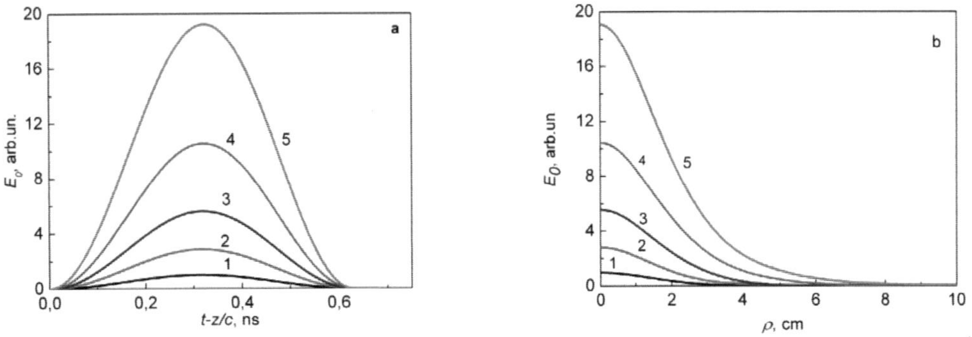

Figure 13. Temporal (1) and radial (2) profiles of the electric field envelope of the amplifying pulse at the different propagation distances z: 1 – 0 cm, 2 – 30 cm, 3 – 60 cm, 4 – 90 cm, 5 – 120 cm. Initial peak intensity is 0.1 W/cm^2.

It can be seen that more significant amplification takes place for the weak initial intensity. For rather intense RF pulses the transported radiation contributed to the EEDF evolution and results in reduction of the gain factor. We see that the trailing edge of the pulse eventually is not amplified.

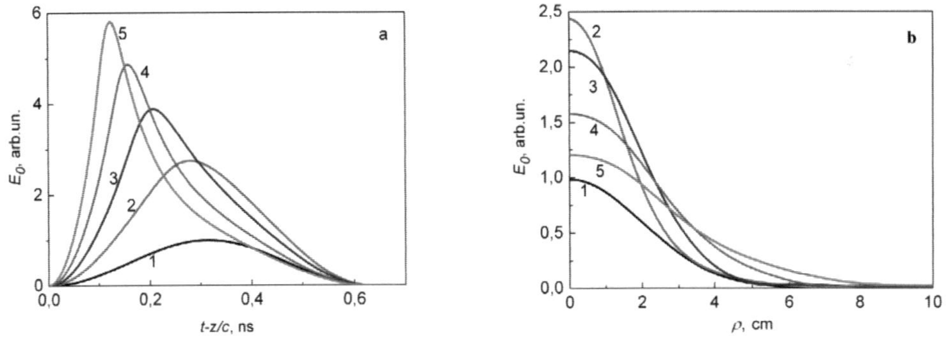

Figure 14. The same, as at figure13, but for the initial peak intensity 1 kW/cm^2.

Conclusions

Thus, it was shown that a plasma channel created in the noble gases by the high intense KrF laser pulse is characterized by the strongly nonequlibrium EEDF which leads to the possibility to use such a channel for amplification of RF radiation. It was demonstrated that xenon plasma has some advantages as the amplifying medium in comparison with other rare gases. Temporal evolution of the gain factor in xenon was investigated. A plasma channel with strongly nonequlibrium EEDF is also characterized by the refractive index greater than unity. The wave equation for the propagation of RF pulses in xenon was solved numerically and analytically. It was shown that such a channel can be considered as a plasma waveguide which can be used for both transportation and amplification of the RF radiation.

Acknowledgments
This work was supported by the Russian Foundation for Basic Research (project no. 15-02-00373). Anna Bogatskaya also thanks the Dynasty Foundation (program for support of students) and the LPI Educational Complex and Special Program of RAS Presidium for young scientist support. Numerical modelling was performed on the SKIF-MSU Chebyshev supercomputer.

References

[1] Agostini P and Di Mauro L F 2004 *Rep. Prog. Phys.* **67** 813
[2] Krausz F and Ivanov M 2009 *Rev. Mod. Phys.* **81** 163
[3] Zhao X M, Wang Y C, Diels J-C and Elizondo J 1995 *IEEE J. Quantum Electron.* **31** 599
[4] Ginzburg V L, Gurevich A V 1960 *Sov. Phys. Usp.* **3** 115
[5] Tzortzakis S, Franco M A, Andre Y-B et al 1999 *Phys. Rev. E* **60**R3505
[6] Rodriguez M, Sauerbrey R, Wille H et al 2002 *Opt. Lett.* **27** 772
[7] Ionin A A, Kudryashov S V, Levchenko A O et al **2012** *Appl. Phys. Lett.* **100** 104105
[8] Penano J, Sprangle P, Hafiz B et al 2012 *J. Appl. Phys.* **111** 033105
[9] Chateauneuf M, Payeur S, Dubois J and Kieffer J-C 2008 *Appl. Phys. Lett.* **92** 091104
[10] Zvorykin V D, Levchenko A O, Smetanin I V and Ustinovski N N 2010 *JETP Lett.* **91** 226
[11] Bogatskaya A V and Popov A M 2013 *JETP Lett.* **97** 388
[12] Bekefi G, Hirshfield Y L and Brown S C 1961 *Phys. Fluids* **4** 173
[13] Bunkin F V, Kazakov A E and Fedorov M V 1973 *Sov. Phys. Usp.* **15** 416
[14] Bogatskaya A V, Volkova E A and Popov A M 2013 *Quantum Electronics* **43** 1110
[15] Fedorov M V, Poluektov N P, Popov A M et al 2012 *IEEE J. Selected Topics in Quantum Electronics* **18** 42

23rd International Laser Physics Workshop (LPHYS'14) IOP Publishing
Journal of Physics: Conference Series **594** (2015) 012018 doi:10.1088/1742-6596/594/1/012018

Particle beams in ultrastrong laser fields: direct laser acceleration and radiation reaction effects

Yousef I. Salamin
Max-Planck-Institut für Kernphysik, Saupfercheckweg 1, 69117 Heidelberg, Germany
Department of Physics, American University of Sharjah, POB 26666,
Sharjah, United Arab Emirates

Jian-Xing Li
Max-Planck-Institut für Kernphysik, Saupfercheckweg 1, 69117 Heidelberg, Germany

Karen Z. Hatsagortsyan
Max-Planck-Institut für Kernphysik, Saupfercheckweg 1, 69117 Heidelberg, Germany

Matteo Tamburini
Max-Planck-Institut für Kernphysik, Saupfercheckweg 1, 69117 Heidelberg, Germany

Antonino Di Piazza
Max-Planck-Institut für Kernphysik, Saupfercheckweg 1,69117 Heidelberg, Germany

Christoph H. Keitel
Max-Planck-Institut für Kernphysik, Saupfercheckweg 1, 69117 Heidelberg, Germany

Abstract. Several aspects of the interaction of particle beams with ultrastrong laser fields are discussed. Firstly, we consider regimes when radiation reaction is not essential and it is demonstrated that employing chirped laser pulses, significant improvement of the direct acceleration of particles can be achieved. Results from single- and many-particle calculations of the particle acceleration, in vacuum, by plane-wave fields, as well as in tightly-focused laser beams, show that the mean energies and their spreads qualify them for important applications. Secondly, we investigate the effect of radiation reaction in electron-laser-beam interactions. Signatures of the quantum radiation reaction during the interaction of an electron bunch with a focused superstrong ultrashort laser pulse can be observed in a characteristic behavior of the spectral bandwidth, and the angular spread of the nonlinear Compton radiation on the laser pulse duration. Furthermore, it is shown that the radiation reaction effects can be employed to control the electron dynamics via the nonlinear interplay between the Lorentz and radiation reaction forces. In particular, it is shown that an ultrarelativistic electron bunch colliding head-on with a strong bichromatic laser pulse can be deflected in a controllable way, by changing either the relative phase or the relative amplitude between the two frequency components of the bichromatic field.

1. Introduction

Recent advances in laser techniques have opened up new avenues for the generation of high energy photons, electrons and ions [1]. When strong laser radiation interacts with an atomic gas jet, extreme ultraviolet and x-ray radiation can be generated via high-order harmonic generation processes [2, 3] which, in turn, has led to the field of attoscience [4]. Gamma rays can be created employing Compton scattering of ultrastrong laser radiation by a relativistic electron beam [5]. Moreover, a laser beam can be used to generate relativistic electron beams up to giga-electronvolt energies via wakefield acceleration [6], allowing for the realization of all-optical setups for Compton radiation sources. In strong laser fields the electron radiation can be so strong that it may disturb the electron dynamics. Then, the radiation reaction effects become nonperturbative and the so-called radiation dominated regime sets in [1]. Currently, a lot of efforts are invested also in accelerating ions with ultrastrong laser fields. Thus, for example, with the target normal sheath acceleration mechanism quasi-monoenergetic ion beams are produced with about MeV energy per nucleon [7]. The main aim of the laser acceleration of ions is to develop more cost effective solutions for medical applications.

In this work several aspects of the interaction of particle beams with strong laser fields are discussed. In the first part of the paper we consider the possibility of accelerating the particles by laser means, directly and in the absence of any background like a plasma or boundaries. The main motivation for laser-acceleration is the desire to replace the large and expensive conventional accelerators with compact and cheap laser-based alternatives. Our work on laser acceleration of electrons and ions to energies that may make them useful for various applications involves unchirped as well as chirped laser pulses. In particular, it has been concluded from our unchirped pulse calculations of ion acceleration for tumour therapy [8, 9] that a laser power of 10 PW would be needed. The laser pulse or beam needs to be focused to a sub-wavelength waist radius ($w_0 < \lambda$), which would correspond to the intensity of 10^{23} W/cm^2. By appropriately chirping the frequency of the laser pulse, our simulations [10, 11, 12, 13, 14, 15] have shown that both the power and intensity may be lowered by one order of magnitude and that the beam/pulse need not be focused to a sub-wavelength spot radius ($w_0 > \lambda$). In this paper results from our chirped-pulse laser acceleration will be discussed.

The second part of the paper is devoted to the radiation reaction effects which take place during an electron beam interaction with laser fields. Firstly, we review our results on signatures of radiation reaction in the radiation dominated regime and in a focused laser beam of short duration [16]. We consider a setup where the relativistic electron beam counterpropagates with the laser beam, and investigate the radiation frequency and angle resolved spectra during multiphoton Compton scattering. In particular, the analysis of the dependence of the spectral bandwidth and the angular spread of the radiation on the laser pulse duration, reveals that their behaviour is very specific in the case of the radiation dominated regime in a tightly focused laser beam.

Finally, we demonstrate that the radiation reaction effects can provide a route to the control of the electron dynamics in the interaction with a strong few-cycle [17] and bichromatic laser pulse [18]. In particular, an ultrarelativistic electron bunch initially counterpropagating with respect to a bichromatic laser pulse, can be deflected in a controllable way, by changing the relative phase or the relative amplitude of the two frequency components of the laser pulse field [18]. In addition, the effect is independent of the initial electron energy, as long as the quantum corrections remain suppressed because a larger initial electron energy, in turn, implies a larger radiation reaction force and, thus, a larger energy loss.

2. Direct laser-acceleration of particles

The focus of our earlier work on direct laser acceleration was on electrons, employing plane-wave and tightly-focused laser beams and pulses. In more recent investigations, attention was

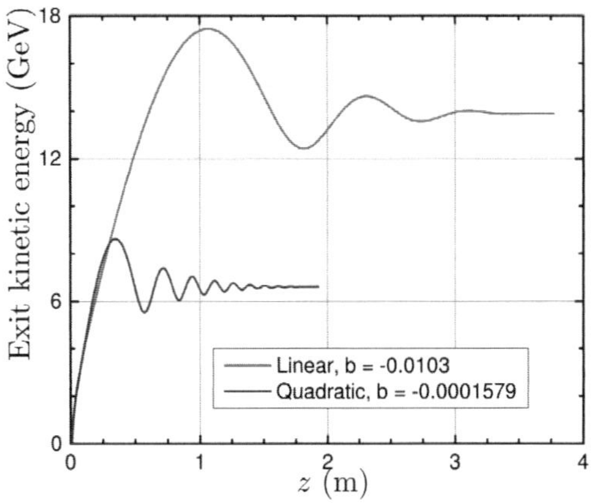

Figure 1. (Color online) Exit electron kinetic energy $K_{exit} = mc^2[\gamma_f - 1]$, where $\gamma_f = \gamma(\eta_f)$, vs. the excursion distance z, following interaction with linear and quadratic chirps, $\omega = \omega_0(1 + b\eta)$ and $\omega = \omega_0(1 + b\eta^2)$, respectively, in which $\eta = \omega_0(t - z/c)$. The pulse is modeled by a finite-duration pulse of $\cos^2$ shape (in place of the Gaussian in Eq. (3), see [20]). In both cases, the dimensionless intensity parameter $a = eE_0/(m\omega c) = 3$ (which corresponds to the intensity $I \sim 1.23 \times 10^{19}$ W/cm^2), the scaled injection energy $\gamma_0 = 10$ ($K_0 \sim 4.6$ MeV), $\lambda_0 = 1$ μm, and the initial phase $\psi_0 = 0$. Temporal width of the $\cos^2$ pulse is $\tau = 50$ fs.

directed at the use of chirped laser pulses to accelerate electrons [19] and then protons. A further shift from linearly- to quadratically-chirped pulses has also been the subject of a more recent publication [20].

The single-particle interaction may be handled by solving the relativistic equations of motion in the electric and magnetic fields of the pulse, $\boldsymbol{E}$ and $\boldsymbol{B}$, respectively, subject to the appropriate initial conditions. With the relativistic momentum $\boldsymbol{p} = \gamma m\boldsymbol{v}$ and relativistic energy $\varepsilon = \gamma mc^2$, where m is the particle's mass and $\gamma = (1 - \beta^2)^{-1/2}$ its Lorentz factor, the equations of motion read

$$\frac{d\boldsymbol{p}}{dt} = q[\boldsymbol{E} + \boldsymbol{\beta} \times \boldsymbol{B}] \quad (1)$$

$$\frac{d\varepsilon}{dt} = q\boldsymbol{v} \cdot \boldsymbol{E}. \quad (2)$$

In the above equations, q is the charge of the particle and $\boldsymbol{\beta} = \boldsymbol{v}/c$ is its velocity, scaled by the speed of light.

2.1. Direct laser acceleration of electrons

With m and q in the equations of motion replaced by those of an electron, dynamics of single and many electrons have been studied within the context of many schemes involving a single laser beam, as well as crossed-beam and beat-wave configurations. In those investigations, unchirped linearly- and radially-polarized laser systems were employed. Later work [20, 21] also employed linearly- and quadratically-chirped laser pulses.

From a purely theoretical point of view, the basic idea of a chirped pulse is to make its frequency a function of the time. In a linear chirp, the central frequency ω_0 would be replaced by $\omega \approx \omega_0(1 + b\eta)$, where $\eta = \omega_0 t - k_0 z$, with $ck_0 = \omega_0$ and c the speed of light in vacuum, for a pulse propagating along z. The constant b, here, is a dimensionless chirp parameter. With this

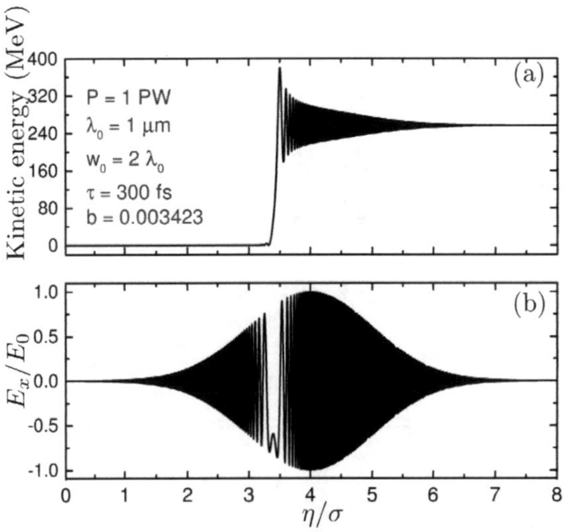

Figure 2. (a) Evolution of the kinetic energy $K = (\gamma - 1)mc^2$ of a single proton during interaction with a single chirped *plane-wave* pulse whose parameters are given. (b) Electric field of the same pulse seen by the proton.

transformation, the electric field of a *plane-wave* Gaussian pulse $E = E_0 \cos(\eta) \exp(-\eta^2/2\sigma^2)$ becomes

$$E = E_0 \cos(\eta + b\eta^2) \exp\left[-\frac{\eta^2}{2\sigma^2}\right], \quad \sigma = \frac{\omega_0 \tau}{\sqrt{2 \ln 2}}, \tag{3}$$

where τ is the temporal width of the pulse.

Note what chirping the frequency typically does to the scaled electric field. The pulse shape and field oscillations get distorted and the electric field appears to develop a low-frequency (almost *quasi-static*) portion. Interaction of a charged particle with the unchirped pulse results in no net gain or loss, because each field cycle consists of two almost identical half-cycles, one positive and the other negative. Interaction with the quasi-static part of the chirped pulse, on the other hand, allows for an extended period of energy gain without loss. Eventually, the particle is left behind the pulse, retaining some of the energy it has gained from the field.

Employing a plane-wave model for the pulse fields, the equations of motion admit analytic solutions in closed form. While the details may be found in the references, it suffices here to display some of the most recent results graphically. In Fig. 1 we show variation of the exit kinetic energy of a single electron subjected to plane-wave linearly- and quadratically-chirped pulses, as functions of the excursion distance of travel along the pulse's propagation direction. It is evident from these model calculations that the same set of parameters (apart from the chirp parameter b) lead to energy gains in the case of a linear chirp that are almost twice what would be gained from a quadratic chirp. This result may be model dependent, though.

2.2. Direct laser acceleration of protons

Figure 2 (a) shows evolution of the kinetic energy of a single proton, from rest at the origin of coordinates, during interaction with a plane-wave 300 fs, 1 PW Gaussian laser pulse of central wavelength $\lambda_0 = 1$ μm. Fig. 2 (b) shows evolution of the scaled electric field during interaction with the proton. The kinetic energy evolution exhibits synchronization with the field oscillations. Interaction with the highly-symmetrical parts of the field (on the left and right) results in no net gain, whereas interaction with the quasi-static part (to the left of the center) does lead to substantial gain. In this example the figure shows clearly that the particle is left behind the pulse with a net gain of approximately 256.4 MeV.

In the discussion above the laser fields of the pulse were modeled in terms of plane-waves with a Gaussian temporal envelope. But a plane-wave is only an idealization, suitable for modeling the fields of a low-power, low-intensity, and sufficiently long laser pulse. The extremely high-power lasers needed for laser acceleration ought to be represented by tightly-focused fields. To this end, we use the fields developed recently by Salamin [14] from a generalized Lax series [22]. In all our calculations, the following field components will be employed, with each multiplied by a Gaussian temporal envelope $\exp(-\eta^2/2\sigma^2)$. First, the electric field components are

$$E_x = E_0 e^{-r^2/w^2}\left\{S_1 + \epsilon^2\left[\xi^2 S_3 - \frac{\rho^4 S_4}{4}\right] + \mathcal{O}(\epsilon^4)\right\}, \tag{4}$$

$$E_y = E_0 e^{-r^2/w^2}\xi\upsilon\epsilon^2\left\{[S_3] + \mathcal{O}(\epsilon^4)\right\}, \tag{5}$$

$$E_z = E_0 e^{-r^2/w^2}\xi\left\{\epsilon\,[C_2] + \mathcal{O}(\epsilon^3)\right\}. \tag{6}$$

In these equations, E_0 is a constant amplitude, $r^2 = x^2 + y^2$, $w = w_0\sqrt{1+\zeta^2}$, $\zeta = z/z_r$, $z_r = k_0 w_0^2/2$ is the Rayleigh length, $\rho = r/w_0$, $\xi = x/w_0$, $\upsilon = y/w_0$, $C_n = (w_0/w)^n \cos(\psi + n\psi_G)$; $n = 1, 2, 3, ...,$ $S_n = (w_0/w)^n \sin(\psi + n\psi_G)$, $\psi_G = \tan^{-1}\zeta$ is the Gouy phase, and $\psi = \psi_0 + \eta - k_0 r^2/R$; $R = z + z_r^2/z$, and ψ_0 is a constant initial phase. also, $\epsilon = w_0/z_r$ is the diffraction angle. To the same order in the diffraction angle, the magnetic field components read

$$B_x = 0, \tag{7}$$

$$B_y = \frac{E_0}{c} e^{-r^2/w^2}\left\{S_1 + \epsilon^2\left[\frac{\rho^2 S_3}{2} - \frac{\rho^4 S_4}{4}\right] + \mathcal{O}(\epsilon^4)\right\}, \tag{8}$$

$$B_z = \frac{E_0}{c} e^{-r^2/w^2}\upsilon\left\{\epsilon\,[C_2] + \mathcal{O}(\epsilon^3)\right\}. \tag{9}$$

Analytic solutions to the equations of motion are, unfortunately, not possible when the above field components are used. Inevitably, one resorts to numerical simulations [10, 11] based on Eqs. (1) and (2) instead. Some results of simulations are displayed in Fig. 3.

For pulse parameters the same as those used in the plane-wave case of Fig. 2, the results of Fig. 3 are almost the same. Exit kinetic energy in Fig. 3 (a) is $\sim$ 257.2 MeV, compared with $\sim$ 256.4 MeV of Fig. 2 (a) of the plane-wave case. Similar synchronous interaction features are displayed also in Figs. 3 (b) and 2 (b). Figures 3 (c) and (d) show how the particle kinetic energy and field oscillations evolve with the excursion distance along the pulse's propagation direction, z. Let us define the *energy gradient* as the energy gained, by the particle from the pulse, per unit forward excursion distance. Evaluated, for our purposes here, by dividing the exit energy gained ($\sim$ 257 MeV) by the excursion along the $z-$axis during effective interaction ($\sim 4\mu$m), the average gradient is over 60 TeV/m. This is more than 6×10^5 times the performance limit on linear conventional accelerators ($\sim$ 100 MeV/m, for electrons).

As was pointed out in the Introduction, particle bunches containing $10^7 - 10^{11}$ particles/bunch are needed for medical and other applications. In recent simulations [10, 11], dynamics of an ensemble of 3000 non-interacting protons was investigated, whose initial positions were distributed randomly within a cylinder centered at the origin of coordinates and of radius $R = 0.3\ \lambda_0$ and height $L = 0.2\ \lambda_0$. On the other hand, the initial conditions on the velocities of the particles were derived from a normal distribution of initial kinetic energies of mean value $\bar{K}_0 = 10$ keV and spread $\Delta K_0 = 0.5$ keV. A code based on the single-particle equations of motion has been used to solve for the particle dynamics in the ensemble. In further calculations, the 3000 protons were assumed to be taken from a hydrogen gas jet or an expanding hydrogen cluster. The initial conditions, otherwise, were assumed to be the same as for the ensemble

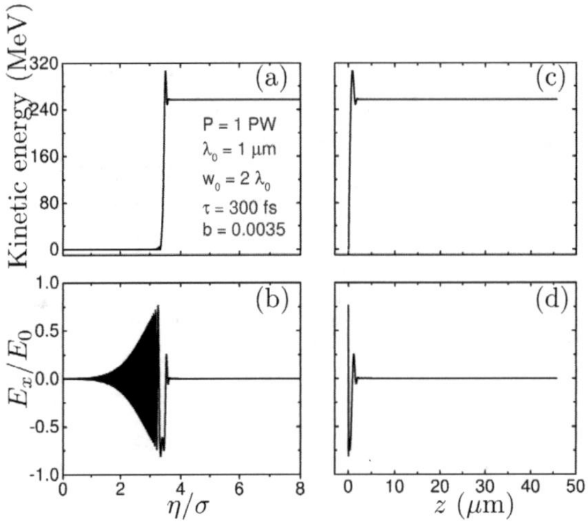

Figure 3. (a) and (b) Evolution of the kinetic energy $K = (\gamma - 1)mc^2$ of a single proton during interaction with a single chirped *tightly-focused* pulse whose parameters are given. (c) and (d) Evolution of the main electric field component of the same pulse, as seen by the proton.

of non-interacting protons. When the pulse impinges upon the target, an under-dense plasma of electrons and protons is generated quickly. The electrons get swiftly accelerated and, being the lighter component, they immediately leave the scene and drag the much more massive, and sluggish, protons behind them. The protons will also be accelerated further by interaction with the pulse. A particle-in-cell (pic) code has been used to solve the equations of motion self-consistently for the particle dynamics in the plasma. Unlike the proton ensemble calculations, in the pic simulations all of the Coulomb particle-particle interactions are taken into account.

The main results from the many particle calculations, employing pulse parameters the same as in Fig. 3, apart from $w_0 = 2.2\lambda_0$ and $\tau = 350$ fs, are as follows [10]. The two calculations (interacting and non-interacting) return exit energy gains within 5% of each other (245.3 MeV and 258.3 MeV, respectively). In both cases, the spreads in the exit kinetic energies are about 1%, which qualifies the protons accelerated in this fashion for use in tumor therapy. The spread in the distribution of the ensemble of non-interacting protons is larger than that of the protons from the plasma. This should come as no surprise in light of the fact that the pic code takes into account motion of, and all effects associated with, the electrons. Attraction of the electrons prevents the protons from spreading out spatially and pulls further on them in the direction of their collective motion, which results ultimately in less exit energy spread.

3. radiation reaction effects in ultrastrong laser fields

3.1. Signatures of radiation reaction in the interaction of electron and laser beams

We consider a relativistic electron beam moving in the field of an ultrastrong laser field opposite to the laser field propagation direction. Then, in the electron rest frame the laser field is Lorentz-boosted and the radiation dominated regime is achievable when the parameter $R \equiv \alpha\xi\chi$ is of order of one, where α is the fine structure constant, $\xi = eE_0/mc\omega_0$ is the relativistic invariant parameter of the laser field strength, $\chi = E'/E_{cr}$ is the quantum strong field parameter, $E_{cr} = m^2c^3/e\hbar$ is the QED critical field, $E' \approx 2\gamma E_0$ is the laser field in the rest frame of the electron and γ is the Lorentz-factor. When $\chi \gtrsim 1$, the photon emission recoil is not negligible and the radiation dynamics is quantum mechanical. If additionally $\alpha\xi \gtrsim 1$, then radiated energy during the electron motion in a single laser period is comparable with the electron total energy and the interaction is in the quantum radiation dominated regime. We calculated in [23] the Compton radiation spectra in the quantum radiation reaction regime taking into account the effects of the radiation reaction. The radiation reaction effect is exhibited in the radiation spectrum as decreasing the photon piling effect at the maximal photon energy, shifting the peak of the spectra towards lower photon energies [23] and, consequently, decreasing the emitted photon numbers in the high energy region. However, detection of these effects in an experiment requires accurate measurement of the emitted photon numbers which is not an easy task. That is why in [24] it was proposed to measure the electron parameters (energy) after the end of interaction as a measure of the radiation reaction in the experiment. Yet it is disappointing, because the electron energy decrease is a trivial signature of the radiation reaction. Then, we asked ourselves if one can find non-trivial signatures of the radiation reaction in the spectra of emitted radiation which have qualitative character and do not require very accurate measurement.

We calculated angle resolved spectra of the electron radiation during the Compton scattering and investigated its dependence on the laser pulse duration. For the calculation of the radiation spectra we assumed that the electron dynamics in the laser field is classical. This is justified because for the ultrarelativistic motion of the electron in the infrared laser field, the electron de-Broglie wavelength is much smaller than the laser wavelength. For the calculation of the radiation spectra the full quantum mechanical formulas have been used in the asymptotic limit $\xi \gg 1$ [25]. In this limit the coherence length of the radiation $l_{coh} \sim \lambda_0/\xi$ is much smaller than the laser wavelength λ_0 and the emission intensity depends only on the local field values, i.e., one can assume that the radiation is emitted from a local point on the electron trajectory. At each point of the trajectory one can calculate the emitted radiation and deduct from the electron energy and the momentum the energy and momentum carried by the emitted Compton photons (and by the absorbed laser photons). One should take into account that the emission of the Compton photons is possible only when the electron absorbs from the laser wave a certain number of laser photons (as required by the energy-momentum conservation). The accurate accounting of the electron momentum change due to the radiation yields an additional radiation reaction term in the classical equation of motion which is derived in [26, 27]:

$$\frac{dp^\alpha}{d\tau} = \frac{e}{mc}F^{\alpha\beta}p_\beta - \frac{\mathcal{I}}{mc^2}p^\alpha + \tau_c\frac{\mathcal{I}}{\mathcal{I}_c}F^{\alpha\beta}F_{\beta\gamma}p^\gamma, \tag{10}$$

where τ is the proper time, $\tau_c \equiv 2e^2/(3mc^3)$ and $\mathcal{I}_c = 2e^2\omega^2\xi^2/(3c)$ is the classical radiation loss rate and $\mathcal{I}$ is the total quantum mechanical radiation loss rate.

The driving laser beam is circularly polarized. We have applied laser pulses of different duration from 0.5 to 9 cycles. The circular polarization has an advantage that the spectra have no dependence on the carrier-to-envelope phase of the short laser pulse. The first panel in Fig. 4 shows the angle resolved total radiation intensity (color coded), θ and ϕ are the polar and azimuthal angles, respectively. For each azimuthal angle there is contribution from different polar angles which correspond to different laser cycles with different magnitudes of the laser field amplitude. We choose the azimuthal angle where the emission intensity is maximal and plot the corresponding spectrum, resolved on the polar angle, in the right column of Fig. 4. The radiation intensity averaged by the azimuthal angle is shown in the middle column of Fig. 4. We inspect the spectral and angular bandwidths of the radiation. The latter is defined as the spectral and angular region where the emission intensity drops to one-half its peak value.

First of all we note from the last column of Fig. 4 that the emission is not uniform with respect to the polar angle: there is a spike at a polar angle far from 180° (direction of the electron motion). The main emission region near 180^0 arises at the peak of the parameter χ, which takes place before the electron reflection in the laser field. This can be deduced from Fig. 5 (the bottom panel). This figure shows the electron velocity direction, which coincides with the emission direction (as the relativistic electron radiates forward). At those moments η

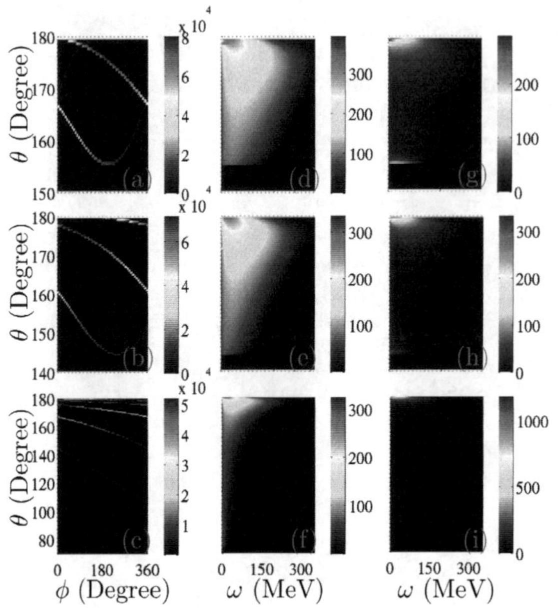

Figure 4. (Color online) The radiation spectrum: (left column) the angle resolved total radiation energy (color coded), θ is the polar and ϕ azimuthal angles; (middle column) the radiation spectrum averaged by the azimuthal angle; (right column) the radiation spectrum at the azimuthal angle where the emission energy is maximal. $\xi = 230$, $\gamma = 1000$, the laser pulse duration is 1 cycle (first row), 1.5 cycle (second row), and 5 cycles (third row).

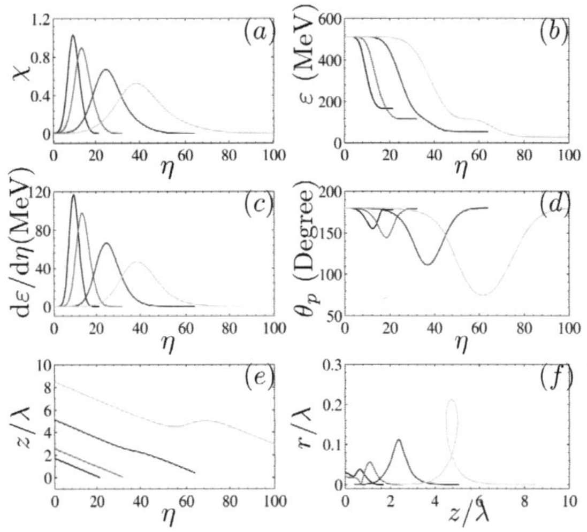

Figure 5. (Color online). The electron dynamics in counterpropagating laser pulses of various durations: (a) The parameter χ; (b) The electron energy, (c) the radiation loss, (d) the emission angle, (e) the electron coordinate along the propagation direction vs the laser phase; (f) The electron trajectory. The black, red, blue, and green curves (from left to right) correspond to the laser pulse durations $\tau_0 = T_0$, $\tau_0 = 1.5T_0$, $\tau_0 = 3T_0$, and $\tau_0 = 5T_0$, respectively.

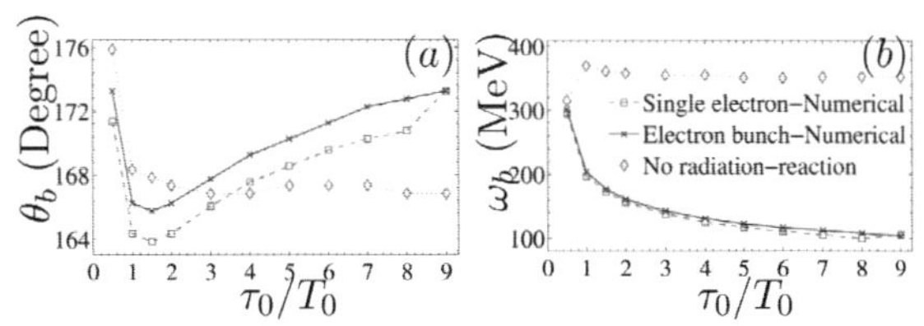

Figure 6. (Color online) The spectral bandwidth (left) and the angular spread (right) of the radiation vs the laser pulse duration. The laser and electron parameters are the same as in Fig. 4.

when χ is maximal, the electron direction θ is still close to 180°. The electron reflection point corresponds to the minimum of the angle θ after which the radiation is negligible. Although near the minimum θ the parameter χ is small and the radiation rate is also small, the total emitted energy is not small because it is accumulated along a longer time interval than in the case of the peak of the parameter χ. The radiation emitted at the minimum of θ is seen as the spike in the spectral distribution of the last column of Fig. 4.

Concentrating on the main emission region and neglecting the spike of the radiation, we plot in Fig. 6 the dependence of the radiation angular and spectral bandwidth on the laser pulse duration. While the spectral bandwidth is decreasing monotonously with increasing laser pulse duration, the angular spread of the radiation shows non-monotonous behavior. The monotonous decrease of the spectral bandwidth is due to the decrease of the maximum of the χ-parameter with increasing laser pulse duration. The latter is due to more decrease of the electron energy because of radiation in a longer laser pulse. The non-monotonous behavior of the angular spread is due to two competing effects: radiation reaction and focusing. The radiation angle is determined by the electron velocity direction at the point where the χ-parameter is maximal. Note that $\chi \sim \xi\gamma$ is proportional to the laser field and the electron energy. If there were no focusing of the laser pulse then in the longer laser pulse the largest χ is reached at a lower field strength ξ (the energy loss due to radiation dominates) and the emission angle is closer to 180°. While when there were no radiation reaction effects, in the longer laser pulse the larger χ is reached at the larger ξ value and the emission angle is more remote from 180°.

3.2. Electron dynamics controlled via radiation reaction

3.2.1. Analytical analysis In the classical framework, radiation reaction effects can be taken into account by the so called Landau-Lifshitz (LL) equation [28]:

$$\frac{dp^\mu}{d\tau} = eF^{\mu\nu}\frac{p_\nu}{mc} + \frac{2e^4}{3m^3c^5}\left[F^{\mu\nu}F_{\nu\alpha}p^\alpha + (F^{\nu\beta}p_\beta)^2\frac{p^\mu}{m^2c^2}\right]. \tag{11}$$

The LL equation is equivalent to the Sokolov equation (10) in the realm of classical electrodynamics, the difference between the two equations being smaller than than quantum effects. Also, the term of the RR force containing the derivatives of the field tensor $F^{\mu\nu}$ [28] has been dropped from Eq. (11) since its contribution is negligible [29].

Let us consider an ultrarelativistic electron with initial momentum $\mathbf{p}_0 = (0, 0, p_{z_0})$ interacting with a counterpropagating laser pulse which, in first approximation, is modeled as a plane-wave pulse. Since the exact analytical solution of Eq. (11) for a plane-wave pulse with arbitrary shape, frequency and polarization is known [30], the final momentum $\mathbf{p}$ after the electron passes through a pulse propagating along the positive z-axis and polarized along the x-axis is [18, 30]:

$$p_x = \frac{\mathcal{I}}{h}, \tag{12}$$

$$p_y = 0, \tag{13}$$

$$p_z = \frac{p_{z0}}{h} + \frac{(h^2-1)mc + \mathcal{I}^2/mc}{2\rho_0 h}, \tag{14}$$

with:

$$h = 1 + \frac{2e^4\rho_0}{3m^3c^5}\int_{-\infty}^{+\infty} d\varphi\, E_x^2(\varphi), \tag{15}$$

$$\mathcal{I} = \frac{2e^5\rho_0}{3m^3c^5}\int_{-\infty}^{+\infty} d\varphi\, E_x(\varphi)\int_{-\infty}^{\varphi} d\vartheta\, E_x^2(\vartheta), \tag{16}$$

where $E_x(\varphi)$ is the plane-wave electric field as a function of the phase $\varphi = (t - z/c)$, $\rho_0 = (\gamma_0 - p_{z0}/mc)$ is the initial Doppler factor, and $\gamma_0 = \sqrt{1 + p_{z0}^2/m^2c^2}$ is the initial relativistic factor. Note that the quantity h accounts for the energy losses associated with the emission of high frequency radiation and appears in the denominator in Eqs. (12)-(14), while the quantity $\mathcal{I}$ appears in the numerator and accounts for the work performed by the Lorentz force in the presence of the radiation reaction force, i.e. along the radiation-reaction-altered electron trajectory [18].

In the following we consider the effect induced by the interaction of an electron bunch with a second-harmonic enriched laser pulse with electric field: $E_x(\varphi) = g(\varphi)[E_1\sin(\omega\varphi + \theta_1) + E_2\sin(2\omega\varphi + \theta_2)]$, where ω is the lower frequency component, $g(\varphi)$ is the pulse envelope, E_1, E_2 are the field amplitudes of each frequency component, and θ_1 and θ_2 are two constant initial phases.

For the sake of comparison, first we briefly discuss the case of a quasimonochromatic plane-wave pulse. In this case ($E_2 = 0$ and duration of several periods) the integrand of Eq. (16) basically contains only odd frequencies, which arise from the cubic nonlinearity with respect to the plane-wave field. Hence, $\mathcal{I}$ averages out to zero and no transverse momentum gain is possible [see Eqs. (12) and (16)]. On the contrary, if both frequencies are simultaneously present with the higher frequency being twice the lower frequency mode then a zero-frequency term arises in the integrand of Eq. (16). As a consequence, $\mathcal{I}$ diverges in the limit of infinite pulse duration. In fact, from Eq. (16) and in the limit of long pulse duration, we obtain $\mathcal{I} \propto E_1^2E_2\Delta t\cos(\theta_2 - 2\theta_1)$, where Δt is the laser pulse duration. Hence, in the interaction with a bichromatic laser pulse the electron can gain a transverse momentum along the polarization axis [see Eq. (12)], which can be controlled by changing either E_1, E_2 or the relative phase $\theta_2 - 2\theta_1$. In fact, it was shown that for an ultrarelativistic electron the deflection angle with respect to the initial propagation direction is [18]:

$$\zeta \approx \arctan\left(\frac{2\mathcal{I}/mc\rho_0}{1 - \mathcal{I}^2/m^2c^2\rho_0^2}\right) \tag{17}$$

if $|\mathcal{I}|/mc\rho_0 < 1$, $\zeta - \pi$ if $\mathcal{I}/mc\rho_0 < -1$ and $\zeta + \pi$ if $\mathcal{I}/mc\rho_0 > 1$. Note that $\mathcal{I}/mc\rho_0$ and the deflection angle ζ are independent of the initial electron energy because higher initial energies imply larger RR effects. As a result, for $|\mathcal{I}|/mc\rho_0 > 1$ all electrons with arbitrarily high initial energy are back reflected as long as quantum effects do not noticeably alter the predictions of classical electrodynamics. We mention that such controlled deflection is also achievable by employing carrier-envelope-phase (CEP) tunable few-cycle laser pulses, due to the very broad bandwidth and the consequent nonlinear frequency combination analogous to the case of a bichromatic pulse. In this latter case, the deflection angle can be controlled by changing the

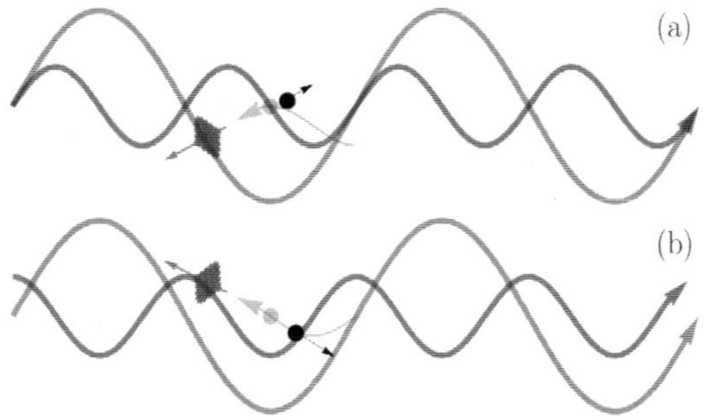

Figure 7. (Color online) Illustation of the emission process in the presence of a bichromatic laser pulse by changing the relative phase between the two frequency components. (a) An electron emits a photon and the recoil reduces its momentum and shifts its position inside the laser field. (b) A change in the relative phase between the two frequency components alters the direction of photon emission. In addition, in (b) the total field is stronger at the electron position because the two frequency components interfere constructively. As a consequence, the radiation reaction force, i.e. the recoil due to the photon emission, is larger in (b) compared to (a). In turn, this results in a larger shift with respect to the laser pulse field.

CEP of the few-cycle pulse [18]. Such ultraintense and CEP-tunable few-cycle laser pulses can be generated, e.g., via the relativistic mirror technique [17]

Physically, the energy and momentum losses associated with the emission of high-frequency radiation modulate the position of the electron within the plane-wave pulse. In the case of a quasimonochromatic plane-wave pulse there are no free parameters to control the effect of the emission and the consequent modulation. On the contrary, if a second higher frequency component is present, its frequency and relative phase can be chosen in such a way that the resulting modulation is resonant with the lower frequency component of the bichromatic field. Hence, the electron emits radiation prevalently along the positive (or negative) direction of the polarization axis, and the effect can be controlled by changing the relative phase between the two frequency components (see Fig. 7 for an illustration).

3.2.2. Simulation results In our simulation we considered a focused bichromatic laser pulse which is 100 fs long between its first and last half maximal intensity. Its total peak intensity is 4.2×10^{21} W/cm^2 with 37% contribution from the second harmonic component, 7 μm waist radius and 1.55 eV energy of the lower frequency component. Initially, the electron bunch is distributed according to a six-dimensional Gaussian probability distribution:

$$f_e(\mathbf{x}, \mathbf{p}) = N_e \frac{e^{-\left[\frac{x^2+y^2}{2\sigma_T^2} + \frac{(z-z_0)^2}{2\sigma_L^2}\right] - \left[\frac{p_x^2+p_y^2}{2\sigma_{pT}^2} + \frac{(p_z-p_{z,0})^2}{2\sigma_{pL}^2}\right]}}{(2\pi)^3 \sigma_T^2 \sigma_{pT}^2 \sigma_L \sigma_{pL}}, \tag{18}$$

with $N_e = 400$ being the total number of electrons and $\sigma_T = 0.2$ μm and $\sigma_L = 0.5$ μm ($\sigma_{pT} = 1$ mc and $\sigma_{pL} = 12$ mc) being the transverse and the longitudinal position (momentum) widths, respectively. For the sake of simplicity, we set the constant phase $\theta_1 = 0$ so the deflection is controlled by $\cos(\theta_2)$ at fixed field amplitudes E_1, E_2.

Figure 8 displays the electron density distribution $n_e(p_z, p_x)$ as a function of the longitudinal p_z and transverse p_x momentum for the interaction of 400 electrons with the focused bichromatic laser pulse both for $\cos(\theta_2) = 0$ and $\cos(\theta_2) = 1$, with and without RR. No appreciable difference

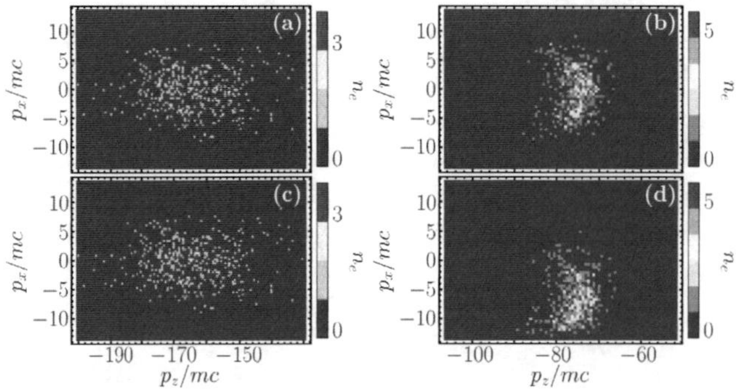

Figure 8. (Color online) Electron density distribution $n_e(p_z, p_x)$ as a function of the longitudinal p_z and transverse p_x momentum after the interaction of 400 electrons with a focused bichromatic laser pulse with 4.2×10^{21} W/cm^2 peak intensity (37% contribution from the second harmonic component) and 7 μm waist radius. Panel (a): $\cos(\theta_2) = 0$ without RR. Panel (b): $\cos(\theta_2) = 0$ with RR. Panel (c): $\cos(\theta_2) = 1$ without RR. Panel (d): $\cos(\theta_2) = 1$ with RR.

between $\cos(\theta_2) = 0$ and $\cos(\theta_2) = 1$ is found if only the Lorentz force is taken into account. Furthermore, if the RR force is neglected, the mean of the momentum distribution remains unaltered after the electron bunch has passed through the laser pulse $\bar{p}_x \approx 0$ and $\bar{p}_z \approx -165$ mc [see Figs. 8(a), 8(c)]. However, if the RR force is taken into account, for $\cos(\theta_2) = 0$ the electrons still move along their initial propagation direction and are distributed symmetrically in the transverse momentum space with $\bar{p}_x \approx 0$ and $\bar{p}_z \approx -76$ mc [see Fig. 8(b)] in good agreement with the plane wave prediction $p_{x,f} \approx 0$ and $p_{z,f} \approx -73$ mc. On the other hand, for $\cos(\theta_2) = 1$ all the electrons are deflected in the transverse direction independently of their initial energy, the mean of the momentum distributions being $\bar{p}_x \approx -6.6$ mc and $\bar{p}_z \approx -76$ mc [see Fig. 8(d)] which implies a deflection angle of about 5°. For the corresponding plane-wave pulse, we obtain $p_{x,f} \approx -5.6$ mc and $p_{z,f} \approx -73$ mc and 4.4° deflection, in fair agreement with the above mentioned focused pulse results.

4. Conclusions

Firstly, we have presented in the paper our results on laser acceleration in vacuum of electrons and protons by chirped pulses, for energies that make them useful for applications. Results from single-particle calculations and many-particle simulations have been discussed. In particular, the work has demonstrated feasibility of the process of vacuum laser acceleration of protons for the tumor therapy, including mean exit energies within the required range of 20 – 580 MeV and spread of about 1%. The calculations have also demonstrated utility of the plane-wave model (in representing the accelerating fields of the pulses employed) for most cases considered. Effect of the particle-particle interactions, when taken into account within the context of a many-particle calculation, on the exit energy spread, has been shown to be small but must be considered seriously in any medical applications.

Secondly, we have discussed the signatures of the radiation reaction during interaction of the relativistic electron beam with a counterpropagating focused short laser pulse. The spectral bandwidth decreases monotonously with increasing laser pulse duration caused by radiation losses. The angular spread of radiation shows different dependence on the pulse duration for short and long laser pulses which results from competition of the radiation reaction effect (in the quantum radiation dominated regime) and the laser field focusing effect. While due to the former the angular spread decreases with increase of the laser pulse duration, due to the latter,

Journal of Physics: Conference Series **594** (2015) 012018 doi:10.1088/1742-6596/594/1/012018

inversely, the angular spread increases with increase of the pulse duration.

Thirdly, we have shown that radiation reaction effects provide a route to the control of the electron dynamics in the interaction of an ultrarelativistic electron bunch colliding head-on with a bichromatic laser pulse. In this case the electron bunch is deflected with respect to the initial propagation direction, and the deflection angle can be tuned by changing the relative amplitude or the relative phase between the two frequency components of the bichromatic field. In addition, the deflection depends on the properties of the laser field only. Indeed, for ultrastrong bichromatic pulses the deflection angle can be so large that all electrons are back-reflected independently of their initial energy as long as quantum corrections are relatively small.

References

[1] A. Di Piazza, C. Müller, K. Z. Hatsagortsyan and C. H. Keitel, Rev. Mod. Phys. **84**, 1177 (2012).
[2] P. Agostini and L. F. Di Mauro, Rep. Prog. Phys. **67**, 813 (2004).
[3] T. Popmintchev, M.-C. Chen, P. Arpin, M. M. Murnane, and H. C. Kapteyn, Nature Photon.**4**, 822 (2010).
[4] F. Krausz, and M. Ivanov, Rev. Mod. Phys.**81**, 163 (2009).
[5] N. D. Powers, I. Ghebregziabher, G. Golovin, C. Liu, S. Chen, S. Banerjee, J. Zhang, and D. P. Umstadter, Nature Photon. **8**, 28 (2014).
[6] E. Esarey, C. B. Schroeder, and W. P. Leemans, Rev. Mod. Phys. **81**, 1229 (2009).
[7] B. M. Hegelich, B. J. Albright, J. Cobble, K. Flippo, S. Letzring, M. Paffett, H. Ruhl, J. Schreiber, R. K. Schulze, and J. C. Fernandez, Nature **439**, 441 (2006).
[8] Y. Salamin, Z. Harman, and C. H. Keitel, Phys. Rev. Lett. **100**, 155004 (2008).
[9] Z. Harman, Y. Salamin, B. Galow, and C. H. Keitel, Phys. Rev. A **84**, 053814 (2011)
[10] B. Galow, Y. Salamin, T. V. Liseykina, Z. Harman, and C. H. Keitel, Phys. Rev. Lett. **107**, 185002 (2011).
[11] Y. Salamin, J. Li, B. Galow, Z. Harman, and C. H. Keitel, Phys. Rev. A **85**, 063831 (2012).
[12] J. Li, Y. Salamin, B. Galow, and C. H. Keitel, Phys. Rev. A **85**, 063832 (2012).
[13] Y. I. Salamin, Phys. Lett. A **376**, 2442 (2012).
[14] Y. I. Salamin, Appl. Phys. B **86**, 319 (2007).
[15] Y. I. Salamin and N. M. Jisrawi, J. Phys. B **47**, 025601 (2014).
[16] Jian-Xing Li, K. Z. Hatsagortsyan, and C. H. Keitel, Phys. Rev. Lett. **113**, 044801 (2014).
[17] M. Tamburini, A. Di Piazza, T.V. Liseykina, and C.H. Keitel, Phys. Rev. Lett. **113**, 025005 (2014).
[18] M. Tamburini, C. H. Keitel, and A. Di Piazza, Phys. Rev. E **89**, 021201(R) (2014).
[19] B. J. Galow, J.-X. Li, Y. I. Salamin, Z. Harman, and C. H. Keitel, Phys. Rev. ST-AB **16**, 081302 (2013).
[20] Y. I. salamin and N. M. Jisrawi, J. Phys. B **47**, 025601 (2014).
[21] N. M. Jisrawi, B. J. Galow, and Y. I. Salamin, Laser Part. Beams **32**, 671 (2014).
[22] M. Lax, W. H. Louisell, and W. B. McKnight, Phys. Rev. A **11**, 1365 (1975).
[23] A. Di Piazza, K.Z. Hatsagortsyan, and C.H. Keitel, Phys. Rev. Lett. **105**, 220403 (2010).
[24] A. G. R. Thomas, C. P. Ridgers, S. S. Bulanov, B. J. Griffin, and S. P. D. Mangles, Phys. Rev. X **2**, 041004 (2012).
[25] V. I. Ritus, J. Sov. Laser Res. **6**, 497 (1985).
[26] I. V. Sokolov, N. M. Naumova, J. A. Nees, and G. A. Mourou, Phys. Rev. Lett. **105**, 195005 (2010).
[27] I. V. Sokolov, J. A. Nees, V. P. Yanovsky, N. M. Naumova, and G. A. Mourou, Phys. Rev. E **81**, 036412 (2010).
[28] L. D. Landau, E. M. Lifshitz, *The classical theory of fields*, 3rd ed. (Pergamon Press, Oxford, 1971), Chaps. 76 and 78.
[29] M. Tamburini, F. Pegoraro, A. Di Piazza, C. H. Keitel, and A. Macchi, New J. Phys. **12**, 123005 (2010).
[30] A. Di Piazza, Lett. Math. Phys. **83**, 305 (2008).

Journal of Physics: Conference Series **594** (2015) 012019 doi:10.1088/1742-6596/594/1/012019

Tunneling time, what is its meaning?

C R McDonald, G Orlando, G Vampa and T Brabec
Department of Physics, University of Ottawa, Ottawa, ON K1N 6N5, Canada
E-mail: cmcdo059@uottawa.ca

Abstract. The tunnel time ionization dynamics for bound systems in laser fields are investigated. Numerical analysis for a step function switch-on of the field allows for the tunnel time to be defined as the time it takes the ground state to develop the under-barrier wavefunction components necessary to achieve the static field ionization rate. A relation between the tunnel time and the Keldysh time is established. The definition of the tunnel time is extended to time varying fields and experimental possibilities for measuring the tunnel time are discussed.

1. Introduction

With the development of quantum mechanics in the first few decades of the 20th century came many phenomena that possessed no classical analogue. The most ubiquitous of these is the phenomenon of tunneling [1] — the ability for a particle to surmount a potential barrier that it would otherwise be prevented from overcoming classically. Tunneling is now described in most introductory quantum mechanics texts [2, 3] and is relevant in many areas of pure and applied science, such as tunnel diodes [4], scanning tunneling microscopy [5] and, in the last few decades, attosecond science [6–9].

In his seminal work, Keldysh developed a dynamical theory of optical field ionization of atoms for a time varying laser field [10]. Ionization was characterized as either multiphoton or tunnel ionization. The regime to which ionization belonged was determined by the Keldysh parameter γ — multiphoton ionizaton for $\gamma \gg 1$ and tunnel ionization for $\gamma \ll 1$. The Keldysh parameter can be defined as the ratio of the Keldysh tunnel time $\tau_k = l/|v|$ and the laser oscillation period $T_0 = 2\pi/\omega$ for a laser with frequency ω. Here the tunnel time can be interpreted as the the time it takes an electron with velocity $v = i\sqrt{2E_0}$ to cross a barrier of width $l = E_0/F$, where E_0 is the electron binding energy and F is the laser electric field. The velocity v is imaginary due to the fact the electron in classically forbidden from travel underneath the barrier. Nonetheless, τ_k displays the characteristics that are expected of a tunnel time. When the laser field changes too fast for the electron to escape under the barrier, tunneling is restricted.

Recent attosecond angular streaking experiments were done to measure the tunnel response time [11]. The measured time was much shorter than those predicted by most theoretical definitions, such as the Büttiker-Landauer or Keldysh times. These experiments raise the question as to the meaning of the tunnel time for ionization from bound systems. So far a more detailed theory of tunnel time for bound electrons is needed to clarify the meaning of the tunnel time.

To answer this question we examine 1D quantum systems subjected to an instantaneous switch-on of a static electric field. When the field is switched on, the ground state transforms into a dynamic resonance state decaying with a time-dependent ionization rate $w(t)$. When the

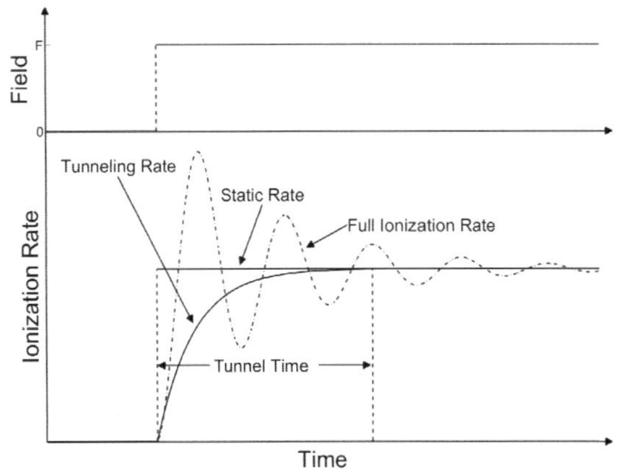

Figure 1. Schematic of the tunneling process for a step function switch-on. The dotted line represents the full ionization rate $w(t)$. The filtered ionization rate (labeled "Tunneling Rate") is represented by the solid curve. The constant line is the static ionization rate $w(F)$.

wavefunction in and around the bound region approaches the static resonance, determined by the solution of the time-independent Schrödinger equation for constant field F, the ionization rate takes on a constant value $w(F)$.

For an unambiguous definition of the tunnel time of bound systems two obstacles must be overcome. First, tunnel ionization is usually measured by the outgoing current at some reference point. However, when this is done the temporal evolution of the ionization rate depends on the reference point. Second, the tunneling signal from the ground state is masked by a contribution due to the sudden switch-on resulting in population and ionization of excited states. The first obstacle can be over come by using a time-dependent generalization of Fano-resonance theory. The ionization rate is determined by the imaginary part of the resonance making this method reference point free. The second obstacle is overcome by pairing this generalized Fano approach with a spectral filtering method that removes the contributions from the excited states. The tunnel time can then be defined as the response time it takes the part of the wavefunction tunneling directly from the ground state to reach the static ionization rate. A schematic of this process is depicted in Fig. 1.

2. Solution of the time-dependent Schrödinger equation

Consider a 1D quantum system initially in its ground state. At time $t = 0$, a static field of strength F is switched on instantaneously. The Hamiltonian in dipole approximation and length gauge is given by,

$$H = \begin{cases} H_0 & \text{for } t < 0 \\ H_0 - Fx & \text{for } t \geq 0 \end{cases} \tag{1}$$

where $H_0 = T + V$ with T the kinetic energy and V the potential. Unless otherwise indicated, atomic units are used throughout. We then express the wavefunction as,

$$|\Psi(t)\rangle = \mathrm{a}(t)|0\rangle + |\phi(t)\rangle \tag{2}$$

where $|0\rangle$ is the unperturbed ground state and $\mathrm{a(t)} = \langle 0|\Psi(\mathrm{t})\rangle$. The last term in Eq. (2) is the part of the wavefunction that develops in response to the external field for which $\langle 0|\phi(t)\rangle = 0$ is enforced; it is zero in the absence of the external field. Inserting Eq. (2) into the time-dependent Schrödinger equation (TDSE) with the Hamiltonian given in Eq. (1) and projecting $\langle 0|$, yields an equation of motion for $\mathrm{a}(t)$,

$$i\dot{\mathrm{a}}(t) = E_0\mathrm{a}(\mathrm{t}) + \langle 0|H_0 - Fx|\phi(t)\rangle, \tag{3}$$

where $E_0 = \langle 0|H|0\rangle$. Choosing the *ansatz* $|\phi(t)\rangle = \mathrm{a}(t)|\varphi(t)\rangle$ we arrive at the expression

$$\Delta E(t) = i\frac{\dot{\mathrm{a}}(t)}{\mathrm{a}(t)} - E_0 \tag{4}$$

where $\Delta E(t) \equiv \langle 0|H|\varphi(t)\rangle$. The rate at which probability leaves the ground state and escapes into the continuum is given by $\mathrm{w}(t) = -2\mathrm{Im}\{\Delta E(t)\}$. To completely determine $|\Psi\rangle$, a corresponding equation for the time evolution of $\phi(t)$ can be obtained by projecting $1 - |0\rangle\langle 0|$ on the TDSE. The resulting equation is not given, as it is not used here. Eq. (4) is the time-dependent generalization of Fano resonance theory [12]. By using the time-independent instead of the time-dependent Schrödinger equation in our derivation, the regular Fano resonance theory would be obtained yielding the static tunneling rate for ΔE.

The above equations were developed for diagnostic purposes. Time propagation is handled by expanding the total wavefunction on the eigenstates of the field-dressed Hamiltonian. This spectral method is more advantageous as it will allow for the separation of the tunneling contribution from the rest of the wavefunction. For more details about the expansion see Ref. [13].

3. Results for the instantaneous switch-on

We begin by solving the TDSE for the 1D quantum dot (QD) potential

$$V(x) = \begin{cases} \frac{1}{2}\alpha^2x^2 - \beta & \text{for } |x| < \sqrt{2\beta}/\alpha \\ 0 & \text{otherwise} \end{cases} \tag{5}$$

with $\alpha = 1/2$ and $\beta = 3/4$. This system contains only the ground state and a first excited state at the continuum threshold. Initially the system is in the state $|0\rangle$; at $t = 0$ the static field is switched on.

Fig. 2 shows the spectral density as a function of energy for the system at $F = 0.08$. We display the results for this field strength as they are representative of the results for all field strengths investigated. Note that the laser field turns discrete bound states into mixed bound-continuum resonance states with a finite spectral width. The spectrum has a minimum at around an energy of -0.35, which separates two distinct features of the step response: (i) The sharp peak to the left represents tunnel ionization from the ground state directly. (ii) The broad peak to the right corresponds to photon absorption and ionization via the excited state resonance — due to the infinite bandwidth of the step function, photon absorption is dominated by single photon transitions. Both channels (i) and (ii) coexist in near infrared laser pulses as well. However, as a single photon cannot reach excited states or the continuum directly, channel (ii) occurs via multi-photon transitions. As long as the minimum is outside the spectral widths of both features, tunneling and excited state channels can be separated by multiplying a spectral filter on the probability amplitudes; see the (red) dashed line in Fig. 2. The choice of the filter function is discussed in the supplemental material of Ref. [13].

The tunnel ionization peak in Fig. 2 reveals the spectral signature of the tunnel ionization dynamics. An exponential decay in time corresponds to a Lorentzian spectrum, the width of which determines the tunnel ionization rate. Our spectrum is Lorentzian with a small

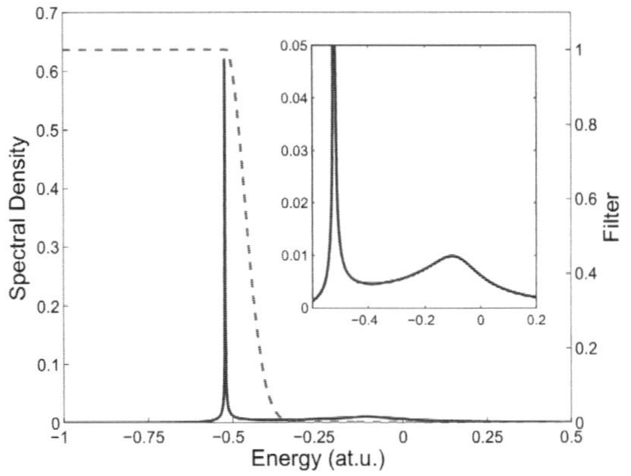

Figure 2. Spectral density (solid blue line) for the 1D quantum dot potential at $F = 0.08$. The spectral filter is shown by the red dashed line. A magnified section of the spectral density is shown in the inset.

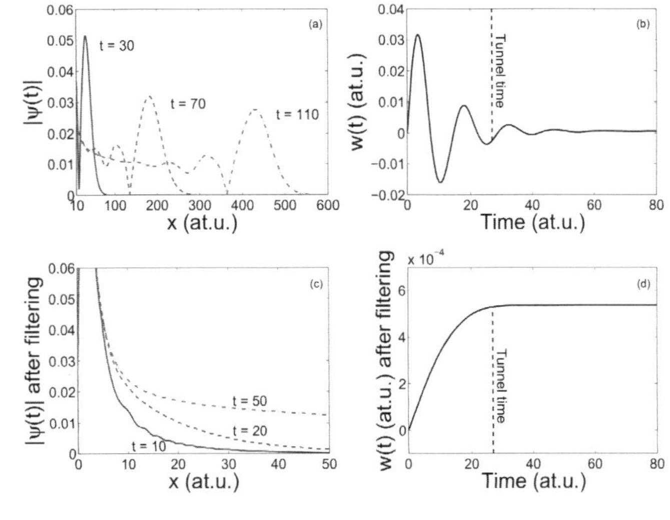

Figure 3. (a) Magnitude of $\Psi(x)$ for $t = 30$ (solid blue), $t = 70$ (dash green) and $t = 110$ (dash-dot red); to illustrate the outgoing wave, the bound region is not shown. (b) Ionization rate $\mathrm{w}(t)$ corresponding to (a). (c) Magnitude of the filtered wavefunction (ground state resonance only) for $t = 10$ (solid blue), $t = 20$ (dash green) and $t = 50$ (dash-dot red); filter is given in Fig. 1. (d) $\mathrm{w}(t)$ corresponding to (c). 1D quantum dot potential and $F = 0.08$ for (a)-(d).

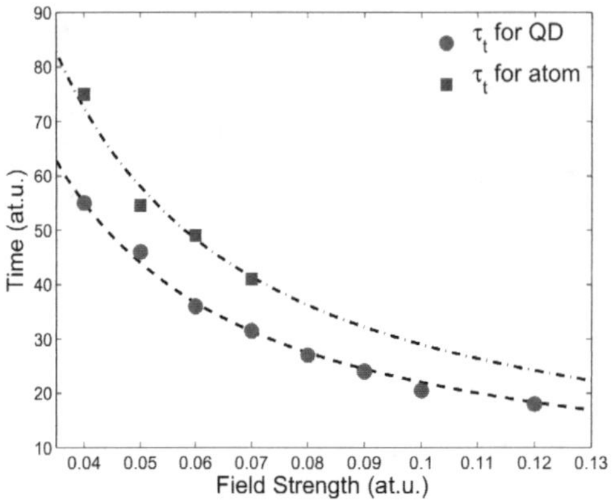

Figure 4. Calculated $\tau_{\rm t}$ for the QD (●) and atomic (■) potentials. The lines in the plot represent the Keldysh time multiplied by a constant to match $\tau_{\rm t}$.

asymmetric distortion and a phase, which are responsible for the switch-on of the tunneling process. As the functional form of the spectral distortion is unknown, extraction of the tunneling time is easier in the time domain.

Figure 3 shows how the system responds to the instantaneous switch-on of the field for both the total (unfiltered) and filtered wavefunction. Figure 3(a) shows that for the total wavefunction an outward propagating wave is created from the excited state ionization channel. In its wake this wave leaves a steady flow of probability from the ground state into the continuum representing the static resonance with constant tunneling rate. The corresponding ionization rate in Fig. 3(b) shows strong oscillations that mask the tunneling signal. Eventually the oscillations decrease and the ionization rate reaches the static rate.

When the filter is applied, the initial outgoing wave ceases to exist and the wavefunction monotonically rises towards a steady-state; see Fig. 3(c). The corresponding ionization rate in Fig. 3(d) shows the resulting tunneling rate rise monotonically and approaches the steady state value. It is this rise time that we define as the electron tunneling time $\tau_{\rm t}$. For times much shorter than the tunneling time, the wavefunction components under the barrier that are necessary for tunneling cannot be developed and tunnel ionization does not occur. For all investigated field strengths, the tunnel time is considerably shorter than the time it takes for the excited state ionization signal to subside. Therefore, filtering is indispensable for extracting the tunnel time.

The tunneling time $\tau_{\rm t}$ is calculated by using $w(t)$ after filtering and determining when the ionization is within 1% of the static value. Results for the QD potential and for a 1D softened Coulomb potential $V(x) = 1/\sqrt{x^2+b^2}$ are shown in Fig. 4. The shielding parameter b is chosen so that the ionization potential is ≈ 0.5 (the same as the QD potential). The spectral features for the Coulomb potential are similar to those of the QD except that the two peaks merge at lower field strength — around $F = 0.07$ — than for the QD. The results of our calculations show that the tunnel time scales with the Keldysh time as $\tau_{\rm t} \approx 2.2\tau_k$ for the QD and as $\tau_{\rm t} \approx 2.9\tau_k$ for the Coulomb potential. In addition to this scaling, it has been shown that the Keldysh time is in fact a lower limit for the tunneling time [14].

4. Sinusoidal switch-on

Up to this point the tunnel time has been defined for a step-function response. In this section the definition for the tunnel time will be generalized to time varying fields. To define the duration of any physical process, it is necessary to have an initial and final reference point. So far we have used the ground state and the static resonance between which the tunnel time can be measured. Tunneling cannot manifest for $t \ll \tau_t$ and maximum tunneling with the static ionization rate is reached for $t \gtrsim \tau_t$.

To extend the definition of τ_t to time varying fields, consider an initially bound electron interacting with a monochromatic, linearly polarized laser field $F = F_0 \sin(\omega t)$ where F_0 is the peak strength and ω the frequency.

To understand what happens in a sinusoidal laser field, first decompose it into small step functions. The field changes from F to $F + \Delta F$ and remains constant over a time Δt; the step size $\Delta F \propto \omega \Delta t$. In the tunneling regime, $\gamma \lesssim 1$, the initial state at F is no longer the unperturbed ground state, but a state that has already developed tunneling components. The dynamics that takes place during Δt is the wavefunction attempting to develop components to realize the static resonance state for $F + \Delta F$. The time needed for realizing the static resonance for one step can be viewed as a local tunnel response time that will vary throughout the laser cycle. However, a meaningful tunnel time cannot be defined, as both initial and final states are tunnel ionizing.

For a sinusoidal field with peak value F_0 and period T_0, a natural way to generalize the tunnel time is to use a quarter cycle; this resembles a softened switch-on function. In contrast to the step function the time of measurement is not free but determined by $T_0/4$; it can only be varied by changing ω. Initial and final reference states are taken at the field minimum and its subsequent maximum. This applies to linearly as well as elliptically polarized fields. It is assumed that the initial state at the field minimum is close to the ground state where tunnel ionization is negligible compared to the field maximum. As a result, the tunnel time is determined by the quarter oscillation period for which the ground state transforms into the static resonance for F_0.

5. Considerations for experimental measurement of the tunnel time

From our analysis it is clear that experiments based on the notion of a tunnel time required for a particle to cross the barrier with fail. There are no time dynamics visible in the wavefunction that indicate such a behaviour. Instead, the tunnel time must be seen as the time required to establish a tunnel through the potential by which the particle can escape. This is achieved by the wavefunction developing the under-the-barrier components of the (quasi-static) resonance.

For a sinusoidal laser field the tunnel time cannot be measured directly because the field changes. However, it is possible to measure it indirectly. In the tunneling limit, for laser frequencies for which $\gamma \leq 1$, tunnel ionization is fully established and will be maximum at and around the peak of an optical half cycle. In the transition regime, tunnel ionization at the peak will decrease as a function of increasing laser frequency. In the multi-photon regime, tunnel ionization is expected to be negligible. As a result, by measuring tunnel ionzation at the peak of the laser half-cycle as a function of laser frequency ω, we expect the tunneling yield to increase with decreasing frequency until it saturates when $\gamma \leq 1$ is reached.

Experimentally, methods have already been developed that can measure tunnel ionization with sub-cycle resolution [11, 15]. Ionization is achieved with nearly circularly polarized laser pulses and the momentum of the ionized wavepacket is recorded. A unique relationship exists between the ionization time and the final momentum of the wavepacket. Wavepackets created at different times along an optical cycle carry different final momenta. As a result, there exists a relation between the time at which an electron wavepacket was born and its final momentum recorded in the momentum spectrum. The ionization dynamics can then be recovered with sub-cycle accuracy from the momentum spectrum.

These experiments measure the *tunnel delay time* — the time difference between the ionization maximum and the peak of the optical half cycle. The presence of a non-zero delay time indicates that tunneling does not have sufficient time to develop over a quarter cycle and

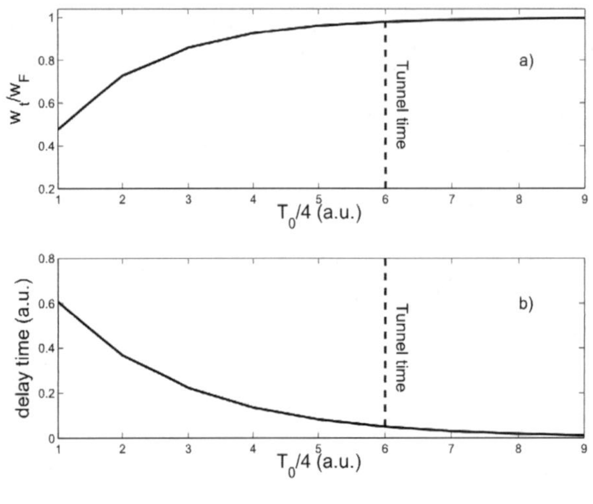

Figure 5. Schematic of tunnel time measurement. (a) Ratio w_t/w_F as a function of a quarter laser period. w_t is the tunneling rate at the half cycle peak; the field F at the half cycle peak being kept constant; w_F is the tunneling rate of the static resonance. (b) The tunnel delay time (difference between half cycle peak and maximum ionization rate) as a function of a quarter laser period. The dashed line in (a) and (b) represents the time at which the quarter cycle period is equal to the tunnel time. At this time the ionization rate reaches a chosen fraction of the static rate and correspondingly the the tunnel delay time approaches zero within some prescribed precision.

that the tunneling time is longer than the quarter cycle. Thus, the delay time obtained from experiment can be viewed as a measure of how far the wavefunction is from establishing the static resonance eigenstate at the half cycle peak. To extract the tunnel time from these experiments that is in agreement with the Keldysh tunnel time, the ionization at the half-cycle needs to measured as a function of ω. The ionization rate saturates to a frequency independent value for $\omega < 1/\tau_t$ and correspondingly the tunnel delay time converges to zero. The tunnel time can be experimentally determined as $\tau_t^{\exp} = 1/\omega_t$ with ω_t being the smallest frequency at which the ionization rate saturates and the tunnel delay time drops to zero (within some prescribed accuracy). A schematic of this idea is depicted in Fig. 5.

6. Conclusion

Nearly five decades ago Keldysh published his seminal work treating the ionization of atoms in strong laser fields. His work presents the basis on which the field of intense laser physics has been constructed. The advent of attosecond science has opened the opportunity to test and explore more deeply some of the predictions of Keldysh's original work. This has rekindled the interest in understanding the Keldysh tunnel time, and tunnel times in general, which there has remained some discrepancy.

Here we have presented a definition of the tunnel time for bound systems. In the case of an instantaneous switch-on of a field with strength F, the tunnel time is the response time for the wavefunction to develop from the ground state to the static resonance for the field value F. We presented a numerical method that allows us to remove the contributions from excited states and determine the tunnel time. It was shown that the calculated tunnel times as a function of F scale with the Keldysh time. This reveals the connection between our definition of tunnel time and that presented by Keldysh nearly five decades ago. Finally, we have extended our definition to time varying fields and discussed how this can be applied to the experimental observation of the tunnel time.

References

[1] Merzbacher E 2002 *Phys. Today* **55** 44
[2] Landau L D and Lifshitz E M 1977 *Quantum Mechanics* (Oxford: Pergamon)
[3] Griffiths D J 1995 *Introduction to Quantum Mechanics* (Englewood Cliffs, NJ: Prentice-Hall)
[4] Esaki L 1958 *Phys. Rev.* **109** 603
[5] Binning G, Gerber C and Weibel E 1982 *Phys. Rev.* **49** 57
[6] Tong X M, Zhao Z X and Lin C D 2003 *Phys. Rev. Lett.* **91** 233203
[7] Baker S, Robinson J S, Haworth C A, Teng T, Smith R A, Chirilă C C, Lein M, Tisch J W G and Marangos J P 2006 *Science* **312**, 424
[8] Meckel M *et al.* 2008 *Science* **320** 1478
[9] Wörner H J, Bertrand J B, Kartashov D V, Corkum P B and Villeneuve D M 2010 *Nature* **466** 604
[10] Keldysh L V 1965 *Sov. Phys. JETP 20* 1307
[11] Eckle P, Pfeiffer A N, Cirelli C, Staudte A, Dörner R, Muller H G, Büttiker M and Keller U 2008 *Science* **322** 1525
[12] Fano U 1961 *Phys. Rev.* **124** 1866
[13] McDonald C R, Orlando G, Vampa G and Brabec T 2013 *Phys. Rev. Lett* **111** 090405
[14] Orlando G, McDonald C R, Protik N H and Brabec T 2014 *Phys. Rev. A* **89** 014102
[15] Shafir D, Soifer H, Bruner D B, Dagan M, Mairesse Y, Patchkovskii S, Ivanov M Y, Smirnova O and Dudovich N 2012 *Nature* **485** 343

Analytical study of residual-current excitation during gas ionization by two-color laser pulse

A. A. Silaev, V. A. Kostin, I. D. Laryushin, and N. V. Vvedenskii

Institute of Applied Physics, Russian Academy of Sciences, Nizhny Novgorod 603950, Russia

University of Nizhny Novgorod, Nizhny Novgorod 603950, Russia

E-mail: silaev@appl.sci-nnov.ru, vk1@appl.sci-nnov.ru, vved@appl.sci-nnov.ru

Abstract. This work is focused on the analytical investigation of the excitation of low-frequency residual current during gas ionization by two-color laser pulses. Such excitation is of great interest since it may be used to convert e ciently laser pulses into low-frequency radiation, in particular, into radiation of the terahertz frequency band. We derive analytical expressions for the residual current density for the cases when two-color laser pulse contains the main strong fundamental-frequency field and additional frequency-tunable field with the frequency close to the double or half frequency of the main field. Derived formulas are used to analyze and compare main features of residual-current excitation in the cases when additional field is second-or half-harmonic of the main field.

1. Introduction

At present, a great attention is given to the terahertz (THz) generation methods based on the use of laser-induced plasmas. These methods are usually realized under the ionization of different gases, including the ambient air, by femtosecond laser pulses. Great interest in these methods is due to the capabilities of obtaining coherent, intense, broadband, and highly directional THz pulses (see, e.g., review [1]). The THz spectrum usually contains a low-frequency core in the range on few THz (typically, 1–3 THz), where the main energy of the THz pulse is concentrated, and a long high-frequency tail, which can spread to tens of THz [2–7]. The latter can be attributed to the currents varying rapidly under the laser pulse action. The main contribution to the low-frequency THz radiation is made by the plasma currents in the long wakefield of the laser pulse. The amplitude of the low-frequency currents is determined by the free-electron residual current density (RCD) [8–15], which is equal to the zero-frequency spectrum of the time derivative of the current density excited during ionization.

The most popular and efficient laser-plasma schemes of THz generation are the so-called two-color schemes, in which the laser pulse at the fundamental frequency is supplemented by its second harmonic obtained using a nonlinear crystal [2–4, 16–18]. In the recent works [19, 20] THz radiation was generated in the inverted scheme, where two-color femtosecond laser pulse contain, besides the fundamental-frequency main field, a weak additional field tunable near the frequency of the half-harmonic. In order to explain the experimental results, the analytical approach to RCD calculation was developed. The approach is based on calculation of the ionization-induced electron-density harmonics at the frequencies of the main and additional fields [19].

In this work we apply the developed analytical approach to derive formulas for the time derivative of low-frequency current and residual current density excited during gas ionization by two-color laser pulses containing the strong fundamental-frequency main field and weak frequency-tunable second-harmonic field ($\omega + 2\omega$ case). For this case, some analytical expressions were derived in [21, 22]. However in the present paper, we for the first time take into account the finiteness of the laser pulse duration, and consider the laser pulses with arbitrary angles between the main and additional field vectors and nonzero frequency detuning of the additional field. The obtained formulas are used to determine the main differences of RCD excitation in commonly used $\omega + 2\omega$ case and inverted $\omega + \omega/2$ case. It is also shown that the frequency detuning of the additional field from the doubled or halved fundamental frequency results in upshift of the low-frequency current spectrum.

2. Analytical model

In order to calculate the density $N(t)$ of free electrons and their current density $\mathbf{j}(t)$, we use the *semiclassical* approach, which is valid in the tunneling regime of ionization [11, 12]. This approach consists in the solution of the following equations [4, 7–12, 23]:

$$\frac{\partial \mathbf{j}}{\partial t} + \nu \mathbf{j} = \frac{e^2}{m} N\mathbf{E}, \quad \frac{\partial N}{\partial t} = (N_g - N)w(E). \tag{1}$$

Here, N_g is the undisturbed gas density, ν is the electron collision frequency, e and m are the electron charge and mass, respectively, and $w(E)$ is the tunneling ionization probability per unit time. We assume here for simplicity that ionization degree and the collision frequency are small enough, $N \ll N_g$ and $\nu \ll \tau_p^{-1}$, where τ_p is the laser pulse duration. Under these assumptions, the RCD $\mathbf{j}_{\mathrm{RCD}}$ can be found from the solution of equations (1) for $\nu = 0$ and $t \to \infty$,

$$\mathbf{j}_{\mathrm{RCD}} = \frac{e^2}{m} \int_{-\infty}^{\infty} N\mathbf{E}\, dt, \quad N(t) = N_g \int_{-\infty}^{t} w\, dt. \tag{2}$$

In what follows, we use equations (2) to find approximate analytical formulas for RCD by using the approach based on calculation of the harmonics of N at the frequencies of the main and additional fields.

2.1. Strong fundamental and weak second-harmonic field

In this section, we consider that the two-color laser pulse is a superposition of the strong field on the fundamental frequency ω_1 and weak additional second-harmonic field on the frequency $\omega_2 = 2\omega_1 + \Delta\omega$ and is given as

$$\mathbf{E}(t) = \mathbf{E}_1(t)\cos(\omega_1 t) + \mathbf{E}_2(t)\cos(\omega_2 t + \varphi)\,. \tag{3}$$

Here, φ is the constant part of phase shift between carriers of fundamental and additional field, $\Delta\omega$ is the small detuning of the second-harmonic frequency, $\mathbf{E}_1(t) = E_1(t)\hat{\mathbf{x}}$ and $\mathbf{E}_2(t) = E_2(t)(\cos\theta\,\hat{\mathbf{x}} + \sin\theta\,\hat{\mathbf{y}})$ are the slowly time-dependent amplitudes of the fundamental- and second-harmonic fields, respectively, θ is the angle between the fundamental- and second-harmonic field vectors, $\hat{\mathbf{x}}$ and $\hat{\mathbf{y}}$ are unit vectors along the x-axis and y-axis respectively.

For small enough frequency detuning, $\Delta\omega \ll \omega_1$, the spectrum of free-electron density is localized near the multiples of the fundamental frequency ω_1. The change of $\Delta\omega$ results in shifts of peaks in this spectrum. Thus, one may present free-electron density as

$$N(t) = [N_1(t)e^{i\omega_1 t} + N_2(t)e^{2i\omega_1 t} + \text{c.c.}] + \delta N(t). \tag{4}$$

Here, $N_1(t)$ and $N_2(t)$ are the complex amplitudes (slow envelopes) of the density oscillations at the fundamental and second-harmonic frequencies, $\delta N(t)$ is the rest part of $N(t)$ which includes low-frequency part located near the zero frequency, and high-frequency part localized near the frequencies $k\omega_1$ with integer k. In what follows we are interested only in fundamental and second-harmonics of $N(t)$ since mixing of these harmonics with the laser field leads to the generation of the low-frequency current density. Assuming that the characteristic time τ_i of density rise is much greater than the fundamental period $T = 2\pi/\omega_1$, we can represent the RCD as

$$\mathbf{j}_{\text{RCD}} \approx \frac{e^2}{m}\int_{-\infty}^{\infty} \overline{N\mathbf{E}}\,dt, \tag{5}$$

where the slowly time-dependent period-averaged product of N and $\mathbf{E}$ is

$$\overline{N\mathbf{E}} = \text{Re}\left[N_1(t)\mathbf{E}_1(t) + N_2(t)\mathbf{E}_2(t)e^{-i\varphi-i\Delta\omega t}\right]. \tag{6}$$

In order to find approximate expressions for $N_1(t)$ and $N_2(t)$, one can use the method presented in [19]. In this method the density harmonics are calculated under the assumption that the pulse envelope is constant and the frequency detuning is zero. We assume that $w(E)$ is a sharp function, i.e., $Ew'(E)/w(E) \gg 1$, where the prime denotes a derivative with respect to the argument. In this case, on the time interval T, the function $w(E(t))$ has two sharp maxima which are located near the maxima of $|\cos(\omega_1 t)|$ (for small enough E_2). This allows one to use Laplace's method and find that

$$N_1(t) \approx 2N_g \frac{\overline{w}(E_1)}{\omega_1}\frac{(\mathbf{E}_1, \mathbf{E}_2)}{E_1^2}\sin(\varphi + \Delta\omega t), \tag{7}$$

$$N_2(t) \approx -\frac{iN_g}{2}\frac{\overline{w}(E_1)}{\omega_1}, \tag{8}$$

where $\overline{w}(E_1) = [2w^3(E_1)/\pi E_1 w'(E_1)]^{1/2}$ is the average ionization rate for sharp function $w(E)$. Using (7) and (8), we obtain the expression for the low-frequency current $\bar{\mathbf{j}}$,

$$\frac{\partial\bar{\mathbf{j}}}{\partial t} \approx \frac{1}{2}j_{\text{osc}}f_1(t)\overline{w}(E_1)\sin(\varphi + \Delta\omega t)\left[4\frac{(\mathbf{E}_1, \mathbf{E}_2)}{E_1^3}\mathbf{E}_1 - \frac{\mathbf{E}_2}{E_1}\right]. \tag{9}$$

Here, $j_{\rm osc} = e^2 N_g E_{01}/m\omega_1$ is the maximum oscillatory current density in the main field, where E_{01} is the maximum fundamental-harmonic amplitude, $f_1(t)$ is is the slowly varying envelope of the fundamental field, i.e., $E_1(t) = E_{01} f_1(t)$.

Taking into account that $\overline{w}(E_1(t))$ is a sharp function of t near its maximum with time width of τ_i, we can write that $\overline{w}(E_1(t)) \approx \overline{w}(E_{01})\exp(-2t^2/\tau_i^2)$, where $\tau_i = 2[d^2E_1/dt^2|_{t=0}\overline{w}'(E_{01})/\overline{w}(E_{01})]^{-1/2}$. Since $\tau_i \ll \tau_p$, the calculation of the integral (9) gives

$$\dot{\mathbf{j}}_{\rm RCD} \approx \frac{\sqrt{2\pi}}{4} j_{\rm osc}\overline{w}(E_{01})\tau_i e^{-(\tau_i\Delta\omega)^2/8}\frac{E_{02}}{E_{01}}\mathbf{\Phi}(\varphi,\theta), \tag{10}$$

$$\mathbf{\Phi}(\varphi,\theta) = \sin\varphi\left[3\cos\theta\,\hat{\mathbf{x}} - \sin\theta\hat{\mathbf{y}}\right], \tag{11}$$

where E_{02} is the maximum second-harmonic amplitude.

2.2. *Strong fundamental and weak half-harmonic field*

Let us consider now the case when the field on the higher frequency ω_2 has stronger intensity [19]. In this case the field with the frequency $\omega_2 = 2\omega_1 + \Delta\omega$ plays the role of the main field, and the field with the frequency ω_1 is the additional half-harmonic field. The electric field $\mathbf{E}(t)$ is specified as follows:

$$\mathbf{E}(t) = \mathbf{E}_1(t)\cos(\omega_1 t + \varphi) + \mathbf{E}_2(t)\cos(\omega_2 t). \tag{12}$$

We assume that the fundamental field is polarized along the x axis, $\mathbf{E}_2(t) = E_2(t)\hat{\mathbf{x}}$ and the half-harmonic field is $\mathbf{E}_1(t) = E_1(t)(\cos\theta\,\hat{\mathbf{x}} + \sin\theta\,\hat{\mathbf{y}})$, where θ is the angle between the fundamental- and half-harmonic fields. In this case the free-electron density can be written as

$$N(t) = [N_1(t)e^{i\omega_2/2t} + N_2(t)e^{i\omega_2 t} + {\rm c.c.}] + \delta N(t). \tag{13}$$

Here, $N_1(t)$ and $N_2(t)$ are the complex amplitudes of the density oscillations at the half-harmonic and fundamental-harmonic frequencies and $\delta N(t)$ is the rest part of $N(t)$. The excitation of low-frequency current density corresponds to mixing the laser field and the fundamental- and half-harmonics of $N(t)$. The low-frequency part of pruduct $N(t)\mathbf{E}(t)$ is

$$\overline{N\mathbf{E}} = {\rm Re}\left[N_1(t)\mathbf{E}_1(t)e^{-i\varphi+i\Delta\omega t/2} + N_2(t)\mathbf{E}_2(t)\right]. \tag{14}$$

The envelopes $N_1(t)$ and $N_2(t)$ are [19]

$$N_1(t) \approx -iN_g\frac{\overline{w}(E_2)}{\omega_2}\frac{(\mathbf{E}_{\omega_1},\mathbf{E}_{\omega_2})}{E_{\omega_2}}e^{-i\varphi+i\Delta\omega t/2}, \tag{15}$$

$${\rm Re}[N_2(t)] \approx \frac{N_g}{4}\frac{\overline{w}(E_2)}{\omega_2}\frac{(\mathbf{E}_{\omega_1},\mathbf{E}_{\omega_2})^2}{E_{\omega_2}^3}\sin(2\varphi - \Delta\omega t), \tag{16}$$

and the time derivative of low-frequency current density is

$$\frac{\partial\overline{\mathbf{j}}}{\partial t} = j_{\rm osc}f_2(t)\frac{\overline{w}'(E_2)}{4}\sin(2\varphi - \Delta\omega t)\left[\mathbf{E}_2\frac{(\mathbf{E}_1,\mathbf{E}_2)^2}{E_2^4} - 4\mathbf{E}_1\frac{(\mathbf{E}_1,\mathbf{E}_2)}{E_2^2}\right], \tag{17}$$

where $j_{\rm osc} = e^2 N_g E_{02}/m\omega_2$ is the maximum oscillatory current density in the field of fundamental harmonic, and $f_2(t)$ is the envelope of fundamental field, $E_2(t) = E_{02} f_2(t)$. The residual current density is

$$\mathbf{j}_{\rm RCD} \approx j_{\rm osc} \frac{\sqrt{2\pi}}{8} \overline{w}'(E_{02}) E_{02} \tau_i \left(\frac{E_{01}}{E_{02}} \right)^2 e^{-(\Delta\omega\tau_i)^2/8} \mathbf{\Phi}(\varphi, \theta), \tag{18}$$

where

$$\mathbf{\Phi}(\varphi, \theta) = -\sin(2\varphi)(3\cos^2\theta \hat{\mathbf{x}} + 2\sin(2\theta)\hat{\mathbf{y}}). \tag{19}$$

3. Discussion

According to analytical formulas (10) and (18) the RCD depends on the intensities $I_1 \propto E_{01}^2$ and $I_2 \propto E_{02}^2$ of the field components, the angle θ, and the phase shift φ as the product of separate functions. Moreover, the dependence of the RCD on the intensity of the additional field, the angle θ and phase shift φ are independent of a specific form of the function $w(E)$. This allows one to explain very clearly the results of experiments on the generation of terahertz radiation using two-color laser pulses. In the recent work [19] analytical formula (18) was used to explain the results of some experiment where fundamental and frequency-tunable half-harmonic field were used to generate THz radiation.

Further we discuss some peculiarities of two mechanisms of RCD excitation in which additional field is the second- and half-harmonics of fundamental field. First of all, it should be noted that formulas (10) and (18) describe the high sensitivity of the RCD to a frequency detuning in both the $\omega+2\omega$ and $\omega+\omega/2$ cases. With the detuning, the phase shift between the carriers of fundamental and additional fields becomes time dependent. This leads to destructive superposition of the contributions to average drift velocity from the electrons born at different values of the phase shift. Equations (9) and (17) are exactly an indications that if this shift changes drastically over the ionization time τ_i , the total RCD is significantly attenuated. The dependence of RCD on $\Delta\omega$ has the form of a resonantlike curve with FWHM $\omega_d = 4\sqrt{\ln 4}/\tau_i$ which is equal in both the $\omega + 2\omega$ and $\omega + \omega/2$ cases. Thus, for incommensurate two-color pulses, the key parameter that determines the efficiency of low-frequency THz generation is $\Delta\omega\tau_i$.

The derived expression (18) for RCD for the $\omega+\omega/2$ case predicts that RCD is equal to zero at $\theta = \pi/2$. It can be explained by the disappearance of the density harmonics N_1 and N_2, since they are proportional to the dot product $(\mathbf{E}_1, \mathbf{E}_2)$ and its square, respectively, due to the Taylor approximation of $w(|\mathbf{E}(t)|)$ when complex amplitudes of density harmonics are calculated. In the $\omega + 2\omega$ case the RCD does not vanish for any angle θ. It is connected by the fact that in the $\omega + 2\omega$ case the excitation of RCD is due to appearance of first and second harmonics of free-electron density. The second harmonic of density does not depend on θ and the term $\overline{N_2 \mathbf{E}_2}$ gives the contribution to RCD at any angle θ.

The derived analytical formulas show that RCD dependence on the intensity I_{add} of additional field is different in the $\omega + 2\omega$ and $\omega + \omega/2$ cases. For the $\omega + 2\omega$ case,

$|\mathbf{j}_{\mathrm{RCD}}| \propto \sqrt{I_{add}}$, while in the $\omega + \omega/2$ case, $|\mathbf{j}_{\mathrm{RCD}}| \propto I_{add}$. This gives the opportunity to get sufficiently high efficiency of generation of RCD in the $\omega+\omega/2$ case, which efficiency can be as high as in the $\omega + 2\omega$ case at some laser pulse parameters.

The equations (9), (17) show that the introduction of the frequency detuning $\Delta\omega$ results in upshift of the low-frequency spectrum of current density. At sufficiently high $\Delta\omega$ the low-frequency component of free-electron current can have central frequency in the mid-infrared range. This gives an opportunity to use such incommensurate two-color pulses to generate ultrashort pulses in mid-infrared range with controlled carrier-envelope phase.

4. Conclusions

In this work we have investigated analytically the excitation of residual current density (RCD) during gas ionization by two-color laser pulses. We derived simple closed-form analytical formulas for RCD in cases when two-color pulse contains the strong fundamental field and weak frequency-tunable second- or half-harmonic field. The obtained analytical expressions (10) and (18) gives the dependences of RCD on the parameters of two-color laser pulses. It is shown that introduction of the frequency detuning of the additional field from the resonance value (which is doubled- or half-fundamental frequency) results in upshift of the low-frequency spectrum of free-electron current density. This gives an opportunity to use incommensurate two-color pulses to generate ultrashort pulses in mid-infrared range.

This work was supported by the Government of the Russian Federation (Agreement No. 14.B25.31.0008), and the Russian Foundation for Basic Research (Grants No. 13-02-00964, No. 14-02-31044, and No. 14-02-00847).

References

[1] Clough B, Dai J and Zhang X C 2012 *Mater. Today* **15** 50–58
[2] Kim K Y, Glownia J H, Taylor A J and Rodriguez G 2007 *Opt. Express* **15** 4577–4584
[3] Kim K Y, Taylor A J, Glownia J H and Rodriguez G 2008 *Nat. Photonics* **2** 605–609
[4] Clerici M, Peccianti M, Schmidt B E, Caspani L, Shalaby M, Giguére M, Lotti A, Couairon A, Légaré F, Ozaki T, Faccio D and Morandotti R 2013 *Phys. Rev. Lett.* **110** 253901
[5] Borodin A V, Panov N A, Kosareva O G, Andreeva V A, Esaulkov M N, Makarov V A, Shkurinov A P, Chin S L and Zhang X C 2013 *Opt. Lett.* **38** 1906–1908
[6] Manceau J M, Averchi A, Bonaretti F, Faccio D, Di Trapani P, Couairon A and Tzortzakis S 2009 *Opt. Lett.* **34** 2165–2167
[7] Babushkin I, Skupin S, Husakou A, Köhler C, Cabrera-Granado E, Bergé L and Herrmann J 2011 *New J. Phys.* **13** 123029
[8] Gildenburg V B and Vvedenskii N V 2007 *Phys. Rev. Lett.* **98** 245002
[9] Wu H C, Meyer-ter Vehn J and Sheng Z M 2008 *New J. Phys.* **10** 043001
[10] Liu C S and Tripathi V K 2009 *J. Appl. Phys.* **105** 013313
[11] Silaev A A and Vvedenskii N V 2009 *Phys. Rev. Lett.* **102** 115005
[12] Silaev A A and Vvedenskii N V 2009 *Phys. Scr.* **T135** 014024
[13] Silaev A A, Ryabikin M Y and Vvedenskii N V 2010 *Phys. Rev. A* **82** 033416

[14] Kostin V A and Vvedenskii N V 2010 *Opt. Lett.* **35** 247–249

[15] Alexandrov L N, Emelin M Y and Ryabikin M Y 2013 *Phys. Rev. A* **87** 013414

[16] Thomson M D, Blank V and Roskos H G 2010 *Opt. Express* **18** 23173–23182

[17] Oh T I, You Y S, Jha N, Rosenthal E W, Milchberg H M and Kim K Y 2013 *New J. Phys.* **15** 075002

[18] Minami Y, Kurihara T, Yamaguchi K, Naka ima M and Suemoto T 2013 *Appl. Phys. Lett.* **102** 041105

[19] Vvedenskii N V, Korytin A I, Kostin V A, Murzanev A A, Silaev A A and Stepanov A N 2014 *Phys. Rev. Lett.* **112** 055004

[20] Balčiūnas T *et al.* 2010 Tunability of thz emission originating from sub-cycle electron bursts in a laser induced plasma *International Conference on Ultrafast Phenomena* (Optical Society of America) p PDP4

[21] Kim K Y 2009 *Phys. Plasmas* **16** 056706

[22] Akhmedzhanov R A, Ilyakov I E, Mironov V A, Suvorov E V, Fadeev D A and Shishkin B V 2009 *Journal of Experimental and Theoretical Physics* **109** 370–378

[23] Rodriguez G and Dakovski G L 2010 *Opt. Express* **18** 15130–15143

Theory of high-harmonic generation in solids

C R McDonald, G Vampa, G Orlando, P B Corkum and T Brabec
Department of Physics, University of Ottawa, Ottawa, ON K1N 6N5, Canada
E-mail: cmcdo059@uottawa.ca

Abstract. High-harmonic generation (HHG) in bulk crystals exposed to intense mid-infrared lasers with photon energies below the bandgap is investigated theoretically. A three dimensional, two-band model that considers both interband and intraband currents is used. It is shown that the interband current is the dominant mechanism for HHG in solids. A physical interpretation of interband HHG — similar to atomic HHG — is provided by saddle point analysis. The effects of dephasing time and driving field wavelength on the harmonic specrum are investigated.

1. Introduction

The most prominent processes involving the interaction of intense laser fields with matter resulting in nonperturbative, nonlinear dynamics is high-harmonic generation (HHG) [1]. This process, mostly studied in atomic and molecular gases, has resulted in the birth of attosecond science [2], as swell as, new ultrafast imaging methods [3,4]. Much less focus had been afforded to corresponding processes in the condensed matter phase. Recent experiments have demonstrated that intense laser-solid interactions offer a wide range of phenomena and applications, including HHG [5–8].

In atoms, the electrons occupy a discrete set bound states; when an electron is ionized it is promoted to the continuum. By contrast, in solids, due to the large number of atoms that make up the lattice, the energy levels become dense and are organized into energy bands. Electrons that are "bound" are contained in the valence band separated from the conduction band by a finite bandgap. When an electron is "ionized" it is promoted to the conduction band leaving behind a hole in the valence band; see Fig. 1. In atoms the hole has a mass that is orders of magnitude greater than the electron and thus its motion can be neglected. In solids, however, electron and hole masses are comparable and thus hole dynamics cannot be ignored.

High-harmonic generation in solids can originate from two channels, an intraband current from nonlinear currents in the individual bands and an interband current due to polarization buildup between the valence and conduction bands; see Fig. 1 for a schematic representation. Theoretically, HHG in solid state systems has been investigated in the single active electron approximation for 1D and 2D band models [9–13]. The focus of the work here is to understand the main driving mechanism of HHG in solids in the underresonant limit. All previous work in this limit has not relied on a full solution of the two-band equations and has exclusively investigated the intraband current. So far the interband current has not been considered.

2. Two-band model for HHG in solids

Our analysis of HHG in solids is based on a two-band tight-binding model coupled to a laser field $\mathbf{F}(t) = \hat{\mathbf{r}}F_0 f(t)$; the dipole approximation is used, as the laser wavevector is much smaller

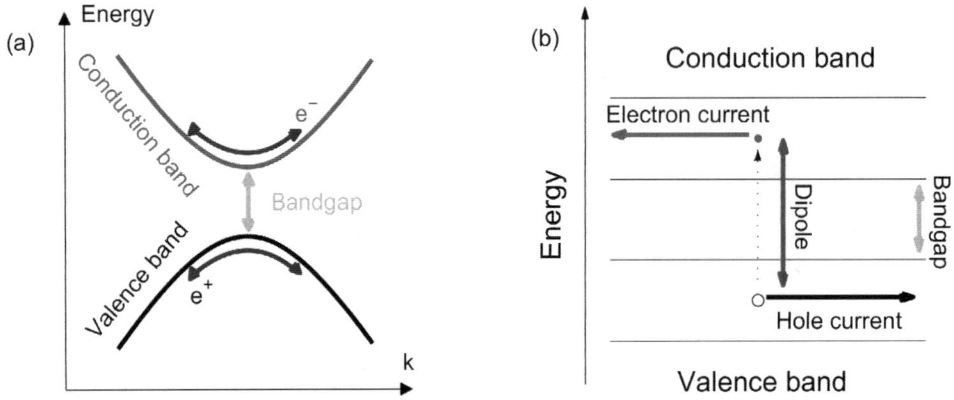

Figure 1. (a) Schematic of the crystal momentum representation of the band structure. (b) Schematic of the energy band structure in space. The intraband current arises from nonlinear currents in the individual bands and the interband current arises from a polarization buildup between the valence and conduction bands.

than the reciprocal lattice vectors. Further, atomic units are used throughout the paper, unless stated otherwise; the laser is characterized by field amplitude $f(t)$. The valence band (holes), the conduction band (electrons) and the bad gap are determined by $E_{\rm v} = -\Delta E_{\rm v}$, $E_{\rm c} = E_{\rm g} + \Delta E_{\rm c}$, and $\varepsilon_g = E_{\rm c} - E_{\rm v} = E_{\rm g} + \Delta E_{\rm g}$, respectively. Intense laser solid interaction is modeled using the two level density matrix equations,

$$\dot{\pi}(\mathbf{K},t) = -\frac{\pi(\mathbf{K},t)}{T_2} - \frac{i}{2}\Omega(\mathbf{K},t)\, w(\mathbf{K},t)\, e^{-iS(\mathbf{K},t)} \tag{1a}$$

$$\dot{n}_{\rm m}(\mathbf{K},t) = i\frac{s_m}{2}\,\Omega^*(\mathbf{K},t)\,\pi(\mathbf{K},t)\,e^{iS(\mathbf{K},t)} + \text{c.c.} \tag{1b}$$

where $n_{\rm m}$ is the probability amplitude of valence (m = v) and conduction (m = c) band, $w = n_{\rm v} - n_{\rm c}$ is the population difference between valence and conduction band. Initially, when all electrons are in the valence band, $n_{\rm v} = w = 1$; $s_m = -1, 1$ for valence and conduction band, respectively. The crystal momentum $\mathbf{k}$ has been transformed into a frame moving with the vector potential $\mathbf{A} = -d\mathbf{F}/dt$, $\mathbf{K} = \mathbf{k} - \mathbf{A}(t)$. As a result, the first Brillouin zone (BZ) is also shifted to $\overline{\rm BZ} = {\rm BZ} - \mathbf{A}(t)$. Further, T_2 accounts for dephasing within the relaxation time approximation, $S(\mathbf{K},t) = \int_{-\infty}^{t} \varepsilon_{\rm g}(\mathbf{K}+\mathbf{A}(t'))dt'$ is the classical action, $\Omega(\mathbf{K},t) = 2\mathbf{F}(t)\,\mathbf{d}(\mathbf{K}+\mathbf{A}(t))$ is the Rabi frequency, and $\mathbf{d}(\mathbf{k}) = i\int d^3\mathbf{x}\, u^*_{{\rm v},\mathbf{k}}(\mathbf{x})\nabla_{\mathbf{k}} u_{{\rm c},\mathbf{k}}(\mathbf{x})$ is the dipole moment. Finally, π is connected to the polarization by

$$\mathbf{p}(\mathbf{K},t) = \mathbf{d}[\mathbf{K}+\mathbf{A}(t)]\pi(\mathbf{K},t)e^{iS(\mathbf{K},t)} + \text{c.c.} \tag{2}$$

The above equations are identical to the two-band semiconductor equations [14] in the single active electron limit. The main difference is the frame of transformation which removes differentiation $\nabla_{\mathbf{k}}$ from the standard two-band equations. This allows for more efficient numerical integration and also allows for generalization of the analytical tools developed for atomic HHG [15] to solids; more detail is given in Ref. [16]. Finally, HHG in solids is determined by an intraband $\mathbf{j}_{\rm ra}$ and interband $\mathbf{j}_{\rm er}$ contribution which are respectively given by [9, 14]

$$\mathbf{j}_{\rm ra}(t) = \sum_{\rm m=c,v} \int_{\rm BZ} \mathbf{v}_{\rm m}[\mathbf{K}+\mathbf{A}(t)]\, n_{\rm m}(\mathbf{K},t)\, d^3\mathbf{K} \tag{3a}$$

$$\mathbf{j}_{\rm er}(t) = \frac{d}{dt}\int_{\rm BZ} \mathbf{p}(\mathbf{K},t)\, d^3\mathbf{K}, \tag{3b}$$

where the band velocity is defined by $\mathbf{v}_{\rm m}(\mathbf{k}) = \nabla_{\mathbf{k}} E_{\rm m}(\mathbf{k})$. The high-harmonic spectrum is obtained from the Fourier transform (FT) of the total current, $\mathbf{j}_{\rm t} = \mathbf{j}_{\rm ra} + \mathbf{j}_{\rm er}$, as $|\mathrm{FT}\{\mathbf{j}_{\rm t}\}|^2$.

In order to better understand the physical processes driving HHG in solids, it is useful to explore Eqs. (1) by using the Keldysh approximation [17], $w(t) \approx 1$ in Eq. (1a). This decouples Eqs. (1) so that they can be formally integrated. In Sec. 3, we will show that the interband term dominates HHG for mid-ir frequencies in ZnO. As such, here we will only focus on the interband term; the intraband term will be considered in future work. Inserting the result into Eq. (3b), we find the interband contribution along laser polarization to be,

$$j_{\rm er}(\omega) = \omega \int_{\rm BZ} d^3\mathbf{k}\, d(\mathbf{k}) \int_{-\infty}^{\infty} dt e^{-i\omega t} \int_{-\infty}^{t} dt' F(t') d^*(\boldsymbol{\kappa}_{t'}) \times e^{-iS(\mathbf{k},t',t)-(t-t')/T_2} + \text{c.c.}, \tag{4}$$

where $\boldsymbol{\kappa}_{t'} = \mathbf{k} + \mathbf{A}(t') - \mathbf{A}(t)$, v_m and d are the band velocity and dipole moment along laser polarization. Further, we have transformed the crystal momentum back to the initial frame $\mathbf{k} = \mathbf{K} + \mathbf{A}(t)$, and $S(\mathbf{k}, t', t) = \int_{t'}^{t} \varepsilon_g(\boldsymbol{\kappa}_\tau) d\tau$. Note that deriving Eq. (4) directly from the Schrödinger equation would yield the same result except for the dephasing term.

Following the semiclassical method developed for atomic HHG [15], saddle point integration with regard to the three integrals $d^3\mathbf{k}, dt', dt$ in Eq. (4) gives three saddle point conditions,

$$\nabla_{\mathbf{k}} S = \Delta\mathbf{x}_{\rm c} - \Delta\mathbf{x}_{\rm v} = 0 \tag{5a}$$

$$\frac{dS}{dt'} = \Delta E(\mathbf{k} - \mathbf{A}(t) + \mathbf{A}(t')) = 0 \tag{5b}$$

$$\frac{dS}{dt} = \Delta E(\mathbf{k}) = \omega. \tag{5c}$$

The above equations reveal a physical picture for interband HHG. In Eq. (5a), $\Delta\mathbf{x}_{\rm m} = \int_{t'}^{t} \mathbf{v}_{\rm m} dt'' = \mathbf{x}_{\rm m}(t) - \mathbf{x}_{\rm m}(t')$ is the distance propagated by the electron/hole in the conduction/valence band ($\rm m = c, v$) between time of birth t' and time of observation t; the velocity is $\mathbf{v}_{\rm m} = \nabla_{\mathbf{k}} E_{\rm m}$. This equation reveals a similar process to the acceleration of an electron in the laser field followed by its subsequent re-encounter with its parent ion in atomic HHG. However, in solids the masses of the electron and hole are comparable and thus crystal hole dynamics cannot be neglected. Eq. (5a) then states that, in solids, after having been accelerated and separated by the laser field, HHG takes place when the electron and hole re-encounter one another. Eq. (5b) states that electrons are born with zero momentum at time t', $\mathbf{k} = \mathbf{A}(t = t') - \mathbf{A}(t') = 0$. The finite bandgap energy results in a complex birth time, which describes tunnel ionization. Finally, Eq. (5c) represents conservation of energy — the electron-hole pair recombines and emits a photon ω with energy equal to the bandgap at the momentum of recombination, $\mathbf{k} = \mathbf{A}(t) - \mathbf{A}(t')$. As a result, the cut-off for interband HHG in a two-band model is limited to the maximum bandgap. Higher harmonics are possible, but require excitation to a higher band.

3. Numerical Results

Our theory is applied using a 3D band structure for ZnO (wurtzite structure). The orientation of the reciprocal lattice is chosen so that $\hat{\mathbf{x}} \parallel \Gamma{-}M$, $\hat{\mathbf{y}} \parallel \Gamma{-}K$, and $\hat{\mathbf{z}} \parallel \Gamma{-}A$ (optical axis).

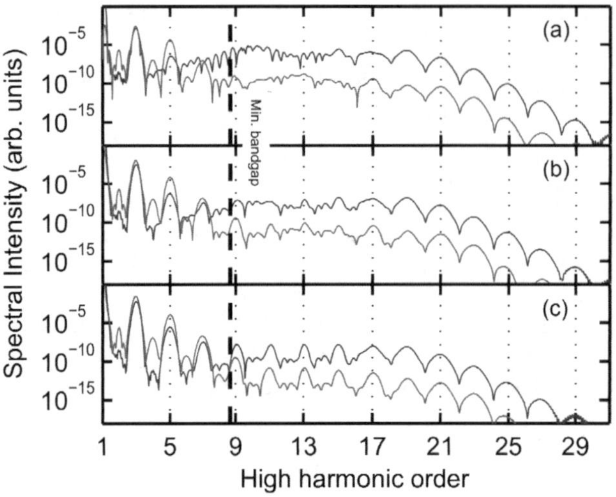

Figure 2. Harmonic spectrum from interband ($|\mathrm{FT}\{\mathrm{j_{er}}\}|^2$, blue line) and intraband ($|\mathrm{FT}\{\mathrm{j_{ra}}\}|^2$, red line) currents for field strength $F_0 = 0.003$; we use a temporal Gaussian envelope with FWHM equal to 10 cycles and cos-carrier with frequency $\omega_0 = 0.014$, corresponding to a laser period $T_0 = 2\pi/\omega_0 = 10.9$ fs; (a), (b), and (c) show dephasing times of $T_2 = \infty$, $T_0/2$, $T_0/4$, respectively. The dashed black vertical lines represent the (minimum) bandgap at the center Brillouin zone.

Valence, conduction band, and bandgap are determined by $E_\mathrm{v} = -\Delta E_\mathrm{v}$, $E_\mathrm{c} = E_\mathrm{g} + \Delta E_\mathrm{c}$, and $\varepsilon_g = E_\mathrm{c} - E_\mathrm{v} = E_\mathrm{g} + \Delta E_\mathrm{g}$, respectively. The bands along Γ–M, K, A were obtained by density functional theory in the local density approximation (LDA-ABINIT). The complete 3D band in the first BZ is approximated as the sum over the three 1D bands. The linearly polarized pulse $\mathbf{F}(t) = \hat{\mathbf{x}}F_0 f(t)$ is characterized by a field amplitude $f(t)$ consisting of a temporal Gaussian envelope with a FWHM of 10 cycles of the fundamental cosine carrier frequency $\omega_0 = 0.014$ ($\lambda = 3.25\,\mu$m). For integration of Eqs. (1) we use 600 points (full BZ) along Γ–M and 200 points (half BZ) along the each of the remaining two directions.

3.1. Effect of dephasing time

Our analysis of HHG begins with a comparison of the two contributions to HHG in Fig. 2. The interband (blue) and intraband (red) currents are shown for dephasing times $T_2 = \infty$, $T_0/2$, $T_0/4$ with T_0 being the laser oscillation period. Non-perturbative HHG takes place above the dashed line ($N = 9$), indicating the minimum bandgap at the center of the BZ. Both contributions show a plateau in the non-perturbative regime, whereas in experiments the harmonic yield drops with increasing harmonic order. This difference is due to the neglect of propagation effects, in particular reabsorption of harmonic photons which increases with growing harmonic orders. The spectra in Fig. 2 were calculated with peak field strength $F = 0.003$. The calculations were repeated with a range of field strengths used in experiments [5]; all calculations yielded similar results. In the plateau region, interband HHG is always dominant by at least two orders of magnitude.

In Fig. 2(a), both spectra are very noisy and do not show a clear odd harmonic structure. The reason for this noise becomes clear by looking at Figs. 2(b) and (c). The odd harmonic spectrum starts to manifest for dephasing times of the order of a half cycle, see Fig. 2(b).

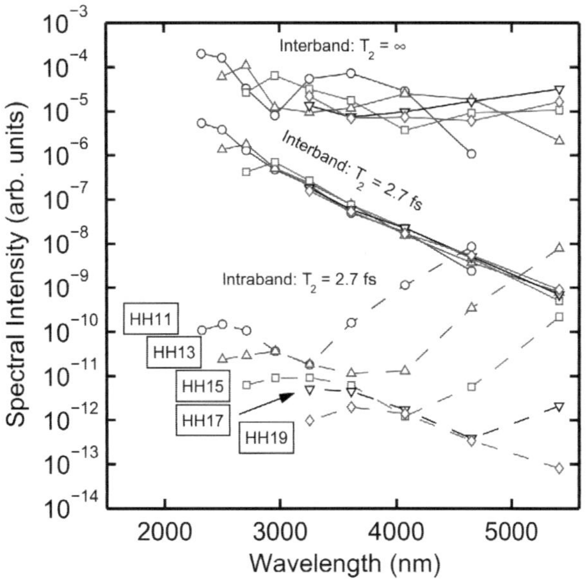

Figure 3. Interband (circles) and intraband (triangles) contributions to harmonic intensity vs laser wavelength λ_0 for high-harmonics (HH) 11 – 19 at field strength $F_0 = 0.003$.

For even shorter dephasing times, the signal to background ratio is further improved in 2(c) and a clean harmonic spectrum emerges. This can be understood as follows. In atomic HHG, propagation effects introducing a phase mismatch between the fundamental laser frequency and its second harmonics can also contribute to attenuating second and higher returns [18]. This is not the case in solids, as absorption dominates phase mismatch for above band gap plateau harmonics [13]. This leaves dephasing as the only mechanism for obtaining a clean harmonic spectrum in agreement with experiments.

3.2. Wavelength dependence of the harmonics

Here we describe how the interband and intraband contributions scale with the wavelength of the driving field. Figure 3 shows this scaling for the interband (solid lines) and the intraband (dashed lines) with $T_2 = 2.7$ fs. When the wavelength is reduced, at some point a given harmonic will move beyond the plateau into the cutoff region and the yield will drop. On the other hand, for increasing wavelength a given harmonic order will drop below the band gap and move from the perturbative to the nonperturbative realm. Each harmonic has only been plotted between these two limits — as long as it stays in the plateau region.

The interband contribution drops exponentially with increasing wavelength, which can be attributed to the dephasing term in Eq. (4). By contrast, the wavelength dependence of the intraband yield is more complex and even increases for longer wavelengths. For most of the investigated wavelength range, the interband HHG clearly dominates. Only around wavelengths, where a given harmonic transits from the non-perturbative into the perturbative regime, the two contributions become comparable. The calculations have been repeated for a range of field strengths between $0.002 \leq F_0 \leq 0.006$, which show a very similar picture with small differences in where inter- and intraband current intersect. Our calculations indicate the dominance of interband HHG over the part of the mid-infrared spectrum where non-perturbative

Journal of Physics: Conference Series **594** (2015) 012021 doi:10.1088/1742-6596/594/1/012021

HHG occurs. Ultimately, the importance of interband versus intraband HHG will have to be settled by experiment. Figure 3 reveals a first suggestion of how this can be done. The opposite gradients of inter- and intraband yield at higher wavelengths present a strong signature for experimentally discerning the two mechanisms.

4. Conclusion

We have presented a 3D, two-band, density matrix analysis for HHG in solids using a single active electron. Numerical results suggest interband HHG, which has been neglected so far, to be the dominant channel. Saddle point integration in the Keldysh/Lewenstein limit reveal the following mechanism responsible for HHG in solids: after the creation of an electron hole pair by tunnel ionization, electrons and holes are accelerated and driven apart by the laser field. When electron and hole re-encounter each other, recombination can occur resulting in the emission of a harmonic photon; the wavevector at the time of recombination determines the energy difference (gap) between the electron-hole pair and therewith the emitted photon energy. By comparison, in atoms an electron is ionized and accelerated by the laser field and ultimately driven back to the core which is stationary due to its weight. Upon recombination it releases its kinetic energy as a high energy photon.

Similar to atomic HHG higher returns and recollisions occur. The influence of higher returns is much stronger in solids than in atoms. In the absence of dephasing, higher returns create a noisy continuum-like structure, which masks the odd harmonic spectrum. Dephasing times of the order of a half cycle are necessary to suppress higher returns and to obtain agreement with the clean odd harmonic structure observed in experiments.

The interband current drops exponentially with increasing laser wavelength. This opens a novel way for measuring dephasing times without the need of ultrashort laser sources. For long wavelengths ($> 4\mu$m) the intraband current grows for increasing wavelengths. The different signs of interband and intraband wavelength dependence present a strong signature allowing to experimentally settle the importance of interband versus intraband HHG.

References

[1] Corkum P B 1993 *Phys. Rev. Lett.* **71** 1994
[2] Corkum P B and Krauz F 2007 *Nat. Phys.* **3** 381
[3] Itatani J, Levesque J, Zeidler D, Niikura H, Pépin H, Kieffer J C, Corkum P B and Villeneuve D M 2004 *Nature* **432** 867
[4] Wörner H J *et al.* 2011 *Science* **334** 208
[5] Ghimire S, DiChiara A D, Sistrunk E, Agostini P, DiMauro L F and Reis D A 2011 *Nat. Phys.* **7** 138
[6] Ghimire S, DiChiara A D, Sistrunk E, Szafruga U B, Agostini P, DiMauro L F and Reis D A 2011 *Phy. Rev. Lett.* **107** 167407
[7] Zaks B, Liu R B and Sherwin M S 2012 *Nature* **483** 580
[8] Schubert O *et al.* 2014 *Nat. Photonics* **8** 119
[9] Golde D, Meier T and Koch S W 2008 *Phys. Rev. B* **77** 075330
[10] Kemper A F, Moritz B, Freericks J K and Deveraux T P 2013 *New J. Phys.* **15** 023003
[11] Pronin K A, Bnadrauk A D and Ovchinnikov A A 1994 *Phys. Rev. B* **50** 3473
[12] Hawkins P G and Ivanov M Yu 2013 *Phys. Rev. A* **87** 063842
[13] Ghimire S, DiChiara A D, Sistrunk E, Ndabashimiye G, Szafruga U B, Mohammad A, Agostini P, DiMauro L F and Reis D A 2012 *Phy. Rev. A* **85** 043836
[14] Haug H and Koch S W 2004 *Quantum Theory of Optical and Electronic Properties of Semiconductors* (World Scientific, Singapore)
[15] Lewenstein M, Balcou P, Ivanov M Y, L'Huillier A and Corkum P B 1994 *Phys. Rev. A* **49** 2117
[16] Vampa G, McDonald C R, Orlando G, Klug D D, Corkum P B and Brabec T 2014 *Phys. Rev. Lett.* **113** 073901
[17] Keldysh L 1965 *Sov. Phys. JETP* **20** 1307
[18] Balcou Ph, Salieres P, L'Huillier A and Lewenstein M 1997 *Phys. Rev. A* **55** 3204

Low-frequency dipole response of a diatomic heteronuclear molecule on an intense ultrashort laser pulse

V. Yu. Kharin and O. V. Tikhonova

Skobeltsyn Institute of Nuclear Physics, Moscow State University, Moscow, 119234 Russia

Department of Physics, Moscow State University, Moscow, 119991 Russia

Email: w.kharin@list.ru

Abstract. In the present work we analyze the dipole response of a diatomic heteronuclear molecule on the nonperturbative action of the external low-frequency laser field. Main results are obtained by numerical solution of the time-dependent Schrodinger equation for the nuclear subsystem of the molecule. We study the spectrum of polarization response and the influence of the coherent repopulation of the ro-vibrational states on it. Generally, we consider two cases: repopulation due to Raman-type transitions leading to the interference stabilization of the molecule and repopulation arising as a result of extremely short pulse action. We also study the transformation of a weak probe pulse propagating through the highly excited medium of the studied molecules.

1. Introduction

Laser research during last decades, both theoretical and experimental, revealed a number of new remarkable features of molecular and atomic media exposed to external electromagnetic fields. Vast part of the new fundamental properties is discovered by virtue of using laser setups with extremely high intensities of generated light [1-7]. From theoretician's point of view, some of them can be described in terms of perturbation theory, but there are also a number of fundamental processes demonstrating essentially nonperturbative quantum-mechanical behavior [1-3], such as tunnel ionization of atoms, plasma formation in clusters, photostabilization[2-6] etc. Molecular media have a lot of properties similar to those of the atomic systems when one deals with electronic subsystem [8] and transitions between electronic states. But once the measurement time is comparable with ro-vibrational evolution times, there are several very important differences between atomic and molecular behavior in the external field. They arise as a result of the nuclear subsystem response on the external field. Molecular-specific phenomena are alignment control [9], optical centrifuging, photostabilization with respect to ro-vibrational transitions [10], and many others. One of the most fascinating effects of the nuclear subsystem dynamics is the interference stabilization with respect to dissociation. It is interesting not only as the pure quantum-mechanical effect that seems counter-intuitive at the first glimpse, but also as an instrument of coherent repopulation of the wide set of ro-vibrational states. Roughly speaking, photostabilization consists in dissociation suppression when the incident field intensity exceeds some critical value. In the case of interference mechanism [2], it results from the multiple coherent Raman-type transitions via lower lying-bound states (so-called V-type transitions), and higher-lying ro-vibrational continuum states (Λ-type transitions). Hence, the additional effect accompanying the interference stabilization regime is the population of the coherent superposition of the molecular states, very wide with respect to initial wave-packet. Looking at the dipole moment of the molecule, one can see that the spectrum of the dipole response gains a variety of frequencies that correspond to different rotational and vibrational transitions and can be very low with respect to incident light frequency.

The theory [2,3], experimental evidence [11], and explanation of the stabilization regime was developed for quasimonochromatic radiation. It can be regarded as the long-pulse approximation. The other limiting case is the case of extremely short pulses [12]. In terms of nuclear subsystem shortness means that the pulse duration is much lower than the vibrational period of the molecule (but still much higher than the inverse frequencies of the electron transitions). The most interesting situation arises when the pulse is characterized by the non-zero integral of electric field over time (so-called half-cycle pulse). The action of such short pulse is remarkable in the sense that the positions of nuclei are nearly unchanged during the pulse, but the nuclear subsystem gains momentum, which determines its further evolution and the dipole response.

The present work is devoted to the numerical study of the dipole response of the molecular media on low-frequency electromagnetic field in these two cases. We intend to analyze spectral characteristics of the response in the non-perturbative regimes of the interference stabilization and extremely short action of monopolar pulses.

2. Numerical model

We consider nuclear subsystem of the heteronuclear molecule within the Born-Oppenheimer approximation using single potential energy surface. The time-dependent Schrödinger equation can be written in the dipole approximation for linearly polarized field as

$$i\hbar\frac{\partial\psi(r,\theta,\varphi,t)}{\partial t}=-\frac{\hbar^2}{2\mu}\Delta\psi(r,\theta,\varphi,t)+\left(V(r)-qrE(t)\cos\theta\right)\psi(r,\theta,\varphi,t). \quad (1)$$

Here r is the internuclear distance, θ and φ are ordinary spherical angles taken from the polarization direction. $V(r)$ denotes the vibrational potential energy of nuclei, $E(t)$ is incident field, q is the effective charge included into the dipole moment of the nuclear subsystem. It is convenient to decompose full wave function over spherical harmonics as

$$\psi(r,\theta,\varphi,t)=\sum_{j=0}^{\infty}\sum_{m=-j}^{j}\frac{u_j^m(r,t)}{r}Y_j^{(m)}(\theta,\varphi). \quad (2)$$

If we assume that only states with $m=0$ are populated initially, we can omit superscript m, since it is always zero, and rewrite TDSE as a system of partial differential equations for u_j:

$$i\hbar\frac{\partial u_j}{\partial t}=-\frac{\hbar^2}{2\mu}\frac{\partial^2 u_j}{\partial r^2}+\left(V(r)+\frac{\hbar^2 j(j+1)}{2\mu r^2}\right)u_j-qrE(t)\left(\beta_{j+1}^j u_{j+1}+\beta_{j-1}^j u_{j-1}\right). \quad (3)$$

Coefficients β origin from the Clebsch-Gordan coefficients and are equal to

$$\beta_{j-1}^j=\frac{j}{\sqrt{(2j-1)(2j+1)}}$$
$$\beta_{j+1}^j=\frac{j+1}{\sqrt{(2j+1)(2j+3)}} \quad (4)$$

In order to provide the possibility of interference stabilization of the molecule we need to consider the molecule with dense vibrational spectrum and effective transitions to the vibrational continuum. Taking it into account we have chosen the system with properties close to the IBr ones. Potential energy surface was approximated by Morse potential curve with bound energy equal to 1.3 eV, steepness parameter $\alpha = 5$, equilibrium internuclear distance of 2.5 angstroms. Initially the molecule is assumed to occupy the highly excited vibrational state. The frequency of the laser radiation (except the case of extremely short action) was taken low enough to provide effective transitions to the vibrational continuum. In our calculations it was 0.1eV. Laser pulse envelope was described using smoothed trapezoidal shape with equal durations of front and tail parts. The system (3) was solved numerically using finite difference method with iterating of the right part of equation. Having the time-dependent wavefunction of the system ψ, we can calculate the vibrational and rotational quantum number distributions over the bound states using ordinary formulae

$$W_j = \sum_v \left|\langle\psi|\varphi_{v,j}\rangle\right|^2,$$

$$W_v = \sum_j \left|\langle\psi|\varphi_{v,j}\rangle\right|^2.$$

Here v is the vibrational quantum number; j is the rotational quantum number, $\varphi_{v,j}$ stands for the corresponding stationary wavefunction. The time-dependent average dipole moment of the system can be obtained as

$$\langle d_z\rangle = \langle\psi|qr\cos\theta|\psi\rangle. \quad (5)$$

3. Results and discussion

While in the weak field regime the polarization response of the molecule to the external field action is perturbative and rather straightforward, in the strong-field regime, when the interference stabilization is present, the perturbation theory-based approach is non-applicable. There are two main characteristics that define discrepancy. First, strong-field dynamics assumes strong repopulation of the large set of ro-vibrational levels during the pulse. It destroys the proportionality between dipole moment of the molecule and electric field, and in the stabilization regime, when multiple Raman-type transitions take place, it also destroys the polynomial dependence usual for nonlinear optics. Second, the post-pulse dipole response is defined by the redistributed population over rotational-vibrational molecular levels. This redistribution is possible due to non-adiabaticity of the incident radiation. So, if the incident pulse is sufficiently long (and has slow-varying envelope), one won't see any post-pulse dynamics, while in the case of short non-adiabatic pulse action the residual polarization signal can be observed in the after-pulse regime.

In our calculations we took a vibrationally excited molecule to provide pure stabilization regime with the single-photon bound state-continuum coupling. The bound state had dissociation energy of 0.082eV, non-excited with respect to rotations. In experiment such excited molecule can be obtained in the field-induced ionization process. If one wants to take into account the thermal effects, he has to deal with non-coherent summation over a set of rotational states, but here we consider highly non-equilibrium system. The time-dependence of the dipole moment induced by the strong pulse interacting with vibrationally excited molecule is presented on the fig. 1.

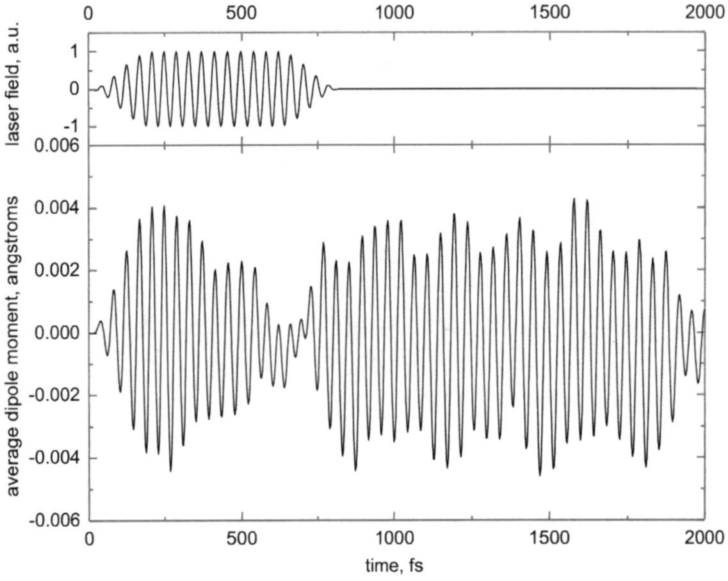

Figure 1. Time dependence of the electric field of the incident pulse (top), and the corresponding dipole response of the molecular system (bottom). Intensity of the incident light was set to 2 TW/cm^2. Initial state of the molecule is described in the text. Dissociation during the pulse has not exceeded the value of 2 percent.

Here one can see that the response is not proportional to the laser field (even in terms of envelope) during the pulse, and moreover, the residual polarization signal is present. It should be noticed that frequencies much lower than the pulse carrier frequency can be distinguished. As it has been said, field-free dynamics is fully determined by the population distribution of differrent ro-vibrational states and the energy differences between populated levels. So we can divide the obtained response into several terms. First, rapidly changing near-resonant response is provided by the vibrational energy gaps. Second, the slowest term, arises as a result of rotational repopulation. Its characteristic time can be estimated as $\tau \sim \hbar / B$ where B is the rotational constant. All these frequencies can be observed in the corresponding spectrum (fig. 2). Intermediate frequencies of the response are supressed. Hence, calculations predict possibility of the non-vanishig response at the rotational molecuar frequencies in the case of multiple Raman-type transitions.

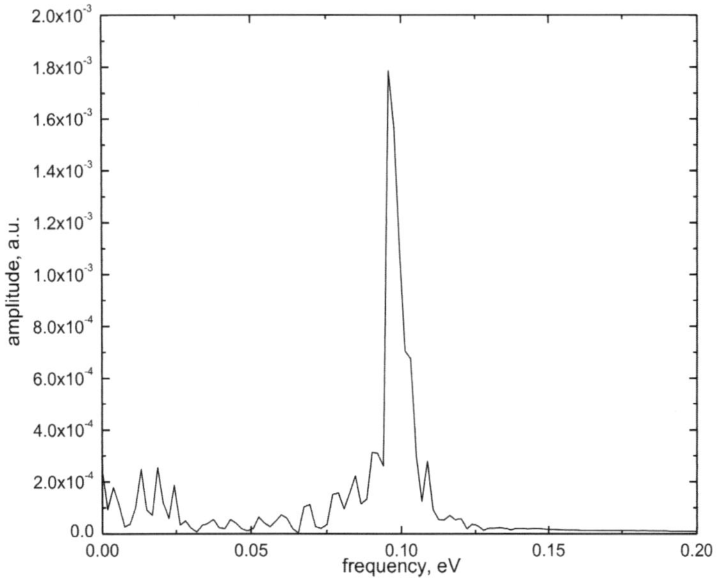

Figure 2. Spectrum of the dipole response shown on the fig. 1.

Now consider the dipole response of the molecule on the action of the extremely short pulse referred to as "half cycle pulse". This means that duration of the pulse is much lower than the vibrational period of the molecule. In addition the only physical property that characterizes the post-pulse evolution of the quantum-mechanical wave packet is the momentum transferred to the nuclear subsystem of the molecule. In order for this quantity to have nonzero values, the corresponding laser pulse must have non-vanishing integral of the electric field over time. It can be achieved experimentally from the single-cycle optical pulse using nonlinear saturating media [13]. Questions of repopulation under action of this pulse are discussed in detail in [14].

The dynamics of the dipole moment of the model molecule under such an action is presented on the fig. 3. We consider two cases, namely relatively weak pulse (fig. 3a), and relatively strong one (fig. 3b). The difference between these two cases consists in the number of ro-vibrational states involved in the post-pulse dynamics. It is seen that the dominating part of the dipole response is provided by the molecular rotations. When the incident pulse is weak, the population distribution over rotational states is not significantly wide. Hence, in the post-pulse dynamics one can mainly see the slow-varying dipole moment, corresponding to the matrix element between $j = 0$ and $j = 1$ states. The vibrational quantum number is nearly unchanged. Low repopulation of vibrational states affects the weak response at relatively high frequency, which can be seen on both pictures. As for the strong field case, the distribution over rotational quantum number is much broader. Since the energy gap between states with rotational quantum numbers $j-1$ and j is roughly proportional to j, the higher excitation of rotations assumes the higher frequencies involved and more rapid evolution of the dipole moment. So in general, we have quasiperiodic response at rotational frequencies which can be controlled by the momentum transferred to the system. One can notice the high-frequency oscillations on the time dependence. They arise from the molecular vibrational motion, and low amplitude is a consequence of low vibrational repopulation during the ultrashort pulse.

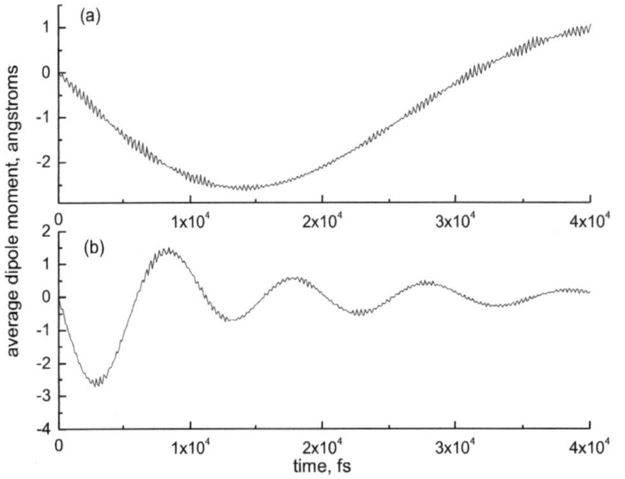

Figure 3. Post-pulse dipole response on the extremely short pulse action. Total momentum transferred to the molecule was set to 1 (a) and 5 (b) inverse angstroms correspondingly. Initial molecular state is chosen to be the higly excited vibrational one (see the text).

We only considered the situation when the initial state is one of the ro-vibrational eigenstates of the molecular Hamiltonian (highly excited one). From the experimental point of view it is more natural to study the behavior of the wide superpositoion of the states. This situation makes the dipole response analysis more complicated. However, the qualitative conclusions about the polarization dynamics obtained earlier remain valid. Such a scheme can be realized as follows. First, the molecule is exposed to the strong laser radiation. Dipole response on the pulse of this sort was discussed above. Strong pulse forms ro-vibrational wave packet and induces residual dipole response. Second, during the evolution of the residual response, the second (probe) pulse comes. It is weak, and the corresponding repopulation of the molecular ro-vibrational levels is rather low. But it causes additional dipole response. This response, in turn, can affect the probe pulse propagation and can have additional components of the spectrum in comparison with the case of the only strong pumpe pulse action.

Dipole response here consists of the part caused by the first pulse, and weak response to the second one which strongly depends on the medium excitation provided by the first pulse. It is interesting to extract the probe-field response. Time dependence extracted from our calculations is presented on fig. 4.

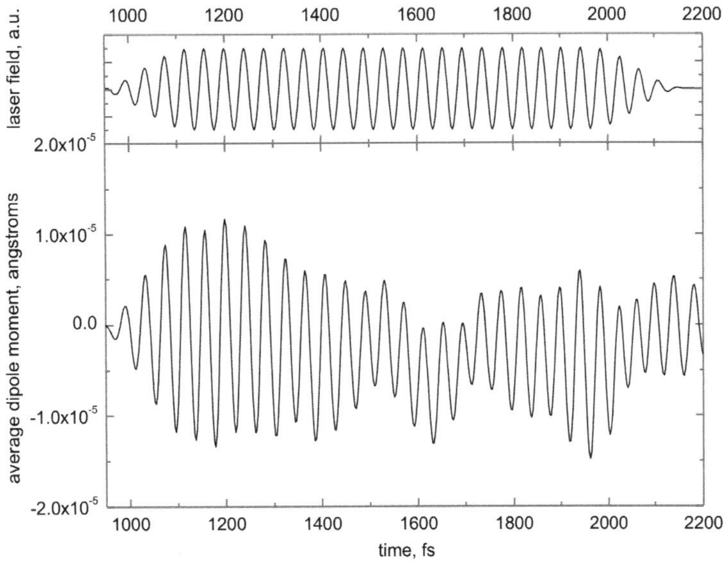

Figure 4. Time dependence of the electric field of the weak probe pulse (top), and the corresponding dipole response of the molecular system (bottom). Pump pulse had intensity 2TW/cm^2 and total duration of 20 optical cycles with the carrier frequency of 0.1eV. The intensity of the probe pulse was 0.2MW/cm^2. Dipole response on the first pulse has been subtracted.

And the fig. 5 presents its spectrum. As it has been expected, the response of the excited molecule strongly differs from ordinary linear response. Again we see the near-resonance component of the vibronic transitions, but at the same time the low frequency components are seen to be strongly enhanced though the probe pulse itself cannot provide neither stabilization, nor the high-order transitions. According to our model, these frequencies can be matched with non-uniform (with respect to j) quasi-resonant population and "de-population" of the rotational states.

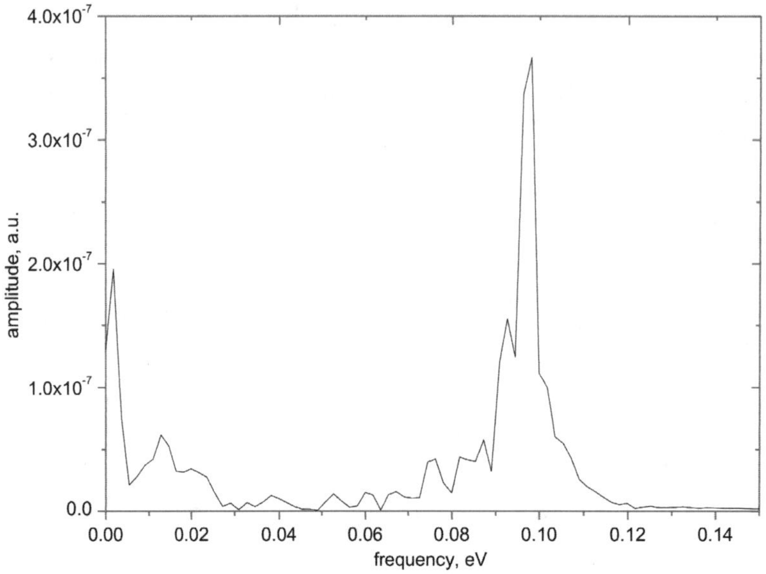

Figure 5. Spectrum of the dipole response shown on the fig.4.

4. Conclusions

We presented numerical data on the dipole response of diatomic heteronuclear molecule on the external laser field action investigated by "ab initio" TDSE calculations. If the incident pulse is strong enough the formation of wide rotational distribution can be achieved due to interference stabilization phenomenon and multiple Raman-type Λ and V transitions. In this case the population response reveals efficient low-frequency spectral components arising from rotational excitation of the molecule. These low-frequency components can be significantly enhanced if the molecular medium is pre-excited by a strong pump pulse providing a wide coherent population of rotational states. Such a possibility of quantum control of the low-frequency dipole response of the molecular medium is of great importance especially for problems of generation and amplification of the terahertz radiation.

Acknowledgements

This work was supported by Russian Foundation for Basic Research grants №14-02-00389_a and №14-02-31114_mol_a.

References

[1] M. V. Fedorov, Atomic and Free Electrons in a Strong Light Field, World Scientific, Singapore (1997).
[2] M. V. Fedorov, A. M. Movsesian, J. Opt. Soc. Am. B, 5(4), 850 (1988).
[3] N. P. Poluektov, M. V. Fedorov, Journal of Experimental and Theoretical Physics, 90, 794 (2000).
[4] M. Gavrila, Journal of Physics B: Atomic, Molecular and Optical Physics, 35(18), R147 (2002).
[5] M. Pont, M. Gavrila, Phys. Rev. Lett., 65(19), 2362 (1990).
[6] M.V.Fedorov, N.P.Poluektov, A.M.Popov, O.V.Tikhonova, V.Yu.Kharin, E.A.Volkova, IEEE J. of Selected Topics in Quantum Electronics, V.18, N1, P. 42-53 (2012)
[7] A. M. Popov, O. V. Tikhonova, E. A. Volkova, Journal of Physics B: Atomic, Molecular and

Optical Physics, 36(10), R125 (2003).
[8] M. V. Fedorov, Journal of Modern Optics, 50(3-4), 513 (2003).
[9] T. Seideman, J. Chem.Phys, 103 (18), 7887-7896 (1995)
[10] V.Yu.Kharin, A.M.Popov, O.V.Tikhonova, Laser Phys, 11, № 11, (2012)
[11] U. Eichmann, T. Nubbemeyer, H. Rottke *et al*, Nature, 461(7268), 1261(2009).
[12] M. Leibscher, I. Sh. Averbukh, P. Rozmej *et al*, Phys. Rev. A, 69(3), 032102 (2004).
[13] D. You, P. H. Bucksbaum, J. Opt. Soc. Am. B, 7, 1651 (1997).
[14] Kharin V.Yu., Tikhonova O.V., Laser Phys. Lett., Vol. 11, 075302 (2014).

Fraunhofer type diffraction of phase-modulated broad-band femtosecond pulses

A. M. Dakova[1], **L. M. Kovachev**[1], **K. L. Kovachev**[1], **D. Y. Dakova**[2]
[1] Institute of Electronics, Bulgarian Academy of Sciences, 72 Tzarigradsko shossee, 1784 Sofia, Bulgaria
[2] Faculty of Physics, University of Plovdiv "Paisii Hilendarski", 24 Tsar Asen Str., 4000 Plovdiv, Bulgaria

E-mail: anelia.dakova@gmail.com
E-mail: lubomirkovach@yahoo.com

Abstract. Attosecond optical pulses with one-two cycles under the envelope diffract in non-paraxial regime on several diffraction lengths. Their intensity profile takes a form similar to Fraunhofer distribution. An analytical theory was developed, where was pointed, that such type of diffraction depends on the spectral width of the optical pulse. In this paper is shown that even for a broad-band phase-modulated femtosecond pulses the diffraction is also of Fraunhofer type.

1. Introduction

In recent years the linear regime of propagation of optical pulses with broad-band spectrum in air and transparent media attracts a considerable attention. The attosecond pulses in UV region as well as femtosecond pulses with time duration *5-15 fs* in optical and IR region admit from one to several optical cycles under their envelopes. For such pulses, as it was pointed in [1], the diffraction is not paraxial and on few diffraction lengths their shape takes parabolic form. In the theory developed in [2,3] it is shown that the non-paraxiality depends on the spectral width $\Delta k_z = \Delta\omega / v_{gr}$ of the pulse. Thus, one additional possibility appears: Is it possible to use phase-modulated broad-band femtosecond pulses with many cycles under their envelope to obtain Fraunhofer type diffraction?

In this paper we will numerically investigate Fraunhofer type diffraction of phase-modulated *25 fs* laser pulses which admit at least 10 cycles at level e^{-1} of the maximum.

This study gives us knowledge about the limits and applicability of paraxiality in the frames of femtosecond optics.

2. Basic theory

The linear scalar non-paraxial amplitude equation, governing the evolution of laser pulses in isotropic dispersive media, is written in Local type coordinates [4]:

$$-2ik_0 \frac{\partial A}{\partial z} = \Delta A_{\perp} + \frac{\partial^2 A}{\partial z^2} - \frac{2}{v}\frac{\partial^2 A}{\partial \tau \partial z} - \frac{\beta}{v^2}\frac{\partial^2 A}{\partial \tau^2} \quad , \tag{1}$$

where $A(\tau, z)$ is the slowly varying amplitude function of the pulse envelope, v is the group velocity, $\beta = k_0 v_{gr}^2 |k''|$ is a parameter characterizing the second order of linear dispersion of the medium, $z = z$ and $\tau = t\text{-}z/v$.

This equation can be solved by using the Fourier method. The Fourier transform is applied to the amplitude function $FFF[A(x,y,\tau,z)] = \hat{A}\left(k_x, k_y, \frac{\Delta\omega}{v}, z\right)$. Here $\Delta\omega = \omega - \omega_0$ and $\Delta k_z = \Delta\omega / v$ are respectively the spectral width and the wave number of the optical pulse.

After a couple of calculations the fundamental solution of equation (1) in $\left(k_x, k_y, \frac{\Delta\omega}{v}, z\right)$ space can be written as follow:

$$A\left(k_x, k_y, \frac{\Delta\omega}{v}, z\right) = \hat{A}(k_x, k_y, \Delta\omega, 0)\exp\left\{ i[(k_0 - \frac{\Delta\omega}{v}) \pm \sqrt{(k_0 - \frac{\Delta\omega}{v})^2 + k_x^2 + k_y^2 - \frac{\beta\Delta\omega^2}{v^2}}]z \right\} \quad (2)$$

The solution of equation (1) in (x,y,z,t) dimensions is simply the inverse Fourier transform of solution (2):

$$A(x,y,\tau,z) = FFF^{-1}\left\langle \hat{A}(k_x, k_y, \Delta\omega, 0)\exp\left\{ i[(k_0 - \frac{\Delta\omega}{v}) \pm \sqrt{(k_0 - \frac{\Delta\omega}{v})^2 + k_x^2 + k_y^2 - \frac{\beta\Delta\omega^2}{v^2}}]z \right\} \right\rangle \quad (3)$$

Here we will investigate more precisely the case of narrow laser pulses when the condition $\Delta k_z << k_0$ is satisfied. Thus, we can assume that $k_0 - \frac{\Delta\omega}{v} \approx k_0$. The solution (3), in this approximation, takes the form:

$$A(x,y,\tau,z) = FFF^{-1}\left\langle \hat{A}(k_x, k_y, \Delta\omega, 0)\exp\left\{ i\left[k_0 \pm k_0\sqrt{1 + \frac{k_x^2 + k_y^2 - \frac{\beta\Delta\omega^2}{v^2}}{k_0^2}} \right] z \right\} \right\rangle \quad (4)$$

On the other hand, the expression under the square root can be presented in Taylor series, because for such type of laser pulses the condition $k_0^2 >> k_x^2 + k_y^2 - \frac{\beta\Delta\omega^2}{v^2}$ is satisfied. If we use the minus in front of the square root, the expression (4) takes the form:

$$A(x,y,\tau,z) = FFF^{-1}\left\langle \hat{A}(k_x, k_y, \Delta\omega, z) = \hat{A}(k_x, k_y, \Delta\omega, 0)\exp\left\{ -i\left[\frac{k_x^2 + k_y^2 - \frac{\beta\Delta\omega^2}{v^2}}{2k_0^2} \right] z \right\} \right\rangle \quad (5)$$

This is the typical spectral kernel of the spatio-temporal equation of the kind:

$$-2ik_0\frac{\partial A}{\partial z}=\Delta A_{\perp}-\frac{\beta}{v^2}\frac{\partial^2 A}{\partial\tau^2} \tag{6}$$

Thus, we proved that the linear evolution of narrow-band light pulses can be described correctly by solving the paraxial spatio-temporal equation (6).

In the context of broad-band optical pulses the condition $\Delta k_z=\frac{\Delta\omega}{v}\approx k_0$ does not allow to present the spectral kernel (2) in Taylor series.

The analyses presented above lead to the next important conclusion: The non-paraxial equation (1) describes correctly the dynamics of narrow-band as well as broad-band optical pulses.

That is why the first numerical experiment with broad-band attosecond laser pulses predicts results quite different from the paraxial diffractive optics [1]. The results obtained in their paper demonstrated semi-spherical Fraunhofer type diffraction at distances of few microns. The initial laser pulse has the form of optical disc, as a superposition of plane waves. In the range of a couple of microns, before to reach the surface where are generated free electrons, the profile is changed into a semi-spherical shape, which is more close to Fraunhofer diffraction.

The problem of non-paraxial diffraction has been discussed in [2] where an analytical solution of the linear amplitude equation is obtained. The normalized non-paraxial linear amplitude equation (1) is derived from the following non-normalized equation written in Laboratory coordinates:

$$-2ik_0\left(\frac{\partial A}{\partial z}+\frac{1}{v}\frac{\partial A}{\partial t}\right)=\Delta A-\frac{1}{v^2}\frac{\partial^2 A}{\partial t^2} \tag{7}$$

In this equation the term connected with dispersion β is neglected because in air $\beta<<1$ and the effects of dispersion on the evolution of laser pulses are very weak.

The method used to solve equation (7) is similar to that of equation (1). The initial condition for Gaussian pulse is: $A_x(x,y,z,t=0)=\exp\left(-\frac{x^2+y^2+z^2}{2r_0^2}\right)$. Using the Fourier method the initial condition problem can be solved and the following exact solution is obtained:

$$\begin{aligned}A_x(x,y,z,t)=&\frac{i}{2\hat{r}}\exp\left(-\frac{k_0^2r_0^2}{2}+ik_0(vt-z)\right)\times\\ &\times\left\{i(vt+\hat{r})\exp\left(-\frac{1}{2r_0^2}(vt+\hat{r})^2\right)erfc\left(\frac{i}{\sqrt{2}r_0}(vt+\hat{r})\right)-i(vt-\hat{r})\exp\left(-\frac{1}{2r_0^2}(vt-\hat{r})^2\right)erfc\left(\frac{i}{\sqrt{2}r_0}(vt-\hat{r})\right)\right\}\end{aligned} \tag{8}$$

where $\hat{r}=\sqrt{x^2+y^2+(z-ir_0^2k_0)^2}$.

This analytical result describes the diffraction of narrow-band pulses as well as that of broad-band laser pulses. For example, when $k_0>>r_0$ (narrow-band case) we have Fresnel diffraction on several diffraction lengths, while for $k_0\geq r_0$ (broad-band case) the solution gives Fraunhofer type diffraction at these distances.

The analytical solution (8) as well as the numerical simulations in [1] are identical and describe very well the semi-spherical diffraction of optical pulses with few cycles under their envelope. On the other hand, the authors in [5] obtained *U*-type diffraction. It is important to mention that this type of

diffraction is described by solving the non-paraxial equation (UPPE equation) [5] in Local time coordinates.
One of the purposes of this work is to show that this *U*-type diffraction in Local time is simply the well-known Fraunhofer diffraction and it is due to the specific nature of Local time transformation.
To find the particularities of the diffraction of narrow and broad-band optical pulses we will investigate the evolution of initial Gaussian pulse in the frames of different types of equations.

a) The non-paraxial equation in Galilean frame:

$$-i\frac{2k_0}{v}\frac{\partial A}{\partial t} = \Delta A_{\perp} - \frac{1+\beta}{v^2}\left(\frac{\partial^2 A}{\partial t^2} - 2v\frac{\partial^2 A}{\partial t \partial z}\right) - \beta\frac{\partial^2 A}{\partial z^2} \tag{9}$$

b) The same type of equation written in Local time frames (1).
c) Paraxial spatio-temporal equation (6).

It is important to be pointed here that similar analytical solution as (8) is obtained for equation (9). As it can be expected, the diffraction is the same for both equations. There is only one difference: the translation of the pulse in Laboratory coordinate system and the stationarity of the pulse in Galilean.

3. Numerical results

To find differences in the diffraction picture we will solve numerically the spatio-temporal paraxial equation (6), the non-paraxial equation (1) written in Local time coordinate system and equation (9) written in Galilean frame for broad-band laser pulses with more than 10 optical cycles under the envelope.
For this purpose we studied the evolution of phase-modulated 25 *fs* laser pulses with a Gaussian spatio-temporal profile in air. The initial condition for equations (1) and (6) is:

$$A(x, y, \tau, z=0) = A_0 \exp\left(-\frac{x^2+y^2}{2r_0} - b\frac{\tau^2}{2\tau_0^2}\right), \tag{10}$$

where r_0 is the transverse size (the spot of the pulse), τ_0 is the temporal size and $b=1+ia$ is a complex number where a is the linear frequency modulation (chirp of the pulse).
The initial condition for equation (9) is:

$$A(x, y, z, t=0) = A_0 \exp\left(-\frac{x^2+y^2}{2r_0} - b\frac{z^2}{2z_0^2}\right), \tag{11}$$

where again r_0 is the transverse size (the spot of the pulse), $z_0=v_{gv}\tau_0$ is the spatio-longitudinal size and a is the same linear frequency chirp.
The connection between parameter a, the real time duration of the pulse t_{FWHM} and its spectral broadening $\Delta\omega_{FWHM}$ can be obtained by using the equation [6]:

$$t_{FWHM}\Delta\omega_{FWHM} = 2\pi c_B\sqrt{1+a^2} > 2\pi c_B \tag{12}$$

The parameter c_B is a constant. In our work the linear frequency modulation changes from 0 to 10.
In the beginning we will investigate the evolution of *25 fs* spectrally-limited optical pulse (chirp parameter $a=0$) governed in the frames of paraxial equation (6) under the initial condition (10). The

numerical results are shown in Figure 1. Here, as well as on the next fiqures, the side projection of the intensity profile is presented trought a distances of one diffractional length.

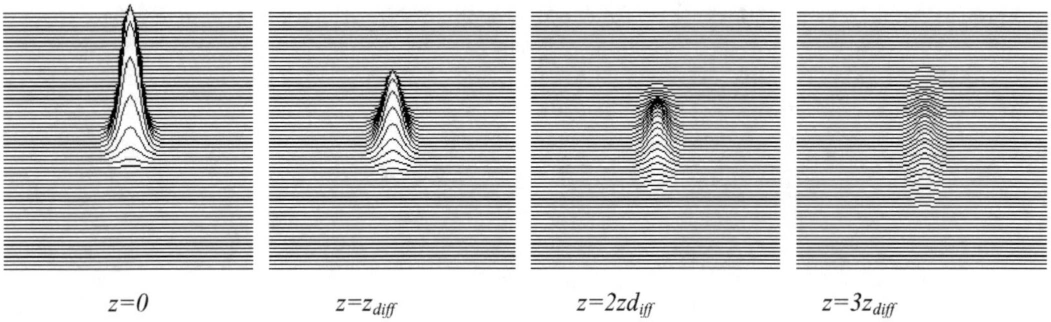

Figure 1. Plot of the side *(x-τ)* projection of the intensity profile of *25 fs* spectrally-limited optical pulse. The numerical result is obtained by solving the paraxial Eq. (6) and initial condition (10). The parameter $a=0$. The typical Fresnel diffraction is clearly seen. At one diffraction length the transverse size of the intensity enlarges tice and the initial amplitude decreases by factor $\sqrt{2}$.

The form of the pulse corresponds to the typical diffractive profile on a several diffractive lengths in the context of paraxial optics.
In Figure 2 are presented the results obtained by using the initial condition (10) and the equation (6). The chirp parameter is $a=10$.

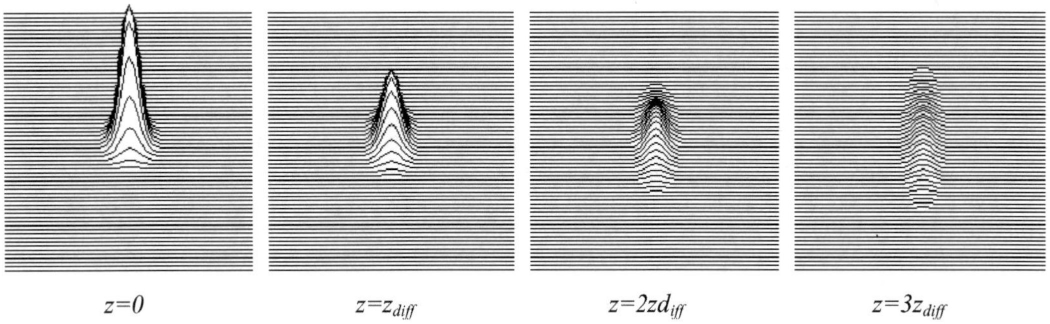

Figure 2. Plot of the side *(x-τ)* projection of the intensity profile of *25 fs* phase-modulated optical pulse. The results are obtained by solving the paraxial Eq. (6) and the same initial condition (10). The pulse is chirped: $a=10$. The phase-modulated pulse governed by the paraxial Eq.(6) again gives Fresnel diffraction. The result correspond to the simple fact that in the frames of paraxial optics the chirp and the spectral band of the pulse do not influence on the diffraction.

The form of the pulse corresponds to the typical diffractive profile on a several diffractive lengths in the context of paraxial optics. The figure shows that in the frames of paraxial optics the chirp and the spectral band of the pulse do not influence on the diffraction.
The same result is obtained by using the non-paraxial spatio-temporal equation (1) for unchirped pulse under the same initial condition (10).

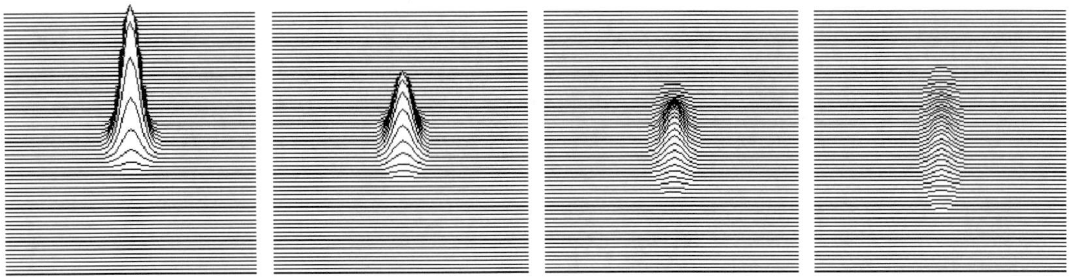

Figure 3. Plot of the side *(x-τ)* projection of the intensity profile of *25 fs* spectrally-limited optical pulse with parameter $a=0$. This result is obtained by solving the spatio-temporal non-paraxial Eq. (1) under the same initial condition (10). In its nature the non-modulated *25 fs* pulse is narrow-band, we see again Fresnel diffraction.

The form of the pulse is not different from the typical diffractive profile obtained in the frames of paraxial optics.
In Figure 4 are presented the results obtained under the same initial condition (10), equation (1) and chirp parameter $a=10$.

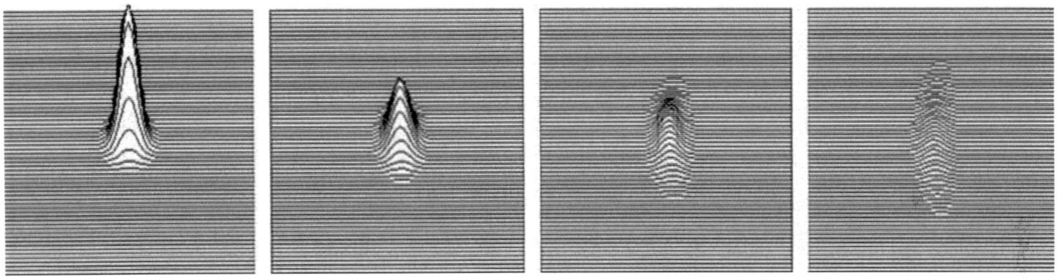

Figure 4. Plot of the side *(x-τ)* projection of the intensity profile of *25 fs* phase-modulated optical pulse with parameter $a=10$ under the same initial condition (10) and the spatio-temporal non-paraxial Eq. (1). The phase-modulated pulse is broad-band and its diffraction is of *U*-type (inverted semi-sphere) as in the case of attosecond pulses.

The form of the pulse significantly differs from the typical diffractive profile obtained in the frames of paraxial optics (Figure 2). The semi-spherical shape of the pulse is clearly seen. The parabolic form of the phase-modulated femtosecond laser pulse is observed on a several diffractive lengths. The Local coordinate system gives an inverted image of the real pulse deformation with respect to the axis *Oz*.
The question is: Is it possible to observe the real Fraunhofer type deformation obtained in the experiments [1]?
The real parabolic shape of the pulse is observed when we use equation (9) in Galilean frame under the initial condition of the kind (11) and chirp parameter $a=10$. The results are shown in Figure 5.

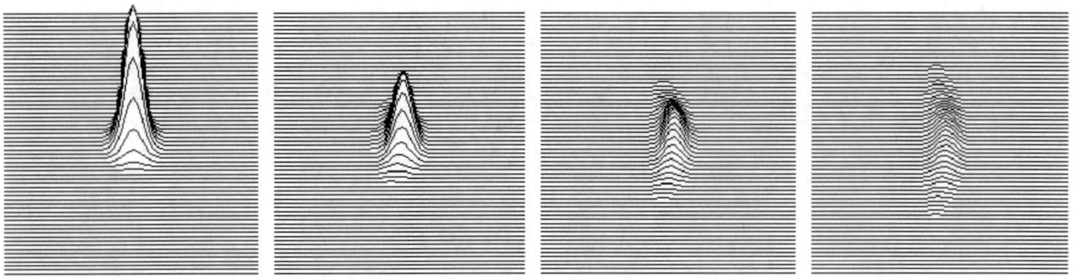

$z=0$ $z=z_{diff}$ $z=2zd_{iff}$ $z=3z_{diff}$

Figure 5. Plot of the side *(x-z)* projection of the intensity profile of *25 fs* phase-modulated optical pulse. The real spatial Fraunhofer type semi-spherical deformation is clearly seen. The initial condition (11), the non-paraxial Eq. (9) in Galilean frame and chirp parameter $a=10$ are used. It is obvious that the Local coordinate system gives an inverted image (with respect to axis *Oz*) of the real pulse deformation.

The intensity profile of the laser pulses in Figure 4 and 5 is a typical example of Fraunhofer diffraction for broad-band optical pulses which is the same as the diffraction of laser pulses with one-two cycles under the envelope (attosecond optical pulses). The solutions in Galilean fames give a real diffractive deformation of the pulses in the space. This is the reason why it is more appropriate to use this coordinate system instead of the Local time.

4. Conclusion

In the present work the propagation of optical pulses with narrow and broad-band spectrum in linear regime is investigated in the frames of three different evolution equations: paraxial spatio-temporal equation (6), non-paraxial amplitude equation in Local tame coordinate system (1) and non-paraxial amplitude equation in Galilean frames (9). The Fraunhofer type diffraction of phase-modulated *25 fs* laser pulses can be observed only in the frames of non-paraxial optics: equations (1) and (9). The paraxial optics is not sensitive to the spectral band of the pulses and cannot present this effect.
Our numerical results have shown that:

- The non-paraxial equation in Local coordinate system (1) gives the correct profile deformation of optical pulses but inverted with respect to the axis *Oz*;
- The real spatial Fraunhofer diffraction of femtosecond and attosecond laser pulses is obtained when the non-paraxial equation is solved in Galilean (9) or Laboratory coordinate system (1).

In this paper we have shown that the Fraunhofer type diffraction is not a characteristic only to attosecond or femtosecond pulses with one or few optical cycles under the envelope. There are different methods to obtain broad-band laser pulses with many cycles under the envelope. For examples: nonlinear self-phase modulation, optical grating etc.
We demonstrated that even in the case of phase-modulated femtosecond pulses with many cycles under the envelope Fraunhofer type diffraction can be observed.

References

[1] N. M. Naumova, J. A. Nees, I. V. Sokolov, B. Hou, and G. A. Mourou 2004 Relativistic Generation of Isolated Attosecond Pulses in $\lambda(3)$ Focal Volume *Phys. Rev. Lett* **92** 063902.

[2] Lubomir M. Kovachev and Kamen Kovachev 2008 Diffraction of femtosecond pulses: nonparaxial regime *JOSA A* **25** 2232-2243.

[3] Lubomir M. Kovachev and Kamen Kovachev 2008 Diffraction of femtosecond pulses: nonparaxial regime: erratum *JOSA A* **25** 3097-3079.

[4] Lubomir M. Kovachev and Kamen Kovachev 2011 *Linear and Nonlinear Femtosecond Optics in Isotropic Media: Ionization-free Filamentation* (InTech Open Access Publisher) ISBN 978-953-307-429-0.

[5] Couairon A., Mysyrowicz A. 2007 Femtosecond filamentation in transparent media *Physics Report* **441** 47-189.

[6] J.-C. Diels, W. Rudolph 2006 *Ultrashort Laser Pulse Phenomena* (New York: Academic Press) 2nd edition.

Journal of Physics: Conference Series **594** (2015) 012024 doi:10.1088/1742-6596/594/1/012024

Circular dichroism in nonlinear electron-positron pair creation

K. Krajewska and J. Z. Kamiński

Institute of Theoretical h sics, Facult of h sics, ni ersit of arsaw, asteura 5, 2- 3 ars awa, oland

-mail katarzyna.krajewska@fuw.edu.pl

Abstract. The trident pair production b -ra photons in the presence of a powerful laser eld is considered in the framewor of strong- eld uantum electrod namics. The dependence of di erential probabilit rates of the process on the handedness of circular polari ations of -ra and laser photons is illustrated. It is re ealed that circular dichroism e ects are most pronounced for short pulse durations of the dri ing elds.

1. Introduction

The possibility of converting energy into mass is one of the most fundamental issues of contemporary physics, as it provides a test of nonlinear quantum electrodynamics (ED). It led Sauter and Schwinger [1, 2] to their prediction of electron-positron (e^-e^+) pair creation from vacuum in the presence of a static electric field. The following investigation of e^-e^+ pair production by means of laser radiation started in the 1960s [3, 4, 5] (for recent reviews, see, Refs. [6, 7]). While the problem was of purely theoretical interest at that time, currently there is a clear perspective for corresponding experimental studies due to a remarkable and still ongoing progress in high-power laser technology. The reason is that field intensities well beyond 10^{20} W cm^2 are available today in many laboratories (see, e.g. [8, 9]). Note that in the near future laser intensities exceeding 10^{22} W cm^2 are expected to become also available (see, for instance, ELI and XCELS proposals [10, 11]).

Current technologies allow, in principle, to study experimentally e^-e^+ pair creation in powerful laser fields. The reason is that relevant field intensities are Doppler upshifted when a highly relativistic particle beam counterpropagates an intense laser pulse (in the rest frame of target particles the laser field intensity is enhanced by $4\gamma^2$ where $\gamma \gg 1$ is the Lorentz factor). This has led to the first, and until recently, only experimental observation of laser-induced e^-e^+ pair production [12]. In the experiment performed at the Stanford Linear Accelerator Center (SLAC) [12], the detected pairs were attributed to a nonlinear Breit-Wheeler process in which a high-energy photon generated by Compton backscattering collided with multiple laser photons. Another scenario of laser-induced pair creation exploiting the relativistic particle beams is the nonlinear Bethe-Heitler process, where the pairs are created directly in collisions of a laser and relativistic particle beams. Even though the latter process appears as a promising tool for e cient generation of electron-positron pairs, it is still awaiting experimental demonstration.

In our previous papers on nonlinear Bethe-Heitler process, we anticipated that in order to treat this process in strong laser fields it is important to account for a finite mass of the colliding target [13, 14, 15, 16, 17]. This would be particularly important when we consider the pair creation by an impact of a strong laser field on a beam of electrons – even lighter particles than nuclei. Let us note that in the famous SLAC experiment [12], electrons carrying energy of almost 50 GeV were colliding with an intense laser pulse. For such energetic electrons, the laser field strength that they experience is enhanced by a factor of 10^5, which would make it possible to perform the respective Bethe-Heitler–type experiment with presently available laser sources. We find this crucial to consider the situation similar to [13, 14, 15, 16, 17] but with an electron beam instead, i.e., the so-called *trident pair production* [18, 19, 20]. In contrast to aforementioned studies, we shall model the driving field as a two-color pulse train. Each color field will consist of a sequence of pulses (see, our previous papers on strong-field ED processes [21, 22, 23] as well): a high-frequency field will mimic X-rays while a low-frequency field will describe the laser beam, both being circularly polarized. This will allow us to study circular dichroism e ects in the trident pair creation.

Regarding convention, in the theoretical formulas we keep $\hbar \quad 1$. However, we present our numerical results in relativistic units such that $\hbar \quad c \quad m_{\rm e} \quad 1$, where $m_{\rm e}$ is the electron mass.

2. Theory

We describe the external electromagnetic field by the following four-vector potential,

$$A^\mu(k\cdot x) \quad A^\mu_{\rm L}(k\cdot x) \quad A^\mu_{\rm X}(k\cdot x), \tag{1}$$

which represents the laser field, $A^\mu_{\rm L}(k\cdot x)$, and the X-ray field, $A^\mu_{\rm X}(k\cdot x)$, both corresponding to pulse trains. More precisely,

$$A^\mu_i(k\cdot x) \quad A_{0i}f_i(k\cdot x)\Big[\varepsilon_1^\mu \sin(N_{i,\rm osc}k\cdot x)\cos\delta_i - \varepsilon_2^\mu\cos(N_{i,\rm osc}k\cdot x)\sin\delta_i\Big] \tag{2}$$

for $i\in\{{\rm L},{\rm X}\}$. A_{0i} defines the field strength, $N_{i,\rm osc}$ is the number of cycles within one pulse, while $f_i(k\cdot x)$ defines the wave envelope. In the following, we shall assume that

$$f_{\rm L}(k\cdot x) \qquad \sin^2\Big(\frac{k\cdot x}{2}\Big), \tag{3}$$

$$f_{\rm X}(k\cdot x) \qquad \sin^2\Big(\frac{N_{\rm X,osc}k\cdot x}{2N_{\rm L,osc}}\Big). \tag{4}$$

$\varepsilon_{1,2} \quad (0,\boldsymbol{\varepsilon}_{1,2})$, present in Eq. (2), are linearly polarized four-vectors satisfying the conditions: $\varepsilon_1^2 \quad \varepsilon_2^2 \quad -\boldsymbol{\varepsilon}_1^2 \quad -\boldsymbol{\varepsilon}_2^2 \quad -1$ and $\varepsilon_1\cdot\varepsilon_2 \quad -\boldsymbol{\varepsilon}_1\cdot\boldsymbol{\varepsilon}_2 \quad 0$. In addition, δ_i describes the ellipticity of the i-th field. For our further purpose, we note that for $\delta_i \quad -\pi/4$ the electromagnetic field (2) becomes left-handed circularly polarized while for $\delta_i \quad \pi/4$ it becomes right-handed circularly polarized. If ω is the fundamental field frequency and $\boldsymbol{n}$ is the field propagation direction, the wave four-vector becomes $k \quad (\omega/c)(1,\boldsymbol{n})$.

Note that for the vector potential defined by Eqs. (1) and (2) it holds that $k\cdot A_i \quad 0$ and $k\cdot k \quad 0$. Based on these properties, one can derive exact solutions of the Dirac equation coupled to the electromagnetic field the so-called Volkov states [24]. These laser-dressed states labeled by four-momentum p and spin pro ection on a quantization axis λ read

$$\psi^{(\beta)}_{p\lambda}(x) \quad \sqrt{\frac{m_{\rm e}c^2}{VE_{\boldsymbol{p}}}}\Big(1-\frac{\beta e}{2k\cdot p}\not{A}\not{k}\Big)u^{(\beta)}_{p\lambda}\,{\rm e}^{-{\rm i}\beta S^{(\beta)}_p(k\cdot x)}, \tag{5}$$

after being normalized in volume V. Here, the electron charge is $e<0$ whereas $\beta \quad 1$ stands for an electron and $\beta \quad -1$ for a positron. In Eq. (5), $u^{(\beta)}_{p\lambda}$ is a four-spinor satisfying the field-free

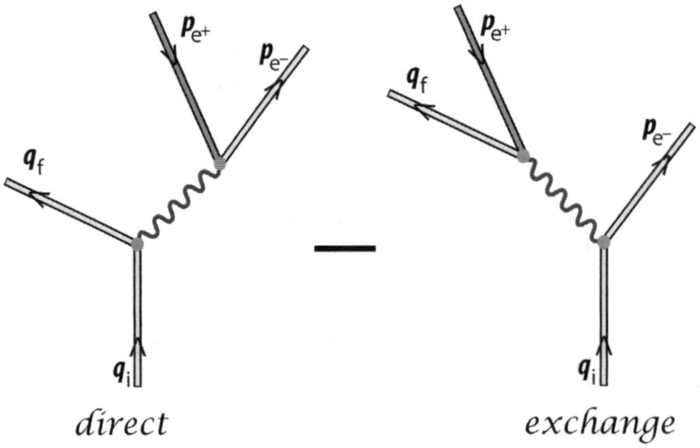

Figure 1. (Color online) The Feynman diagrams for the trident pair production. The fermion lines (double solid lines) represent the Volkov states for electrons and positrons in the background laser field.

equations, $(\not{p} - \beta mc)u^{(\beta)}_{p\lambda} \quad 0$, while the phase

$$S^{(\beta)}_p \quad p\cdot x \quad \int^{k\cdot x} \mathrm{d}\phi\Big[\beta e\frac{A(\phi)\cdot p}{k\cdot p} - \frac{(eA)^2(\phi)}{2k\cdot p}\Big] \tag{6}$$

has a meaning of the classical action of a particle (antiparticle) in the electromagnetic field.

The probability amplitude of e^-e^+ pair production by the impact of a powerful laser field on a relativistic electron, to the leading order in fine structure constant α, has the form

$$S_{\mathrm{fi}} \quad \mathcal{A}(q_{\mathrm{i}}, q_{\mathrm{f}}, p_{\mathrm{e}^+}, p_{\mathrm{e}^-}) - \mathcal{A}(q_{\mathrm{i}}, p_{\mathrm{e}^-}, p_{\mathrm{e}^+}, q_{\mathrm{f}}), \tag{7}$$

where the first term relates to the *direct* diagram whereas the second term relates to the *exchange* diagram, as shown in Fig. 1. Both terms can be written in a compact form as

$$\mathcal{A}(q_{\mathrm{in}}, q_{\mathrm{out}}, p_{\mathrm{in}}, p_{\mathrm{out}}) \quad -4\pi\mathrm{i}\alpha\int\mathrm{d}^4x\int\mathrm{d}^4y\, j^{(++)\mu}_{q_{\mathrm{out}},q_{\mathrm{in}}}(x)\mathcal{D}_{\mu\nu}(x-y)j^{(+-)\nu}_{p_{\mathrm{out}},p_{\mathrm{in}}}(y), \tag{8}$$

where

$$\mathcal{D}_{\mu\nu}(x-y) \quad \int\frac{\mathrm{d}^4\mathcal{K}}{(2\pi)^4}\tilde{\mathcal{D}}_{\mu\nu}(\mathcal{K})\,\mathrm{e}^{-\mathrm{i}\mathcal{K}\cdot(x-y)}, \tag{9}$$

$$\tilde{\mathcal{D}}_{\mu\nu}(\mathcal{K}) \quad -\frac{1}{\mathcal{K}^2}\Big(g_{\mu\nu} - \xi_{\mathrm{G}}\frac{\mathcal{K}_\mu\mathcal{K}_\nu}{\mathcal{K}^2}\Big), \tag{10}$$

is the photon propagator, with the gauge fixing constant ξ_{G}. Eq. (8) contains also four-currents $j^{(++)\mu}_{q_{\mathrm{out}},q_{\mathrm{in}}}(x)$ and $j^{(+-)\nu}_{p_{\mathrm{out}},p_{\mathrm{in}}}(y)$ which are expressed in terms of Volkov solutions (5),

$$j^{(++)\mu}_{q_{\mathrm{out}},q_{\mathrm{in}}}(x) \quad \bar{\psi}^{(+)}_{q_{\mathrm{out}}\lambda_{\mathrm{out}}}(x)\gamma^\mu\psi^{(+)}_{q_{\mathrm{in}}\lambda_{\mathrm{in}}}(x), \tag{11}$$

$$j^{(+-)\nu}_{p_{\mathrm{out}},p_{\mathrm{in}}}(y) \quad \bar{\psi}^{(+)}_{p_{\mathrm{out}}\chi_{\mathrm{out}}}(y)\gamma^\nu\psi^{(-)}_{p_{\mathrm{in}}\chi_{\mathrm{in}}}(y). \tag{12}$$

Journal of Physics: Conference Series **594** (2015) 012024 doi:10.1088/1742-6596/594/1/012024

Next, we introduce the Fourier expansion of both four-currents,

$$j^{(++)\mu}_{q_{\rm out},q_{\rm in}}(x) \quad \frac{m_{\rm e}c^2}{V\sqrt{E_{\boldsymbol{q}_{\rm out}}E_{\boldsymbol{q}_{\rm in}}}}\sum_{M=-\infty}^{\infty} \mathcal{C}^{\mu}_{M}(q_{\rm out}\lambda_{\rm out}, q_{\rm in}\lambda_{\rm in})\mathrm{e}^{-\mathrm{i}(Mk+\bar{q}_{\rm in}-\bar{q}_{\rm out})\cdot x}, \tag{13}$$

$$j^{(+-)\nu}_{p_{\rm out},p_{\rm in}}(y) \quad \frac{m_{\rm e}c^2}{V\sqrt{E_{\boldsymbol{p}_{\rm out}}E_{\boldsymbol{p}_{\rm in}}}}\sum_{L=-\infty}^{\infty} \mathcal{F}^{\nu}_{L}(p_{\rm out}\chi_{\rm out}, p_{\rm in}\chi_{\rm in})\mathrm{e}^{-\mathrm{i}(Lk-\bar{p}_{\rm in}-\bar{p}_{\rm out})\cdot y}, \tag{14}$$

where $\mathcal{C}^{\mu}_{M}(q_{\rm out}\lambda_{\rm out}, q_{\rm in}\lambda_{\rm in})$ and $\mathcal{F}^{\nu}_{L}(p_{\rm out}\chi_{\rm out}, p_{\rm in}\chi_{\rm in})$ are the corresponding Fourier coe cients and where the four-momenta of particles dressed by the laser field are defined, in general, as

$$\bar{p} \quad p - \frac{e^2\langle(A_{\rm L} \quad A_{\rm X})^2\rangle}{2k\cdot p}k. \tag{15}$$

Here, $\langle ...\rangle$ means the average over $k\cdot x$ from 0 to 2π. Plugging these expansions and the definition of the photon propagator into Eq. (7), we obtain

$$\mathcal{A}(q_{\rm in}, q_{\rm out},p_{\rm in}, p_{\rm out}) \quad \frac{-4\pi\mathrm{i}\alpha}{\sqrt{E_{\boldsymbol{q}_{\rm out}}E_{\boldsymbol{q}_{\rm in}}E_{\boldsymbol{p}_{\rm out}}E_{\boldsymbol{p}_{\rm in}}}}\frac{(m_{\rm e}c^2)^2}{V^2}(2\pi)^4\sum_{M,L}\int \mathrm{d}^4\mathcal{K}\,\delta(Mk \quad \mathcal{K} \quad \bar{q}_{\rm in}-\bar{q}_{\rm out})$$
$$\times\delta(Lk-\mathcal{K}-\bar{p}_{\rm in}-\bar{p}_{\rm out})\mathcal{C}^{\mu}_{M}(q_{\rm out}\lambda_{\rm out}, q_{\rm in}\lambda_{\rm in})\,\tilde{\mathcal{D}}_{\mu\nu}(\mathcal{K})\,\mathcal{F}^{\nu}_{L}(p_{\rm out}\chi_{\rm out}, p_{\rm in}\chi_{\rm in}), \tag{16}$$

where the remaining integral can be performed exactly. This leads to the following expansion,

$$\mathcal{A}(q_{\rm in}, q_{\rm out}, p_{\rm in}, p_{\rm out}) \quad \sum_{N=-\infty}^{\infty}\delta(\bar{q}_{\rm in}-\bar{q}_{\rm out}-\bar{p}_{\rm in}-\bar{p}_{\rm out} \quad Nk)\mathcal{A}_N(q_{\rm in}, q_{\rm out}, p_{\rm in}, p_{\rm out}), \tag{17}$$

where $M \quad L \quad N$ can be interpreted as a net number of photons exchanged with the laser field by target and product particles. The delta function in Eq. (17) expresses the four-momentum conservation condition, while $\mathcal{A}_N(q_{\rm in}, q_{\rm out}, p_{\rm in}, p_{\rm out})$ is the probability amplitude of the trident process by N-photon absorption:

$$\mathcal{A}_N(q_{\rm in}, q_{\rm out}, p_{\rm in}, p_{\rm out}) \quad \frac{-4\pi\mathrm{i}\alpha}{\sqrt{E_{\boldsymbol{q}_{\rm out}}E_{\boldsymbol{q}_{\rm in}}E_{\boldsymbol{p}_{\rm out}}E_{\boldsymbol{p}_{\rm in}}}}\frac{(m_{\rm e}c^2)^2}{V^2}(2\pi)^4\,t_N(q_{\rm in}, q_{\rm out}, p_{\rm in}, p_{\rm out}), \tag{18}$$

where $t_N(q_{\rm in}, q_{\rm out}, p_{\rm in}, p_{\rm out}) \quad \sum_L \mathcal{C}^{\mu}_{N-L}(q_{\rm out}\lambda_{\rm out}, q_{\rm in}\lambda_{\rm in})\tilde{\mathcal{D}}_{\mu\nu}(Lk-\bar{p}_{\rm in}-\bar{p}_{\rm out})\mathcal{F}^{\nu}_{L}(p_{\rm out}\chi_{\rm out}, p_{\rm in}\chi_{\rm in})$.

Let us define the total probability rate of trident pair production by a two-color pulse train:

$$W_N \quad \frac{1}{2}\sum_{\lambda_{\rm i},\lambda_{\rm f},\lambda_{\rm e^+},\lambda_{\rm e^-}=\pm}\int\frac{V\mathrm{d}^3q_{\rm f}}{(2\pi)^3}\frac{V\mathrm{d}^3p_{\rm e^+}}{(2\pi)^3}\frac{V\mathrm{d}^3p_{\rm e^-}}{(2\pi)^3}\delta(\bar{q}_{\rm i}-\bar{q}_{\rm f}-\bar{p}_{\rm e^+}-\bar{p}_{\rm e^-} \quad Nk)$$
$$\times \quad (2\pi)^4V|\mathcal{A}_N(q_{\rm i}, q_{\rm f}, p_{\rm e^+}, p_{\rm e^-})-\mathcal{A}_N(q_{\rm i}, p_{\rm e^-}, p_{\rm e^+}, q_{\rm f})|^2, \tag{19}$$

where the coe cient in front of the integral stands for an average over the initial and summation over the final spin degrees of freedom. We also define the probability rates related to the direct and exchange diagrams:

$$W_N^{(\rm direct)} \quad \frac{1}{2}\sum_{\lambda_{\rm i},\lambda_{\rm f},\lambda_{\rm e^+},\lambda_{\rm e^-}=\pm}\int\frac{V\mathrm{d}^3q_{\rm f}}{(2\pi)^3}\frac{V\mathrm{d}^3p_{\rm e^+}}{(2\pi)^3}\frac{V\mathrm{d}^3p_{\rm e^-}}{(2\pi)^3}\delta(\bar{q}_{\rm i}-\bar{q}_{\rm f}-\bar{p}_{\rm e^+}-\bar{p}_{\rm e^-} \quad Nk)$$
$$\times \quad (2\pi)^4V|\mathcal{A}_N(q_{\rm i}, q_{\rm f}, p_{\rm e^+}, p_{\rm e^-})|^2 \tag{20}$$

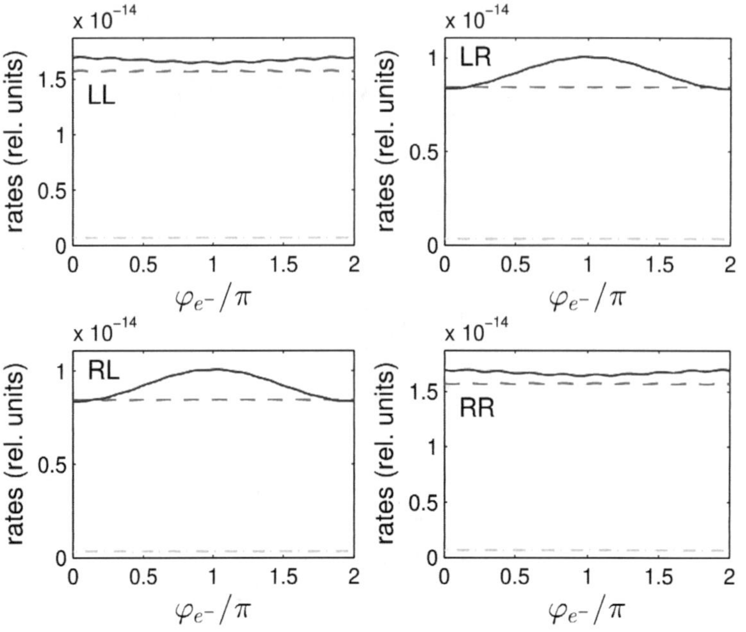

Figure 2. (Color online) Dependence of the di erential probability rates of trident pair creation by a two-color pulse train, as defined implicitly by (22), on the azimuthal angle of the product electron (solid blue line). The contributions from each diagram represented in Fig. 1 are shown: dashed red line corresponds to the direct diagram while the dotted-dashed green line corresponds to the exchange diagram. Each panel is for a di erent polarization setup for instance, LL which marks the top left panel relates to the results when both driving fields are left-handed circularly polarized, while RL in the bottom left panel relates to the results when the laser field is right-handed- and the X-ray field is left-handed circularly polarized. The results correspond to the case when both fields are composed of very long pulses, namely, $N_{\mathrm{L,osc}}$ 16 and $N_{\mathrm{X,osc}}$ 160. The remaining parameters are defined in the text.

and

$$W_N^{(\mathrm{exchange})} \quad \frac{1}{2} \sum_{\lambda_{\mathrm{i}},\lambda_{\mathrm{f}},\lambda_{\mathrm{e}^+},\lambda_{\mathrm{e}^-}=\pm} \int \frac{V\mathrm{d}^3 q_{\mathrm{f}}}{(2\pi)^3} \frac{V\mathrm{d}^3 p_{\mathrm{e}^+}}{(2\pi)^3} \frac{V\mathrm{d}^3 p_{\mathrm{e}^-}}{(2\pi)^3} \delta(\bar{q}_{\mathrm{i}} - \bar{q}_{\mathrm{f}} - \bar{p}_{\mathrm{e}^+} - \bar{p}_{\mathrm{e}^-} \quad Nk) \times (2\pi)^4 V |\mathcal{A}_N(q_{\mathrm{i}}, p_{\mathrm{e}^-}, p_{\mathrm{e}^+}, q_{\mathrm{f}})|^2. \tag{21}$$

Following Ref. [13], one can partially perform integrals in Eqs. (19), (20), and (21) by using the corresponding four-momentum conservation condition. As a result, one can represent W_N in Eq. (19) as

$$W_N \quad \sum_{\ell} \int \mathrm{d}E_{\boldsymbol{q}_{\mathrm{f}}} \mathrm{d}^2 \quad _{\boldsymbol{q}_{\mathrm{f}}} \mathrm{d}^2 \quad _{\boldsymbol{p}_{\mathrm{e}^-}} \frac{\mathrm{d}^5 W_N^{(\ell)}}{\mathrm{d}E_{\boldsymbol{q}_{\mathrm{f}}} \mathrm{d}^2 \quad _{\boldsymbol{q}_{\mathrm{f}}} \mathrm{d}^2 \quad _{\boldsymbol{p}_{\mathrm{e}^-}}}, \tag{22}$$

which implicitly defines the N-photon di erential probability rate of the trident process. Here, ℓ labels solutions of the four-momentum conservation condition: $\bar{q}_{\mathrm{i}} - \bar{q}_{\mathrm{f}} - \bar{p}_{\mathrm{e}^+} - \bar{p}_{\mathrm{e}^-} \quad Nk \quad 0$ where, in the most general case, $\ell \quad 1, 2, 3, 4$. Similarly one can define the N-photon di erential probability rates for the direct and exchange contributions to the process, which is based on

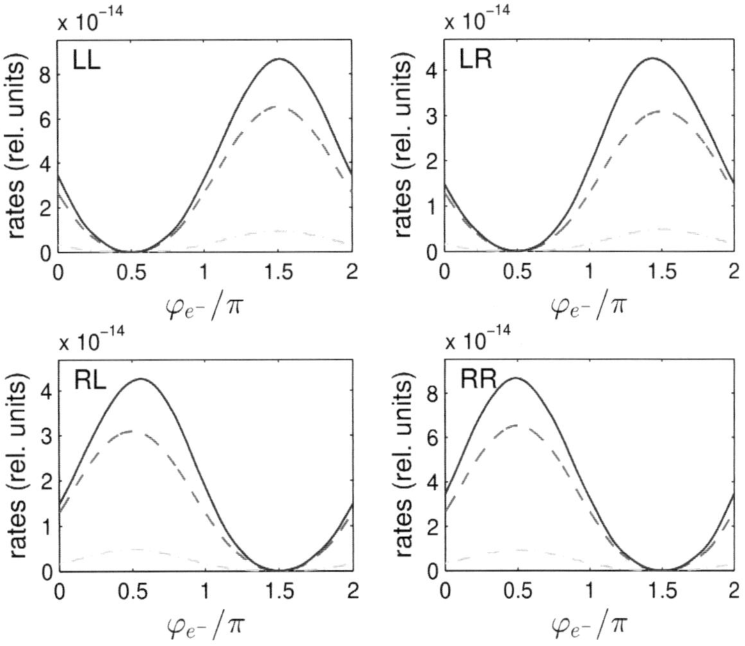

Figure 3. (Color online) The same as in Fig. 2 but both driving fields consist of much shorter pulses such that $N_{\mathrm{L,osc}}$ 2 and $N_{\mathrm{X,osc}}$ 20.

Eqs. (20) and (21). These quantities are plotted below to illustrate a possibility of observing circular dichroism in the trident process.

3. Numerical illustration

We present the results calculated in the reference frame in which the initial electron is at rest, meaning that $\boldsymbol{q}_{\mathrm{i}}$ $\mathbf{0}$. In the considered setup, we also assume that the electromagnetic field propagates against the z-axis, such that $\boldsymbol{k}$ $-\frac{\omega}{c}\boldsymbol{e}_z$, while the final electron is detected with the momentum $\boldsymbol{q}_{\mathrm{f}}$ $-m_{\mathrm{e}}c\boldsymbol{e}_z$. The parameters describing the laser field are ω_{L} $N_{\mathrm{L,osc}}\omega$ $m_{\mathrm{e}}c^2$ and μ_{L} $\frac{|eA_{0\mathrm{L}}|}{m_{\mathrm{e}}c}$ 0.2, whereas for the X-ray field: ω_{X} $N_{\mathrm{X,osc}}\omega$ $10\omega_{\mathrm{L}}$ and μ_{X} $\frac{|eA_{0\mathrm{X}}|}{m_{\mathrm{e}}c}$ 0.01. The results presented below correspond to the case when the energy of $30\omega_{\mathrm{L}}$ is absorbed from the electromagnetic field. Since the e^-e^+ pairs are mainly generated in the direction of the external field propagation (cf., in Refs. [13, 14]), we assume that the polar angle of the produced electron is θ_{e^-} 0.975π. In Fig. 2, we plot the dependence of di erential probability rates on the azimuthal angle of the produced electron φ_{e^-} in the case when a single modulation of the laser field contains 16 cycles ($N_{\mathrm{L,osc}}$ 16) while a single modulation of the X-ray field contains 160 cycles ($N_{\mathrm{X,osc}}$ 160). Each panel in Fig. 2 is for a di erent polarization setup. The upper left panel, labeled as LL , is for the case when both fields are left-handed circularly polarized. The upper right panel, labeled as LR , is for the case when the laser field is left-handed- while the X-ray field is right-handed circularly polarized. The same labeling scheme relates to the lower panels. Moreover, in each panel we present the results for the direct process (dashed red line), for the exchange process (dotted-dashed green line), and for the complete process with both contributions included (solid blue line). One can conclude from Fig. 2 that for the given parameters the exchange diagram contributes very little to the trident pair creation.

Nevertheless, the probability amplitude of the exchange process interfere constructively with the probability amplitude of the direct process, which holds for any polarization of the driving fields. It follows from Fig. 2 that the distributions for "LL" and "RR" polarizations, as well as for "LR" and "RL" polarizations, are the same. In other words, the probability rates do not change under a simultaneous change of handedness of both polarized fields. Let us compare the distributions corresponding to only one different handedness of either of the polarized fields, for instance, for "LL" and "LR" polarizations. One can see a qualitative difference between these distributions, despite a similar magnitude. It happens that more pronounced circular dichroism effects are observed if both external fields are composed of much shorter pulses. This is illustrated in Fig. 3 for $N_{\rm L,osc} = 2$ and $N_{\rm X,osc} = 20$ with the remaining parameters the same as in Fig. 2. In this case, the rates with different handedness of the polarized X-ray field are similar in shape but they differ in magnitude substantially. What is more striking, however, is that the rates seems to depend significantly on the polarization of the laser field (compare, for instance, the results for "LL" and "RL" polarizations). This is the main reason for differences between the rates for "LL" and "RR", as well as "LR" and "RL", polarizations.

4. Conclusions

We have presented preliminary results on the trident pair production in the presence of an X-ray field and a laser field, both modeled as circularly polarized pulse trains. We have demonstrated a sensitivity of the differential probability rates of pair creation to the handedness of the polarized driving fields. As we have shown, such circular dichroism is stronger for shorter pulses consisting the trains. A more complete analysis of this effect will be presented elsewhere.

Acknowledgments

This work was supported by the Polish National Science Center (NCN) under Grant No. 2011/01/B/ST2/00381.

References

[1] F. Sauter, Z. Phys. **69**, 742 (1931).
[2] J. Schwinger, Phys. Rev. **82**, 664 (1951).
[3] H. R. Reiss, J. Math. Phys. **3**, 59 (1962).
[4] A. I. Nikishov and V. I. Ritus, Sov. Zh. Eksp. Teor. Fiz. **46**, 776 (1964) [Phys. JETP **19**, 529 (1964)].
[5] V. P. Yakovlev, Zh. Eksp. Teor. Fiz. 49, **318** (1965) [Sov. Phys. JETP **22**, 223 (1966)].
[6] F. Ehlotzky, K. Krajewska, and J. Z. Kamiński, Rep. Prog. Phys. **72**, 046401 (2009).
[7] A. Di Piazza, C. Müller, K. Z. Hatsagortsyan, and C. H. Keitel, Rev. Mod. Phys. **84**, 1177 (2012).
[8] http://cuos.engin.umich.edu/researchgroups/hfs/facilities/hercules-petawatt-laser/
[9] http://www.unl.edu/diocles/
[10] http://www.eli-beams.eu/
[11] http://www.xcels.iapras.ru/
[12] D. Burke *et al.*, Phys. Rev. Lett. **79**, 1626 (1997).
[13] K. Krajewska and J. Z. Kamiński, Phys. Rev. A **82**, 013420 (2010).
[14] K. Krajewska and J. Z. Kamiński, Phys. Rev. A **84**, 033416 (2011).
[15] K. Krajewska, Laser Phys. **21**, 1275 (2011).
[16] K. Krajewska and J. Z. Kamiński, Phys. Rev. A **85**, 043404 (2012).
[17] K. Krajewska and J. Z. Kamiński, Phys. Rev. A **86**, 021402(R) (2012).
[18] H. Hu, C. Müller, and C. H. Keitel, Phys. Rev. Lett. **105**, 080401 (2010).
[19] A. Ilderton, Phys. Rev. Lett. **106**, 020404 (2011).
[20] B. King and H. Ruhl, Phys. Rev. D **88**, 013005 (2013).
[21] K. Krajewska and J. Z. Kamiński, Phys. Rev. A **85**, 062102 (2012).
[22] K. Krajewska and J. Z. Kamiński, Phys. Rev. A **86**, 052104 (2012).
[23] K. Krajewska, C. Müller, and J. Z. Kamiński, Phys. Rev. A **87**, 062107 (2013).
[24] D. M. Volkov, Z. Phys. **94**, 250 (1935).

Adaptive technique for matching the spectral response in skin lesions' images

P Pavlova[1], E Borisova[2], E Pavlova[3], L Avramov[2]
[1] Technical University Sofia Sofia branch of Plovdiv
[2] Institutes of Electronics, Bulgarian Academy of Sciences, Sofia, Bulgaria
[3]University hospital "Tsaritsa Ioanna-ISUL", Sofia, Bulgaria

E-mail:p_pavlova@gbg.bg

Abstract. The suggested technique is a subsequent stage for data obtaining from diffuse reflectance spectra and images of diseased tissue with a final aim of skin cancer diagnostics. Our previous work allows us to extract patterns for some types of skin cancer, as a ratio between spectra, obtained from healthy and diseased tissue in the range of 380 – 780 nm region. The authenticity of the patterns depends on the tested point into the area of lesion, and the resulting diagnose could also be fixed with some probability. In this work, two adaptations are implemented to localize pixels of the image lesion, where the reflectance spectrum corresponds to pattern. First adapts the standard to the personal patient and second – translates the spectrum white point basis to the relative white point of the image. Since the reflectance spectra and the image pixels are regarding to different white points, a correction of the compared colours is needed. The latest is done using a standard method for chromatic adaptation. The technique follows the steps below:
-Calculation the colorimetric XYZ parameters for the initial white point, fixed by reflectance spectrum from healthy tissue;
-Calculation the XYZ parameters for the distant white point on the base of image of non-diseased tissue;
-Transformation the XYZ parameters for the test-spectrum by obtained matrix;
-Finding the RGB values of the XYZ parameters for the test-spectrum according sRGB;
Finally, the pixels of the lesion's image, corresponding to colour from the test-spectrum and particular diagnostic pattern are marked with a specific colour.

1. Introduction
The main goal of our investigations is a development of non-invasive technique for skin cancer diagnostics using optical spectra in the range of wavelength from 380 to 780 nm, where are the most significant changes between healthy and diseased skin.

In a previous investigation we achieved the pattern distributions based on reflected spectra, corresponding to some types of skin damage [1]. The patterns are normalized ratio between spectra of healthy and diseased tissue. The main problem in this investigation – the lack of patients in the set of every kind of cutaneous pathology that is patterned, is attended with other one - variations of the spectra in dependence on position of the tested area used for spectrum, obtaining inside of one lesion and stage of cancer progress [2]. Developing the presented technique we want to support the solving these problems combining images and spectra, and ensure the fidelity of the patterns.

23rd International Laser Physics Workshop (LPHYS'14) IOP Publishing
Journal of Physics: Conference Series **594** (2015) 012025 doi:10.1088/1742-6596/594/1/012025

The current investigations in the area give much information about human skin properties [3] and relation between illness and different features of optical signals [4, 5]. However the process of illness recognition is still unstable and depends on variable factors. The analysis of the all actual optical methods applied in practice for non-invasive skin cancer diagnostics [6] shows that "further studies are needed in finding the appropriate method or combination of methods that can have wide clinical applications".

Searching for answer of the question: "which areas of the healthy skin and tissue inside of the lesion produce particular spectra?", we defined three tasks for realization as a part of the technique:

1. The spectra obtained from the pigmented skin match the pixels of the same areas in the image
2. The spectra of the healthy skin corresponds to the pixels of healthy skin in the image
3. Division the healthy spectrum into pattern distribution, the result will be a spectrum for pigmented area/ lesion, adapted to particular subject, if there is available "patterned" disease.

For this work we use well developed mathematical techniques for operating with the signals that are part of colorimetry. The present standards in that area establish two methods for finding the parameters of one colour: by spectral distribution in the range of visible light and by integrated signals from three detectors of visible light [7]. The both type of obtaining leads to 6 parameters description: 3 basic - XYZ and 3 relative -xyz that are primary in CIE system [7]. Two of them – x,y, form chromaticity, and are presented as a graphical diagram named Locus. The all other used parameters are result from transformation of the XYZ [8,9].

The use of three detectors for R(red),G(green) and B(blue) reproducing signals is the base of image registration by cameras. Development of computer technologies brought to establishment of special standard (ICC) for exchanging the information about colour parameters between different digital devices [10]. This standard also describes translations, needed for adaptation the colours measured in different conditions.

There are described results from our first attempt to apply the ICC standard in achievement of the aim to combine spectral data with the registered digital images.

2. Methods and materials

According to the ICC standard cameras are calibrated in sRGB space. It means generation of RGB reproducing signals that gives chromaticity coordinates inside of triangle area in Locus diagram with relative white point to illuminant –D65 – averaged daylight. Chromaticity coordinates –x,y,z of basic stimulus – Rp,Gp,Bp and white point for the sRGB are given in table 1. Figure 1 shows the position of basic triangle inside of Locus.

Table 1. sRGB and white source coordinates

	Rp	Gp	Bp	white point D_{65}
x	0,6400	0,3000	0,1500	0,3127
y	0,3300	0,6000	0,0600	0,3290
z	0,0300	0,1000	0,7900	0,3583

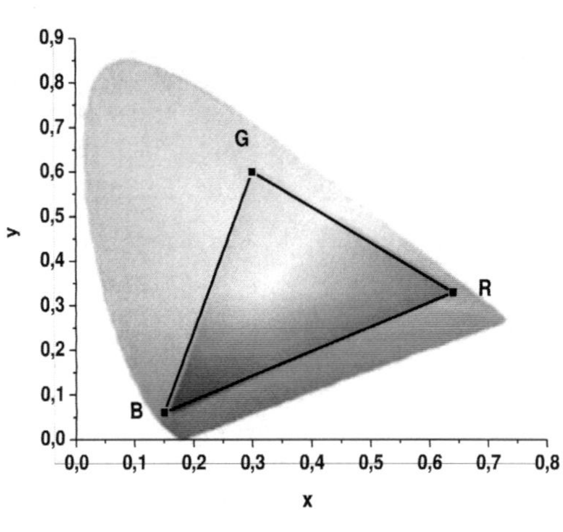

Figure1. The sRGB triangle area inside of Locus

The cameras are not colorimeters, because they transform the colours outside of the area of triangle to points inside of it, and the standard [8] fixes the common used transformations about it. However, colorimetry gives calculation /measurement of colour lightness and chromaticity by:

- Integrated signal from three detectors corresponding to three stimuli
- Spectral distribution

That is why we can find a colour, once from registered spectral distribution and again – as a relation between reproducing signals for pixels of the image of the same area. These two colours could be directly compared at the level of chromaticity, if the both coordinates are inside of the sRGB triangle. Specially developed software, appropriate to rules of registration, has been used for calculation the chromaticity coordinated for spectra and pixels of the images. Conditions of registration and calibration of devices in both cases are described below.

2.1 Spectral registration

System USB4000, product of "Ocean Optics", Inc. equipped with a fibre and a high-sensitivity linear CCD detects the reflected light dispersed by a grating with 600 lines/mm, and a total spectral resolution of the micro-spectrometer approximately 1.5 nm. The portion of light reflected from a sample is expressed as percentage R (%), related to a signal reflected from a standard white surface (Spectralon®). The software calculates R by the following equation:

$$R_\lambda = \frac{S_\lambda - D_\lambda}{SR_\lambda - D_\lambda} \times 100\% \qquad (1)$$

Where R_λ – reflectance (%), S_λ – intensity of the light reflected from the sample, D_λ – the dark currant correction, SR_λ – intensity of the light reflected from the standard white surface, index λ – wavelength. Calibration procedure:

- Measurement of the light reflected from the white surface and the dark current measurement (without any external signal). The calibration procedure of the spectrometer fix the white point to E- illuminant [7], that means chromaticity coordinates x=0.3333 y=0.3333.

2.2 Image registration

For the test of our software, we used images, registered by •7.1 Megapixels camera – Olimpus, initial calibration - sRGB and daylight to illuminate the tested skin

The CIE parameters of XYZ systems were calculated for the both cases – spectrum and pixels of the image. The coordinates are transformed according to standard colour adaptive procedure –CAT, for adaptation in case of different white points. The coordinates of chromaticity are compared by the difference between them. Finally – the pixels in the image have been marked by white colour in case of coincidence of coordinates.

In this work, two adaptations are implemented to localize pixels of the image lesion where the reflectance spectrum corresponds to the pattern. First adaptation of the pattern to the personal patient by multiplying to healthy spectra:

$$fd_p(\lambda) = fh_p(\lambda).fr(\lambda) \qquad (2)$$

where $fd_p(\lambda)$ is tested spectra for particular patient, $fh_p(\lambda)$ -healthy spectra for particular patient, $fr(\lambda)$ -pattern distribution for particular skin damage;

And second – translation of the spectrum white point basis to the relative white point of the image.

2.3 Main stages of the software processing

- CIEXYZ parameters calculation using spectra from skin areas and from adapted patterns;
- CIEXYZ parameters calculation for each image pixel;

23rd International Laser Physics Workshop (LPHYS'14) IOP Publishing
Journal of Physics: Conference Series **594** (2015) 012025 doi:10.1088/1742-6596/594/1/012025

- CAT (colour adaptation transformation)- using matrix M with coefficients defined by Bradford [11]

$$\begin{bmatrix} X_d \\ Y_d \\ Z_d \end{bmatrix} = [M].\begin{bmatrix} X_s \\ Y_s \\ Z_s \end{bmatrix} \qquad [M] = \begin{bmatrix} 1.0209581 & 0.0132500 & 0.0107183 \\ 0.0216398 & 0.9773060 & 0.0002158 \\ 0.0040457 & -0.0079939 & 1.0896730 \end{bmatrix} \quad (3)$$

Where XYZ – colour parameters in CIEXYZ; Index d- illuminant E; Index s – illuminant D_{65}.

- Obtaining the similarity by comparing the distance between coordinates of the points from spectrum and pixels and fixing the pixels with Δx and Δy under a limiting value.

.

$$\Delta x = [(x_{sprectrum} - x_{pixel})^2]^{1/2}$$
$$\Delta y = [(y_{sprectrum} - y_{pixel})^2]^{1/2} \quad (4)$$

The test data include spectra obtained from
• pigmented skin areas
• 4 different areas of healthy skin: elbow areas- medial and lateral, hand-palm, and surrounded pigmented area and also images of the same areas
- six patterns for skin damages: Basal-Cell Carcinoma (BCC); Compound Nevus; Derma Nevus; Dysplastic Nevus; Malignant Melanoma (two patterns).

3. Experimental results

Figure 2 gives the graphical view of the point corresponding to particular test –spectra and their position as chromaticity.

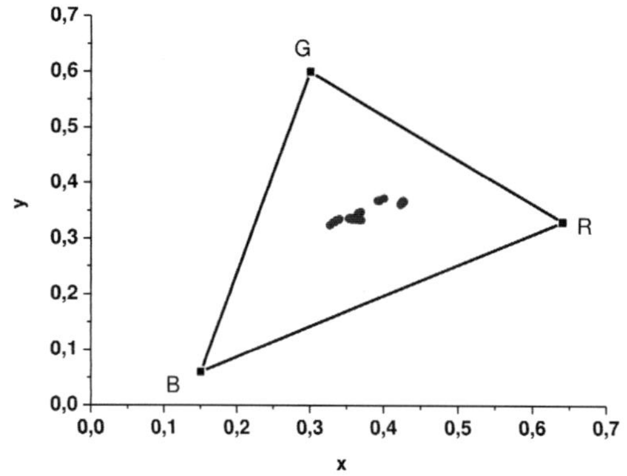

Figure 2. Chromaticity coordinates of some test-spectra and sRGB triangle

The figures 3 and 4 show matched pixels in the images in limiting distance between chromaticity coordinates of pigmented and healthy areas and spectra, obtained from the same areas.

Figure 5 shows results of matching pixels using different "healthy" spectra applied to pigmented areas of one patient.

Figure 6 shows matching with use of three types of "healthy" spectra.

At figure 7 are given spectrum of surrounded "healthy" spectrum and results of matching pixels with different limiting values.

Figures 8 and 9 are presented the test – spectra, results of calculation using patterns and matched pixels in the image for limiting value 0.003.

23rd International Laser Physics Workshop (LPHYS'14) IOP Publishing
Journal of Physics: Conference Series **594** (2015) 012025 doi:10.1088/1742-6596/594/1/012025

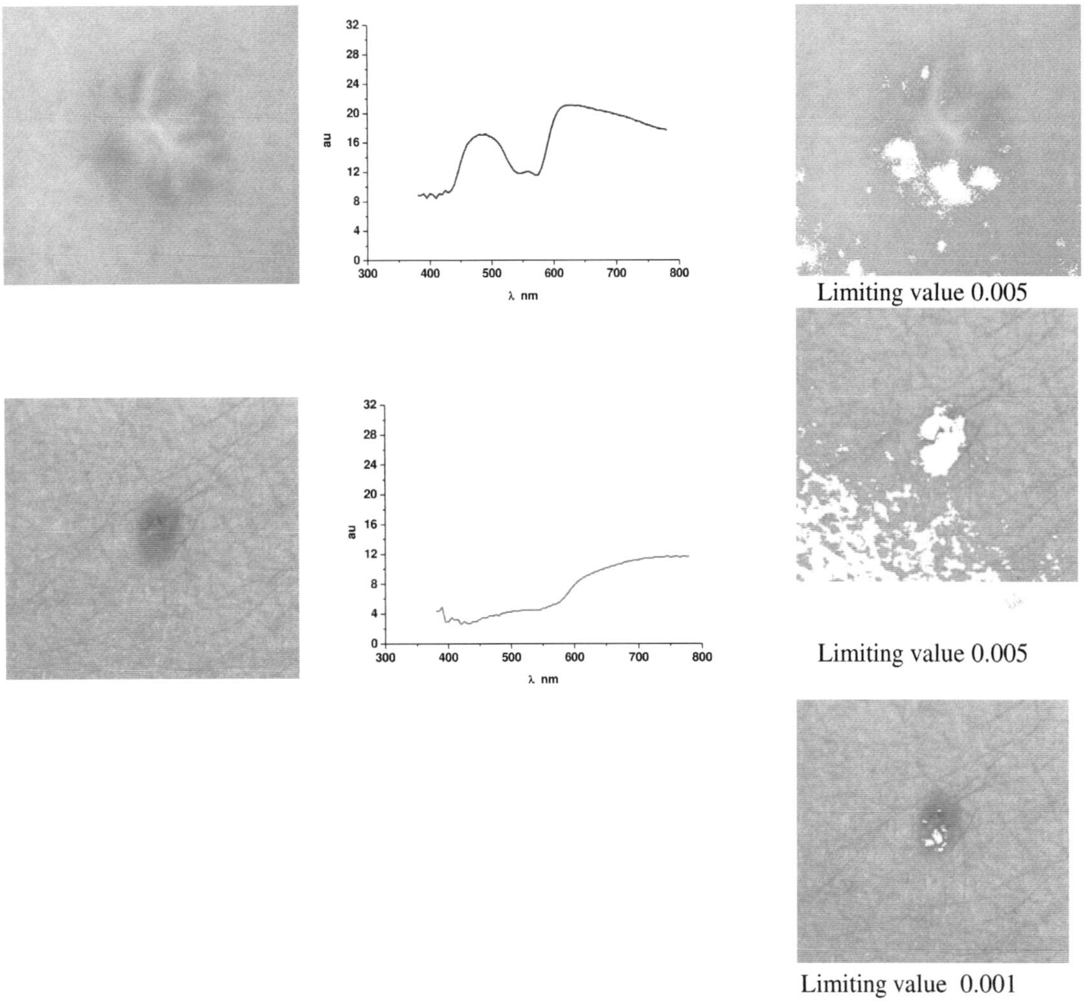

a) b) c)

Figure 3. Original images - a), registered spectra -b), images with matched pixels for pigmented areas – c)

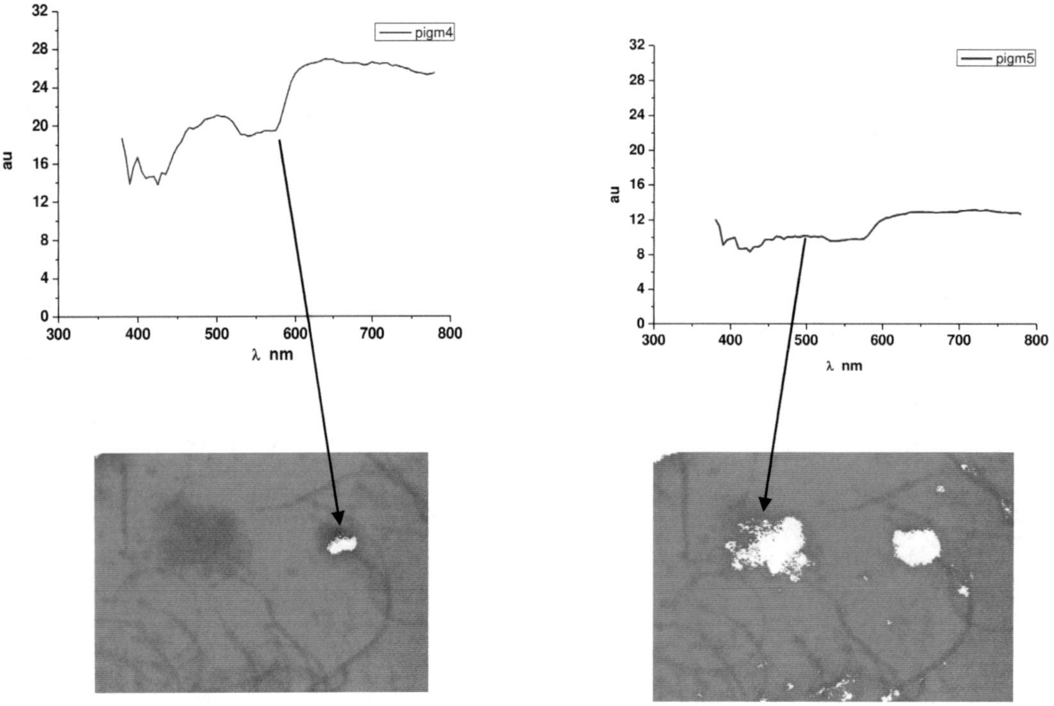

Figure 4. Spectra and matched pixels for two different areas of one patient

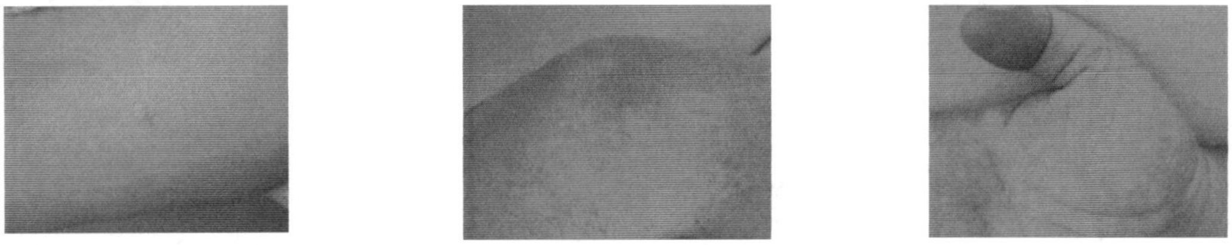

a)

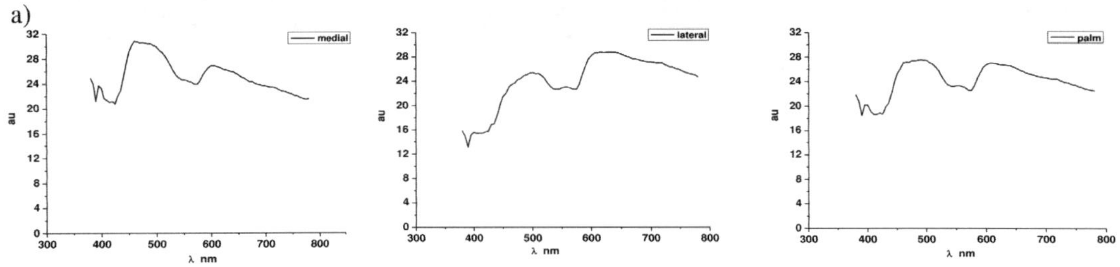

b)

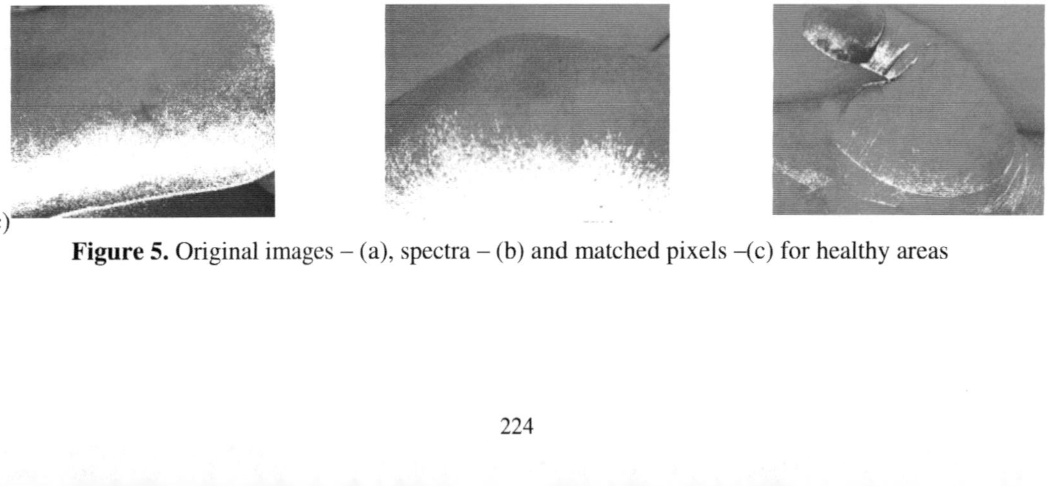

c)

Figure 5. Original images – (a), spectra – (b) and matched pixels –(c) for healthy areas

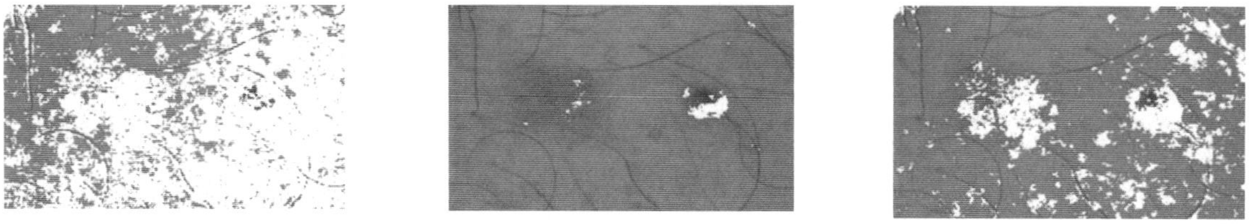

a) b) c)

Figure 6. Matched pixels from: a)- medial area healthy spectrum, b)- lateral area healthy spectrum, c)- palm area healthy spectrum

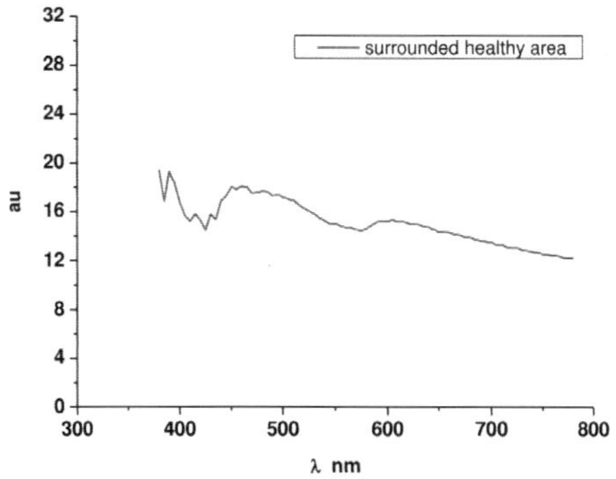

a)

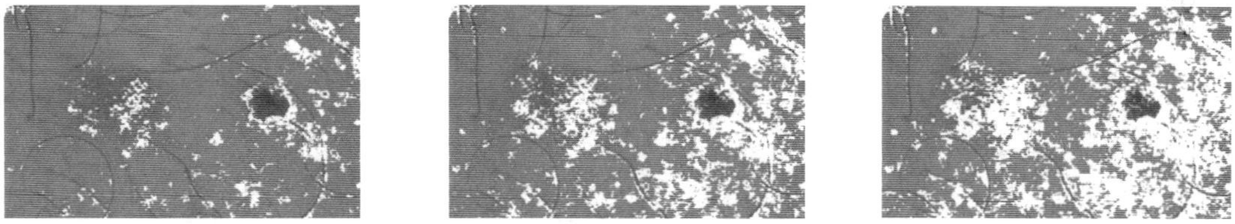

Limiting value 0.0002 Limiting value 0.0003 Limiting value 0.0004

b)

Fiure7. Spectrum from surrounded healthy area- a), matched pixels with different limiting values – b)

Journal of Physics: Conference Series **594** (2015) 012025 doi:10.1088/1742-6596/594/1/012025

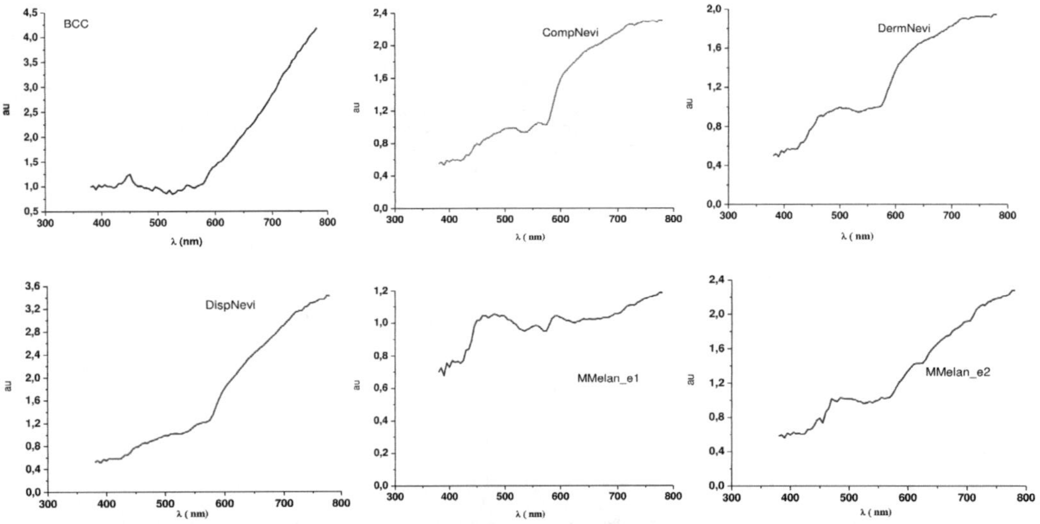

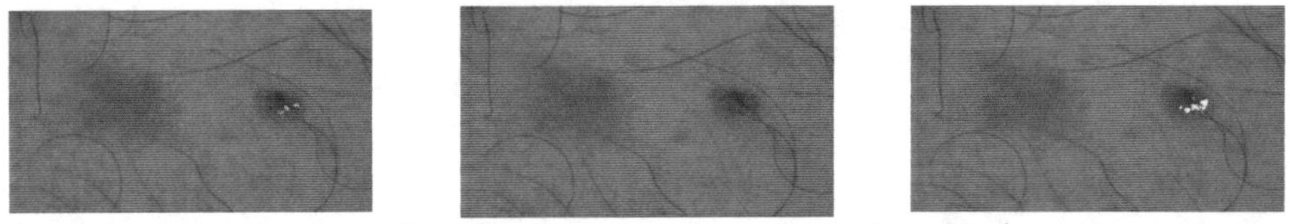

BCC CompNevi DermNevi

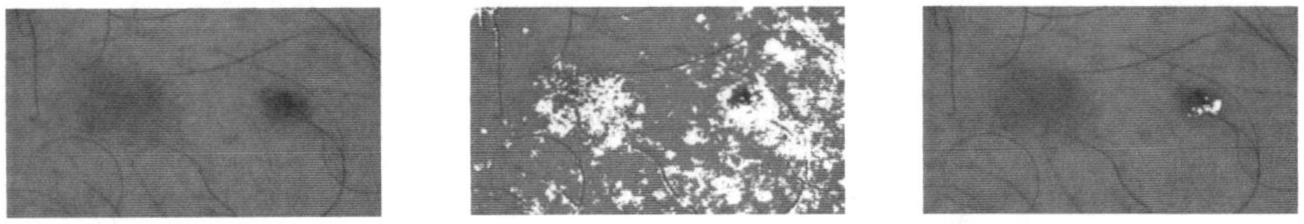

DispNevi MMel_e1 MMel_e2

Figure 8. Spectra for different patterns based on lateral healthy spectrum and corresponding matched pixels in the images

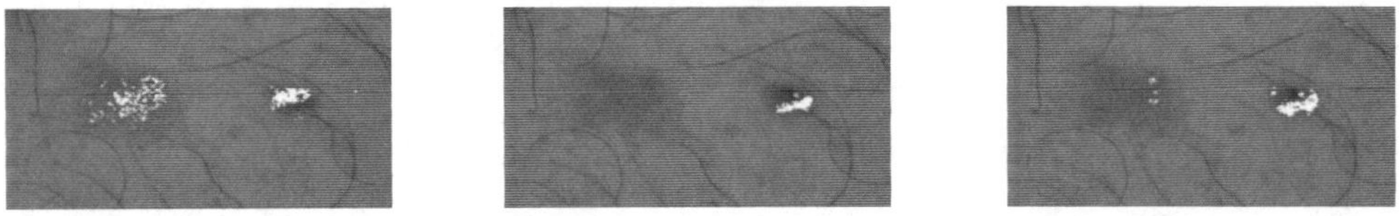

BCC CompNevi DermNevi

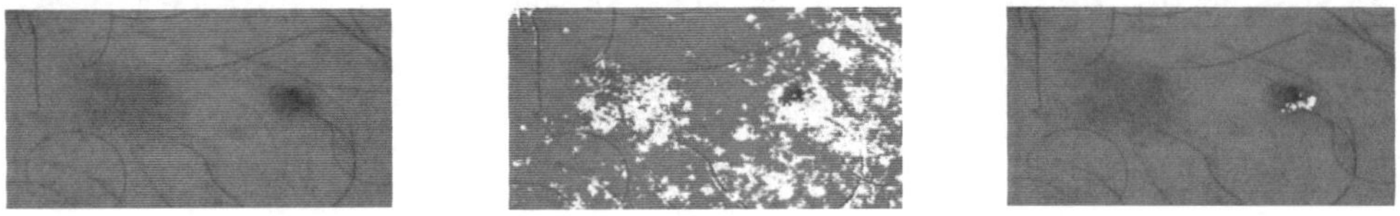

DispNevi MMel_e1 MMel_e2

Figure 9. Corresponding matched pixels in the images results from different patterns based on surrounded area healthy spectrum

4. Discussion

The points of chromaticity coordinates of the spectra are inside the sRGB triangle (fig. 2) it means that the spectral chromaticity could be directly compared with the points from the image.

Analysing the results from pixel matching we extracted needed conditions for registration:

For the spectra:

- The position of obtaining healthy spectra is very important for the extraction of the adapted spectra. The spectrum from surrounded lesion area seems to be most suitable for this aim.
- The numeric aperture of the fibre – if it is too high the signal is averaged from healthy and diseased area

For the images:

- Equal lightening of the whole area – healthy and diseased.
- If the level of illumination is too low or too high, the sensors work with non-linear transfer of signals. As a result the possible number of chromatical coordinates decreases. It changes the distance for comparison.
- Available blics always give bad results of processing.

Regarding to the fidelity of presented pattern distributions and in respect to results for different limiting values and subjects it could be said that only the different types of nevi are suitable for testing diagnoses using these reflectance spectra transformations. Similar effect of applicability is observed as well, when compare feasibility of diffuse-reflectance spectroscopy for cutaneous lesions diagnosis with other spectral techniques [12]. The melanin-pigmented pathologies, such as nevi, are highly differentiated by sub-type, when their reflectance spectra are compared, but non-melanoma pathologies do not reveal high diagnostic specificity of differentiation, if only their reflectance properties are used as diagnostic indicators.

5. Conclusion

The problem of matching pixels in the image coloured from obtained spectrum give us two main conclusions:

First conclusion must be that at this initial stage of the work we fix more problems than good results. In spite of fact that theoretically the problem could be solved, practical experiments show very strong dependence on used illumination and detection devices.

Second one – application of ICC standard principals has to be adapted to this particular task.

6. References

[1] Pavlova P, Borisova E, Avramov L 2004 Automation of the Skin Cancer Diagnostics on the Bases of Reflectance Spectroscopy Proc. of IX International Conference "Biomedical Physics and Engineering Sofia 260-265

[2] Pavlova P, Borisova E, Avramov L, Petkova El, Troyanova P 2011 Investigation of Relations between Skin Cancer Lesions' Images and Their Reflectance and Fluorescent Spectra ed. by Mandi Murph Melanoma in the Clinic - Diagnostics, Management and Complications of Malignancy InTech Open Access Publisher, DOI: 10.5772/19232.

[3] Lister T, Wright Ph A, Chappell P 2012 Optical properties of human skin, *J. Biomed.Opt.* **17(9)** 090901 doi:10.1117/1

[4] Richards-Kortum R, Sevick-Muraca E 1996 Quantitative optical spectroscopy for tissue diagnosis. *Annu Rev Phys Chem.***47** 555-606

[5] Drakaki E et al. 2009 Laser-induced fluorescence and reflectance spectroscopy for the discrimination of basal cell carcinoma from the surrounding normal skin tissue, *Skin Pharmacol. Physiol.* **22(3)** 158–165

[6] Calin MA, et al. 2013Optical techniques for the noninvasive diagnosis of skin cancer. *J Cancer Res Clin Oncol.* **139(7)** 1083-104. doi: 10.1007/s00432

[7] CIE 15.2/ 1996 - Colorimetry

[8] Judd D B 1979 Contributions to Color Science NBS

23rd International Laser Physics Workshop (LPHYS'14) IOP Publishing
Journal of Physics: Conference Series **594** (2015) 012025 doi:10.1088/1742-6596/594/1/012025

[9] Kuehni R G 2003 Color Space and Its Divisions: Color Order from Antiquity to the Present John Wiley & Sons

[10] ICC.1:2004-10 Image technology colour management — Architecture, profile format, and data structure

[11] www.bruselindbloom.com

[12] Borisova E, Troyanova P, Pavlova P, Avramov L 2008 Diagnostics of pigmented skin tumors based on laser-induced autofluorescence and diffuse reflectance spectroscopy *Quantum Electronics* **38(6)** 597-605

REAL-TIME MONITORING OF CHANGES OF THE PATHOGENS CONCENTRATION IN WATER UNDER INFLUENCE OF NANOSILVER PARTICLES BY RESONANCE LASER SPECTROSCOPY TECHNIQUES

T.Yu. Moguilnaya[1], P.V. Bobkov[1], V. I. Tomilin[1], and A. B. Kononenko[2]

1 - JSC "SERVET-M", 115404, Russia, Moscow, ul. 8th Radial,

2 - Veterinary Sanitation, Hygiene and Ecology Institute 123022, Russia, Moscow, Zvenigorodskoe Highway 5

E-mail: mogilnay@mail.ru

Abstract We investigated the nonlinear scattering in solutions of water containing E. coli and the mixture of E. coli and silver nanoparticles. As a result of these investigations, we developed a method of controlling the concentration of pathogenic organisms in the water and the concentration of the nanoparticles influences these organisms. For the study we selected pathogens - goldish staphylococcus, E-coli and coliphage MS-2. Standard biological method is qualitatively in agreement with our results.

1 INTRODUCTION

The optical phenomena of the stimulated scattering in water solutions, which contain DNA, are investigated insufficiently. However, these researches are great interest for the development of automatic control systems of potable water. Content of pathogenic microorganisms in water can be varied from 10^{-8} to 10^{5} mg / m^{3}. Hygienic standards of quality of potable water require a continuous of monitoring absence pathogenic microorganisms directly in a water flow.

Despite of a great number of laboratory devices for potable water's quality, there is no express analyzer for monitoring of pathogenic organisms, which could be directly embedded into the automatic checkout systems. The reasons are low concentrations of pollutions and additional effects (superluminescence, stimulated Brillouin scattering (SBS), which impede automatic data-processing.

Laser methods are widely applied analysis of structure of multicomponent liquids and the unique express tool for low concentration's determination.

In our previous papers, we showed different pathogens [1-3], that stimulated radiations (superluminescence in water solutions, which contain DNA) can appear at enough low power of exciting radiation [1]. Frequencies of these stimulated radiations depend on a DNA type and, accordingly, on a type of pathogenic organism. [3]. The radiation density in this solution allows nonlinear phenomena, such as stimulated Brillouin scattering (SBS)- [1]. This density is reached due to combining of the superluminescence fields and laser radiation [2]. In result, characteristic peak occurs. Similar peaks are well interpreted by methods of the theory of pattern recognition. The purpose of our research is showing that the residual impurity in potable water don't give superluminescence and don't capable to form significant peak and thus to complicate automatic recognition. In addition, we are interested in the opportunity of optical controlling method of destruction pathogenic organisms in the stream on the same hardware. We are developed a method for verifying the effectiveness silver's influence to pathogens. We investigated the possibility of realizing this method Also, we investigated the possibility of destruction pathogenic organisms by adding a solution of silver nanoparticles.

2 EXPERIMENTAL SETUP

We use a made-self stand in our experiments. The Fig. 1 shows the scheme of the experimental stand. The device contains the block with sources of radiation in optical IR-range. The receiver laser radiation with a high accuracy of spectral resolvability (Δ<0.02 nm) is chosen. As a result the developed experimental stand included the following elements (fig.1):

The input waveguide brings one of the exciting radiations (with wavelength λ1=1017 nm, λ2=810 nm λ= 670 nm) to the cuvette, containing the sample. The radiation, which passed through the cuvette and output waveguide, was analyzed by spectrum analyzer "Agilent"

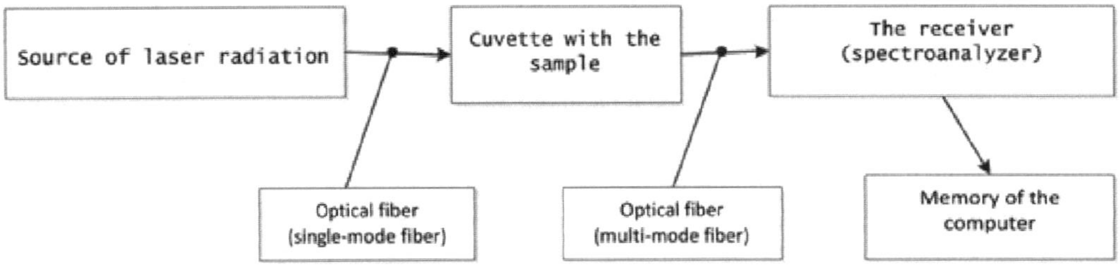

Fig.1. Scheme of the experimental stand

3 THE STUDY OF WATER

In order to reveal effects, which can impede automatic data-processing, we examined probes of potable water (without E- coli), which was given from five different areas of Russia. The creation of express analyzers, allowing full automatic signal processing, requires a study of additional signal characteristics apart from analysis of main spectral characteristics (peak position, intensity and shape of spectral bands) of radiation. For instance, a choice recognition algorithm depends on strong additional signal's characteristics, such as a signal/noise ratio, an averaging, stability of exciting and emitted radiations, absence of overlapping spectral bands from analyzed substance (E-coli) and other substances (water and impurities).

To study stability of exciting radiations from our laser sources, we study spectra of their radiations, passing through the empty cuvette. The fig. 2 shows the spectra of radiations from the laser sources with wavelengths 1017, 810, respectively. The laser sources with wavelengths 1017, 810 nm demonstrated unimode laser radiations with deviation of wave about 3 nm. However, we observed multimode laser radiation from the laser source with λ3 = 670 nm (fig.4). The visible spectral region is inconvenient for automatic recognition [3]. For this reason, we didn't use laser source with λ3 = 670 nm our further experiments. Spectra of the laser sources which have passed through object of research have qualitatively identical appearance. Therefore we give selective results for one lasers wavelength. However, we used all results for developing the software. It is made to provide signal recognition with probability 0.95.

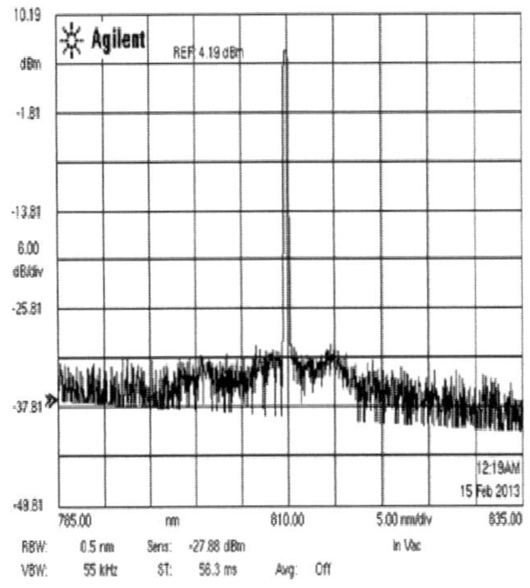

Fig.2. The spectrum of laser radiation with wavelength λ2 = 810 nmpassing through the empty cuvette with waterat averaging degrees 10

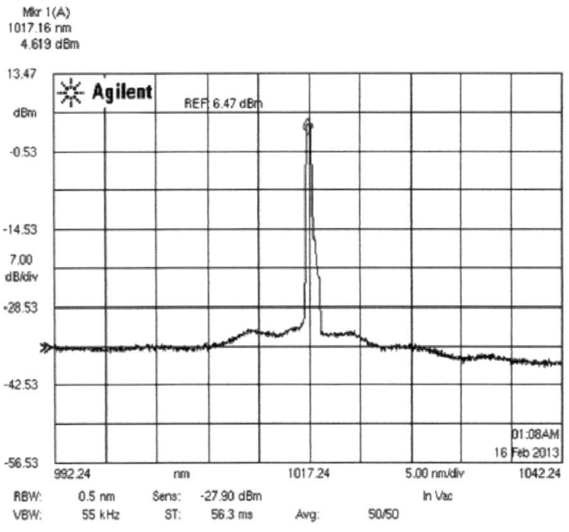

Fig.3. The spectrum of laser radiation with wavelength λ1 = 1017 nm, passing through the empty cuvette with waterat averaging degrees 10

Also we determined optimal averaging degrees of signal characteristics. The fig. 3 shows spectra of laser radiation with wavelength λ1 = 1017 nm. We found out, that the increase of averaging degrees from 10 to 50 did not change signal/noise ratio. Thus, we preferred averaging degree 10. We received the same results for the laser with a wavelength 810 nm.

The analysis of spectra of water from the different regions revealed, that additional spectral bands from water impurities doesn't appear in wavelength regions from 750 to 1500 nm, which we selected for diagnostic of E-coli. However, we observed deviations in intensity and in wavelength of exciting radiations. We investigated water for Mosrentgen's region and the region Pechatniki, water from the well, and water from the Moskva-river. Fig. 4 shows the results of statistical analysis of the intensity from exciting radiations, passing through the samples of water from these regions.

	Descriptive Statistics (OSA in Mosrentgen.stw)				
Variable	Valid N	Mean	Minimum	Maximum	Std.Dev.
Amplitude	60	5,242	3,536	7,695	1,110093
WaveLength	60	16,933	1015,850	1017,650	0,362410

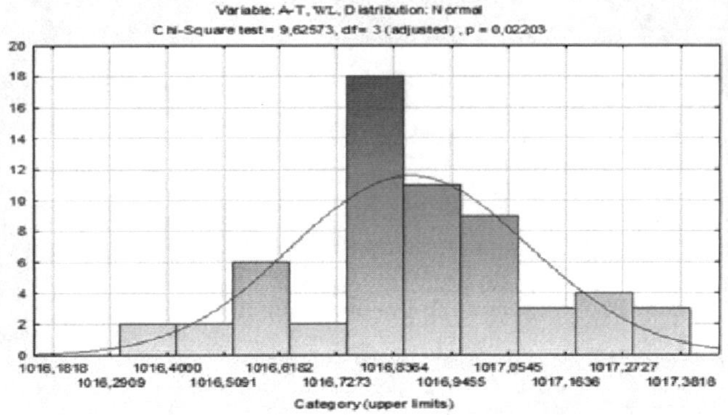

Fig.4. Results of statistical processing statistics (above) and distribution of deviation of parameters of the laser peaks (below) for various, impurities of water

Fig. 5 represents an example of statistical processing of intensity of laser radiation for different samples from one region. Figure 6 shows to intensity of peaks laser and impurity from frequency for various samples from different regions.

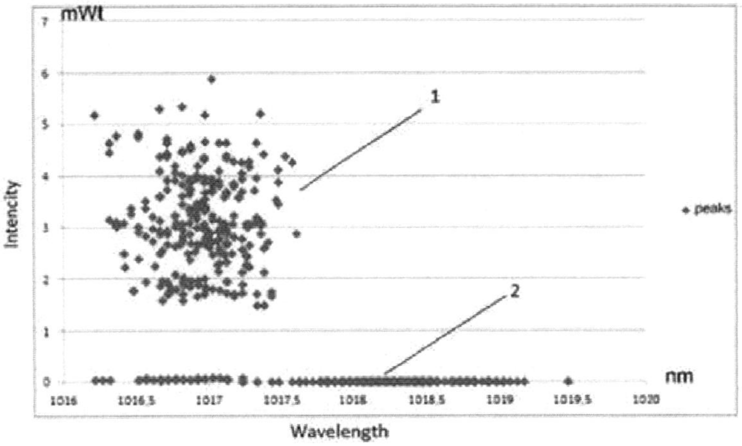

Fig.5. The distributions of the peak position: 1 - distribution of the maximum of the spectrum of the laser radiation passing through the water, taken in different regions 2- distribution of maxima of the spectra of impurities.

The statistical processing of the portable water spectra from different region shows, that the standard deviations for the characteristics of the exciting radiations doesn't exceed 20% for intensity and 0,03% for wavelength. The analysis of water spectra from different regions has shown, that in the chosen range of frequencies there are no parasitic peaks of a luminescence of water impurity. The mistake caused by own noise, does not exceed 0.001%. Therefore, the frequency maximum of intensity peak can serve as informative parameter at recognition of an image.

4 THE STUDY OF WATER SOLUTION OF E-COLI

We researched water solution of E- coli (strain K12) with concentrations 10, 10^2, 10^3, 10^4, 10^5, 10^6, 10^7, 10^8, 10^9 cells/ml (lg tds50).. The water solutions of E- coli (strain K12) were prepared by the following procedure. The culture was grown on the meat-peptonic agar during 24 hours, after produced washing off sterile physiological solution, set on the standard absorbance concentration 10^9 microbial mews/ml. The other solutions were prepared by the tenfold breeding prepared suspensions containing from 10^8 to 10 microbial/ml).

We investigated E-coli solutions of that what to receive informative parameters for automatic recognition of existence it in water. We analyzed about 700 spectra of the E- coli for exciting wavelength 810 nm, about 100 spectra of the E- coli for wavelengths 670 and 1017 nm, and 1400 spectra of the water (without E- coli) for each the areas.

The fig. 6 shows the example of spectra for two concentration 10 and 10^3 cells/ml. We found, that if concentration of E-coli is less than 10^3 cells\ml, we could see the peak in Stokes area. And if concentration of E-coli changes from 10^3 to 10^7 cells\ml, we could see the peak is in anti-Stokes area. The peak is in Stokes area if concentration of E-coli exceeds 10^8. The fig. 7 shows dependence of the logarithm of intensity on concentration E-coli at averaging 10.We found out that the logarithm of intensity depend on concentration same at environment excitement by the laser with a length of wave of 1017 nm.

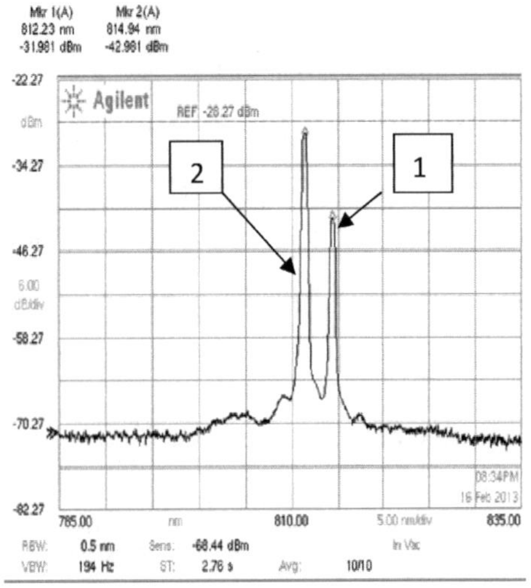

Fig.6. Example of spectra of radiation with concentration E.coli 10^3 o/l: 1 –laser; 2 E – coli

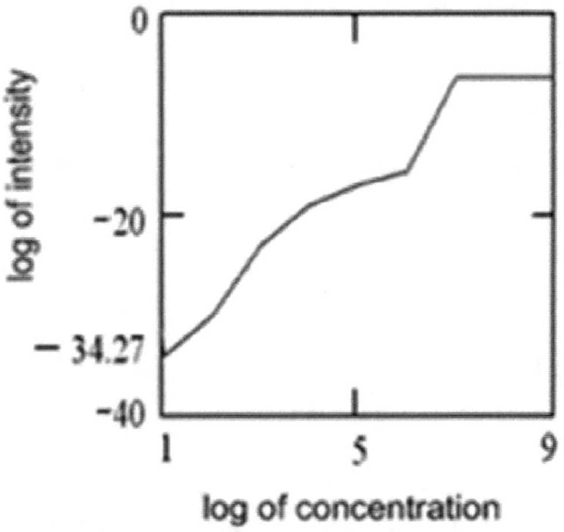

Fig.7. The dependence of the logarithm of intensity on concentration E.coli at averaging 10

We studied change peak dynamics of the pathogens. Dynamics change of intensity distribution of E-coli differs from laser radiation dynamics. The peak arises after some delay. Time of the delay varies from shares of seconds up to ten seconds. The E-coli peak grows, and the laser peak decreases. The laser's peak becomes less than the luminescence peak, if E-coli concentration of more than, 10^3. Both radiations are polarized and coherence [1]. The fig. 7 shows the dependence of the logarithm on intensity concentration E-coli (dBm) at time (μs) at virus concentration $10^{\,2}$ cells\ml.

We found out, that exciting radiations with wavelengths 1017 and 810 nm induce the stimulated Brillouin scattering in spectra of the water, containing pathogen DNAs. Threshold power level for appearance the stimulated Brillouin scattering was achieved, by electromagnetic fields from the exciting radiation and the stimulated fluorescence. We found out, that peak positions and widths are "fingerprints" for pathogens under study, and optical densities of these bands is proportional pathogen content, if this content is less 15%. Thus, Stokes and anti-Stockes bands of the stimulated Brillouin scattering can be used to recognize the pathogens under studies.

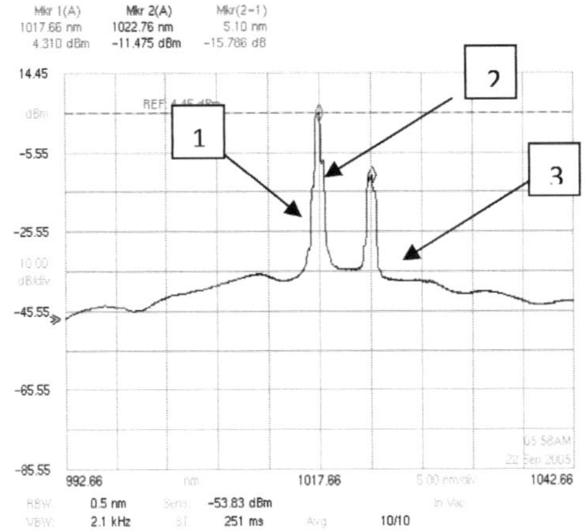

Fig. 8. The emission spectra : 1 – of the laser and 2-silver in the water content of0.0001mg silver\cell water excitation wavelength λ2.=1017nm; 3 – of the pathogen7lgn tcd50

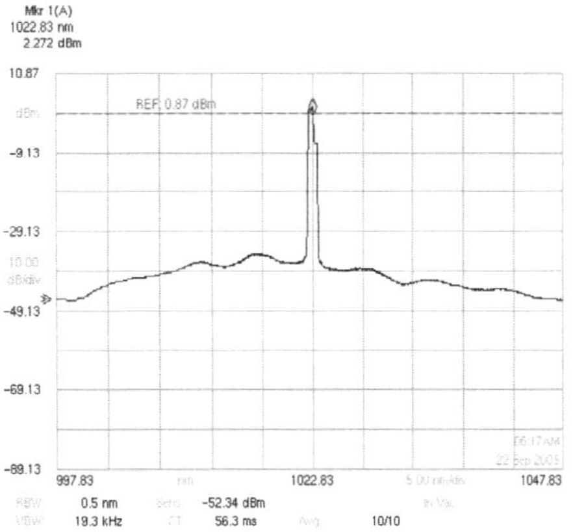

Fig. 9. The emission spectra : – of the laser and the silver content of silver in water0.001mg\cell water excitation wavelength λ2.=1017nm; 2 – of the pathogen content7lgn tcd50

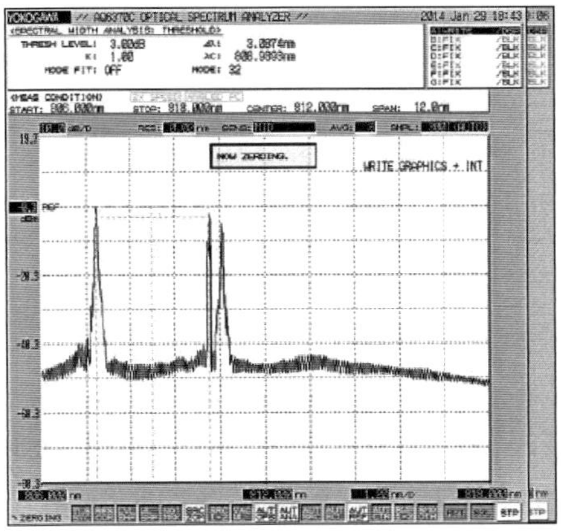

Fig. 10. *The emission spectra: 1 – of the laser and silver in the water content of 0.0001mg silver\cell water excitation wavelength λ2.=1017nm; 2 – of the pathogen (goldish staphilococcus) 7lgn tcd50*

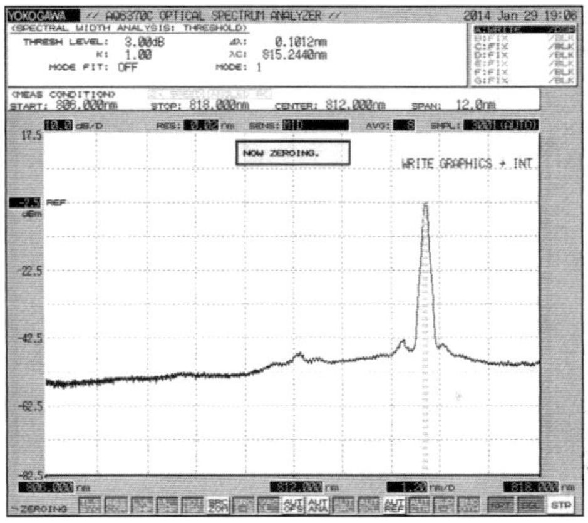

Fig. 11. *The emission spectra: 1 – of the laser and the silver content of silver in water0.001mg\cell water excitation wavelength λ2.=1017nm;2 – of the pathogen content (goldish staphilococcus)7lgn tcd50*

Based on these studies, we have developed a working model of the device for the automated control of pathogens in the water directly.

5 THE STUDY OF WATER SOLUTION OF NANSILVER AND PATHOGENS

We studied the influence of nanosilver on changes on the concentration of the live pathogens. For the study we selected E-coli, Salmonella and goldish staphilococcus.

We added to solutions containing pathogens, solutions of silver nanoparticles in water. Fig 8 and 9 show typical spectra of the E-coli solutions. Figure 10 and 11 shows the emission spectra of the laser and silver in the water content of 0.001 mg silver \ l cell water excitation wavelength λ2. = 1017 nm goldish staphilococcus 7 lgn tcd 50 (10^7 cells\ ml) a (Fig 9). The emission spectra of the laser and the silver content silver in water 0.0001 mg o\ ml cell water excitation wavelength λ2. = 1017 nm patogens content lgn tcd 7 50 (10^7 cells \ ml). The intensity of this peak decreases with the concentration of the Bouguer's law. Comparative analysis of the spectral distribution of the solution in water, silver and silver virus at a wavelength of 1017 nm showed (Figure 8 9-), that at silver concentrations up to 0.01 mg \ L, reaction occurs protein with silver. That leads to the emission peak of silver at a wavelength of 1022. Peaks, corresponding to E-coli disappears completely. Whereby, this emission spectrum has a width of distribution lower than the spectral width of the laser in the air. The intensity of these peaks is depend on the content of silver in the solution, indicating complete separation of the protein from the nucleic acid in solution and complete destroying of the virus. Other pathogens studies showed similar results.

5 CONCLUSION

We developed a new method that combines the method of forced luminescence and stimulated Brillouin scattering. We carried out the statistical spectral analysis of water from different regions in order to determine the statistical errors of the method.

The analysis of spectra of water from different regions showed, that there are not parasitic peaks of a luminescence of water impurity in the chosen range of wavelength (750-1400 nm). The error of signal caused by own noise, doesn't exceed the 0.01%. Therefore the frequency of maximum of peak of intensity can serve as informative parameter at recognition of an image.

We researched spectral characteristics of E-coli in water an amount necessary for the initial identification with 95% probability.

We found out that exciting radiations with wavelengths 1017 and 810 nm induces the stimulated Brillouin scattering in spectra of the water, containing pathogen DNAs. Threshold power level for appearance the stimulated Brillouin scattering was achieved, by electromagnetic fields from the exciting radiation and the stimulated fluorescence. We found out, that peak positions and widths of are "fingerprints" for pathogens under study, and optical densities of these bands is proportional pathogen content, if this content is less 15%.

Thus, Stokes and anti-Stockes bands of the stimulated Brillouin scattering can be used to recognize the pathogens under studies.

Based on these studies, we have developed a working model of the device for the automated control of pathogens in the water directly.

We studied the nanosilver influence on the pathogens concentration. For the study we selected pathogens - goldish staphylococcus, E-coli and coliphage MS-2. Standard biological method is qualitatively in agreement with our results.

REFERENCES

[1] T.Yu. Moguilnaya., Botikov .A. G., Saguitova EA Influence of instability of laser radiation on accuracy of record and reading of the information of diagnostic complex " Intest1" " ; Proceeding of Spie 31 .07 -4.07 2005, USA Volume 5908-41; 12 cnh.;

[2] T.Yu. Moguilnaya E.A. Saguitova, K.A. Prokhorov, , E.I. Grebenuk, N.A. Grebenuk Laser instrumentation for express diagnostic of toxin in liquid biological systems Fall Meeting 2000: Optical Technologies in Physics and Medicine II, Proceeding of SPIE, Vol.4241 (2001), p.504-506 –

[3] T.Yu. Moguilnaya A. Saguitova, K.A. Prokhorov, E.I. Grebenuk, N.A. Grebenuk Laser instrumentation for express diagnostics of impurities and toxins in liquid food Photonic Detection and Intervention Technologies for Safe Food, Proceeding of SPIE, Vol. 4206 (2001), p. 256-265

XeCl laser treatment of polyethersulfone membrane in the air and water

A Hashemi Panah[1], H pazokian[2], M Mollabashi[1], J Barzin[3]

[1] Department of Physics, Iran University of Science and Technology, Tehran, Iran
[2] Laser and Optics Research School, Tehran, Iran
[3] Biomaterial Department, Iran Polymer and Petrochemical Institute

E-mail: h_pazokian@iust.ac.ir

Abstract. XeCl laser irradiation of Polyethersulfone membranes in air and water were done. The irradiated surface were modified chemically or morphologically depends on the laser parameters and the mediums in which irradiation is done. The results in improving the surface hydrophilicity and biocompatibility for the biological applications were compared.

1. Introduction

When a biomaterial contacts tissue fluid or blood, its surface comes in to contact with the physiological environment. Since, many implantable biomaterials are constructed from polymeric materials, the biocompatibility of these materials is critical in their biological applications such as blood and cell contact works. The surface morphology and the surface chemistry specially wetting properties of the surface can significantly affect the extent of these interactions [1].

Nowadays laser irradiation is known as an appropriate method for both chemical and morphological modification of the materials used in different industries, especially medical demands [2-4].

Laser processing can perform in ambient air, vacuum or in the presence of a gas or liquid. Laser interaction in the presence of liquid mostly consists of conversion of part of the light energy into a mechanical impulse, which transports material and induces shock waves. Also in some cases dissociation products react chemically with materials. Most of these effects can be achieved using neutral liquids, but water is the most common, cheap and safe medium which used for wet etching.
Laser interaction in water has been studied since the 1970s and the commercial applications of water-assisted laser cutting, shock processing and surface cleaning from particulate contamination is now commercially available [5].

Polyethersulfone (PES) is among polymeric materials for medical applications. It is a derivative of high temperature engineered thermoplastic polysulfone polymer. It is frequently used for various medical applications such as membranes for hemodialysis and medical implants [5-7]. Unfortunately, the low hydrophilicity of Polysulfones has undesirable effects on its blood contact works [6, 8].

In this paper XeCl laser treatment of polyethersulfone membranes in air and water is investigated and the results of improving the surface hydrophilicity and biocompatibility for the biological applications are compared.

2. Materials and Methods

2.1. Materials preparation
Polyethersulfone (PES, Ultrason E6020, MW = 58,000, flakes) supplied by BASF Co. was employed as main polymer and polyvinylpyrrolidone (PVP, K90) provided by Fluka Co. was used as additive. Dimethylacetamide (DMAc) (Aldrich) was used as the solvent and pure water as nonsolvent.
The phase inversion method [9] was used for the preparation of membranes. The polymer solutions of membranes containing 18% (wt) PES and 3% (wt) PVP in DMAc were used.

2.2. Laser irradiations
XeCl laser (Lambda Physics, 308 nm, τ = 25 ns) was used as irradiation source. The laser was operated at repetition rate of 1 Hz. The central part of the beam was selected by an aperture. The samples were irradiated at various fluences in two different environments (air and water).

2.3. Sample characterization
The morphology and hydrophilicity of the surface before and after irradiation was examined by scanning electron microscopy (TESCAN, Vega series) and statistic water drop contact angle measurements (contact angle goniometer (Krüss G10)), respectively. Attenuated total reflectance (ATR) spectra for determination of chemical changes on PES membrane surfaces were recorded using a FTIR spectrophotometer (Perkin Elmer, Model: Spectrum-100) in the region 650 to 4000cm^{-1} with a resolution of 4 cm^{-1}.

Cell culture study was done using L929 mouse cells for surface biocompatibility measurement.

3. Results and discussions

3.1. The morphology of membranes
Figure1 shows the SEM images of the surface morphology and the cross section of the non-irradiated polyethersulfone membrane. As it can be seen in figure, the membrane consists of a very thin and dense skin layer overlaying on a thick and highly porous sub layer.

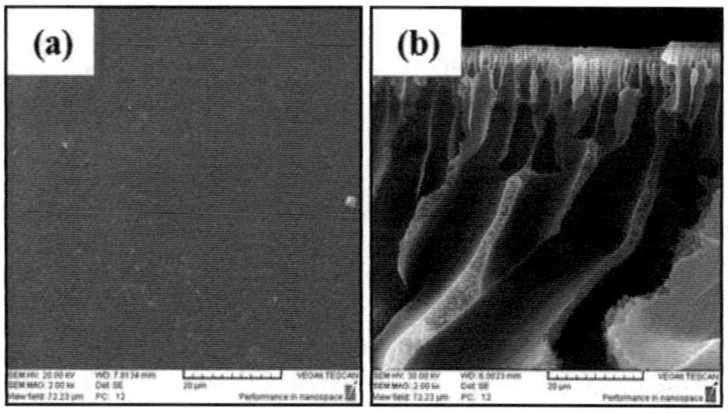

Figure 1. SEM micrographs of (a) the surface morphology and (b) cross section of Polyethersulfone membrane

3.1.1. Irradiation in the air. When PES membrane is irradiated in the air with 45 pulses at a fluence of 50 mJ/cm^2 (Figure 2a), the skin morphology changed and a wavy pattern formed on the surface. This deformation of the skin could be a result of thermal interaction of the laser and the membrane skin. With increasing the laser fluence the skin is removed from the surface at local positions and it is ablated completely at a fluence of 85 mJ/cm^2 (Figure 2b). More increase in the laser fluence leads to removing the sub layers and observation of the inner structure of the membrane. Figure 2c shows sample irradiated at a fluence of 120 mJ/cm^2 and 45 pulses.

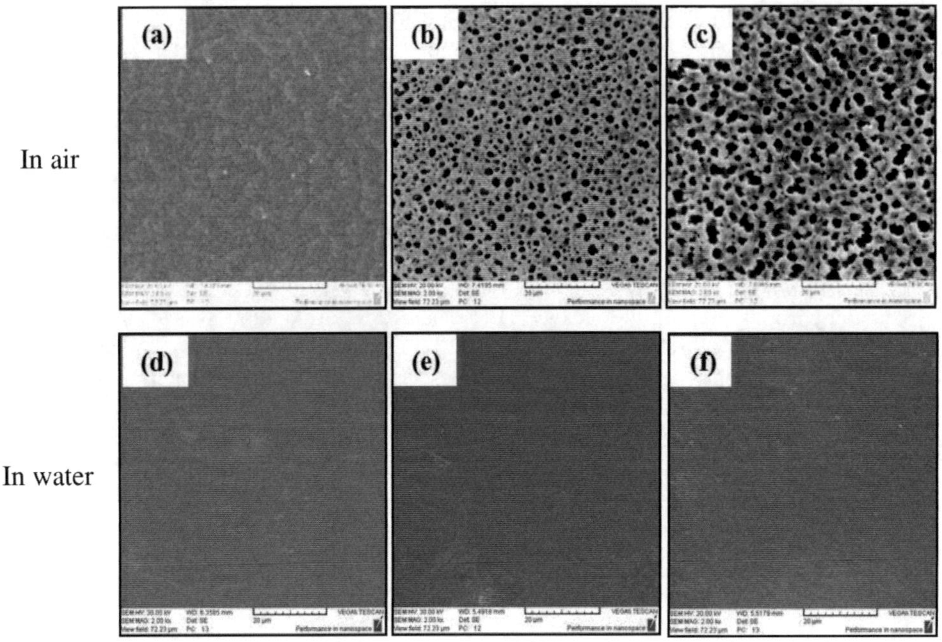

Figure 2. SEM micrographs of the samples irradiated with 45 pulses in air at fluences of; (a), (d) 50 mJ/cm^2, (b), (e) 85 mJ/cm^2 and (c), (f) 120 mJ/cm^2

3.1.2. Irradiation in the water. The samples were putted in distillate water and the laser beam was focused through the water on to the target surface. The height of water above the membrane was about 3-4 mm. Since water has low absorption at 308 nm, the same fluences as above were applied for irradiation of the samples in water. This helps us to avoid the results may happen due to large difference in laser fluences.

Here the laser pulse interacts with both the water and the membrane which causes vaporisation of the solid target and a small amount of water. Irradiation at all fluences (50, 85 and 120 mJ/cm^2 (Figure 2, (d, e and f))) has no effect on the surface morphology.

3.2. Chemical modifications

High power laser interaction with the sample both in air and water produce plums of the particles due to the material ablation. In the initial process the interaction of light with the sample surface causes to vaporization of the surface material and a small amount of surrounding liquid. Then chemical reactions between the ablated species and molecules in the liquid can subsequently occur. The ablated plum of the particles expands more freely in air than water. This lead to formation of molecules contained atoms from the target material and the water.

Figure 3 shows FTIR spectra of PES membranes before and after irradiation at fluences of 50 and 120 mJ/cm^2 in air and water. A new peak at 1734 cm^{-1} assigned to C-O bond, appeared following irradiation of the samples at a fluence of 50 and 120mJ/cm^2 in air. The peak intensity increases with increasing the laser fluence and then decreases at higher laser fluences (not shown). Formation of this peak is due to interaction of the laser with the O_2 molecules of the air and the carbon atoms of the membrane. There are also two bands located at 1665 cm^{-1} and 3430 cm^{-1} that are associated with the C=O vibration for PVP molecules and the OH stretching vibration of water molecules. The presence of OH group is related to the membranes preparation method which used a water bath to form a porous membrane. Since porous materials may hold in their pores small amounts of water that in practical terms, are almost impossible to remove, this peak is observed in the IR spectra. The intensity of this peak decreases during irradiation due to material removal from the surface and vaporization of free water.

The peak intensity is higher for samples irradiated with the similar conditions in the water. This probably is related to the confinement of the ablated products with water and then increasing the number of reactions with water molecules near the surface of membrane.

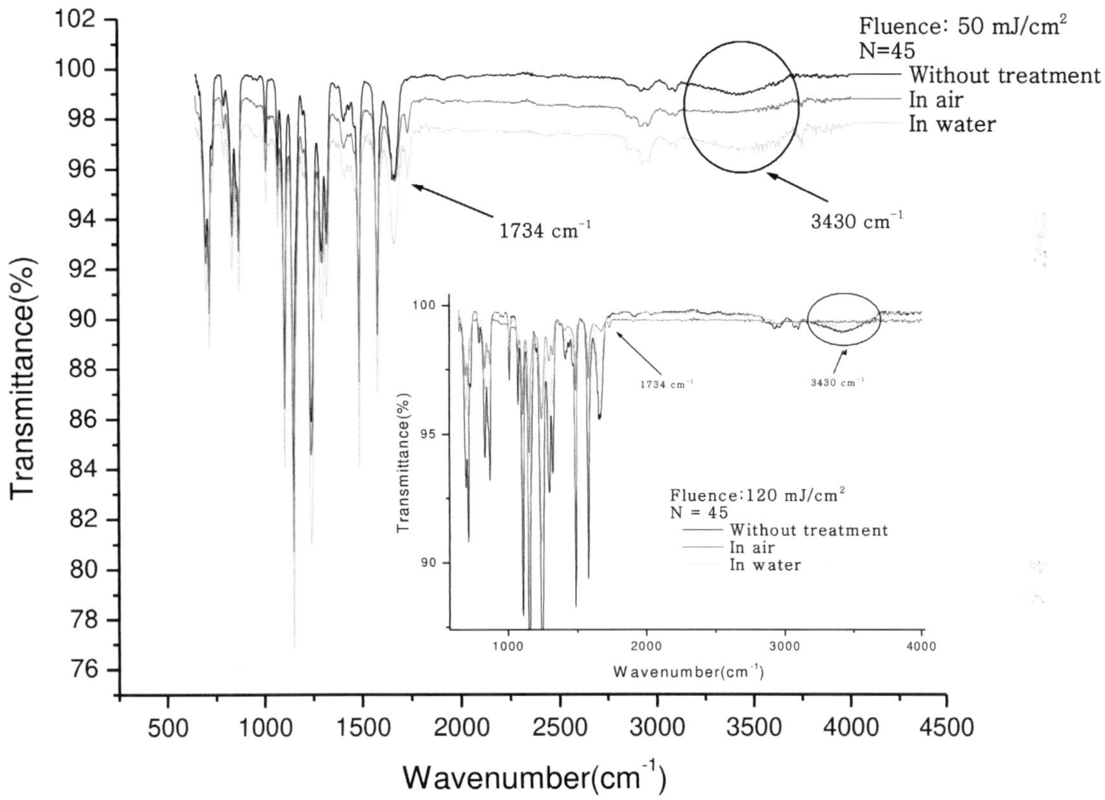

Figure 3. ATR-FTIR spectrum of untreated and laser-treated surface with different fluences (50, 120mJ/cm^2) in the air and water

3.3. Contact angle measurements and surface hydrophilicity

Table1 shows contact angle of water droplet on the surface before and after laser in air and water. The contact angle of water droplet on untreated surface is about 88°. Laser irradiation of surface with 50 mJ/cm^2 and 45 pulses in air leads to a decrease in water droplet contact angle to 79°. With increasing the fluence to 120 mJ/cm^2 contact angle increases to about 121.3°. The results are consistent with decreasing the peak intensity of the polar groups in IR spectrum. Removing the skin and interaction of the water with the porous surface also leads to increasing the water contact angle due to the presence of the air pockets which repel the water droplet from the surface.

For the samples irradiated in water the contact angle changed within a smaller range and a different manner. It increased slightly with irradiation at 50 mJ/cm^2 and then decreased at higher fluences. These results can be explained by difference in OH and CO peak intensities at various fluences since no morphological changes were observed for sample irradiated in water.

Table 1. The contact angles of water droplet on the samples

Fluence(mJ/cm^2)		0	50	85	120
Contact angle (°)	In air	88	79.6	84.2	121.3
	In water	88	90.2	86.7	84.6

3.4. Cell Culture Study and surface biocompatibility

Figure 4 shows the optical micrographs of cells cultured on the untreated and laser treated surfaces in the air and water. in our last paper we observed that the surface roughness is very important in cell adhesion to the PES film surface. The result of the cell culture in this experiment showed no significant increase on adhesion of the cell to the surface following irradiation of the membrane in air or water in spite of the increasing the surface roughness for samples irradiated at a fluence of 120 mJ/cm^2 in air. The number of adhered cells even decreases for membrane irradiated at a fluence of 50 mJ/cm^2 in water.

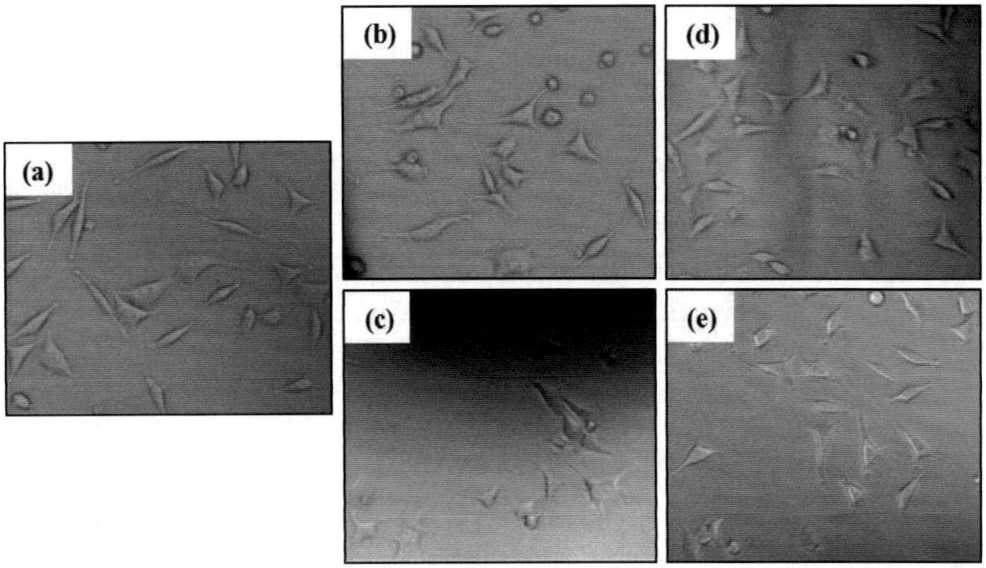

Figure 4. Optical micrographs of the cell cultured on (a) the untreated, and laser treated surfaces at fluence of 50 mJ/cm^2 (b) in air , (c) in water and at fluence of 120 mJ/cm^2 (d) in air, (e) in water

In summary the cell adhesion is more for membranes irradiated in the air than in the water. However, the lack of toxicity of both of the samples to evaluate the biocompatibility in contact with the living environment is evident.

4. Conclusions

Laser processing in water, except to cool the material or to increase plasma pressure, it can also act as a chemical reagent. In this study, the surface of PES membrane was irradiated with XeCl laser in air and distilled water in the same condition. The interaction of light with the sample surface causes to vaporization of the surface material and a small amount of surrounding the target (in air or water). Then chemical reactions between the ablated species and molecules in the environment can occur. The link between environmental segregated particles such as O_2 molecules of the air or the OH stretching vibration of water molecules on the surface of the sample occurs. These peaks are observed in the IR spectra of both of them.

The irradiated samples in the water have higher peak intensity in the OH stretching vibration. The contact angle of the water droplet is decreased more for samples irradiated in water. The cell adhesion is also more for membranes irradiated in the air than in the water.

References

[1] Hunt J A, Williams R L, Tavakoli S M and Riches S T 1995 Laser surface modification of polymers to improve biocompatibility *J. Mater. Sci: Mater. Med.* **6** 813-817

[2] Brayfield C A, Marra K G, Leonard J P, Cui X T and Gerlach J C 2008 Excimer laser channel creation in polyethersulfone hollow fibers for compartmentalized in vitro neuronal cell culture scaffolds *Acta. Biomater.* **4** 244-255

[3] Pazokian H, Mollabashi M, Selimis A, Stratakis E, Barzin J and Jelvani S 2013 Effect of pulse duration on KrF laser treatment of a polyethersulfone film: cell culture study *Appl. Phys. A* **110** 633-637

[4] Pazokian H, Selimis A, Stratakis E, Mollabashi M, Barzin J and Jelvani S 2011 KrF laser ablation of a polyethersulfone film: Effect of pulse duration on structure formation *Appl. Surf. Sci.* **258** 169-175

[5] Kruusing A 2004 Underwater and water-assisted laser processing: Part 1—general features, steam cleaning and shock processing *Opt. Laser. Eng.* **41** 307-327

[6] Belfer S, Fainchtain R, Purinson Yand Kedem O 2000 Surface characterization by FTIR-ATR spectroscopy of polyethersulfone membranes-unmodified, modified and protein fouled *J. Membr. Sci.* **172** 113-124

[7] Pazokian H, Barzin J, Mollabashi M, Jelvani S and Abolhosseini S 2012 Exploring the Influence of a XeCl Laser Treatment on Biocompatibility of Polyethersulfone Film *Laser. Phys.* **22** 922-929

[8] Watanabe K, Torikai H, Yang Q S, Sasoh A, Sano Y and Mukai N 2005 Shock wave phenomena in underwater laser peening *Shock Waves (Proceedings of the 24th International Symposium on Shock Waves Beijing, China July 11–16, 2004)* ed Professor Z Jiang (Berlin: Springer) pp 1039-1042

[9] Barzin J, Madaeni S S and Mirzadeh H 2005 Effect of Preparation Conditions on Morphology and Performance of Hemodialysis Membranes Prepared from Polyether Sulphone and Polyvinylpyrrolidone *Iran. Polym. J.* **14** (4) 353-360

23rd International Laser Physics Workshop (LPHYS'14) IOP Publishing
Journal of Physics: Conference Series **594** (2015) 012028 doi:10.1088/1742-6596/594/1/012028

Change of phase composition of metals by the radiation of the fiber laser for management of their magnetic characteristics

A O Andreyev[1], V D Mironov[1], V N Petrovskiy[1], E A Protasov[1], N M Prokopova[1], P S Dzhumayev[1], V I Polsky[1], E V Tret'yakov[2]

[1]National Research Nuclear University "MEPhI", Moscow, Russia
[2]NPO CNIITMASH, Moscow, Russia

E-mail: VNPetrovskij@mephi.ru

Abstract. Main technological parameters of heat treatment of ferromagnetic alloy samples of special stainless steel were experimentally determined in order to create the paramagnetic zone of a given size there.

1. Introduction

One of problems of modern materials science is receiving functional materials with pre-set physical characteristics. Such materials after being correspondingly processed (as a rule, thermal) possess high dynamic magnetic characteristics that allow using them for production of special equipment details.

Now sharply the quantity of products applied in the industry from ferromagnetic materials increased greatly. The main indicators of quality of such materials are specific magnetic losses and magnetic induction. One of promising methods of heat treatment for receiving materials with adjustable magnetization can be their radiation by laser radiation [1–6].

At the same time, despite prospects of such method of processing, there are no data on influence of various modes of laser processing on magnetic properties of special alloys. The large number of the factors influencing magnetic properties of materials in the course of their processing is the reason of it. Among these factors, besides density of power of laser radiation, an operating mode of the laser (continuous or pulse), the speed and orientation of movement of a laser bunch on sample surfaces, the size and geometry of processed sites, etc. should be mentioned.

With the advent of in recent years powerful fiber lasers - technological lasers of new generation, ample opportunities for improvement of existing and creations of new laser technologies on the basis of lasers of this type open.

In this paper the results on development of technology of heat treatment of samples from a special alloy of stainless steel by the radiation of the fiber laser, intended for production of rotor plates with the set phase structure, for the purpose of reduction of influence of losses by vortex currents are presented. Techniques and schemes for supervision and registration of magnetization distribution and other characteristics of the ferromagnetic material which was affected by laser radiation are presented.

2. Experimental setup

Heat treatment of samples consisted of heating of certain zones of a sample to the temperature about of 800 °C. At such temperature there is a transition from ferromagnetic to a paramagnetic phase for stainless steel of the chosen chemical composition.

Working off of modes of heat treatment of samples of the alloy intended for production of rotor plates, was carried out on the robotized installation equipped with the robot of AVV firm, the fiber LS-5 laser with a power of 5 kW and the scanner of a laser beam (fig. 1). The initial

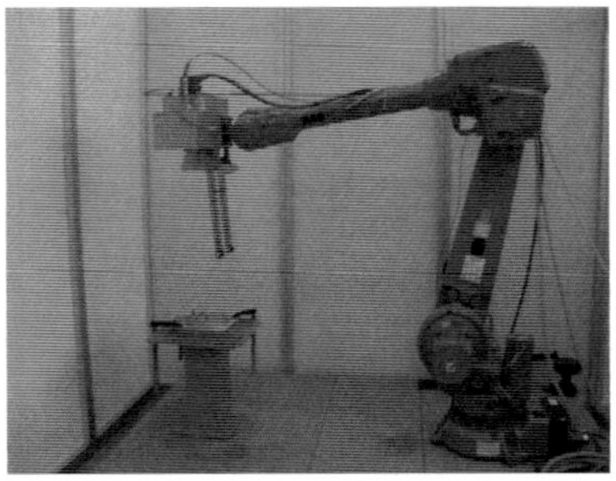

Figure 1. General view of experimental installation.

material used in experiments has in the structure 100% a martensite phase. Experimental strips of 8×200 mm in size on which the technology of heat treatment, and rotor plates with a diameter of 200 mm was fulfilled were cut out from this material. During experiments on selection of modes of heat treatment processing by the laser radiation of separate zones of metal strips was carried out. On rotor plates then according to the picked-up modes the zones which have been set by the specification on production of details were processed. The experiment scheme on selection of modes of heat treatment is submitted in figure 2.

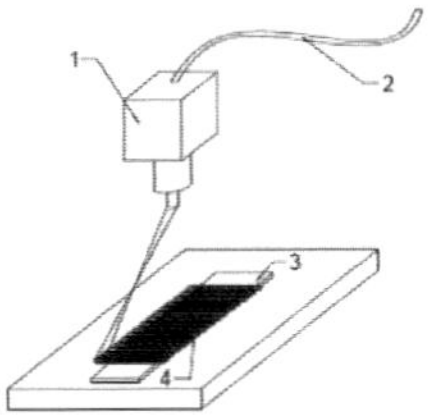

Figure 2. Experiment scheme: 1 - scanners of laser radiation, 2 - transport fiber from a source of laser radiation, 3 - sample, 4 - out-of-focus radiation scanning a surface of a sample.

In figure 3 the typical photo of the plate processed by laser radiation in various modes is submitted.

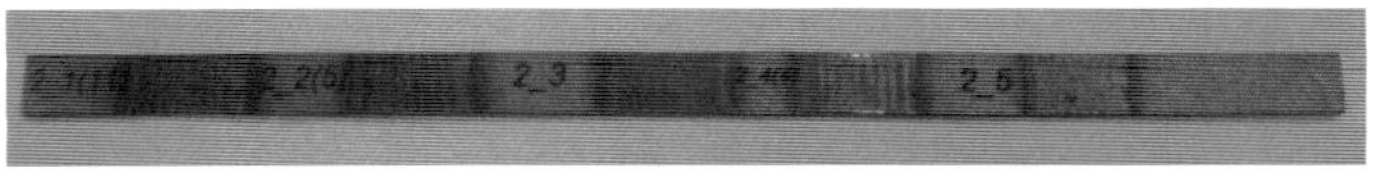

Figure 3. Sample of a plate of 8×200 mm processed in various regimes.

In figure 4 fragments of rotor plates samples are presented, and numbers designate areas of the samples processed by laser radiation.

3. Experimental results and discussion

At present there are some techniques of definition of magnetic characteristics of ferromagnetic products, such as magnetization distribution, the sizes of domains, thickness of domain borders

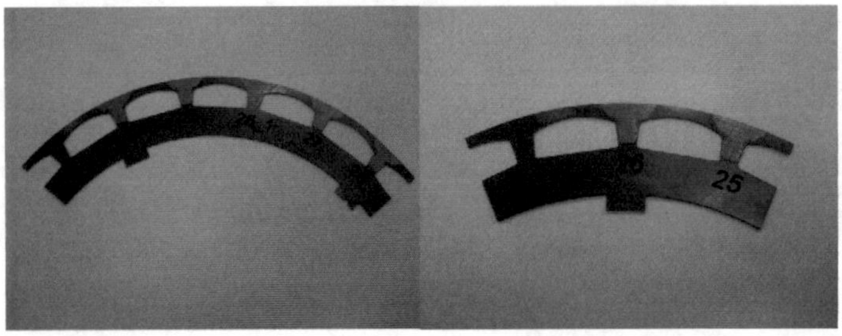

Figure 4. Fragments of samples of the rotor plates, processed by laser radiation.

and others. Two methods are rather evident: method of powder figures and magnetooptical method. The magnetooptical supervision method of domains by means of Kerr or Faraday's magnetooptical effects doesn't demand use of the intermediate suspension environment therefore it is convenient for research. Also magnetooptical method possesses high resolution, allowing to define local distribution of magnetization in ferromagnetic samples [7, 8].
For carrying out pilot studies of magnetization in the irradiated samples the special magnetooptical installation was developed and used.
As a source of radiation the lighter of OCB1 with a power of 100 W was used. Also two lenses of Gelios-44 with focal length of f=58 were used; two polarizers: at the exit from a source there was PF-40.5 polarizer, as the analyzer PF-49 polarizer was used. Registration of the received images was carried out by a digital camera. On value of illumination we can estimate an angle of rotation of the plane of polarization, and, therefore, and magnetization of a sample.
Light from the source, linearly polarized after passing through a polarizer, got on a glass prism and then to a magnetooptical film at which passing there was a turn of the plane of polarization. Having reflected from it, light left other side of a prism, passed through the analyzer and a lens and got on the digital camera.
Studied samples represented long narrow ferromagnetic plates (fig. 3) and the fragments of rotor plates (fig. 4) processed in certain places by laser radiation. Each site was processed with a certain speed of pass of a beam and with a certain value of a defocusing of radiation. In figure 5 the characteristic photo of the irradiated site of a plate is submitted.
Figure 5 presents a clear boundary of transition of metal from a condition of a martensitny phase (ferromagnetic) in an austenitic phase (paramagnetic).

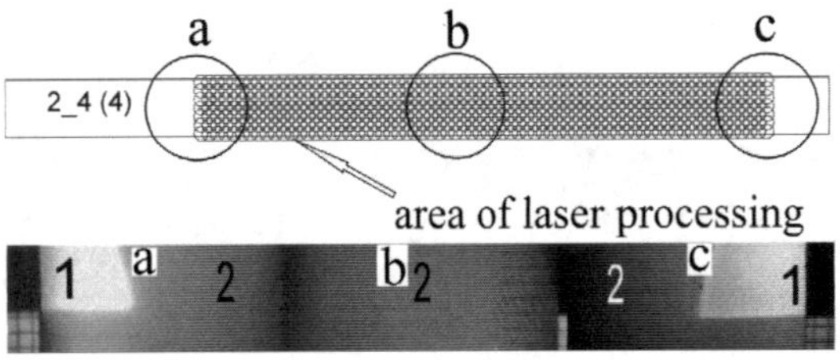

Figure 5. Magnetooptical research of a plate (regime 2_4 (4)): and - the left edge, - the central part, in - the right edge; 1-ferromagnetic areas, 2-paramagnetic areas.

Figure 6 clearly shows (1) difficult area of a plate processed by laser radiation allocated with a dark background in form, and a light background (2), plate area not processed by laser radiation. Areas (3) correspond to the windows which have been cut out in a rotor plate.

For the analysis of transition area size between ferromagnetic and paramagnetic phases, and also estimates of change of chemical composition of metal in various areas of the irradiated sample, the metalgraphic analysis (fig. 7-9) was carried out.

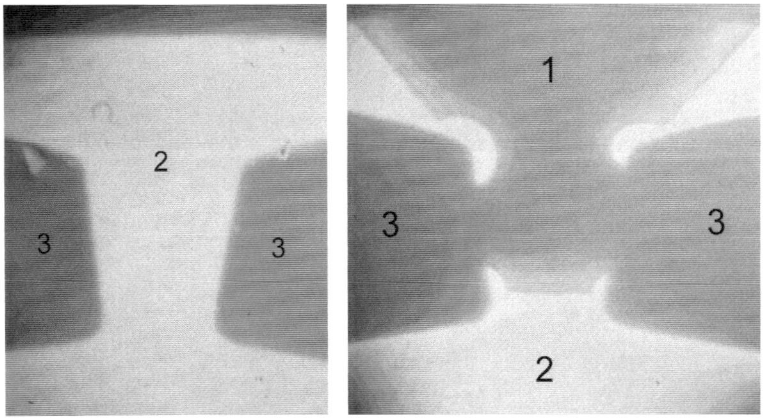

Figure 6. The typical photo of a fragment of a plate of a rotor before processing by laser radiation.

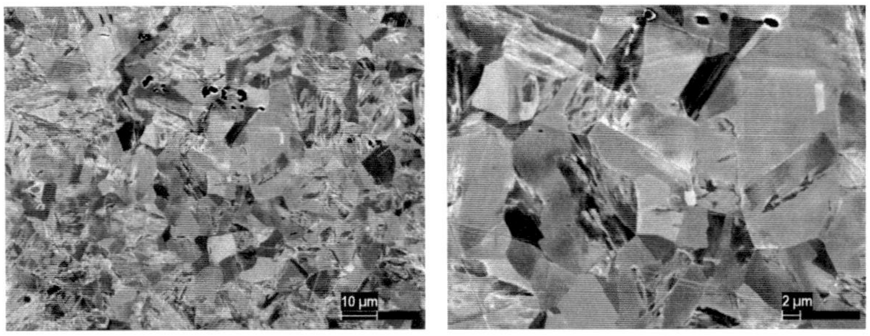

(a) (b)

Figure 7. Structure of the irradiated part of a steel strip (a and b- different increases).

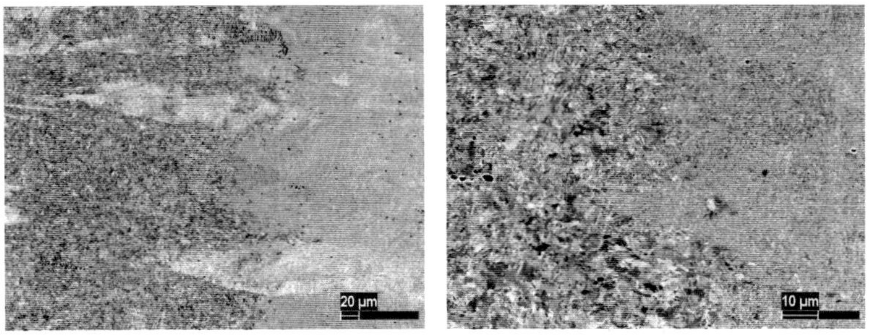

(a) (b)

Figure 8. Structure of transitional area of a steel strip: at the left the irradiated area, on the right unirradiated (a and b- different increases).

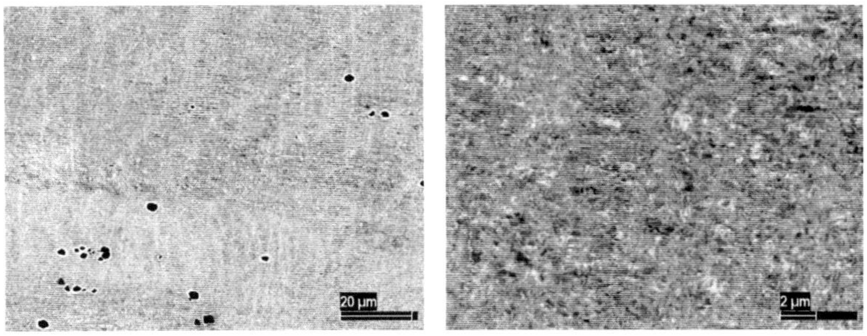

(a) (b)

Figure 9. Structure of the irradiated area (a and b- different increases).

Dependence of microhardness near limit of the section of the irradiated and unirradiated areas of a plate (fig. 10) was defined also.

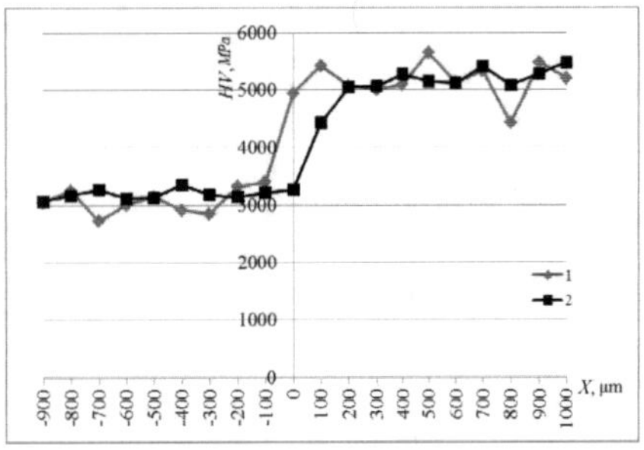

Figure 10. Results of measurements of microhardness on border of the irradiated and unirradiated areas of a plate (1, 2 measurements in various sections of a boundary zone).

4. Conclusion

The technology of heat treatment of plates made of gradient material is developed for correction of magnetic properties of separate plate sites. The scheme on the basis of the magnetooptical effects is chosen, allowing to define with high resolution local magnetic characteristics of the ferromagnetic materials which were affected by laser radiation. The experimental installation, allowing to observe, photograph and measure magnetization distribution in ferromagnetic samples is developed. By means of a magnetooptical method measurements of distribution of magnetization of the ferromagnetic samples processed by laser radiation are carried out. The metalgraphic analysis of the irradiated samples is carried out. It is established that the transitional area from ferromagnetic to a paramagnetic phase makes about 400 microns. It is established that in processing places, at optimum modes of laser radiation, magnetization is absent.

Acknowledgments

This work was supported by the RFBR (project number 14-02-00369-a) and the Ministry of Education and Science of the Russian Federation (unique identifier PNIER RFMEFI58214X0004).

References

[1] Semenov A, Gavrilyuk A, Gavrilyuk A, Semirov A, Malov A, Gavrilyuk B, Turik N and Mokhovikov A 2010 *Inorganic Materials* **46** 617–623
[2] Kayukov S V, Hares E and IA D 2003 *News of the Samara Russian Academy of Sciences scientific center* **5** 66
[3] Bukhanova I, Divinsky V V and Zhuravel V M 2004 *Electrical equipment* 39
[4] Voicu M, Dinu S Oros C, Dima G and Patrol E 2011 *U.P.B. Sci. Bull. Series A* **73** 213
[5] Gorzhon V V, Smolyakov A, Babich N G and Semenko M 2009 *Physics of metals and metallurgical science* **108** 133
[6] Mohaman S, Diebolder R and Herr U 2008 *Journal of Applied Physics* **103**
[7] Tikadzumi S 1987 *Physics of ferromagnetics. Magnetic characteristics and practical applications* (M.)
[8] Protasov E 2009 *Fundamentals of magneto-optics* (MEPhI - M.)

DETECTION OF SINGLE BACTERIA – CAUSATIVE AGENTS OF MENINGITIS USING RAMAN MICROSCOPY

[1]T.V. Baikova, [2]S.A. Minaeva, [3]A.V. Sundukov, [3]T.S. Svistunova, [2]V.N. Bagratashvili, [1]M.V. Alushin, and [1]S.A. Gonchukov

[1]National Research Nuclear University MEPhI, Moscow, Russia
[2]Institute of Laser and Information Technologies, Russian Academy of Sciences, Troitsk, Russia [3]Infectious Clinical Hospital No 2, Moscow, Russia
gonchukov@mephi.ru

Abstract.
Early diagnostics of meningitis is a very topical problem as it is a fulminant disease with a high level of mortality. The progress of this disease is, as a rule, accompanied by the appearance of bacteria in the cerebrospinal fluid (CSF) composition. The examination of the CSF is well known to be the only reliable approach to the identification of meningitis. However, the traditional biochemical analyses are time consuming and not always reliable, simple, and inexpensive, whereas the optical methods are poorly developed. This work is devoted to the study of Raman spectra of several bacterial cultures which are mainly present during meningitis. Raman microscopy is a prompt and noninvasive technique capable of providing reliable information about molecular-level alterations of biological objects at their minimal quantity and size. It was shown that there are characteristic lines in Raman spectra which can be the reliable markers for determination of bacterial form of meningitis at a level of a single bacterium.

Journal of Physics: Conference Series **594** (2015) 012029 doi:10.1088/1742-6596/594/1/012029

Introduction
Bacteria invasion into the cerebral spinal fluids (CSF) is a serious problem leading to the development of meningitis – a dangerous bacterial form of the central nervous system (CNS) illness with a high level of mortality. The early diagnosis and the cure started of this illness are the actual problems of medicine. The examination of CSF has been well known to be the only reliable approach to the identification of meningitis [1, 2]. The current tools available for detecting the bacteria penetration into CSF are cytological, bacteriological, virological and serological analysis techniques. They are time-consuming and not always reliable, simple, and inexpensive. At present the optical methods of CFS analysis are poorly developed unfortunately. The Raman spectroscopy undoubtedly holds promise for CSF analysis, because it is a prompt and noninvasive technique capable of providing reliable information about molecular changes. In so doing the detection of pathological bodies impurities in CSF can be realized at a level of a single bacterium or cell [3-6]. It is very important as first bacterial identification is necessary for early diagnosis of bacterial meningitis. As it was shown earlier on [3, 5-7], a single bacterium is well detected in CSF by optical microscope. As regards the problem of different bacterial culture identification using Raman spectrometer the presence of the CSF matrix is not required. This work is devoted to the study of Raman spectra of several pure bacterial cultures which are mainly present during meningitis.

2. Materials and Methods

2.1. Sample preparation
Cultivation and preparation of in vitro cultured bacteria of four different species of Staphylococcus aureus, Streptococcus dysgalactiae, Streptococcus pyogenes and Enterococcus faecalis were performed by the bacteriological laboratory of the Moscow Infectious Clinical Hospital No 2. The different bacterial species were cultured in nutrition medium and their colonies suspended afterwards in phosphate buffered saline solution. About 1 ml of the suspension were spotted onto the clean polished surface of a substrate and dried at room temperature. 70% ethanol solution is used for bacteria inactivation.

2.2. Spectroscopic instrumentation
The spectral measurements were fulfilled with the help of Nicolet Almega XR spectrometer with a 532-nm laser excitation source (a Nd:YAG CW laser generating a 20-mW second-harmonic radiation). The spectrometer was equipped with a microscope capable of an X10 and X50 magnification. In microscopic investigations, the diameter of the probe laser beam spot was 1 mkm. The Stokes shift components in the range of 400–3100 cm^{-1} at a spectral resolution of 2 cm^{-1} were used to analysis. Wave number calibration was performed by using the characteristic silicon reference peak at 521 cm^{-1}. To reduce the effect of fluorescence the phenomenon of photobleaching for 30 seconds prior to the recording of the Raman spectra was used. The spectra obtained were computer processed using the OMNIK software.

Journal of Physics: Conference Series **594** (2015) 012029 doi:10.1088/1742-6596/594/1/012029

3. Results nd Discussion

Our investigations showed that a bacterium sample in the form of a droplet dried at room temperature retains the main characteristic features of its Raman spectrum for at least one month. The sample in this form is convenient to store and transport. In so doing the computer processing of its Raman spectrum is simplified because of the absence of water. However the choice of the substrate can essentially influence on the detection quality of Raman spectrum as own substrate spectral lines and informative lines can be overlapped. In literature enough attention is not usually given to this problem. As was shown, the base informative lines of bacteria Raman spectra are concentrated in the range of 700–1700 cm^{-1}. About 20 substrates were examined in this work. Raman spectra of several substrates are presented in Fig. 1.

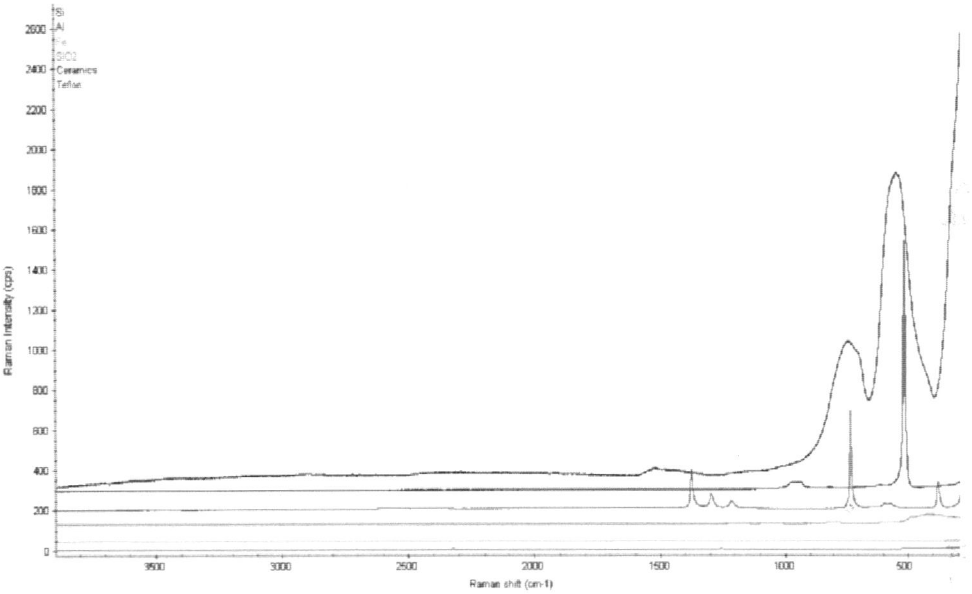

Figure 1. Raman spectra of six substrates. Bottom up: Al-mirror, steel, quartz, teflon, silicon, PZT ceramics. Excitation wavelength is equal to 532 nm.

It can be seen, that the substrates from metal, semiconductor, dielectric and even organic can be used for measurements in a large spectral range. However among them the metal substrates are more preferable for wide-range spectral investigations.

Raman spectra of four bacterial cultures spotted onto Al-mirror and fused quartz which are mainly present during meningitis are presented in Fig. 2. As it can be seen, there is the intensive signal around the wavenumber of 3000 cm^{-1}. This band is constituted by the closely spaced several lines associated with the CH_2 and CH_3 vibrations of proteins and lipids. This signal is peculiar to all bacteria and therefore can not be of interest for differential analysis. The fingerprint region between the wavenumbers of 700-1700 cm^{-1} can be used for the identification of different bacteria. The spectra contain the information on the molecular composition of the whole bacteria, such as fatty acids, carbohydrates, proteins and nucleic acids [8, 9]. Raman spectra depend, obviously, on the measuring position within the bacteria along of different molecular composition [7]. However their basic features are retained for the bacterium as a whole. Characteristics lines of four investigated bacteria can be determined in Fig. 2.

As for Enterococcus faecalis spectral identification there are three lines at wavenumbers of 1000, 1360 and 1600 cm^{-1} which represent both proteins and Amide I bands. In Raman spectrum of Streptococcus dysgalactiae there are several strong lines in range of 1000-1700 cm^{-1}. Three lines at 1451, 1580 and 1662 cm^{-1} stand out against others. They correspond to the vibrations of CH groups in lipids, amino acids and carbohydrates. Two Raman lines at 1582 cm^{-1} and 1636 cm^{-1} which represent both C=C vibrations of phenylalanine and Amide I band can be used for Streptococcus pyogenes identification. As respects Staphylococcus aureus there are two strong lines at 1152 cm^{-1} and 1520 cm^{-1} for its identification. These lines in Raman spectrum indicate the presence of carotenoids in molecular structure of Staphylococcus aureus and can serve as the valid dominant of this bacterium detection.

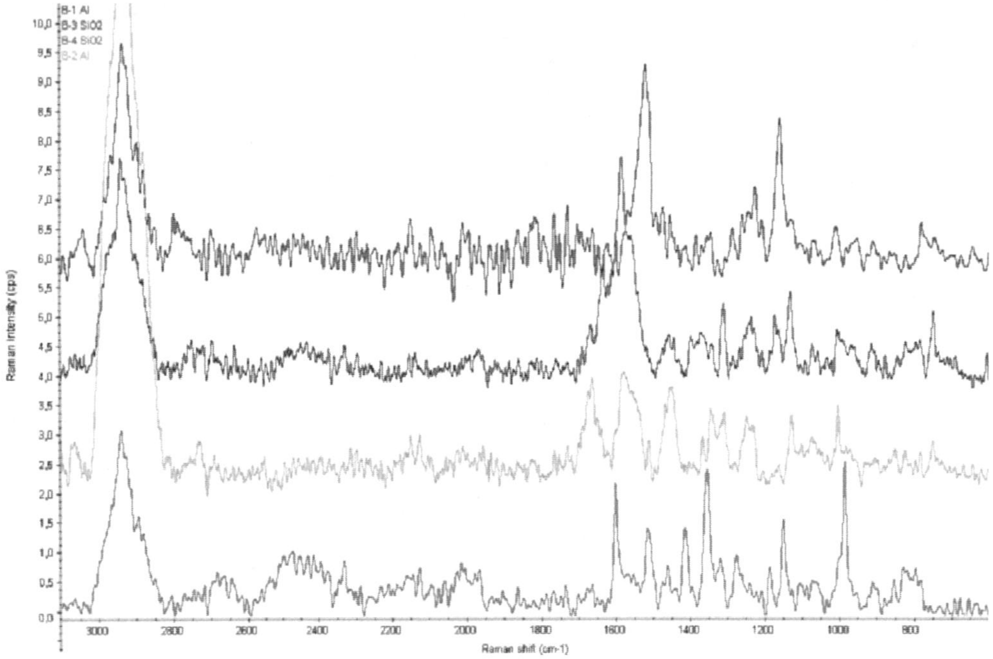

Figure 2. Raman spectra from (bottom up) Enterococcus faecalis, Streptococcus dysgalactiae, Streptococcus pyogenes and Staphylococcus aureus. Excitation wavelength is equal to 532 nm.

As it is known, the wide-spread carotenoids beta-carotene and lycopene strongly absorb in the green–blue area of the spectrum [10-12]. Therefore, using corresponding excitation wavelengths in this area, resonance Raman spectroscopy can take place and more intensive carotenoid lines can be acquired. As a result, noninvasive and fast detection of this bacterium can be realized. An argon laser operating at wavelengths of 488 and 514.5 nm could be a good excitation source in such a case.

4. Conclusion

In view of its potential for the fast and noninvasive analysis at a molecular level the Raman spectroscopy has been the method of precise research of biological objects for the last years. Unfortunately in the domain of meningitis diagnosis the Raman spectroscopy is poorly developed. In the frames of this work the Raman spectra of four pure bacterial cultures which are mainly present during meningitis were investigated. It was shown that a bacterium sample in the form of a droplet spotted onto metal mirror and dried at room temperature retains the main characteristic features of its Raman spectrum for at least one month. It was demonstrated that there are characteristic lines in Raman spectra which can be the reliable markers for the determination of a kind of bacteria – causative agents of meningitis. Further work in this direction has to lead to database forming for early bacteria identification.

References

[1] 2008 *Methods for the Laboratory Investigation of the Cerebrospinal Fluid* (Moscow: MONIKI), in Russian
[2] Zenkov L R and Ronkin M A 1991 *Functional Diagnostics of Nervous Diseases. A Doctor's Guide* (Moscow: Meditsina), in Russian
[3] Harz M., Kiehntopf M., Stöckel S, Rösch P, Straube E, Deufel T, Popp J, Direct Analysis of Clinical Relevant Single Bacterial Cells from Cerebrospinal Fluid During Bacterial Meningitis by Means of Micro-Raman spectroscopy *J. Biophoton,* **2,** 70 (2009)
[4] R. Sathyavathi, N.C. Dingari, I. Barman, P.S.R. Prasad, S. Prabhakar, D.N. Rao, R.R. Dasari, J. Undamatla, Raman spectroscopy provides a powerful, rapid diagnostic tool for the detection of tuberculous meningitis in ex vivo cerebrospinal fluid samples, *J. Biophoton,* **1,** 6 (2012)
[5] S.A. Gonchukov, T.V. Lonkina, V.N. Bagratashvili, S.A. Minaeva, A.V. Sundukov, T.E. Migmanov, J. Lademann, M.E. Darvin, Raman spectroscopy of pathological cells of cerebrospinal fluid, *Meditsinskaya Fizika,* No. **2,** 97 (2013), in Russian
[6] S.A. Gonchukov, T.V. Lonkina, S.A. Minaeva, A.V. Sundukov, T.E. Migmanov, J. Lademann, M.E. Darvin, V.N. Bagratashvili, Confocal Raman microscopy of pathologic cells in cerebrospinal fluid, *Laser Physics Letters,* **11,** № 015602 (2014)
[7] M. Harz, M. Kiehntopf, S. Stöckel, P. Rösch, T, Deufel, J. Popp, Analysis of single blood cells for CSF diagnostics via a combination of fluorescence staining and micro-Raman spectroscopy *Analyst,* **133,** 1416 (2008)
[8] Z. Movasaghi, S. Rehman, I.U. Rehman, Raman Spectroscopy of biological tissues, *Applied Spectroscopy Reviews,* **42,** 493 (2007)
[9] O. Samek, H.H. Telle, L.G. Harris, M. Bloomfield, D. Mack, Raman spectroscopy for rapid discrimination of *Staphylococcus epidermis* clones related to medical device-associated infections, *Laser Physics Letters,* **5,** 465 (2014)
[10] M.E. Darvin, I. Gersonde, S. Ey, N.N. Brandt, H. Albrecht, S.A. Gonchukov, W. Sterry, J. Lademann, Noninvasive detection of beta-carotene and lycopene in human skin using Raman spectroscopy, *Laser Phys.* **14,** 231 (2004)
[11] Darvin M.E., Zastrov L., Gonchukov S.A., Lademann J., Influence of IR Radiation on the Carotenoid Content in Human Skin, *Opt. Spectrosc.* **107,** 917 (2009)
[12] S. Gonchukov, A. Sukhinina, D. Bakhmutov, T. Biryukova, M. Tsvetkov, V. Bagratashvily, Periodontitis diagnostics using resonance Raman spectroscopy on saliva, *Laser Physics Letters,* **10,** № 075610 (2013)

This work was supported by the Russian Foundation for Basic Research (Grant No. 12-02-00672-a).

Turbid Media Extinction Coefficient for Near-Infrared Laser Radiation

T Dreischuh, L Gurdev, O Vankov, D Stoyanov, L Avramov

Institute of Electronics, Bulgarian Academy of Sciences
72 Tzarigradsko chaussee Blvd., 1784 Sofia, Bulgaria

E-mail: tanjad@ie.bas.bg, lugurdev@ie.bas.bg

Abstract. In this work, extended investigations are performed of the extinction coefficient of Intralipid-20% dilutions in distilled water depending on the Intralipid concentration, for laser radiation wavelengths in the red and near-infrared regions covering the so-called tissue optical window. The extinction is measured by using an approach we have developed recently based on the features of the spatial intensity distribution of laser-radiation beams propagating through semi-infinite turbid media. The measurements are conducted using separately two dilution-containing plexiglass boxes of different sizes and volumes, in order to prove the appropriateness of the assumption of semi-infinite turbid medium. The experimental results for the extinction are in agreement with our previous results and with empiric formulae found by other authors concerning the wavelength dependence of the scattering coefficient of Intralipid - 10% and Intralipid - 20%. They are also in agreement with known data of the water absorptance. It is estimated as well that the wavelengths around 1320 nm would be advantageous for deep harmless sensing and diagnostics of tissues.

1. Introduction

The investigations in the fields of optical diagnostics and therapy [1-3] are based on the knowledge of the optical characteristics (parameters) of tissue and tissue-like phantoms. The main such characteristics from biomedical point of view are: the absorption coefficient μ_a describing the optical - energy dissipation and the thermal effect on tissues; the integral scattering, μ_s, backscattering, μ_{bs}, and reduced-scattering, $\mu_{rs} = \mu_s(1-g)$, coefficients describing the optical-energy dispersion; the extinction (total attenuation) coefficient $\mu_e = \mu_a+\mu_s$; and the scattering anisotropy factor, g. The backscattering and extinction coefficients determine the informative depth of light penetration in tissue, from where the (single-sided tomography) signal is strong enough to be accurately detected.

The use of phantoms is necessary to avoid tedious and complicated in vivo experimental procedures for testing and calibration of novel diagnostic and therapeutic methods and instruments. The dilutions of Intralipid (IL) fat emulsions in distilled water [4-8] are an important class of liquid phantoms having numerous advantages. Certainly, the optical properties of tissues and phantoms determine the radiative transfer inside such media. Moreover, they condition the therapeutic effects in tissues as well as the possibility of specifying characteristic inhomogeneities in tissue diagnostics. Therefore, the determination of the optical characteristics and the development of methods for this purpose are of primary importance for the biomedical photonics of tissues [9,10]. The optical radiation wavelengths λ employed in the biomedical photonics cover the interval from 600 nm to 1300 nm, where the absorption in tissues (mainly in water) is minimal: from 0.002 cm^{-1} at λ = 600 nm to ~1.35

cm^{-1} at $\lambda = 1300$ nm [11-13]. This wavelength interval is often called the tissue optical (or therapeutic) window [14]. The scattering coefficient of tissues for radiation from the optical window is orders of magnitude larger than the corresponding absorption coefficient. Thus, the tissues are turbid media for such radiation.

The main goals of the present work are as follows. First, using an approach we have developed recently [15,16], to determine experimentally the extinction coefficient of Intralipid dilutions in distilled water, for laser radiation of different wavelengths within the tissue therapeutic window. Another aim is to perform a comparison and mutual validation of the results for μ_e obtained here and corresponding results for μ_s obtained using empiric formulae derived by other authors [6-8]. The third goal is to prove experimentally the assumption that the plexiglass boxes employed containing the IL dilutions are sufficiently large to be considered as semi-infinite turbid media. At last, the work is also intended to estimate and illustrate the eventual advantage of using long-wavelength laser radiation (around the upper limit of the therapeutic window) for deeper harmless diagnostics of tissues.

The theoretical basis of the method of measuring the extinction, the experimental arrangements and procedures, and the materials employed are described in the following section 2. In section 3, the experimental results are analysed and discussed and compared with results obtained by other authors. In section 4, possible diagnostic advantages of longer-wavelength laser radiation are evaluated and discussed. The main results of the work are summarized in section 5.

2. Methods, experimental procedures and phantoms

2.1. Extinction measurement approach

The extinction (total attenuation) coefficient of the Intralipid dilutions of interest, occupying large plexiglass containers, has been determined by measuring the longitudinal, in-depth profile of the on-axis intensity ($\propto$ the detected small-receiving-aperture gated light power) of forward propagating inside laser radiation beams. The depth in the IL dilutions is given by the coordinate z along the beam axis $0z$ that is perpendicular to and begins from the internal frontal wall of the container (figure 1). Like tissues, the IL dilutions produce forward-peaked scattering of light, mainly within a $\sim\gamma_m = [2(1-g)]^{1/2}$ – wide angular interval. Then, the spatial (transversal and in-depth) distribution of the forward-propagating light intensity is analytically obtainable as a small-angle solution of the radiative-transfer equation [16,17]. This solution is valid to depths several times as large as the transport mean free path of the photon μ_{rs}^{-1} in the investigated medium. Simple analytical expressions of the on-axis detected power $J(z)$ are obtained [16] for the low-scattering zone, where $\mu_s z \sim 1$ or $\mu_s z < 1$ and the single-scattered and unscattered light is prevailing, and the developed-scattering zone, where $\mu_s z >> 1$ and the multiple-scattering light is prevailing. The expression of $J(z)$ in the low-scattering zone have different forms, depending on the relation between γ_m and the angle of acceptance γ of the receiving optical system. When $\gamma >> \gamma_m$, $J(z)$ drops down exponentially with z, with decay constant μ_a, when $z < (w^2 + E^2)^{1/2}$, and with decay constant μ_e, when $z > (w^2 + E^2)^{1/2}$; w and E are the radii of the sensing (collimated) laser beam at z=0 and of the receiving aperture, respectively. When $\gamma << \gamma_m$, $J(z)$ decreases exponentially with only one decay constant μ_e. In the developed-scattering zone, $J(z)$ behaves $\propto z^{-3}\exp(-\mu_a z)$ or $z^{-4}\exp(-\mu_a z)$ when $\gamma^2 >> \mu_{rs} z$ or $\gamma^2 << \mu_{rs} z$, respectively. In the experiments performed, the conditions $\gamma << \gamma_m$ and $\gamma^2 << \mu_{rs} z$ are in power. Under these conditions, $J(z) \propto z^{-4}\exp(-\mu_a z)$, in the developed-scattering zone, and

$$J(z) = J_0 \exp(-\mu_e z)[1 + \delta(z, \mu_s, w, \gamma, E, g)] \quad (1)$$

in the low-scattering zone, where $J_0 = P_t(E^2/w^2)$, P_t is the incident beam power, and δ is a term that is negligible to some depth z_n decreasing with the increase of the turbidity ($\propto \mu_s$) of the medium under investigation. According to (1), a log-linear fit of the experimental data obtained for $J(z)$ in the low-scattering zone, at $\gamma << \gamma_m$ and $\delta <<1$, allows one to determine the extinction coefficient.

2.2. Experimental

The experimental setup for measuring the forward-propagating light power $J(z)$ is schematically shown in figure 1. The light sources employed are laser diodes emitting nearly collimated continuous-wave optical beams of about 1 mm radius and wavelengths λ = 672 nm, 847 nm, and 1326 nm. The Intralipid dilutions under investigation have been placed in two plexiglass boxes having parallelepiped and cubic forms with sizes of 12×12×22 cm and 25 cm side, respectively. The axis of the incident laser beam is perpendicular to the entrance wall of the container and is considered as coincident with the axis $0z$ beginning from the internal entry wall of the container and oriented forward, in the direction of incidence of the beam (figure 1). The forward-propagating light power inside the container is measured by using a scanning optical fiber of $2E$ = 0.1 mm core diameter and $\gamma \sim 9°$ (~ 0.16 rad) half angle of acceptance. The fiber is oriented antiparallel to the beam axis and is connected with an optical radiometer Rk-5100 (Laser precision corp., USA) in external locking regime, with a RqP-546 silicon probe for 672 and 847 nm radiation, and a RkP-545 pyroelectric probe for 1326 nm radiation, a 14bits ADC and a computer for appropriate data processing. The noise equivalent power (NEP) of the radiometer is 2.10^{-12} W, with RqP-546, and 2.10^{-7} W, with RkP-545. Its averaging (low-pass filtering) time constant τ_a is chosen to be 1 s. The on-axis detected-power distribution (on-axis intensity profile) is measured by a longitudinal scan of the fiber along the beam axis, implemented by using a linear translation stage with an integrated stepper motor and controller LTS 300/M (Thorlabs, Inc., USA). In the experiments, the sampling step is varied from 0.1 mm to 10 μm. They have been performed in a dark environment at practically entirely damped stray light influence. The least measured power values of interest exceed considerably the NEP of the radiometer. Thus, the data fluctuations should result from the signal-conditioned shot noise, digitizing noise, laser power fluctuations, and scintillations due to the random walk of the scattering particles. By the low-pass filtering performed with time constant 1 s, along with 400 measurements per point, the fluctuation level is reduced to about 1% for all the wavelengths at the least sampling steps employed.

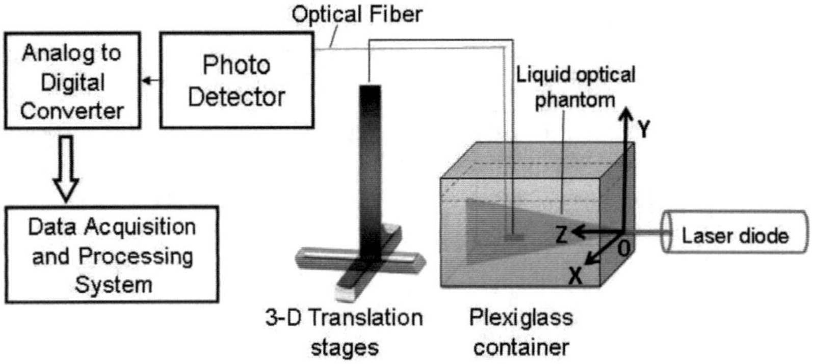

Figure 1. Experimental setup.

2.3. Intralipid phantoms

The turbid media investigated in the work are dilutions in distilled water of different amounts (volume parts) of Intralipid-20% (Fresenius Kabi AB, Sweden). The IL concentration has usually been defined as the volume fraction of soybean oil and egg lecithin forming the scattering submicron pellets in the dilution [6]. The volume fraction of these components is 22.74% in stock IL-20% and 11.95%, in stock IL-10%. The dilutions (phantoms) are prepared just before the measurements performed at room temperature. The IL concentration varies from 0.02 to 3.2 % in volume for different wavelengths. The

Intralipid bags employed in the experiments described below are from the same production batch (#10GH5382).

3. Results and discussion

The measured on-axis light power profiles $J(z)$ involve both the low scattering and developed scattering zones. An illustration of such recorded profiles at different IL concentrations is given in figure 2. The exponential fall-off regions in the low-scattering zone are well distinguishable. As expected, the extent of these regions decreases with the increase of the IL concentration [15]. The measurements of the profiles $J(z)$ have been performed repeatedly by using both the above-described plexiglass containers, the larger cubic box and the smaller one having parallelepiped form.

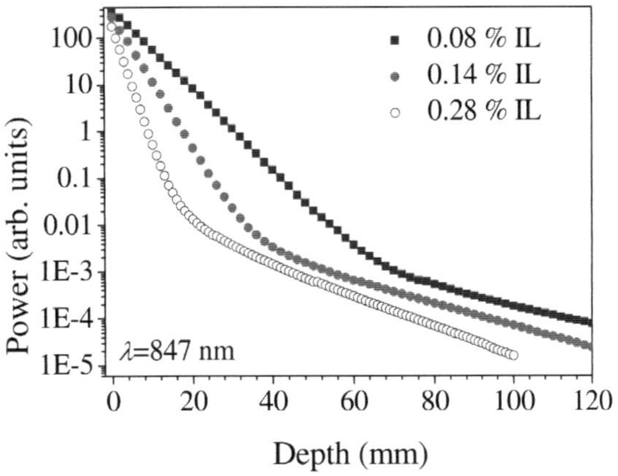

Figure 2. Measured in-depth light power profiles $J(z)$ for different Intralipid concentrations.

The log-linear fit of the results obtained for $J(z)$ in the low-scattering zone provides the values of the extinction coefficient μ_e for the wavelengths of interest. The dependence of the results for μ_e on the IL concentration C is illustrated in figure 3 for $\lambda = 847$ nm and $\lambda = 1326$ nm. The values of μ_e obtained using both the containers are close to each other within the statistical accuracy. This allows one to assume that the investigated dilutions occupy large-enough volumes to be considered as semi-infinite turbid media. The dependence $\mu_e(C)$ indicates a linear run in general at relatively low IL concentrations, which is in agreement with the results obtained previously by us and other researchers [15,16,18,19]. At $\lambda = 1326$ nm a linear dependence is observed up to the maximum IL concentrations concerned in the experiments with the small box. The experimental points obtained in this case can also be well fitted by a quadratic function but with negligibly small nonlinearity. The analogous experiments conducted with the large box lead to analogous results covering the same fitting dependences up to 2% IL concentration (figure 3b). The further increase of the concentration, and respectively of the amount of liquid in the container, leads to already noticeable deformations of it and respectively to noticeable distortions (apparent nonlinearity) of the results for $\mu_e(C)$. At $\lambda = 847$ nm, the experimental points obtained with both the containers follow the same nonlinear-in-general dependence $\mu_e(C)$ up to 2% IL concentration again (figure 3a). Above this concentration, the experimental points obtained with the large box deviate down due to the above-mentioned reason. In this case, the (quadratic, found by fitting) nonlinearity is noticeable for IL concentrations above 0.3-0.5%. (The experimental points obtained for $C \leq 0.5$ % obey a linear good-fit dependence $\mu_e(C)$.) Such a nonlinear behaviour has also been observed previously [18] in experiments with low-concentration dilutions of milk. It is perhaps a stage of a more general nonlinear dependence of μ_s on C at higher IL-20% concentrations (up to 25%) observed in [20] and analyzed theoretically in [21]. The nonlinear run of the extinction coefficient needs a profound and adequate physical explanation and mathematical description.

The experimental results show as well that with the increase of λ, the extinction coefficient μ_e decreases due to decreasing the light scattering (figure 4), and the interval of linearity at low IL concentrations extends.

The intercepts (μ_e at zero IL concentration) of the fitting curves in figure 4 are close to the published data for the absorption coefficients of pure water μ_{aw} (e.g., [11-13]). For instance, at λ = 1326 nm, the curve fitting all the experimental points has intercept $\mu_e(0)$ ~ 2.07 cm^{-1} that practically coincides with the literature data. Thus, the method of measuring $\mu_e(C)$ and the results obtained can be assumed as reliable.

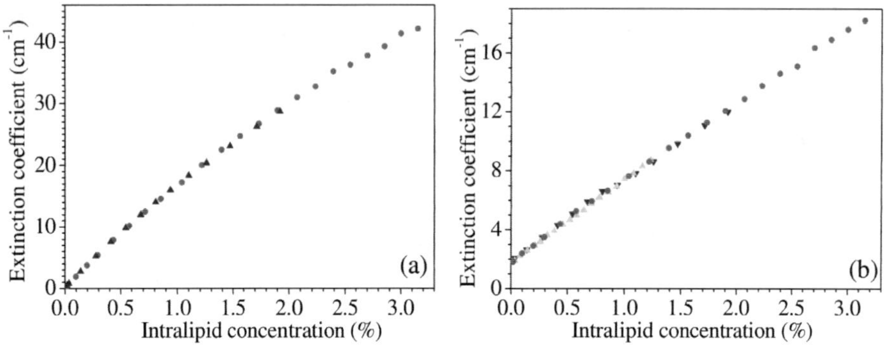

Figure 3. Extinction coefficients of Intralipid-20% dilutions, depending on the IL concentration, determined by using 12x12x22 cm box (circles) and 25x25x25 cm box (triangles) at λ = 847 nm (a) and λ = 1326 nm (b).

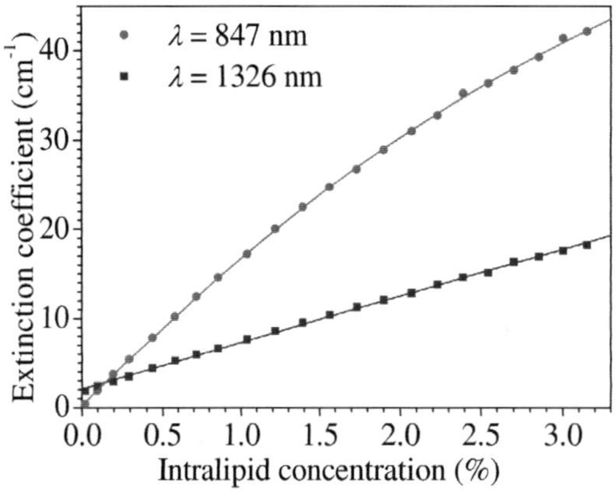

Figure 4. Extinction coefficients of dilutions with different Intralipid concentrations determined by using 12x12x22 cm box, at λ=847 nm (dots) and λ=1326 nm (squares), and the corresponding fitting curves.

There are derived in the literature [6-8] some empiric analytical expressions of the scattering coefficient μ_s of dilutions of IL-10% and IL-20%, as a function of λ. Such expressions have been derived from experimental data and Mie-theory calculations and help the estimation of the values of

μ_s, at different radiation wavelengths λ, depending on the IL concentration C. All the known empiric dependences $\mu_s(\lambda)$ can be written in a common form:

$$\mu_s(\lambda) = a\lambda^{b} \qquad (2)$$

where λ is given in [μm], and μ_s in [mL^{-1}Lcm^{-1}], implying that the volume fraction of IL-10% or IL-20% (having 11.952% and 22.74% volume concentration of scattering particles, respectively) in distilled water is in units of [ml/l]. A first such expression concerning dilutions of IL-10% and the wavelength interval 0.4 μm ≤ λ ≤ 1.1μm, with a = 0.16 and b = -2.4, have been established by van Staveren et al. [6]. Two more similar expressions concerning dilutions of IL-20%, with $\lambda \in$(0.4 μm, 1 μm), a = 0.249 and b = -2.397, and $\lambda \in$(0.5 μm, 2.25 μm), a = 0.317 and b = -2.59, are given by Michels et al [7] and Aernouts et al [8].

In figure 5, the experimentally obtained results for $\mu_e(C)$ corrected for the absorption of water are compared with linearly extrapolated data $\mu_s(C)$ obtained on the basis of the above-discussed expressions (2). It is seen that at lower concentrations (say, C < 0.5 %), where the scattering is assumed independent and the extinction coefficient should depend linearly on the concentration, the experimental results for $\mu_e(C)$ and the extrapolated results for $\mu_s(C)$ nearly coincide. At larger concentrations, two pairs of near results are outlined. The first pair includes the extrapolated results, following from the formulae derived in [6] and [8]. The second pair includes the experimental results obtained here and those following from the formula of Michels et al. [7]. Such a proximity, especially of the experimental and extrapolated results, is a confirmation of the reliability, for different intervals of concentrations, of both the method for measuring μ_e and the empiric formulae.

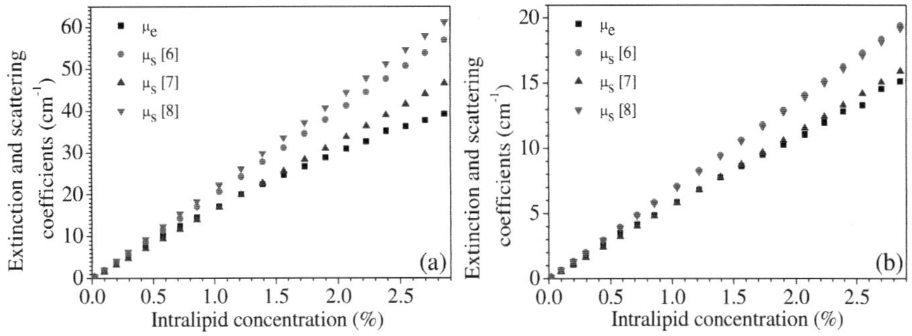

Figure 5. Experimentally-determined extinction coefficient μ_e (corrected for the absorption of water), as a function of the Intralipid concentration, compared with the scattering coefficient μ_s evaluated using the empiric formulae derived in [6-8] at λ = 847 nm (a) and 1326 nm (b).

4. Single-sided lidar-type sensing of turbid media

The smaller extinction and higher backscattering in tissue-like turbid media of laser radiation with longer wavelength λ would allow one to use such radiation for deeper sensing and diagnostics of tissues and other similar turbid media. A simple approximate illustration of such a possibility is given in figure 6, where, based on the single-scattering lidar equation [22], the normalized time-to-range resolved detected return power from a turbid medium, irradiated by laser pulses of wavelength λ = 672 nm, 847 nm, or 1326 nm, is evaluated as a function of the depth of sensing z. It is seen in the figure that the longer the wavelength the stronger the return signals from all the depths in the medium. The medium is chosen to be a 0.2% IL dilution with μ_e determined here (see also figure 4), and $\mu_s \sim \mu_e$ and g evaluated using the corresponding formulae of van Staveren et al. [6]. At an IL concentration of 0.2%, the low-scattering zone, where the lidar equation should still be valid, may achieve depths of the order of centimeters (see also in [23]). For more clarity, let us write explicitly the lidar equation in the form [22]

$$S(z,\lambda) = \mu_{bs}(\lambda)\exp[-2\mu_e(\lambda)z] , \quad (3)$$

where $S(z,\lambda) = P_b(z,\lambda)/(E_p cA/2z^2)$ is the so-called lidar S function, $P_b(z,\lambda)$ is the detected, time-to-range resolved return power, c is the speed of light, E_p is the sensing pulse energy, and $A=\pi E^2$ is the receiving aperture area. For Henyey-Greenstein indicatrix, the backscattering coefficient

$$\mu_{bs} = [(1-g)/4\pi](1+g)^{-2}\mu_s , \quad (4)$$

and

$$g(\lambda) = 1.1\text{-}0.58\lambda\ [\mu\text{m}]. \quad (5)$$

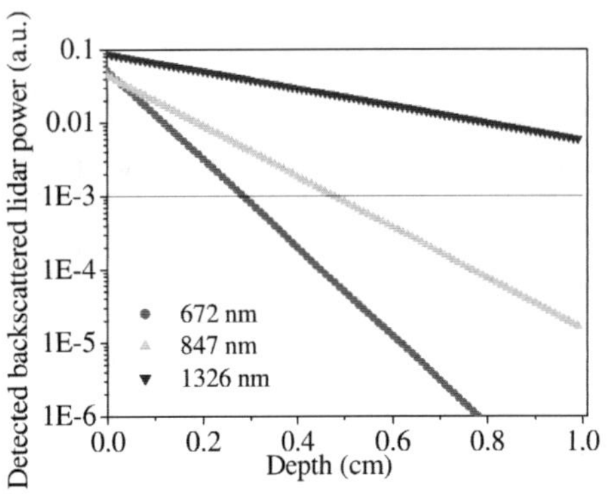

Figure 6. Dependence on depth of the return-signal power at single-sided lidar-type diagnostics of Intralipid dilution of 0.2% concentration for different radiation wavelengths.

The results represented in figure 6 are calculated by using (2)-(5). It is evident that the longer wavelength would ensure deeper informative sensing and diagnostics. From another point of view, however, the longer-wavelength radiations within the therapeutic window undergo stronger absorption in water, which may lead to undesirable increase of the temperature of the diagnosed in vivo tissues. When a laser beam/pulse of effective duration τ_p is propagating through a watery turbid medium, in the small-angle-approximation zone [16], the heat deposition and the temperature rise T along the beam axis can be estimated, using the expression (see also in [9,24])

$$T(z) = \mu_a P_t \tau_p [\pi w^2(z)\rho c_t]^{-1}\exp(-\mu_a z) , \quad (6)$$

where P_t is the peak pulse power, $w(z)$ is the beam radius at depth z, and ρ and c_t are, respectively, the mass density and the specific thermal capacity of water. When z decreases, $w(z)$ tends to the initial beam radius w. On the contrary, when z increases and approaches the developed scattering zone, $w^2(z) = \mu_{rs} z^3 /6 >> w$ [16]. So, it is clear [see (6)] that the temperature increment T is maximum at the entrance into the medium, when $z \rightarrow 0$. Then we have

$$T(z{=}0) = \mu_a P_t \tau_p [\pi w^2 \rho c_t]^{-1} . \quad (7)$$

Assume further, taking into account the recommendations of the American National Standards Institute and the International Electrotechnical Commission about the safe, maximum permissible exposures to laser radiation [25], that $\lambda = 1326$ nm, $\mu_a = 2.16$ cm^{-1} [12], the pulse energy $E_p = P_t\tau_p = 8\mu$J, $P_t = 38$ mW, $w = 1$ mm, $\rho = 1000$ kg/m^3, and $c_t = 4186$ Jkg^{-1}K^{-1}. Under such conditions, we obtain that T(0) ~ 1.314x10^{-4} K per a pulse. The thermal relaxation time τ_r of the heated region (of length μ_a^{-1} and radius w) is [9,26]

$$\tau_r \sim \min [(4\mu_a)^{-2}, (w/2)^2]/k_d , \quad (8)$$

where $k_d = k_c /(\rho c_t)$ is thermal diffusivity, and k_c is thermoconductivity. For water, $k_c \cong 0.58$ W/(mK) and respectively $k_d \cong 1.38\text{x}10^{-7}$ m^2/s. Then, using (8) we obtain that $\tau_r \sim 2$ s. The lidar type sensing of tissues would require many laser shots along one line of sight in order to accumulate a sufficient statistical volume for accurate signal estimation [23]. The pulse repetition rate of sensing however will be restricted by the thermal relaxation time τ_r. So, if we do not like that T exceeds ~ 0.5 K during the sensing procedure, the pulse repetition rate should not exceed 2000 Hz, that is 4000 laser shots and 32 mJ deposited light energy during every 2 s long time interval. Then, the maximum temperature rise, near the entrance of the laser beam into the investigated medium, will be 0.523 K. If one should deposit, for instance, 3.2 J radiative energy in order to achieve a prescribed signal-measurement accuracy to a given depth, the sensing procedure should evidently last 200 s.

In general, a careful selection is necessary of optimum sensing conditions ensuring optimum diagnostic safety and convenience, accuracy and resolution, and, respectively, depth of reliable detection of characteristic more or less contrasting inhomogeneities.

5. Conclusion

In this work, we have investigated experimentally the extinction coefficient of IL-20% dilutions in distilled water, depending on the IL concentration, for several laser radiation wavelengths in the red and near-infrared regions. The approach to measuring the extinction coefficient has been developed by us recently. It is based on the specific behavior in the low-scattering zone of the on-axis intensity of laser radiation beams propagating through the investigated media. The measurements have been conducted, using separately two dilution-containing plexiglass boxes of different volumes, in order to prove the appropriateness of the assumption of semi-infinite turbid medium.

The experimental results for the extinction are in agreement with our previous results concerning the dependence of μ_e on C. They are also in agreement with empiric formulae found by other authors [6-8] concerning the wavelength dependence of μ_s of IL-10% and IL-20% dilutions. It has been shown as well that the values obtained for the extinction coefficient do not depend on the choice of the plexiglass container and tend to the water absorptance at vanishing IL concentration.

As a whole, the results obtained in the work confirm the consideration of the experimental phantoms as semi-infinite media. They also confirm and extend theoretical and experimental results obtained previously, and reveal advantages of using relatively long wavelengths (around the upper limit of the therapeutic window) for deeper sensing and diagnostics of tissues and mimicking turbid media.

Acknowledgments

This work has been supported in part by the Bulgarian National Science Fund under the project "Development of biophotonics methods as a basis of oncology theranostics".

References

[1] Gibson A P, Hebden, J C and Arridge S R 2005 *Phys. Med. Biol.* **50** R1–R43
[2] Drexler W and Fujimoto J G 2008 Optical *Coherence Tomography: Technology and Applications* (Berlin: Springer)
[3] Taroni P et al. 2010 *J. Biomed. Opt.* **15** 060501
[4] Flock S T, Jacques S L, Wilson B C, Star W M and van Gemert M J C 1992 *Lasers Surg. Med.* **2** 510–9
[5] Ninni P D, Martelli F and Zaccanti G 2011 *Phys. Med. Biol.* **56** N21
[6] van Staveren H J, Moes, C J M, van Marie J, Prahl S A and van Gemert M J C 1991 *Appl. Opt.* **30** 4507-14
[7] Michels R, Foschum F and Kienle A 2008 *Opt. Express* **16** 5907-25
[8] Aernouts B, Zamora-Rojas E, Van Beers R, Watté R, Wang L, Tsuta M, Lammertyn J and Saeys W 2013 *Opt. Express* **21** 32450
[9] Kim A and Wilson B C 2011 Measurement of *ex vivo* and *in vivo* tissue optical properties:

methods and theories *Optical-Thermal Response of Laser-Irradiated Tissue* 2nd edn ed A J Welch and M J C van Gemert (Berlin: Springer) chapter 8, p.267

[10] Jacques S L 2013 *Phys. Med. Biol.* **58** R37–R61

[11] Palmer K F and Williams D 1974 *JOSA* **64** 1107-10

[12] Wieliczka D M, Weng S and Querry M R 1989 *Appl. Opt.* **28** 1714-19

[13] Pope R M and Fry E S 1997 *Appl. Opt.* **36** 8710-23

[14] Anderson R R and Parrish J A 1981 *J. Invest. Dermatol.* **77** 13–19

[15] Gurdev L, Dreischuh T, Bliznakova I, Vankov O, Avramov L and Stoyanov D 2012 *Phys. Scripta* **T149** 014074

[16] Gurdev L. Dreischuh T, Vankov O, Bliznakova I, Avramov L and Stoyanov D 2014 *Appl. Phys. B* **115** 427-41

[17] Ishimaru A 1978 *Wave Propagation and Scattering in Random Media* Vol. 1 (New York: Academic Press)

[18] Campbell S D, Menon S, Rutherford G, Su Q and Grobe R 2007 *Laser Physics* **17** 117-23

[19] Marchesini R, Bertoni A, Andreola S, Melloni E and Sichirollo A 1989 *Appl. Opt.* **28** 2318-24

[20] Zaccanti G, Bianco, S and Martelli, F 2003 *Appl. Opt.* **42** 4023–30

[21] Giusto A, Saija R, Iatì M A, Denti P, Borghese F and Sindoni O I 2003 *Appl. Opt.* **42** 4375-80

[22] Gurdev L, Dreischuh T and Stoyanov D 2011 Deconvolution of long-pulse lidar profiles *Lasers – Applications in Science and Industry* ed K Jakubczak (Vienna: Intech) chapter 13 pp 249-276

[23] Gurdev L, Dreischuh T and Stoyanov D 2007 *Proc. SPIE* **6604** 66042I

[24] Walsh J T 2011 Basic interactions of light with tissue *Optical-Thermal Response of Laser-Irradiated Tissue* 2nd edn ed A J Welch and M J C van Gemert (Berlin: Springer) chapter 2, p.13

[25] Gould T, Wang Q, Pfefer T J 2014 *Biomed. Opt. Express* **5** 832-847

[26] Cox, B 2013 *Introduction to Laser-Tissue Interactions* Lecture notes MPHY3886/M866/G886: Optics in Medicine (London: University College London, UK)

Polarization sensitive excitation-emission matrices for detection of colorectal tumours – initial investigations

E. Borisova[1], Ts. Genova[1], Al. Zhelyazkova[1], L. Angelova[1], M. Keremedchiev[2], N. Penkov[2], B. Vladimirov[3], O. Semyachkina-Glushkovskaya[4], L. Avramov[1]

[1]Biophotonics lab., Institute of Electronics, Bulgarian Academy of Sciences, 72, Tsarigradsko chaussee Blvd., 1784 Sofia, Bulgaria;
[2]Surgery Departament, University Hospital "Tsaritsa Yoanna-ISUL", 8 Bialo more str., 1527 Sofia, Bulgaria;
[3]Gastroenterology Departament, University Hospital "Tsaritsa Yoanna-ISUL", 8 Bialo more str., 1527 Sofia, Bulgaria;
[4]Physiology of Human and Animals lab., Biology Department, Saratov State University, 83 Astrakhanskaya str., Saratov, Russia

E-mail: borisova@ie.bas.bg

Abstract. Excitation-emission matrices were used for detection of the autofluorescence properties of surgically excised normal and cancerous mucosa of lower gastrointestinal tract tissues (colon and rectum carcinoma). Linear polarization of the excitation and emission fluorescence light was additionally applied to evaluate the influence of anisotropic fluorophores presented in the tissues investigated. Excitation applied was in the region of 280-440 nm, using step of 10 nm for the scanning, fluorescence emission was detected in the region of 300-800 nm, with scanning step of 1 nm. Excitation and emission light were investigated in parallel and perpendicular linear polarization modes respectively. These investigations are part of the concept to proof the feasibility of autofluorescence system for a real clinical application. Autofluorescence detection could make the entire diagnostics procedure more personal, patient friendly and effective and will help for further understanding of tumors nature and to improve patients' lives. In the current investigation major spectral features without and with linear polarization applied are addressed and lower GIT lesions' emission properties are evaluated.

1. Introduction

The limitations of standard endoscopy for detection and evaluation of cancerous changes in gastrointestinal tract (GIT) are significant challenge and initiate development of novel diagnostic modalities. Therefore many spectral and optical techniques are applied recently into the clinical practice for obtaining qualitatively and quantitatively new data from gastrointestinal neoplasia with a different level of clinical applicability and diagnostic success. Fluorescence imaging has been one of the most promising technologies in this area. The technique is very topical with its practical application in intra-operative, image-guided resection of tumors, because it permits minimal surgery intervention and friendly observation conditions. It could be based on endogenous fluorophores response to an excitation in the end of UV-blue spectral range, or on exogenous fluorescence, when fluorescent markers are applied to improve the detection sensitivity and the contrast ratio of the signals obtained from normal and abnormal tissue sites.

The first gastrointestinal laser-induced fluorescence spectroscopy (LIFS) of gastrointestinal tract was performed by Kapadia et al. in 1990 [1]. In that *ex vivo* study normal colonic tissue and adenomatous polyps were investigated, using an excitation at 325 nm and the emission was evaluated in the 350 - 600 nm range. Very promising results were reported due to the applied multivariate linear regression analysis of the emission spectra, which resulted 94% accuracy in differentiating hyperplasia and 100% accuracy in differentiating adenoma from normal colonic tissues.

Similar prospective results were observed during *in vivo* autofluorescence investigations of normal, hyperplastic and adenomatous polyps. [2] In that study excitation wavelength at 370 nm was chosen. Considering fluorescence characteristics of normal and cancerous tissue in the spectral range from 400 nm to 700 nm, the research group had developed probability-based algorithm for tissue classification with a sensitivity of 90% and a specificity of 82 %.

Due to the first promising results, many groups started to work in the next two decades in the field of autofluorescence detection of gastrointestinal neoplasia and dysplasia with different level of success, depending from the excitation wavelength applied and specificity of the autofluorescence spectra received.

About ten years later were proposed the endoscopic systems for clinical GIT autofluorescence detection. Several feasibility studies are reported based on several endoscope models produced by the specialized companies, working in the field of endoscopic equipment development.

Few autofluorescence endoscopic systems had been implemented for diagnostics of the GIT tumours, such as the LIFE-GI system and its second generation *LIFE II*; Xillix Technology Co.,Canada, D-Light system of Karl Storz GmbH, DAFE system (Diagnostic AutoFluorescence Endoscope) of Richard Wolf GmbH, XCLV-260HP and XCF-Q240FAI; Olympus, Tokyo, Japan. The feasibility of those systems in diagnostic endoscopy of the esophagus, stomach and colon cancer diagnostic has been investigated by several research groups. [3-5]

However, to improve the diagnostic accuracy, multiple excitation wavelengths have to be applied, to cover the range of all endogenous fluorophores, which could add to the diagnostic value of the autofluorescence spectra detected from normal and cancerous GIT tissues.

For a first time, an excitation – emission mapping using several excitation wavelengths for gastrointestinal tissues autofluorescence detection was performed by the group of Richard-Kortum at al. [6], in the beginning of 1990s. The group performed investigation of the autofluorescence of colonic tissue *ex vivo* using excitation - emission matrices, for identifying excitation wavelengths, which would result an emission spectra with diagnostic meaning. They stated that excitation wavelengths at 330 nm, 370 nm and 430 nm were the most suitable for autofluorescence differentiation between normal and cancerous colonic tissues and 404, 480, and 680 nm were found to be the most useful emission wavelengths for diagnose.

Recent investigations in this field are related to the searching of some optimal excitation wavelengths and revealing of the fluorescence peculiarities, which could have diagnostic value for better differentiation between normal, dysplastic and malignant GIT mucosa.

Imaizumi et al., 2011. had investigated the fluorescence of NADH in human colonic tissue samples [7]. Diagnostic meaning of the fluorescence spectra in this study was evaluated by observing ratio images created by dividing the fluorescence spectra obtained with excitation wavelengths of 365 nm and 405 nm. The ratio images showed a higher signal intensity in the adenomas and hyperplastic polyps, respectively, than in the adjacent normal mucosa. This study showed that metabolic alteration of colonic adenomas could be detected by imaging mucosal autofluorescence in the wavelength range of 450 to 490 nm under excitation at 365 and 405 nm.

Another group applied excitation wavelength of 355, 361, 370, 380, 384, 390 and 410 nm to biopsied or surgically resected colorectal tissue samples from 50 subjects with different degrees of colorectal dysplasia [8]. Autofluorescence from the samples was observed in the spectral region of 400-570 nm. The most significant difference in the autofluorescence signal of normal, benign and malignant tissue observed was the intensity at 460 nm of the emission, related to the NADH, which is lower for malignant and benign tissues. The obtained sensitivity, specificity and accuracy in differentiating between normal, benign and malignant tissue was found to be 100%, 90% and 96,7% respectively, which is an excellent perspective for further development of the autofluorescence excitation-emission mapping of the GIT mucosa for the needs of early tumor diagnosis.

We plan to gain more significant data base for the main spectral characteristics of lower GI neoplasia using excitation in the range of 280-440 nm, with a step of 10 nm. Additionally applied linear polarization for excitation and emission in two positions – parallel and perpendicular polarization for the light beams from the excitation source and from the fluorescent tissue sample will allow us to make some conclusions about the influence of structural proteins and other polarization-sensitive endogenous fluorophores in the GIT mucosa – normal and tumour. We foresee as well to develop appropriate algorithms for benign/malignant tissue differentiation, based on the spectral features, obtained for normal mucosa and colorectal pathologies with and without polarizing of the excitation/emission light pairs. Initial results obtained are promising that the specific differences are observed between normal/tumour, as well as for dysplasia/tumour pairs.

2. Methods and materials

Our investigations here are based on *ex vivo* measurements of excitation-emission matrices (EEM) for lower gastrointestinal (GIT) neoplasia to evaluate the applicability of autofluorescence technique for clinical observations of lower GIT tumours *in vivo*. Fluorescence signal from the surgically excised samples was obtained in three modalities – EEM – excitation –emission matrix without polarization optics, and with linear polarizers applied on excitation and emission pathway of the light in two positions – parallel and perpendicular for the excitation and emission light beams.

The EEMs are graphically represented with excitation wavelength on one axis, emission wavelength on the second, and fluorescence intensity forms the third axis. This method for three-dimensional fluorescence spectroscopy provides enough information about the fluorescence spectra of biological tissue samples for determining excitation wavelengths that gave emission fluorescence spectra containing the most diagnostic meaning for clinical diagnostic analysis. EEMs were performed over pairs of cancerous tissue and healthy tissue from the lower GIT from 9 different patients. The procedure of obtaining the investigated samples includes their excision during surgery for removal of GIT neoplasia lesions. After the surgical removal biological samples are transported in isothermal conditions and safe-keeping solution from the hospital to the spectral laboratory, where their fluorescence is investigated. All patients received and signed written informed consent and this research is approved by Ethics committee of University Hospital "Tsaritsa Yoanna", Sofia.

Spectrofluorimeter FluoroLog 3 (HORIBA Jobin Yvon, France) was used for the measurements. This system's light source is Xenon lamp with power 300 W, performance range of 200-650 nm and PMT detector with performance range of 220-800 nm for fluorescence detection. Since our samples vary in shape and dimensions, their fluorescence was investigated with additional module F - 3000 of Fluorolog 3, which allows investigation of samples outside of the sample chamber.

Measurements of the fluorescence signals of the different tissue samples obtained in EEMs were performed with applied excitation in 280-440 nm spectral region with step of 10 nm and emission observed between 300 nm and 800 nm, with step of scanning - 1 nm. Polarization measurements were performed in the same spectral ranges for excitation and emission light, but on the way of the light beams were applied linear polarizers in two positions - parallel and perpendicular to each other and polarized EEMs were detected. After the performing of both types of spectroscopic measurements for healthy and cancerous tissue, the samples were stored in formalin solution. Histological analysis of the tissues detected by the autofluorescence EEM approach was used as a "gold" standard for the clinical diagnosis of the pathology investigated.

3. Results and Discussion

Main differences observed between the fluorescence spectra of healthy and cancerous tissue are in the intensity of the fluorescence originating from the amino acids –tyrosine and tryptophan, the enzymes and coenzymes NADH and FAD, and from the structural proteins elastin and collagen.

On Figure 1. are presented the results in a form of spectral curves (1a and 1b) and contour maps for EEMs (1c and 1d) of a healthy mucosa and a colon carcinoma. No polarization elements are applied.

The differences observed arise from the different metabolic rate and structural characteristic of cancerous cells to healthy cells. The higher metabolic rate of the cancerous cells results in intensive production of the amino acids tyrosine and tryptophan, hence we observe higher intensity of their fluorescence in comparison with other fluorophores in these tissues in the UV spectral region – 320-350 nm peaks. Cancer cells undergo aerobic glycolysis which results in elevated NADH:NAD+ ratios, where NAD+ is the non-fluorescent oxidized form of NADH. This may be one of the reasons for the observed lower intensity of the fluorescence maxima of NADH in cancerous cells (max at 440-460 nm region).

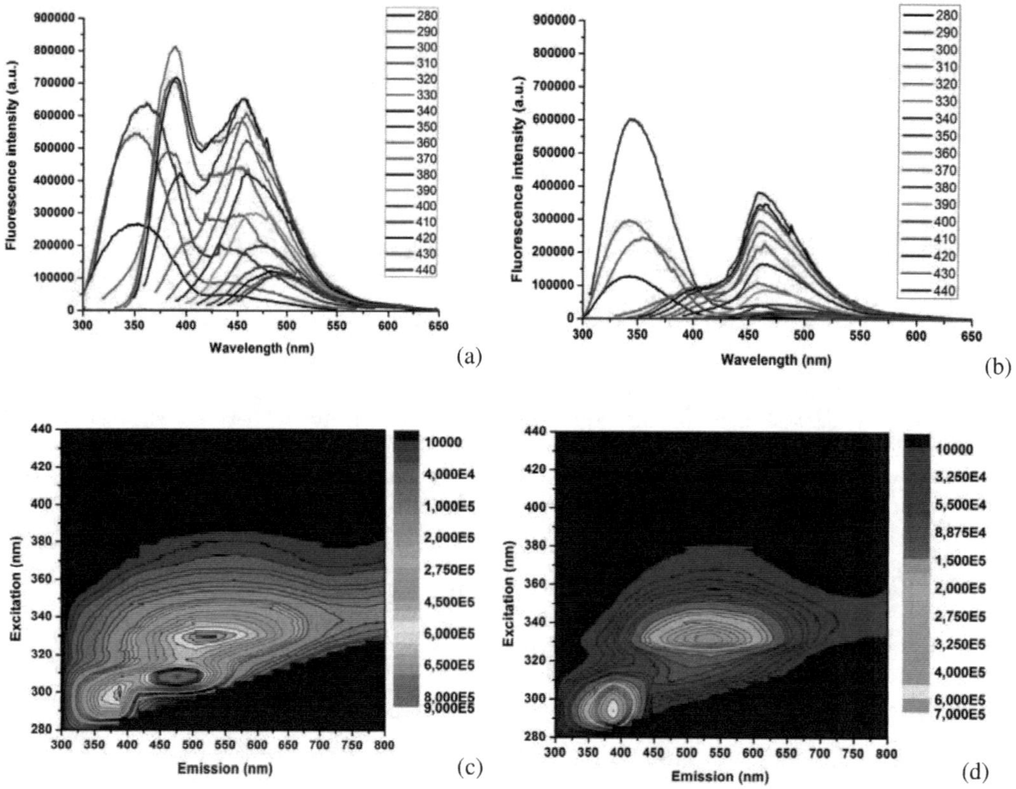

Figure 1. Comparison between (a) autofluorescence spectra of normal colon mucosa and (b) autofluorescence spectra of colon carcinoma using excitation wavelengths in the region of 280-440 nm; (c) EEM of normal colon mucosa (d) EEM of colon carcinoma

Journal of Physics: Conference Series **594** (2015) 012031 doi:10.1088/1742-6596/594/1/012031

Abnormal oversized growths of cancerous cells results in lack of structural proteins in volume unite of the cancerous tissues. This reduction of the quantity of the structural proteins affects the fluorescence spectra of cancerous tissue by lowering the intensity of fluorescence maxima of structural proteins, in comparison with the same maxima in the spectra of healthy tissue. The maximum of collagen type I fluorescence at 400-410 nm is significantly higher for normal mucosa in comparison with tumour site. The partial demolition of the extracellular matrix due the tumour growth also affected fluorescence signal resulted from protein cross-links in the region of 480-500 nm. This signal is about two times lower for the tumour vs. normal GIT mucosa.

Excitation-emission matrices were used for defining most suitable excitation wavelength for fluorescence diagnostic of colonic tissue ex vivo, also by Li et al. [9]. This group performed the measurement for excitation in the range of 260-540 nm with increment of 20 nm and emission observed in a range starting 20 nm above the excitation wavelength and extending to 800 nm at 5-nm intervals. They observe four maxima in the EEMs of the investigated normal and cancerous colonic tissue. The fluorophores, those maxima are arising from, are addressed as tryptophan and tyrhosyne for excitation of 280 nm and emission of 330 nm, NADH with excitation maxima of 350nm and emission at 480 nm and 460-605 nm and FAD with excitation maxima of 460 nm and emission at 520 nm. These authors observed also additional peaks in the fluorescence spectra of adenomatous colonic tissue, at 635 nm and 710 nm, addressed to the endogenous porphyrins in cancerous tissues. In our investigations we did not observed yet significant level of endogenous porphyrins fluorescence in the GIT tumours investigated. However, with the broadening of the spectral database and new samples investigated we could expect such fluorescence to be detected.

The feasibility of different wavelengths (namely 337, 375, 405 and 460 nm) for an excitation of autofluorescence of colonic cancer is investigated recently [10]. This group acknowledged that the autofluorescence spectra obtained under 337 nm excitation could provide more diagnostic information. In our own investigations, we observe significant changes in the level of amino acids autofluorescence in the region of 320-350 nm. The autofluorescence intensity for normal mucosa is about 2,5 times higher vs. tumour, when excitation at 280 nm is applied and about 2 times higher for the 290-300 nm excitation. Significant changes in the spectral shape of the emission for normal and cancerous mucosa are observed for all cases investigated when 310-330 nm excitations is applied. This region is appropriate to be used for diagnostic differentiation not only due to intensity changes occurred, but due to rapid changes in the spectral shape for both tissue states.

The results obtained from polarization autofluorescence measurements for the normal colon and carcinoma reveal the influence of collagen anisotropy in the autofluorescence features of the normal and abnormal tissues. On fig.2 are compared EEMs of normal colon mucosa (2a and 2b) and carcinoma (2b and 2d) for both modalities – parallel (2a and 2c) and perpendicular (2b and 2d) linear polarization of the excitation and emission light from the tissue samples detected.

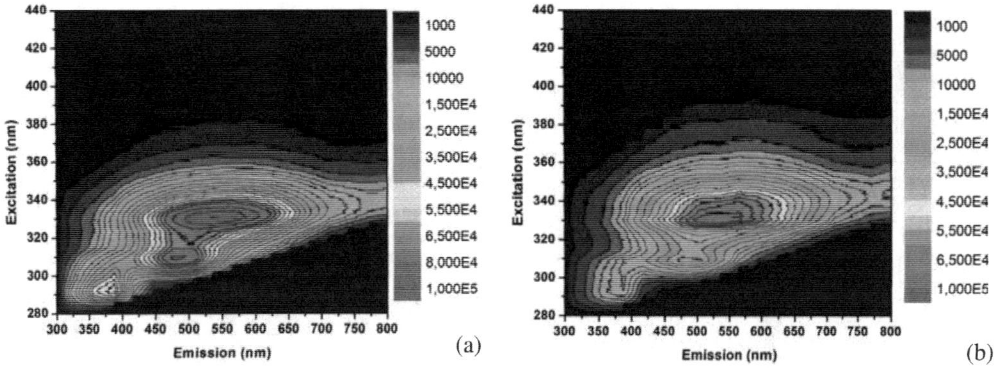

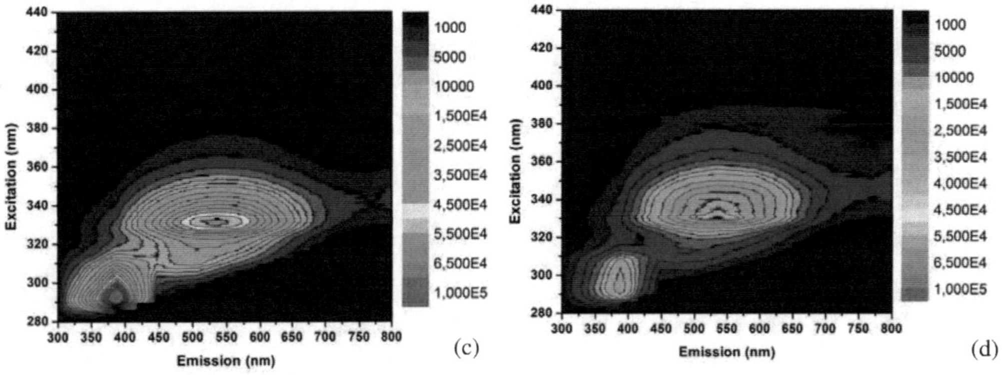

Figure 2. Comparison between EEMs of normal colon mucosa with parallel polarization (a) and perpendicular polarization (b) of the excitation and emission light beams; Comparison between EEMs of colon carcinoma with parallel polarization (c) and perpendicular polarization (d) of the excitation and emission light beams. Excitation is in the region of 280-440 nm, autofluorescence emission detected is in the region 300-800 nm.

Some of the biologically important fluorophores are anisotropic by their nature, due to the specific structure of chirality. Typical example of scattering and fluorescence anisotropic molecule in the human tissues is its major structural compound – collagen, due to its fiber structure. The structure of collagen fibers could be responsible for the higher degree of polarization effects observed for the fluorescence signal obtained from this fluorophore as well.

The most significant changes are observed in the region of 380-420 nm, where the collagen type I fluorescence maximum is situated, as well as in the region of 480-520 nm, where the protein cross-links fluorescence has place.

In the longer wavelength spectral region, for excitation range longer than 360 nm, when the co-enzimes, such as NADH and flavins are excited to fluoresce, no significant spectral shape changes for the parallel and perpendicular positions of polarizer/analyzer couple were observed. Only the fluorescence maxima related to structural proteins and their cross links reveal spectral shape differences in that comparison.

The slight spectral shift of the maxima between the parallel and cross-polarized fluorescence spectra observed for thick tissue layers ($\approx$ 2 mm), or as in our case of *ex vivo* study, could be associated to the absorption properties of the tissue investigated. [11] The observed spectral peculiarities could be used as indicators for a discrimination of the tissue health condition.

Other investigators observed that the main difference between fluorescence spectra of normal and cancerous tissue is the lower intensity in the maxima of 380-400 nm, which is due to collagen fluorescence. [10] The lower intensity of this maximum is discussed to be caused by increasing of the mucosa resulting from neoplastic transformation so that lower the fluorescence quantum yield of collagen mainly distributing in the submucosa. Other researchers comment that the most pronounced difference in the autofluorescence properties of normal colon vs. colorectal carcinoma is related to the NADH signal. [8] Its maximum at 460 nm is observed to be significantly lower for the pathologic tissue. We also observe such decrease of the collagen and NADH autofluorescence, which allow us to compare our results with these of other research groups and to develop summarized picture of the endogenous fluorophores observed in the normal and malignant GIT mucosa.

The reduction of the quantity of the structural proteins affects the fluorescence spectra of cancerous tissue by lowering the intensity of fluorescence maxima of structural proteins, in comparison with the same maxima in the spectra of healthy tissue. In the autofluorescence spectra of cancerous tissue, additional difference, in comparison with the spectra of normal tissue is the spectral shift of the NADH and FAD maxima in the range of the FAD fluorescence, which we interpreted like increasing of FAD contained in the cancerous tissue.

4. Conclusions

GIT cancer diagnostics could be developed based on the difference in metabolism and structure for normal and diseased tissues. The fluorescence anisotropy measurement of the tissues shows its high sensitivity to the structural and morphological changes, related to health condition. The correlation between such small changes, related to the extracellular matrix decrease of structure and integrity could be more pronounced in the case of tumour lesions, where such integrity and structural matrix are partially demolished due to the lesion growth. The polarization autofluorescence measurements may provide a noninvasive method for cancer detection. In order to further demonstrate the feasibility of this method, studies need to be carried out to measure the fluorescence anisotropy of GIT benign and malignant lesions. Our preliminary investigations on measurements of healthy mucosa and colon carcinoma *ex vivo* allow finding the most appropriate excitation wavelengths and the specific fluorophores, which could contribute to the diagnostic discrimination of normal and cancerous GIT.

5. Acknowledgements

This work is supported in part by the National Science Fund of Bulgarian Ministry of Education, Youth and Science under grants #DMU-03-46/2011 "Development and introduction of optical biopsy for early diagnostics of malignant tumours", and "Development of biophotonics methods as a basis of oncology theranostics", and by the Ministry of Education and Science of Russia under Grant № 17.488.2014/K. E. Borisova acknowledged as well her personal fellowship from Foundation UNESCO/L'Oreal "For young women in life sciences", Bulgaria – 2014, whence the polarization optics was purchased.

References

[1] Kapadia C, Cutruzzola F, O'Brien K, Stetz M, Enriquez R, Deckelbaum L 1990 *Gastroenterology* **99** 150

[2] Cothren R, Sivak M,Van Dam J, Petras R, Fitzmaurice M, Crawford J, Wu J, Brennan J, Rava R, Manoharan R, Feld M, 1996 *Gastrointest Endosc* **44** 168.

[3] Haringsma J, Tytgat G, Yano H, Iishi H, Tatsuta M, Ogihara T, Watanabe H, Sato N, Marcon N, Wilson B, Cline R, 2001 *Gastrointest Endosc* **53** 642

[4] Tajiri H 2007 Proc. Jpn. Acad., Ser. B **83** 248.

[5] Kwon Y, Cho Y, Yoon T, Kim H, Choi M, 2012 *World J Gastrointest Endosc* **4** 57

[6] Richards-Kortum R, Rava R, Petras R, Fitzmaurice M, Sivak M, Feld M, 1991 *Photochem Photobiol* **53** 777

[7] Imaizumi K, Harada Y, Wakabayashi N, Yamaoka Y, Konishi H, Dai P, Tanaka H, Takamatsu T 2012 *Gastrointest endosc* **75** 110

[8] Luo X, Zhang B, Li J, Luo X, Yang L 2012 *Z. Med. Phys.* **22** 40

[9] Li B, Xie S 2005 *World J Gastroenterol* **11** 3931

[10] Liu, L, Nie, Y, Lin, L, Li, W, Huang, Z, Xiea S, Li, B 2012 *Photodiagn Photodyn* **10** 111

[11] Tuchin V, Wang L, Zimnyakov D eds., [Optical Polarization in Biomedical Applications], Springer-Verlag Berlin Heidelberg, 218-223 (2006).

Influence of additions of nanoparticles TaC on a microstructure laser cladding

M A Murzakov[1], V N Petrovskiy[1], V I Polski[1], V D Mironov[1], N M Prokopova[1], E V Tret'yakov[2]

[1]National Research Nuclear University MEPhI, Moscow, Russia
[2]NPO CNIITMASH, Moscow, Russia

E-mail: VNPetrovskij@mephi.ru

Abstract. The features of a laser cladding of nickel-based powders with TaC nanopowder additives have been experimentally investigated. The minimum depth of pro-melting of a basis, microhardness distribution over the cross section of the substrate, and the saturation of the metal of the cladding with basis components has been determined in the experiments.

1. Introduction

Laser cladding is applied to the surface of the workpiece coating by melting the filler material. Moreover, since the base melts minimally coating properties mainly depend on the properties of the filler material [1].

Nanostructured coatings are the following step in the development of functional coatings. The modern mechanical engineering sets tasks of decrease in coefficient of friction at preservation of wear resistance, increase of coating viscosity together with preservation of anticorrosive properties. And answers to these questions can be found in application of nanocoating. Nanostructured coatings in a thickness from a few to hundreds of microns have properties of superior wear resistance, corrosion resistance not only in relation to the main material, but also to traditional coating [2-3]. Due to its unique properties, radically different from the properties of the macro- and micropowder of identical composition, nanopowder injected into the molten pool contributes to its modification, causing redistribution of harmful impurities between grain boundaries and reducing their size. Refractory nanoparticles having a melting point of over 1600°C can act as centers of crystallization.

2. Experimental setup

The experimental setup for cladding structurally consists of the following major parts:

- Robot ABB;
- ABB robot control unit;
- Fiber laser;
- Chiller;
- Scanning laser head;
- Control unit of the scanning laser head.

In the course of this work fiber laser LS-5 power of 5 kW and scanning laser head has been used.

The microstructure of deposited coatings was investigated using an optical metallographic microscope Neofot 30 manufactured by Carl Zeiss Jena (Germany) with magnification of the microscope 500x. Determination of the elemental composition of the samples was carried out by X-ray microanalysis in the analytical scanning electron microscope at an accelerating voltage EVO-50 10-20 kV (probe current 5-50 nA) using:

- energy-dispersive spectrometer of the INCA x-ACT with a resolution 133 eV produced by Oxford Instruments (UK);

- wave spectrometer INCA Wave-500 with a resolution of 5 eV and a sensitivity of 0.01 wt. % manufactured by Oxford Instruments (UK).

3. Technique of the experiment

Experiments were made with use of standard powder on the basis of nickel 1350 made by, chemical composition of which is given in Table 1 and also when using powder 1350 with additives of nanopowder tantalum carbide (TaC).

Table 1. Chemical composition of the powder based on nickel 1350 and substrate.

	Ni	C	Fe	Cr	B	Si
Powder 1350	base	0.45	2.9	11.0	2.3	3.6
Substrate	0.12	0.17	98.54	0.14	-	0.42

Laser cladding of standard powder on the basis of nickel and powder with additives of TaC nanopowder was carried out on flat surfaces of substrates from low-carbonaceous steel, chemical composition of which is given in Table 1.

Before the experiments on a laser cladding both with standard metal powder on the basis of nickel, and standard powder with TaC nanopowder additives, the powder about 1 mm thick layer was put on a flat surface of samples beforehand with use water solution of cellulose as binding substrate.

When carrying out experiments on a cladding with additives of nanopowder mixes with concentration of nanoparticles of 10% and 20% were used. During the experiments the following parameters of a mode of processing were changed:

- Radiation power;
- Linear speed;
- The size of a laser radiation spot on a surface of a built-up sample.

Power of radiation varied in the range from 1500 to 3000 W with a step of 200-300 W, the speed of a cladding changed from 5 mm/s to 20 mm/s, with a step of 2-3 mm/s. The defocusing made +230 mm and +150 mm.

4. Microstructural researches of obtained laser claddings

Figure 1 presents microsection of laser cladding of standard powder on the basis of nickel 1350 with the indication of points of detailed shooting of sample structure. Cladding has sharp border with a substrate. The material of a cladding does not get deeply into a substrate, but has good coupling with a substrate.

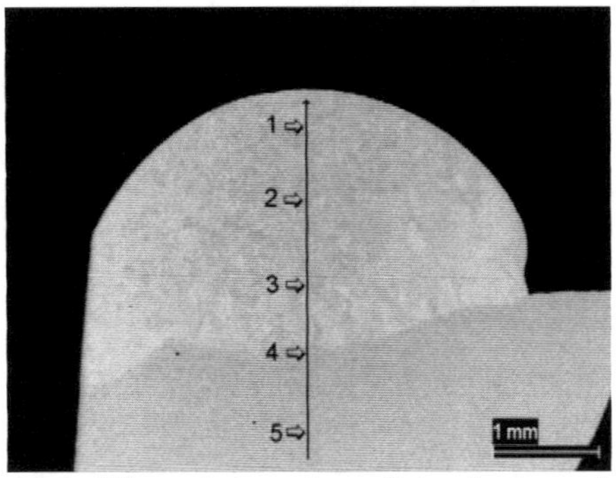

Figure 1. General view of laser cladding microsection of standard powder on the basis of nickel 1350.

Figure 2 presents detailed photos of a microstructure of a cladding in the points designated by numbers 1-5 in figure 1.

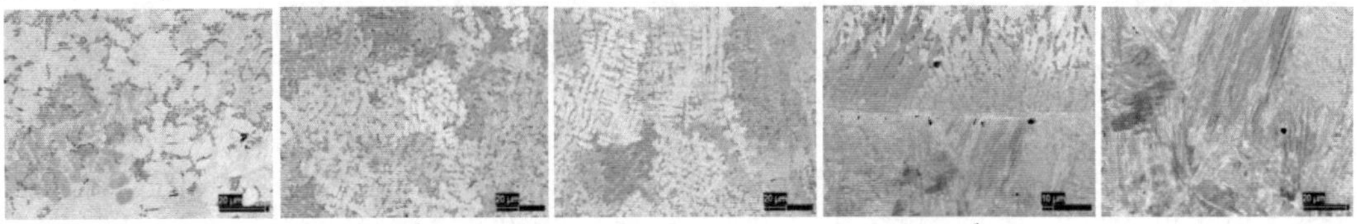

a b c d e

Figure 2. Detailed structure of a cladding of standard powder 1350: a) point 1; b) point 2; c) point 3; d) point 4; e) point 5 in Figure 1.

Structure of the top part of a cladding (point 1) – large equiaxed grains with early stages of dendrites. The central part of a cladding (points 2 and 3) – dendritic structure, characteristic for the congealed melt. Point 4 is sharp transition from structure of the stiffened fusion of a cladding to structure of the zone of thermal influence (ZTI). Point 5 (ZTI) is structure, typical for needle martensite. Substrate structure is typical hypoeutectic steel: rather equiaxed grains of ferrite and perlite. Size of ZTI is insignificant (~ 300 microns).

In figure 3 results of microhardness measurement of a standard powder cladding on the basis of nickel 1350 are presented.

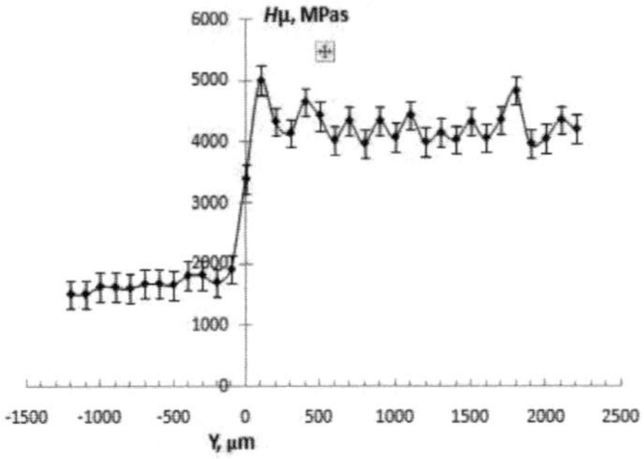

Figure 3. Microhardness change on cladding height; zero point corresponds to a point of 4 (Figure 1).

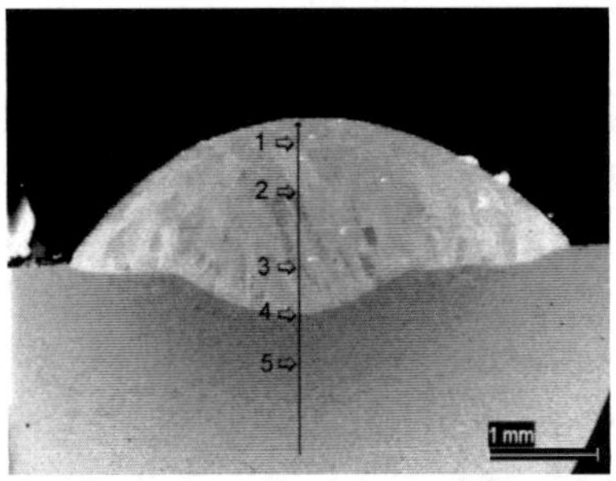

Figure 4. General view of a laser cladding microslice of standard powder on the basis of nickel 1350 with TaC nanopowder additives.

In figure 4 the microstructure of a cladding of standard powder 1350 with addition of TaC nanoparticles with concentration of 20% is shown.

a b c d e

Figure 5. Detailed structure of a cladding of standard powder 1350 with nanopowder TaC additives. a) point 1; b) point 2; c) point 3; d) point 4; e) point 5 in Figure 4.

The main regularities of structure change are the same, as on the previous cladding with standard powder on the basis of nickel. Differences are in the following:

- in a body of a cladding a large number of finely divided inclusions (Figure 5a, c) is observed;
- the element analysis showed that black impregnations are chrome carbides, white impregnations are tantalum carbides;
- the ZTI area from a cladding decreased to 10 microns;
- the ZTI general area also decreased to ~ 200 microns.

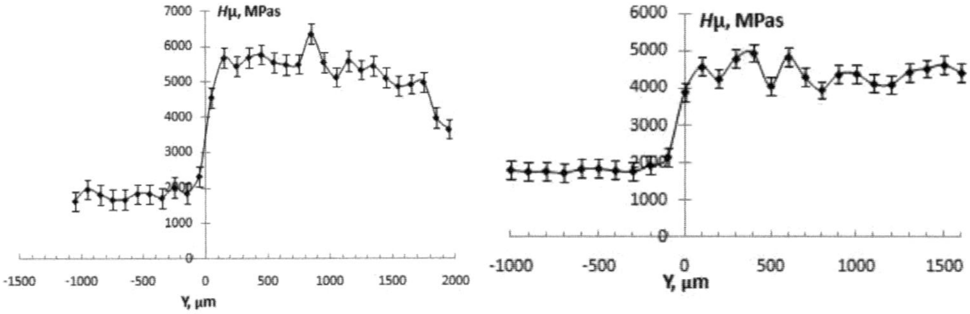

Figure 6. Microhardness change on height of a cladding of standard powder mix on the basis of nickel and a TaC nanopowder additive (concentration of TaC of 10% left side and concentration of TaC of 20% right side).

The maximum microhardness is in a cladding (~ 4500 MPas), minimum is in a steel substrate (~ 1800 MPas).

As the conducted researches showed, microhardness of a cladding with additives of nanopowder of carbide of tantalum differs from the microhardness of a cladding of standard powder on the basis of nickel. And the tendency of increase in microhardness with reduction of concentration of TaC nanoparticles is shown.

The specified result is explained by the fact that in the course of cladding melted main powder and particles of nanopowder of a refractory material mix up. As nanoparticles of a refractory material have melting temperature higher, than temperature of a molten pool, they remain in a molten pool in a firm state and at the subsequent crystallization of metal of a molten pool are the centers of crystallization of forming crystallites and strengthening particles in the nanodimensional range.

5. Conclusion

Comparison of laser cladding structure obtained with use of standard powder on the basis of nickel and standard powder on the basis of nickel with additives of carbide of tantalum nanopowder with different concentration is carried out.

It is shown that microhardness of a cladding with carbide of tantalum nanopowder additives differs from the microhardness of a cladding of standard powder on the basis of nickel. And the tendency of increase in microhardness with reduction of nanoparticles of TaC concentration is shown. Increase of microhardness shows advantages of developing of surface layer with high operational characteristics with use for a cladding of standard metal powders with nanopowders of refractory materials additives.

Acknowledgments

The study was supported by the RFBR (project no. 14-02-00369-a) and the Ministry of the Education and Science (unique identifier PNIER RFMEFI58214X0004).

References

[1] Grygoryants A G, Shiganov I N, Misyurov A I 2008 *Technological processes of laser processing* M.:MGTU, 664.

[2] Chen C, Zhang M, Chang Q 2008 *Laser in Engineering.* **18** 85

[3] Xue-song G, Zong-jun T, Zhi-dong L 2012 *Transactions of Nonferrous Metals Society of China* **22** 2498

Journal of Physics: Conference Series **594** (2015) 012033 doi:10.1088/1742-6596/594/1/012033

Spectral diagnostics of a vapor-plasma plume produced during welding titanium with a high-power ytterbium fiber laser

S A Uspenskiy[1], V N Petrovskiy[1], D P Bykovskiy[1], V D Mironov[1], N M Prokopova[1], E V Tret'yakov[2]

[1]National Research Nuclear University "MEPhI", Moscow, Russia
[2]NPO CNIITMASH, Moscow, Russia

E-mail: VNPetrovskij@mephi.ru

Abstract. This work is devoted to the research of welding plume during high power ytterbium fiber laser welding of a titanium alloy in the Ar shielding gas environment. High speed video observation of a vapor-plasma plume for visualization of processes occurring at laser welding was carried out. The coefficient of the inverse Bremsstrahlung absorption of laser radiation is calculated for a plasma welding plume by results of spectrometer researches. The conclusion deals with the impact of plasma on a high-power fiber laser radiation.

1. Introduction

Ytterbium fiber laser production technology develops high- and ultra high-power laser radiation at a wavelength of 1.07 μm with high beam quality. Therefore the fiber lasers are of considerable interest as an energy source in deep penetration metal welding processes, when the heat input power density is the major parameter setting the weld depth/width ratio and the process velocity. In welding with traditionally used 10.6 μm wavelength CO_2 laser radiation this parameter was limited by the plasma shielding effect, i.e. by the absorption and refraction of the laser radiation in the welding plasma plume. As the wavelength of ytterbium fiber laser radiation is about 10 times shorter, the plasma absorption, being inversely proportional to λ^2, should be negligibly small. However, the diffraction effects, which worsen the power stability and the beam focusing quality, can limit the technical advantages of the fiber laser [1].

2. Experimental setup

Welding experiments were carried out in a robot welding cell (fig. 1) based on 10 kW ytterbium fiber laser (1). Radiation was delivered by a transport fiber (2) (200 μm core dia.) to the optical processing head (3) and focused using a lens having a focus distance of 200 mm. The diameter of the focused beam on the metal surface (4) was 0.35 mm. A robotic arm (5), holding the welding optical head (3), provided the independent discrete horizontal shift of the high power fiber laser beam focus area in two mutually perpendicular directions with the accuracy 0.1 mm. The positioning stage supplied also a possibility of the movement of the specimen near the chosen focus position with a xed speed as well as the vertical shift of the specimen, when the distance between the metal surface (4) and the focusing lens is kept constant with the accuracy

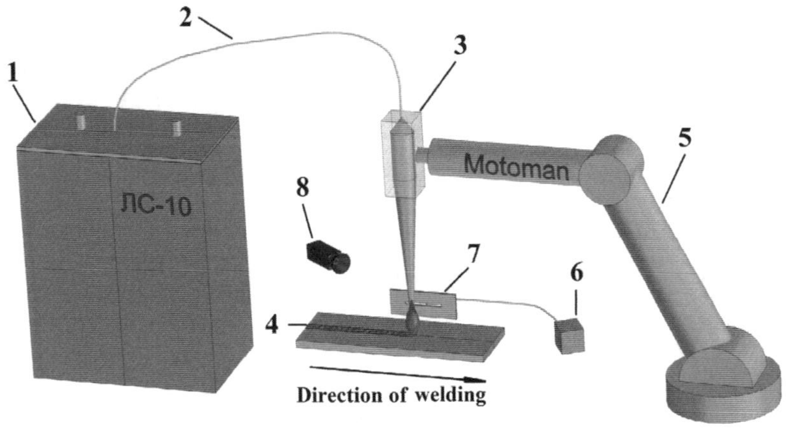

Figure 1. Experimental scheme for the observation of the welding plume.

0.1 mm. To determine the parameters of the welding plasma spectrometer ASP-150TF (6) was used with a fiber-optical input with the possibility of tuning the spectral range and spectral resolution of the spectrometer 0.09 nm. Measurements were carried out at a distance of 5 mm with aperture (7) above the sample surface (4) of titanium 6 mm thick. We used a high speed camera (8) for visualization processes during laser welding.

3. Results of high-speed video observation

The weld plume was observed with a high-speed video camera at frame rate 17000 fps. It was shown that besides the weld plasma there was a stream of small particles of condensed vapor, which almost agreed geometrically with the laser beam caustic and accelerated up to 10 m/s (fig. 2a). When we were using cross-jet the plume, this stream of particles was seen too. Cross-jet is not effective, because vapor reflect at robot, wall and another part of welded details back to caustic (fig. 2b). It is necessary to use additional cross-jet, venting for reducing the amount of steam in the caustic.

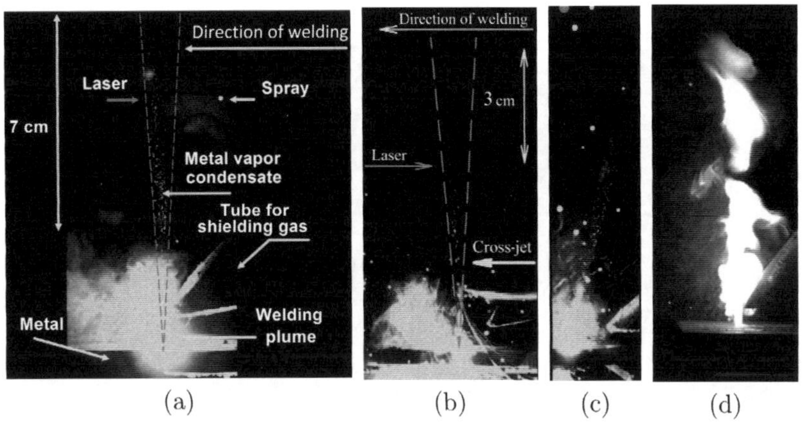

Figure 2. The frame of high-speed video observation of welding titanium without (a) and using cross-jet (b), during welding with the angle 15° to normal (c), during welding of steel (for comparison) (d).

The stream of metal vapor particles spreads along caustic of laser radiation also during welding with the angle 15° to normal (fig. 2c). We state, that after vaporization particles of metal fly to caustic laser and heat on side of laser radiation. It explains sloped direction of stream particles upper welding plume. There is a strong stream of metal particles in caustic beam, but in the case of steel [2](fig. 2d), there was burning of vapor. It is explained due to different thermal properties of these metals. There are no vapors and metal particles in caustic laser beam during welding of steel and using cross-jet is more effective in this case, in contrast with welding of titanium.

4. Result of spectrometric research

To study the welding plume using the spectral pyrometry method [3] a grating spectrometer with a fiber-optic input and a silicon CCD array was used to record the spectrum in the range of 200-1160 nm. The lineup includes 2048 pixels. The spectral resolution of the spectrometer is 1.4 nm.

Experimental spectral intensity $I(\lambda, T)$ of thermal radiation (fig. 3a), i.e. power radiated by a unit surface in unit spectral range is:

$$I(\lambda, T) = \frac{\varepsilon C_1 \lambda^{-5}}{\exp\left(\frac{C_2}{\lambda T}\right) - 1} \tag{1}$$

where $C_1 = 37418\ W \cdot \mu m^4/cm^2$, $C_2 = 14388\ \mu m \cdot K$, ε - emission coefficient, wavelength expressed in micrometers, the intensity in $W/cm^2 \cdot \mu m$. For the Wines region ($C_2/\lambda T \gg 1$) the previous expression after the transformation takes the form

$$\ln\left(\lambda^5 I\right) - \ln\left(\varepsilon C_1\right) = -C_2/\lambda T \tag{2}$$

If $\varepsilon = const$, a section of the spectrum in this regions linear in the coordinate plane (X, Y), where $X = \ln(\lambda^5 I)$ and $Y = C_2/\lambda$, and the slope depends on the temperature of the plume (fig. 3b). The dependence $\varepsilon(\lambda)$ appears only logarithmically. If ε changes in the selected range permanently or only slightly, its influence is manifested in a parallel shift straight along the X axis, but the slope and the desired temperature remain unchanged. Detailed description of the technique for measuring the temperature of the flame is represented in the works [3, 2]

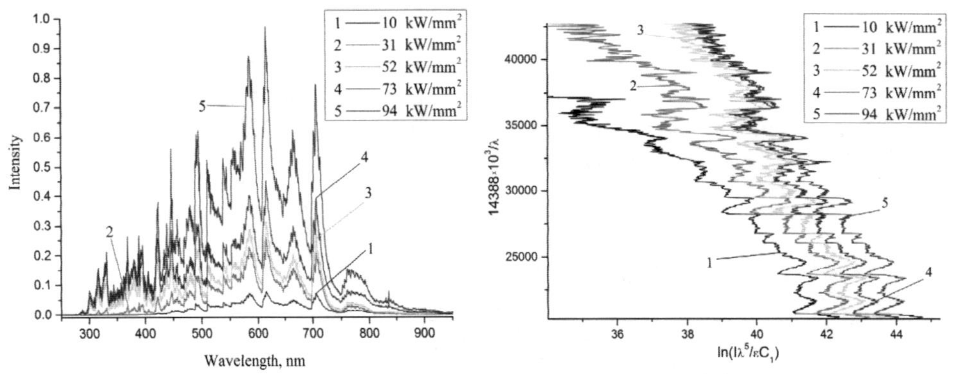

Figure 3. Panoramic emission spectrum of the plasma plume using shielding gas argon (a); spectra of plasma plume rebuilt in the new coordinates (b).

Based on the spectroscopic data to calculate the degree of ionization (fig. 4b) using Saha equation taking into account single ionization at a gas pressure of 1 atm:

Journal of Physics: Conference Series **594** (2015) 012033 doi:10.1088/1742-6596/594/1/012033

$$\frac{\alpha^2}{1-\alpha} = \left(\frac{2\pi m}{h}\right)^{3/2} \frac{(kT)^{5/2}}{p} \exp\left(-\frac{eV_i}{kT}\right) \tag{3}$$

where, α is ionization degree, m is mass of electron, V_i is ionized voltage, e is electronic charge, and p is gas pressure.

Electron density (fig. 4c) was calculated for these temperatures, using the equation of state of ideal gas, according to the equation

$$n_e = \frac{\alpha p}{kT} \tag{4}$$

From the above given parameters of the welding plasma state the inverse Bremsstrahlung coefficient (fig. 4d) can be calculated for the ytterbium fiber laser wavelength [4, 5] (1.07 μm) by equation

$$k_{ib} \approx \frac{\nu_c n_e e^2}{c m_e \omega} \approx 3.3 \cdot 10^{-41} \frac{n_e^2}{T^{3/2}} \tag{5}$$

where ν_c is the collision frequency, c - speed of light, ω - laser radiation frequency.

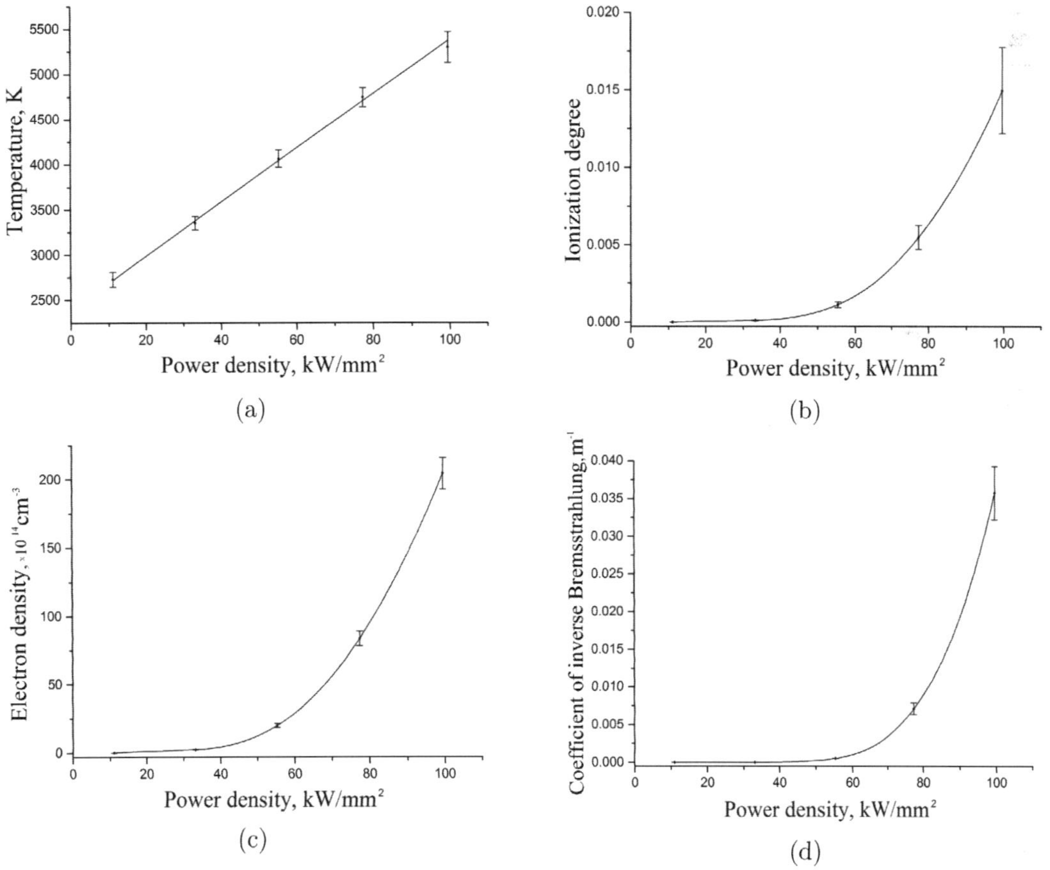

Figure 4. Temperature dependence on power density (a), ionization degree dependence (b), electron density dependence (c), coefficient of inverse Bremsstrahlung absorption dependence (d).

From the obtained absorption coefficient value it can be concluded that the welding plasma during the deep penetration metal welding with high power and high brightness solid-state lasers has practically no influence on the laser radiation and on the weld quality, even with a laser power density of up to 50 kW/mm^2. Temperature of boiling titanium (3560 K) [6] is achieved when power density is $\sim$20 kW/mm^2. It leads to sharp increasing of ionization degree, electron density and coefficient of inverse Bremsstrahlung absorption. In case of welding steel listed parameters except the last one have values of the same order, but coefficient of the inverse Bremsstrahlung is less at two orders.

5. Conclusion

The features of the vapor-plasma plume formed by welding titanium alloy VT-23 with a high power fiber laser were determined. On frame of high-speed video it is shown that 3 cm high welding plume is formed on the metal surface, and it varies periodically in the welding direction. Stream of the fine particle size fractions of a millimetre is observed in the caustic of the laser beam, moving down at a speed of 15 m/s. System of cross-jet in the direction of welding cannot effectively deal with the negative impact of the metal vapor radiation. Boiling titanium is achieved when the laser power density is higher than 40 kW/mm^2, which leads to a sharp increase in the density of free electrons in the plasma, the degree of ionization and an inverse Bremsstrahlung absorption. Plasma plume is in a weekly ionized state (degree of ionization $\alpha \leq 2{\cdot}10^{-2}$) and negligible absorbs radiation (coefficient of inverse Bremsstrahlung $k_{ib} \leq 4{\cdot}10^{-2}$ m^{-1}). The total absorption by plasma is less than 1%.

Acknowledgments

This work was supported by the RFBR (project number 14-02-00369-a) and the Ministry of Education and Science of the Russian Federation (unique identifier PNIER RFMEFI58214X0004).

References

[1] Scheglov P Y 2012 *Study of Vapour-plasma Plume During High Power Fiber Laser Beam Influence on Metals* (Bundesanstalt für Materialforschung und -prüfung (BAM))
[2] Petrovskiy V N, Uspenskiy S A, Shcheglov P Y, Gumenyuk A V and Rethmeier M 2011 *Nuclear physics and engineering* **2** 159
[3] Kopev V A, Kossyi I A, Magunov A N and Tarasova N M 2006 *Instruments and Experimental Techniques* 1
[4] Kawahito Y , Matsumoto N, Mizutani M and Katayama S 2008 *Science and Technology of Welding and Joining* **13** 744
[5] Kim K R and Farson D F 2001 *Journal of Applied Physics* **89** 681
[6] Zefirov N S 1995 *Chemical encyclopedia* vol 4

Self-similar Shape Mode of Optical Pulse Propagation in Medium with non-stationary Absorption

Vycheslav A. Trofimov, Tatyana M. Lysak, Mihail V. Fedotov, Alexander S. Prokopenko

Lomonosov Moscow State University, Moscow 119992, Russia

e-mail vatro@cs.msu.ru

Abstract. We discuss laser pulse propagation with the self-similar shape in a medium with instantaneous nonlinear absorption. We consider two cases of the laser pulse propagation. First one corresponds to problem of laser-induced plasma generation in silica under action of TW laser pulse. The second one corresponds to femtosecond laser pulse propagation in medium with nanoparticles of noble metals. In both cases the mode of the self-similar shape of pulse is of interest. We discuss also a physical mechanism of non-linear acceleration or slowing-down for laser pulse propagation in a medium with nanoparticles. The last phenomena are important, in particular, for a problem of data processing of all optical method. We used analytical approach for considered problem as well as computer simulation.

1. Introduction

Soliton or soliton-like profile of laser beam or shape of laser pulse plays big role in modern nonlinear optics and attaches great attention of various scientist (see, for example [1-20]). We want to stress that in our opinion, investigation of a laser beam propagation mode with self-similar profile (or shape) is actual problem if such investigation is made for a medium with non-linear absorption because powerful laser system produces high intensive laser pulse, for which a multi-photon absorption takes place due to plasma formation under its propagation.

In our previous papers [21-26] we have investigated the possibility of laser pulse propagation with self-similar profile in a medium with two-photon absorption (TPA) or multi-photon absorption and also taking into account a presence of either cubic nonlinear response of a medium (self-focusing or de-focusing) or resonant nonlinearity or both nonlinearity together. In all cases under consideration we have found out a propagation mode with the self-similar profile of laser beam. To achieve this aim we formulate the problem as corresponding eigenvalue nonlinear problem. Obviously, certain simplification has to be made because of continuous decreasing of maximal intensity of the laser beam due to nonlinear absorption. Earlier we have applied this approach for finding the soliton solution of various problems of nonlinear optics [27-31].

It should be stress, that in contrast to previous papers we consider below a possibility of self-similar shape mode of laser pulse propagation if a medium response is non-instantaneous.

Consideration is provided for two problems. The fist one is laser pulse propagation in medium with TPA and taking into account self-action of laser pulse due to time-dependent generation of free electrons. The second problem is a propagation of laser pulse in a medium with nanorods. In both cases we show that there is some distance of laser pulse propagation along which the shape of laser pulse changes in self-similar mode.

2. Laser pulse propagation in a medium with nanorods

Below we consider a femtosecond laser pulse propagation in a medium with nanorods, taking into account the aspect ratio (ellipticity) of nanorods changing due to their melting because of multi-photon absorption (MPA) of optical radiation. In the framework of slowly varying envelope of wave packet, this process can be described by the following dimensionless nonlinear equations

$$\frac{\partial A}{\partial z}+iD\frac{\partial^2 A}{\partial t^2}+\frac{\delta_0+i\xi\Delta}{1+\Delta^2}(\varepsilon-1)A|A|^{2(k-1)}=0,\ 0<z\le L_z, 0<t<L_t.$$
$$\frac{\partial \varepsilon}{\partial t}=-\frac{\tilde{\delta}}{1+\Delta^2}(\varepsilon-1)|A|^{2k}. \qquad (1)$$

with initial and boundary conditions for complex amplitude and aspect ratio of nanorods

$$A(z,0)=A(z,L_t)=0,\quad 0\le z\le L_z,\ A(0,t)=A_0(t),\quad 0\le t\le L_t,\ \varepsilon(z,t=0)=\varepsilon_0,\ 0\le z\le L_z.$$

Above A is dimensionless slowly varying envelope of electric field pulse, normalized on the maximum value of square root from the incident pulse intensity ($z=0$). The nanorods are melting under the action of laser light. As a consequence, their aspect ratio ($\varepsilon=a/b$) is varied, where a and b are major and minor axes of nanorods, ε_0 is its initial value. Parameter k is the number of photons, involved in the absorption. E.g., $k=1$ is one-photon absorption (OPA), $k=2$ is two-photon absorption (TPA) and so on. Variable z is a dimensionless longitudinal coordinate, along which the optical radiation propagates; t is a dimensionless time in the system of coordinates moving with a pulse, time is measured in the units of τ_{pulse} - duration of the incident pulse; the dimensionless duration of laser pulse is denoted as τ_p; L_t is a dimensionless time interval, during which the laser pulse interaction with nanorods is analyzed, L_z is a dimensionless length of nonlinear medium. Parameter D characterizes the group velocity dispersion (GVD). Parameters δ_0 and $\tilde{\delta}$ characterize the absorption of laser light. Coefficient ξ characterizes self-action of laser pulse due to detuning of the carrier frequency of wave packet from the central frequency of the absorption spectrum. The case of $\Delta=0$ corresponds to optical pulse propagation at pure amplitude grating of the medium. It means an influence only of light energy absorption on the laser pulse propagation. In opposite case ($\Delta=\pm1$), the phase gratings are also induced by laser radiation. It should be mentioned that under consideration the positive sign of parameter Δ ($\Delta=1$, this case is named by us as positive grating) corresponds to pulse compression and at negative sign of this parameter ($\Delta=-1$, this case is named by us as negative grating) the decompression of laser pulse occurs.

We consider the mode of laser pulse propagation with self-similar shape [26]. This mode takes place, if the following equality is valid

$$\frac{\delta_0\tau_p{}^2}{D(1+\Delta^2)}\frac{(k-1)^2}{2k}\left(\frac{\xi}{\delta_0}\Delta+\sqrt{\left(\frac{\xi}{\delta_0}\Delta\right)^2+\frac{4k}{(k+1)^2}}\right)=1. \qquad (2)$$

Obviously, the parameter k must be greater than 1. In this case the complex amplitude of incident light is defined as

$$A(z=0,t)=A_0(t)=ch^{-1/(k-1)}\left(\frac{t-L_t/2}{\tau_p}\right)\cdot$$
$$\cdot\exp\left(-i\frac{k+1}{2(k-1)}\left(\sqrt{\left(\frac{\xi}{\delta_0}\Delta\right)^2+\frac{4k}{(k+1)^2}}-\frac{\xi}{\delta_0}\Delta\right)\ln\left(ch\left(\frac{t-L_t/2}{\tau_p}\right)\right)\right),0\le t\le L_t. \quad (3)$$

To verify our assumption about a validity of the self-similar shape of laser pulse under its propagation in a medium with nanorods, we use computer simulation based on the conservative finite-difference scheme. For qualitative estimation of the laser pulse propagation we calculate a deviation τ_c of pulse center from the initial one as follows

$$\tau_c(z)=\int_0^{L_t}(t-L_t/2)|A(z,t)|^2\,dt\bigg/\int_0^{L_t}|A(z,t)|^2\,dt \quad (4)$$

and full width at half maximum (FWHM) pulse duration $\Delta\tau$, measured at half maximum intensity level.

Below we consider only pure amplitude grating ($\Delta=0$), which means that we do not take into account the laser pulse self-action due to detuning between the carrier frequency of the wave packet and the central frequency of the absorption spectrum. Hence, the value of ξ does not matter. Additionally, we choose the initial aspect ratio of nanorods being equal to $\varepsilon_0=2$.

While the light propagates along the medium, its intensity decreases due to nonlinear absorption. Moreover, in the case of significant melting ($\tilde{\delta}>1$), the center of the pulse shifts into the area with less aspect ratio of nanorods. So, to estimate self-similarity of the pulse shape at various sections we use the following procedure. First, we renormalize pulse shape at maximum intensity (this procedure is not a rule). Then, we are comparing a calculated intensity shape of the pulse whith the intensity distribution of the incident pulse, taking into account the pulse center shifting:

$$A_{app}^2(z,t)=ch^{-2/(k-1)}\left((t-L_t/2-t_z(z))/\tau_z(z)\right),\ \ 0\le t\le L_t, \quad (5)$$

We determined parameters t_z and τ_z, using minimization of the functional:

$$J(z)=\int_0^{L_t}\left(I(z,t_i)-\left(A_{app}(z,t)\right)^2\right)^2 dt\,. \quad (6)$$

Here $I(z,t_i)=|A(z,t_i)|^2/|A_{max}(z,t)|^2$ is intensity of the laser pulse normalized on its maximum value in the section under consideration.

Fig. 1 shows evolution of laser pulse, having the self-similar shape (3) at the input section (z=0), along the coordinate z for parameters $k=2$, $\tau_p=1$, $\delta_0=3\sqrt{2}$ and $D=1$. For clarification of nanorods influence on laser pulse propagation, in Fig. 1a the evolution of the pulse shape in a homogeneous medium ($\tilde{\delta}=0$) without nanorods is shown. Figs. 1b,c illustrate laser pulse propagation features under moderate melting ($\tilde{\delta}=10$) or strong melting ($\tilde{\delta}=50$), correspondingly. Let us note, that these values of parameter $\tilde{\delta}$ satisfy the relation (2), so the laser pulse shape changes in a self-similar mode up to section $z=0.6$ for a homogeneous medium ($\tilde{\delta}=0$), for example. Then, a distortion of the pulse shape takes place near the pulse edges. Nevertheless, up to the section $z=1.0$

the pulse shape can be approximated by the inverse hyperbolic cosecant of the second degree (6) with high accuracy: the value of the functional J does not exceed $5.8\cdot10^{-3}$ (see Fig. 2 d).

Opposite to the homogeneous medium, the soliton–like shape of a laser pulse undergoes a compression up to 1.5 times at the initial distance of propagation in the medium with nanorods (Fig. 2a, b). In this case, the pulse center shifts towards the area of time increasing. However, the pulse center shifting becomes negligible if the absorption ($\tilde{\delta}$) becomes strong, or, equivalently, if strong increasing of incident light intensity takes place. For the medium values of parameter $\tilde{\delta}$ ($\tilde{\delta}=10$), a slight asymmetry of the pulse shape is well seen, which becomes negligible for high value of $\tilde{\delta}$ ($\tilde{\delta}\geq 50$) (let us compare with Fig. 1b,c). Nevertheless, similar to the pulse propagation in a homogeneous medium, the pulse preserves its shape up to the section $z=1.0$ with enough high accuracy – the value of the functional J does not exceed $6.6\cdot10^{-3}$ (see Fig. 2d). Of course, this accuracy is worse than for in the case of the homogeneous medium, especially at the beginning of the pulse propagation distance, due to the pulse asymmetry.

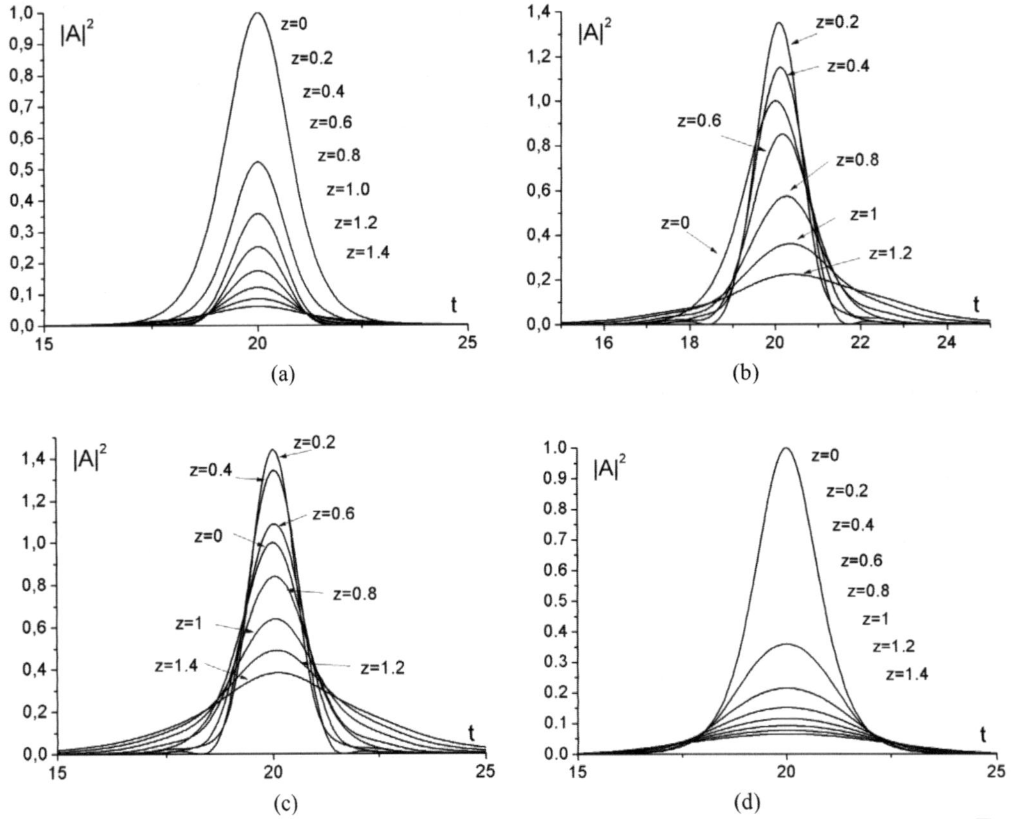

Fig.1. Pulse shape at different sections for $\tilde{\delta}$ = 0 (a,d); 10 (b); 50 (c) and $k=2$, $\tau_p=1$,$\delta_0=3\sqrt{2}$ and $D=1$. Fig. 1d shows evolution of the pulse shape for unchirped incident pulse, propagating in a homogeneous medium ($\tilde{\delta}$ = 0).

To emphasize an influence of the pulse chirp, in Fig. 1d the evolution of the incident unchirped pulse

$$A(z=0,t)=A_0(t)=ch^{-1/(k-1)}\left(\frac{t-L_t/2}{\tau_p}\right),\ k=2$$

in a homogeneous medium ($\tilde{\delta}=0$) is shown. As it is well seen, the pulse duration increases and its maximal intensity decreases much faster than it happens in the case of initially chirped pulse (compare corresponding curves in Fig. 2a,b). Preservation of the pulse shape is also much worse: the value of the functional (6) in some sections up to ten times exceeds its value for the incident chirped pulse (Fig.2e).

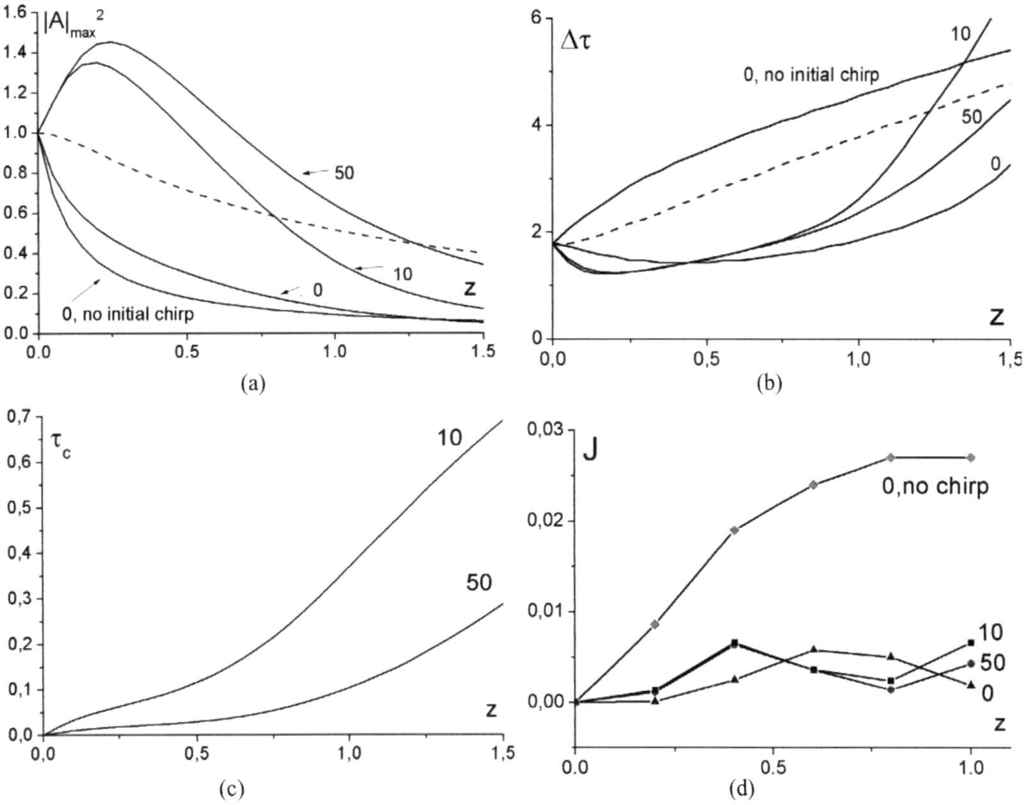

Fig. 2. Evolution of the maximal intensity (a), FWHM pulse duration (b), deviation of the pulse center from its initial position (c), functional J (d) along z-coordinate for various values of parameter $\tilde{\delta}$ (for solid lines the values of $\tilde{\delta}$ are depicted in the Figs.) and $k=2$, $\tau_p=1$, $\delta_0=3\sqrt{2}$ and $D=1$. Dashed lines correspond to a propagation of the incident unchirped pulse in a homogeneous medium.

Computer simulation have shown that a self-similar mode of the pulse propagation under three- and four-photon absorption is similar to the case of TPA (let us compare Fig.2,3,4). Nevertheless, some qualitative differences take place. Because we chose the same absorption coefficient ($\delta_0=3\sqrt{2}$) as it was for TPA, then we have to decrease the pulse duration in accordance to the equality (2). Therefore, we chose $\tau_p=0.6389$ for three-photon absorption and $\tau_p=0.5118$ for four-photon absorption. Together with $D=1$, these values allow self-similar propagation of the incident pulse in the form (3) if we deal with a homogeneous medium. One can stress, that in all cases the FWHM pulse duration remains approximately the same. This fact allows revealing the influence of photon number involved in the process of light absorption on the self-similar propagation of laser radiation.

In comparison to the process of TPA, increasing the number of absorbed photons leads to a less weak decreasing of the light intensity while the pulse propagates in a homogeneous medium (Fig.3a,4a). This is obviously a consequence of increasing the number of photons involved in absorption. Indeed, if the intensity decreases less than 1, the multy-photon absorption results in smaller absorption. In a like manner to TPA, the presence of nanorods in the medium leads to focusing of the chirped pulse at the initial part of propagation because of the reflection of the pulse from the gradient of absorption grating induced by changing of the nanorod aspect ratio. In the case of MPA, focusing of the laser pulse becomes slightly stronger with the growth of the number of photons.

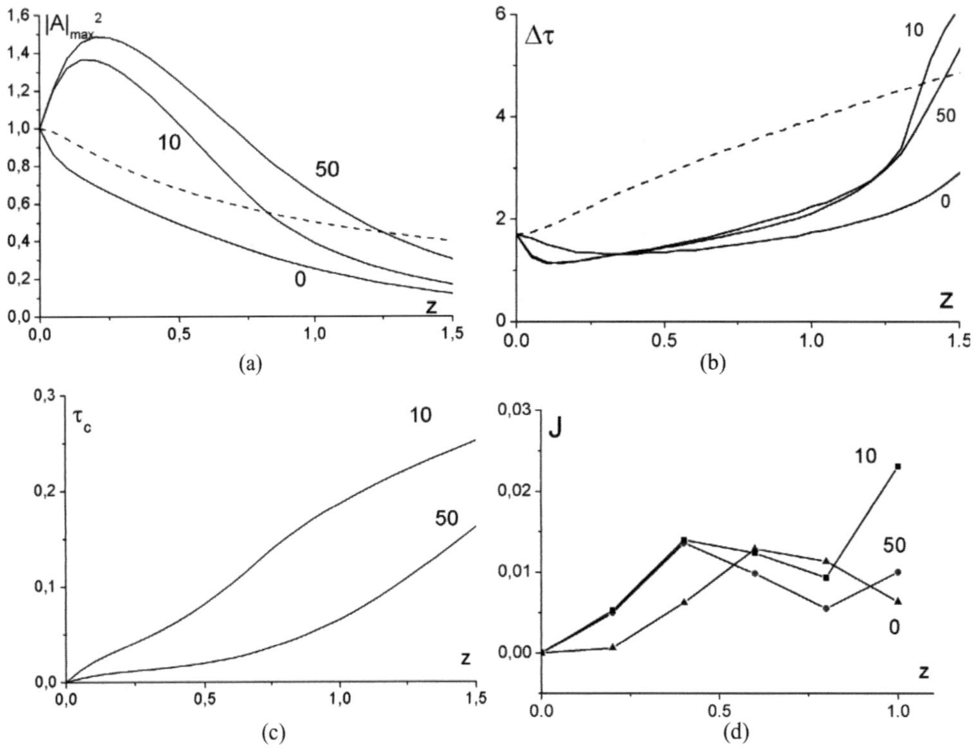

Fig. 3. Evolution of maximal intensity (a); FWHM pulse duration (b); deviation of pulse center from its initial position (c); functional J (d) along z-coordinate for various values of $\tilde{\delta}$ (solid lines, value of $\tilde{\delta}$ is shown in the Fig) and three photon absorption (k=3), $\delta_0 = 3\sqrt{2}$, $\tau_p = 0.6389$ and $D = 1$. Dashed lines show a propagation of the incident unchirped pulse in the medium without absorption ($\delta_0 = \tilde{\delta} = 0$).

The most remarkable fact is a significant decreasing of deviation of the pulse center from its initial value. As it is well seen from Fig. 2c, 3c, 4c, the deviation at the medium exit is up to 5 times less for $\tilde{\delta} = 10$ and up to 2.5 less for $\tilde{\delta} = 50$. This is also because of the weaker absorption due to increasing the photon number under the absorption of laser energy, if the maximal intensity decreases less than 1. Nevertheless, similar to the case of TPA, the pulse center deviation from its initial position is greater for $\tilde{\delta} = 10$ than for $\tilde{\delta} = 50$. Unfortunately, multy-photon absorption leads to distortion of self-similarity shape of the pulse. Remember, that for medium wuth TPA, the functional value does not exceeds 0.01

despite the strength of nanorod melting (Fig.2d). Approximation of the pulse shape, calculated from computer simulation, by the inverse hyperbolic cosecant of the corresponding degree in the case of strong melting is better than for medium strength melting. In the case of three-photon absorption, the functional value exceeds 0.01 for some sections of z-coordinate and it exceeds 0.02 somewhere in the case of four-photon absorbtion. For all the cases, approximation of the laser pulse shape along the initial part of its propagation is much better for homogeneous medium. Nevertheless, along the last part of laser pulse propagation, the approximaton of the laser pulse shape is the best for the strong melting ($\tilde{\delta} = 50$).

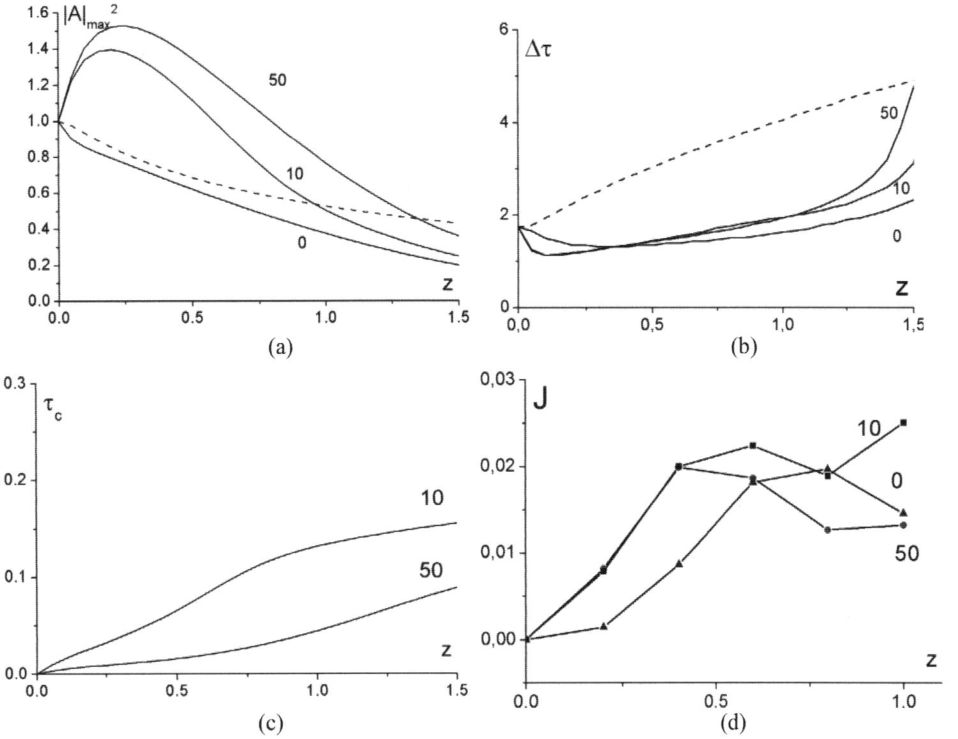

Fig.4. Evolution of maximal intensity (a); FWHM pulse duration (b); deviation of pulse center from its initial position (c); functional J (d) along z-coordinate for various values of $\tilde{\delta}$ (solid lines, value of $\tilde{\delta}$ is shown in the Fig.) and four-photon absorption (k=4), $\delta_0 = 3\sqrt{2}$, $\tau_p = 0.5118$, $D = 1$. Dashed lines show a propagation of the incident unchirped pulse in the medium without absorption ($\delta_0 = \tilde{\delta} = 0$).

3. Laser pulse propagation in medium with time-dependent free electron concentration.

Below we consider laser pulse propagation in a medium with time-dependent free electron concentration under the TPA of laser energy. This process is governed by the following set of dimensionless equations:

$$\frac{\partial A}{\partial z} + iD\frac{\partial^2 A}{\partial t^2} + i\alpha\rho A + \delta_0 |A|^2 A = 0, \quad 0 < z \le L_z, \quad 0 < t < L_t,$$
$$\frac{\partial \rho}{\partial t} + \frac{\rho}{\tau} = q|A|^4. \qquad (7)$$

These equations are solved with zero-value boundary conditions

$$A(z,0) = A(z,L_t) = 0, \quad 0 \le z \le L_z \tag{8}$$

and inital condition for free electron concentration

$$\rho(z,0) = 0, \quad 0 \le z \le L_z,$$

and initial distribution of complex amplitude

$$A(0,t) = A_0(t), \quad 0 \le t \le L_t. \tag{9}$$

Above, $\rho(z,t)$ is a free electron concentration. Parameter τ characterizes a free electron relaxion time; parameter q is proportional to squared intensity of incident laser pulse.

As it is well-known, sign of the parameter D can be positive or negative in dependence of laser pulse wavelength. We will take this into account by changing of the sign for parameter α, which describes changing of a medium refractive index. In our notation its positive sign corresponds to laser pulse compression at positive value of parameter D. We choose below this sign of D for definiteness.

We follow the propagation of incident laser pulse with self-similar shape. In the case of a medium with TPA (pure absorption grating) and without free electron generation, such shape is defined as

$$A_0(t) = e^{i\sqrt{2}\cdot\ln(ch(ct))}/ch(ct), \quad c = \sqrt{\frac{\delta}{3\sqrt{2}D}}. \tag{10}$$

The main question under consideration is the following: what distance of the pulse propagation with self-similar shape is? It depends on the problems parameters. We fixed value of the following parameters: D=0.25, q=1 and changed other parameters: τ = 0.1, 0.5, 0.75, 1.0 , $\alpha = 0, \pm 5, \pm 10, \pm 20, \pm 30, \pm 40$, $\delta_0 = 3, 4$. It should be emphasized that the absorption coefficient influences on occuring the self-similar mode of the laser pulse propagation in a strong way. So, if the

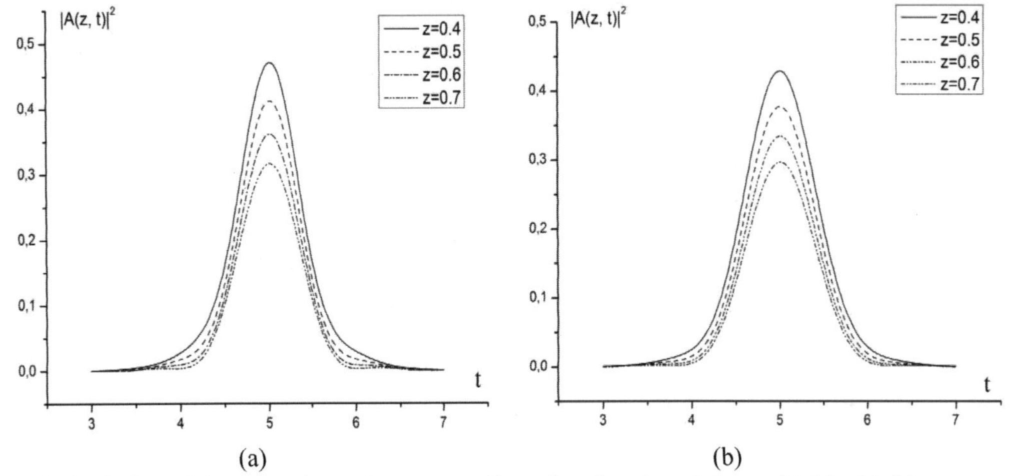

Fig.5. Evolution of the pulse shape along z-coordinate for $\delta_0 = 3$, τ=0.1, α = 10 (a), -10 (b).

absorption coefficient is equal to 4, then the amplitude grating affects mainly on the pulse popagation. If this parameter is equal to 3, then self-action of the laser pulse due to refractive grating influences essentially on taking place of the self-similar mode. As an example, in Fig. 5 we show evolution of the laser shape for its propagation in medium with nonlinear response induced self-compression of optical pulse (Fig. 5a) or its self-decompression (Fig.5b). We see that in Fig. 5b the pulse shape remains the same up to distance that is equal to 0.7 dimensionless units. This Figure illustrates also a positive role which plays self-decompression of laser pulse if nonlinear absorption takes place. As it is well-known, without nonlinear absorption such regime of the laser pulse propagaion can not appear.

4. Conclusion

We showed that the self-similar mode of laser pulse propagation takes place in a medium with its time-dependent response. The distance, along which the laser pulse shape preserves its self-similarity, depends on the interaction parameters. If laser pulse propagates in a medium with nanorods, under certain conditions the soliton apears in the area of the front of nanorod melting.

The mode of laser pulse propagation with its self-similar shape takes place also for laser radiation interaction with plasma.

5. Acknowledgmets

This work was financially supported in part by the Russian Foundation for Basic Research (grant number 13-01-92695-Ind-a).

References

[1] Akhmanov S A, Vysloukh V A and Chirkin A S 1992 *Optics of femtosecond laser pulses* (NY: American Institute of Physics)
[2] Shen Y R 1984 *The Principles of Nonlinear Optics* (NY: J. Wiley Press)
[3] Agrawal G P 1995 *Nonlinear fiber optics* (NY: Academic Press)
[4] Sukhorukov A P 1988 *Nonlinear Wave Interactions in Optics and Radiophysics* (Moscow: Nauka)
[5] Kivshar Y S and Agrawal G P 2003 *Optical Solitons: From Fibers to Photonic Crystals* (San Diego: Academic Press)
[6] Akhmediev N N and Ankiewicz A 1997 *Solitons: Nonlinear Pulses and beams* (London: Chapman and Hall Press)
[7] Brull L and Lange H 1986 Stationary, oscillatory and solitary wave type solution of singular nonlinear Schrödinger equations *Math. Meth. in the Appl. Sci.* **8** (4) 559–75
[8] Malomed B A 2006 *Soliton management in periodic systems* (NY: Springer)
[9] Berge L, Mezentsev V K, Rasmussen J J et al 2000 Self-guiding light in layered nonlinear media *Opt. Lett.* **25** (14) 1037–39
[10] Tower I and Malomed B A 2002 Stable (2+1)-dimensional solitons in a layered medium with sign-alternating Kerr nonlinearity *JOSA* B **19** (3) 537–43
[11] Yannopapas V, Modinos A and Stefonou N 2003 Anderson localization of light in inverted opals *Phys. Rev.* B **68** 193205
[12] Goodman R H, Holmes Ph J and Weinstein M I 2004 Strong NLS soliton-defect interactions *Physica* D **192** 215–42
[13] Driben R, Oz Y, Malomed B A et al. 2007 Mismatch management for optical and matter-wave quadratic solitons *Phys. Rev.* E **75** 026612
[14] Serkin V N and Hasegawa A 2002 Exactly integrable nonlinear Schrodinger equation models with varying dispersion, nonlinearity and gain: application for soliton dispersion *IEEE Journal of Selected Topics in Quantum Electronics* **8** (3) 418
[15] Serkin V N, Hasegawa A and Belyaeva T L 2007 Nonautonomous solitons in external potentials *Physics Review Letters* **98** 074102
[16] Baizakov B B, Malomed B A and Salerno M 2003 Multidimensional solitons in periodic

potentials *Europhysics Letters* **63** (5) 642–48

[17] Al Khawaja U 2009 Integrability of a general Gross–Pitaevskii equation and exact solitonic solutions of a Bose–Einstein condensate in a periodic potential *Phys. Letters* A **373** 2710–16

[18] Xu Q Q, Yao D Z, Liu X N,Zhou Q, Xiong G G 2012 Solitary propagation effect of a well-defined chirped femtosecond laser pulse in a resonance-absorbing medium Physical *Review A* **86** (2) 86.023853

[19] Mahalingam A, Porsezian K, Mani Rajan M S and Uthayakumar A 2009 Propagation of dispersion–nonlinearity-managed solitons in an inhomogeneous erbium-doped fiber syste *J. Phys. A: Math. Theor.* **42** 165101

[20] Loomba Sh, Gupta R, Kaur H, Mani Rajan M S 2014 Self-similar rogue waves in an inhomogeneous generalized nonlinearSchrödinger equation *Physics Letters A* 378 2137

[21] Trofimov V A, Matysevich O V and Smotrov D A 2011 Mode of propagation of optical radiation with self-similar pulse shape in layered medium with nonlinear absorption *Proc. of SPIE* **8095** 80951K

[22] Trofimov V A, Matysevich O V and Smotrov D A 2011 Mode of propagation of optical radiation with self-similar pulse shape in layered medium with nonlinear absorption *Proc. of SPIE* **8095** 80951K

[23] Trofimov V A, Rozantsev A V 2012 2D soliton formation of BEC at its interaction with external potential *Proc. of SPIE* **8497** 84970F

[24] Trofimov V A, Zakharova I G, Smotrov D A, Lan Sh 2013 Self-similar pulse shape mode for femtosecond pulse propagation in optical fiber with multi-photon absorption and nonlinear refraction *Proc. of SPIE* **8785** 87750K

[25] Trofimov V A, Zakharova I G, Konar S 2014 Self-similar pulse-shape mode for femtosecond pulse propagation in medium with resonant nonlinearity *Proc. of SPIE* **9136** 91360Z

[26] Trofimov V, Zakharova I, Fedotov M 2014 Self-similar Shape Mode of Optical Pulse Propagation in de-focusing Medium with two-photon Absorption *Journal of Physics: Conference Series (22nd International Laser Physics Worksop (LPHYS 13))* **497** 012023.

[27] Trofimov V A and Matusevich O V 2008 Numerical method for 2D soliton solution at SHG in media with time-dependent combined nonlinearity *Mathematical Modelling and Analysis* **13** (1) 123–34

[28] Trofimov V A, Lysak T M and Matysevich O V 2010 Influence of transverse perturbation of soliton propagation direction on laser radiation evolution along the layered medium *Proc. of SPIE* **7714** 77140K

[29] Trofimov V A, Lysak T M, Matysevich O V and Lan Sh 2010 Parameter control of optical soliton in one-dimensional photonic crystal *Mathematical modeling and analysis* **15** (4) 517–532

[30] Matusevich O V, Trofimov V A, Yudina E A and Malomed B A 2009 The evolution of two-frequency solitons in an optical fiber with a longitudinally nonuniform nonlinearity *Optics and Spectroscopy* **106** (1) 99–107

[31] Trofimov V A, Lysak T M and Matysevich O V 2010 Influence of transverse perturbation of soliton propagation direction on laser radiation evolution along the layered medium *Proc. of SPIE* **7714** 77140K

Spectroscopy of laser-induced autoionizing nonlinear resonances in atomic systems

L I Pavlov, V V Kovachev
Department of Physics, Faculty of Mathematics and Natural Sciences /FMNS/, South-West University, BG-2700, Bulgaria

E-mail: l_pavlov2004@yahoo.com

Abstract. Efficient harmonic generation at laser-induced autoionizing state in continuum, were observed for the first time by our laboratory [1,2,3]. In this paper, nonlinear frequency mixing in alkali metal vapors at induced continuum structure in Na, are studied in detail. Besides, measurement of nonlinear optical susceptibility at induced autoionizing resonance, is performed. Tunable ultraviolet radiations by four-photon processes near induced resonances are obtained. Estimations of the real and imaginary parts of the nonlinearities, are estimated. We will specially note, that the resonant photoabsorption of ions is experimentally investigated since it gives information about the autoionizing states, which is impossible to obtain on the base only of the emission spectra. We use also laser inducing for obtaining of powerful radiation source in VUV. Thus, many actual problems as the selective impact on matter, laser photochemistry and laser isotope separation are connected directly with the induced autoionizing nonlinear resonances. The main interest is to control the spectral characteristics of continuum. The discrete level from the continuum structure leads to appearance of asymmetrical autoionizing resonance onto the background of the broad-band absorption line.

1. Introduction

First observation of quite efficient nonlinear frequency conversion when the levels of Fano type are laser-induced in continuum, are reported by our laboratory [1,2,3]. We should note, that this possibility of inducing a continuum structure, by laser field coupling with a lower lying bound state, was theoretically predicted in [4,5,6]. If the continuum structure is induced [5], then the photoexcitation rate can be greatly enhanced and such resonant structure becomes important [5] under strong laser field excitation. According to the experimental study [1], a gain of the output THG (third harmonic generation) is much more times efficient due to the induced nonlinear optical resonances in continuum.

It is important, that the bound state coupled to the continuum by the intense radiation field, behaves as an autoionizing state [5], when modifying with a narrow structure the photoionization spectrum. According to theoretical predictions [4,5], the atomic excitations of the photoionization spectra leads to quantum interference of transitions to the continuum from various Rydberg states, which gives highly excited atoms above the ionization potential, with the enhancement of laser-inducing field intensity. The induced autoionizing states in atoms can precisely control the photoionization rate by an external electromagnetic coherent field, which intensity and frequency can be varied. When the continuum is structured by induced autoionizing state, it is possible to obtain high efficiency of the nonlinear optical processes in atomic systems. Concerning the first experimental evidence of induced autoionizing state, we can characterize this process of efficient tripling [1] – as resonant third harmonic generation in sodium. The experimental research of autoionizing-like states can also require a measurement of other atomic properties such as the birefringence, the dichroism, and especially the efficiency of harmonics generation, which can be tested experimentally. Measurement of the optical polarization rotation in cesium at the induced autoionizing state was obtained in [7], Cs. Considerable efforts were devoted to the direct observation of induced autoionizing states in alkali metal atoms, even due to photoionization of molecules (Na_2) [8]. Experimental evidence of induced continuum structure was obtained also in the vicinity of an autoionizing state in rare gases [9].

We can consider, that after the first experimental confirmation of high efficient third harmonic generation [1] near laser-induced autoionizing (AI) state, such strongly resonant phenomena were consequently obtained in a number of coherently excited atomic systems. Our conclusion, is that due to the first experimental efforts – it was created indeed a new quite active scientific field as Resonant Autoionization Laser Spectroscopy. Recently, the experimental study in this new field continue for the continuum structure in different gases [9], atoms etc. An observation of induced autoionizing state in [9] was possible when using the process of multiphoton ionization. Obtaining of the induced autoionizing structure via coherent electromagnetic field was realized by extensive experimental study in various atoms during the last years. These experimental attempts demonstrate also the interference character of laser-induced autoionizing state. As a result, the line shape becomes dependent on the temporal and spatial intensity distribution – of the inducing laser. Further, following the results of experiments in [10], we consider that we can perform coherent amplification without inversion of population. Thus, near induced continuum structure, the inversionless amplification and a lasing is efficient [10], and it is for example favorable for short-wavelength lasing systems in ultraviolet (UV) and VUV region.

In this paper, we investigate four-photon frequency mixing in sodium vapor $2\omega_d + \omega_d` = \omega_s$ near laser-induced AI resonance. The optical mixing efficiency increases, when the dye laser frequency ω_d corresponds to the two-photon 3s-5s transition in sodium.

2. Experimental setup

We used in our experiment a dye laser with two output frequencies ω_d and $\omega_d`$, pumping by second harmonic of Ng^{3+}:glass laser. Powerful fundamental laser radiation of ω_1 frequency couples the 10s level (Fig. 1) to the continuum of Na. The experimental setup is given in Fig. 1.

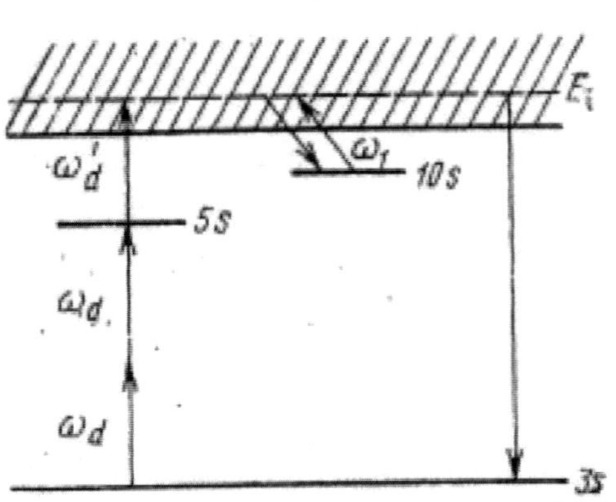

Figure 1. Energy diagram of optical frequency mixing in sodium at induced resonance from 10s level to E_i state in the continuum.

We use a mode-selector within the cavity of mode-locked Nd^{3+}:glass laser and ensures narrowing of the output linewidth to a magnitude of $\Delta\nu_l \leq 0.1$ cm^{-1}. The train consists of 10 laser pulses of 2ns duration each. Second harmonic generation ω_2 is of 30% efficiency in KDP crystal and it is used to pump the dye laser. Dye laser contains a solution of Rhodamine 6GDN, B,F3B in ethanol. The dye laser has two output frequencies ω_d and ω_d` which are tunable independently using TF-17 grating and two mirrors $M_d$ , $M_d$`. The output pulses of dye laser are 2 ns long. The dye laser linewidths are $\Delta\nu_d$, $\Delta\nu_d$` $\leq 0.1$ cm$^{-1}$. As it is shown in Fig. 2, part of the fundamental $\omega_l$ radiation enters in Na cell collinearly with both dye laser $\omega_d$, $\omega_d$` beams.

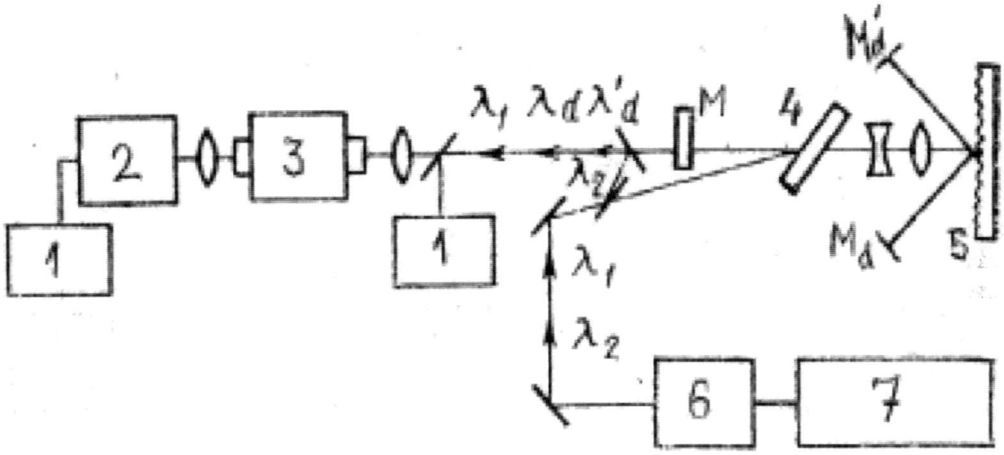

Figure 2. Experimental setup.
1 – photodetectors; 2 – DMR-4 monochromator; 3 – Na cell; M, M_d, M_d` - mirrors; 4 – dye cell; 5 – grating TF-17; 6 – KDP crystal; 7 – Nd^{3+}:glass laser.

The radiations ω_l, ω_d, ω_d` are focused into the Na cell by a 30 cm focal length lens. A sodium vapor pressure of 5 Torr is used at T = 441°C temperature in order to avoid influence of the absorption coefficient. The signal ω_s wave of the optical mixing is obtained in cell of sodium vapor and xenon. The ω_s radiation is separated by double monochromator DMR-4 and it is registered by photodetector system.

3.Results

The effect of the AI state induced by ω_l beam is studied in this work. The results are compared in case of the ω_l radiation absence (i.e., no inducing of AIS), and in case of collinear irradiation of Na cell by all frequencies ω_d, ω_d` and $\omega_l$. When the dye wavelength are $\lambda_d$ = 602,3 nm and $\lambda_d$` = 615,8 nm, then we have precise coincidence of $2\omega_d$ with the two-photon transition, and ω_d` corresponds to the transition into continuum. The dependence of the signal intensity I_s versus ν_s frequency is presented in Fig. 3. Measurements are made when the phase-matching is fulfilled, and the ω_l frequency corresponds to the transition from 10s state to E_i level in continuum (Fig. 1). Near the induced AIS, the I_s intensity increases of the order of magnitude. When the nonlinear interaction is not phase-matched, the efficiency is $\eta = P_s/ P_d \sim 10^{-6}$. When the dye laser frequency ω_d exactly coincides with the two-photon transition in sodium, the efficiency increases up to 10^{-5}. If we have a phase-matching in a mixture of Na vapor and Xe, the measured maximum efficiency is 2×10^{-4}.

We used in experiment maximum intensities I_d, I_d` , and then we can reach also maximum for the I_s signal , but under these conditions we observed the saturation effect.

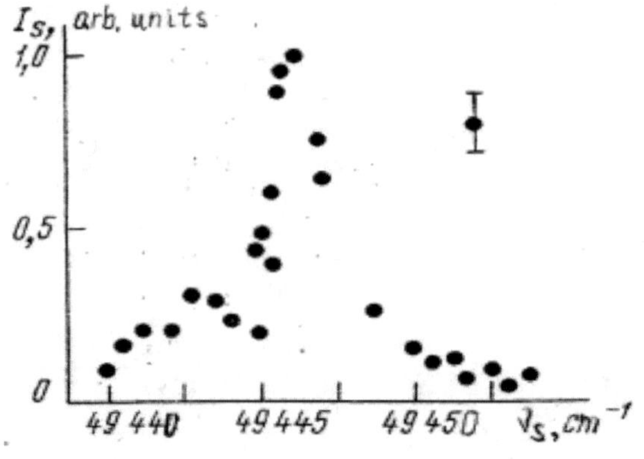

Figure 3. Signal intensity I_s in frequency mixing process as a function of ν_s , at induced autoionizing resonance.

Dependence of I_s upon I_d and I_d` , is presented in Fig. 4. Concerning the fundamental radiation intensity $I_1$ , we keep it constant, and the saturation is evident for signal $I_s$ - versus both dye laser output intensities $I_d$ , $I_d$` (Fig.4).

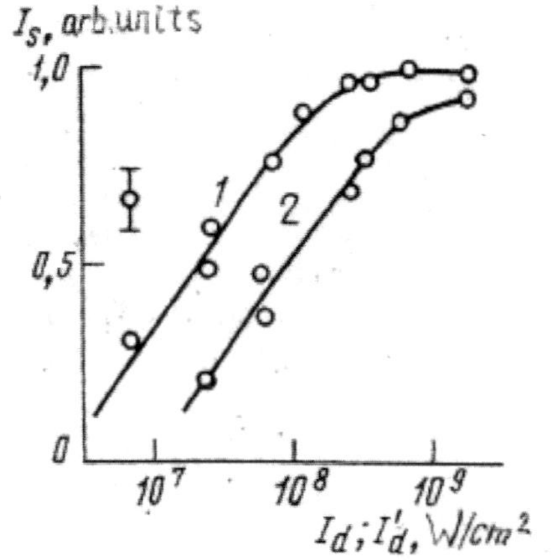

Figure 4. Dependence of signal I_s upon the dye laser outputs I_d (curve 1) , I_d` (curve 2,) and observation of the saturation effect.

Laser-induced autoionizing state leads to great efficiency of harmonic generation and of nonlinear conversion processes, but it is also important to examine the ionization and to measure the ionizatioin yield when scanning the tunable frequency via resonance. Then, the experimental setup consists of a time-of-flight mass-spectrometer, in which the Na atoms are introduced in the form of an effusive beam. The beams of two tunable dye lasers ω_d , ω_d` of 0,2 cm^{-1} bandwidth are combined in a dichroic mirror and cross the atomic beam, focused with a 15 cm achromatic lens. The ionized species are extracted with a dc electric field at right angles with respect to the atomic and laser beams through a 2-mm-diameter aperture in front of which the two laser beams are focused, so that only ions produced at the focus could reach the microchannel plate detector.

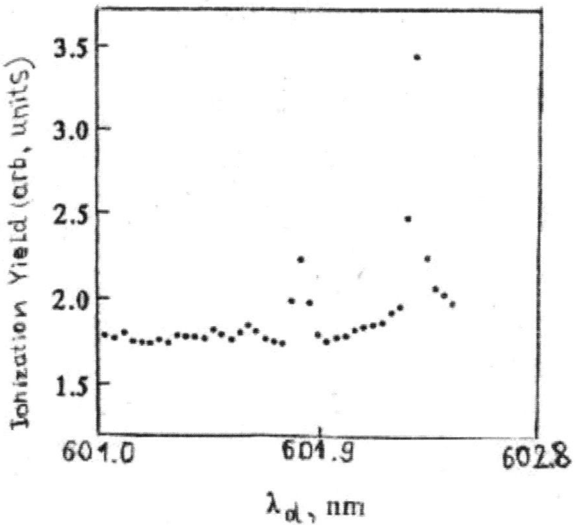

Figure 5. Ionization spectrum of sodium scanning λ_d in the range 601.0 – 602.8 nm and exactly on laser-induced autoionizing resonance coupling 10s level to the continuum.

A summary of the main experimental results is given in Fig. 5 and Fig. 6. The Fig. 5 – 6 show a scan of the wavelength λ_d between 601,0 nm and 602,8 nm, and two resonances are detected, correspondingly.

The peak at 602,3 nm corresponds to the two-photon resonant ionization of Na when $2\omega_d$ coincides precisely with 3s – 5s two-photon transition in sodium. The second peak of the intensity in Fig. 5 appears only when the fundamental ω_l laser beam is introduced and we have induction of autoionizing resonance as a continuum structure near 601.9 nm (Fig. 5).

As it is evident from Fig. 6, following our experimental results with mass analysis of the ions, we have only one peak of the ionization yield at the two-photon 3s – 5s transition ; i.e., there is no in this case an inducing of autoionizing resonance and the dependence in Fig. 6 is obtained for the absence of ω_l radiation in front of Na atoms. As it is demonstrated on Fig. 5, the asymmetry is evident in the observed laser-induced continuum structure when measuring the ionization yield spectra.

The ability to introduce additional tunable resonance in the nonlinear optical susceptibility of sodium (Fig. 5) by the injection of a radiation ω_l field is an attractive possibility in order to obtain an efficient optical frequency mixing process. Such effective nonlinear conversion at an induced autoionizing state is possible with an enhancement more than order of magnitude of the signal ω_s wave within the frequency mixing $2\omega_d + \omega_d` = \omega_s$ process in sodium. At the same time, the dye laser $2\omega_d$ frequency corresponds to two-photon 3s – 5s transition in the atomic Na system. Since we use the induced autoionizing resonance and we involve two-photon 3s – 5s transition as well (Fig. 5), then we apply two dye laser ω_d , $\omega_d`$ frequencies and the fundamental ω_l beam in Na atoms. The result of inducing the autoionizing state verify the significance of modification of the atomic system with a continuum structure, and we obtain greater ω_s efficiency.

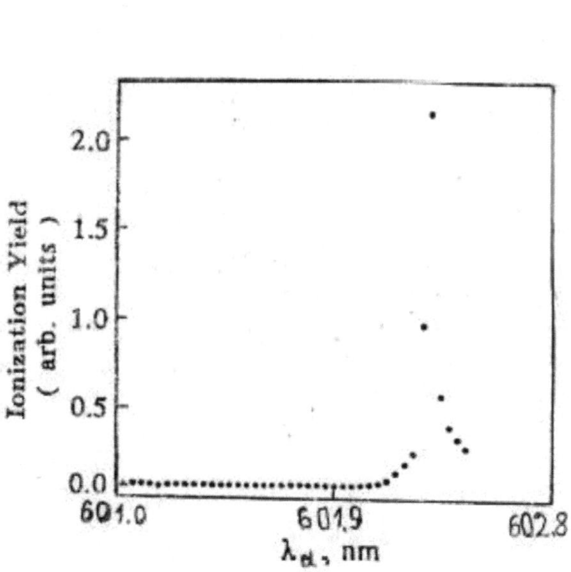

Figure 6. Ionization yield of Na when tuning the λ_d within the region 601.0 – 602.8 nm, but without using of ω_1 fundamental beam and there is no induced autoionizing resonance under these conditions.

We study frequency mixing when using collinearly dye laser ω_d , $\omega_d`$ and the ω_1 beams for enhancement of the signal ω_s wave by a direct fifth-order nonlinear mixing process. We will establish conclusively that the increasing of frequency mixing efficiency arises first of all from a resonance in the nonlinear optical susceptibility. The optical frequency mixing is described by direct and cascade six-photon processes. The direct process corresponds to the

$$\chi^{(5)} (\omega_s ; \omega_d , \omega_d , \omega_d` , \omega_1 , - \omega_1)$$

nonlinear optical susceptibility. We consider, that the cascade six-photon interaction consists of two successive four-photon processes. The interaction $2\omega_d + \omega_d` = \omega_s$ determines the frequency mixing. Second cascade process is of the two-photon Raman amplification type which corresponds to the $\chi^{(3)} (\omega_s ; \omega_1 , \omega_s , - \omega_1)$ optical nonlinearity. Consequently, the nonlinear processes up to fifth-order can produce contributions to the optical frequency mixing, as well as those of cascade third-order processes. The laser-induced autoionizing tunable resonances are mostly preferable for a substantial enhancement of the optical frequency mixing efficiency and for other nonlinear conversion processes. The efficiency increasing at autoionizing resonances is substantial and the continuum structuring can be of practical interest. The induced resonance in the atomic continuum may be quite valuable on harmonic generation in UV and VUV range.

Important feature of the expected structure is the obtaining of an asymmetric profile, similar to that of Fano resonance, providing by quantum-mechanical interference between the two-photon 3s – 5s transition in sodium and the induced autoionizing state to the continuum structure. The efficient frequency mixing and observing of powerful signal wave requires a combination of phase-matching, larger focal spot to avoid breakdown, and longer Na cell length, which also contributes to the efficiency enhancement. Phase-matching can be accomplished in rare gases. Our measurements were made with phase-matched mixtures of Na vapor and xenon at a given density of Xe gas.

We can get more information about temporal and spatial line shape, if we use experimental technique with higher resolution in scanning the ionization spectrum. In order to obtain high resolution in scanning the ionization enhancement in sodium, we use amplification stage for dye laser, which provides 50-mJ pulses. Besides, we apply linearly polarized radiation tunable around λ_d = 602,3 nm and λ_d coincides with two-photon resonance 3s – 5s in Na. The dye lasers ω_d, ω_d` and the coherent inducing ω_1 beam, are focused by 20- and 30-cm focal length lenses, and overlap in the center of the heat pipe in a counterpropagating geometry. We should specially note, that in counterpropagating configuration – the output spectrum of the nonlinear optical mixing is free of Doppler broadening. Thus, further narrowing of the bandwidths of signal ω_s wave and of dye laser outputs is possible. The beams temporal nanosecond duration contribute to better resolution of the frequency scanning system, as well. The optimum tuning is provide by an optical delay between the interacting beams. Control of the polarized rotation of these beams, is also performed.

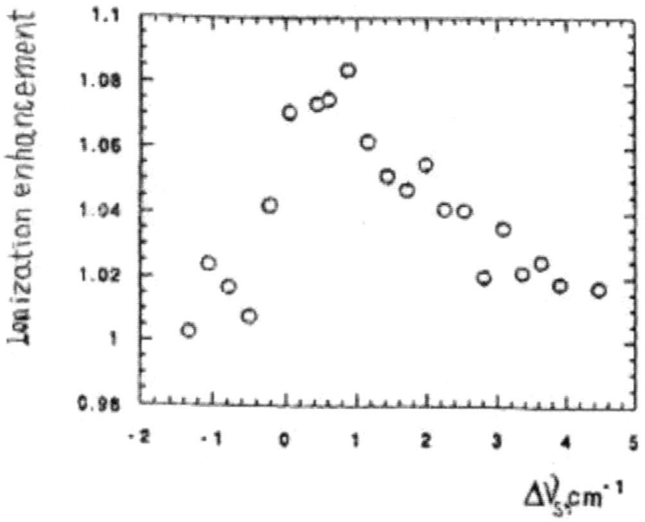

Figure 7. Ionization enhancement versus the laser tuning near the induced autoionizing resonance when scanning the signal linewidth $\Delta\nu_s$, cm^{-1}.

In Fig. 7, the ionization enhancement is plotted upon the laser tuning near the induced autoionizing resonance. The ionization signal is averaged over 100 laser pulses. The heat pipe produced a sodium density ~ 2 x 10^{14} atoms/cm^3. A sensitivity of about 10^{-2} in the measurement of the ionization enhancement was possible only by a careful optimization of experimental conditions, performed in order to minimize the background. The ionization enhancement is determined as a ratio of the ionization signal measured at induced autoionizing resonance, and this is compared with the ionization measured in the absence of inducing ω_1 laser, i.e. when the continuum is not structured by fundamental ω_1 beam. The line shape is asymmetric, following the spatial and temporal electromagnetic field distribution. The accuracy of this measurement provides clear evidence of an asymmetric line, which character of the line shape is typical - for the laser-induced autoionizing state in atomic systems. The cross section of photoionization of the two-photon 3s – 5s transition also increases, since we use in experiment powerful dye laser I_d, I_d` intensities at a constant fundamental I_1 beam.

4.Discussion

In order to obtain the correct physical interpretation of the experimental results, we need to include the four-photon as well as six-photon processes of nonlinear frequency mixing – within the processes studied here. Such simultaneous action of several direct and cascade processes was well-known for the higher harmonic generation [11,12]. It turned out that these processes are quite important also for the laser-induced AI resonances. The direct six-photon process at AIS is determined by fifth order nonlinearity:

$$I_s \sim |\chi^{(5)}(\omega_s ; \omega_d , \omega_d , \omega_d` , \omega_l , - \omega_l )|^2 \; N^2 \; I_d^2 \; I_l^2 \; I_d`$$

where N is the atomic density. Both consequent cascade four-photon processes are described by third order nonlinearities

$$\chi^{(3)} (\omega_s ; \omega_d , \omega_d , \omega_d`)$$

and

$$\chi^{(3)`} (\omega_s ; \omega_l , \omega_s , - \omega_l) .$$

The efficiency enhancement of the nonlinear process at induced AIS can be explained by the resonant increasing of the nonlinear optical susceptibility. As it is evident in Fig. 4, the saturation effect arises with the increasing of the intensities I_d , $I_d`$ . When $\omega_d$ precisely corresponds to the two-photon transition 3s - 5s , the saturation can be connected with a broadening of the resonant line after fundamental power decreasing, because of two-photon absorption. The enhancement of $I_d`$ leads to a broadening of the induced autoionizing resonance and limits the output signal intensity I_s . Another reason for saturation could be the dispersion of the excited atoms, when the processes of self-focusing and defocusing take place.

Depending on the duration of the pulses, the ionization may be complete in the central part of the focused region, and we can consider the parameters over the spatial distribution of the laser intensities. Recently, studies of induced autoionizing states include experimental results on nonlinear optical processes and photoionization in alkali, in alkaline – earth atoms etc. Experimental investigations are very useful in clarifying the role of the induced autoionizing resonances in nonlinear interactions of atoms when applying laser radiation ω_l field.

It is important to perform the experimental study both in the strong- and weak-field regime when referring to the fundamental ω_l intensity coupled the bound state to the continuum. At the increasing intensity, the width of the structure becomes dependent on the photoionization rate from the excited state. As a result, the line shape becomes dependent on the temporal and spatial intensity distribution of the fundamental ω_l beam. The main purpose of the study near the induced autoionizing nonlinear resonance is an interpretation of the line shape, taking into account the laser bandwidths and the hyperfine structure of atomic levels. The results provide a complementary information to that given by photoionization experiments. Besides, measurements performed at different vapor densities show a clear evidence of line-broadening effects.

Considering the accurate determination of the atomic parameters, the process of structuring the continuum in the conditions of experiment is found to be quantitatively consistent with the theoretical predictions. A careful optimization of the interaction conditions and of the sensitivity of the apparatus allow a measurement of dispersive profile of the autoionizing-like resonance suitable to an interpretation. According to the results, the signal ω_s wave of the frequency mixing can be enhanced by introducing the laser ω_l beam and structuring the continuum.

The obtained results here involve a detailed study of the ionization process. In addition, experiments are carried out to control the ionization rate and at the same time – the efficiency of the optical frequency mixing. The ω_l radiation is used to couple an autoionizing state to the continuum at the required position to affect the induced nonlinear resonance. The ionization rate is found to be dependent on the frequency ω_l . Results of detailed study of this process include also its effect on the efficiency of the frequency-mixing process.

5. Conclusion

The main obtained result in this experiment is an optimization of optical frequency mixing at phase-matching and using two-photon transition near autoionizing-like resonance. These processes can be applied for creating of high efficient conversion into UV and VUV range. It is also interesting to study the processes via induced AI states at various temporal delays between the pulses of intensities I_d , I_d` , I_l . Using tunable laser beams, we can induce narrow nonlinear resonances in an arbitrary field of continuum, which realize both the ionization and the effective nonlinear frequency conversion. The induced autoionizing resonances have much advantages in order to optimize laser isotope separation (LIS) and the higher order harmonics generation into VUV and soft X-ray region.

6. References

[1] Pavlov L I, Dimov S S, Metchkov D I, Mileva G M, Stamenov K V and Altshuller G B 1982 *Phys. Letters* A **89** 441-443
[2] Dimov S S, Pavlov L I, Stamenov K V, Heller Yu I and Popov A K 1982 *Appl. Physics* B **30** 35
[3] Dimov S S, Pavlov L I, Stamenov K V and Altshuller G B 1983 *Opt. and Quant. Electronics* **15** 305
[4] Armstrong L Jr, Beers B L and Feveuille S 1975 *Phys.Rev.* A **12** 1903
[5] Knight P L, Milonni P W 1980 *Phys. Rep.* **66** 21
[6] Heller Yu I and Popov A K 1976 *Opt. Commun.* **18** 449
[7] Heller Yu I, Lukinykh V F and Popov A K 1981 *Phys. Letters* A **82** 4
[8] Snitman A, Sofer I, Golub I, Yogev A, Shapiro M, Chen Z and Brumer P 1996 *Phys.Rev.Letters* **76** 2886
[9] Hutchinson M H R and Ness K M 1988 *Phys. Rev. Letters* **60** 105
[10] Harris S E 1989 *Phys. Rev. Letters* **62** 1033
[11] Akhmanov S A, Martinov V A, Saltiel S M and Tunkin V G 1976 *JETP Letters* **22** 143
[12] Mitev V M, Pavlov L I and Stamenov K V 1979 *Opt. and Quant. Electronics* **11** 229

Self-induced diffraction patterns in nonlinear Fabry-Perot interferometer on lithium niobate

V. Shandarov, A. Perin, V. Ryabchenok

Tomsk State University of Control Systems and Radioelectronics, Lenin Prospect 40, Tomsk, 634050, Russia

E-mail: shandarovvm@svch.rk.tusur.ru

Abstract. The effects of self-action of light fields within Fabry-Perot interferometers based on photorefractive lithium niobate plates both, of X- and Z- cuts are experimentally investigated. Formation of one-dimensional practically regular diffractive elements within the interferometers with slightly non-parallel input and output surfaces has been demonstrated at laser light wavelength of 532 nm. It has been found that these diffractive elements are formed due to the photorefractive response of the crystal to the steady-state optical interference patterns appeared within the interferometer at multiple reflections of light beam. From the Raman-Nath far field diffraction patterns the ordinary refractive index change within the interferometer on Z-cut crystal has been estimated.

1. Introduction

Light propagation in optically nonlinear medium may result in the conversion of the spatial and spectral structures of light fields due to the self-action effects. At the photorefractive nature of nonlinear response the spatial self-action of light may occur in the medium at optical powers of a microwatt level [1] that is very useful for the experimental studies. Very high optical nonlinearity results in the slow nonlinear response of photorefractive media that also brings the advantage to the time evolution study of light field self-action using simple and not too expensive tools. The most interesting among the spatial conversion effects for optical beams is the light localization in the form of spatial solitons that makes possible design of all-optical photonic elements and circuits with light controlled by light. Another result of similar effects relates to generation inside the nonlinear medium regular diffractive elements which also can be used for the light control needs. It should be noted that the boundaries of the nonlinear medium or interfaces between various materials may sufficiently vary the self-action characteristics of light fields. For example, generation of regular light patterns becomes possible in nonlinear optical cavities where the optical feedback comes into play [2 - 5]. Similar effects have been observed in photorefractive crystals and other nonlinear media with both, self-focusing and self-defocusing nonlinear response [2, 5].

In this work, we study the effect of self-formation of one-dimensional diffractive structures in photorefractive Fabry-Perot interferometers (FPI) with slightly non-parallel optical surfaces based on photorefractive lithium niobate ($LiNbO_3$) plates of two cuts.

2. Experimental setup and conditions

The X- and Z-cut $LiNbO_3$ plates with polished to optical grade largest opposite sides are used as Fabry-Perot interferometers in experiments. The plate dimensions are 1×15×10 mm^3 along X, Y, Z axes for X-cut sample and 10×10×2 mm^3 along X, Y, Z axes for Z-cut one. The Z-cut sample is doped with Fe ions (0.05%) during the crystal growth. Photorefractive properties of the X–cut sample result from the doping of its surface layer with iron (Fe) and copper (Cu) ions using the thermal diffusion of these impurities [6]. For this sample the thickness of Fe-doped and Cu-doped layers, where the photorefractive nonlinearity is high enough, make up 50 and 150 μm that is much less than the total thickness of the sample itself. The basic schematic of the experimental setup used to study the features of light field self-action in FPI is shown in Figure 1. Here a frequency-doubled continuous wave Nd^{3+}:YAG laser, operating on a single longitudinal mode (light wavelength λ=532 nm, output power up to 50 mW), is used as a light source. The laser beam is collimated to necessary diameter (10 and 20 mm in various experiments) with a collimator built with two spherical lenses (CL). The central part of it with almost uniform intensity field is separated from the broad beam with a diaphragm to expose FPI. Using square shaped diaphragms, the transverse dimensions of light fields at the input surface of the crystal samples (LN) are provided as 3mm × 3mm or 5mm ×5mm. The light polarization corresponds to the extraordinary wave in the X-cut plate and to the ordinary wave in the Z-cut one. The light intensity of the input light field in most experiments on self-formation of diffractive structures in FPI is ~35 mW/cm^2. The FPI samples are placed over the linear positioning holders with micrometric control of their location. The light patterns at the output surface of FPI are imaged with the spherical imaging lens (SL) onto the sensitive plane of the laser beam analyzer BS-FW-FX33 (LBA) connected to the personal computer (PC). In experiments we study the temporal evolution of spatial distributions of light field intensity at the output surface of FPI and check the self-induced refractive index changes within the samples using the optical probing with both, broad and narrow light beams.

The optical probing method with a broad light beam exploits the effect of light diffraction on the diffractive structure. The analysis of far field diffraction patterns can confirm at this approach the occurrence of diffractive structures within the crystal sample and estimate the modulation depth of the sample refractive index by the nonlinearly generated diffractive element. The second approach is based on the optical probing of the diffractive element using light beam with its diameter less than spatial period of diffractive structure, as in the one of configurations of optical probing of surface acoustic waves [7].

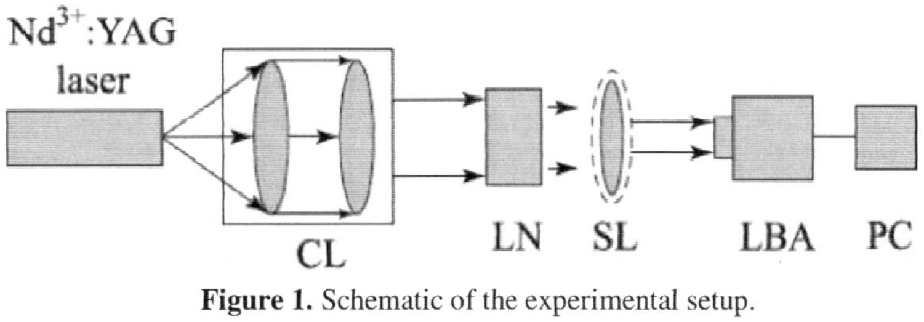

Figure 1. Schematic of the experimental setup.

3. Experimental results and discussions

As the result of FPI exposure with a laser beam, we observe generation of almost regular one-dimensional light patterns at its output surface. These patterns appear in near to 10 minutes for Z-cut sample and 30 minutes for X-cut one after the exposure start. They develop within the initial almost uniform light images. These light patterns point to the stationary spatial modulation of the material refractive index due to the photorefractive effect.

3.1. Light patterns at the output surface of FPI

To study the evolution of light patterns at the output surface of FPI's, we use light field with almost uniform intensity over its cross-section with dimensions 3 mm × 3 mm. A part of light image at the input surface of FPI is shown in Figure 2a. At the initial stage (t=0) the light pattern at the output surface shows slight one-dimensional periodic modulation of its intensity because of the nonparallel input and output surfaces of the crystal plate (Figure 2b). But optical probe experiments do not find any periodic diffractive objects within FPI for this initial stage. In some minutes after the exposure start the increase of the intensity modulation depth in the light field at the crystal output surface is observed. It reaches some maximal depth in 50 – 60 minutes for the Z-cut sample and in 70 – 80 minutes for the X-cut FPI at input light intensity 35 mW/cm^2. The example of such light patterns observed at the output surfaces for the Z-cut FPI is shown in Figure 2c. The spatial period of the periodic light pattern is ~260 μm here and it does not depend on the incidence angle of the light beam to the input surface.

a b c

Figure 2. Light fields at the input and output surfaces of Z-cut FPI: a – the initial pattern at the input surface; b – light pattern at the output surface at t=0; c – light pattern at the output surface after FPI exposure for 25 minutes (input light field intensity 35 mW/cm^2).

To reveal the mechanism of the light modulation depth increase, the optical probe of the FPI with narrow light beam (beam diameter at the input surface is 23 μm, λ=532 nm) is used. The exposed area of FPI is scanned with this probe beam along direction parallel to the "wave vector" of the light pattern. The crystal sample is moved with the micrometric accuracy with respect to the light beam in that case. As the intensity distribution within the cross-section of the probe beam as its peak intensity are investigated with LBA. Some results of this experiment are illustrated by Figure 3. The step of FPI movement with respect to the probe beam is 25 μm and the optical power of this beam is less than 0,01 mW to prevent photorefractive distortions of its intensity profile. The results demonstrate the dependence of the peak intensity of the probe beam on its location at scanning the single period of the diffractive structure within this FPI. The light patterns of the beam cross-sections for the uniform (unexposed) area of the crystal and for its different locations at the exposed area probing are shown here. It may be concluded that the beam peak intensity increases for its propagation along the central part of the region with increased refractive index that results in its self-focusing and the compression of its intensity cross-section in direction of light pattern "wave vector" (image "a"). Propagation of the probe beam in the slope part of this region results in its refraction with peak intensity decreasing and distortions of the intensity distribution within the probe beam cross-section (images "b" and "c"). The scanning of the total exposed area with narrow probe beam demonstrates practically periodic dependence of its peak intensity on the location within this area. The same results are observed for FPI based on X-cut $LiNbO_3$ crystal with some distinctions only in the quantitative characteristics of these effects. Thus, the results of experiments performed evidently show that practically periodic diffractive structure is generated within photorefractive FPI during its exposure with light field of uniform intensity distribution.

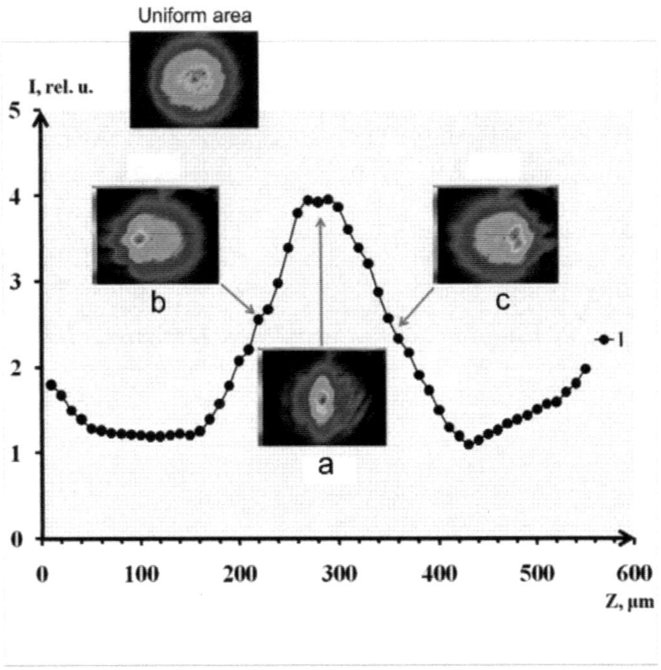

Figure 3. Optical probing of induced diffractive pattern with the narrow (focused) light beam. Dependence I(z) illustrates the beam peak intensity change at scanning single period of the diffractive structure; a – cross-section of the probe beam located in the area of maximal refractive index change; b and c – its cross-sections at probing opposite slopes of optical inhomogeneity; image "Uniform area" shows cross-section of this beam with no diffractive element.

3.2. Far field diffraction of light on the diffractive structures within FPI

We also use another approach to study the diffractive structures induced within FPI. It is based on the diffraction of the broad light beam on the induced phase diffractive structure. The light patterns in the far field can inform in this case on spatial period of the structure and refractive index change within it. For this purpose the light beam of He-Ne laser (λ=633 nm) with diameter of 2 mm and optical power of 0,5 mW is used. Its polarization corresponds to the ordinary wave of the crystal. Far field of diffracted light is studied with LBA and intensities of diffraction maxima are measured with a silicon photodiode.

The diffraction patterns observed in experiments are illustrated by images in Figure 4. Here the left image corresponds to the light beam passed through unexposed Z-cut $LiNbO_3$ based FPI. After FPI exposure for 30 minutes with light of λ=532 nm we observe in the far field the pattern with many diffraction maxima typical for Raman-Nath diffraction. The diffraction angle corresponds to the spatial period of the diffractive gratings of 270 μm. The ratio of the first and zero diffraction maxima intensities η_m depends on the refractive index change within the diffractive structure. This change Δn_0 is estimated using the known expression [8]:

$$\eta_m = J_m^2(\delta) = J_m^2\left(\frac{2\pi d\Delta n_0}{\lambda\cos(\theta)}\right)$$

where J_m is a Bessel function, d is a grating thickness, Δn_0 is a refractive index change, λ is wavelength, and θ is the diffraction angle. For measured η_m=8% and 2 mm FPI thickness we estimate the ordinary refractive index change value of induced within FPI grating as $\Delta n_0 \approx 3.7\cdot 10^{-5}$.

Similar effects of light diffraction we also observe for X-cut $LiNbO_3$ based FPI, but with slightly different diffraction angles.

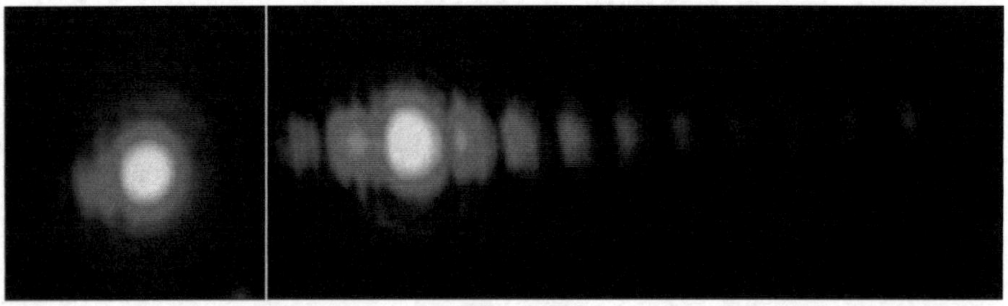

Figure 4. Far field pattern of He-Ne laser beam diffraction on nonlinearly induced diffractive element in FPI. Left – light pattern before FPI exposure; right – the pattern after FPI exposure.

3.3. Discussions

The analysis of experimental results shows that diffractive structures are formed within FPI`s due to multiple reflections of light beam from its boundaries together with not ideally parallel input and output surfaces of FPI. The angles between these surfaces are measured as 46±5 arc seconds and 40±5 arc seconds for Z-cut and X-cut $LiNbO_3$ plates. The interference of almost co-propagating and almost counter-propagating coherent optical waves results in the formation of complicated three-dimensional (3D) steady-state light patterns within FPI. With taking into account the reflection coefficient value of light at the interface "lithium niobate - air", the 3D interference pattern may be considered as the superposition of two 1D interference patterns. The first pattern is formed by two almost counter-propagating waves and the second one is formed by two almost co-propagating waves. In turn, these steady state light patterns may result in arising within FPI the photorefractive gratings. The estimations of spatial periods of similar gratings in FPI with non-parallel surfaces (β=46 arc seconds) give values of ~118 nm for reflecting grating and ~0.3 mm for transmitting grating. Because of the slow photorefractive nonlinear response of $LiNbO_3$ (more than 10 minutes in our experiments with optical powers used), formation of reflecting gratings requires experimental setup very stable mechanically and thermally, and their formation is not observed in the experiments. From the other side, the spatial period Λ for transmitting photorefractive gratings is determined by the expression:

$$\Lambda = \frac{\lambda}{2n \cdot sin(2\beta)}$$

where n, the refractive index of $LiNbO_3$; β, the angle between input and output surfaces. For $\beta<1'$ this period $\Lambda\approx0.3$ mm and it corresponds to the spatial period value observed for the self-induced diffractive structures in our experiments.

Conclusion

The formation of 1D diffractive structures within photorefractive Fabry-Perot interferometers based on lithium niobate has been demonstrated at light wavelength of 532 nm. It has been revealed that these structures are formed due to the photorefractive response of lithium niobate crystal to the three-dimensional steady-state light interference patterns which appear within the crystal plates because of their not ideally parallel optical surfaces. The effect observed may be useful for design of new nonlinear optical elements for the laser devices and for all-optical photonic components.

Acknowledgments

This study was carried out with the financial support of Ministry of Education and Science of Russia (the task N 3.878.2014/K of the project part).

References

[1] Crosignani B, Segev M, Engin D, DiPorto P and Yariv A 1993 *Journal of Optical Society of America B* **10** 446
[2] Buljan H, Soljacic M, Carmon T, and Segev M 2003 *Phys. Rev. E* **68** 016616
[3] Smirnov E, Stepic M, Shandarov V, and Kip D 2006 *Appl. Phys. B* **85** 135
[4] Perin A S, Kozlova S M, and Shandarov V M 2011 *Phys. Wave Phenom.* **19** 6
[5] Zhu Nan, Guo Ru, Liu Simin, Liu Zhaohong and Song Tao 2006 *J. Opt. A: Pure Appl. Opt.* **8** 149
[6] Kip D 1998 Appl. Phys. B **67** 131
[7] Whitman R L and Korpel A 1969 *Applied Optics* **8** 1567
[8] Korpel A 1981 *Proc. IEEE* **69** 48

Analysis of nonlinear optical materials properties by simple powder technique

O A Morozov, A K Naumov, A V Lovchev and M R Garipov
Kazan Federal University, 420008, Kremlevskaja str., 18, Kazan, Russian Federation

E-mail: uh4rad@rambler.ru

Abstract. The article describes a simple technique for comparative analysis of the second harmonic generation properties of new samples and etalon materials by their powders. The effectiveness of the method was tested and and measuring of nonlinear coefficient and damage threshold of the well known materials: KDP, LiIO3, m-nitroaniline was demonstrated. The parameters of the new promising nonlinear material DNPAP were measured.

1. Introduction

The first experiments for obtaining the second harmonic generation (SHG) of laser oscillation has been successfully implemented by Franken et al. on crystalline quartz [1]. These experiments were carried out in the same year, when the first laser was created. Thus researchers have found a simple way to get new laser frequencies. At present about fifty crystals have possibility of laser radiation frequency doubling but only few crystals are used in the lab. It is because a number of additional demands on the properties of crystals are associated with using their in practice. These properties are transparency region, processability of large-sized crystal growing, cost and etc. A number of chemical and physical properties such as hygroscopicity, optical homogeneity, radiation resistance, hardness, and thermal characteristics are further narrowing the list of the crystals. Therefore, a new nonlinear materials searching still is a very urgent task.

At the initial stage search for new crystals was very difficult. It is because it takes a lot of time to grow good optical quality new promising crystals of sufficient size for studies. This is required the development of crystal growth technology, which in turn requires a great financial and time resources. Nonlinear properties of these materials can be predicted using available modern theories. Their results are often confirmed by experimental measurements, despite their lack of accounting in the theory of other properties of studying materials, which influence on their nonlinearity. However, the actual values of nonlinear coefficients of new crystals ultimately are determined experimentally. But for these measurements single crystals samples require sufficient size. To solve this problem it is necessary to improve growth technology of these crystals, which is quite difficult and costly. It will be turned out especially more costly when the new crystal, for which growth technology was developed, after examining its nonlinear properties would be not promising. Therefore, along with the new SHG crystals creation the experimental techniques have been developed and are being developed. Some techniques allow measuring the nonlinear parameters of materials with minimal time and resources without using for these single-crystal samples. The development of such techniques is necessary for modern quantum electronics.

One of the ways that meet these requirements is the method of the nonlinear properties materials study proposed by S.K. Kurtz and T.T. Perry [2]. They first proposed systematic experimental study of nonlinear materials characteristics by powder method. This method has been widely used by experimentalists for a long time up to now. In this paper authors described the methodology of research with a well-developed at that time experimental setup. Method for sample preparation and technology of measurement was also described in detail. The experimental setup consisted of Q-switched laser and a photomultiplier with oscilloscope that determined the sufficient sensitivity for detecting radiation of SHG excited in the powders. The properties of a large number of new nonlinear materials, which presented in the paper, were measured in comparison with the intensity of the quartz SHG. In a later paper [3] Kurtz et al adduce a design description of analyzer, which is made in form of device, and its opportunity of studies of SHG and non-central symmetry powder materials properties. This device is based on experimental setup described in [2]. Created analyzer can detect materials SHG at the level of 1/1000 with respect to the quartz SHG. Paper [4] describes a method for the study of SHG in powders using inexpensive equipment. YAG:Nd^{3+} laser operating in free-running mode was used to excite the SHG, which was cheaper, compared with Q-switched lasers for that time. It is well known that the pulse of free-running YAG:Nd^{3+} laser is a train of microsecond pulses with a total duration equal to duration of the glow pumping lamp. Peak power of every microsecond pulse was enough for SHG and its registration by PMT. Signal was split and came to the photomultipliers, one of which was placed in the pump channel - the other one in the SHG test channel. Author applied integrator for registration of this signal. Concluding about SHG efficiency can be done from the ratio of PMT signals. In papers [5-6] authors suggested another powder method for studying properties of materials including the SHG parameters. Experimental setup scheme has the same dual-channel scheme with the PMT in each channel. Measure of the SHG efficiency of new materials by powder technique carried out as well with respect to quartz crystal SHG, which sample was installed in one of the channels. Feature of their setup was using of crystalline semi-spherical rutile (TiO_2) prism in the test channel. Powder of studied material was pressed by base of this prism. The powder sample was excited by evanescent wave (EW). It was partially emerged from the prism under total internal reflection from the surface to which the analyzed powder was pressed. As a result, authors solved the problem of the penetration depth of the radiation in the powder. Also with help of this prism by polarization way the SHG radiation was separated out from the main radiation. In paper [7] the idea of using EW in powder technique for test SHG properties of new materials got development in both directions as theoretical as experimental way. As a result they have shown the efficiency of using EW method for study properties of new materials in powder state. In their setup, powder was held tightly against the semi-cylindrical prism, which was placed on a double goniometer with angular movement controlled by computer. By this method, they studied and explored a number of new organic materials. It should be noted that the values of the effective nonlinear coefficient and refractive indices of the sample are obtained as fitting parameters of measured dependences. According to the authors, SHG signal from the EW is less sensitive to the phase-matching conditions for the material.

Thus, the powder method allows us to investigate the SHG efficiency of completely new synthesized crystalline materials for which single crystals growth technology is not known yet. First of all this method is necessary for chemists to intensify search for new effective nonlinear materials. But they continue to apply the setups built under the scheme of Kurtz and Perry that was designed in 1968. In this paper we describe a simple way, which is based on Kurtz and Perry method, for measuring parameters of new nonlinear materials by powder technique with using available modern equipment.

2. Basis of proposed powder technique

SHG coefficient determination by powder technique in the proposed method is based on a method of S.K. Kurtz and T.T. Perry [2] as mentioned above. In this method the SHG parameters of an unknown powder are compared with SHG of quartz powder. These parameters of nonlinearity for quartz were well known and also they previously have been determined for a single crystal.

In [2] the formula relating the intensity of the second harmonic $I_{total}^{2\omega}$ to intensity of the incident fundamental harmonic of the laser radiation I_{ext}^{ω} and the substance nonlinear optical coefficient $d^{2\omega}$ is presented. It is important to take into account that the grain size of powder of the substance - r is much more than coherence length $l_c = \frac{\lambda}{4(n_{2\omega}-n_{\omega})}$.

Let's consider the equation of the second harmonic intensity from the intensity of incident fundamental radiation.

$$I_{total}^{2\omega} = \frac{32\pi}{c}(d^{2\omega})^2\left[\frac{64\pi I_{ext}^{\omega}}{\lambda(n_\omega+1)^2(n_{2\omega}+1)}\right]^2 L\frac{l_c^2}{r}sin^2(\frac{\pi r}{2l_c}),$$

where L is a thickness of plane-parallel slab, λ is a wavelength of fundamental optical signal. We divide it by the square of the fundamental radiation intensity:

$$\frac{I_{total}^{2\omega}}{(I_{ext}^{\omega})^2} = \frac{32\pi}{c}(d^{2\omega})^2\left(\frac{64\pi}{\lambda(n_\omega+1)^2(n_{2\omega}+1)}\right)^2 L\frac{l_c^2}{r}sin^2(\frac{\pi r}{2l_c}).$$

This ratio denote by the coefficient k.

$$\frac{I_{total}^{2\omega}}{(I_{ext}^{\omega})^2} = k$$

Then, we take the coefficients ratio of the sample and the standard.

$$\frac{k_{sample}}{k_{standard}} = \left(\frac{d_{sample}^{2\omega}}{d_{standard}^{2\omega}}\right)^2\left(\frac{(n_{\omega\,standard}+1)^2(n_{2\omega\,standard}+1)}{(n_{\omega\,sample}+1)^2(n_{2\omega\,sample}+1)}\right)^2\frac{l_{c\,sample}^2}{l_{c\,standard}^2}\frac{sin^2(\frac{\pi r}{2l_{c\,sample}})}{sin^2(\frac{\pi r}{2l_{c\,standard}})}.$$

After this express from it nonlinear optical coefficient and then we obtain:

$$d_{sample}^{2\omega} \approx d_{standard}^{2\omega}\frac{(n_{\omega\,sample}+1)^2(n_{2\omega\,sample}+1)}{(n_{\omega\,standard}+1)^2(n_{2\omega\,standard}+1)}\left(\frac{k_{sample}}{k_{standard}}\right)^{\frac{1}{2}}\frac{l_{c\,standard}}{l_{c\,sample}}\left|\frac{sin(\frac{\pi r}{2l_{c\,standard}})}{sin(\frac{\pi r}{2l_{c\,sample}})}\right|.$$

It is evident that if the samples are equally prepared for determining this value we must know the corresponding coefficient for the standard, intensities of the incident and doubled frequency light and the refractive index of materials at corresponding wavelengths.

In practice, measurement of the nonlinear coefficient of new material in powder form reduces to the preparing powders with the same grain size as for powder of comparison crystal. Also the comparison crystals must have well known values of the nonlinear coefficients, for example KDP and $LiIO_3$. For each powder sample measurement of SHG intensity dependences from the energy density of the incident laser radiation is need to register as much as possible in the identical conditions. Such dependencies have the form of the parabolas: $E_{shg} = a + bx^2$, where b contains the nonlinear coefficient peculiar for the crystal powder for which this relationship is built. If we take into account the refractive indices for the incident and SHG waves in comparable materials, it is easy to directly obtain the ratio of their conversion factors from the ratio of the coefficients b. In presence of the tabulated value of the corresponding coefficient for one of them - a reference sample, the required coefficient for another sample is easily calculated.

3. Samples

For demonstrating the performance of the proposed method we took the powders of KDP, $LiIO_3$ and m-nitroaniline crystals with known values of nonlinearity from the literature. As an additional sample we took powder of a new synthesized material on which effect of SHG was visually observed, but the value of nonlinear coefficients was not known yet. This material was synthesized at the Institute of Chemistry, Kazan Federal University of Yu.G. Shtyrlin's group. Abbreviation of this material is DNPAP [8].

Powders of the materials were prepared by mechanical milling of the crystals and sifting through calibrated sieves. For the experiments we took each powder sample of the crystals with the same grain size from 90 to 120 μm.

4. Technique of experiment

For determination of nonlinear coefficients of new materials in powder form, which are due to SHG, we used experimental setup, which concept is displayed in figure 1.

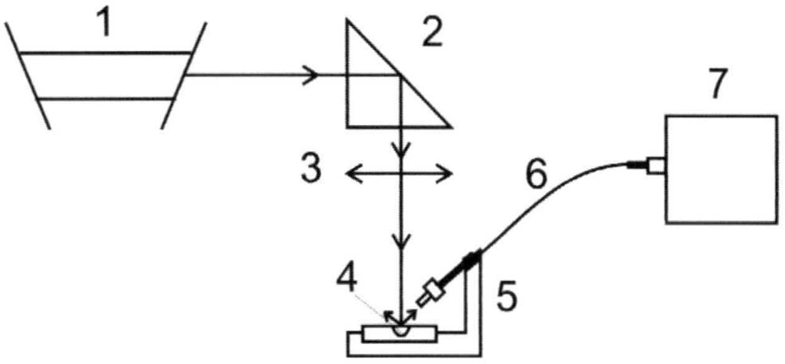

Figure 1. The scheme of the experimental setup for determining nonlinear coefficients of materials in powder form by SHG: 1 - YAG:Nd^{3+} laser, 2 – 90-degree prism, 3 - focusing lens, 4 - sample of powder on the substrate, 5 - a table with fiber mount, 6 - quartz optical fiber, 7 - «StellarNet» spectrometer.

Radiation of the fundamental harmonic of YAG:Nd^{3+} laser (1), was turned via prism (3) so the beam fell on a sample vertically. Then, the radiation focused by the lens (2) on the sample of the investigated powder (4). Sample of the powder prepared in a special way on a special substrate (see details below). It was placed on a special table (5) with a fiber which was attached to it (6). The other end of the fiber was connected to the spectrometer StellarNet (7), which was used for recording the emission spectrum of SHG of the powder.

For the most accurate determination of the b coefficients of comparable materials in powder form it was needed to perform two conditions. Firstly, it was needed an equal preparation and identical installation of each powder sample on the table of experimental setup. Secondly, for the r>>lc condition implementation it was needed to choose the identical spot diameter of pumping radiation for both samples taking into account the particle size of this powder materials. To perform first condition mentioned above we used substrate with a special groove. This groove had depth satisfying the condition $h >> l_h$ for laser penetration depth (3-4 mm more than enough). The sample of powder freely without compaction was poured into the groove. Its excess was equalized by smooth straight edge of a plate against edge of the groove. After that, the substrate was mounted on a horizontal table (5) with special detents for its uniform installation. Thus, the experimental conditions to excite the SHG for each powder sample were reproduced identically as closely as possible. Diameter of the spot of the incident radiation was not less than 1 mm, that at a particle size of 100 μm is fully satisfy to the second condition.

Incident and converted radiation in the form of two spectral lines were recorded on spectrometer (7). This spectrum included the following components: specular and diffuse scattering of fundamental harmonic of the laser and the conversion signal due to the SHG of researched substances. Operational control of radiation intensity changes of the fundamental wave and of second harmonic by maxima on the spectrum was carried out. Additionally the energy of the laser fundamental harmonic was controlled by "Ophir" power meter. Its measuring head (on the figure is not shown) was set in place of a table (5) and during main experiment it was removed.

Also, in our experiments we performed measurements of damage threshold of researched samples under the influence of multiple laser pulses. For this, the samples were exposed to several tens of thousands pulses of fundamental harmonic of YAG:Nd^{3+} laser. The damage threshold of the samples was controlled by several methods. Firstly, it has determined by dislocation of quadratic dependence of the SHG intensity on the density of the incident radiation. Secondly, it was defined with a constant energy density of the laser pulses by decreasing of SHG intensity on the number of laser pulses. Thirdly, the ultimate control of samples destruction presence was performed by analysis of them under a microscope. Thus we obtained sufficiently reliable data on the damage thresholds of the studied samples.

5. Results and analysis of them

During the work the measurements were made and relationships of changes in the SGH intensity for all four samples – $LiIO_3$, KDP, m-nitroaniline, and DNPAP from the energy density of the laser incident radiation were described. The results of measurement were plotted as graphs which are shown in figure 2.

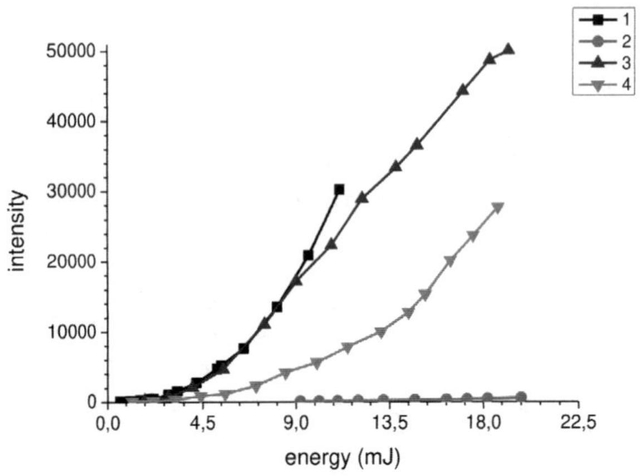

Figure 2. SHG intensity of samples (1 - $LiIO_3$; 2 - KDP; 3 - m-nitroaniline; 4 - DNPAP) versus incident radiation energy density of a pulsed YAG:Nd^{3+} laser.

The obtained dependences are well approximated by a quadratic function, as it is shown for $LiIO_3$ sample in figure 3.

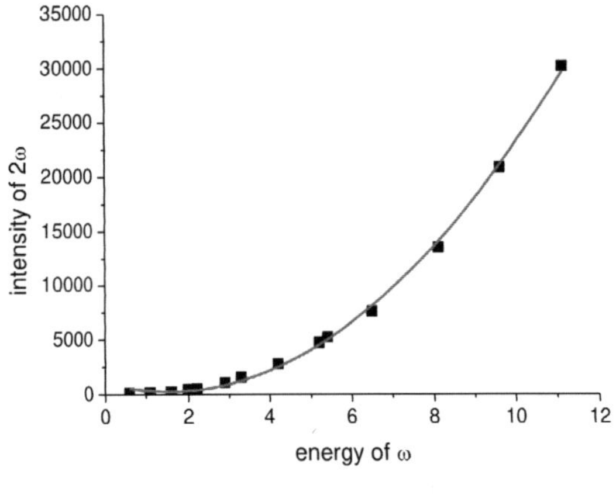

Figure 3. SHG beam intensity dependence data for $LiIO_3$ sample vs the incident radiation energy density of a pulsed YAG:Nd^{3+} laser, which was approximated by quadratic function.

This is due to the fact that the dependence of frequency conversion has a view of quadratic function as a rule. The resulting parameters of approximation of experimental data are the parameters of frequency conversion of studied samples. Namely, the coefficient of the quadratic part data is due to the nonlinear coefficient of sample. This approximation has been made for each graph data, which is shown in figure 2, and thus b coefficients were determined for each sample. From the ratio of the b coefficients for samples with unknown nonlinear coefficient to the b coefficient for sample with known one, in accordance with their refractive index data, the values of nonlinear coefficient $d^{2\omega}$ were calculated for each sample. The corresponding results are shown in table 1.

Table 1. The data of laser-optical properties of the studied crystalline materials.

Crystal	Normalized nonlinear coefficient	Absolute nonlinear coefficient (m/V)	Damage thresholds (multiple pulses) (W/cm^2)
KDP[a]	1	$6.3\cdot10^{-13}$	$1\cdot10^{10}$
$LiIO_3$	11.16	$7.03\cdot10^{-12}$	$2.1\cdot10^{10}$
m-nitroaniline	9.08	$5.72\cdot10^{-12}$	$2.3\cdot10^{10}$
DNPAP	6.44	$4.06\cdot10^{-12}$	$4.4\cdot10^{10}$

[a] The data were taken from literature [9]

Thus in the table 1 we gave data of the nonlinear coefficient for each studied sample normalized with respect to KDP; the absolute nonlinear coefficient values calculated from the relative coefficients, with using data known from the literature for the KDP; the damage thresholds against multiple laser pulses.

The obtained results were compared with reference data for the widely known crystals described in the literature [9]. In this article value of $d^{2\omega}$ coefficient at a wavelength of 1.06 µm is equal $7.02\cdot10^{-12}$ m/V for $LiIO_3$ crystal. As we can see, the data from this article is in a very good agreement with data from table 1. The optical damage thresholds data of the studied materials have a good agreement also [9]. This proves the correctness of the methodic, which is described in this paper.

So, the date that we received for the new material DNPAP is also correct. The above data allow us to tell about the promising of DNPAP material as an SHG material. It is due to the rather large value of the nonlinear optical coefficient comparable to $LiIO_3$. Also, as we can see from table, this new material has large optical damage threshold – about two time more than for materials, which taken for comparison in the work.

Thus, by the work it was demonstrated simple and effective method of measurement of nonlinear properties of a new material.

6. Conclusion

Simple technique for analyzing of new nonlinear optical materials properties has been developed and demonstrated on well-known crystalline materials. Nonlinear coefficient and damage threshold of KDP, $LiIO_3$, m-nitroaniline of the samples were measured. The data coincided with data known from literature.

Nonlinear-optical parameters for new promising nonlinear material - DNPAP were measured. The values of these parameters allow us to hope to apply of this material in practice.

Acknowledgments

This work supported by RFBR, research project No. 14-02-31299 mol_a.

References

[1] Franken P A, Hill A E, Peters C W and Weinreich G 1961 *Phys. Rev. Lett.* **7** 118
[2] Kurtz S K and Perry T T 1968 *J. Appl. Phys.* **39** 3798

[3] Dougherty J P and Kurtz S K 1976 *J.Appl Cryst.* **9** 145
[4] Hirsch R and Stuhmer G 1976 *Phys. Stat. Sol.* **37** 105
[5] Kiguchi M, Kato M, Okunaka M and Taniguchi Y 1992 *Appl. Phys. Lett.* **60** 1933
[6] Kiguchi M, Kato M, Kumegawa N and Taniguchi Y 1993 *Mol. Crysr. Liq. Crysr.* **227** 133
[7] Kremer R, Boudrioua A and Loulergue J C 1999 *J. Opt. Soc. Am. B* **16** 83
[8] Strelnik A D, Garipov M R, Petukhov A S, Shtyrlin N V, Lodochnikova O A, Litvinov I A, Naumov A K, Morozov O A, Klimovitskii A E and Shtyrlin Yu G 2014 *Spec. Acta A–Mol. Biomol. Spec.* **117** 793
[9] Choy M M and Byer R L 1976 *Phys. Rev. B.* **14** 1693

Optimization of the nonlinear regime of self-compression at femtosecond laser pulses in silica and air

V. Slavchev, L. Kovachev , L. M. Ivanov

Institute of Electronics, Bulgarian Academy of Sciences
72 Tzarigradsko shossee,1784 Sofia, Bulgaria

e-mail: lubomirkovach@yahoo.com

Abstract. In the present work it is demonstrated two efficient methods of self-compression of femtosecond pulses based on suitable selection of optical elements, parameters of the medium and laser radiation. The basic idea is that the phase modulated pulses more efficiently can be compressed trough nonlinear mechanisms. The first method can be applied for mediums with significant dispersion like fused silica, where the sign of the dispersion of the group velocity is important. We show that the combination of focusing by optical lens and a balance between anomalous dispersion and nonlinearity lead to significant compression from 100fs to~20-30fs of optical pulse. The second method for self-compression is by using only one optical diffraction grating to obtain broadband pulses and the following self-compression in nonlinear regime. In the second case in addition is observed generation of X wave.

1. Introduction

One of the main tasks of modern laser physics is to obtain ultra short laser pulses. There are different optical schemes for self-compression based on pair diffraction grating or by other optical elements when the pulses propagate in linear regime [1]. The shortening of pulse can be obtained also by using nonlinear mechanisms such as self-phase modulation, when a pulse reaches critical power for self-focusing or self-modulation. Another modern method for compressing of pulses in time is by using of hollow optical fibers filled with noble gas. During propagation of negative chirped pulses in hollow optical fiber the self-phase modulation leads to compression. These effects have been studied in different fiber types, including single-mode fibers [1, 2], hollow fibers filled with noble gas [3, 4]. Problem of this method is in the damage of the fibers. There is also one limitation of the energy connected with the recompressing of the pulse due to plasma ionization of the gas. The most efficient schemes for compression require careful analyze for collection of possible optical elements, linear and nonlinear effects to obtain maximal compression. This work investigates two different mechanisms for self-compression of femtosecond pulses - in fused silica and in air.

In the first case we compare the self-compression of laser pulses during their nonlinear propagation in the normal and anomalous dispersion region of a glass. The pulses are spatial phase modulated initially by long focused lens. The numerical experiment is performed for 100 fs pulses at wavelength 800 nm and 1400 nm correspondingly.

In the second case we perform comparing of the nonlinear compression of no-chirped initial 25 fs pulse and the chirped by grating ones, during their propagation in air. Our numerical results show, that after only one pass trough optical diffraction grating the chirped pulse self-compress more efficiently in nonlinear regime and in addition obtain X wave deformation.

2. Nonlinear regime of narrow-band pulse in silica

The no-modulated femtosecond pulses admit relatively narrow band spectrums $\Delta k_z << k_0$ and their evolution can be describe into the frame of nonlinear spatio-temporal nonlinear optics. The main problem in this regime is to use proper combination of optical lens and to by find proper balance between dispersion and nonlinearity [5].

2.1 Basic equation

The normalized scalar paraxial (3D+1) amplitude equation describing the propagation of narrow-band optical pulse in nonlinear dispersive medium has the form:

$$i\frac{\partial A}{\partial z} = \frac{1}{2}\Delta_\perp A - \beta\frac{\partial^2 A}{\partial t^2} + \gamma|A|^2 A \qquad (1)$$

Where *A (x,y,z,t)* is the scalar amplitude function, characterizing the pulse envelope. $\beta = {z_{diff}}/{z_{disp}}$ is a dimensionless parameter giving the ration between the dispersion and the diffraction length of the pulse, $z_{disp} = t_0^2/k''$ is the dispersion length, $z_{diff} = k_0 r_\perp^2$ - the diffraction length. The nonlinearity of the medium is of Kerr type. It is determined by the expression $\gamma = k_0^2 r_\perp^2 n_2 |A_0|^2$, k_0 - is the wave number, $\Delta_\perp = \partial^2/\partial x^2 + \partial^2/\partial y^2$ is the transverse Laplace operator. This equation counts effects of paraxial diffraction, second order linear dispersion and Kerr type nonlinearity. We use the initial conditions for the numerical experiments Gaussian pulse modulated by lens:

$$A(x,y,t,0) = \exp\left\{-\frac{x^2+y^2}{2} - \delta\frac{t^2}{2}\right\}\exp\left\{if'\left(x^2+y^2\right)\right\} \qquad (2)$$

where $\delta = \frac{r_\perp^2}{t_0^2}$ is the relation between spatial transverse and longitudinal time durations, and $f' = \frac{\pi r_\perp^2}{\lambda_0 f}$ is normalized length factor.

2.2 Nonlinear pulse propagation in the positive dispersion region of silica

We investigate numerically the propagation of laser pulse in the positive dispersion regions of fused silica by using the following parameters of the laser source: time duration $t_0 = 100$ fs; spot $r_\perp = 140\mu m$ at carrying wave wavelength $\lambda = 800$ nm. The dispersion of the group velocity in the positive dispersion region is k"=3.3x10^{-26} s^2/cm and the focus distance of the lens is *f = 30cm*. The parameters of the pulse of the medium are selected to obtain the relation between diffraction and dispersion length to be *β = 0.5*. The nonlinear coefficient γ =2.2 is chosen to be slightly above the critical for self-focusing. The numerical result shown in Fig. 1 is obtained by solving equations (1) under initial conditions (2).

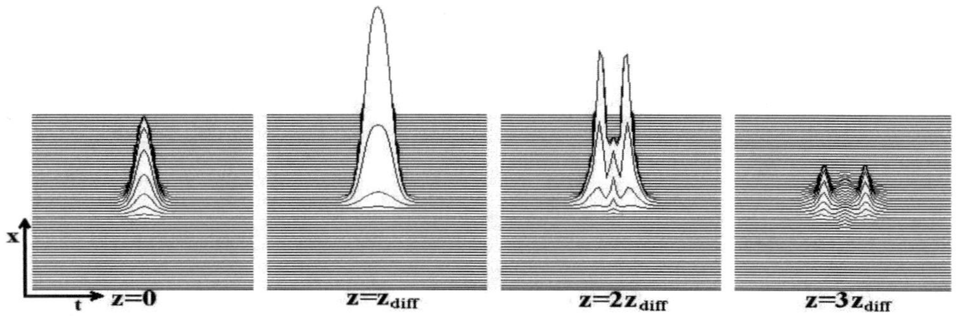

Figure 1: Nonlinear propagation of initial modulated by lens (f=30 cm) optical pulse in medium with normal dispersion β= 0.5 and coefficient of nonlinearity γ =2.2. The side (x, t) projection of the intensity profile is plotted. Numerical result is obtained by solving equation (1) under initial condition (2).

The numerical result shown in Fig.1 presents side (x,t) projection of the intensity profile at four distances $z = 0$, $z = z_{diff}$, $z = 2z_{diff}$, and $z = 3z_{diff}$. The evolution of the pulse describes well the observed in the experiments and numerical simulations separation of the temporal profile into two maximums and the significant spectral enlarging.

2.3 Nonlinear pulse propagation in the negative dispersion region of silica

Two main tasks are placed when we investigate the propagation of optical pulse in the negative region of dispersion in fused silica. The first purpose is by one long focusing optical lens in nonlinear regime to minimize the energy in the pedestal of the self-focused pulse. The second purpose is the negative frequency modulation (chirp) of the dispersion to compensate the positive phase modulation of the nonlinearity. In this way significant part of the energy will go in the nonlinear focus, which will lead to natural self-compression and self-focusing of the pulse. On Fig.2 is presented the nonlinear propagation of a pulse, spatially phase modulated by the same lens with *f=30 cm*, and dynamics governed by numerical solution of equation (1) under initial condition (2). The dimensionless parameters are β=-0.5 and γ=2.2. As can be expected from the performed above qualitative analysis of the physical parameters, significant self-focusing without loss of energy in the pedestal is observed. The compression of the pulse is from 100 fs up to ~20-30 fs.

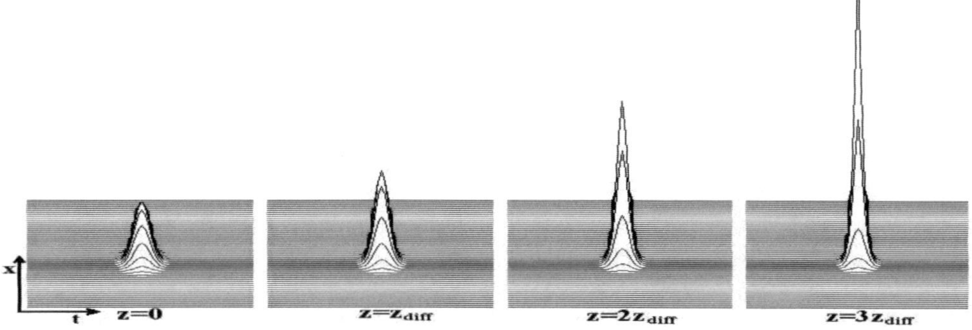

Figure 2: Propagation of initially modulated by lens (f = 30cm) optical pulse in medium with negative dispersion β= -0.5 and coefficient of nonlinearity γ =2.2. These parameters correspond to propagation of a laser pulse on carrying wavelength λ=1400nm, with time duration t_0 = 100 fs and

spot $r_{\perp} = 180\mu m$. The corresponding at λ=1400nm negative group velocity dispersion is k"=-3.3x10^{-26} s^2/cm.

3. Non-paraxial nonlinear regime of broad–band pulses in air modulated by optical grating

The spatio- temporal equation (1) can't govern the propagation of broad-band pulses [6]. This is the reason, if we investigate nonlinear regime of propagation of pulses with time duration of order of 20-30 fs, to use non-paraxial envelope equation [6]. The pulses with such duration at short distances enlarge significantly their spectrum by self-phase modulation and reaches spectral widths of the order of $\Delta k_z \approx k_0$. Other reason to investigate phase modulation by grating in gases is that the group velocity dispersion in air is very small (k"=2.5x10^{-31} s^2/cm) and practically we can't modulate the phase of the pulse by dispersion on few diffraction lengths.
The scalar non-linear non-paraxial envelope equation in Galilean frame is given by:

$$2i\alpha\delta^2 \frac{\partial A}{\partial t'} = \Delta_{\perp} A - \beta \frac{\partial^2 A}{\partial z'^2} + \left(\delta^2 + \beta\right)\left(\frac{\partial^2 A}{\partial t'^2} - \frac{\partial^2 A}{\partial t' dz'}\right) + \gamma\left(|A|^2 A + A^3 e^{(2i\alpha(z - \hat{v}_n t))}\right) , \quad (4)$$

where β and γ are the same normalized numbers and in addition the following normalized constants appear: $\alpha = k_0 z_0$; $z_0 = v_{gr} t_0$; $\delta = r_{\perp} / z_0$ and $\hat{v}_n = \left(v_{ph} - v_{gr}\right)/ v_{gr}$ is the normalized group- phase velocity delay. The number α with precise 2π counts the number of cycles under the envelope and in this way we can control the spatial (and temporal) shape of the initial pulse in respect to optical period λ_0. In gases the standard way to obtain large phase modulated spectrum is trough a diffraction grating. In our numerical experiment we will modulate the chirp parameter by the complex number ia in the initial conditions [1]. That why our initial Gaussian pulse takes the form:

$$A(x, y, z, 0) = A_0 e^{\left(\frac{x^2 + y^2 + (1 + ia) z^2}{2}\right)} \quad (5)$$

In this paper we investigate the following two basic cases:

1) White (no-phase modulated) initial Gaussian pulse ($a = 0$).
2) Broad-band phase modulated Gaussian pulse with chirp parameter $a = 8$.

In the both cases the initial width of the pulse is selected to by 25 femtoseconds.

3.1 Propagation of white (no-phase modulated) initial Gaussian pulse (a=0).

To investigate the nonlinear propagation of no-modulated 25 fs pulse we solve numerically the envelope equation (4) under the initial condition (5) with chirp parameter $a = 0$. As the group velocity dispersion in air is of order of k"=2.5x10^{-31} s^2/cm the relation between the diffraction and the dispersion lengths is a very small number - $\beta \approx 10^{-5}$. Thus, practically the spatial evolution of the pulse appears as balance between non-paraxial diffraction and nonlinearity. The nonlinear coefficient γ =1.8 again is chosen to be slightly above the critical for self-focusing. The numerical result shown in Fig. 1 presents side (x,z) projection of the intensity profile at four distances $z = t = 0$, $z = z_{diff}$, $z = 2z_{diff}$, and $z = 3z_{diff}$. As it can be expected, the typical separation of the pulse on two maximums with significant enlarging of the spectral width is observed. The main part of the energy is situated in the pedestal and the self-compression is not so effective.

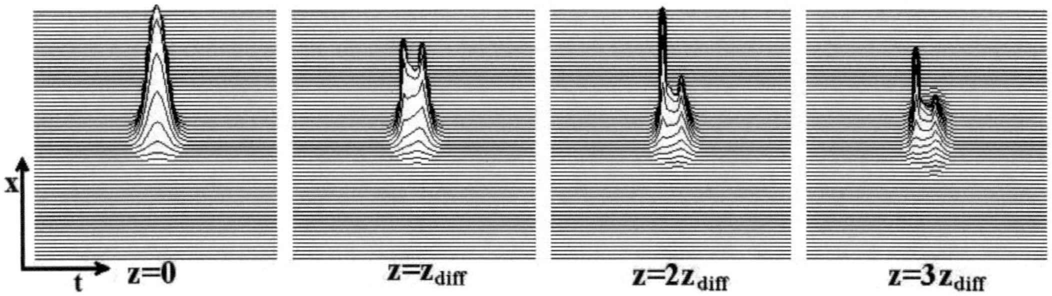

Fig.3.Propagation of spectrally limited (narrow-band) 25fs pulse (a=0) governed by Eq.(4). The typical self-compression with two maximums is observed. Side (x,z) projection of the intensity profile at four distances $z = t = 0$; $z = z_{diff}$; $z = 2z_{diff}$; $z = 3z_{diff}$ is plotted. The main part of the energy is situated in the pedestal and the self-compression is not so effective.

3.2 Propagation of broad- band pulse with chirp parameter a=8;

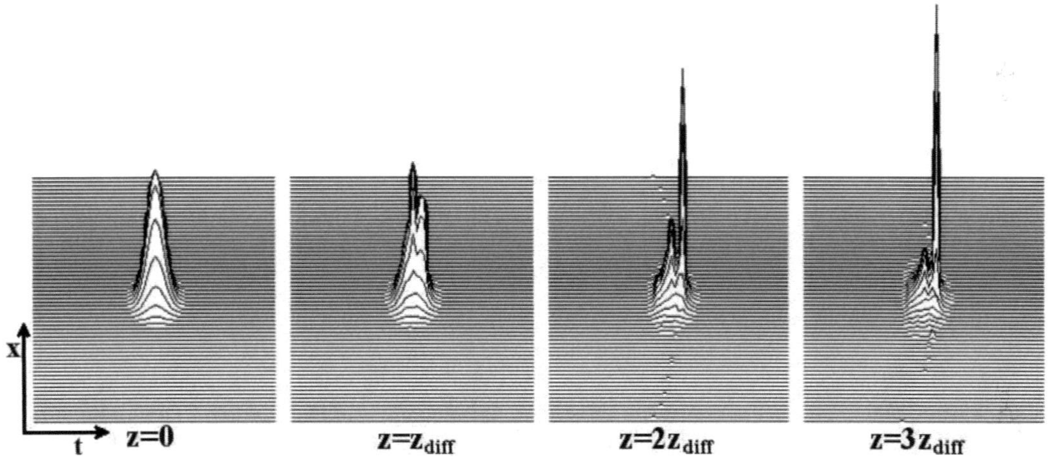

Fig.4. Propagation of broad-band 25fs pulse (with chirp parameter a=8) governed by Eq. (4).

On Fig 4 self-compression with decreasing maximums is clearly seen. In addition, typical X-wave formation in the nonlinear focus is observed. In Galilean frame we have a real deformation of the X-wave, while in earlier works performing in Local time coordinate system the inversed pictures of the deformation are presented.

4. Conclusions

In the present work two efficient methods for self-compression of femtosecond pulses based on suitable selection of optical elements, parameters of the medium and laser radiation are demonstrated. We explore the basic idea, that the phase modulated pulses more efficiently can be compressed trough nonlinear mechanisms. The first method is applied for mediums with significant dispersion like fused silica, where the sign of the dispersion of the group velocity is important. We show that the combination of focusing by optical lens for relative reduction of energy in the pedestal and balance between anomalous dispersion and nonlinearity lead to significant compression from 100fs to~20-30fs of optical pulse in silica. The second method for self-compression is applied to gases, where the dispersion is negligible. In this case we use optical diffraction grating to obtain broadband pulses and the following self-compression in nonlinear regime. In the second case in addition is observed generation of X wave.

References

[1] J.-C.Diels, W. Rudolph, "Ultrashort Laser Pulse Phenomena" 2nd edition, (Academic Press, 2006).

[2] S. A. Planas, N. L. Pires, C. H. Brito and H. L. Fragnito, "Spectral narrowing in the propagation of chirped pulses in single-mode fibers," Opt. Lett. **18**, 699 (1993).

[3] B. R. Washburn, J. A. Buck and S. E. Ralph, "Transform-limited spectral compression due to self-phase modulation in fibers," Opt. Lett. 25, 445 (2000).

[4] Krzysztof Iwaszczuk, Binbin Zhou, Morten Bache and Peter Uhd Jepsen" Spectral Compression of Intense Femtosecond Pulses by Self Phase Modulation in Silica Glass" CLEO Technical Digest. Optical Society of America, 2012. p. JW4A.61.

[5] A. E. Dokurina, E. O. Smetanina and V. O. Kompanets, "Femtosecond filamentation of Bessel–Gaussian beams under conditions of anomalous group-velocity dispersion", Journal of Optical Technology, 81, 08 (2014).

[6] Lubomir M. Kovachev and Kamen Kovachev, "Linear and Nonlinear Femtosecond Optics in Isotropic Media. Ionization-free Filamentation", .Laser Systems for Applications / Book 1, chapter, InTech, (2011), DOI: 10.5772/23213.

Many-body excitation spectra of trapped bosons with general interaction by linear response

Ofir E. Alon

Department of h sics, ni ersit of aifa at Oranim, Ti on 36 6, Israel

-mail ofir@research.haifa.ac.il

Abstract. The linear-response theor of the multicon gurational time-dependent artree for bosons method R-MCTD B for computing man -bod excitations of trapped Bose- instein condensates h s. Re . **88**, 236 6 2 13 ; . Chem. h s. **140**, 341 2 14 is implemented, for the rst time, for s stems with general interparticle interaction. This allows us to in estigate the man -bod excitation spectrum of interacting bosons with, for instance, a long-range interaction. Illustrati e numerical examples for repulsi e and attracti e bosons are pro ided. The R-MCTD B theor is capable of identif ing all excitations, including the excitations which are not unra eled within Bogoliubo –de Gennes e uations. The theor is herewith benchmar ed against the exactl -sol able one-dimensional harmonic-interaction model. s a complementar result, using a complex transformation, we represent R-MCTD B in a compact bloc -diagonal form, opening up thereb an a enue for treating larger man -bod s stems. e expect the R-MCTD B theor and its implementation for general interparticle interaction to pro ide a pro ed probe into the man -bod excitations in ol ed in the out-of-e uilibrium d namics of trapped interacting bosons.

1. Introduction

The standard and most popular avenue to compute the dynamics (and excitations) of a Bose-Einstein condensate (BEC) is the Gross-Pitaevskii equation, which assumes all bosons to occupy a single one-particle state [1-4]. Clearly, whenever the system under investigation is not fully condensed one has to go beyond Gross-Pitaevskii theory in order to faithfully account for the system s dynamics, a matter which is well documented in the literature, see, e.g., the recent book [5] and references therein.

The usual way to account for excitations in a BEC is to build them atop the Gross-Pitaevskii ground state, taking a particle out from the condensate to an excited one-particle state. Formally, this leads to the Bogoliubov–de Gennes equations which often are also referred to as linear response of the Gross-Pitaevskii equation [1-10]. It turns out, as we have recently shown for contact interaction [11], that even if the ground state is well described by the fully-condensed Gross-Pitaevskii wavefunction, many excitations are missing by this standard treatment. For fully-fragmented ground states [12], more excitations appear [13]. Thus, to describe systems whose ground states are not fully condensed and, importantly, to identify new classes of excitations which cannot be resolved by the standard tools, suitable methods are in need.

Consider N trapped bosons, interacting by a generic interparticle interaction. The many-particle Hamiltonian is written as follows

$$H(\mathbf{r}_1,\ldots,\mathbf{r}_N) \quad \sum_{j=1}^{N} h(\mathbf{r}_j) \quad \sum_{k>j=1}^{N} W(\mathbf{r}_j,\mathbf{r}_k). \tag{1}$$

Here $h(\mathbf{r})$ is the one-body Hamiltonian which consists of kinetic and potential (trap) terms and $W(\mathbf{r},\mathbf{r}')$ is a generic interparticle interaction which is symmetric to permutation of the particles coordinates. In the familiar case for ultracold bosonic atoms it takes the form $W(\mathbf{r},\mathbf{r}') \quad \lambda_0\delta(\mathbf{r}-\mathbf{r}')$ where the interaction parameter λ_0 is proportional to the s-wave scattering length.

We recall that by approximating the system s time-dependent wavefunction by the product state

$$_{\rm GP}(\mathbf{r}_1,\ldots,\mathbf{r}_N \; t) \quad \phi(\mathbf{r}_1,t)\cdots\phi(\mathbf{r}_N,t) \equiv |N \; t\rangle \tag{2}$$

and utilizing the Dirac-Frenkel variational principle, the Gross-Pitaevskii equation, $i\dot{\phi}(\mathbf{r},t)$ $[h(\mathbf{r}) \quad \lambda|\phi(\mathbf{r},t)|^2]\phi(\mathbf{r},t)$, where $\lambda \quad \lambda_0(N-1)$, is obtained.

The linear-response theory atop the Gross-Pitaevskii mean-field wavefunction (2) is formally obtained by linearizing the Gross-Pitaevskii equation around the ground-state solution. It results in the Bogoliubov–de Gennes equations which take on the matrix form

$$\boldsymbol{\mathcal{L}}_{\rm BdG}\begin{pmatrix} u^k \\ v^k \end{pmatrix} \quad \omega_k \begin{pmatrix} u^k \\ v^k \end{pmatrix}, \qquad \boldsymbol{\mathcal{L}}_{\rm BdG} \quad \begin{pmatrix} h \quad 2\lambda|\phi_0|^2 - \mu & \lambda(\phi_0)^2 \\ -\lambda(\phi_0^*)^2 & -(h^* \quad 2\lambda|\phi_0|^2 - \mu) \end{pmatrix}. \tag{3}$$

The linear-response matrix $\boldsymbol{\mathcal{L}}_{\rm BdG}$ depends explicitly on the ground-state orbital ϕ_0. μ is the chemical potential. The excitation spectrum ω_k of the BEC as well as the so-called response amplitudes u^k and v^k are obtained by solving the eigenvalue system (3).

2. Many-body theory

In many situations as mentioned above the ground state of the BEC cannot be described well by the wavefunction Eq. (2). Importantly, even when the ground state is well approximated by the Gross-Pitaevskii wavefunction (2), the standard linear-response atop misses many excitations [11], also see the benchmarks in Sec. 3 below. The natural idea was to use a more extended ansatz for the system s wavefunction, and then to perform linear response atop [11, 14]. We recall that linear response atop the exact ground state gives rise to the exact many-body excitation spectrum, see, e.g., [15, 16].

Let us brie y describe the underlying many-body theory for BECs used in the present work before we proceed to its linear response. In the multiconfigurational time-dependent Hartree for bosons (MCTDHB) method [17, 18] the bosons are allowed to occupy not one but $j \quad 1,\ldots,M$ one-particle functions (modes) ϕ_j. The many-body wavefunction is assembled by distributing the N bosons over the M one-particle functions

$$(t) \quad \sum_{\vec{n}} C_{\vec{n}}(t)|\vec{n} \; t\rangle, \tag{4}$$

where $C_{\vec{n}}(t)$ are the expansion coe cients and $|\vec{n} \; t\rangle$ are permanents (Fock states) with $\vec{n} \equiv (n_1,\ldots,n_M)$, $n_1 \quad \ldots \quad n_M \quad N$. Utilizing the Dirac-Frenkel variational principle, the one-particle functions ϕ_j as well as the expansion coe cients $C_{\vec{n}}$ are determined self-consistently. This leads to a system of coupled equations which has been coined in the literature the MCTDHB method [17, 18]. The MCTDHB method has been used for unveiling many-body phenomena with repulsive and attractive BECs in one-dimensional setups, see, e.g., [19-26], and benchmarked against an exactly-solvable model [27]. Most recently, MCTDHB was extended to two and three spatial dimensions, and employed to establish the mechanism of fragmentation and generic regimes of dynamics in repulsive BECs with strong, finite-range interparticle interactions [28, 29].

The successes of the MCTDHB method to accurately compute many-body out-of-equilibrium dynamics of BECs stem from the employment of the self-consistent (time-adaptive) multiconfigurational wavefunction (4). The wavefunction (4) contains a substantially larger number of variational parameters (i.e., modes, and Fock states and their expansion coe cients) in comparison with the standard Gross-Pitaevskii wavefunction (2). Clearly, if the obtained many-body dynamics is accurate then MCTDHB manages to resolve the excitation spectra of BECs. This has motivated us to pursue its linear response, as a venue to research on the many-body level excitations of trapped BECs directly, i.e., without propagation.

The derivation of the linear-response (LR) theory atop the wavefunction (4) is rather lengthly but otherwise straightforward [11, 14]. We will not repeat it here and begin from the final result for the resulting LR-MCTDHB theory, which takes on the form of the eigenvalue equation [11, 14]

$$\boldsymbol{\mathcal{L}} \begin{pmatrix} \mathbf{u}^k \\ \mathbf{v}^k \\ \mathbf{C}_u^k \\ \mathbf{C}_v^k \end{pmatrix} \quad \omega_k \begin{pmatrix} \mathbf{u}^k \\ \mathbf{v}^k \\ \mathbf{C}_u^k \\ \mathbf{C}_v^k \end{pmatrix}. \tag{5}$$

The linear-response matrix $\boldsymbol{\mathcal{L}}$ of the many-boson wavefunction is more involved than the Bogoliubov–de Gennes linear-response matrix (3). We will discuss its structure shortly. Physically, the response amplitudes of all modes, $\mathbf{u}^k$ and $\mathbf{v}^k$, and of all expansion coe cients, $\mathbf{C}_u^k$ and $\mathbf{C}_v^k$, combine to give the many-body excitation spectrum ω_k. For comparison, in the Bogoliubov–de Gennes linear-response matrix (3) there is only a single block representing the sole one-particle function used to describe the BEC within Gross-Pitaevskii theory. In Ref. [11] we have successfully managed to explicitly construct $\boldsymbol{\mathcal{L}}$ for bosons interacting by contact potential and obtained the many-body excitation spectrum.

There is another way to group the linear-response matrix which we are now going to exploit. Namely, first to list the orbitals and coe cients 'u blocks and then the respective 'v blocks. In this way each of the new blocks has the same dimension, see below. The spectrum, of course, does not change. Hence, reshuffling the blocks of $\boldsymbol{\mathcal{L}}$ the final result can be written as

$$\boldsymbol{\mathcal{L}} \begin{pmatrix} \mathbf{u}^k \\ \mathbf{C}_u^k \\ \hline \mathbf{v}^k \\ \mathbf{C}_v^k \end{pmatrix} \quad w_k \begin{pmatrix} \mathbf{u}^k \\ \mathbf{C}_u^k \\ \hline \mathbf{v}^k \\ \mathbf{C}_v^k \end{pmatrix}, \qquad \boldsymbol{\mathcal{L}} \quad \left(\begin{array}{c|c} \boldsymbol{\mathcal{L}}^u & \boldsymbol{\mathcal{L}}^v \\ \hline -(\boldsymbol{\mathcal{L}}^v)^* & -(\boldsymbol{\mathcal{L}}^u)^* \end{array} \right), \tag{6}$$

where details of the blocks of $\boldsymbol{\mathcal{L}}$ are collected in [30].

The above general relation between the 'u and 'v blocks of $\boldsymbol{\mathcal{L}}$ is appealing since we may mix them and eventually block diagonalize $\boldsymbol{\mathcal{L}}$. For this, consider the transformation

$$\mathbf{Q} \quad \frac{1}{\sqrt{2}} \begin{pmatrix} \mathbf{1} & \mathbf{1}\theta \\ \mathbf{1} & -\mathbf{1}\theta \end{pmatrix}, \tag{7}$$

where θ is the operation of complex con ugation, for example, $\theta \mathbf{v}^k \quad (\mathbf{v}^k)^*$. It is not di cult to show that $\mathbf{Q}$ block diagonalizes $\boldsymbol{\mathcal{L}}$ Consult the appendix for additional details. The final result reads

$$\begin{aligned} \boldsymbol{\mathcal{L}}_f^{(2)} \begin{pmatrix} \mathbf{f}^k \\ \mathbf{C}_f^k \end{pmatrix} \quad w_k^2 \begin{pmatrix} \mathbf{f}^k \\ \mathbf{C}_f^k \end{pmatrix}, &\qquad \boldsymbol{\mathcal{L}}_f^{(2)} \quad (\boldsymbol{\mathcal{L}}^u - \boldsymbol{\mathcal{L}}^v\theta)(\boldsymbol{\mathcal{L}}^u \quad \boldsymbol{\mathcal{L}}^u\theta), \\ \boldsymbol{\mathcal{L}}_g^{(2)} \begin{pmatrix} \mathbf{g}^k \\ \mathbf{C}_g^k \end{pmatrix} \quad w_k^2 \begin{pmatrix} \mathbf{g}^k \\ \mathbf{C}_g^k \end{pmatrix}, &\qquad \boldsymbol{\mathcal{L}}_g^{(2)} \quad (\boldsymbol{\mathcal{L}}^u \quad \boldsymbol{\mathcal{L}}^u\theta)(\boldsymbol{\mathcal{L}}^u - \boldsymbol{\mathcal{L}}^v\theta), \end{aligned} \tag{8}$$

Journal of Physics: Conference Series **594** (2015) 012039 doi:10.1088/1742-6596/594/1/012039

with the relations between the eigenvectors blocks $\mathbf{f}^k, \mathbf{g}^k \quad \frac{1}{\sqrt{2}}[\mathbf{u}^k \pm (\mathbf{v}^k)^*]$ and $\mathbf{C}^k_f, \mathbf{C}^k_g \quad \frac{1}{\sqrt{2}}[\mathbf{C}^k_u \pm (\mathbf{C}^k_v)^*]$.

Eq. (8) is the main result of this work on the theory side We compactly represent the LR-MCTDHB theory in a block-diagonal form. Because the resulting (square of the) excitation spectrum is bound from below, this will open up an avenue for treating larger systems. First, this in particular would allow one to use standard diagonalization techniques for matrices with bound spectra. Here, Eq. (8) defines the basic operation of matrix-to-vector multiplication used in such techniques. Second, it reduces the size of the linear-response matrix to half its size. Furthermore, since $\mathbf{Q}$ in Eq. (7) is a complex transformation, the block diagonalization can be used with complex wavefunctions and with complex hermitian Hamiltonians. This situation is encountered, e.g., when a system is described in the rotating frame. Thus, the block diagonalization derived in the present work for the many-body theory also generalizes the literature block-diagonalization treatment of the Bogoliubov–de Gennes equations, see, e.g., [31-33], which only used a real transformation. This concludes our block diagonalization of the LR-MCTDHB theory.

3. Illustrative numerical examples and benchmarks

In the following section we would like to report, for the first time, the implementation and application of LR-MCTDHB with general interparticle interaction. As an illustrative system we have chosen the harmonic-interaction model [34-37]. This is an analytically-solvable model, yet, it is not at all trivial to be treated numerically. First, because the interaction is non-contact the construction of various matrix elements for computing the ground-state wavefunction and its response matrix are much more involved. We have now successfully coped with this demanding task. Second, the solution of the problem on the computer is done in the laboratory frame, where all the bosons are indistinguishable. This is unlike the analytical solution which exploits the separability of the center-of-mass and relative coordinates degrees-of-freedom. As mentioned above, the harmonic-interaction model and a time-dependent extension of which have been used to benchmark MCTDHB [27]. Hence, we expect the model to be instrumental in benchmarking the excitation spectrum computed by the many-body linear-response theory.

In Eq. (1) the one-body Hamiltonian is now $h(x) \quad -\frac{1}{2}\frac{\partial^2}{\partial x^2} \quad \frac{1}{2} \; ^2x^2$. The two-body interaction reads $W(x-x') \quad K(x-x')^2$, representing thereby long-range interaction. The parameter $K < 0$ ($K > 0$) indicates repulsion (attraction) between the bosons. In what follows we set without loss of generality $\quad 1$. The exact excitation energies of the one-dimensional harmonic-interaction model are known and given by [34, 36]

$$\omega[n_{CM}, n_{rel}] \quad n_{CM} \quad n_{rel}\delta_N, \qquad \delta_N \quad \sqrt{1 \quad 2NK}, \tag{9}$$

with the center-of-mass (CM) and relative coordinates (rel) quantum numbers $n_{CM} \quad 1, 2, 3, \ldots$ and $n_{rel} \quad 2, 3, \ldots$.

We consider $N \quad 1000$ weakly-interacting repulsive bosons ($K \quad -0.0001$) and compare in Table 1 below the LR-MCTDHB results with the analytical formula (9). The many-body theory, computed at the level of $M \quad 2$ orbitals in the expansion (4), improves as one would expect the accuracy of excitations unveiled by Bogoliubov–de Gennes theory. More important, it unravels additional excitations. We see the capability of the LR-MCTDHB theory to describe numerically-exactly the center-of-mass and relative coordinates excitations. The physical distinction here for the excitations missing in the Bogoliubov–de Gennes theory is that all consist of one quantum of center-of-mass excitation plus at least one additional quantum of excitation. The latter can be either another center-of-mass or a relative coordinates excitation, see the rightmost column of Table 1. We stress again that the computation is done in the

	M 1	M 2	Exact analytical	n_{CM}, n_{rel}
E_{GS}	447.26$\underline{949371}$	447.2663819$\underline{4}$	447.26638190	0, 0
ω_1	1.00000000	1.00000000	1.00000000	1, 0
ω_2	1.78$\underline{907797}$	1.788854$\underline{43}$	1.78885438	0, 2
ω_3	n a	2.0000$\underline{4476}$	2.00000000	2, 0
ω_4	2.683$\underline{61696}$	2.683281$\underline{68}$	2.68328157	0, 3
ω_5	n a	2.7888$\underline{8891}$	2.78885438	1, 2
ω_6	n a	3.0000$\underline{7751}$	3.00000000	3, 0
ω_7	3.57$\underline{815595}$	3.5777$\underline{1028}$	3.57770876	0, 4
ω_8	n a	3.683$\underline{97387}$	3.68328157	1, 3

Table 1. Spectrum of the one-dimensional harmonic-interaction model with N 1000 bosons and repulsion K -0.0001. Comparisons of LR-MCTDHB and the exact results for the ground, E_{GS}, and excited states, ω_k $E_k - E_{GS}$. The last column assigns the excitations in terms of center-of-mass and relative coordinates quantum numbers. Some excitations are first uncovered at the M 2 level of theory, i.e., they are not available (n a) within Bogoliubov–de Gennes theory (M 1). Convergence with the number or orbitals M to the exact results is clearly seen. Underlined digits indicate the di erence to the exact result. All quantities are dimensionless.

	M 1	M 2	Exact analytical	n_{CM}, n_{rel}
E_{GS}	547.67$\underline{691206}$	547.6748349$\underline{7}$	547.67483495	0, 0
ω_1	1.00000000	1.00000000	1.00000000	1, 0
ω_2	n a	2.0000$\underline{3648}$	2.00000000	2, 0
ω_3	2.190$\underline{70765}$	2.1908902$\underline{6}$	2.19089023	0, 2
ω_4	n a	3.0000$\underline{6435}$	3.00000000	3, 0
ω_5	n a	3.190$\underline{91687}$	3.19089023	1, 2
ω_6	3.286$\underline{06147}$	3.286335$\underline{42}$	3.28633535	0, 3
ω_7	n a	4.00$\underline{129579}$	4.00000000	4, 0
ω_8	n a	4.28$\underline{580642}$	4.28633535	1, 3

Table 2. Same as Table 1 but for N 1000 bosons with attraction K 0.0001. Some excitations are first uncovered at the M 2 level of theory, i.e., they are not available (n a) within Bogoliubov–de Gennes theory (M 1). Convergence with M to the exact results is clearly seen. Note the interchange of order of some center-of-mass and relative coordinates excitations in comparison with the repulsive system. All quantities are dimensionless.

laboratory frame, where all bosons are equivalent, i.e., the separability to center-of-mass and relative coordinates, which is special for harmonic traps, is not exploited.

We now move to attractive interaction and compute the excitation spectrum of N 1000 bosons with K 0.0001. The results are collected in Table 2 and show, as above, the capability of LR-MCTDHB to uncover the missing excitations and to numerically-exactly converge to the analytical results. The physical distinction of the missing excitations having one quantum of center-of-mass excitation plus at least one additional quantum of excitation is readily seen in Table 2. When comparing the repulsive and attractive systems, please note the interchange of order of some center-of-mass and relative coordinates excitations.

4. Concluding remarks

The linear-response theory of the multiconfigurational time-dependent Hartree for bosons method for computing many-body excitations of trapped Bose-Einstein condensates has been implemented for systems with general interparticle interaction. This allows us to investigate the excitation spectrum of interacting bosons with, for instance, long-range interaction. As illustrative examples we considered, separately, repulsive and attractive bosons within the one-dimensional harmonic-interaction model. The many-body theory improves the accuracy of excitations unveiled within Bogoliubov–de Gennes theory. Chie y, the many-body linear-response theory is capable of identifying all excitations, including the excitations which are not unraveled within Bogoliubov–de Gennes equations, and to numerically converge to their exact values. The results of the present work serve to benchmark the LR-MCTDHB method.

As a complementary result, we compactly represent the theory in a block-diagonal form. This is made possible by identifying a complex transformation, applicable also for complex wavefunctions and complex hermitian Hamiltonians. This will open up an avenue for treating larger systems. We expect the LR-MCTDHB theory and its implementation for general interparticle interaction to provide an important and proved probe into the many-body excitations involved in the out-of-equilibrium dynamics of trapped BECs.

Acknowledgments

I am grateful to Alexe Streltsov and Lorenz Cederbaum for stimulating and continuous discussions.

Appendix A. Block diagonalization by a complex transformation

Consider the eigenvalue system

$$\begin{pmatrix} \mathbf{A} & \mathbf{B} \\ -\mathbf{B}^* & -\mathbf{A}^* \end{pmatrix} \begin{pmatrix} \mathbf{u} \\ \mathbf{v} \end{pmatrix} \quad \omega \begin{pmatrix} \mathbf{u} \\ \mathbf{v} \end{pmatrix}, \tag{A.1}$$

where $\mathbf{A}, \mathbf{B}$ and $\mathbf{u}, \mathbf{v}$ are the blocks of a square matrix and its eigenvector, respectively. The eigenvalue ω is assumed to be real.

Let us examine the transformation matrix $\mathbf{Q} \quad \frac{1}{\sqrt{2}} \begin{pmatrix} 1 & 1\theta \\ 1 & -1\theta \end{pmatrix}$, Eq. (7) of the main text. Interestingly, the transformation matrix $\mathbf{Q}$ is neither a unitary nor an anti-unitary operator. Its inverse exists and reads

$$\mathbf{Q}^{-1} \quad \frac{1}{\sqrt{2}} \begin{pmatrix} 1 & 1 \\ 1\theta & -1\theta \end{pmatrix}, \qquad \mathbf{Q}\mathbf{Q}^{-1} \quad \mathbf{Q}^{-1}\mathbf{Q} \quad \begin{pmatrix} 1 & 0 \\ 0 & 1 \end{pmatrix}. \tag{A.2}$$

Multiplying with $\mathbf{Q}$ from the left on both sides of Eq. (A.1) we find

$$\left\{\mathbf{Q} \begin{pmatrix} \mathbf{A} & \mathbf{B} \\ -\mathbf{B}^* & -\mathbf{A}^* \end{pmatrix} \mathbf{Q}^{-1}\right\} \left\{\mathbf{Q} \begin{pmatrix} \mathbf{u} \\ \mathbf{v} \end{pmatrix}\right\} \quad \begin{pmatrix} \mathbf{0} & \mathbf{A} - \mathbf{B}\theta \\ \mathbf{A} \quad \mathbf{B}\theta & \mathbf{0} \end{pmatrix} \begin{pmatrix} \mathbf{f} \\ \mathbf{g} \end{pmatrix} \quad \omega \begin{pmatrix} \mathbf{f} \\ \mathbf{g} \end{pmatrix}, \tag{A.3}$$

where $\begin{pmatrix} \mathbf{f} \\ \mathbf{g} \end{pmatrix} \quad \begin{pmatrix} \frac{1}{\sqrt{2}}(\mathbf{u} \quad \mathbf{v}^*) \\ \frac{1}{\sqrt{2}}(\mathbf{u} - \mathbf{v}^*) \end{pmatrix}$. Multiplying now Eq. (A.3) from the left with the transformed matrix we obtain the desired result in a block-diagonal form

$$\begin{pmatrix} (\mathbf{A} - \mathbf{B}\theta)(\mathbf{A} \quad \mathbf{B}\theta) & \mathbf{0} \\ \mathbf{0} & (\mathbf{A} \quad \mathbf{B}\theta)(\mathbf{A} - \mathbf{B}\theta) \end{pmatrix} \begin{pmatrix} \mathbf{f} \\ \mathbf{g} \end{pmatrix} \quad \omega^2 \begin{pmatrix} \mathbf{f} \\ \mathbf{g} \end{pmatrix}. \tag{A.4}$$

This concludes our derivation.

Using the operation of complex con ugation θ above becomes redundant if $\mathbf{A}$ and $\mathbf{B}$ are real quantities. In this specific case, which is the only one considered within Bogoliubov–de Gennes equations in the literature, see, e.g., [31-33], $\mathbf{Q}$ becomes a unitary and real transformation.

References

[1] Pitaevskii L and Stringari S 2003 *Bose-Einstein Condensation* (Oxford University Press, New York)
[2] Leggett A J 2008 *Quantum Liquids: Bose Condensation and Cooper Pairing in Condensed-Matter Systems* (Oxford University Press, New York)
[3] Pethick C J and Smith H 2008 *Bose-Einstein Condensation in Dilute Gases, 2nd ed.* (Cambridge University Press, New York)
[4] Yukalov V I 2011 *Phys. Part. Nuc.* **42** 460
[5] Proukakis N P, Gardiner S A, Davis M J and Szymanska M H (eds) 2013 *Quantum Gases: Finite Temperature and Non-Equilibrium Dynamics, Vol. 1 Cold Atoms Series* (Imperial College Press, London)
[6] Bogoliubov N N 1947 *J. Phys. USSR* **11** 23, reprinted in Pines D 1961 *The Many-body Problem* (W. A. Benjamin, New York)
[7] Ruprecht P A, Edwards M, Burnett K and Clark C W 1996 *Phys. Rev. A* **54** 4178
[8] Esry B D 1997 *Phys. Rev. A* **55** 1147
[9] Gardiner C W 1997 *Phys. Rev. A* **56** 1414
[10] Castin Y and Dum R 1998 *Phys. Rev. A* **57** 3008
[11] Grond J, Streltsov A I, Lode A U J, Sakmann K, Cederbaum L S and Alon O E 2013 *Phys. Rev. A* **88** 023606
[12] Cederbaum L S and Streltsov A I 2003 *Phys. Lett. A* **318** 564
[13] Grond J, Streltsov A I, Cederbaum L S and Alon O E 2012 *Phys. Rev. A* **86** 063607
[14] Alon O E, Streltsov A I and Cederbaum L S 2014 *J. Chem. Phys.* **140** 034108
[15] Fetter A L and Walecka J D 1971 *Quantum Theory of Many-Particle Systems* (McGraw-Hill, New York)
[16] Olsen J and Jørgensen P 1985 *J. Chem. Phys.* **82** 3235
[17] Streltsov A I, Alon O E and Cederbaum L S 2007 *Phys. Rev. Lett.* **99** 030402
[18] Alon O E, Streltsov A I and Cederbaum L S 2008 *Phys. Rev. A* **77** 033613
[19] Sakmann K, Streltsov A I, Alon O E and Cederbaum L S 2009 *Phys. Rev. Lett.* **103** 220601
[20] Grond J, Schmiedmayer J and Hohenester U 2009 *Phys. Rev. A* **79** 021603(R)
[21] Streltsov A I, Alon O E and Cederbaum L S 2009 *J. Phys. B* **42** 091004
[22] Streltsov A I, Alon O E and Cederbaum L S 2011 *Phys. Rev. Lett.* **106** 240401
[23] Březinová I, Lode A U J, Streltsov A I, Alon O E, Cederbaum L S and Burgdörfer J 2012 *Phys. Rev. A* **86** 013630
[24] Heimsoth M, Hochstuhl D, Creffield C E, Carr L D and Sols F 2013 *New J. Phys.* **15** 103006
[25] Hochstuhl D, Hinz C M and Bonitz M 2014 *Eur. Phys. J. Special Topics* **223** 177
[26] Březinová I, Burgdörfer J, Lode A U J, Streltsov A I, Cederbaum L S, Alon O E, Collins L A and Schneider B I 2014 *J. Phys.: Conf. Ser.* **488** 012032
[27] Lode A U J, Sakmann K, Alon O E, Cederbaum L S and Streltsov A I 2012 *Phys. Rev. A* **86** 063606
[28] Streltsov A I 2013 *Phys. Rev. A* **88** 041602(R)
[29] Streltsova O I, Alon O E, Cederbaum L S and Streltsov A I 2014 *Phys. Rev. A* **89** 061602(R)
[30] The blocks of the linear-response matrix (6) are given by

$$\boldsymbol{\mathcal{L}}^u = \begin{pmatrix} \mathbf{P}\boldsymbol{\rho}^{+\frac{1}{2}}\boldsymbol{\mathcal{L}}^u_{oo}\boldsymbol{\rho}^{+\frac{1}{2}}\mathbf{P} & \mathbf{P}\boldsymbol{\rho}^{+\frac{1}{2}}\boldsymbol{\mathcal{L}}^u_{oc}\mathbf{P}_C \\ \mathbf{P}_C(\boldsymbol{\mathcal{L}}^u_{oc})^\dagger\boldsymbol{\rho}^{+\frac{1}{2}}\mathbf{P} & \mathbf{P}_C(\cdot)^{\hat{H}-\varepsilon}\mathbf{P}_C \end{pmatrix}, \quad \boldsymbol{\mathcal{L}}^v = \begin{pmatrix} \mathbf{P}\boldsymbol{\rho}^{+\frac{1}{2}}\boldsymbol{\mathcal{L}}^v_{oo}(\boldsymbol{\rho}^*)^{+\frac{1}{2}}\mathbf{P}^* & \mathbf{P}\boldsymbol{\rho}^{+\frac{1}{2}}\boldsymbol{\mathcal{L}}^v_{oc}\mathbf{P}^*_C \\ \mathbf{P}_C(\boldsymbol{\mathcal{L}}^v_{oc})^t(\boldsymbol{\rho}^*)^{+\frac{1}{2}}\mathbf{P}^* & \mathbf{0} \end{pmatrix}.$$

The meaning of the ingredients is as follows: $\mathbf{P}$ is a projector matrix on the subspace of one-particle functions orthogonal to the ground-state modes ϕ_j; $\mathbf{P}_C$ is a projector matrix on the subspace of permanents' coefficients orthogonal to the ground-state coefficients $C_{\vec{n}}$; $\boldsymbol{\rho}$ is the reduced one-particle density matrix [38] of the ground state; and $(\cdot)^{\hat{H}-\varepsilon}$ marks the operation of the Hamiltonian in Fock space utilizing mapping of permanents [39] and ε is the ground-state energy. Finally, the sub-blocks $\boldsymbol{\mathcal{L}}^u_{oo}$, $\boldsymbol{\mathcal{L}}^v_{oo}$ and $\boldsymbol{\mathcal{L}}^u_{oc}$, $\boldsymbol{\mathcal{L}}^v_{oc}$ collect, respectively, the couplings between the modes and the couplings between the modes and coefficients. Together with further details, they are given in [11, 14].
[31] Hutchinson D A W, Zaremba E and Griffin A 1997 *Phys. Rev. Lett.* **78** 1842
[32] Huepe C, Tuckerman L S, Métens S and Brachet M E 2003 *Phys. Rev. A* **68** 023609
[33] Ronen S, Bortolotti D C E and Bohn J L 2006 *Phys. Rev. A* **74** 013623
[34] Cohen L and Lee C 1985 *J. Math. Phys.* **26** 3105
[35] Gajda M, Załuska-Kotur M A and Mostowski J 2000 *J. Phys. B* **33** 4003
[36] Yan J 2003 *J. Stat. Phys.* **113** 623
[37] Gajda M 2006 *Phys. Rev. A* **73** 023603
[38] Coleman A J and Yukalov V I 2000 *Reduced Density Matrices: Coulson's Challenge, Vol. 72 Lectures Notes in Chemistry* (Springer, Berlin)
[39] Streltsov A I, Alon O E and Cederbaum L S 2010 *Phys. Rev. A* **81** 022124

Numerical simulation of nonequilibrium states in a trapped Bose-Einstein condensate

A N Novikov[1,2], V I Yukalov[1,2] and V S Bagnato[2]

[1] Bogolubov Laboratory of Theoretical Physics, Joint Institute for Nuclear Research, Dubna 141980, Russia

[2] Instituto de Fisica de São Carlos, Universidade de São Paulo, CP 369, 13560-970 São Carlos, São Paulo, Brazil

Abstract. In this work we present numerical study of a trapped Bose-Einstein condensate perturbed by an alternating potential. The relevant physical situation has been recently realized in experiment, where the trapped condensate of ^{87}Rb, being strongly perturbed, exhibits the set of spatial structures. Firstly, regular vortices are detected. Further, increasing either the excitation amplitude or modulation time results in the transition to quantum vortex turbulence, followed by a granular state. Numerical simulation of the nonequilibrium Bose-condensed system is based on the solution of the time-dependent 3D nonlinear Schrödinger equation within the static and dynamical algorithms. The damped gradient step and time split-step Fourier transform methods are employed. We demonstrate that computer simulations qualitatively reproduce the experimental picture, and describe well the main experimental observables.

1. Introduction

As is well known, trapped Bose atoms in equilibrium at zero temperature form Bose-Einstein condensate (BEC) in the ground state. This system has been thoroughly studied theoretically as well as experimentally. For details we refer to the review articles and books [1, 2, 3, 4] and references therein. If the condensate is slightly moved from equilibrium it usually exhibits collective excitations. Such dynamics is typical not only for BEC, but for the variety of quantum systems [5, 6, 7].

Nonequilibrium BEC can be created by injecting into the system energy, which can be realized by applying an external modulation field. Two main ways of moving trapped BEC far from equilibrium are usually considered. The first one is a resonant excitation of coherent modes [8, 9]. In this case the applied external perturbation is weak, but the modulation frequency of the alternating field is in resonance with one of the transition frequencies between topological modes. This method is a bit difficult for experimental realization, requiring fine tuning to resonance. The other way is to employ a strong external alternating field superimposed onto the stationary trap [10], where either the modulation time or excitation amplitude determine the energy injected into the system. In both these cases, BEC transfers from its ground to excited states, demonstrating a set of dynamical regimes. The second approach has been recently realized in the experiment [11, 12].

The experimental realization of nonequilibrium condensates stimulates interest to its numerical modeling and computer simulation [13]. Dealing with a trapped Bose-Einstein condensate, the nonlinear Schröedinger equation (NLSE), which, in BEC physics sometimes is called the Gross-Pitaevskii equation [1], is the main tool for the theoretical investigation. Nowadays, different numerical schemes [13, 14] have been developed and successfully applied for the solution of second order partial differential equations, like NLSE. The methods differ in their stability, accuracy, computational cost, and ability of keeping the NLSE original properties. In general, all such methods could be classified as explicit, implicit and finite Fourier transform or pseudospectral methods. Each numerical scheme has its own merits and disadvantages, determining the accuracy of the NLSE solution. The presence of either random forces or potentials, being a prerequisit of nonequilibrium systems, complicates the numerical procedure, since the relevant NLSE solution is not smooth.

In our recent paper [15] we presented the results of numerical simulation of a trapped condensate, under external perturbation, showing the appearance of a granular state, analogous to that observed in the experiment. The simulations were based on the solution of the NLSE, with the parameters corresponding to the experimental setup. In the present paper we give the details of the accomplished numerical modeling. Also, we demonstrate that the chosen numerical methods qualitatively describe the typical physical picture, observed in the experiment.

The paper is organized as follows. In Sec. 1, we briefly review the experiment on the strong perturbation of a trapped BEC by external modulating field. Details of computer simulation are given in Sec. 2. In Sec. 3, we present the numerically simulated time-amplitude diagram.

2. Experiment on perturbation of ^{87}Rb

In the experiment [11, 12], the ^{87}Rb atoms (atomic mass $m = 1.444^{-22}$ g, scattering length $a_s = 0.557 \times 10^{-6}$ cm), being cooled down to the temperature much lower than $T_c = 276$ nK, form the ground state Bose-Einstein condensate. The total number of atoms in the BEC fraction is $N \approx 1.5 \times 10^5$. Stationary trapping potential has cylindrical symmetry, with the trap frequencies: $\omega_r = 2\pi \times 210$ Hz, $\omega_z = 2\pi \times 23$ Hz. After the stationary BEC state, obeying the Thomas-Fermi shape, is achieved, the trapping potential is modulated by an additional external alternating field, oscillating with the frequency $\Omega_0 = 2\pi \times 200$ Hz. As a result, the total trap potential V_{ext} can be approximated as [12]:

$$\begin{aligned} V_{ext} = \frac{1}{2}(\lambda^2[x\cos\theta_1 + y\sin\theta_1 - z\sin\theta_2 - \delta_1(1-\cos\Omega_0 t)]^2 + \\ +[y\cos\theta_1 - x\sin\theta_1 - \delta_2(1-\cos\Omega_0 t)]^2 + \\ +[z\cos\theta_2 + x\sin\theta_2 - \delta_3(1-\cos\Omega_0 t)]^2), \end{aligned} \tag{1}$$

where $\lambda = \omega_r/\omega_z$, $\theta_i = A_i(1-\cos\Omega_0 t)$, $A_1 = \frac{\pi}{60}$, $A_2 = \frac{\pi}{120}$. The trapping frequency ω_r determines the oscillator length l_r that defines the modulation parameters of the potential: $(\delta_1, \delta_2, \delta_3) = \alpha(2,5,3)$ μm $/l_r$, where α is an effective amplitude of pumping.

If the external perturbation imposes a certain angular momentum onto the system, like in the case of the rotating [16] or laser-stirred condensate [17], the formed vortices are aligned along the rotation axis. The number of vortices depends on the modulation frequency. The aligned vortices can form a completely ordered lattice.

But in our case, as is seen from the expression (1), the modulation field used in the experiment does not impose a certain rotation axis. This feature makes the main difference of our experiment, as compared to the case of rotating BEC. Contrary to the rotated BEC, the external perturbation, we use, just shakes the condensate cloud, injecting the energy into the trap and thus generating the diverse coherent topological modes [10].

Mathematically, coherent topological modes are stationary solutions of NLSE, including the ground state and other excited states. Different modes correspond to rather different spatial distributions of the condensate density, forming a variety of spatial shapes. As a result, different structures can be generated in a trapped condensate [8, 9].

First modes, generated in the experiment, are quantum vortices. Since the external field does not inject a rotation moment, the vortices appear as vortex-antivortex pairs with the winding numbers plus/minus one These modes are energetically the most stable. The number of created vortices depends on the injected energy, hence can be controlled by either the pumping amplitude or by the modulation time. At a fixed amplitude, almost a linear dependence of the vortex number from time is observed, until the distance between the vortex centers is less than four coherence lengths, when the interaction energy of two vortices is much smaller than the vortex energy. When the number of vortices becomes sufficiently large, the transition to random vortex turbulence occurs.

The continuous energy pumping into the BEC cloud creates more vortices than the trap can host, as a result of which the vortices start strongly interacting and colliding, which destroys the regime of vortex turbulence. Then the system passes to the next dynamical regime, where the condensate forms high-density grains surrounded by sufficiently rarified gas. It can be demonstrated [15] that these grains are coherent objects.

3. Numerical simulation

Experimental characteristics are found to be in good agreement with analytical estimates [10, 15]. Now we aim at accomplishing numerical calculations for the same setup as in the experiment. In this section, we focus on the techniques of numerical modeling as applied to the solution of NLSE describing the condensate dynamics driven by external perturbations. The corresponding numerical simulation is realized in two steps: first, we calculate the parameters of the initial conditions describing the stationary state of the BEC in the trap, and then we consider time propagation. The latter is performed via the solution of the time-dependent NLSE for the desired time interval. In both these cases, the full 3D geometry is considered, and the wave function is determined on the 3D space grid. Below, the main points of the accomplished numerical simulations are illustrated.

3.1. Initial conditions

First, the wave function, describing the condensate stationary state, has to be computed. To this end, we consider the eigenproblem:

$$H(\varphi)\varphi_n = E_n\varphi_n, \tag{2}$$

where the index n enumerates quantum states and the corresponding wave functions φ_n and energies E_n. In equation (2), the nonlinear Hamiltonian $H(\varphi)$ is typical for a trapped BEC:

$$H(\varphi) = -\frac{\hbar^2}{2m}\nabla^2 + V_{ext} + gN|\varphi|^2 \,, \tag{3}$$

where $g = \frac{4\pi\hbar^2 a_s}{m}$ is the interaction strength depending on the scattering length a_s and atomic mass m, V_{ext} is the trapping potential. The parameters of the Hamiltonian (3) exactly correspond to the experimental setup. The BEC of ^{87}Rb atoms with the related a_s and m are considered. The total number of condensed particles is $N = 1.5 \times 10^5$. The external field V_{ext} is represented by the total trap potential in form (1). Since we interested in stationary solutions, the external perturbation is switched off, so that $V_{ext} \equiv V_{ext}(\mathbf{r}, t = 0)$.

The eigenproblem (2) can be solved by a variety of numerical schemes, but iterative methods are the most convenient. We used the damped gradient step method [18, 19], where the initial guess ϕ_n is improved by the iteration steps. Mathematically, this can be represented in the form:

$$\phi_n^{(i+1)} = \mathcal{O}[\phi_n^{(i)} - \mathcal{D}(H(\phi_n^i) - < H(\phi_n^i >)\phi_n^i] \,, \tag{4}$$

where $\mathcal{D}$ is a damping operator, the symbol $\mathcal{O}$ labels the orthonormalization of the set of functions ϕ_n at each iteration step marked as i. The Schmidt orthonormalization is usually employed. Hamiltonian (3) is recalculated after the each iteration step. The main problem of the numerical scheme (4) is the proper choice of the damping operator D providing the convergence of the iteration procedure. The damped gradient step method operates with the damping operator in the form

$$\mathcal{D} = \frac{\delta}{T + E_0} , \tag{5}$$

where δ is a parameter controlling the iteration step size, $T = -\frac{\hbar^2}{2m}\nabla^2$ is the operator of kinetic energy. Since kinetic energy can happen to be close to zero, parameter E_0 stabilizes the iteration process.

If the procedure converges, the final set of ϕ_n, obtained after a number of iteration steps, corresponds to the eigenfunctions $\varphi_n(\mathbf{r})$ with the spectra E_n.

After the eigenproblem (2) is solved, the total wave function $\Psi(\mathbf{r})$ can be written as a superposition of the states

$$\Psi(\mathbf{r}) = \sum_n C_n \varphi_n(\mathbf{r}) , \tag{6}$$

and then normalized to the total number of condensed atoms

$$N = \int |\Psi(\mathbf{r})|^2 d\mathbf{r} . \tag{7}$$

At $t = 0$, there is no external perturbation, thus the condensate is in equilibrium state corresponding to the lowest energy E_0. To take this into account, we apply the condition

$$C_0 >> C_{n>0} \tag{8}$$

to the notation (6).

Representation (6) of the wave function corresponds to its expansion over the basis of coherent topological modes [9], which allows one to treat non-ground-state condensates [8].

3.2. *Time evolution*

The static solution is followed by the time propagation of the condensate wave function $\psi(\mathbf{r})$. For this purpose, we solve numerically the nonlinear 3D time-dependent NLSE

$$i\hbar\frac{\partial}{\partial t}\Psi(\mathbf{r},t) = [-\frac{\hbar^2}{2m}\nabla^2 + V_{ext}(\mathbf{r}) + g|\Psi(\mathbf{r},t)|^2]\Psi(\mathbf{r},t) , \tag{9}$$

using the obtained $\Psi(\mathbf{r})$ as the initial condition. Similarly to the previous case, the parameters of NLSE exactly correspond to the experimental setup. To solve NLSE, the time split-step Fourier transform (TSSFT) method [20, 21] is chosen. The advantage of this numerical scheme [13] is that it is unconditionally stable and keeps the main properties of NLSE, such as mass conservation, time reversibility, i.e., time-inverse invariance. Also, the method is second-order accurate in both time and space.

The main idea of the TSSFT is in splitting the time-propagation between the kinetic and potential components of the total energy. Both these components can be approximated by exponents related to the solutions of the linear Schrödinger equation (see Ref. [21] for details). Mathematically, the mechanism of time propagation has the following form:

$$\Psi(\mathbf{r}, t + \Delta t) \approx \exp(iT\Delta t/2)\exp(-iV\Delta t)\exp(iT\Delta t/2)\Psi(\mathbf{r},t) , \tag{10}$$

where $T = -\frac{\hbar^2}{2m}\nabla^2$ and $\widehat{V} = V_{ext} + g|\Psi(\mathbf{r},t)|^2$ are the operators of kinetic and potential energy, respectively.

The relation between the exponential operators obeys the Baker-Cambell-Hausdorf theorem, which states that $e^A e^B = e^C$ is valid if $C = A + B + [A, B] + ...$ Since the TSSFT method skips the commutation relations of the kinetic and potential operators, numerical error grows with time. The symmetrical form of (10), with respect to the operators, allows to remove the first commutator term, increasing the accuracy of time propagation [22].

The exponential derivative operator is easily calculated in the momentum space rather than in the coordinate space, because of the well known relation

$$\exp\left(\frac{d^2}{d\mathbf{q}^2}\right)\Psi(\mathbf{q},t) = \exp\left(-k^2\right)\mathcal{F}[\Psi(\mathbf{q},t)] \ , \tag{11}$$

where symbol $\mathcal{F}$ means the Fourier transform, k is a variable conjugated to q.

Taking into account expressions (10) and (11), the full time propagation of the condensate wave function for the time step Δt is performed in three stages:

(i) The Fourier transform of the initial function $\Psi(\mathbf{r},t)$ is followed by the half time-step $\Delta t/2$, with the operator of kinetic energy in the momentum space. The first stage is completed by the inverse Fourier transform that results in the new function $\Psi(\mathbf{r},t) \rightarrow \tilde{\Psi}(\mathbf{r},t)$.

(ii) Recalculation of the potential operator V with the function $\tilde{\Psi}(\mathbf{r},t)$ and further full-time step propagation of $\tilde{\Psi}(\mathbf{r},t)$, with the recalculated term V. As a result, the transformation $\tilde{\Psi}(\mathbf{r},t) \rightarrow \hat{\Psi}(\mathbf{r},t)$ is obtained.

(iii) The final stage equals the first step, but with the function $\hat{\Psi}(\mathbf{r},t)$. This completes the time propagation, since $\hat{\Psi}(\mathbf{r},t) \rightarrow \Psi(\mathbf{r},t+\Delta t)$.

The time step Δt should be small enough to keep the accuracy of time propagation. The increment Δt fulfills the condition [23]

$$< \Psi|H|\Psi > \ << \frac{2\pi}{\Delta t} \ . \tag{12}$$

3.3. Other numerical aspects

Dealing with a grid-based technique, the parameters of the computation box are extremely important. The size of the computational box strictly depends on the trapping frequencies, total number of condensed particles and the kind of boundary conditions. The proper size of the 3D grid avoids the unphysical truncation of the operated wave function and its reflection from the grid boundaries during the perturbation. The later is important, if reflecting boundary conditions are employed. In the present simulation, we used the reflecting boundary conditions for both, the static and dynamical simulations. In the experiment, the applied external perturbation heats up the BEC leading to the transition from condensed to thermal phase for a part of atoms. This process is almost regular in time. Consequently, the number of condensed particles in the trap permanently decreases. To follow the experiment, we introduced into the system a small dissipation replacing the term $i\hbar\frac{\partial}{\partial t}\Psi(\mathbf{r},t)$ in the NLSE (9) by $(i-\gamma)\hbar\frac{\partial}{\partial t}\Psi(\mathbf{r},t)$, where γ is a phenomenological term. The same approach is also used by the other authors [24]. In simulations, during the 60 ms of perturbation, approximately 15 percents of the initial condensed particles are lost, which is in good agreement with the experiment. Of course, introducing the dissipation term γ does not conserve the initial norm of the wave function (7), which implies the diminishing number of condensed atoms.

4. Results and discussion

Computer simulations reproduce well the experimental results. The applied external perturbation, modelled by (1) creates, first, dipole oscillations of the condensate cloud, followed by vortex structures corresponding to vortex-antivortex pairs. The vortex nature of the observed structures is confirmed by quantized circulation having the values plus or minus one. The circulation is calculated by the method proposed in Ref [25]. Increasing either the amplitude of pumping or the excitation time produces more and more vortices, creating the random vortex tangle, representing, according to Feynman [26], quantum turbulence. The next regime observed in the simulations is grain turbulence, where continuous trap modulation destroys all the vortices, forming high-density grains surrounded by rarified gas. The typical physical picture, obtained in numerical simulations, is represented by the time-amplitude diagram shown in Fig. 1. The experimental diagram, published in Ref [12], is in qualitative agreement with that obtained in the numerical simulation. The area inside the rectangle reproduces the experiment in the best way.

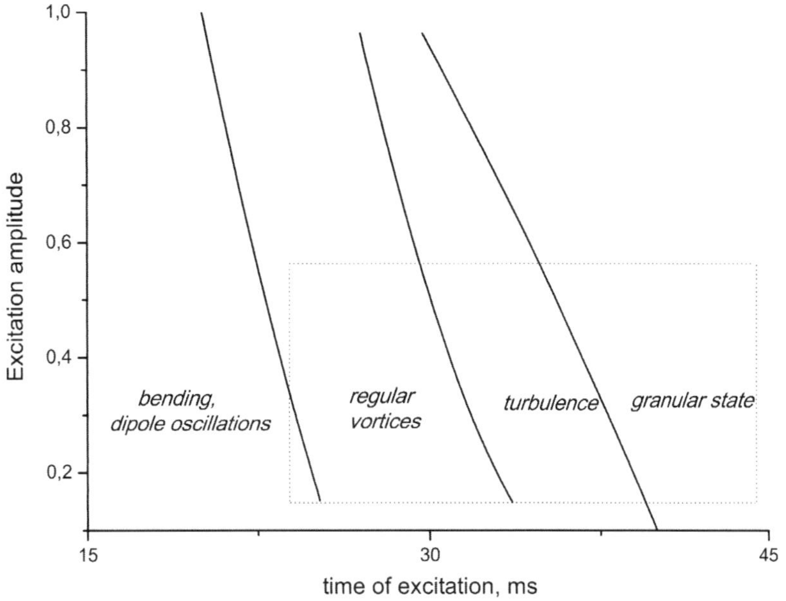

Figure 1. The time-amplitude diagram, as found in numerical simulations. The area inside the rectangle is in the best agreement with the experiment.

Numerical simulations give very good quantitative values for the main characteristics observed in the experiment and also found from theoretical arguments. Thus, the critical number of vortices $N_{cr} \approx 25$, required for creating the regime of quantum turbulence, is the same in the experiment, numerical simulations, as well as in theoretical estimates. The grains can be considered as coherent objects, since their typical linear size practically coincides with the coherence length, as is found in the experiment and in the simulation.

In numerical simulations, we also observe one more regime [15], arising after the granular state, the regime of wave turbulence, when all BEC is destroyed, and turbulence is formed by small-amplitude waves. Here we compare the experimental and numerical results, because of which the regime of wave turbulence is not included in the diagram, since it has not yet been reached in experiment.

5. Acknowledgments
Financial support from FAPESP and CNPq (Brazil) is appreciated. Two of the authors (V.I.Y. and A.N.N.) acknowledge financial support from the Russian Foundation for Basic Research, under project 14-02-00723.

6. References
[1] Pitaevskii L and Stringari S 2003 *Bose-Einstein Condensation* (Oxford: Clarendon)
[2] Pethick C J and Smith H 2008 *Bose-Einstein Condensation in Dilute Gases* (Cambridge: Cambridge University)
[3] Courteille P W, Bagnato V S and Yukalov V I 2001 *Laser Phys.* **11** 659
[4] Yukalov V I 2011 *Phys. Part. Nucl.* **42** 460
[5] Birman J L, Nazmitdinov R G and Yukalov V I 2013 *Phys. Rep* **526** 1
[6] Yannouleas C and Landman U 2007 *Rep. Prog. Phys.* **70** 2067
[7] Saarikoski H, Reimann S M, Harju A and Manninen M 2010 Rev. Mod. Phys. **82** 2785
[8] Yukalov V I, Yukalova E P and Bagnato V S 1997 *Phys. Rev. A* **56** 4845
[9] Yukalov V I, Yukalova E P and Bagnato V S 2002 *Phys. Rev. A* **66** 043602
[10] Bagnato V S and Yukalov V I 2013 *Prog. Opt. Sci. Photon.* **1** 377
[11] Shiozaki R F, Telles G D, Yukalov V I and Bagnato V S 2011 *Laser Phys. Lett.* **8** 393
[12] Seman J A, Henn E A L, Shiozaki R F, Roati G, Poveda-Cuevas F J, Magalhães K M F, Yukalov V I, Tsubota M, Kobayashi M, Kasamatsu K and Bagnato V S 2011 *Laser Phys. Lett.* **8** 691
[13] Antoine X, Bao W and Besse C 2013 *Comput. Phys. Comm.* **184** 2621
[14] Taha T R and Ablowitz M J 1984 *J. Comput. Phys.* **55** 203
[15] Yukalov V I, Novikov A N and Bagnato V S 2014 *Laser Phys. Lett.* **11** 095501
[16] Fetter A L 2009 *Rev. Mod. Phys.* **81** 647
[17] Madison K W, Chevy F, Wohlleben W and Dalibard J 2000 *Phys. Rev. Lett.* **84** 806
[18] Blum V, Lauritsch G, Maruhn J A and Reinhard P G 1992 *J. Comput. Phys.* **100** 364
[19] Reinhard P G and Cusson R Y 1982 *Nucl. Phys. A* **378** 418
[20] Hardin R H and Tappert F D 1973 *SIAM Rev. Chronicle* **15** 423
[21] DeVries P L 1987 *AIP Conf. Proc.* **160** 269
[22] Fleck J, Morris J K and Feit M J 1976 *Appl. Phys.* **10** 129
[23] Feit M D, Fleck J A and Steiger A 1982 *J. Comput. Phys.* **47** 412
[24] Tsubota M, Kobayashi M and Takeushi H 2013 *Phys. Rep.* **522**, 191
[25] Foster C J, Blakie P B and Davis M J 2010 *Phys. Rev. A* **81** 023632
[26] Feynman R P 1955 *Porg. Low Temp. Phys.* **1**, 17

Rydberg excitation of a Bose-Einstein condensate

N Malossi[a,b,1], **M M Valado**[a,b], **E Arimondo**[a,b,c], **O Morsch**[a,b], **and D Ciampini**[a,b,c]

[a] Dipartimento di Fisica 'E. Fermi', Università di Pisa, Largo Pontecorvo 3, 56127 Pisa, IT
[b] INO-CNR, Via G. Moruzzi 1, 56124 Pisa, IT
[c] CNISM UdR Pisa, Dipartimento di Fisica 'E. Fermi', Università di Pisa, Largo Pontecorvo 3, 56127 Pisa, IT

E-mail: ciampini@df.unipi.it

Abstract. The excitation to high-lying Rydberg states is investigated in a sample of 87-Rb atoms undergoing the Bose-Einstein condensation phase transition, in the limit that the spatial dimensions of the condensed cloud are smaller than the (Wan-der-Waals) dipole blockade radius. We measure the number of excitations by single ion counting as a function of the dimensions of the condensed and thermal fraction. We observed a distinct decrease in the number of excitations due to the change of the atomic spatial density distribution, from the thermal cloud, through a bimodal distribution, to the condensed cloud. When only the condensed part is present, the average number of excitations measured levels off at around one which is compatible with having a single collective excitation in the condensed cloud.

1. Introduction

In recent years the excitation of atoms to high lying Rydberg states in quantum gases has been widely investigated both in theory and experiments. When atoms are excited to high-lying Rydberg states, a strong interaction arises between them via the dipole-dipole or van-der-Waals potential [1, 2, 3] that suppresses the resonant optical excitation of more than one Rydberg atom inside a certain blockade volume (dipole or Wan-der-Waals blockade) [3, 4]. In a simplified model the Rydberg excitation is shared by the atoms inside the blockade volume, generating a collective excitation of the atoms. The Rydberg blockade is at the heart of many quantum information processing schemes [5, 6] including quantum gates between individual atoms [7, 8, 9, 10] or between collective excitation in many-body systems [11]. The feasibility of such schemes relies on the coherent manipulation of a single collective excitation in a cold atomic cloud [12] and on the scalability of the system, which is related to creating arrays of excitations like in arrays of single atoms [13] (e.g. excitations in a Mott-insulator) or in arrays of many-body systems.

In this context the studies of Rydberg excitation of a paradigmatic many-body system like Bose-Einstein Condensates (BECs) in different geometries (3D clouds, cigar shapes and optical lattices) is of primary importance. The Rydberg excitation of an atomic gas undergoing the BEC phase transition was studied for the first time in [14] where it was demonstrated that due to the blockade effect the change in the atomic density distribution during the phase transition is reflected in the change of the fraction of atoms excited to a Rydberg state. Our group performed in [15] the Rydberg excitation in a Rb-BEC in a quasi-one-dimensional geometry and

[1] Present address: Scuola di Scienze e Tecnologie, Sezione di Fisica, Università di Camerino, I-62032 (MC), IT.

in one-dimensional optical lattices, demonstrating the formation of a chain of collective Rydberg excitations compatible with the blockade radius and the spatial length of the one-dimensional cloud. The Rydberg excitation of a condensate in the Mott-insulator phase has been examined by [13] without exploring the condensate transition phase. More recently, the excitation of a single collective state to a high-lying Rydberg state inside a BEC [16] was used to demonstrate the strong coupling between a single Rydberg electron and the whole condensate.
In the present work the excitation to a high-lying Rydberg state is investigated in an atomic sample undergoing the BEC phase transition, composed of thermal and condensed clouds, in a regime where the dimensions of the condensed cloud are smaller than the Rydberg blockade sphere. We measure the number of excitations by field ionizing the Rydberg atoms and counting the resulting ions as a function of the dimensions of the thermal and condensed clouds, which are controlled by evaporative cooling. We observe a distinct decrease in the number of Rydberg excitations across the phase transition when the spatial density is bimodal (thermal and condensed fraction); well below the critical temperature, when only the condensed fraction is detectable, we measure around one excitation, which is compatible with the expected average of a single collective excitation in the condensed cloud.

2. Experimental realization

In our experiment we create ultra-cold and condensed clouds of 87-Rb atoms in an optical dipole trap. The apparatus and the experimental procedure are described in [17]. Evaporative cooling is performed by ramping down linearly the power of the two trapping beams ($w = 70\,\mu$m gaussian waist, crossing at right angle). The number of atoms is measured after a free fall of $t_{tof} = 18.3\,\mu$s by absorption imaging and the density distribution is fitted by a bimodal gaussian distribution in order to obtain both the thermal and the condensate fractions. The number of atoms in the cloud decreases from $N^{at} \simeq 2.5 \times 10^5$ thermal atoms when the mean trap frequency is $\langle\nu\rangle \approx 80\,$Hz to $N^{at} \simeq 8 \times 10^4$ condensed atoms, with negligible thermal fraction, for $\langle\nu\rangle \approx 65\,$Hz. Decreasing further the dipole trap power leads to a reduction of the number of atoms in the BEC. The mean size of the condensate $\langle\sigma_C\rangle$ and of the thermal component $\langle\sigma_T\rangle$ inside the optical trapping potential have been measured as a function of the average trap frequency. The average in-trap size has been obtained from the relations $\langle\sigma_{C,T}\rangle = (\sigma_{C,T\,x}(0)^2\sigma_{C,T\,z}(0))^{1/3}$ for our elliptical condensate cloud and

$$\sigma_{C,T\,i}(0) = \frac{\sigma_{C,T\,i}(t_{tof})}{\sqrt{1+\omega_i^2 t_{tof}^2}} \tag{1}$$

where $\omega_i = 2\pi\nu_i$ is the trap frequency and $\sigma_{C,T\,i}(t)$ is the Gaussian width (along the $i-$direction) of the density distribution $n_{C,T}(t)$ of the condensate or the thermal cloud, measured after time of flight

$$n_{C,T}(t) = n_0\, e^{-\left[\frac{x^2+y^2}{\sigma^2_{C,T\,x}}+\frac{z^2}{\sigma^2_{C,T\,z}}\right]} \tag{2}$$

The volumes of the condensed and the thermal fraction inside the optical trap are then calculated as $V_{C,T} = \frac{2}{3}\pi\langle\sigma_{C,T}\rangle^3$ and normalized to the volume of a superatom $V_{sa} = \frac{2}{3}\pi R_{db}^3$ with the dipole blockade radius $R_{db} \simeq 10\,\mu$m of the $71S$ state as derived from the data of ref. [4]. The results for the volumes of the thermal and condensate fractions (open circles) are plotted in Fig. 1(a). Since the BEC volume is always smaller than the thermal cloud volume, its volume becomes relevant only for $\langle\nu\rangle \leq 65\,$Hz, when the thermal fraction is negligible. As shown in Fig. 1(a), the volume of the condensate remains approximately constant in the explored frequency range.

The ground state atoms are resonantly excited to the $71S$ Rydberg state using a two-photon excitation process with detuning via the intermediate $6P$ level (see [18, 19, 20] for details).

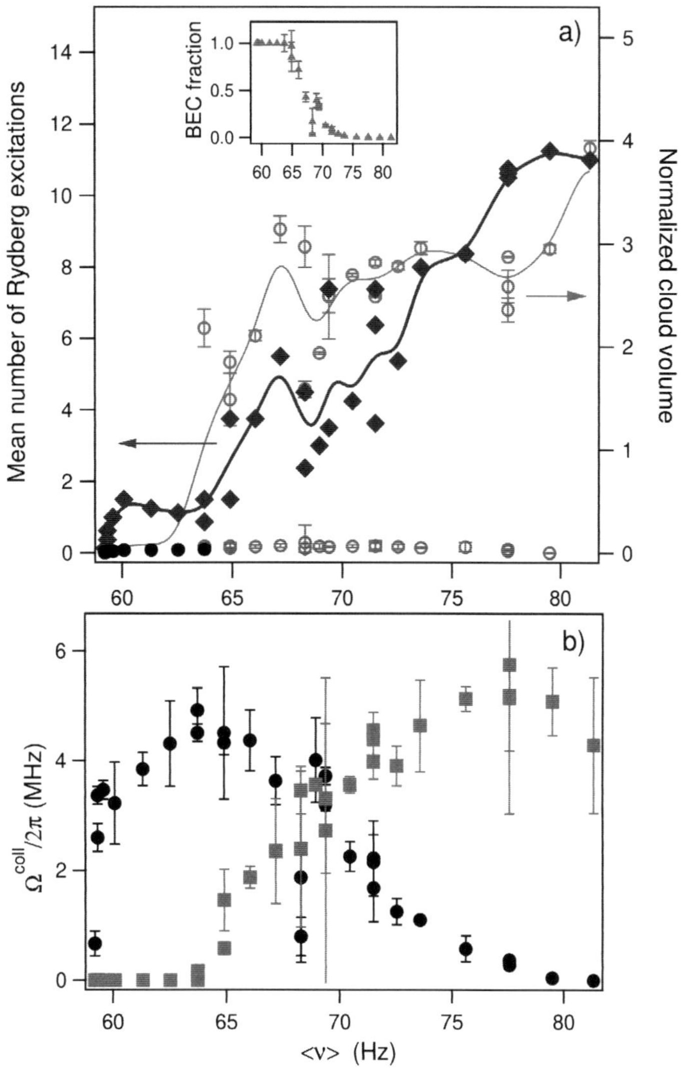

Figure 1. In (a) the measured mean number of Rydberg excitations (full blue diamonds and left scale), calculated normalized volume (right scale) of the thermal fraction (open red circles) and of the condensate fraction (open gray circles and closed black circles) as a function of the mean trapping frequency. At low trapping frequency, where the thermal fraction is negligible, the total cloud volume coincides with the volume of the condensate (closed black circles). The continuous blue line is obtained by smoothing the mean number of excitations, the continuous red line is obtained by smoothing the total cloud volume (open red circles and closed black circles) and both serve as a guide for the eyes. The insert shows the BEC fraction as a function of mean trapping frequency. In (b) collective Rabi frequencies Ω^{coll} vs mean trapping frequency calculated for the atoms in the condensate fraction (full black circles) and the thermal fraction (full red squares), including the correction for the gravitational sag.

Rubidium atoms in the $71S$ state have repulsive interactions, suppressing unwanted collisions and ionization. In the Rydberg excitation scheme the excitation lasers are detuned by $\Delta = 1.9\,\mathrm{GHz}$ from the $6P$ state, which can therefore be adiabatically eliminated and the laser interaction will be described by a two-level effective system with Rabi frequency Ω. The first step laser at 420 nm is sent through a single mode optical fiber and the Gaussian beam waist when reaching the atoms is $100\,\mu$m. The second step laser at 1013 nm has a Gaussian waist of $110\,\mu$m. Both beams are aligned with respect to the minimum of the combined optical trap potential and gravitational potential when the trap frequency is high and the gravitational sag is negligible. For the lowest trap frequencies explored in this experiment, the gravitational sag is approximately $z_{sag} = -40\,\mu$m, comparable to the Gaussian waists of the Rydberg lasers. The Rabi frequency of the Rydberg excitation has been corrected taking into account the decrease of the local laser intensity at the position of the atoms. For the atoms in the highest laser intensity region of the Rydberg lasers $\Omega/2\pi$ is 20 kHz. After the excitation laser pulse of duration $\tau_{exc} = 0.5\,\mu$s an electric field is applied for $2\,\mu$s in order to field ionize the Rydberg atoms and to accelerate the resulting ions towards a channeltron where they are individually counted. The overall detection efficiency is $\eta \approx 40\%$ [21]. In the following, the number of Rydberg excitations is derived from the measured number of detected ions taking into account the finite detection efficiency. The excitation-detection cycle is applied 20 times to the same atomic sample followed by absorption imaging of the remaining ground state atoms. Since the excitation time is shorter than the coherence time of the laser (limited by the finite linewidth of the excitation laser of approximately 0.5 MHz), the full atomic dynamics takes place in the coherent regime.

3. Results

Because of dipole blockade effect, the ensemble of N_{db}^{at} atoms within the $V_{sa} \approx 2 \times 10^3\,\mu\mathrm{m}^3$ $71S$ superatom volume characterized by a collective atomic excitation, is driven by the collective Rabi frequency $\Omega^{coll} = \sqrt{N_{db}^{at}}\Omega$ [3, 12]. The resonant excitation timescale is given by Ω^{coll}, and depends on the Rydberg laser intensity and on the atom density. Since the size of the condensate cloud is smaller than the dipole blockade radius, all the condensate atoms belong to a single superatom and the collective Rabi frequency $\Omega_C^{coll} = \sqrt{N_C^{at}}\Omega$ is in the MHz range, as shown in Fig. 1 (b). Since the size of the thermal cloud is larger than the dipole blockade radius, within the superatom picture more than one excitation can be accomodated in the thermal cloud. The collective Rabi frequency for the excitations of the thermal cloud is plotted in Fig. 1(b) and is calculated considering the average density of thermal atoms, including corrections for the gravitational sag.

We measured the full counting statistics of the Rydberg excitations as in [21], and their mean number is shown in Fig. 1(a) as a funcion of the mean trapping frequency. The smoothing lines reported in Fig. 1(a) as a guide for the eyes for the number of the Rydberg excitations and for the normalised volume evidence the correlations between the two data sets. Below 65 Hz the number of Rydberg excitations reaches a plateau even though the number of atoms in the condensate decreases by a factor of ~ 3 from 65 Hz to 60 Hz trap frequency. We interpret this as the onset of the single excitation regime in which a single superatom contained inside the condensed fraction (which is smaller than the blockade radius by a factor of 2.5) undergoes Rabi flopping between the ground and Rydberg states. The fact that we measure around one excitation rather than 0.5 could be due to the presence of an undetectable thermal fraction (i.e. fewer than 2×10^3 atoms) that can accommodate on average another excitation. For very low values of the trapping frequency, below 60 Hz, the number of Rydberg excitations goes to zero, since for such low number of BEC atoms the collective Rabi frequency becomes small and $\Omega_C^{coll} \cdot \tau_{exc} \ll 1$. As expected, above 65 Hz the increasing volume of the thermal fraction and the increase in Ω_T^{coll} lead to a number of Rydberg excitations exceeding the plateau value of one.

4. Conclusions

In the present work we investigated the excitation to high-lying Rydberg in an atomic sample undergoing BEC phase transition and in the condition that the dimension of the condensed cloud is smaller than the blockade radius. We measured the number of excitations by single ion counting as a function of the thermal and condensed cloud dimensions which are controlled by evaporative cooling. We observed a distinct decrease in Rydberg excitations across the phase transition when the spatial density is bimodal (thermal and condensed fraction). In the regime where only the condensed fraction is detectable, the number of Rydberg excitations is compatible with a single collective excitation inside the condensed cloud. In order to explain the observed deviation from the simple single-excitation picture, further theoretical work taking into account the full atomic density distribution across the phase transition will be necessary.

Acknowledgments

This work was supported by the E.U. through Grant No. 265031-ITN-COHERENCE.

References

[1] Gallagher T F 1994 *Rydberg Atoms* (Cambridge: Cambridge University Press)
[2] Gallagher T F and Pillet P 2008 *Adv. At. Mol. Opt. Phys.* **56** 161
[3] Comparat D and Pillet P 2010 *J. Opt. Soc. Am.* B **27** A208
[4] Saffman M, Walker T G, Mølmer K 2010 *Rev. Mod. Phys.* **82** 2313
[5] Jaksch D, Cirac J I, Zoller P, Rolston S L, Côté R and Lukin M D 2000 *Phys. Rev. Lett.* **85** 2208
[6] Lukin M D, Fleischhauer M, Côté R, Duan L M, Jaksch D, Cirac J I and Zoller P 2001 *Phys. Rev. Lett.* **87** 037901
[7] Urban E, Johnson T A, Henage T, Isenhower L, Yavuz D D, Walker T G and Saffman M 2009 *Nature Phys.* **5** 110
[8] Isenhower L, Urban E, Zhang X L, Gill A T, Henage T, Johnson T A, Walker T G and Saffman M 2010 *Phys. Rev. Lett.* **104** 010503
[9] Gaëtan A, Miroshnychenko Y, Wilk T, Chotia A, Viteau M, Comparat D, Pillet P, Browaeys A and Grangier P 2009 *Nature Phys.* **5** 115
[10] Wilk T, Gaëtan A, Evellin C, Wolters J, Miroshnychenko Y, Grangier P and Browaeys A 2010 *Phys. Rev. Lett.* **104** 010502
[11] Pohl T, Demler E and Lukin MD 2010 *Phys. Rev. Lett.* **104** 043002
[12] Dudin Y O, Li L, Bariani F and Kuzmich A 2012 *Nature Phys.* **8** 790
[13] Schauß P, Cheneau M, Endres M, Fukuhara T, Hild S, Omran A, Pohl T, Gross C, Kuhr S and Bloch I 2012 *Nature* **491** 11596
[14] Heidemann R, Raitzsch U, Bendkowsky V, Butscher B, Löw R and Pfau T 2008 *Phys. Rev. Lett.***100** 033601
[15] Viteau M, Bason M G, Radogostowicz J, Malossi N, Ciampini C, Morsch O and Arimondo E 2011 *Phys. Rev. Lett.* **107** 060402
[16] Balewski J B, Krupp A T, Gaj A, Peter D, Buchler H P, Low R, Hofferberth S and Pfau T 2013 *Nature* **502** 12592
[17] Zenesini A, Sias C, Lignier H, Singh Y, Ciampini D, Morsch O, Mannella R, Arimondo E, Tomadin A and Wimberger S 2008 *New J. Phys* **10** 053038
[18] Viteau M, Radogostowicz J, Chotia A, Malossi N, Fuso F, Ciampini D, Morsch O, Ryabtsev I and Arimondo E 2010 *J. Phys.* B **43** 155301
[19] Viteau M, Radogostowicz J, Chotia A, Malossi N, Fuso F, Ciampini D, Morsch O, Ryabtsev I and Arimondo E 2011 *J. Phys.* B **44** 079802 (erratum)
[20] Viteau M, Bason M, Radogostowicz J, Malossi N, Morsch O, Ciampini D and Arimondo E 2013 *Laser Phys.* **23** 015502
[21] Malossi N, Valado M M, Scotto S, Huillery P, Pillet P, Ciampini D, Arimondo E and Morsch O 2014, *Phys. Rev. Lett.* **113**, 023006

All-optical production of ^{6}Li quantum gases

A Burchianti[1,2], J A Seman[1,2,3], G Valtolina[1,2,4], A Morales[5], M Inguscio[2,6], M Zaccanti[1,2] and G Roati[1,2]

[1]INO-CNR, via Nello Carrara 1, 50019 Sesto Fiorentino, Italy
[2]LENS and Università di Firenze, Via Nello Carrara 1, 50019 Sesto Fiorentino, Italy
[4]Scuola Normale Superiore, Piazza dei Cavalieri, 7, 56126 Pisa, Italy
[5]Department of Physics, ETH Zürich, 8093 Zürich, Switzerland
[6]INRIM, Strada delle Cacce 91, 10135 Torino, Italy

E-mail: burchianti@lens.unifi.it

Abstract. We report efficient production of quantum gases of ^{6}Li using a sub-Doppler cooling scheme based on the D_1 transition. After loading in a standard magneto-optical trap, an atomic sample of 10^9 atoms is cooled at a temperature of 40 μK by a bichromatic D_1 gray-molasses. More than 2×10^7 atoms are then transferred into a high-intensity optical dipole trap, where a two-spin state mixture is evaporatively cooled down to quantum degeneracy. We observe that D_1 cooling remains effective in the deep trapping potential, allowing an effective increase of the atomic phase-space density before starting the evaporation. In a total experimental cycle of 11 s, we produce weakly-interacting degenerate Fermi gases of 7×10^5 atoms at $T/T_F < 0.1$ and molecular Bose-Einstein condensates of up 5×10^5 molecules. We further describe a simple and compact optical system both for high-resolution imaging and for imprinting a thin optical barrier on the atomic cloud; this represents a first step towards the study of quantum tunneling in strongly interacting superfluid Fermi gases.

1. Introduction

Ultracold atoms, due to their high tunability and control, represent an ideal test ground for many-body physics [1]. In particular, the advent of ultracold atomic Fermi gases has proved the possibility to address, with unprecedented clarity, open questions in condensed matter which arise in strongly correlated electron systems [2]. Indeed, contrary to real electronic materials, dilute degenerate Fermi gases confined in optical traps offer a wealth of experimentally accessible parameters, including the interaction strength between particles, the confinement and the dimensionality. Continuous efforts have been thus devoted to the development of experimental methods for fast and efficient production of large and highly degenerate fermionic samples which are the starting point for more complex experiments. In this context, all-optical schemes are preferred to magnetic ones, due to their higher flexibility [3]. One of the main advantages is the possibility of trapping any internal state, eventually in the presence of a magnetic field. This allows the implementation of efficient forced evaporation close to a magnetically-tuned Feshbach resonance [4, 5]. The feasibility of this approach requires cold and dense atomic clouds in order to match the optical trap volume and depth. This guarantees that a large fraction of atoms is

[3] Present address: Instituto de Física, Universidad Nacional Autónoma de México, Apartado Postal 20-364, 01000 México Distrito Federal, Mexico.

transferred from the magneto-optical trap (MOT) into the optical potential, where the sample is then evaporatively cooled to quantum degeneracy.

Atomic samples at sub-Doppler temperatures are generally produced by exploiting Sisyphus cooling in optical molasses [6]. However, for lithium and potassium, which are the only alkali-metal atoms with stable fermionic isotopes, the standard Sisyphus mechanism, working on the D_2 transition, is complicated by the unresolved hyperfine splitting of the excited state. Lately, several groups have proved efficient sub-Doppler cooling of bosonic ^{7}Li and two potassium isotopes using gray-molasses [7–10], where Sisyphus cooling occurs on the blue side of D_1 transition [11–14]. Recently, following the pioneering work in Ref. [7] on ^{40}K, we have successfully extended the D_1 gray-molasses cooling technique to the case of ^{6}Li [15]. Similar results have also been reported in Ref. [16], where simultaneous sub-Doppler laser cooling of ^{6}Li and ^{40}K has been achieved using the D_1 transitions for both atomic species. We recall that atomic samples of both ^{6}Li and ^{40}K at temperatures of a few tens of μK have also been produced in MOTs, working on narrow dipole transitions lying in the near-UV region [17–19]. Despite its effectiveness, the latter method has the drawback to require special optics and quite expensive laser sources, when compared with the previous one.

In this paper, we describe sub-Doppler cooling of ^{6}Li using the D_1 transition and demonstrate how the introduction of this novel cooling stage allows for efficient production of quantum gases of ^{6}Li. After collecting an atomic ^{6}Li sample in a standard MOT, we apply bichromatic cooling in a D_1 gray-molasses. A laser field, blue-detuned from the $2S_{1/2}(F=3/2) \rightarrow 2P_{1/2}(F'=3/2)$ transition, produces Sisyphus cooling, while a weaker laser field allows phase-coherent addressing of the $2S_{1/2}(F=1/2) \rightarrow 2P_{1/2}(F'=3/2)$ transition, creating velocity-selective coherent population trapping (VSCPT), as soon as the two lasers are in Raman configuration [8]. The emergence of quantum interference at the Raman resonance provides an extra-cooling effect. After 2 ms in the D_1 molasses, the temperature of the atomic cloud drops from 500 μK to 40 μK, without any significant changes in the number of captured atoms ($\sim 10^9$ atoms). Then, more than 2×10^7 atoms are transferred into a far-detuned optical dipole trap (ODT) at 1070 nm. We observe that the D_1 cooling works even in the presence of a deep trapping potential. This implies a 40-fold increase of the phase-space density of the sample before starting the evaporation. The scattering lengths between two of the lowest spin states of ^{6}Li are magnetically tuned via their Feshbach resonance allowing us to explore both the BEC and BCS side of the resonance. As a proof of the feasibility of the developed technique, we produce, in a total cycle time of 11 s, either molecular Bose-Einstein condensates (mBECs) of 5×10^5 molecules or binary mixtures of degenerate Fermi gases of 3.5×10^5 atoms per spin state at $T/T_F < 0.1$, where T_F is the Fermi temperature. Furthermore, we discuss the realization in our setup of a simple and inexpensive optical system for both high-resolution imaging and for imprinting a thin optical barrier on the atomic sample. This repulsive potential, superimposed to the one produced by the optical trap, creates a double-well potential. We plan to use this trap geometry to investigate coherent tunneling of ^{6}Li quantum gases throughout different interaction regimes.

2. Experimental setup

Our experimental setup (see Fig. 1) consists of two main parts: the oven assembly, where a enriched ^{6}Li sample (isotopic abundance>99%) is heated at a temperature of 440°C, and the science chamber, where we perform the experiments. These two regions are connected by the Zeeman slower tube and by a differential pumping stage section. The residual pressure in the chamber is 10^{-11} Torr. The chamber is a spherical octagon with a large optical access. In particular, two reentrant CF100 viewports allow us to set up a high-resolution imaging system along the vertical direction (z-axis). The atoms from the thermal beam, slowed down by the Zeeman slower, enter in the science chamber, where they are captured by a MOT operating on D_2 ($2S_{1/2} \rightarrow 2P_{3/2}$) optical transitions (see Fig. 2). The MOT light configuration consists

Journal of Physics: Conference Series **594** (2015) 012042 doi:10.1088/1742-6596/594/1/012042

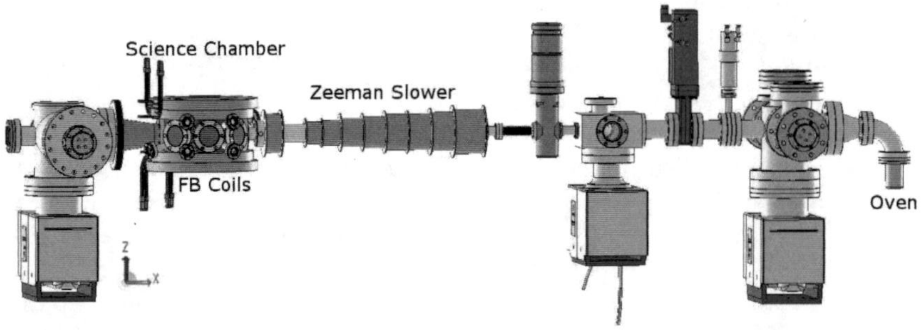

Figure 1. Sketch of the experimental setup. A thermal atomic beam is produced by evaporating a ^{6}Li sample in a oven kept at 440°C. The atoms slowed by a Zeeman slower enter inside the science chamber where they are collected by a MOT. The Feshbach (FB) coils produce a suitable magnetic field for tuning the atomic scattering length.

of three mutually orthogonal retro-reflected laser beams with circular polarization (σ^+/σ^-), crossing in the center of the chamber (see Fig. 3). The beams have a $1/e^2$ radius of about 1.5 cm, and peak intensity of about 7 I_S, where I_S=2.54 mW/cm^2 is the saturation intensity of the D_2 transition. Each beam contains both cooling (-9 Γ detuned from the F=3/2$\rightarrow$F'=5/2 transition, where $\Gamma = 2\pi \times 5.87$ MHz) and repumper (-6 Γ detuned from the F=1/2$\rightarrow$F'=3/2 transition) light. The cooling power is a factor 1.5 higher than the repumper one. In 5 s, 2×10^9 atoms are loaded into the MOT and cooled at a temperature of 2.5 mK. Then, the atomic beam, coming from the oven, is blocked with a mechanical shutter and the Zeeman slower light and magnetic field are turned off. It follows a cooling and compression phase, where the intensity and the detuning of both cooling and repumper light is reduced. After this phase lasting 2 ms, the temperature of the atomic cloud drops to 500 μK and 1.6×10^9 atoms remain trapped. At this point, we turn off the D_2 lights and the MOT magnetic quadrupole field, while turning on the D_1 molasses. The laser beams forming the D_1 molasses follow the same optical path of those of the D_2 MOT (see Fig. 3). Each beam contains the cooling (F=3/2$\rightarrow$F'=3/2) and repumper lights (F=1/2$\rightarrow$F'=3/2), both blue detuned with respect to the resonances (see Fig. 2). The maximum available peak intensity for beam is 3 I_S and the ratio between cooling and repumper power is set close to 5, the value that experimentally gives the maximum D_1 cooling efficiency. The D_2 and D_1 lights are derived from two amplified laser diodes which are controlled with two acusto-optic modulators (AOMs), acting as fast light switches. The two lasers inject the same tapered amplifiers, so the same optical components are used to realize the D_2 MOT and D_1 molasses. This scheme has the advantage that no further adjustment of the molasses alignment is needed.

During the D_1 cooling phase the atoms are transferred into a ODT consisting of one single focused laser beam derived from a 200 W multi-mode ytterbium fiber laser with a central wavelength of 1073 nm. The trapping beam forms an angle of about 15° with one of the horizontal MOT/molasses beam, as shown in Fig. 3, and it is focused on the atoms with a waist of 42 μm. In order to enlarge the spatial overlap between the molasses and the ODT, the trapping volume is increased by modulating the frequency and amplitude of the ODT's control AOM at a frequency (1.2 MHz and 600 kHz, respectively) larger than the trapping frequencies.

This produces an effective elliptic Gaussian-shaped beam with waists of about 42 μm (along the z-axis)$\times$85 μm. The initial trap power is set to 120 Watt which corresponds to a trap depth of $\sim$1.3 mK.

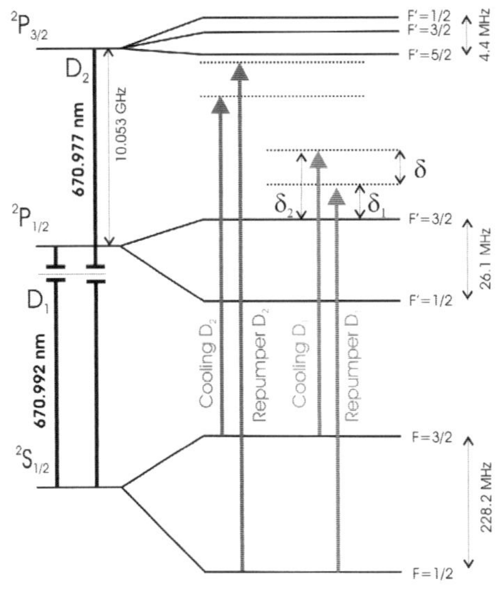

Figure 2. ^{6}Li level scheme (not to scale) showing the D_2 and D_1 hyperfine structure and the transitions used for the D_2 MOT and the D_1 molasses. The laser detuning from the repumper and cooling transitions of the D_1 molasses are δ_1 and δ_2, respectively, while their relative detuning is δ.

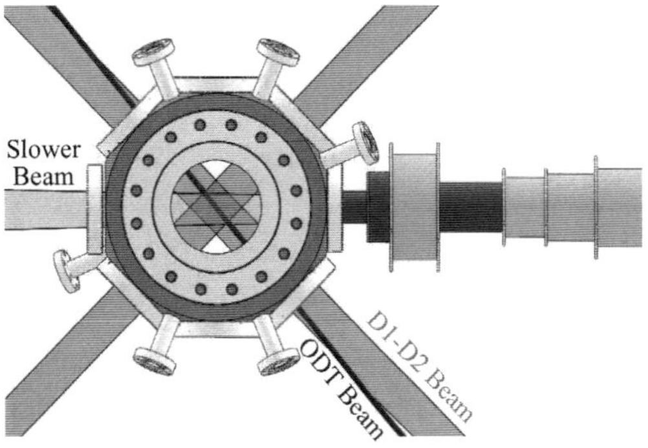

Figure 3. Top view of the science chamber, showing the configuration of the D_2 MOT/D_1 molasses beams, the Slower beam and the ODT beam.

3. D_1 gray-molasses cooling

We characterize the D_1 gray-molasses by measuring the temperature T and and the number of atoms N as a function of different parameters. T and N, after 2 ms in the molasses, are determined by time-of-flight (TOF) measurements, using absorption imaging resonant with the D_2 transition. In Fig. 4 (a) the values of T and the molasses capture efficiency, defined as N/N_0, where $N_0 = 1.6\times10^9$ is the initial atom number, are reported as a function of the relative detuning $\delta = \delta_1 - \delta_2$, where δ_1 and δ_2 are the detuning of the D_1 repumper and cooling lasers from the $F = 1/2 \to F' = 3/2$ and $F = 3/2 \to F' = 3/2$ transitions, respectively (see Fig. 2). The detuning of the cooling light δ_2 is fixed at 5.4 Γ, and $I_{rep} = 0.18\ I_{cool}$, with $I_{cool} = 2.7\ I_S$. At the Raman resonance, i.e. $\delta = 0$, the temperature reaches its minimum value $T_{min} = 40.5(1.0)$ μK with a capture efficiency N/N_0 of 75 %, as a consequence of both Sisyphus cooling on the blue of the $F \to F' = F$ transition and the formation of a coherent dark state in a three level Λ-system [7–10]. Furthermore, the temperature dependence on δ, in a narrow range around the Raman resonance, exhibits an asymmetric Fano-profile, which is a signature of the occurrence of quantum interference [20]. For δ slightly blue-detuned from the resonance, a strong heating accompanied by a decrease of the capture efficiency is instead observed, as already discussed in Refs. [8, 9]. Out of the resonance, both T and N/N_0 reach stationary values due to the Sisyphus

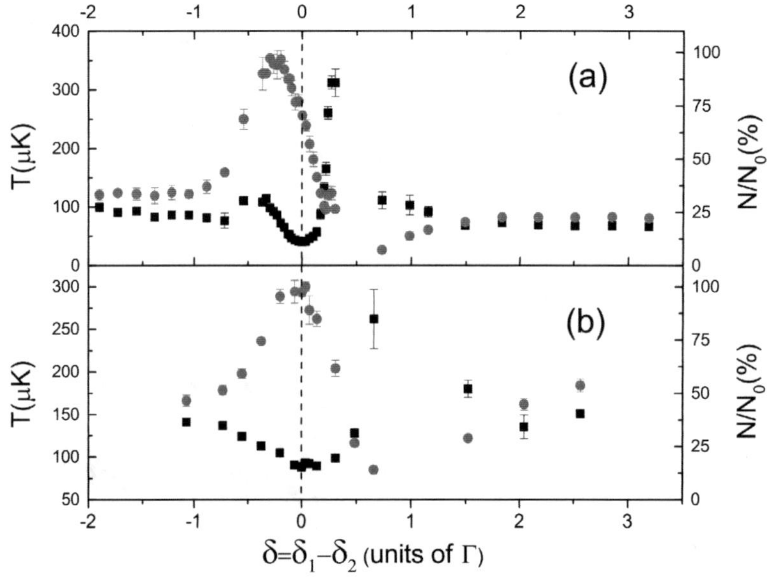

Figure 4. (a) Temperature T (black squares) and capture effciency N/N_0 (gray circles), with $N_0 = 1.6\times10^9$, as a function of the relative detuning δ. The laser intesities are: $I_{cool} = 2.7\ I_S$, $I_{rep} = 0.5\ I_S$ and $\delta_2 = 5.4\ \Gamma$. The molasses time is 2 ms. (b) Temperature T (black squares) and capture efficiency N/N_0 (gray circles), with $N_0 = 2\times10^7$, as a function of the relative detuning δ. The laser intesities are $I_{cool} = 2.7\ I_S$, $I_{rep} = 0.5\ I_S$ and $\delta_2 = 5.4\ \Gamma$. The molasses time is 300 μs and the ODT power is 120 Watt. In (a) and (b) the dotted vertical line indicates the value $\delta = 0$. Figures taken form Ref. [15].

effect alone. As the Raman condition is fulfilled, we find that the molasses efficiency is almost insensitive to δ_2 from a broad range of values from 3.5 to 8 Γ. This observation anticipates that the spatially varying AC Stark shift, due to the presence of the ODT, should not affect the D1 cooling mechanism, provided that the absolute detuning remains in this range. This is confirmed by our measurements of the D_1 light-shifts as a function of the trap intensity. Indeed, we find a differential Stark shift of about +2 Γ at the initial trap depth.

We directly verify the effectiveness of the gray-molasses in the presence of the ODT potential. The dipole trapping laser is turned on during the D_2 cooling stage, about 6 ms before switching to the D_1 light, and its power is ramped up to the maximum value in 5 ms. Mode-matching between the MOT and the optical trap is optimized by unbalancing the relative intensity of the MOT beams, in such a way to produce a cigar-shaped cloud perpendicular to the z-axis and elongated in the direction of the ODT beam. Despite this anisotropy, the performance of the gray-molasses is not substantially modified with respect to the balanced configuration. The number of atoms transferred into the ODT is maximized by applying a D_1 molasses cooling phase, lasting 2 ms, with δ=-0.2 Γ. This value corresponds to the maximum capture efficiency of the gray-molasses. In this way, we load into the optical potential about 2×10^7atoms at T=135(5) μK. In Fig. 4 (b), we show the Fano-profile recorded in the trap, 25 ms after the end of the ODT loading, by applying a further D_1 cooling stage lasting 300 μs. The molasses parameters are the same used in Fig. 4 (a). The dependence of T and N/N_0 on δ exhibits a behavior similar to the one observed in Fig. 4 (a). At δ=0, the temperature reaches its minimum value, T_{min} = 80(5) μK, which is a factor 2 higher than that found in the absence of the trapping potential. This is accompanied by a broadening of the Fano-profile. This effect can be ascribed to the large atomic density in the trap ($\sim 2\times 10^{13}$ atoms/cm^3) which limits the D_1 cooling efficiency [14]. We observe that the molasses capture efficiency, corresponding to the minimum temperature, is 100 %. This is not surprising since now the initial temperature is low enough that all the atoms are cooled by the molasses. The optimal cooling strategy thus consists in a first D_1 phase, lasting 2 ms at δ=-0.2 Γ, followed by a second one, lasting 300 μs at δ=0. In this way, we first maximize the number of trapped atoms and then minimize their temperature. The last D_1 phase is followed by a 25 μs hyperfine pumping to the $|F = 1/2\rangle$ manifold, realized by turning off the D_1 repumper light before the cooling one. The hyperfine pumping increases the atomic temperature by about 10%.

4. Production of ^{6}Li quantum gases

To reach the quantum degenerate regime, we start the evaporative cooling of the two spin states $|F = 1/2, m_F = \pm 1/2\rangle$, labeled as $|1\rangle$ and $|2\rangle$, at 840 G, close to the center of the Feshbach resonance [21]. After the hyperfine pumping stage, we ramp the magnetic field to the final value in about 30 ms and we decrease the trap laser power from 120 to 30 W in 500 ms. Multiple resonant radio-frequency sweeps create an incoherent balanced mixture of the $|1\rangle$ and $|2\rangle$ states. After this first evaporation ramp, we typically have 1×10^7 atoms per spin state at T$\simeq$ 30 μK. Molecular BECs are then produced by continuing the evaporation of $|1\rangle$-$|2\rangle$ mixture at 800 G, where the s-wave scattering length a_{12} is of the order of 11000 a_0 [21], with a_0 being the Bohr radius. The temperature T and the total number of atoms N as a function of the evaporation time are shown in Fig. 5 (a).

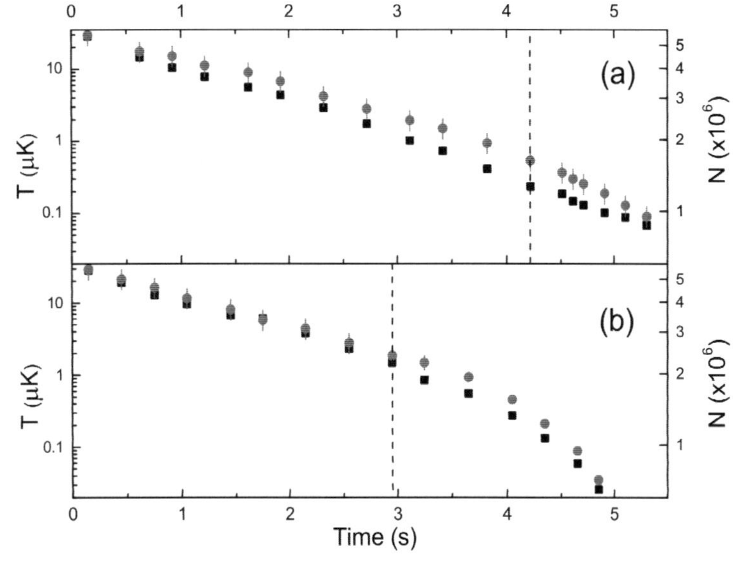

Figure 5. (a) Temperature T (black squares) and total atom number N (gray circles) as function of time during forced evaporation in the ODT at B=800 G. The dotted vertical line indicates the evaporation time for which T/T$_c$=1. (b) Temperature T (black squares) and total atom number N (gray circles) as function of time during forced evaporation in the ODT at B=300 G. The dotted vertical line indicates the evaporation time for which T/T$_F$=1. Figures taken form Ref. [15].

At a magnetic field of 800 G, molecules are formed via three-body recombination when the temperature of the sample becomes lower than the molecular binding energy. At $T=T_c=210(20)$ nK, we observe the onset of condensation for $N_{mol} \simeq 1 \times 10^6$ molecules. To image the molecular cloud in time of flight, we adiabatically sweep the magnetic field to 690 G to reduce the inter-particle interaction. At T_c, the radial trap frequencies are $2\pi\times239(2)$ Hz (along the z-axis) and $2\pi\times111(3)$ Hz and the axial frequency, given by the magnetic curvature of our Feshbach coils, is $2\pi\times8.2(1)$ Hz. By further reducing the trap depth, we observe the formation of an almost pure mBEC of about $N_{mol} \simeq 5 \times 10^5$. The same method is used to produce a unitary Fermi Gas at the Feshbach resonance. We observe ultracold clouds of about 2×10^6 particles at a temperature corresponding to T_c, when sweeping to the BEC side.

We follow a slightly different strategy for producing weakly-interacting Fermi gases. In this case, we evaporate the $|1\rangle$-$|3\rangle$ spin-mixture, where the $|3\rangle$ state corresponds to the $|F=3/2, m_F=-3/2\rangle$ state at low magnetic fields. The mixture is created at $T\simeq30$ μK by transferring 100% of the atoms from the state $|2\rangle$ to the $|3\rangle$ by a radio-frequency sweep. The mixture is then evaporated at 300 G, where the $|1\rangle$-$|3\rangle$ s-wave scattering length is $a_{13} \simeq -880a_0$, about three times larger than $a_{12} \simeq -290a_0$ [21], strongly enhancing the evaporation efficency. The axial confinement, at this magnetic field, is provided by a magnetic curvature generated by an additional pair of coils. In Fig. 5 (b), we report the evaporation trajectory for the $|1\rangle$-$|3\rangle$ mixture at 300 G. This trajectory is similar to the one reported in Fig. 5 (a), showing the efficient thermalization of the $|1\rangle$-$|3\rangle$ mixture. After 3 seconds of forced evaporation, the system enters the degenerate regime with $N_{|1\rangle}=N_{|3\rangle}=2\times10^6$ atoms. After a further 2 seconds, we produce a highly degenerate Fermi gas of $N_{|1\rangle}=N_{|3\rangle}=3.5\times10^5$ atoms at $T/T_F \simeq 0.06(1)$, where T_F is the Fermi temperature defined as $k_BT_F = \hbar\varpi(6N_i)^{1/3}$, where ϖ is the mean trap frequency. Here, we measure radial trapping frequencies of $2\pi\times231(3)$ Hz (along the z-axis) and $2\pi\times111.5(2)$ Hz, and an axial trapping frequency of $2\pi\times12.4(1)$ Hz.

5. High-resolution imaging and production of a thin optical barrier

We implement an optical setup, along the vertical direction, both for high-resolution imaging and for producing engineered optical potentials. In particular, we describe how to create and imprint on the atomic cloud a green sheet of light. The main component of our system is an aspheric lens [22] with a diameter of 45 mm and a focal length of 32 mm, placed at a few millimeters from the upper reentrant viewport (see Fig. 6 (a)). The lens has a numerical aperture of 0.61 which corresponds to a theoretical resolution of $\sim\lambda$, thus approaching the diffraction limit. The absorption imaging and the vertical MOT beams, which have linear orthogonal polarizations, are superimposed, using a polarization beam splitter, then they are circularly polarized with opposite sign by a $\lambda/4$ waveplate and sent through the chamber via the lower reentrant viewport. In order to retro-reflect back the incoming MOT beam, we place after the aspheric lens in close sequence a $\lambda/4$ waveplate and a wire-grid polarizer (WGP). The relative optical axis between the $\lambda/4$ waveplate and the WGP is adjusted in a such way that the MOT beam is reflected back whereas the imaging beam is transmitted. All the optical elements are assembled together in a custom-designed holder, made with PEEK polymer, which is, in turn, aligned using a five-axis optical positioner. The laser light used to produce the optical barrier is provided by a 532 nm Nd:YVO4 laser. A 532 nm beam, after crossing a cylindrical lens with a focal length of 200 mm, is overlapped to the imaging beam, coming from the bottom, by means of a dichroic mirror. Then, the green beam, which is properly linearly polarized, is transmitted by the WGP and is focused by the aspheric lens on the atomic sample. This configuration produces a repulsive sheet of light with waists of about 2 μm (along the longitudinal axis of the ODT)x850 μm (see Fig. 6 (b)). The minimum value of the axial waist is limited by residual aberrations as well as diffraction effects introduced by the optical components. The green barrier added to the optical dipole trap potential produces a double-well potential landscape for the atoms. As preliminary result, we show in Fig. 7 the in-situ absorption image, recorded by a CCD camera, and the

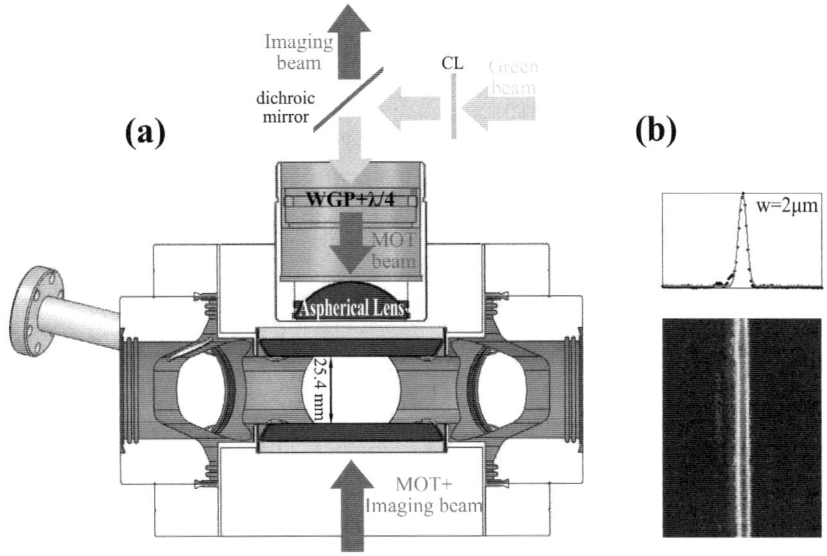

Figure 6. (a) Optical system for high-resolution imaging and production of a repulsive like-sheet potential. The vertical MOT beam is retro-reflected back by a wire-grid polarizer (WGP) and a $\lambda/4$ waveplate. The 670 nm imaging beam and the 532 nm beam are superimposed using a dichroic mirror. By combining a cylindrical lens (CL) and the objective (aspheric lens), we create a thin optical barrier for the atoms. (b) Intensity profile of the 532 nm beam, after the cylinrical lens, projected by the obective on a CCD camera.

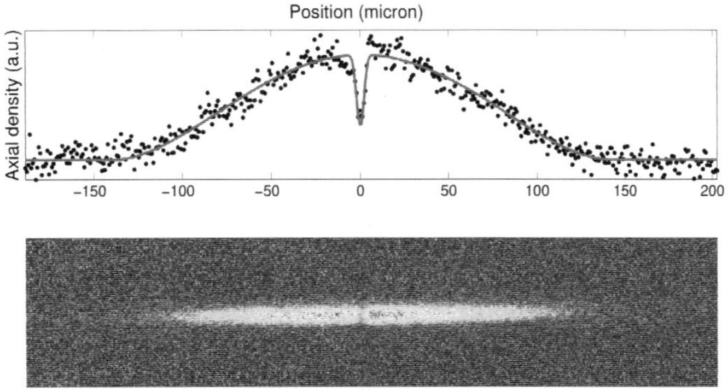

Figure 7. In-situ absorption image of a mBEC at 690 G (bottom panel) and its axial density profile (top panel) in a double-well potential formed by the superposition of the ODT trap potential and a green barrier with a size of about 2 μm. The density profile has been fitted by a sum of a Thomas-Fermi function and a Gaussian peak (red curve).

relative axial density profile of a mBEC (produced following the procedure described in Sec. 4) of about 1.2×10^4 molecules at 690 G in the double-well trap. The atomic density distribution has been determined using the absorption imaging along the z-axis. The power of the green beam on the atomic cloud is about 2 mW.

In the future, we plan to investigate the coherent oscillations between two fermionic superfluids across the barrier. Although the Josephson effect in weakly linked Bose-Einstein condensates has been theoretically described [23] and experimentally observed [24–26], much less is known about its fermionic counterpart. Recently, several theoretical papers have investigated the tunneling dynamics of both fermionic superfluids [27–29] and spin-polarized Fermi gases [30, 31] through a barrier. Our setup allows to address all these regimes by simply tuning the atomic scattering length across the Feshbach resonance. The case of a Fermi gas at unitarity is of special interest since all scales are set by the inverse of the Fermi wave vector k_F. In our typical experimental conditions k_F^{-1} is of the order of 0.5 μm; therefore, the coherence length $l_{coh} \sim k_F^{-1}$ is still comparable with the barrier size, allows for testing of different theoretical approches.

6. Conclusion

In conclusion, we have demonstrated an all-optical method for fast and efficient production of quantum gases of ^{6}Li, without the need of any additional coolant atomic species [32]. Our scheme is based on a D_1 gray-molasses phase which provides sub-Doppler cooling with a high capture efficiency. This ensures an efficient transfer into an ODT, where we evaporate down to the degenerate regime. We observe that the D_1 cooling also works in the presence of a deep optical dipole potential. This result suggests the possibility to exploit this scheme to image ^{6}Li atoms in optical lattices with single-site resolution [33, 34]. In an experimental run, lasting 11 s, we produce either mBECs and crossover superfluids of 5×10^5 dimers/pairs or binary mixtures of degenerate Fermi gases of 3.5×10^5 atoms per spin state at $T/T_F < 0.1$.

We have also developed, and integrated in our setup, a simple optical system for both high-resolution imaging and for creating a thin optical barrier. We have then verified that this system allows us to produce and imprint on the atomic cloud a green barrier with a size of about 2 μm. As this repulsive well is added to the harmonic potential, generated by our optical trap, an effective double-well potential is created. We plan to study the tunneling dynamics of two quantum gases of ^{6}Li in the double-well trap, starting from the BEC regime of composite molecules reaching the BEC-BCS crossover. This represents the first step towards the experimental realization of a Josephson junction with ultracold Fermi gases.

Acknowledgments

We would like to thank the technical staff of LENS for assistance. Special acknowledgments to the LENS Quantum Gases group. This research was supported under the European Research Council Grant No. 307032 QuFerm2D and the European Council Advanced Grant No. 247371 DISQUA.

References

[1] I. Bloch, J. Dalibard, and W. Zwerger, Rev. Mod. Phys. **80**, 885 (2008).
[2] Proceedings of the International School of Physics Enrico Fermi, Course CLXIV, Edited by M. Inguscio, W. Ketterle, and C. Salomon, 2007, IOS Press Amsterdam.
[3] M. D. Barrett, J. A. Sauer, and M. S. Chapman, Phys. Rev. Lett. **87**, 010404 (2001).
[4] K. M. O'Hara, S. L. Hemmer, M. E. Gehm, S. R. Granade, and J. E. Thomas, Science **298**, 2179 (2002).
[5] S. Jochim, M. Bartenstein, A. Altmeyer, G. Hendl, S. Riedl, C. Chin, J. Hecker Denschlag, and R. Grimm, Science **302**, 2101 (2003).
[6] J. Dalibard and C. Cohen-Tannoudjii, J. Opt. Soc. Am. B **6**, 2023 (1989).
[7] D. Rio Fernandes, F. Sievers, N. Kretzschmar, S. Wu, C. Salomon, and F. Chevy, Europhys. Lett. **100**, 63001 (2012).
[8] A. T. Grier, I. Ferrier-Barbut, B. S. Rem, M. Delehaye, L. Khaykovich, F. Chevy, and C. Salomon, Phys. Rev. A **87**, 063411 (2013).
[9] D. Nath, R. K. Easwaran, G. Rajalakshmi, and C. S. Unnikrishnan, Phys. Rev. A **88**, 053407 (2013).
[10] G. Salomon, L. Fouch, P. Wang, A. Aspect, P. Bouyer, and T. Bourdel, Europhys. Lett. **104**, 63002 (2013).
[11] G. Grynberg and J. -Y. Courtois, Europhys. Lett. **27**, 41 (1994).

[12] M. Weidemüller, T. Essingler, M. A. Ol'shanii, A. Hemmerich, and T. W. Hänsch, Europhys. Lett. **27**, 109 (1994).
[13] D. Boiron, C. Triché, D. R. Meacher, P. Verkerk, and G. Grynberg, Phys. Rev. A **52**, R3425 (1995).
[14] D. Boiron, A. Michaud, J. M. Fournier, L. Simard, M. Sprenger, G. Grynberg, and C. Salomon, Phys. Rev. A **57**, R4106 (1998).
[15] A. Burchianti, G. Valtolina, J. A. Seman, E. Pace, M. De Pas, M. Inguscio, M. Zaccanti, and G. Roati, Phys. Rev. A **90**, 043408 (2014).
[16] F Sievers, S Wu, N Kretzschmar, DR Fernandes, D Suchet, M Rabinovic,C V Parker, L Khaykovich, C Salomon, F Chevy, arXiv:1410.8545 (2014).
[17] P. M. Duarte, R. A. Hart, J. M. Hitchcock, T. A. Corcovilos, T.-L. Yang, A. Reed, and R. G. Hulet, Phys. Rev. A **84** ,061406R (2011).
[18] D. C. McKay, D. Jervis, D. J. Fine, J. W. Simpson-Porco, G. J. A. Edge, and J. H. Thywissen, Phys. Rev. A **84**, 063420 (2011).
[19] J. Sebastian, C Gross, K Li, H. C. J. Gan, W Li and K. Dieckmann, Phys. Rev. A **90**, 033417 (2014).
[20] B. Lounis and C. Cohen-Tannoudji, J. Phys. II France **2**, 579 (1992).
[21] G. Zürn, T. Lompe, A. N. Wenz, S. Jochim, P. S. Julienne, and J. M. Hutson, Phys. Rev. Lett. **110**, 135301 (2013).
[22] A similar optical scheme has been developed by the Jochim group in Heidelberg.
[23] A. Smerzi, S. Fantoni, S. Giovanazzi, and S. Shenoy, Phys. Rev. Lett. **79**, 4950 (1997).
[24] F. S. Cataliotti, S. Burger, C. Fort, P. Maddaloni, F. Minardi, A. Trombettoni, A. Smerzi, and M. Inguscio, Science **293**, 843 (2011).
[25] M. Albiez, R. Gati, J. Fölling, S. Hunsmann, M. Cristiani, and M. K. Oberthaler, Phys. Rev. Lett. **95**, 4950 (2005).
[26] S. Levy, E. Lahoud, I. Shomroni, and J. Steinhauer, Nature **449**, 579 (2007).
[27] A. Spuntarelli, P. Pieri, and G.C. Strinati, Phys. Rev. Lett. **99**, 040401 (2007).
[28] P. Zou and F. Dalfovo, arXiv:1401.2007v1 (2014).
[29] F. Ancilotto, L. Salasnich, and F. Toigo, Phys. Rev. A **79**, 033627 (2009).
[30] L. Salasnich, G. Mazzarella, M. Salerno, and F. Toigo, Phys. Rev. A **81**, 023614 (2010).
[31] T. Macrì and A. Trombettoni, Laser Phys. A **81** 23, 095501 (2013).
[32] Z. Hadzibabic, S. Gupta, C. A. Stan, C. H. Schunck, M. W. Zwierlein, K. Dieckmann, and W. Ketterle, Phys. Rev. Lett. **91**, 160401 (2003).
[33] W. S. Bakr, J. I. Gillen, A. Peng, S. Fölling, and M. Greiner, Nature **462**,74 (2009).
[34] J. F. Sherson, C. Weitenberg, M. Endres, M. Cheneau, I. Bloch, and S. Kuhr, Nature **467**, 68 (2010).

Solitons in atomic condensates, with optical lattices and field-induced dipole moments

Lauro Tomio[1,2], H F da Luz[1], A Gammal[3] and F Kh Abdullaev[4]

[1]Centro de Ciências Naturais e Humanas (CCNH), Universidade Federal do ABC, 09210-170, Santo André, Brazil

[2]Instituto de Física Teórica, UNESP, 01140-070, São Paulo, Brazil

[3]Instituto de Física, Universidade de São Paulo, 05508-090, São Paulo, Brazil

[4]Department of Physics, Kulliyyah of Science, International Islamic University of Malaysia, 25200, Kuantan, Malaysia

E-mail: tomio@ift.unesp.br

Abstract. We report some investigations on the existence of matter-wave solitons when considering cross-combined optical lattices (OL) in three dimensions (3D), where we have a nonlinear OL in one of the directions, which is perpendicular to a two-dimensional (2D) plane with linear OLs in either one or both directions. This study can be useful to manage 3D solitons through spatial modulations of the scattering length in one of the OL directions. Another independent study is reported, by considering bright solitons manifested in a bosonic condensate gas carrying collinear dipole moments, which induces an external polarizing field with strength periodically modulated along one of the coordinates. This leads to an effective nonlocal nonlinear lattice, with solitonic solutions. Their dynamics and mobility can be investigated by an effective one-dimensional (1D) model. Interactions between solitons are also reported within this 1D model. In all the cases, we consider full numerical and variational approaches.

1. Introduction

In the present contribution we review some recent works on nonlinear optical lattices that we have been investigating in Refs. [1, 2]. By considering Ref. [1], we show results obtained on the existence of three-dimensional (3D) matter-wave solitons when considering cross-combined optical lattices (OL), with a nonlinear optical lattice (NOL) in one of the directions, perpendicular to a plane where we have linear OL in either one or both directions. Our section II is dedicated to this case. First, we noticed that one cannot obtain stable 3D solitons when in one of the perpendicular directions there is no optical lattice or trapping interaction. However, in the full 3D OL cross-combined case, we can obtain families of 3D solitons, which can be stable for both attractive and repulsive interactions. Sample results are presented in this case, analyzing the existence of soliton solutions, as well as their time evolution and stability.

Next, in section III, by considering a bosonic gas of particles carrying collinear dipole moments, following [2], it is shown that one can obtain an effective nonlocal nonlinear lattice in a condensate, where bright solitons can be manifested. Such dipole moments are induced by an external polarizing field with the strength periodically modulated along the coordinate. Our aim, in this case, is to investigate the existence and stability of bright solitons in a system controlled by an effective induced dipole-dipole interaction (DDI) in nonlocal nonlinear lattice. Some results on the soliton stability and dynamics are also presented. The model we consider is a quasi-1D dipolar Bose-Einstein condensate (BEC) with electric dipole moments of particles induced by the dc field with the local strength periodically varying in space, while its direction is uniform, being oriented along the system's axis, thus giving rise to the attractive DDI.

In both sections the reported results were obtained from exact numerical simulations, as well as from variational procedures, as detailed in [1, 2]. In section III we also add sample results on the dynamics and collisions of solitons. In section IV we have some final considerations.

2. Solitons in cross-combined linear and nonlinear optical lattices

When we have cross-combined OL, with linear ones in the $x-$ and $z-$direction, and a nonlinear one in the remaining $y-$direction, it is shown in [1] that one can generate 3D matter-wave solitons for some parametric regions. In this section we report some of the results obtained in this work. In case of a 2D BEC, it was shown in [3] that, for conservative systems, a NOL in one direction by itself cannot give stable localized solutions in the case of attractive interactions. Experimentally, one can generate a NOL by a periodic spatial modulation of the scattering length on an optically induced Feshbach resonance [4]. In [5] it was verified that a crossed linear and nonlinear OL allows to stabilize 2D solitons against collapse or decay with either attractive or repulsive mean interaction.

The existence of 3D matter-wave solitons were demonstrated, analytically by using a multi-Gaussian variation approach (VA); and, numerically, by the relaxation method detailed in [6], as well as by direct time integrations of the Gross-Pitaevskii equation (GPE), which also allows us to study the dynamics of the corresponding matter waves. In order to investigate the stability of the solutions, it was considered the Vakhitov-Kolokolov (VK) [7] necessary criterion.

Due to the lattice anisotropy, the solitons display elliptical cross sections, which in our VA are accounted by a multi-Gaussian *ansatz* with different parameters for each one of the perpendicular spatial directions. The VA results are shown in terms of chemical potentials (μ) as functions of the number of particles (N), with the VK criterion being used to investigate the stability of the solutions. Such results are compared with direct numerical integrations of the full GPE.

The basic mean-field formalism for the 3D GPE, in dimensionless units ($\hbar = m = 1$), follows below. The GPE with cross-combined linear and nonlinear OLs, can be given by

$$\mathrm{i}\frac{\partial u}{\partial t} = -\nabla^2 u - Vu - \Gamma|u|^2 u, \tag{1}$$

where the linear V and nonlinear Γ interactions are given, respectively, by

$$V \equiv V(x,z) = \varepsilon_x \cos(2x) + \varepsilon_z \cos(2z) \tag{2}$$

and

$$\Gamma \equiv \Gamma(y) = \chi + \gamma\cos(\lambda\, y). \tag{3}$$

The parameter χ is linearly proportional to the two-body scattering length, with the optical lattices parametrized by ϵ_x and ϵ_z, in the linear case; and by γ and λ in the nonlinear case. The variational *ansatz* is a combination of three Gaussian functions, with parameters A, a, b and c:

$$U(x,y,z) = A\exp\left[-\frac{(ax^2 + by^2 + cz^2)}{2}\right]. \tag{4}$$

By defining μ as the chemical potential, we look for variational solutions of the form

$$u(x,y,z,t) = U(x,y,z)\exp(-\mathrm{i}\mu t). \tag{5}$$

2.1. Results - 3D Solitons with quasi-2D OLs

We start this 3D investigation by considering a cross-combined quasi-2D OLs, where we have a nonlinear OL in the $y-$ direction, but no trap or OL in the $z-$ direction, such that $\varepsilon_z = 0$ and $\varepsilon_x \equiv \varepsilon$. The corresponding variational results are being exemplified by some curves given in Fig. 1, in the $\mu - N$ plane, obtained for different values of the parameter ε, when we fix the parameters of the nonlinear interaction to $\chi = 1$, $\gamma = 0.5$ and $\lambda = 2$. As verified, no stable 3D solitons are predicted to exist, according to the VK criterion, which predicts stability for $d\mu/dN < 0$ in the attractive case.

The variational approach gives in general a good agreement with full numerical PDE simulations. It is also verified that PDE simulations confirm the prediction of the VK criterion about the instability of 3D solitons. The agreement between VA and PDE numerical simulations is exemplified by the case with $\varepsilon_x = 3$, as shown in Fig. 1.

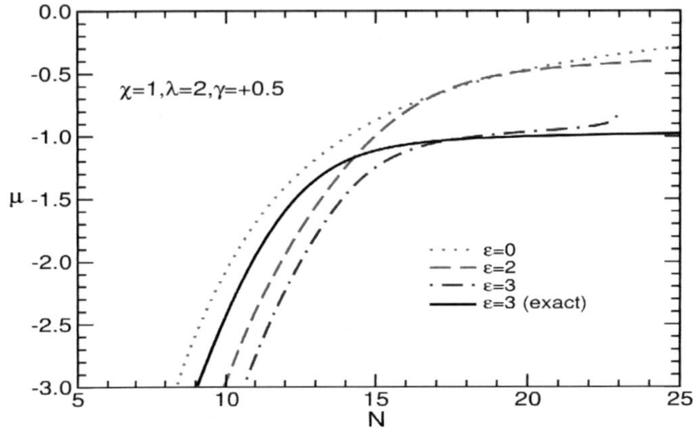

Figure 1. Variational results obtained for μ in terms of N, are shown in case of 3D solitons, for crossed 1D LOL ($x-$direction) with 1D NLOL ($y-$direction), without any constraint in the third $z-$direction ($\varepsilon_z = 0$), considering attractive mean two-body interactions ($\chi = 1$). The results are presented for $\varepsilon_x \equiv \varepsilon$, with other parameters fixed as shown inside the frame. For $\varepsilon = 3$, we compare the VA (dot-dashed line) with direct numerical solutions (solid line).

2.2. Results - 3D Solitons for two linear with one nonlinear OLs

By considering OL in three perpendicular directions, with a nonlinear OL in one of them, some sample results are displayed in Figs. 2 and 3, for attractive and repulsive cases, respectively. As we observe, in both the cases there are branches of the curves for which one can predict stable 3D solitons, by considering the VK criterion ($d\mu/dN < 0$). For comparison with specific VA solutions (solid lines), in each of the figures we add one dotted line with direct PDE calculations.

2.3. Results - Time evolution and stability

In Fig. 4, we present some results related to time evolution and stability of the solitons, considering the evolution of a profile obtained from direct numerical time integrations of the GP equation. The profile is for one specific point of the dotted curve shown in Fig. 3 for the repulsive case, with $\mu = -2.5$. This point is close to the limit that defines the stable or unstable regions according to the VK criterion. As shown in Fig. 4, by reducing ε_z, the instability occurs in a very short time interval.

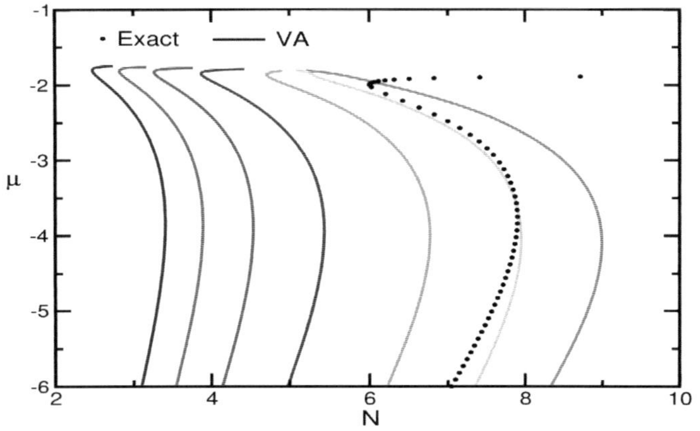

Figure 2. In cross-combined 2D LOL with 1D NOL, VA results (solid lines) for 3D solitons are shown for μ in terms of N. We fix $\varepsilon_x = \varepsilon_z \equiv \varepsilon = 3$, with attractive interaction $\chi = 1$, varying the NOL strength, such that, from left to right, $\gamma = 3.0, 2.5, 2.0, 1.5, 1.0, 0.7, 0.5$, respectively. For comparison with PDE simulations, a sample result with $\gamma = 0.5$ is given by the dotted curve.

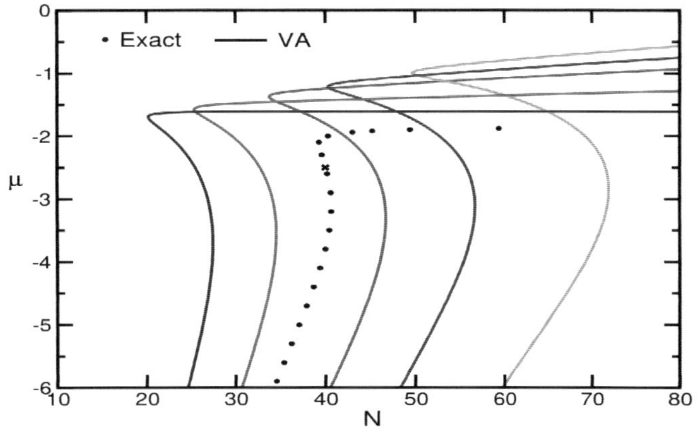

Figure 3. VA results (solid lines) for μ versus N are shown for fixed $\varepsilon = 3$ and $\gamma = 0.5$, with several repulsive mean nonlinearity interactions, $\chi =$ 0.0, -0.1, -0.2, -0.25, -0.3 (from left to right, respectively). For comparison with VA, it is included PDE results for $\chi = -0.2$.

This section is concluded by some remarks evidenced by the results. While in the 2D OL cross-combined case, 3D solitons are always unstable, in the full 3D OL cross-combined case, we can obtain families of 3D solitons, which can be stable for both attractive and repulsive interactions. In perspective of potential applications, we should point out that, besides providing an alternate way to create stable solitons in 3D, with cross-combined linear and nonlinear optical lattices (both in BEC and nonlinear optics), these results can also be useful in practical applications, opening the possibility to manage stable 3D solitons through spatial modulations of the scattering length in one of the optical lattice directions.

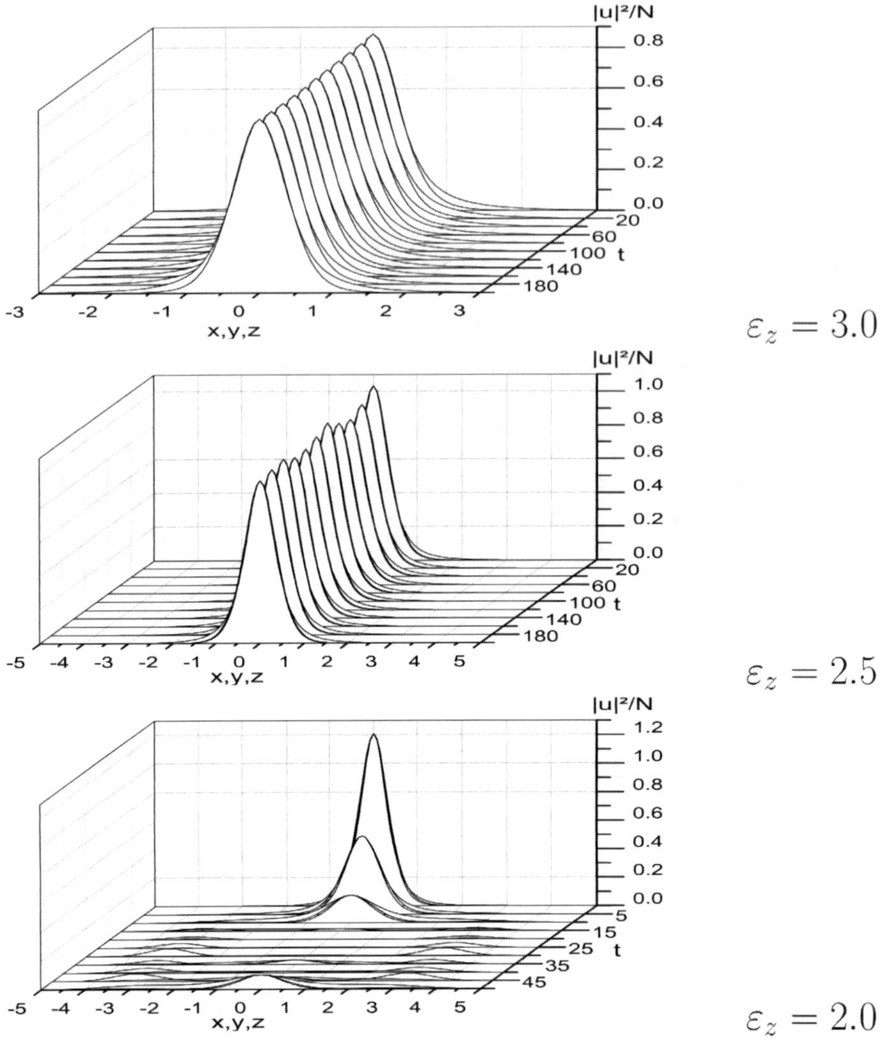

Figure 4. The soliton dynamics, for the repulsive case given in Fig. 3, for $\mu = -2.5$, with ε_x =3.0, $\gamma = 0.5$, and $\chi = -0.2$. The loss of stability is verified by the reduction of the linear parameter in one direction, from $\varepsilon_z = 3.0$ (top panel) to $\varepsilon_z = 2.5$ (middle panel) and $\varepsilon_z = 2.0$ (lower panel). The profiles in the $y-$direction are slightly wider.

3. Nonlocal nonlinear lattice from induced dipole-dipole interactions

In this section we report some results that have been investigated in Ref. [2], considering a condensate elongated along axis x, with dipole moments of polarizable molecules or atoms induced by an external field directed along x too. The local strength of the polarizing field also varies along x. Accordingly, one can use an effectively 1D GPE, with the DDI term derived from the underlying 3D GPE. The necessary spatially modulated dc electric and/or magnetic field can be imposed by ferroelectric or ferromagnetic lattices. Besides the fact that one could consider magnetic dipole moments induced by a solenoid, within our formalism we have considered local dipole moment induced by a polarizing electric field.

In addition to the ferroelectric lattice, the periodic modulation of the strength of the electric field oriented *perpendicular* to the system's axis (x) can be provided by a capacitor with the separation between its electrodes modulated in x periodically, as discussed in Ref. [8].

3.1. 1D GPE formalism with DDI interaction

From the original 3D GPE, as shown in Ref. [2], by considering DDI interaction, we can derive the following reduced 1D equation:

$$\mathrm{i}\frac{\partial \phi}{\partial t} = -\frac{1}{2}\frac{\partial^2 \phi}{\partial x^2} + \beta f^2(x)\phi + g|\phi|^2\phi - f(x)\phi \int_{-\infty}^{+\infty} f(x')|\phi'|^2 R(x-x')dx', \tag{6}$$

where $\phi \equiv \phi(x,t)$ and $\phi' \equiv \phi(x',t)$, with the effective 1D kernel given by

$$R(x) = \sigma\frac{10}{\pi}\left[\left(1+2x^2\right)\exp\left(x^2\right)\mathrm{erfc}(|x|) - \frac{2}{\sqrt{\pi}}|x|\right]. \tag{7}$$

Here, the parameter σ takes values $\sigma = +1$ for the attractive DDI, and $\sigma = -1/2$ for the repulsive DDI between dipoles oriented perpendicular to x, which makes it possible to consider the latter case too. However, for the perpendicular orientation of the dipoles, we have shown by means of numerical results for different values of μ, that the wave-function profiles are delocalized, i.e., they do not build bright solitons. Therefore, in the next we restrict our analysis to the attractive DDI, with $\sigma = 1$. Actually, the kernel (7) can be replaced by a simplified expression [9],

$$R(x) = \frac{10}{\pi\sqrt{(\pi\, x^2+1)^3}}. \tag{8}$$

It is very close to (7), but smoother near $x = 0$, while (7) has a cusp at this point.

Note that the DDI can be represented by a pseudopotential that includes a contact-interaction (delta-functional) term [10, 11]. Therefore, the spatially modulation may induce a position-depending part of the contact interactions too. However, in the present setting, we restrict ourselves to a pure nonlinear nonlocal lattice. We assume that the dynamics of the system in the perpendicular directions is completely frozen, i.e., the transverse trapping frequency, $\omega_\perp$, is much larger than the longitudinal one, $\omega_\perp \gg \omega_{||}$. On the other hand, if $\omega_\perp$ is not too large, interesting transverse effects may occur, such as the Einstein-de Haas effect [12, 13, 14, 15], which is beyond the scope of our present analysis.

In Eq. (6), the nonlinear term $g|\phi|^2\phi$ accounts for the usual collisional two-body interaction, with the effective DDI potential being composed by linear and nonlinear terms, as follows:

$$V_{\mathrm{eff}}^{(\mathrm{DDI})}(x;|\phi|^2) = f(x)\left[\beta f(x) - \int_{-\infty}^{+\infty} f(x')|\phi'|^2 R(x-x')dx'\right], \tag{9}$$

where the modulation function is chosen, as said above, in the form of a periodic one:

$$f(x) = f_0 + f_1\cos(kx), \tag{10}$$

with the parameters f_0, f_1, and $k \equiv 2\pi a_\perp/\lambda = 2\pi/\Lambda$. β in Eq.(6) can vary from 1 to 10, under typical physical conditions, if the constant part of the modulation function is fixed as $f_0 \equiv 1$.

The Hamiltonian corresponding to Eq. (6) is

$$\begin{aligned} H &= \int_{-\infty}^{+\infty} dx \left[\frac{1}{2}\left|\frac{\partial\phi}{\partial x}\right|^2 + \frac{g}{2}|\phi|^4 + \beta f^2(x)|\phi|^2\right] \\ &- \frac{1}{2}\int_{-\infty}^{+\infty} dx f(x)|\phi|^2 \int_{-\infty}^{+\infty} dx' f(x')|\phi'|^2 R(x-x'). \end{aligned} \tag{11}$$

Note that it contains not only the spatially modulated nonlinear DDI, but also the additional linear potential, $\beta f^2(x)$, which is derived from the locally induced dipole moment with the polarizing field [8]. The existence of bright-soliton solutions can be investigated by solving the corresponding eigenvalue problem, obtained from Eqs. (6) and (9), with $\phi = |\phi|e^{-i\mu t}$:

$$-\frac{1}{2}\frac{\partial^2\phi}{\partial x^2} + g|\phi|^2\phi + V_{eff}^{DDI}(x;|\phi|^2)\phi = \mu\phi. \qquad (12)$$

3.2. Results - Bright solitons, existence and stability

In our following analysis, we consider full numerical solutions of Eq. (12), as well as the corresponding VA results, for two characteristic cases. The solutions produced by the VA are compared with their numerically counterparts in Figs. 5 and 6, for $g = 0$, $f_0 = 1$, $f_1 = 0.5$, and $\Lambda = 0.5$ ($k = 4\pi$). The corresponding numerical solutions of Eq. (12) were obtained by means of a relaxation technique [6]. The effect of the nonlinearity, β (the scaled polarizability), is illustrated in the left panel of Fig. 5. As shown, VA and the numerical results are in good agreement. The variational and numerical results for the soliton profile corresponding to $\mu = 0$ and $\beta = 6$ are compared in the right panel of the figure. We also note that for large β the contribution of the linear lattice grows, leading to more accurate results for the variational approach [16].

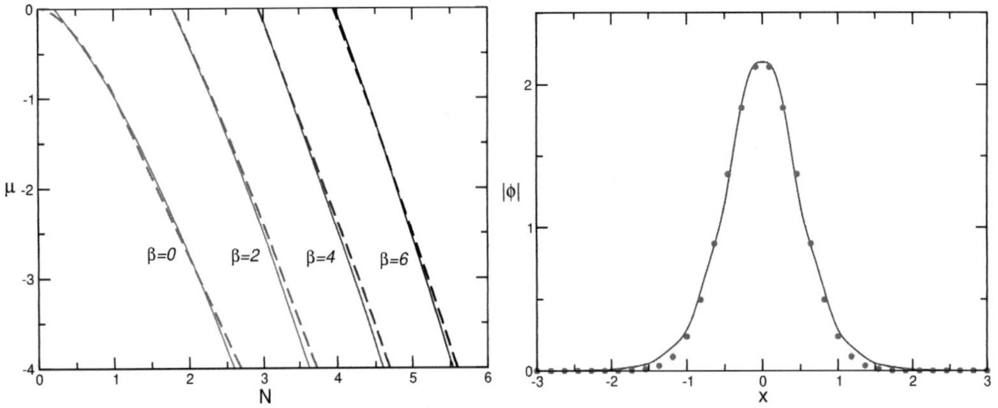

Figure 5. In the left panel, we show μ versus N from direct numerical solutions of Eq. (12) (solid lines), and from the corresponding VA (dashed lines). The parameter β is indicated inside the frame, with the other parameters fixed to $g = 0$, $f_0 = 1$, $f_1 = 0.5$ and $\Lambda = 0.5$ ($k = 4\pi$). In the right panel, we have the profile of the wave function centered in $x = 0$, for $\beta = 6$ and $\mu = 0$. The full-numerical solution (solid line) is compared with VA (dotted line).

In the left panel of Fig. 6, for $k = 2$ ($\Lambda = \pi$) and $\beta = 6$, with the other parameters the same as in Fig. 5, the numerical results for the μ versus N are shown in three distinct regions: two stable (1 and 3) and one unstable (region 2). The profiles, displayed in the right panel, clearly show that the wave function profiles are centered at $x = \pi/2$ in the stable region 1. In the unstable region 2, it is verified a transition from the stable region 1 to the stable region 3, with the center of the profiles moving to $x = 0$. As seen in Fig. 6, the VA gives a perfect agreement with the numerical solutions in the region 1 (where the center of the profile is located at $x = \zeta = \pi/2$), when $\mu > -2$, and in region 3 (where the center of the profile is at $x = \zeta = 0$), when $\mu < -9$. However, the simple Gaussian ansatz we have used cannot follow the behavior presented by the numerical results in region 2.

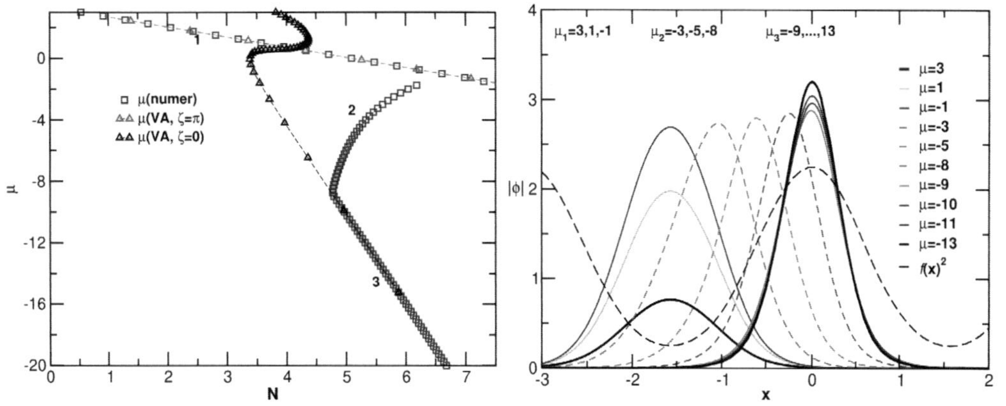

Figure 6. In the left panel, μ is given in terms of N, by considering numerical and variational results, for $\Lambda = \pi$ (i.e., $k = 2$). Other parameters are $g = 0$, $f_0 = 1$, $f_1 = 0.5$ and $\beta = 6$. In this panel, we indicate three regions (1, 2 and 3) of the numerical solutions, following the variation of the chemical potential (μ_1, μ_2 and μ_3), to identify the corresponding profiles in the right panel. The modulation function, $f^2(x)$, is shown by the black-dashed line in the right one.

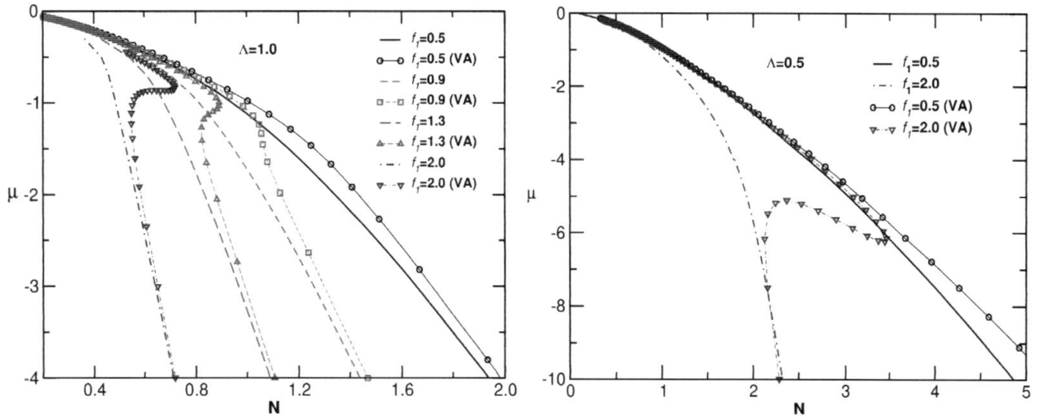

Figure 7. Plots for μ versus N are shown for different values of f_1 (as indicated inside the frames), with $\beta = 0$, $f_0 = 1$, and $\Lambda = 1$ (left panel) or 0.5 (right panel).

Numerical results, for μ versus N compared with VA findings for $\beta = 0$, are shown in Fig. 7, for different values of f_1, with $\Lambda = 1$ (left panel) and 0.5 (right panel). The VK criterion for stability, $\partial\mu/\partial N < 0$ [7], is well verified by our numerical results. Simulations of temporal evolutions that were performed validate the VK criterion in the present model. However, the VA results cannot follow the results to the full extension, besides the fact that they present very good agreement for large negative values of μ. As seen in the left panel of Fig. 7, for $\Lambda = 1$ the VA correctly predicts the stability and converges to numerical results for $\mu < -1.5$ at all values of f_1. In the case of $\Lambda = 0.5$ (right panel), the VA results are equally accurate at $\mu < -6$. On the other hand, for $f_1 = 2$ and $\Lambda = 0.5$, the VA solutions deviate from the full numerical results in an interval of N, where one can obtain more than one μ for the same N.

3.3. Dynamics of bright solitons

In order to conclude this section, we address some results on the mobility of solitons and their collisions. The soliton motion in nonlinear lattices was previously considered in Refs. [17, 18, 19]. First, we present full numerical solutions of the 1D GPE (6), exploring a parameter region for finding stable bright solitons. This is followed by consideration of a dynamical version of the VA, with some results for frequencies of oscillations of perturbed solitons being compared with full numerical calculations.

The propagation of a soliton is presented in the two panels of Fig. 8. In the left panel, for $\mu = -1$ and $N \approx 1.02$, we show the soliton propagation by considering a time interval from 0 to 20, and velocity equal to 1 (dimensionless units). In the right panel, for $\mu = -10$ and $N \approx 4.86$, with profiles separated by time intervals $\Delta t = 2$, we show that the soliton is being trapped at a fixed position. The parameters, in both the cases, are $\beta = g = 0$, $f_0 = 1$, $f_1 = 0.5$, and $\Lambda = 0.5$.

In Fig. 9 it is given a density plot for the interaction of two solitons, with zero phase difference between them, such that they attract each other. The parameters are the same as in Fig. 8. In such a case, we noticed a transition from a bound state to a breather. These results are extracted from figures 7 and 8 of Ref. [2], where more details are presented.

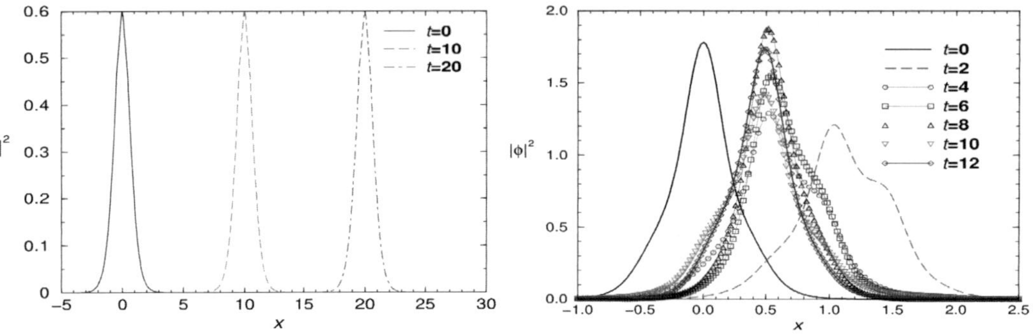

Figure 8. Propagation of a soliton. In the left panel, soliton profiles are shown with time intervals $\Delta t = 10$ (the velocity is 1), for $\mu = -1$ and $N \approx 1.02$. In the right panel, $\mu = -10$, $N \approx 4.86$, we show the soliton being trapped at a fixed position, with the profiles given in intervals $\Delta t = 2$. In both cases, $f_0 = 1$, $f_1 = 0.5$ and $\Lambda = 0.5$ (with $\beta = 0$).

4. General conclusions

We report in section II results obtained on the existence of solitons and the corresponding dynamics from the case of cross-combined linear and nonlinear optical lattices, within a full 3D model. As shown, in the full 3D OL cross-combined case, we can obtain families of 3D solitons, which can be stable for both attractive and repulsive interactions. In perspective of potential applications, we should point out that, besides providing an alternate way to create stable solitons in 3D, with cross combined linear and nonlinear optical lattices (both in BEC and nonlinear optics), these results can also be useful in practical applications, opening the possibility to manage stable 3D solitons through spatial modulations of the scattering length in one of the OL directions. In section III, we report investigations related to another model, considering a 1D reduced case. Within this model, we study the conditions for the existence of bright solitons considering induced dipole-dipole interactions. The results were verified by comparison with numerical solutions of the respective 1D GPE. The stability of the soliton solutions is verified by considering the VK criterion. We also report investigations related to the dynamics of solitons and the interactions between them, including merger into breathers.

Acknowledgments

We thank the Brazilian agencies FAPESP, CNPq and CAPES for partial support.

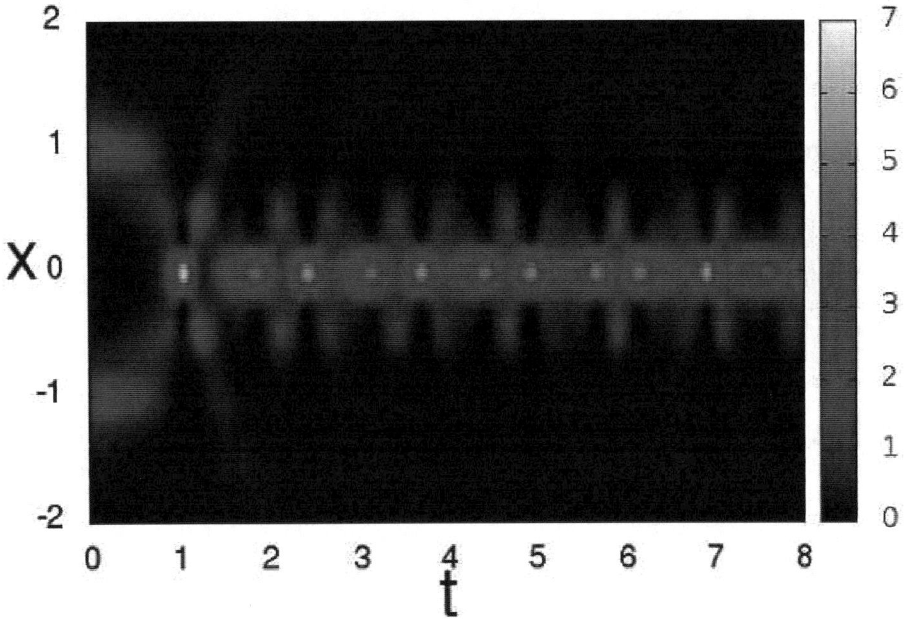

Figure 9. The interaction of two solitons represented in a density plot (scale in the right of the figure), for $\mu = -10$ and $N \approx 4.86$, with zero average velocity. The soliton position x is shown as a function of time t. The parameters are the same as in Fig. 8, with zero phase difference between the solitons. A transition from a bound state to a breather is evidenced.

References

[1] Abdullaev F Kh, Gammal A, da Luz H L F, Salerno M and Tomio L 2012 *J. Phys.* B **45** 115302
[2] Abdullaev F Kh, Gammal A, Malomed B A and Tomio L 2014 *J. Phys.* B **47** 075301
Abdullaev F Kh, Gammal A, Malomed B A and Tomio L 2013 *Phys. Rev.* A **87** 063621
[3] Abdullaev F Kh, Gammal A, da Luz H L F and Tomio L 2007 *Phys. Rev.* A. **76** 043611
[4] Fedichev P O, Kagan Yu, Shlyapnikov G V and Walraven J T M 1996 *Phys. Rev. Lett.* **77** 2913
[5] da Luz H L F, Abdullaev F Kh, Gammal A, Salerno M and Tomio L 2010 *Phys. Rev.* A. **82** 043618
[6] Brtka M, Gammal A and Tomio L 2006 *Phys. Lett.* A **359** 339
[7] Vakhitov N G and Kolokolov A A 1973 *Izv. Vyssh. Uchebn. Zaved. Radiofiz.* **16** 1020 [English translation: 1973 *Radiophys. and Quantum Electron.* **16** 783]
[8] Li Y, Liu J, Pang W and Malomed B A 2013 *Phys. Rev.* A **88** 053630
[9] Cuevas J, Malomed B A, Kevrekidis P G and Frantzeskakis D J 2009 *Phys. Rev.* A **79** 053608
[10] Baranov M 2008 *Phys. Rep.* **464** 71
[11] Bortolotti D C E *et al.* 2006 *Phys. Rev. Lett.* **97** 160402
[12] Kawaguchi Y, Saito H and Ueda M 2006 *Phys. Rev. Lett.* **96** 080405
[13] Swislowski T *et al.* 2011 *Phys. Rev.* A **83** 063617
[14] A de Paz *et al.* 2013 *Phys. Rev.* A **87** 051609
[15] Wall M L, Maeda K and Carr L 2013 *Ann. Phys.* **525** 845
[16] Kartashov Y V, Malomed B A and Torner L 2011 *Rev. Mod. Phys.* **83** 247
[17] Abdullaev F Kh and Garnier J 2005 *Phys. Rev.* A **72** 061605(R)
[18] Abdullaev F Kh, Abdumalikov A A and Galimzyanov R M 2007 *Phys. Rev.* A **367** 149
[19] Kartashov Y V, Vysloukh V A and Torner L 2008 *Opt. Lett.* **33** 1774

Classical-like wakes past elliptical obstacles in atomic Bose-Einstein condensates

G. W. Stagg, A. J. Allen, C. F. Barenghi and N. G. Parker

Joint Quantum Centre Durham–Newcastle, School of Mathematics and Statistics, Newcastle University, Newcastle upon Tyne, NE1 7RU, United Kingdom

E-mail: nick.parker@newcastle.ac.uk

Abstract. We reinvestigate numerically the classic problem of two-dimensional superfluid flow past an obstacle. Taking the obstacle to be elongated (perpendicular to the flow), rather than the usual circular form, is shown to promote the nucleation of quantized vortices, enhance their subsequent interactions, and lead to wakes which bear striking similarity to their classical (viscous) counterparts. Then, focussing on the recent experiment of Kwon *et al.* (arXiv:1403.4658) in a trapped condensate, we show that an elliptical obstacle leads to a cleaner and more efficient means to generate two-dimensional quantum turbulence.

1. Introduction

A prototypical example in classical (viscous) fluid dynamics is the incompressible flow past a cylinder, depicted in Fig. 1. The nature of the flow is determined by the Reynolds number $\mathcal{R} = vd/\nu$, where v is the flow velocity away from the obstacle, d is the obstacle diameter, and ν is the fluid's kinematic viscosity. For $\mathcal{R} \lesssim 5$ the flow is laminar around the obstacle. For $5 \lesssim \mathcal{R} \lesssim 50$ a steady symmetric wake forms downstream of the obstacle. For $10^2 \lesssim \mathcal{R} \lesssim 10^5$ the wake becomes asymmetric and time dependent, forming the famous Bénard–von Kármán vortex street structure. At even higher $\mathcal{R}$, the wake becomes turbulent.

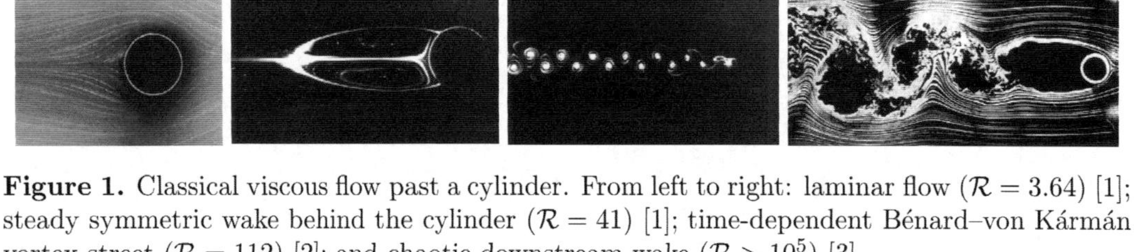

Figure 1. Classical viscous flow past a cylinder. From left to right: laminar flow ($\mathcal{R} = 3.64$) [1]; steady symmetric wake behind the cylinder ($\mathcal{R} = 41$) [1]; time-dependent Bénard–von Kármán vortex street ($\mathcal{R} = 112$) [2]; and chaotic downstream wake ($\mathcal{R} > 10^5$) [3].

In a superfluid the situation is fundamentally different [4]. There is no viscosity so a Reynold's number cannot be defined. Landau's criterion states that dissipation through elementary excitations kicks in only above a critical velocity; for the dilute Bose gas relevant here this critical velocity is approximately the speed of sound. Meanwhile, where eddies do exist in the fluid, they are constrained by quantum mechanics to possess quantized vorticity and circulation.

In their pioneering simulations of the 2D nonlinear Schrödinger equation (NLSE), Frisch *et al.* [5] showed that superflow past a circular obstacle becomes dissipated through the nucleation of quantized vortices at the object, which then get carried downstream.

Atomic Bose-Einstein condensates provide a clean and highly controllable vehicle for studying superfluid phenomena, from persistent currents [6] and Josephson junctions [7] to quantized vorticity [8] and quantum turbulence [9, 10, 11, 12]. Various experiments have probed the response of the condensate to a moving obstacle [10, 11, 13, 14, 15]. The obstacle in these cases takes the form of a repulsive Gaussian-shaped potential created by a blue-detuned laser beam. Of these experiments, [10, 11] employed a highly flattened condensate such that the dynamics were effectively two-dimensional and provided a close analog of the cylindrical problem. Moreover, they employed this scenario to generate two-dimensional quantum turbulence, a dynamical state of the condensate characterized by a statistical distribution of excitations over a large range of lengthscales.

In the mean-field picture, such a 2D condensate can be parameterized by a macroscopic wavefunction $\psi(x, y, t)$ whose behaviour follows the 2D Gross-Pitaevskii equation [16],

$$i\hbar\frac{\partial\psi}{\partial t} = \left(-\frac{\hbar^2}{2m}\nabla^2 + V(x,y,t) + g|\psi|^2 - \mu\right)\psi. \tag{1}$$

Here $g = 4\pi\hbar^2 a_s/\sqrt{2\pi}l_z m$ is a nonlinear coefficient arising from the contact atomic interactions, with a_s being the atomic s-wave scattering length and m the atomic mass, μ is the chemical potential of the condensate, and $V(x, y, t)$ specifies the potential exerted on the condensate.

Whereas Frisch *et al.* [5] considered a "hard" cylinder, it is more natural in the context of atomic condensates to employ a "soft" Gaussian potential of the form $V(x, y) = V_0 e^{-(x^2+y^2)/d^2}$, where d and V_0 parameterize the width and amplitude of the potential. In practice this changes the quantitative, but not the qualitative, behaviour. Sasaki *et al.* [17] recently provided an extensive picture of 2D superflow past such a Gaussian obstacle. The flow regimes are depicted in Fig. 2, based on simulations of the 2D GPE using similar parameters to [17]. At low flow velocity [(a)], the fluid undergoes smooth laminar flow around the obstacle. The streamlines of this flow are symmetric about $x = 0$, as in perfect potential flow. At a critical flow velocity, the local fluid velocity (which is highest at the poles of the obstacle) exceeds the speed of sound, breaking Landau's criterion. Vortex pairs of opposite sign are nucleated periodically and drift downstream, forming a collimated wake of vortex pairs which are widely separated from each other [(b)]. For higher velocities, alternating pairs of like-signed vortices are nucleated [(c)]. At higher velocities, vortex nucleation becomes highly irregular [(d)], forming a chaotic downstream distribution of vortices and sound (density) waves. These quantum fluid flow patterns bear some analogy to the classical flow patterns of Fig. 1, particularly for the laminar flow and chaotic regimes. The alternating like-signed vortices form a somewhat primitive analog of the Bernárd-von Kármán vortex street, while the vortex-antivortex pairs have no obvious classical analog.

In this work we discuss the rich variety of quantum wake regimes, often in close analog to the classical counterparts, which can be obtained via the simple modification of the obstacle to an *elliptical* shape. In other words, we employ an elliptical potential of the form $V(x, y) = V_0 e^{-(\varepsilon^2 x^2+y^2)/d^2}$, where ε is the ellipticity. This is readily achieved experimentally by using cylindrical focussing (rather than spherical focussing) of the laser beam which induces the potential. First, we explore these dynamics in a homogeneous system, which serves to demonstrate the salient behaviour of superfluid flow past an elliptical obstacle, away from boundaries and trap-induced density inhomogeneities which also influence the vortex dynamics. Then we consider the realistic scenario provided by the recent experiment of Kwon *et al.* [11], and investigate how the dynamics therein are modified by an elliptical obstacle.

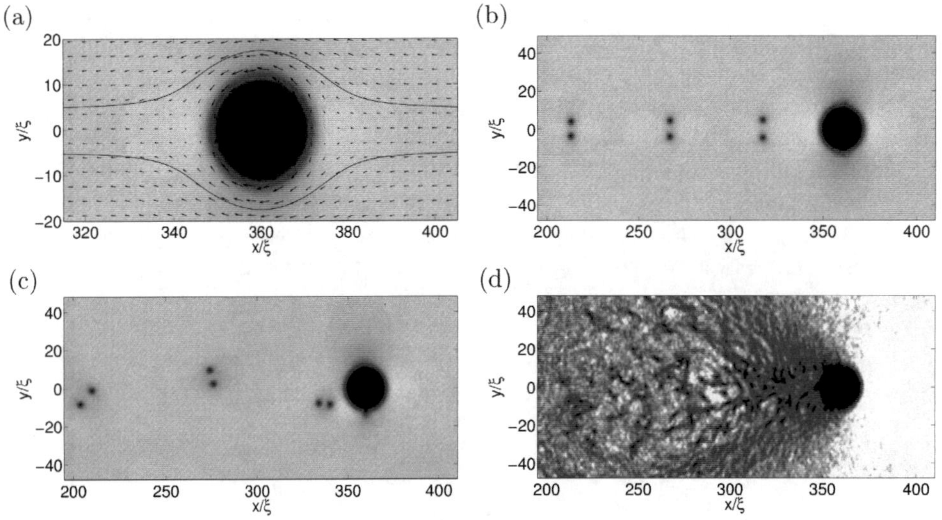

Figure 2. Condensate density during flow (in the positive x-direction) past a circular obstacle ($d = 5\xi$) at various flow speeds. Yellow/black represent high/low density. For reference the critical velocity here is $v_c \approx 0.36c$. (a) Laminar flow at a sub-critical flow speed ($v = 0.3c$). The fluid velocity vector field and two streamlines illustrate the flow pattern. (b) Nucleation of vortex-antivortex pairs ($v = 0.365c$). (c) Nucleation of like-signed vortex pairs ($v = 0.37c$). (d) Chaotic vortex nucleation and generation of strong sound waves, forming a turbulent wake ($v = 0.9c$). Length and speed are expressed in terms of the healing length $\xi = \hbar/\sqrt{mn_0 g}$ and speed of sound $c = \sqrt{n_0 g/m}$.

2. Flow Past an Elliptical Obstacle: Homogeneous Condensate

In the following we simulate the 2D GPE for an untrapped 2D condensate with uniform density n_0, flowing at speed v along x (modelled via the addition of a Galilean term $i\hbar v \frac{\partial}{\partial x}\psi$ to the right-hand side of the GPE (1)). A static elliptical obstacle potential, elongated in y ($\varepsilon > 1$) and with arbitrary size d (expressed in terms of the healing length $\xi = \hbar/\sqrt{mn_0 g}$), pierces the condensate. Further technical details are provided elsewhere [18]. Note that the qualitative behaviour in insensitive to additional noise.

In the following we describe the behaviour at increasing flow speed v. At sub-critical speeds, the condensate undergoes laminar flow around the obstacle, much like for the circular obstacle [Fig. 2(a)]. However, due to the decreased radius of curvature at the poles of the obstacle, the local fluid velocity is higher there than for the circular obstacle. As such the critical velocity is reduced with increasing ε, as revealed by GPE simulations (data points in Fig. 3). Approximating the BEC as an inviscid Euler fluid, the maximum local fluid velocity is $v_{\max} = (1 + \varepsilon)v$, which will exceed the speed of sound c at a critical flow speed $v_c = c/(1 + \varepsilon)$. Despite its crudeness, this approximation (black dashed line in Fig. 3) predicts the correct scaling with ε. Inclusion of higher-order effects, e.g. the reduction of density around the obstacle due to the raised fluid speed (Bernoulli's theorem) [19, 18], bring the model into more precise quantiative agreement (red solid line in Fig. 3).

For $v > v_c$ the ellipticity also increases the rate at which vortices are nucleated. Importantly, a large vortex nucleation rate implies that vortices are close together and thereby interact strongly. (For the circular obstacle, while a higher rate of vortex shedding can be achieved by using a faster flow, this also carries the vortices downstream more quickly, and so the vortex interactions

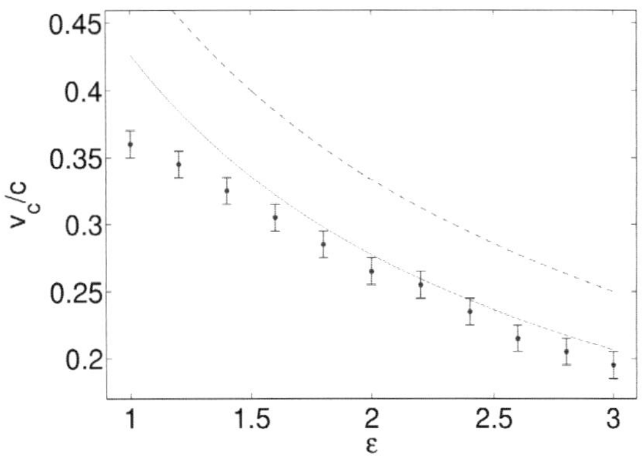

Figure 3. Critical velocity against obstacle ellipticity ε ($d = 10\xi$) according to GPE simulations (blue bars), the Euler prediction $v_c = c/(1+\varepsilon)$ (black dashed), and the corrected Euler prediction [19, 18] (solid red).

are not significantly increased.)

The rapid nucleation of vortex-antivortex pairs leads to a cluster of vortices in the one half-plane and antivortices in the other half-plane [Fig. 4(a)]. In each cluster, the same-sign vortices tend to corotate around each other, such that the cluster approximates a macroscopic vortex. A steady-state is reached whereby the freshly-nucleated vortices entering the clusters is balanced by vortex pairs which escape the cluster downstream. The close analogy to classical flow becomes apparent in plotting the trajectories of the vortices [Fig. 4(c)]; two symmetric lobes of circulating vortices are revealed, in striking similarity to the symmetric classical wake in Fig. 1.

While the above flow pattern is symmetric in y, it is dynamically unstable to the formation of

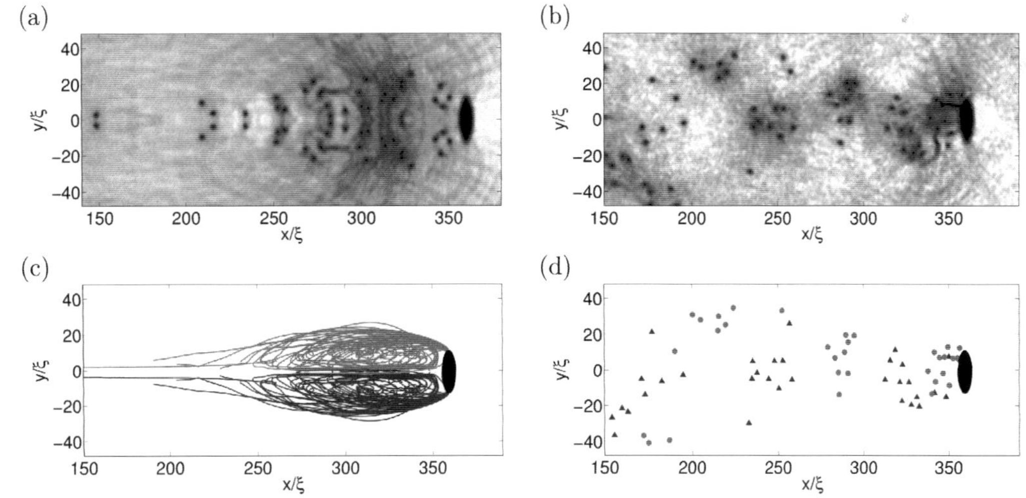

Figure 4. Condensate flow ($v = 0.52c$) past an elliptical obstacle ($d = 5\xi$, $\varepsilon = 3$). (a) Symmetric wake. (b) Asymmetric wake. (c) Time-integrated vortex trajectories for the symmetric wake. (d) Vortex positions for the asymmetric wake shown in (b). Red and blue represent vortices of opposite circulation.

an asymmetric pattern. A typical snapshot of this is shown in Fig. 4(b), with Fig. 4(d) indicating the corresponding position and circulation of the vortices. This pattern is characterised by the nucleation of several like-signed vortices from one pole of the obstacle, followed by nucleation of several vortices of the opposite sign from the opposite pole. The result is a wake consisting of alternating clusters of like-signed vortices. The number of vortices in each cluster increases with increasing v, d and ε. This wake is a strongly analogous to the classical Bénard–von Kármán vortex street shown in Fig. 1(c). Indeed, if one imagines course-graining over the clusters, one obtains a flow pattern of the form of the Bénard–von Kármán vortex street. At even higher flow speeds, the vortex nucleation becomes irregular, forming a chaotic wake like that presented for the circular obstacle in Fig. 2(d).

Overall, we see that the superfluid flow past an elliptical obstacle leads to four flow pattern regimes, each of which is bears striking analogy to the classical viscous cases depicted in Fig. 1. These findings confirm the intuition that a sufficiently large quanta of vorticity/circulation reproduces classical results.

3. Flow Past an Elliptical Obstacle: Trapped Condensate

Next we explore superfluid flow past an obstacle within a trapped condensate. We follow the recent experiment by Kwon *et al.* [11], which employed a highly-oblate and effectively 2D condensate with 1.8×10^6 ^{23}Na atoms. The in-plane harmonic trapping is circularly symmetric $V_{\text{trap}}(x, y) = \frac{1}{2}m\omega_r^2(x^2 + y^2)$ with radial frequency $\omega_r = 2\pi \times 15$ Hz. A blue-detuned laser obstacle ($V_0 = 15\mu$ and $d = 10.6\mu$m) pierces the condensate at the origin, following which the harmonic trap is shifted by 37μm in x at speed v. For reference, the speed of sound at the trap centre is $c_0 = \sqrt{n_0 g/m} \approx 4.6$ mm/s (where the 0-subscript denotes peak values), decreasing further out in accord with the decreasing density as $c(\mathbf{r}) = c_0\sqrt{n(\mathbf{r})/n_0}$. The laser obstacle is then adiabatically removed over 0.4s. As observed experimentally [11], and consistent with GPE simulations [20], vortices become nucleated for $v \gtrsim 0.5$mm/s, with the number of vortices generated increasing as v is further increased.

First we consider the circular obstacle employed in the experiment with $v = 1.4$ mm/s, with the condensate evolution depicted in Fig. 6. The translation of the trap initiates a centre-of-mass mode of the condensate. As the condensate first "sloshes" to the left past the obstacle, a large number of vortices are nucleated in an irregular manner from each pole (53 ms). Significant energy is also driven into sound (density) waves, shock waves and shape excitations, which are visible in these plots. Before the vortices are washed far downstream, the condensate sloshes to the right, generating further vortices and mixing together the old and new vortices (80 ms). This process repeats several times as the condensate sloshes. When the obstacle is removed a disordered arrangement of vortices is apparent (0.5 ms). This is consistent with the experimental observations, and synonymous with a state of two-dimensional quantum turbulence. Following this, there is a slow decay in the vortex number. In our simulations, which are free of dissipation, this occurs through the annihilation of vortices of opposite circulation; experimentally [11], and in the presence of dissipation terms to the GPE [20], vortices will also undergo thermal dissipation whereby they drift to the condensate edge and are lost.

We now consider the case where the circular obstacle is replaced with an elliptical one (with $\varepsilon = 3$) and (for reasons to be made evident shortly) a reduced speed $v = 0.8$ mm/s. The vortices are now nucleated in a *regular* fashion, forming two clusters of like-signed vortices (53 ms), much like in the homogeneous system. No strong sound waves are visible in the condensate. As the condensate changes direction, again two like-signed clusters are formed, but now with alternate polarity (80 ms). Over time, these vortices mix together, and once the obstacle is removed, the condensate and vortex distribution is similar to the circular case. Indeed, the number of vortices in the condensate is essentially identical for both cases ($\sim$ 80 vortices), despite the reduced speed for the ellipse.

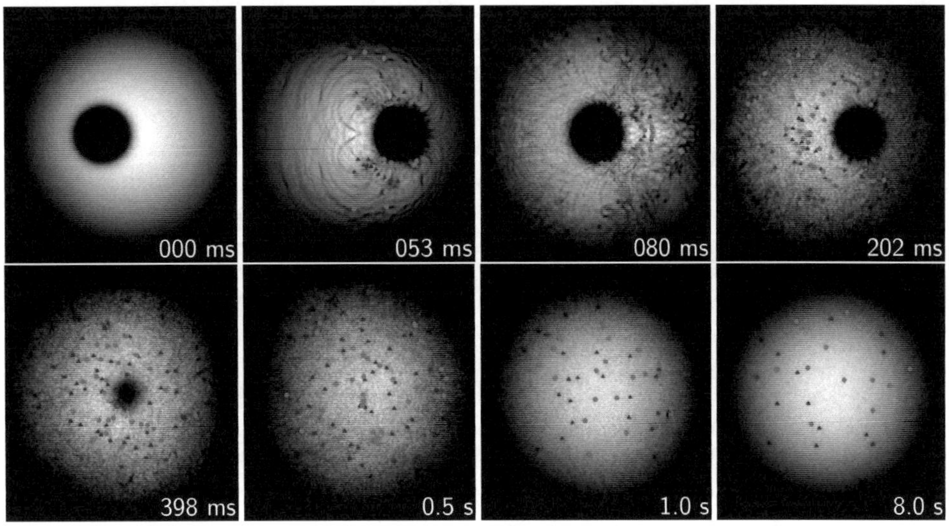

Figure 5. Condensate flow past a circular obstacle with speed $v = 1.4$mm/s, following the experiment of Ref. [11]. Each field of view is of size $[170\mu\text{m}]^2$, centered to best display the condensate. Vortices with positive (negative) circulation are highlighted by red circles (blue triangles).

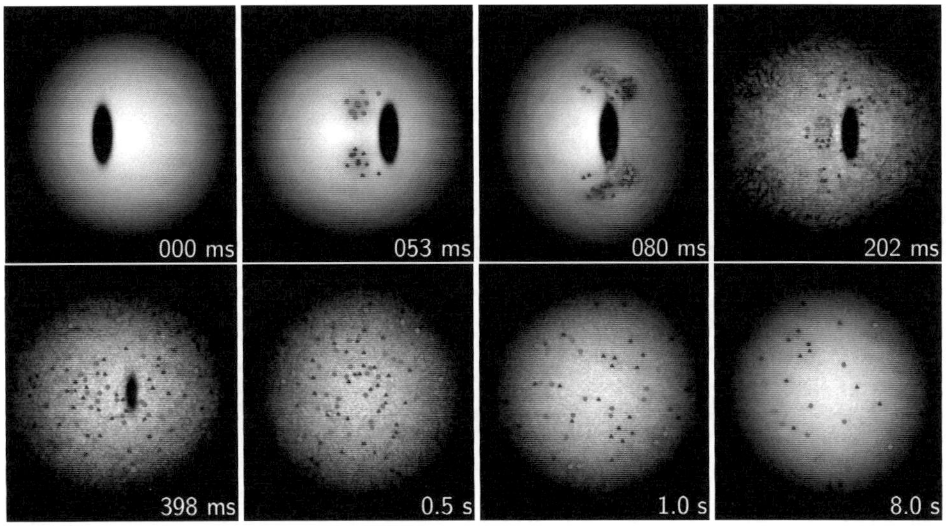

Figure 6. As for Fig. 6 but with an elliptical obstacle ($\varepsilon = 3$) and a speed $v = 0.8$mm/s.

At the same speed v, the elliptical obstacle will typically generate a larger number of vortices than the circular obstacle. For example, for $v = 0.8$ mm/s, the ellipse and circle generate approximately 80 and 60 vortices, respectively (measured following removal of the obstacle). However, for sufficiently high speeds ($v \gtrsim 1$ mm/s) the enhancement due to the elliptical obstacle diminishes; in both cases the number of vortices driven into the condensate saturates. This suggests that for a given condensate there exists a natural maximum vortex density which can be supported. This limit is likely to be imposed by the quadratic increase of vortex annihilation events with vortex density [11], which will act to cancel out the freshly-nucleated vortices.

4. Conclusions

Through simulations of the GPE we have examined two-dimensional superfluid flow past an elliptical, rather than a circular, obstacle. This simple and experimentally feasible modification enables control over the critical speed for vortex nucleation, the vortex nucleation rate, and the ensuing vortex interactions. For an ellipse which is elongated perpendicular to the flow, the enhanced vortex interactions promote the formation of quantum wakes (composed of multiple singly-charged vortices) which closely resemble those of classical (viscous) flow past a cylinder. Our observations [18] suggest the emergence of classical behaviour from the presence of a sufficient number of quanta

When incorporated into the experimental scenario of Ref. [11], such classical-like wakes are formed initially but become disrupted by the limited size and sloshing mode of the system. Employing a condensate which is elongated in the direction of translation would promote the formation of longer-lived vortex clusters, and provide an experimental route to study these wakes and more general coherent structures of quantized vortices [21, 22]. Meanwhile, for the purposes of generating two-dimensional quantum turbulence, the elliptical obstacle allows for more efficient (more vortices for a given speed) and cleaner (reduced disruption due to sound and shock waves) nucleation of vortices.

Acknowledgments

This work made use of the facilities of N8 HPC provided and funded by the N8 consortium and EPSRC (Grant No. EP/K000225/1). The Centre is co-ordinated by the Universities of Leeds and Manchester.

[1] S. Taneda, J. Phys. Soc. Jpn. **11**, No. 3, 302 (1956).
[2] S. Taneda, J. Phys. Soc. Jpn. **50**, No. 4, 1398 (1981).
[3] Photograph reproduced with permission by Thomas Corke and Hassan Nagib.
[4] J. Annett, *Superconductivity, Superfluids and Condensates* (Oxford University Press, Oxford, 2004).
[5] T. Frisch, Y. Pomeau, and S. Rica, Phys. Rev. Lett. **69**, 1644 (1992).
[6] Ryu *et al.*, Phys. Rev. Lett. **99**, 260401 (2007); S. Moulder *et al.*, Phys. Rev. A **86**, 013629 (2012).
[7] R. Gati and M. K. Oberthaler, J. Phys. B **40**, R61 (2007).
[8] A. L. Fetter, J. Low Temp. Phys. **161**, 445 (2010).
[9] E. A. L. Henn *et al.*, Phys. Rev. Lett. **103**, 045301 (2009).
[10] T. W. Neely, *et al.*, Phys. Rev. Lett. **104**, 160401 (2010).
[11] W. J. Kwon *et al.*, arXiv:1403.4658 (2014).
[12] M. Tsubota, J. Phys.: Condens. Matter **21**, 164207 (2009); A. C. White, B. P. Anderson and V. S. Bagnato, Proc. Nat. Acad. Sci. USA **111**, 4683 (2014); A. J. Allen, N. G. Parker, N. P. Proukakis and C. F. Barenghi, J. Phys.: Conf. Ser. **544**, 012023 (2014).
[13] C. Raman, *et al.*, Phys. Rev. Lett. **83**, 2502 (1999).
[14] R. Onofrio *et al.*, Phys. Rev. Lett. **85**, 2228 (2000).
[15] S. Inouye *et al.*, Phys. Rev. Lett. **87**, 080402 (2001).
[16] L. P. Pitaevskii and S. Stringari, *Bose-Einstein Condensation* (Oxford University Press, Oxford, 2003).
[17] K. Sasaki, N. Suzuki, and H. Saito, Phys. Rev. Lett. **104**, 150404 (2010).
[18] G. W. Stagg, N. G. Parker and C. F. Barenghi, J. Phys. B **47**, 095304 (2014).
[19] T. Winiecki, Ph.D. thesis, The University of Durham, 2001.
[20] G. W. Stagg, A. J. Allen, N. G. Parker and C. F. Barenghi, arXiv:1408:3268 (2014).
[21] A.W. Baggaley, J. Laurie, and C.F. Barenghi, Phys. Rev. Lett. **109**, 205304 (2012).
[22] M. T. Reeves, T. P. Billam, B. P. Anderson and A. S. Bradley, Phys. Rev. A **89**, 053631 (2014).

Mutual information–energy inequality for thermal states of a bipartite quantum system

Aleksey Fedorov[1,2] **and Evgeny Kiktenko**[2,3]

[1]Russian Quantum Center, 100 Novaya Street, Skolkovo, Moscow 143025, Russia
[2]Bauman Moscow State Technical University, 5 2nd Baumanskaya, Moscow 105005, Russia
[3]Geoelectromagnetic Research Center of Schmidt Institute of Physics of the Earth, Russian Academy of Sciences, P.O. 30 Troitsk, Moscow Region 142190, Russia.

E-mail: `akf@rqc.ru`; `evgeniy.kiktenko@gmail.com`

Abstract. In this work, we consider an upper bound for the quantum mutual information in thermal states of a bipartite quantum system. This bound is related with the interaction energy and logarithm of the partition function of the system. We demonstrate the connection between this upper bound and the value of the mutual information for the bipartite system realized by two spin-1/2 particles in the external magnetic field with the XY-Heisenberg interaction.

1. Introduction

A paradigm of consideration of quantum systems as a potential basis for applications in various technologies, *e.g.*, information processing, communications, and metrology, inspires a new wave of studies many-body systems with focus on quantum correlation, quantum entanglement, and discord phenomena [1–3]. In particular, outstanding findings are obtained for spin models [4–21]. Remarkable progress in experiments with electronic spins [5], nuclear spins [6], quantum dots [7, 8], and atomic ensembles in optical lattices [9, 10], allows to design quantum systems with effective spin chain Hamiltonians. This fact opens intriguing prospectives for investigation of fundamental concepts in these systems and revealing novel interesting phenomena.

Paramount importance of the XY Hamiltonian spin model [4] has been demonstrated [13–18]. In particular, it has been shown that XY Hamiltonian appears in quantum dot spins and cavity QED systems towards to their applications for quantum information processing [11–15].

Recently, a novel class of so-called entropy–energy inequalities of has been revealed [22]. This class of inequalities connects the mean value of the Hamiltonian and the entropy of an arbitrary state for a given Hamiltonian of the system. In the present work, we follow the strategy of Ref. [22] to construct an upper bound for the quantum mutual information in thermal states of a bipartite quantum system. We consider the relation between this bound and the value of the mutual information for the bipartite system realized by two spin-1/2 particles in the external magnetic field with the XY-Heisenberg interaction.

Our paper is organized as follows. In Section 2, we construct the inequality for the quantum mutual information and energy in bipartite quantum systems. We consider this inequality in the framework of the model of spin-1/2 particles with the XY Hamiltonian in the external magnetic field in Section 3. In Section 4, we give our conclusion and prospectives.

2. Mutual information–energy inequality

Consider a bipartite quantum system AB, which is described by density operator in the Hilbert space $\mathcal{H}_{AB} = \mathcal{H}_A \otimes \mathcal{H}_B$. The Hamiltonian of the system has the following generic form:

$$H = H_A \otimes \mathbb{I}_B + \mathbb{I}_A \otimes H_B + H_{\rm int}, \tag{1}$$

where H_A and H_B describe the subsystems without an interaction, $H_{\rm int}$ describes the coupling between subsystems, $\mathbb{I}_A$ and $\mathbb{I}_B$ are the identity operators in $\mathcal{H}_A$ and $\mathcal{H}_B$ respectively.

We study the state of thermal equilibrium, which is given by the following expression:

$$\rho_{AB} = \frac{1}{Z_{AB}} \exp{(-\beta H)}, \qquad Z_{AB} = {\rm Tr}[\exp{(-\beta H)}], \tag{2}$$

where β is inverse temperature in the corresponding energy units and Z is the partition function. Taking partial traces in Eq. (2), we obtain the states of subsystem:

$$\rho_A = {\rm Tr}_B[\rho_{AB}], \qquad \rho_B = {\rm Tr}_A[\rho_{AB}]. \tag{3}$$

It is clear that the total energy of state (2) could be represented as a sum of three terms:

$$E = E_A + E_B + E_{\rm int}, \quad E_A = {\rm Tr}[\rho_A H_A], \quad E_B = {\rm Tr}[\rho_B H_B], \quad E_{\rm int} = {\rm Tr}[\rho_{AB} H_{\rm int}]. \tag{4}$$

Following the strategy of Ref. [22], to derive an inequality between the mutual information and energy in bipartite thermal state, we start with the notation of the relative entropy [1],

$$S(\rho||\sigma) = {\rm Tr}[\rho \ln \rho - \rho \ln \sigma] \geq 0. \tag{5}$$

This relation is a measure of difference between states with density matrices ρ and σ. We note that the identity $S(\rho||\sigma) = 0$ takes place in the case $\rho = \sigma$ only. Using the von Neumann entropy

$$S(\rho) = -{\rm Tr}[\rho \ln \rho], \tag{6}$$

one can rewrite Eq. (5) as follows:

$$S(\rho||\sigma) = -S(\rho) - {\rm Tr}[\rho \ln \sigma]. \tag{7}$$

By expanding the trivial identity $S(\rho_{AB}||\rho_{AB}) = 0$ and substituting the explicit form of thermal state (2) in second term in the RHS of Eq. (7), we obtain:

$$0 = -S(\rho_{AB}) - {\rm Tr}[\rho_{AB}(-\beta H - \ln Z_{AB})] = -S(\rho_{AB}) + \beta E + \ln Z_{AB}, \tag{8}$$

or, equivalently,

$$S(\rho_{AB}) = \beta E + \ln Z_{AB}. \tag{9}$$

Let us consider the relative entropy between the state ρ_A of the subsystem A and an another thermal state $\tilde{\rho}_A$,

$$\tilde{\rho}_A = \frac{1}{Z_A} \exp{(-\beta H_A)}, \qquad Z_A = {\rm Tr}[\exp{(-\beta H_A)}]. \tag{10}$$

In the general case, we have

$$0 \leq S(\rho_A||\tilde{\rho}_A) = -S(\rho_A) - {\rm Tr}[\rho_A(-\beta H_A - \ln Z_A) = -S(\rho_A) + \beta E_A + \ln Z_A \tag{11}$$

or, equivalently, the following inequality:

$$S(\rho_A) \leq \beta E_A + \ln Z_A. \tag{12}$$

One can see that the same inequality could be obtained for the entropy of ρ_B:

$$S(\rho_B) \le \beta E_B + \ln Z_B, \qquad Z_B = \mathrm{Tr}[\exp(-\beta H_B)]. \tag{13}$$

Finally, by considering the standard definition of the quantum mutual information

$$I(\rho_{AB}) = S(\rho_A) + S(\rho_B) - S(\rho_{AB}) \tag{14}$$

and using Eqs. (9), (12), and (13), we obtain the following inequality:

$$I(\rho_{AB}) \le \beta(E_A + E_B - E_{AB}) + \ln \frac{Z_A Z_B}{Z_{AB}} = -\beta E_{\rm int} + \ln \frac{Z_A Z_B}{Z_{AB}}, \tag{15}$$

which gives the upper bound for mutual information in thermal bipartite state.

For the purposes of convenience, we introduce the notation for the upper bound of the mutual information

$$I_{\rm ub}(\rho_{AB}) = -\beta E_{\rm int} + \ln \frac{Z_A Z_B}{Z_{AB}}. \tag{16}$$

Clearly that in the case of absence of interaction, *i.e.*, at $H_{\rm int} = 0$, the upper bound and mutual information itself goes to zero:

$$Z_{AB} = Z_A Z_B, \qquad I(\rho_{AB}) = I_{\rm ub}(\rho_{AB}) = 0. \tag{17}$$

Inequality (15) is a central object of the present paper.

3. Results for XY-Heisenberg interaction

In this section, we discuss an implementation of inequality (15) for the particular spin model. We consider XY-Heisenberg interaction between two spin-1/2 particles in an inhomogeneous magnetic field acting along Z-axis. The systems is described by the following Hamiltonian:

$$H_A = B_1 \sigma_z, \qquad H_B = B_2 \sigma_z, \qquad H_{\rm int} = g(\sigma_x \otimes \sigma_x + \sigma_y \otimes \sigma_y), \tag{18}$$

where $\sigma_i (i \in (x, y, z)$ stands for the corresponding Pauli matrices, g is the coupling constant, and B_1 and B_2 are magnetic fields acting on particles (magnetic moments of particles are set to be unit).

The ground state of Hamiltonian (18) turns to be the entangled state

$$|\Psi\rangle = \frac{1}{\mathcal{C}} \left(\frac{2g}{\sqrt{(B_1 - B_2)^2 + 2g^2}} |\uparrow\downarrow\rangle - |\downarrow\uparrow\rangle \right), \tag{19}$$

at $B_1 B_2 < g^2$ with $\mathcal{C}$ being the normalization factor, and the separable state at $B_1 B_2 > g^2$ (for positive B_1 and B_2, the state is $|\uparrow\uparrow\rangle$, and it is $|\downarrow\downarrow\rangle$ otherwise).

We start from consideration the case with no external field, *i.e.*, $B_1 = B_2 = 0$. Due to the symmetry of Hamiltonian (18), the reduced states ρ_A and ρ_B of corresponding thermal state ρ_{AB} (2) are proportional to identity operators. On the other hand, we have $Z_A = Z_B = 2$ (as the dimension of the Hilbert space of the subsystems) and $E_A = E_B = 0$. Thus, inequalities (12) and (13) turn into identities and we obtain:

$$I(\rho_{AB}) = I_{\rm ub}(\rho_{AB}) = -\beta E - \ln Z_{AB} + 2\ln 2. \tag{20}$$

Situations of non-zero uniform and nonuniform magnetic fields are presented in Fig. 1. First we note that in Fig. 1a, with entangled ground state, the upper bound (16) explodes at $\beta^{-1} \to 0$

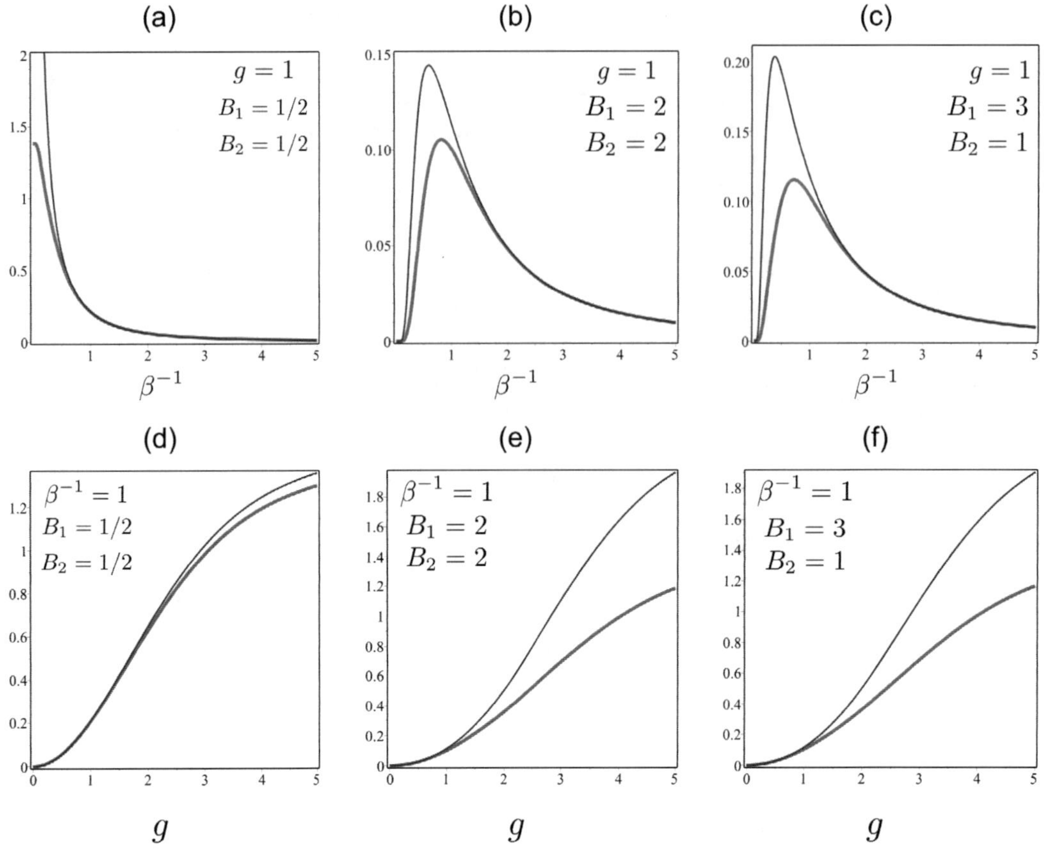

Figure 1. Quantum mutual information (red) given by Eq. (14) and its upper bound (blue) given by Eq. (16): (a)–(c) are shown as functions of temperature β^{-1} at fixed coupling constant $g = 1$; (d)–(f) are shown as functions of the coupling constant g at fixed temperature $\beta^{-1} = 1$. The external magnetic fields are: (a) and (d) $B_1 = B_2 = 1/2$; (b) and (e) $B_1 = B_2 = 2$; (c) and (f) $B_1 = 3, B_2 = 1$.

and becomes even higher than maximal level of mutual information for two-qubit system, *i.e.*, the value $2\ln 2$. However, at high temperatures $\beta^{-1} > g$ the upper bound tends to real value of the mutual information at all considered values of the magnetic field.

The behavior of the quantum mutual information and its upper bound as functions of the coupling constant g at fixed temperature β^{-1}, which is presented in Fig. 1(d-f), confirms with foregoing statement: the difference between $I(\rho_{AB})$ and $I_{\text{ub}}(\rho_{AB})$ grows with increasing of the coupling constant g. It seems like non-uniformity of magnetic field does not bring any significant changes in the general picture.

4. Conclusion

We have introduced the upper bound (16) for the quantum mutual information in thermal states of bipartite quantum systems based on the interaction energy and logarithm of the partition functions. To illustrate our results, we have considered two spin-1/2 particles interacting by XY-Heisenberg model in the inhomogeneous magnetic field. It turned out that in the considered model the introduced the upper bound tends to the value of mutual information for the values of temperatures (in the energy units) higher than the interaction energy. The important open question is whether it is true for all possible models of interactions and dimensions of subsystems.

Acknowledgments
Authors thank V.I. Man'ko and O.V. Man'ko for productive discussions. We thank S.N. Filippov, Y.V. Kurochkin, and Y. Huang for useful comments. We acknowledge the support from the Council for Grants of the President of the Russian Federation (grant SP-961.2013.5, EOK), the Dynasty Foundation (AKF), and RFBR (projects 14-08-00606 & 14-02-00937). We are also grateful to the organizers of the 23rd Laser Physics Workshop (Sofia, July 14–18, 2014) for kind hospitality.

References

[1] Vedral V 2002 *Rev. Mod. Phys.* **74** 197
[2] Amico L, Fazio R, Osterloh A, and Vedral V 2008 *Rev. Mod. Phys.* **80** 517
[3] Modi K, Brodutch A, Cable H, Paterek T, and Vedral V 2012 *Rev. Mod. Phys.* **84** 1655
[4] Lieb E, Schultz T, and Mattis D 1961 *Ann. Phys. (N.Y.)* **16** 407
[5] Vrijen R, Yablonovitch E, Wang K, Jiang H W, Balandin A, Roychowdhury V, Mor T, and DiVincenzo D 2000 *Phys. Rev. A* **62** 012306
[6] Kane B E 1998 *Nature* **393** 133
[7] Loss D and Divincenzo D P 1998 *Phys. Rev. A* **57** 120
[8] Burkard G, Loss D, and DiVincenzo D P 1999 *Phys. Rev. B* **59** 2070
[9] Sørensen A and Mølmer K 1999 *Phys. Rev. Lett.* **83** 2274
[10] Porras D and Cirac J I 2004 *Phys. Rev. Lett.* **92** 207901
[11] Lidar D A, Bacon D, and Whaley K B 1999 *Phys. Rev. Lett.* **82** 4556
[12] DiVincenzo D P, Bacon D, Kempe J, Burkard G, and Whaley K B 2000 *Nature* **408** 339
[13] Imamoglu A, Awschalom D D, Burkard G, DiVincenzo D P, Loss D, Sherwin M, and Small A 1999 *Phys. Rev. Lett.* **83** 4204
[14] Privman V, Vagner I D, and Kventsel G 1998 *Phys. Lett. A* **239** 141
[15] Zheng S B and Guo G C 2000 *Phys. Rev. Lett.* **85** 2392
[16] Wang X 2001 *Phys. Rev. A* **64** 012313
[17] Franchini F, Its A R, Jin B-Q, and Korepin V E 2007 *J. Phys. A: Math. Theor.* **40** 8467
[18] Franchini F, Its A R, and Korepin V E 2007 *J. Phys. A: Math. Theor.* **41** 025302
[19] Safonov E and Lychkovskiy O 2013 *Phys. Rev. A* **87** 042105
[20] Huang Y 2014 *Phys. Rev. B* **89** 054410
[21] Huang Y and Moore J E 2014 *arXiv:1405.1817*
[22] Figueroa A, López J, Castanõs O, López-Peña R, Manko M A, and Manko V I 2014 *arXiv:1408.7099*

Journal of Physics: Conference Series **594** (2015) 012046 doi:10.1088/1742-6596/594/1/012046

High-Fidelity, Weak-Light Polarization Gate Using Room-Temperature Atomic Vapor

L. Deng and E.W. Hagley

National Institute of Standards and Technology, Gaithersburg, Maryland 20899 USA

Runbing Li
State Key Laboratory of Magnetic Resonance and Atomic and Molecular Physics, Wuhan Institute of Physics and Mathematics, Chinese Academy of Sciences, Wuhan 430071, China
Center for Cold Atom Physics, Chinese Academy of Sciences, Wuhan 430071, China

Chengjie Zhu
School of Physics and Engineering, Tongji University, Shanghai 200092 China

Email: lu.deng@nist.gov

Abstract: Using a polarization-selective-Kerr-phase-shift technique we demonstrate a fast, all-optical, high-fidelity polarization gate in a room-temperature atomic medium. By writing a π-phase shift to one selected circularly-polarized component of a linearly-polarized input signal field and by equalizing the gain of both circularly-polarized components we can maintain the original strength of the signal field and yet achieve a perfect 90° rotation of its linear polarization., demonstrating a fast, high-fidelity, dynamically-controlled polarization gate operation. The orthogonal linear polarization switching field intensity can be as low as 2 mW/cm^2 using a warm rubidium vapor, which is equivalent to a 100-nanosecond pulse containing about 200 photons and confined in a typical commercial photonic hollow-core fiber with a 5-μm mode diameter.

1. Introduction

Efficient optical-field gating operations with very low gating control light intensities are critically important to next-generation advanced telecommunication technology. One of such devices is the optical polarization phase gate where a polarization-encoded optical signal field is controlled by a weak light field via nonlinear Kerr effect. Although various weak-light and low power consumption optical phase gate operation schemes have already been proposed [1-4] and a continuously controlled optical phase gate has been recently demonstrated [5], to date no weak-light, dynamic polarization gate has been experimentally demonstrated in room-temperature media. Here, we describe the concept and principles of Polarization-Selective-Kerr-Phase-Shift (PSKPS) based polarization gate operation and we show experimental results of a fast, high fidelity, all-optical polarization gate operation based on this PSKPS protocol. The work presented here opens the door for possible potential applications and implementations of extremely low control light optical logic operations in advanced telecommunication protocols and networks.

2. The concept and principle of PSKPS

The concept of the PSKPS protocol and the all-optical polarization gate operations are based on the principle of "selective and differential Kerr phase engineering of different polarization components of the single linearly polarized signal field". This process can be understood with a lifetime broadened atomic system shown in Fig. 1a. Here, the phase engineering process involves a three-photon transition where the core of the process is a three-state Λ-type Raman gain scheme. A strong pump field couples the fully occupied ground state to an upper electronic state with a large one-photon detuning to suppress spontaneous emission. A linearly polarized signal field is forced, by the atomic transition selection principle, to address two separate channels involving the two circularly polarized components of the linearly polarized input signal field. To complete the three-photon process a circularly-polarized phase-control field writes different nonlinear Kerr phase shifts $\phi^{(\pm)}$ to the different signal polarization branches. Because of the Raman gain in the core Λ-scheme, the different polarization components of the signal field will acquire different nonlinear gain and nonlinear Kerr phase shift. If the differential nonlinear gain $\Delta G = G^{(+)} - G^{(-)}$ can be made negligibly small while simultaneously achieving $\phi^{(+)} = 0$ and $\phi^{(-)} = \pi$, then at the exit of the medium the linear polarization of the signal field will have undergone a perfect 90° rotation. With this perfect polarization gate operation, a CNOT gate and a host of other fast, all-optical logic gates [6] can be easily constructed.

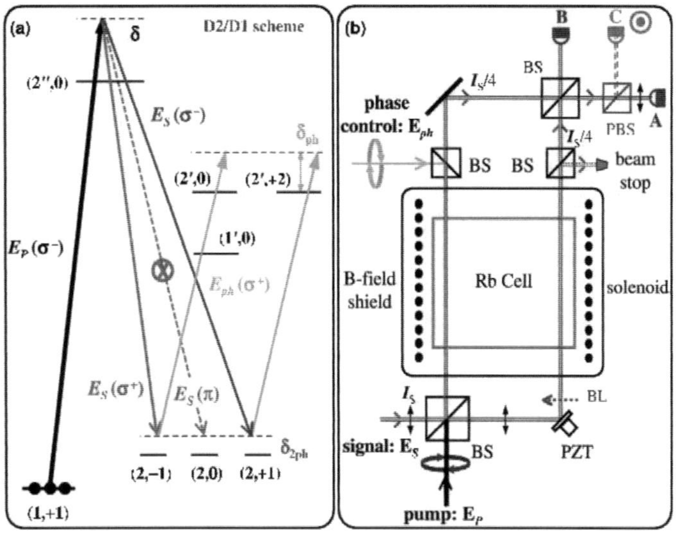

Figure 1: (a) Energy diagram and laser couplings for the photonic polarization gate. δ: one-photon pump field (E_P , thick black arrow) detuning. δ_{2ph}: two-photon detuning. δ_{ph}: the phase-control field detuning. A magic detuning $\delta_{ph} = \delta_{magic}$ is defined as the detuning that results in an orthogonal polarization rotation of the signal field (blue and red arrows) using minimal phase-control field (E_{ph}) and pump field (E_P). (b) Experimental setup. BS: 50-50 beam splitter. PBS: polarization beam splitter. BL: beam block. *A*, *B*, and *C*: detectors. PZT: piezo-controlled mirror.

3. Experiment

Experimentally, we use a warm ^{87}Rb vapor to demonstrate the PSKPS-based polarization gate operation (Fig. 1b). The ^{87}Rb vapor cell has a length of 7.5 cm and a diameter of 2 cm. It is filled with about 933 Pascal Neon buffer gas and it is also shielded from ambient magnetic fields under three layers of μ-metal. During the operation the cell is actively temperature stabilized to 322 K with a number density $n_0 = 6\text{x}10^{11}/\text{cm}^3$.

First, we optically pump the medium using a linearly polarized light field that couples the ($5S_{1/2}$, F=2) hyperfine manifold to the $5P_{3/2}$ manifold, resulting in all atoms initially are in the ground state manifold ($5S_{1/2}$, F=1). Immediately following the optical pumping process, we turn on a strong (8 – 15 mW/cm^2), circularly polarized pump laser coupling the ($5S_{1/2}$, F=1) to ($5P_{3/2}$, F"=2) transition with a one-photon detuning of $\delta/2\pi = 1.2$ GHz (Fig. 1a). A linearly polarized signal field which is derived from the same laser that provides the pump power is also injected into the medium. It has an intensity of < 30 μW/cm^2 and it couples the ($5P_{3/2}$, F"=2) to ($5S_{1/2}$, F=2) transition with a similar large one-photon detuning $\delta/2\pi = 1.2$ GHz. The pump and the signal field form the usual two-photon Raman gain configuration with a typical two-photon detuning of $\delta_{2ph}/2\pi$=700 kHz. The pump and signal fields are overlapped and combined using a 50-50 beam splitter (BS).

We further split this dual-beam into the two arms of a Mach-Zehnder interferometer using a separate beam splitter (Fig. 1b). One arm will later be overlapped with a phase-control laser, and the other arm serves as a reference for phase analysis. After exiting the cell, both arms of the Mach-Zehnder interferometer are joined together using a 50-50 BS so that the phase of the signal field can be analyzed. The output from this Mach-Zehnder interferometer is then further analyzed with a high precision polarization beam splitter (PBS). The signals simultaneously detected at two polarization ports of the PBS yield an assessment of how effective, complete and efficient of the polarization switching process. Because of the frequency proximity of the pump and signal fields these beams must be physically separated at the signal detectors. This is because that even these two fields have orthogonal linear polarizations an extremely small nonlinear-polarization residual in the strong pump could pass the analyzing PBS and enters the detector for the signal field polarization analysis, and therefore dominates the detector response. In our experiment we set up two special filters at a distance of about 1 m from the exit of the vapor cell to improve the signal/leakage-pump discrimination and this proved very effective.

The Kerr phase shift is introduced with a weak circularly polarized phase-control field coupling the ($5S_{1/2}$, F=2) to ($5P_{1/2}$, F'=2) transition. This phase-control field is injected into one arm of the Mach-Zehnder interferometer using another 50-50 BS and it propagates in the opposite direction in comparison with the signal field (Fig. 1b). We first scan the Mach-Zehnder interferometer to make sure that a π-phase shift is written to the arm of the interferometer where the signal and phase control field propagate anti-co-linearly. By adjusting the intensity of the phase control field we can switch the output light field from one polarization port of a polarization beam splitting cube (PBS) to the orthogonal output port of this PBS, demonstrating the fast orthogonal polarization rotation. With the warm rubidium vapor at about 322 K we routinely execute such polarization rotation with a phase control light field of 2 mW/cm^2. To put this phase control-field intensity in perspective, a single 780 nm photon of 100 ns pulse length propagating in a 5 μm mode diameter photonic hollow fiber has an intensity of about 10 μW/cm^2. Therefore, 2 mW/cm^2 would be equivalent to about 200 phase-control photons in such a photonic fiber.

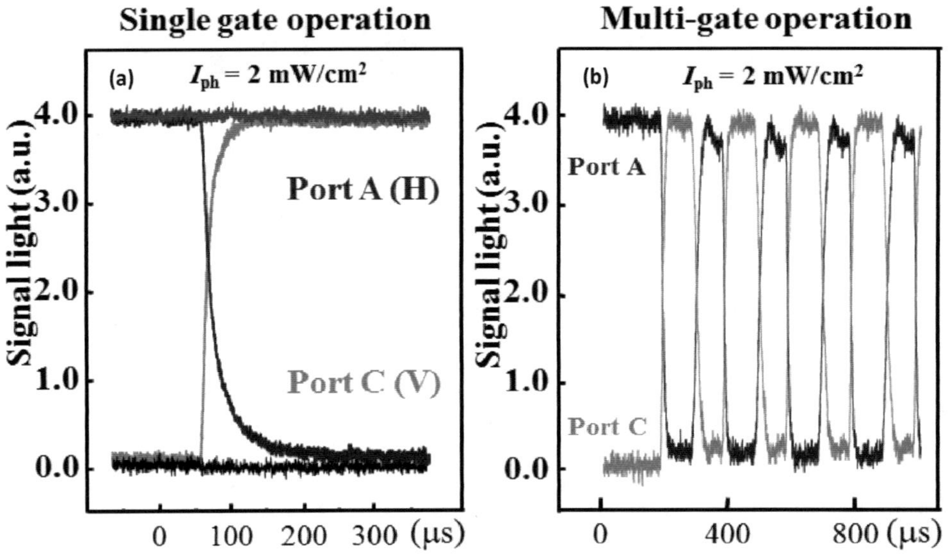

Figure 2: (a) Polarization gate operation. A signal field with horizontal linear polarization is rapidly switched to a signal field with vertical linear polarization by using a phase-control field of 2mW/cm^2. The signal field switching fidelity is typically better than 96%. The switching time can be reduced to 2 μs which is the limit of our system. (b) High fidelity, multiple pulse polarization flips by modulating the phase-control light. The polarization switching can be as fast as MHz which is the limit of our system.

In Fig. 2a we demonstrate signal-light linear polarization switching using a single control field pulse of intensity 2 mW/cm^2. Slight optimization of signal, pump and phase-control light fields timing resulted in a polarization switching at about 2 μs which is roughly the performance limit of our frequency-intensity modulators. Due to the broad Doppler profile of room-temperature atomic medium nearby parasitic transitions, even are very far from resonance, and still contribute to some degree to the loss of the signal and this explains why the signal intensity has not exceeded 100% in this Raman gain scheme. That is, the small gain in this Raman gain configuration is compensated partially by the background absorption of the signal field. In Fig. 2b we show multiple phase-control field pulses by modulating the phase-control light source. This multi-pulsed operation results in high fidelity probe-field polarization switching that can be achieve on the order of MHz (again the limit rate of our control devices). This demonstration proves the viability of fast polarization switching operations using the PSKPS technique developed here.

4. Conclusion

The experimental strategy and methods demonstrated here have significant potential for quantum information processing, and may lead to the development of critical architectural building blocks for quantum computers. Our experiment demonstrates the first high fidelity, all-optical very weak-light polarization gate operations in room-temperature atomic medium. The intensity of the phase-control light can be significantly reduced by using a high density atomic vapor trapped in a photonic hollow fiber. In addition, using of short pulsed phase control field can also reduce the complication of fast relaxation rate resulted from the transverse motion of atoms in the hollow fiber. Further improvements and combinations of the PSKPS protocol and technique demonstrated here with schemes such as electromagnetically induced transparency (EIT) [7, 8] may lead to further reductions in the required phase-control field intensity. Indeed, fully second quantization theoretical calculation has shown that the PSKPS technique is readily applicable in EIT-based schemes at single photon levels where no probe gain is present. This raises an interesting possibility of polarization gate operations at single-signal photon levels without the need to focus the phase-control light field to the diffraction limit.

References

[1] C. Ottaviani, D. Vitali, M. Artoni, F. Cataliotti, and P. Tombesi, Phys. Rev. Lett. **90**, 197902 (2003).

[2]. M.D. Lukin and A. Imamoglu, Nature (London) **413**, 273-276 (2001).

[3]. K.J. Resch, J. S. Lundeen, and A. M. Steinberg, Phys. Rev. Lett. **89**, 037904 (2002).

[4]. S. Rebic, D. Vitali, C. Ottaviani, P. Tombesi, M. Artoni, F. Cataliotti, and R. Corbal´an, Phys. Rev. A **70**, 032317 (2004)

[5] R.B. Li, L. Deng, and E.W. Hagley, Phys. Rev. Lett. **110**, 113902 (2013).

[6] R.B. Li, L. Deng, and E.W. Hagley (submitted)

[7] S.E. Harris Harris, Physics Today 50, 36 (1997).

[8] C.J. Zhu, L. Deng, and E.W. Hagley (submitted).

Density-dependent modifications of the transition spectrum of an atom located inside cold atomic ensemble

A S Kuraptsev
Junior Researcher, St. Petersburg Polytechnical University, Polytechnicheskaya st. 29, Saint-Petersburg, Russia

E-mail: aleksej-kurapcev@yandex.ru

Abstract. On the basis of general theoretical results developed in our group previously [Sokolov I M *et al* 2011 *J. Exp. Theor. Phys.* **112** 246], the transition spectrum of a single atom inside cold atomic ensemble is analyzed under conditions when the averaged interatomic separation is less than or comparable with the wavelength of resonant radiation. Density-dependent shifts of resonance as well as distortion of the spectral shape of the atomic transition are investigated.

1. Introduction

Influence of the environment on atomic transition spectrum has attracted considerable interest both because of its fundamental importance and because of its significance in different applications such as quantum metrology, quantum information science, and lasing in disordered media [1] – [8]. These influences are of special interest for realization of highly stable and accurate optical atomic clocks [9] – [14], particularly neutral-atom-based optical frequency standards [15] – [20].

The question of density-dependent spectral line shifts for cold atomic gases recently gave rise to a wide discussion [21] – [31], and this question is especially important for optical frequency standards [17], [20]. In case when atomic density is quite big (so that average interatomic separation is comparable with wavelength of resonant light) spectral line shift caused by dipole-dipole interaction can be comparable with natural line width. In the vast majority of experiments the external quasi-resonant radiation excites whole atomic ensemble. Therefore, most of theoretical papers are focused on the analysis of this case.

In the present paper we will discuss local excitation inside cold atomic ensemble. It can be experimentally realized by illuminating of the ensemble with two narrow and off-resonant orthogonally propagated light beams. Each beam does not cause single-photon excitation, but their simultaneous interaction with atoms in the crossing region can cause two-photon excitation from the ground S to excited D state if conditions of two-photon resonance are satisfied. The subsequent spontaneous transition from D to P state leads to population of the studied P state. The thereby described method allows obtaining small cluster of excited atoms in the middle of the cloud. For simplicity thereafter in the paper we will consider the case of only one atom excitation. Note that possibilities of two-photon excitation $5s\ S-2(1/2) \rightarrow 5p\ P-2(j) \rightarrow 5d\ D-2\ (j)'$ of rubidium atoms have been already studied experimentally [32]. In our opinion the one-atom spectral response analyzed in this paper can be the foundation of a new practical method of studying of collective effects in dense media.

The main goal of the present work is to study theoretically the transition spectrum of an atom excited in a dense and cold atomic ensemble using quantum microscopic approach developed in our group [33]. A great advantage of this approach is that interatomic correlations are taken into account, including many-body correlations. Correct description of correlations is achieved by taking into account the discrete structure of the medium without the approximation of continuous medium.

2. Basic equations

The calculation of atomic transition spectrum in this paper will be made on the basis of a microscopic quantum approach developed in [33]. Here we will mention only the main approximations of this approach and will show the principal analytical expressions utilized for numerical calculations.

We consider an ensemble consisting of N motionless atoms. All atoms have the ground state with $J = 0$; $m = 0$ and the excited level with $J = 1$; $m = -1, 0, +1$. This model of atomic levels allows us to take into account the polarization properties of electromagnetic field correctly. For detailed comparison of this model and scalar model see [34], [35].

In the framework of quantum microscopic approach [33] we solve the non-stationary Schrodinger equation for the wave function of the joint system consisting of atoms and the field generated in the process of the evolution.

$$i\hbar \frac{\partial \psi}{\partial t} = H\psi \tag{1}$$

The Hamiltonian of the joint system can be presented as $H = H_0 + V$ where H_0 is and the sum of Hamiltonians of free atoms and free field, V is the operator of interaction between atoms and electromagnetic field. In the framework of dipole approximation used here, the operator of interaction V can be written as follows

$$V = -\sum_a \boldsymbol{d}^{(a)} \boldsymbol{E}(\boldsymbol{r}_a) \tag{2}$$

In this formula $\boldsymbol{d}^{(a)}$ is the dipole moment operator of the atom a, $\boldsymbol{E}(\boldsymbol{r})$ is the field operator.

We will seek the wave function ψ as an expansion in a set of eigenstates $\{|l\rangle\}$ of the operator H_0:

$$\psi = \sum_l b_l(t)|l\rangle \tag{3}$$

Here, the subscript l defines the state of all atoms and the field.

An important approximation used in the microscopic approach is in restriction of the total number of states $|l\rangle$ taken into account. We will calculate all radiative correction up to the second order of the fine structure constant. In this case we can consider only the following states (see [36]):

1) $\psi_g = |g, g, \dots, g\rangle|\boldsymbol{k}\alpha\rangle$ all atoms are in the ground state, and 1 photon is in the system
2) $\psi_{e_m} = |g, g, \dots, g, e_m, g, \dots, g\rangle|vac\rangle$ 1 atom is in the excited state, other atoms are in the ground state, and no photon in the system
3) $\psi_{e_m e'_{m'}} = |g, \dots, g, e_m, g, \dots, g, e'_{m'}, g, \dots, g\rangle|\boldsymbol{k}\alpha\rangle$ nonresonant states with two excited atoms and one photon

Here the index e indicates the number of excited atom, $e = 1,2, \dots, N$ and the index m indicates the angular momentum projection, $m = -1,0,+1$. The total number of one-fold excited states ψ_{e_m} is $3N$. The last states $\psi_{e_m e'_{m'}}$ are necessary for a correct description of the dipole-dipole interaction at short interatomic distances.

The restriction of the total number of states allows us to covert the infinite system of equations for quantum amplitudes of states of the joint system b_l to finite system $3N \times 3N$ for the amplitudes of the

Journal of Physics: Conference Series **594** (2015) 012047 doi:10.1088/1742-6596/594/1/012047

one-fold atomic excited states b_{e_m}. The amplitudes of other quantum states b_g and $b_{e_m e'_{m'}}$ can be obtained via b_{e_m}. It allows us to obtain the wave function of the joint system.

In this paper we analyze the transition spectrum of one atom excited inside dense and cold atomic ensemble at the initial time. Let us consider the excited atom located in the center of the ensemble and we denote this atom number 1. For determinacy we will assume that at the initial time $t = 0$ only one substate $m = -1$ of the excited atom is populated. All the other atoms of the ensemble are in their ground state. Electromagnetic field at $t = 0$ is in the vacuum state. So we consider initial conditions as follows: $b_g(0) = 0$; $b_{e_m e'_{m'}}(0) = 0$; $b_{e_m}|_{e=1;m=-1} = 1$; $b_{e_m}|_{e=1;m=0,+1} = 0$; $b_{e_m}|_{e=1,2,\ldots,N;m} = 0$. Hereafter we will associate the Fourier-amplitude $b_0(\omega)$ with Zeeman sublevel $m = -1$ of the central atom $e = 1$.

The transition spectrum $b_0(\omega)$ contains fluctuations connected with random spatial configuration of atoms in the ensemble. In the real experiments the residual thermal motion of atoms makes all these configurations uniformly distributed. Therefore the experimental result is averaged over random spatial configurations of atoms in the ensemble. Accordingly, we perform multiple Monte-Carlo averaging of $b_0(\omega)$ over random spatial configurations of atoms.

3. Results and discussion

Interaction of a radiating atom with its surroundings causes a shift of the atomic transition as well as distortion of the spectral line. The specific type of transition spectrum depends both on the atomic density and on the size of the ensemble, i.e., on the number of atoms in it. We have analyzed how the function $b_0(\omega)$ changes with the number of atoms for the given density and found that for small ensembles when the mean-free path of photon less or comparable with linear size of atomic ensemble these changes are very essential. As the number of atom and linear size increase the changes in $b_0(\omega)$ become more and more weak. This dependence has an evident tendency to saturation. In this section we will discuss sufficiently big atomic ensembles when $b_0(\omega)$ for the given atomic density does not practically depend on the number of atoms. So results obtained in this paper can be used for description of single-atom excitation inside any macroscopic ensemble with reasonable accuracy. The saturation is a simplest scaling law; it is a feature of local excitation. For the excitation of whole atomic ensemble by probe radiation the scaling law is more complex. In [37] we have analyzed this law for cross section of scattered light.

Figure 1 shows the imaginary part of $b_0(\omega)$ (absorption curve) for different atomic densities. It demonstrates that the absorption curve is significantly modified as compared with the classical Lorenz curve typical for a free atom (dot-dashed curve). The maximum of the spectral line is shifted, moreover we observe both red and blue shift depending on the atomic density. The atomic densities $n = 0.05, 0.1, 0.2$ have been analyzed in [31] and we observed red shifts for all these values.

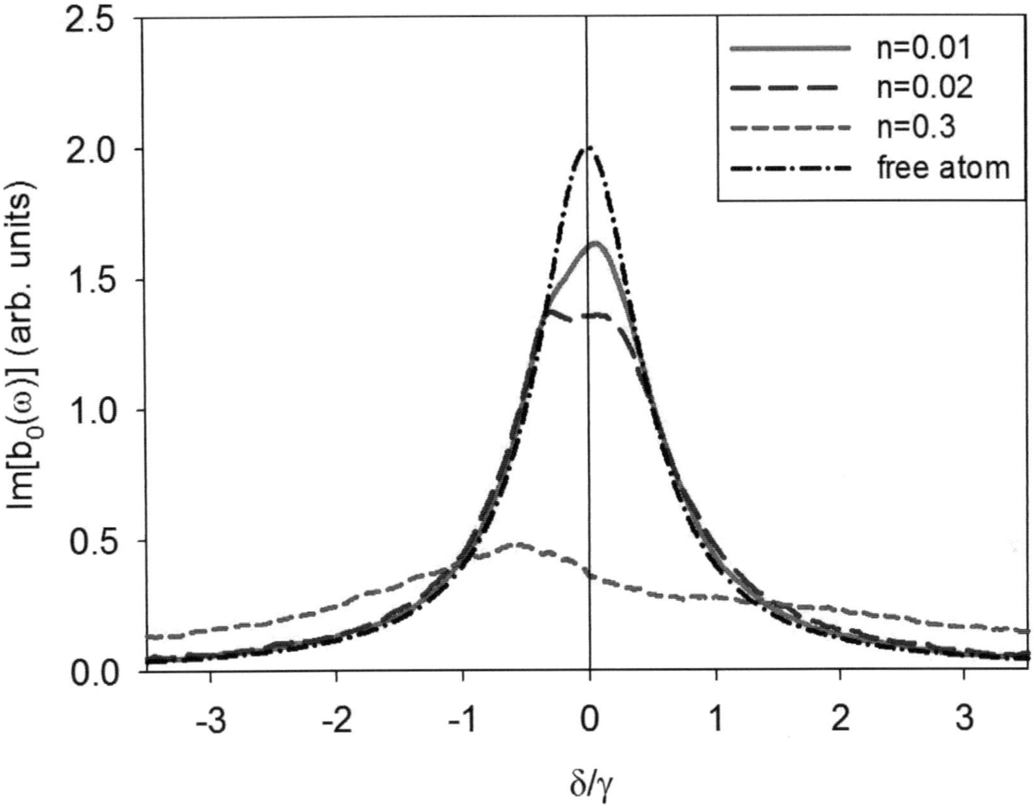

Figure 1. Absorption curve corresponding to an atom in ensemble that is excited at initial time; γ is the natural line width of free atom; $\delta = \omega - \omega_0$, where ω_0 is the resonant transition frequency of free atom; n is the atomic density in units k_0^3 where $k_0 = \omega_0/c$

Increasing the density of surrounding atoms causes enhancement of collective effects and consequently more considerable modification of atomic transition spectrum. It is clearly confirmed by the curve corresponding to the density $n = 0.3$ on the figure 1. The contour has relatively wide wings as compared with a free atom.

Figure 2 shows the real part of $b_0(\omega)$ (dispersion curve) for the same densities as figure 1. It also illustrates the enhancement of modifications of atomic transition spectrum caused by collective effects as density increases.

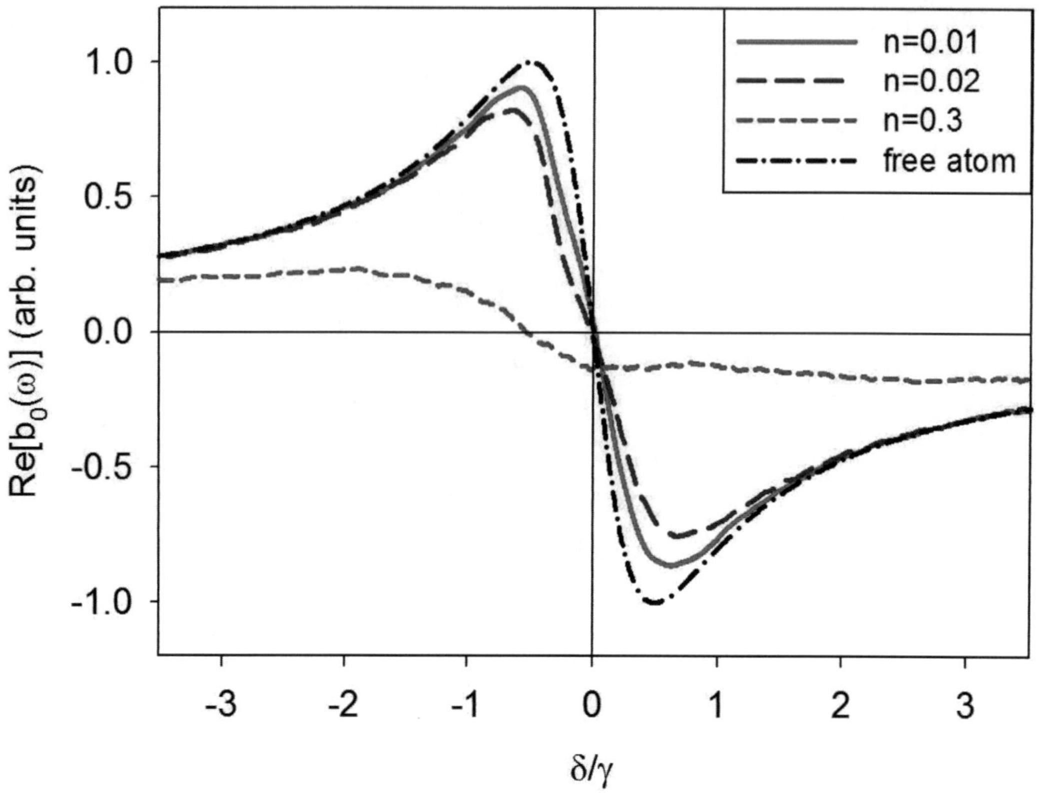

Figure 2. Dispersion curve corresponding to an atom in ensemble that is excited at initial time; all parameters are the same as in figure 1

The exciting feature of collective effects caused by resonant dipole-dipole interaction is that different observables have different spectral dependencies for dense atomic ensembles and these dependencies transform differently as density increases. Thus, the spectrum of the dielectric permittivity of cold and dense atomic clouds has a blue shift [28], [38], [39]. The maximal probability to excite the dense ensemble is observed for a negative detuning of exciting radiation [40]. The calculation of the total scattering cross section shows that the corresponding spectrum has several resonances [33]. Two of them are red shifted, and the third is in the blue region. Spectral dependence of fluorescence also has complicated behavior. It depends both on direction of fluorescence and on its polarization [33]. By now density dependence of spectral properties of cold clouds has been studied in several experiments [17], [41], [42], in which laser-induced fluorescence of atomic ensemble was studied. Experiments showed red shift and some distortion of the excitation spectrum.

4. Conclusion

In this paper we analyze the influence of the dipole-dipole interatomic interaction on the transition spectrum of an excited atom in cold atomic ensemble under conditions when the averaged interatomic separation is less than, or comparable with, the wavelength of resonance radiation. Both red and blue shifts of resonance are obtained depending on the atomic density. The distortion of spectral shape of atomic transition is investigated.

The calculations were made on the basis of a quantum approach taking into account the vector nature of the electromagnetic field and Zeeman structure of atomic sublevels. A continuous medium approximation was not used. It allowed us to take into consideration random inhomogeneity of the atomic system and consequently the existing interatomic correlations.

Together with [31] the results of this paper give sufficiently complete information about the dependence of transition spectrum of an atom in cold atomic ensemble on the atomic density. In our opinion the results obtained in this paper are very important for future improvement in quantum frequency standards based on optical transitions in cold atomic ensembles. For optimization of these devices the density dependence of the main characteristics has to be taken into account.

Note that the model of motionless atoms used here can describe not only cold atoms but impurity centers in solids as well. Therefore the results can be relevant in the area of lasing from randomly disordered medium as well as in the investigation of lasers based on quantum dots. There are also a number of proposals about using cold atomic ensembles and ensembles of impurity centers in solids in quantum-information applications. Large coherence time makes possible to use such systems as elements of quantum memory.

Acknowledgements

The author thanks I. M. Sokolov for useful discussions. The author acknowledges financial support from the Ministry of Education and Science of the Russian Federation (State Assignment 3.1446.2014K) and the Russian Foundation for Basic Research (Grant No. RFBR-14-02-31422)

References

[1] Cao H 2003 *Lasing in disordered media* (Progress in Optics vol 45) ed E Wolf (New York: Elsevier) chapter 6 pp 317-371

[2] Wiersma D S 2008 *Nat. Phys.* **4** 359

[3] Conti C and Fratalocchi A 2008 *Nat. Phys.* **4** 794

[4] Baudouin Q *et al* 2013 *Nat. Phys.* **9** 357

[5] Fleishhauer M *et al* 2005 *Rev. Mod. Phys.* **77** 633

[6] Hau L V 2008 *Nat. Photon.* **2** 451

[7] Ospelkaus S 2008 *Nat. Phys.* **4** 622

[8] Campbell G K *et al* 2008 *Metrologia* **45** 539

[9] Courtillot I *et al* 2003 *Phys. Rev.* A **68** 030501

[10] Esnault F-X *et al* 2013 *Phys. Rev.* A **88** 042120

[11] Falke St *et al* 2011 *Metrologia* **48** 399

[12] Kazakov G A *et al* 2012 *New J. Phys.* **14** 083019

[13] Muller S T *et al* 2011 *J. Opt. Soc. Am.* B **28** 2592

[14] Esnault F-X *et al* 2011 *Adv. Space Res.* **47** 854

[15] Hollberg L *et al* 2001 *IEEE J. Quant. Electron.* **37** 1502

[16] Katori H *et al* 2003 *Phys. Rev. Lett.* **91** 173005

[17] Ido T *et al* 2005 *Phys. Rev. Lett.* **94** 153001

[18] Porsev S G *et al* 2008 *Phys. Rev.* A **78** 032508

[19] Bloom B J *et al* 2014 *Nature* **506** 71

[20] Chang D E *et al* 2004 *Phys. Rev.* A **69** 023810

[21] Friedberg R *et al* 1973 *Phys. Rep.* C **7** 101

[22] Morice O *et al* 1995 *Phys. Rev.* A **51** 3896

[23] Manassah J T 2012 *Adv. Opt. Photonics* **4** 108

[24] Scully M O and Svidzinsky A A 2010 *Science* **328** 1239

[25] Maki J J *et al* 1991 *Phys. Rev. Lett.* **67** 972

[26] van Kampen H *et al* 1999 *Phys. Rev.* A **59** 271

[27] Cundiff S T 2002 *Laser Phys.* **12** 1073

[28] Fofanov Ya A *et al* 2011 *Phys. Rev.* A **84** 053811

[29] Keaveney J *et al* 2012 *Phys. Rev. Lett.* **108** 173601

[30] Javanainen J *et al* 2014 *Phys. Rev. Lett.* **112** 113603

[31] Kuraptsev A S and Sokolov I M 2014 *Phys. Rev.* A **90** 012511

[32] Bayram S B *et al* 2006 *J. Phys.* B **39** 2545

[33] Sokolov I M *et al* 2011 *J. Exp. Theor. Phys.* **112** 246

[34] Sokolov I M *et al* 2010 *J. Mod. Opt.* **57** 1833

[35] Skipetrov S E and Sokolov I M 2014 *Phys. Rev. Lett.* **112** 023905

[36] Stephen M J 1964 *J. Chem. Phys.* **40** 669

[37] Sokolov I M *et al* 2013 J. Mod. Opt. **60** 50

[38] Chomaz L *et al* 2012 New J. Phys. **14** 055001

[39] Kuraptsev A S *et al* 2012 *Optics and Spectroscopy* **112** 401

[40] Fofanov Ya A *et al* 2013 *Phys. Rev.* A **87** 063839

[41] Balik S *et al* 2013 *Phys. Rev.* A **87** 053817

[42] Pellegrino J *et al* 2014 *Phys. Rev. Lett.* **113** 133602

Quantum theory of measurements as quantum decision theory

V I Yukalov[1,2] and D Sornette[2,3]

[1]Bogolubov Laboratory of Theoretical Physics, Joint Institute for Nuclear Research, Dubna 141980, Russia
[2]DMTEC, ETH Zürich, Swiss Federal Institute of Technology, Zürich CH-8092, Switzerland
[3]Swiss Finance Institute, c/o University of Geneva, Geneva CH-1211, Switzerland

E-mail: yukalov@theor.jinr.ru

Abstract. Theory of quantum measurements is often classified as decision theory. An event in decision theory corresponds to the measurement of an observable. This analogy looks clear for operationally testable simple events. However, the situation is essentially more complicated in the case of composite events. The most difficult point is the relation between decisions under uncertainty and measurements under uncertainty. We suggest a unified language for describing the processes of quantum decision making and quantum measurements. The notion of quantum measurements under uncertainty is introduced. We show that the correct mathematical foundation for the theory of measurements under uncertainty, as well as for quantum decision theory dealing with uncertain events, requires the use of positive operator-valued measure that is a generalization of projection-valued measure. The latter is appropriate for operationally testable events, while the former is necessary for characterizing operationally uncertain events. In both decision making and quantum measurements, one has to distinguish composite non-entangled events from composite entangled events. Quantum probability can be essentially different from classical probability only for entangled events. The necessary condition for the appearance of an interference term in the quantum probability is the occurrence of entangled prospects and the existence of an entangled strategic state of a decision maker or of an entangled statistical state of a measuring device.

1. Introduction

Developing the theory of quantum measurements, von Neumann [1] mentioned that the process of quantum measurements is analogous to decision making. The formal analogy between these processes has been described in several mathematical works [2–5]. However, this analogy remained rather formal, without comparing quantum measurements with real decision making, as done by humans. Several important questions have not been answered:

(i) First of all, in what sense human decision making could be characterized by quantum measurements?

(ii) What would be a general scheme for describing measurements under uncertainty and decisions under uncertainty?

(iii) How to correctly define a quantum probability of non-commuting events for both, quantum measurements and quantum decisions?

(iv) What is the role of entanglement in quantum measurements, and in quantum decision making?

(v) When should decision making be treated by quantum rules and when is it sufficient to use classical theory?

In this report, we present a general approach and mathematical techniques common to quantum measurements and decision making that provides natural answers to these questions. The reason for developing a common quantum approach to measurements and to decision making is twofold. First, quantum theory provides tools for taking into account behavioral biases in human decision making [6–8]. Note that the possibility of describing cognition effects by means of quantum theory was suggested by Bohr [9, 10]. Second, formulating a quantum theory of decision making defines the main directions for creating artificial quantum intelligence [11, 12].

2. Operationally testable events

Let us first briefly recall the definition of quantum probabilities for measurements corresponding to operationally testable events and their connection to simple events in decision making [6–8].

The operator of an observable $\hat{A}$ is a self-adjoint operator, whose eigenvectors are given by the eigenproblem

$$\hat{A}|n\rangle = A_n|n\rangle \; , \tag{1}$$

forming a complete basis. The closed linear envelope of this basis composes a Hilbert space $\mathcal{H}_A \equiv span\{|n\rangle\}$. The measurement of an eigenvalue A_n is an event that we denote by the same letter. This event is represented by a projector $\hat{P}_n$ according to the correspondence

$$A_n \rightarrow \hat{P}_n \equiv |n\rangle\langle n| \; . \tag{2}$$

The system statistical state, or the decision maker strategic state, is given by a statistical operator $\hat{\rho}_A$. The probability of an event A_n is

$$p(A_n) \equiv \mathrm{Tr}_A \hat{\rho}_A \hat{P}_n \equiv \langle \hat{P}_n \rangle \; . \tag{3}$$

In the theory of measurements or decision theory, the set $\mathcal{A} \equiv \{\hat{P}_n\}$ of projectors plays the role of the algebra of observables, with the expected value (3) being the event probability.

In eigenproblem (1), a nondegenerate spectrum is tacitly assumed. If the spectrum is degenerate, then in the eigenproblem

$$\hat{A}|n_j\rangle = A_n|n_j\rangle \tag{4}$$

an eigenvalue A_n corresponds to several eigenvectors $|n_j\rangle$, where $j = 1, 2, \ldots$. The Hilbert space can again be composed spanning all these eigenvectors, $\mathcal{H}_A \equiv \{|n_j\rangle\}$. Now an event of measuring A_n is associated with the projector

$$\hat{P}_n \equiv \sum_j \hat{P}_{n_j} \qquad (\hat{P}_{n_j} \equiv |n_j\rangle\langle n_j|) \; . \tag{5}$$

And the event probability is defined as in Eq. (3).

A basis, formed by the eigenvectors of a self-adjoint operator, is complete, because a self-adjoint operator is normal. The set of eigenvectors of any normal operator, such that $\hat{A}^+\hat{A} = \hat{A}\hat{A}^+$, forms a complete basis. Normal operators include self-adjoint and unitary operators. In the case of degenerate spectra, the basis may be not uniquely defined. In quantum measurements, this does not lead to any principal problem, since one can use the summary projector (5). But in decision theory, the question arises: in what sense a single event A_n can correspond to several projectors $\hat{P}_{n_j}$, since an operationally testable event either happens or not?

Fortunately, this problem of nonuniqueness is easily avoidable, both in the theory of quantum measurements and in decision theory. This is done by invoking the von Neumann suggestion of degeneracy lifting [1]. For this purpose, the operator of the observable $\hat{A}$ is slightly shifted by an infinitesimal term,

$$\hat{A} \rightarrow \hat{A} + \nu\hat{\Gamma} \qquad (\nu \rightarrow 0) \; , \tag{6}$$

where $\hat{\Gamma}$ is an operator lifting the symmetry responsible for the spectrum degeneracy. Then the operator spectrum splits into the set

$$A_n \rightarrow A_n + \nu\Gamma_{n_j} \; , \tag{7}$$

thus, removing the degeneracy. Finally, the event probability is defined as

$$p(A_n) = \lim_{\nu\rightarrow 0} p(A_n + \nu\Gamma_{n_j}) \; . \tag{8}$$

In that way, the degeneracy is avoided, and one always deals with a unique correspondence between eigenvalues, representing events, and eigenvectors. This procedure is also analogous to the Bogolubov method of quasi-averages, breaking the symmetry of statistical systems by introducing infinitesimal sources [13, 14].

The union of mutually orthogonal (incompatible) events is represented by the projector sum:

$$\bigcup_n A_n \rightarrow \sum_n \hat{P}_n \qquad (\hat{P}_m\hat{P}_n = \delta_{mn}\hat{P}_n) \; . \tag{9}$$

The probability of such a union is additive:

$$p\left(\bigcup_n A_n\right) = \sum_n p(A_n) \; . \tag{10}$$

In the definition of the quantum probability (3), the averaging is done with a statistical operator $\hat{\rho}_A$, generally, implying a mixed state. In some special cases, a quantum system can be prepared in a pure state described by a wave function $|\psi\rangle$, which corresponds to the statistical operator $|\psi\rangle\langle\psi|$. The setups of experiments with physical systems and with decision makers are quite different. Accomplishing quantum measurements, we may meet two types of systems, open and quasi-isolated.

An *open* system interacts with its surrounding, keeps information on the preparation of its initial state and, generally, can feel past interactions through retardation memory effects. Even if, at the given time, the interactions with its surrounding can be neglected, the system remains open, if it possesses the memory of its preparation and of past interactions. In addition, there can exist quantum statistical correlations making the system entangled with its surrounding.

A physical system is *quasi-isolated* when it is isolated from its surrounding interactions, does not have retardation memory effects, and is not entangled with surrounding through quantum statistical correlations. Such a system can be conditionally treated as isolated for a short instance of time. However, to confirm its isolation, one needs to realize measurements, at least nondestructive, which perturbs the system making it not absolutely isolated, but quasi-isolated [15–17].

Contrary to physical systems, a decision maker, even being isolated from a surrounding society, always keeps memory and information of many previous interactions. Therefore a decision maker has always to be considered as an open system.

3. Composite separable events

When one deals with not just a single event, but with several events, the situation becomes more involved. This especially concerns the measurements of noncommuting observables, whose operators do not commute with each other [18]. There has been an old problem of correctly defining the quantum probability of such events. It has been shown [19] that the Lüders probability [20] of consecutive measurements is a transition probability between two quantum states and cannot be accepted as a quantum extension of the classical conditional probability. Also, the Wigner distribution [21] is just a weighted Lüders probability and it cannot be treated as a quantum extension of the classical joint probability. The correct and most general definition of a quantum joint probability can be done [19] by employing the Choi-Jamiolkowski isomorphism [22, 23] expressing the channel-state duality.

Composite events, that are composed of two or more events, can be separable or entangled. First, we consider separable events.

Let us be interested in the measurements involving two observables associated with two operators, $\hat{A}$ and $\hat{B}$, with the related eigenproblems

$$\hat{A}|n\rangle = A_n|n\rangle , \qquad \hat{B}|\alpha\rangle = B_\alpha|\alpha\rangle . \tag{11}$$

As above, we shall denote the events of measuring the eigenvalues A_n and B_α by the same letters, respectively. To each event, we put into correspondence the appropriate projectors,

$$A_n \rightarrow \hat{P}_n \equiv |n\rangle\langle n| , \qquad B_\alpha \rightarrow \hat{P}_\alpha \equiv |\alpha\rangle\langle\alpha| . \tag{12}$$

Constructing two Hilbert space copies

$$\mathcal{H}_A \equiv \text{span}\{|n\rangle\} , \qquad \mathcal{H}_B \equiv \text{span}\{|\alpha\rangle\} , \tag{13}$$

we define the algebras of observables

$$\mathcal{A} \equiv \{\hat{P}_n\} , \qquad \mathcal{B} \equiv \{\hat{P}_\alpha\} , \tag{14}$$

acting on the corresponding Hilbert spaces. The composite algebra $\mathcal{A} \bigotimes \mathcal{B}$ acts on the tensor-product space $\mathcal{H}_A \bigotimes \mathcal{H}_B$. The system statistical state (decision-maker strategic state) also acts on the space $\mathcal{H}_A \bigotimes \mathcal{H}_B$. The joint probability of events, corresponding to the observables from the algebra $\mathcal{A} \bigotimes \mathcal{B}$, is defined as

$$p\left(\mathcal{A} \bigotimes \mathcal{B}\right) \equiv \text{Tr}_{AB}\, \hat{\rho} \mathcal{A} \bigotimes \mathcal{B} , \tag{15}$$

with the trace over $\mathcal{H}_A \bigotimes \mathcal{H}_B$.

For any two operators from the algebra $\mathcal{A}$, it is possible to introduce the Hilbert-Schmidt scalar product that is a map

$$\sigma_A : \mathcal{A} \times \mathcal{A} \rightarrow \mathbb{C} . \tag{16}$$

Thus, for the operators $\hat{A}_1$ and $\hat{A}_2$ from the algebra $\mathcal{A}$, the scalar product reads as

$$(\hat{A}_1, \hat{A}_2) \equiv \text{Tr}_A \hat{A}_1^+ \hat{A}_2 , \tag{17}$$

with the trace over $\mathcal{H}_A$. The operators from the algebra $\mathcal{A}$, acting on the Hilbert space $\mathcal{H}_A$, and complemented with the scalar product σ_A, form the Hilbert-Schmidt operator space

$$\mathcal{A} \equiv \{\mathcal{A}, \mathcal{H}_A, \sigma_A\} . \tag{18}$$

Similarly, one defines the Hilbert-Schmidt space

$$\widetilde{\mathcal{B}} \equiv \{\mathcal{B}, \mathcal{H}_B, \sigma_B\} \; . \tag{19}$$

The tensor-product of the above Hilbert-Schmidt spaces forms the space

$$\widetilde{\mathcal{C}} \equiv \widetilde{\mathcal{A}} \bigotimes \widetilde{\mathcal{B}} \; . \tag{20}$$

An operator $\hat{C}$ acting on $\widetilde{\mathcal{C}}$ is called *separable* if and only if it can be represented as a sum

$$\hat{C} = \sum_i \hat{A}_i \bigotimes \hat{B}_i \qquad (\hat{A}_i \in \widetilde{\mathcal{A}}, \; \hat{B}_i \in \widetilde{\mathcal{B}}) \; . \tag{21}$$

A composite event is named a prospect. The prospect $A_n \bigotimes B_\alpha$ corresponds to the prospect operator

$$\hat{P}\left(A_n \bigotimes B_\alpha\right) = \hat{P}_n \bigotimes \hat{P}_\alpha \; . \tag{22}$$

The prospect operator (22) is evidently separable. Therefore the corresponding prospect $A_n \bigotimes B_\alpha$ is also called separable. The related prospect probability writes as

$$p\left(A_n \bigotimes B_\alpha\right) = \mathrm{Tr}_{AB} \, \hat{\rho} \hat{P}_n \bigotimes \hat{P}_\alpha \; , \tag{23}$$

with the trace over $\mathcal{H}_A \bigotimes \mathcal{H}_B$.

More generally, the prospect $A_n \bigotimes \bigcup_\alpha B_\alpha$, where $\bigcup_\alpha B_\alpha$ is a union of mutually orthogonal events, corresponds to the prospect operator

$$\hat{P}\left(A_n \bigotimes \bigcup_\alpha B_\alpha\right) = \sum_\alpha \hat{P}_n \bigotimes \hat{P}_\alpha \; . \tag{24}$$

This operator is separable, and the related prospect probability

$$p\left(A_n \bigotimes \bigcup_\alpha B_\alpha\right) = \sum_\alpha p\left(A_n \bigotimes B_\alpha\right) \tag{25}$$

is additive with respect to the events $A_n \bigotimes B_\alpha$.

4. Composite entangled events

An operator $\hat{C}$ from the Hilbert-Schmidt space (20) is termed *entangled*, or non-separable, if it cannot be represented as sum (21), so that

$$\hat{C} \neq \sum_i \hat{A}_i \bigotimes \hat{B}_i \qquad (\hat{A}_i \in \widetilde{\mathcal{A}}, \; \hat{B}_i \in \widetilde{\mathcal{B}}) \; . \tag{26}$$

The appearance of entangled events, corresponding to entangled operators, is connected with the existence of not operationally testable measurements that also are called uncertain measurements, or incomplete measurements, or indecisive measurements, or inconclusive measurements. Respectively, one can keep in mind uncertain events in decision making.

Let us define an uncertain event B as a set of possible events B_α, characterized by weights $|b_\alpha|^2$,

$$B = \{B_\alpha : \; \alpha = 1, 2, \ldots\} \; . \tag{27}$$

Since the uncertain event is not operationally testable, it is not required that the weights $|b_\alpha|^2$ be summed to one. The uncertain-event operator is

$$\hat{P}(B) = |B\rangle\langle B| \; , \tag{28}$$

with the state

$$|B\rangle = \sum_\alpha b_\alpha |\alpha\rangle \tag{29}$$

that is not necessarily normalized to one. The uncertain-event operator (28), which can be written as

$$\hat{P}(B) = \sum_{\alpha\beta} b_\alpha b_\beta^* |\alpha\rangle\langle\beta| \; , \tag{30}$$

is not a projector onto a subspace that would correspond to a degenerate spectrum, because it does not have form (5). Moreover, it is not a projector at all, as far as it is not necessarily idempotent,

$$\hat{P}^2(B) = \langle B|B\rangle \hat{P}(B) \neq \hat{P}(B) \; ,$$

since state (29) is not, generally, normalized to one.

A composite event, formed of an operationally testable event A_n and an uncertain event (27), is the uncertain prospect

$$\pi_n = A_n \bigotimes B \; , \tag{31}$$

whose prospect operator is

$$\hat{P}(\pi_n) \equiv |\pi_n\rangle\langle\pi_n| = \sum_\alpha |b_\alpha|^2 \hat{P}_n \bigotimes \hat{P}_\alpha + \sum_{\alpha\neq\beta} b_\alpha b_\beta^* \hat{P}_n \bigotimes |\alpha\rangle\langle\beta| \; . \tag{32}$$

This operator is not separable in the Hilbert-Schmidt space (20), because the operator $|\alpha\rangle\langle\beta|$ does not pertain to space (19) composed of projectors $\hat{P}_\alpha$. Hence, operator (32) is entangled. The corresponding prospect (31) is termed entangled.

The prospect operators are assumed to satisfy the resolution of unity

$$\sum_n \hat{P}(\pi_n) = \hat{1} \; , \tag{33}$$

where $\hat{1}$ is the identity operator in the space $\mathcal{H}_A \bigotimes \mathcal{H}_B$. But these prospect operators are not necessarily orthogonal, since

$$\hat{P}(\pi_m)\hat{P}(\pi_n) = \langle\pi_m|\pi_n\rangle |\pi_m\rangle\langle\pi_n| \; ,$$

and they are not idempotent, as far as

$$\hat{P}^2(\pi_n) = \langle\pi_n|\pi_n\rangle \hat{P}_n \; .$$

That is, they are not projectors. The family $\{\hat{P}(\pi_n)\}$ of such positive operators, obeying the resolution of unity (33) is named *positive operator-valued measure* [2, 4, 5].

For a given lattice $\{\pi_n\}$ of prospects, the prospect probabilities

$$p(\pi_n) = \mathrm{Tr}\hat{\rho}\hat{P}(\pi_n) \; , \tag{34}$$

where the trace is over $\mathcal{H}_A \bigotimes \mathcal{H}_B$, satisfy the conditions

$$\sum_n p(\pi_n) = 1 \; , \qquad 0 \leq p(\pi_n) \leq 1 \; , \tag{35}$$

Journal of Physics: Conference Series **594** (2015) 012048 doi:10.1088/1742-6596/594/1/012048

making the set $\{p(\pi_n)\}$ a probability measure.

The analysis of the prospect probability (34) results in the following properties [24–26]. The probability can be written in the form

$$p(\pi_n) = f(\pi_n) + q(\pi_n) \; . \tag{36}$$

The first term,

$$f(\pi_n) = \sum_\alpha |b_\alpha|^2 p\left(A_n \bigotimes B_\alpha\right) \; , \tag{37}$$

corresponds to classical probability, possessing the features

$$\sum_n f(\pi_n) = 1 \; , \qquad 0 \le f(\pi_n) \le 1 \; . \tag{38}$$

The classical term (37) is an objective quantity reflecting the given properties of the prospect. The second term,

$$q(\pi_n) = \sum_{\alpha \neq \beta} b_\alpha b_\beta^* \langle n\alpha | \hat{\rho} | n\beta \rangle \; , \tag{39}$$

is purely quantum, caused by interference and coherence effects. This quantum term in the theory of measurements is called interference factor, or coherence factor, and in decision theory, attraction factor, since it characterizes the subjective attractiveness of different prospects to the decision maker.

According to the *quantum-classical correspondence principle* [27], when quantum effects disappear, quantum theory should reduce to classical theory, which implies

$$p(\pi_n) \rightarrow f(\pi_n) \; , \qquad q(\pi_n) \rightarrow 0 \; . \tag{40}$$

In general, the reduction of quantum measurements to their classical counterparts is called decoherence [28].

The quantum factor (39) varies in the range

$$-1 \le q(\pi_n) \le 1 \tag{41}$$

and satisfies the *alternation law*

$$\sum_n q(\pi_n) = 0 \; . \tag{42}$$

For a large number of considered prospects N, we get the *quarter law*

$$\lim_{N\rightarrow\infty} \frac{1}{N} \sum_{n=1}^N |q(\pi_n)| = \frac{1}{4} \; . \tag{43}$$

A known example of the arising interference under measurements is provided by the double-slit experiment [1]. The passage of a particle through one of two slits corresponds to prospect (31). The operationally testable event A_n is the registration of the considered particle by an n-th detector, while the passage through one of the two slits, either B_1 or B_2 is described by the uncertain event $B = \{B_1, B_2\}$.

The quantum term is not always nontrivial [29]. To be nonzero, it requires the validity of two necessary conditions. The first condition is that the considered prospect be entangled, as described in Sec. 4. The second necessary requirement is the entanglement of the statistical operator. The latter is entangled when, e.g.,

$$\hat{\rho} \neq \hat{\rho}_A \bigotimes \hat{\rho}_B \; , \tag{44}$$

where

$$\hat{\rho}_A \equiv \text{Tr}_B \hat{\rho} \; , \qquad \hat{\rho}_B \equiv \text{Tr}_A \hat{\rho} \; ,$$

that is, when the statistical state of a composite system is not a product of its partial subsystem states. More generally, the system state is entangled if it cannot be represented in the form

$$\hat{\rho} \neq \sum_\alpha \lambda_\alpha \hat{\rho}_{\alpha A} \bigotimes \hat{\rho}_{\alpha B} \; , \qquad (45)$$

where

$$\sum_\alpha \lambda_\alpha = 1 \; , \qquad 0 \leq \lambda_\alpha \leq 1 \; .$$

Condition (45) is necessary, but not sufficient for the occurrence of a nonzero term (39).

In quantum theory, entanglement is a well known notion. In decision theory, it corresponds to the correlations between different possible events that are perceived by the decision maker.

5. Conclusion

We have shown that quantum measurements and quantum decision making can be described by the same mathematical tools. In both these cases, the problem of degeneracy can be avoided by employing the von Neumann method of degeneracy lifting, which is analogous to the Bogolubov quasi-averaging procedure. The correct definition of quantum probabilities of composite events, called prospects, is done by using the Choi-Jamiolkowsky isomorphism. This allows us to describe any type of composite events, including those corresponding to noncommutative observables.

The notion of uncertain events and measurements makes it feasible to give a general scheme for describing measurements under uncertainty and decisions under uncertainty. This is done by means of positive operator-valued measures. Prospects are classified into two principally different types, separable and entangled, depending on the structure of the related prospect operators in the Hilbert-Schmidt space.

The appearance of a quantum interference term in the quantum probability of composite events is shown to require two necessary conditions, prospect entanglement and statistical state entanglement.

Classical measurements and classical decision making are particular cases of the corresponding quantum counterparts. The reduction of quantum measurements and decision making to the classical limit occurs when the interference term becomes zero.

The investigation of the analogies between quantum measurements and quantum decision making suggests the ways of creating artificial quantum intelligence [11, 12] and gives keys for better understanding of quantum effects in self-organization of complex systems [30].

[1] von Neumann J 1955 *Mathematical Foundations of Quantum Mechanics* (Princeton: Princeton University)
[2] Benioff P A 1972 *J. Math. Phys.* **13** 908
[3] Holevo A S 1973 *J. Multivar. Anal.* **3** 337
[4] Holevo A S 2011 *Probabilistic and Statistical Aspects of Quantum Theory* (Berlin: Springer)
[5] Holevo A S and Giovanetti V 2012 *Rep. Prog. Phys.* **75** 046001
[6] Yukalov V I and Sornette D 2008 *Phys. Lett. A* **372** 6867
[7] Yukalov V I and Sornette D 2009 *Eur. Phys. J. B* **71** 533
[8] Yukalov V I and Sornette D 2009 *Entropy* **11** 1073
[9] Bohr N 1933 *Nature* **131** 457
[10] Bohr N 1958 *Atomic Physics and Human Knowledge* New York: Wiley

[11] Yukalov V I and Sornette D 2009 *Laser Phys. Lett.* **6** 833
[12] Yukalov V I and Sornette D 2014 *Springer Proc. Phys.* **150** 37
[13] Bogolubov N N 1967 *Lectures on Quantum Statistics*, Vol. 1. New York: Gordon and Breach
[14] Bogolubov N N 1970 *Lectures on Quantum Statistics*, Vol. 2. New York: Gordon and Breach
[15] Yukalov V I 2003 *Phys. Lett. A* **308** 313
[16] Yukalov V I 2012 *Phys. Lett. A* **376** 550
[17] Yukalov V I 2012 *Ann. Phys.* **327** 253
[18] Gudder S 1979 *Stochastic Methods in Quantum Mechanics* New York: North-Holland
[19] Yukalov V I and Sornette D 2013 *Laser Phys.* **23** 105502
[20] Lüders G 1951 *Ann. Phys.* **15** 663
[21] Wigner E 1932 *Phys. Rev.* **40** 749
[22] Choi M D 1972 *Can. J. Math.* **24** 520
[23] Jamiolkowski A 1972 *Rep. Math. Phys.* **3** 275
[24] Yukalov V I and Sornette D 2010 *Adv. Complex Syst.* **13** 659
[25] Yukalov V I and Sornette D 2011 *Theor. Decis.* **70** 283
[26] Yukalov V I and Sornette D 2014 *IEEE Trans. Syst. Man Cybern. Syst.* **44** 1155
[27] Bohr N 1913 *Phil. Mag.* **26** 857
[28] Zurek W H 2003 *Rev. Mod. Phys.* **75** 715
[29] Yukalov V I and Sornette D 2014 *Top. Cogn. Sci.* **6** 79
[30] Yukalov V I and Sornette D 2014 *Adv. Complex Syst.* **17** 1450016

Structural monitoring system with fiber Bragg grating sensors: implementation and software solution

Aleksey Fedorov[1,2], Vladimir Lazarev[2], Ilya Makhrov[2], Nikolay Pozhar[2], Maxim Anufriev[2], Alexey Pnev[2], and Valeriy Karasik[2]

[1]Russian Quantum Center, Novaya str. 100, Skolkovo, Moscow 143025, Russia
[2]Bauman Moscow State Technical University, 2nd Baumanskaya str. 5, Moscow 105005, Russia

E-mail: akf@rqc.ru; sintetaza@mail.ru

Abstract. We present a structural health monitoring system for nondestructive testing of composite materials based on the fiber Bragg grating sensors and specialized software solution. Developed structural monitoring system has potential applications for preliminary tests of novel composite materials as well as real-time structural health monitoring of industrial objects. The software solution realizes control for the system, data processing and alert of an operator.

1. Introduction

The implementation of nondestructive structural health monitoring (SHM) systems for industrial objects is a challenging problem. In general, SHM is a method of integrity analysis of an object via real-time measurements of a class of parameters, which characterize the object's integrity. The main assignment of SHM systems is an alert of the system operator in case of exceeding the measured parameters of a certain critical value, where this value is determined from preliminary tests. Problems of a design of SHM are of current interest for civil and industrial buildings, pipelines, airplanes, water transport, and in many other scopes [1–3].

Composite materials have unique physical and chemical properties [4]. These materials have corrosion resistance, high heat resistance, wear resistance, rigidity and low mass. The properties make them useful in a wide spectrum of technological applications, *e.g.*, in space technologies, aviation engineering, shipbuilding, and etc. The design of SHM systems for composite materials is especially topical [1].

There are a two main approaches to design of a measurement part of SHM systems. The first method of SHM involves mounting sensors directly into the material. The second one is monitoring of the construction by the installation of sensors to certain elements of the structure. In the case of composite material structures, SHM systems are mostly based on embedding the sensors into the material [1].

In this way, the main question is what type of sensors is the most useful for SHM of composite materials. Practically, fiber optic sensors form the most useful class of sensors for SHM systems. There are several types of fiber-based optic sensors. The most common solutions are sensors based on measuring light intensity using changes in fiber curvature or reflected light from a mirrored surface, the Fabry–Perot sensors, and the fiber Bragg grating (FBG) sensors. Using of

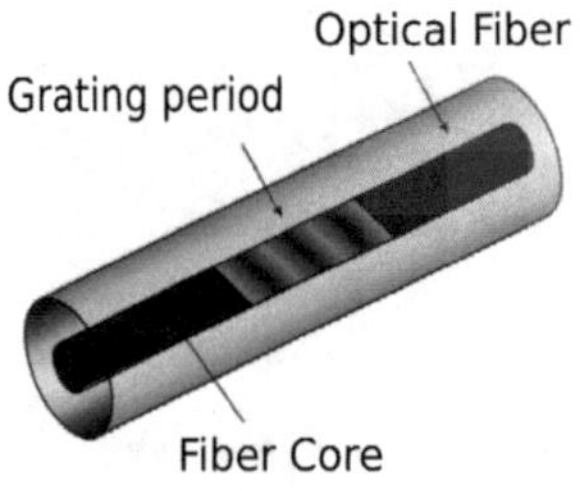

Figure 1. Structure of FBG sensor. Typical size of sensors is 5 − 7 mm with typical grating period being 500 nm.

terahertz vision system for nondestructive testing of composite materials is of current interest as well [5, 6].

Using of an embedded FBG sensors for measurement of distributed stress and strain fields of the composite material has significant advantages [1,7,8]. First of all, FBG sensors can be easily embedded into materials to provide damage detection or internal strain field mapping. The FBG sensors supposed to be located in fiber optic cable mounted in the sample. Furthermore, the key feature of FBGs sensors is that the information about perturbations is encoded in wavelength. Indeed, the main principle of FBG sensors work is transform of spectral or phase characteristics of the test optical signal propagating in the sample according to the perturbations of a measured physical quantity. Due to their high noise resistance to the effects of non-informative influencing factors [9, 10], FBG sensors attract increased attention [1].

FBG is a periodic structure of refractive index which is formed in a fiber core (see, Fig. 1). SHM systems have a measurement part, which typically consists from: FBG sensors, a source of test optical signal, and a measuring device for the signal reflected from FBG. Important problem is the embedding the sensor into the material sample. Useful solution of this problem is proposed in [9–12].

On the other hand, practical application and implementation of SHM systems require using of a software solution for control, fast collection, reliable storage, and effective processing of large data flow from the measurement part in real time. Indeed, the data proceeding is an another kind of problem, which can be solved efficiently only using software solution. The software part is needed for the solution of the following classes of problems: processing of the measurements from FBG sensors, visualization of the measurements results, and "user–measurement part" interaction, and control for the system.

In this work, we present developed SHM system for composite materials based on the fiber Bragg grating sensors. We present the software solution for developed SHM system. The system has two main scopes of applications. The first is preliminary tests system for novel composite materials. The other one is application for real-time SHM with an alert for an operator.

2. Implementation of SHM system

There are two physical reasons for the fluctuation of the FBG sensor central wavelength $\Delta\lambda$: the mechanical stress Δl and the temperature ΔT variations [8]. The central wavelength of Bragg grating as function of the strain and the temperature can be presented as follow

$$\Delta\lambda = 2\left(\Lambda\frac{\partial n}{\partial T} + n\frac{\partial\Lambda}{\partial T}\right)\Delta T + 2\left(\Lambda\frac{\partial n}{\partial l} + n\frac{\partial\Lambda}{\partial l}\right)\Delta l, \qquad (1)$$

where n is the effective refractive index and Λ is the grating period.

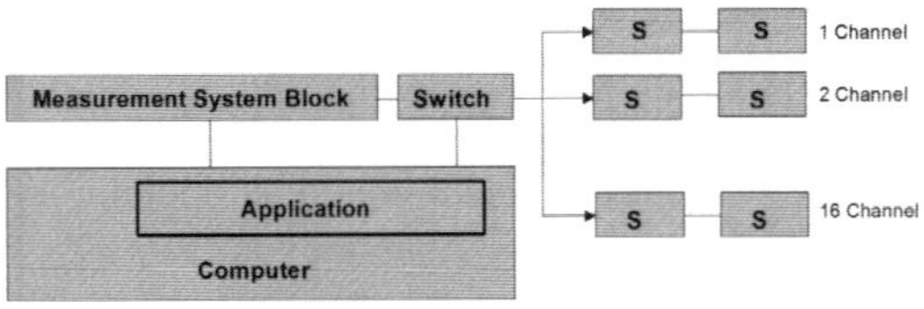

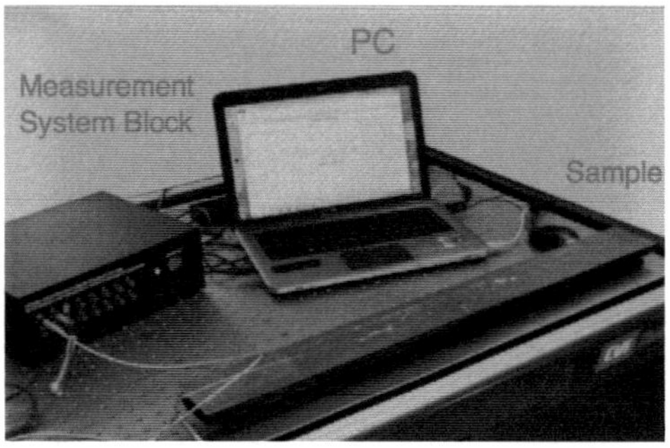

Figure 2. *Left*: Structural monitoring system scheme. Channel of the measurement system is a fiber optic cable. There are 16 channels embedded in the sample. Each channel includes up to 15 sequentially arranged Bragg sensors (S). Each sensor has a central wavelength of the reflected signal. *Right*: Implementation of SHM system.

The first term of the right-hand side of Eq. (1) describes the effect of temperature on the wavelength

$$\Delta\lambda = \lambda(a_\Lambda + a_n)\Delta T, \quad (2)$$

where a_Λ is the thermal expansion coefficient and a_n is the thermo-optic coefficient of the fiber.

In turn, second part of the right-hand side of Eq. (1) describes the nature of the dependence of the central wavelength λ from the applied strain, which can be represented in the form

$$\Delta\lambda = \lambda(1 - p_e)\varepsilon, \qquad p_e = \frac{n^2}{2}\left[p_{12} - \nu(p_{11} + p_{12})\right] \quad (3)$$

where ε is the relative strain, p_e is strain constant for optical fiber, p_{11} and p_{12} are the Pockels coefficients in the optical stress tensor, and ν is the Poisson coefficient.

Functioning of the measurement part of SHM system with FBG sensors is based on the following principles (see Fig. 2). Each channel of the measurement system is a fiber optic cable (see Fig. 2a). There are 16 channels, embedded in the composite material. All of these sensors are sequentially arranged Bragg sensors. Sensor have a unique central wavelength of the reflected signal in channel.

The aim of SHM systems is to measure the strain impact, but for the reliability of the data it is necessary to consider the effect of temperature. In other words, the idea is to isolate temperature and strain effects. For this reason, strain-invariant (called "temperature") sensors in FBG are used for measuring the temperature effect only.

The measurement part (measurement system block, Fig. 1) is intended for interrogating the sensors by applying the test input of each channel of broadband optical signal (the bandwidth is $\Delta\lambda_{\text{test}} \approx 40$ nm). The part makes the measurement of wavelengths reflected from the Bragg grating, *i.e.*, it works as an optical spectrum analyzer. Signal reflected from FBG sensors is the input to an optical switch (switch, Fig. 1), which is recorded after the measuring system. The switch produces a serial connection channel to measurement system block, which performs measurements of the central wavelengths of the reflected radiation from each FBG sensor (see Fig 2). These measurement system block and optical switch are connected to PC via the standard USB-interface (see Fig. 2).

Start our consideration from the simplest case of a one-channel measurement part. Suppose that there are n sensors in the channel, where $n - k$ is a quantity of Bragg sensors and k is a quantity of temperature Bragg sensors. Initial data for the structural monitoring system is the

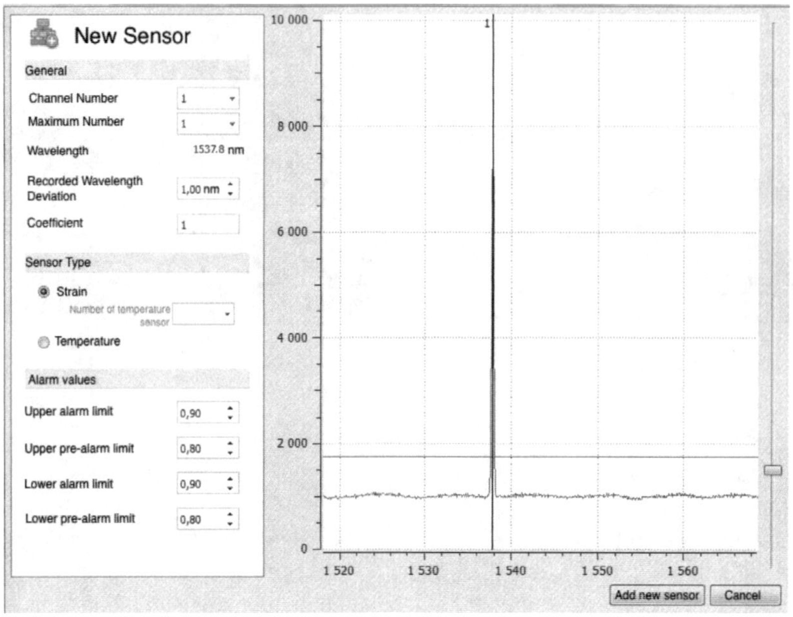

Figure 3. Software interface: adding of a new sensor in the measurement part of SHM system.

set of the reflection wavelength

$$\lambda = (\lambda_1, \ldots, \lambda_n), \tag{4}$$

where λ_i is the reflection wavelength of ith sensor. The result of the measurement in the channel is following set

$$\lambda^r = (\lambda_1^r, \ldots, \lambda_n^r), \tag{5}$$

where λ_i^r is the measured reflecting wavelength of ith sensor.

First, initial reflection wavelength of strain-invariant sensors compared with measured for identifying a temperature effect δT. The set of wavelengths

$$\lambda^r \mapsto \bar{\lambda}^r = (\bar{\lambda}_1^r, \ldots, \bar{\lambda}_n^r) \tag{6}$$

is formed from λ^r for excluding the temperature effect. Second, ith component of the vector $\bar{\lambda}^r$ compared with ith component of the vector λ for temperature sensors. Main aim of this procedure is to identify sensors with non-zero difference

$$\delta_i = \bar{\lambda}^r - \lambda_i. \tag{7}$$

Finally, from grating period Λ, number of sensor and the sample length l it is possible to localize the position of strain effects of the sample.

It is clear that for multi-channel system, *e.g.*, for our system, there is matrix $\lambda = \lambda_{ij}$ where i is the sensor number and j is the channel number. Matrix elements, obviously, λ_{ij} are the reflection wavelengths of the ith sensor in the nth channel. However, general procedure of data processing is the same.

We describe the principles of operation established for that purpose multi-threaded software applications. We note that there is a energy depend memory part in the measurement system

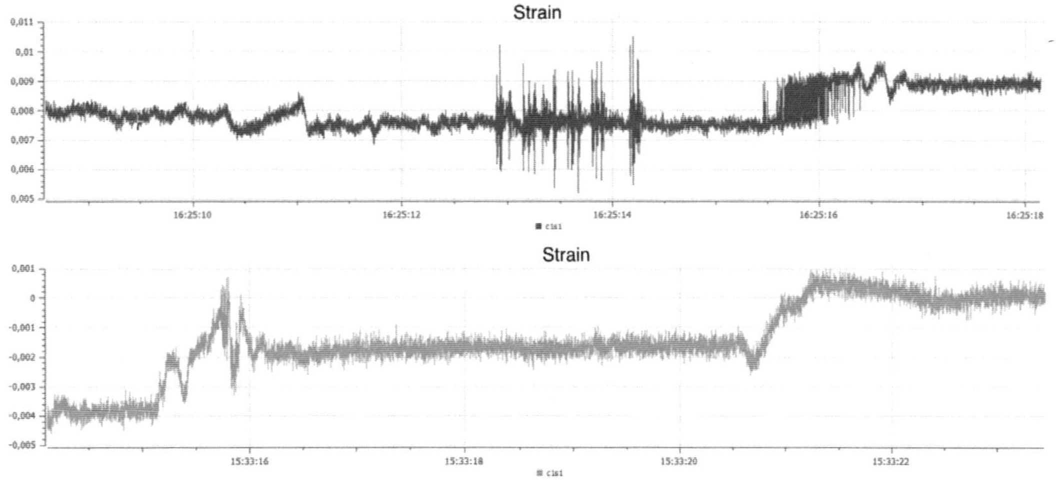

Figure 4. The results of testing of composite materials: strain is shows as function of time.

block and its DLL library used for getting data. The GUI software application is a system of tabs with different settings and for displaying data.

Screen of the main control page of developed software solution is presented on the Fig. 3. From software point of view, every sensor is characterized via set of parameters. As it has been mentioned before, every sensor measures both temperature and mechanical strain. In this way, first parameter of the sensor is a type. In addition, there are channel number, lower and upper limit among them.

On practice, special sensors which are sensitive only to changes in temperature are used. This allows to correct measurement results obtained from other sensors. After the addition of sensors in the channels formed by the summary table, which allows the operator to control the number and types of currently active sensors. The operator at any time is able to move the sensors for tracking in real time.

Via DLL libraries optical switch and measurement system implemented algorithms for channel management in the optical switch and selection the optimum mode of data collection, *i.e.*, a combination of channels, the number of active sensors in the channel sample rate sensors.

Additionally, the software part solves the problem of data visualization. Figure 4 presents our laboratory experimental setup in accordance with the general experimental setup (Fig. 2). Figures 5a and 5b show the results of processing measurement signals from the measurement equipment in the test laboratory strain of the composite material in various load conditions: this is strain as function of time. It should be noted that an important element in developing the software module is a data storage as an integral part of the system when used in practice, is the storage and archiving.

3. Conclusion

Structural monitoring systems are actively used to solve problems of the analysis of the dynamics of aging infrastructure in order to prevent their destruction. Developed systems allows to realize non-destructive structural health monitoring of structures made of composite materials. Software allows to perform the following kinds of problem: control the measurement part (activation of sensors, one of their number, channel selection and polling frequency sensors). Visualization of measurement results in real time is obtained via using a software application.

The use of software and hardware optoelectronic measurement system integrated continuous non-destructive testing is promising for a wide range of engineering problems. The software part of the complex can be easily modified to the case of using another type of sensor or monitoring by the sensors to certain elements of the structure.

Acknowledgments

We thank Scientific-Educational Center "Photonics and Infrared Technology" for support and K.I. Zaytsev for useful comments. AKF is a Dynasty Foundation Fellow. We are grateful to the organizers of the 23rd Laser Physics Workshop (Sofia, July 14–18, 2014) for kind hospitality.

References

[1] Kinet D, Mégret P, Goossen K W, Qiu L, Heider D, and Caucheteur C 2014 *Sensors* **14** 7394.
[2] Tennyson R C, Mufti A A, Rizkalla S, Tadros G, and Benmokrane B 2001 *Smart Mater. Struct.* **10** 560.
[3] Tanaka T, Yamamoto M, Katayama T, Nakahira K, Yamane K, Shimano Y, and Hirayama K 2002 *Earthquake Engineering and Engineering Seismology* **4** 75.
[4] Berthelot J M 1999 Composite Materials: Mechanical Behavior and Structural Analysis; Springer: New York.
[5] Stoik C, Bohn M, and Blackshire J 2008 *Opt. Express* **21** 17039.
[6] Yakovlev E V, Zaytsev K I, Fokina I N, Karasik V E, and Yurchenko S O 2014 *J. Phys. Conf. Ser.* **486** 012008.
[7] Othonos A, Kalli K 1999 Fiber Bragg Gratings: Fundamentals and Applications in Telecommunications and Sensing; Artech House: London.
[8] Vasil'ev S A, Medvedkov O I, Korolev I G, Bozhkov A S, Kurkov A S, and Dianov E M 2005 *Quantum Electron.* **35** 1085.
[9] Karasik V E, Lazarev V A 2011 *Herald of the Bauman Moscow State Technical University. Instrumentation. Special Issue "Modern Problems of Optical Engineering"* 157.
[10] Grigoriev V V, Lazarev V A, Mityurev A K, Pnev A B, Tikhomirov S V, and Hatyrev N P 2009 *Sensors and Systems* **1** 15.
[11] Grigoriev V V, Lazarev V A, Mityurev A K, Pnev A B, and Tikhomirov S V 2009 *Photon Express. Special Issue* **6** 156.
[12] Grigoriev V V, Lazarev V A, Neverova N A, Mityurev A K, Pnev A B, and Tikhomirov S V 2008 *Automation and Remote Control* **71** 2200.
[13] Fedorov A K, Lazarev V A, Makhrov I P, Pozhar N O, Anufriev M N, and Pniov A B 2013 Software solution for a structural monitoring system with fiber Bragg grating sensors, in: *Collection of scientific papers of the Third International Conference "High Performance Computing"*, Kyiv, p. 125.

Journal of Physics: Conference Series **594** (2015) 012050 doi:10.1088/1742-6596/594/1/012050

On Vacuum Polarization and Schwinger Pair Production in Intense Lasers

Sang Pyo Kim

Department of h sics, unsan ational ni ersit , unsan 5 3- 1, orea
Center for Relati istic aser Science, Institute for Basic Science IBS , Gwangju 5 - 12, orea

-mail sangkim@kunsan.ac.kr

Abstract. e re iew and elaborate the complex e ecti e action at one-loop and at ero or nite temperature in the in-out formalism for scalar Q D and the acuum persistence in time-dependent electric elds. sing the gamma-function regulari ation, we nd the e ecti e action in the proper-time integral and the pair-production rate in an exponentiall increasing electric eld. e appl the uantum in ariant theor to the scalar eld in time-dependent electric elds and clarif the meaning of multi-pair states. nd pair-production rates are deri ed when the initial state is the Min ows i acuum or the adiabatic acuum and are compared with two representations for the Vlaso e uation. Finall , the contour integral method gi es the pair-production rate in the exponentiall increasing electric eld.

1. Introduction

The vacuum polarization and the production of electron-positron pairs by background electric fields is one of the most interesting nonperturbative phenomena in quantum field theory. Heisenberg and Euler [1] and Schwinger [2] have shown that the vacuum under the action of a constant electromagnetic field could be polarized as a medium and the e ective action could gain an imaginary part, implying the instability of the Dirac sea (Minkowski vacuum) due to the pair production. With the rapid development of intense lasers, probing the vacuum structure may be attainable by such a facility as Extreme Light Infrastructure (ELI) or International Center for Zetta-Exawatt Science and Technology (IZEST) in the near future.[1]

The coherent state of numerous low energy photons from intense lasers provides an electromagnetic field, which is spatially localized and temporally pulsed. However, computing the one-loop e ective action for the localized pulse has been a theoretical challenge in quantum field theory in strong field physics (for a good review and references, see [3]). Further, electron-positron pairs can be produced in a localized, time-dependent electric field. The vacuum instability implies the one-loop e ective action should be complex in electric fields. In fact, the in-out formalism by Schwinger and DeWitt [4, 5] leads to a complex e ective action, the real part of which is the vacuum polarization and twice the imaginary part of which is the decay rate of the vacuum and is related to the pair production rate. The methods so far introduced for calculating the pair-production rate are the Vlasov equation [6, 7, 8], the worldline instanton [9, 10], the phase integral [11, 12], the contour integral [13, 14], to name a few. The methods for the

[1] I, http //www.eli-laser.eu/ and I ST, http //www.i est.pol techni ue.edu/.

vacuum polarization are the resolvent technique [15], the Feynman worldline path integral [16], the Dirac-Heisenberg-Wigner formalism [17], the Schwinger-DeWitt in-out formalism [18, 19]. There are other useful methods either for the vacuum persistence or for the vacuum polarization, which are not listed here [3].

In this paper, we review and elaborate a recent field theoretic method for the complex e ective action and the quantum invariant method for quantum states for a charged field in time-dependent electric fields. In the in-out formalism, the one-loop action integrated over the four-volume is the scattering matrix $e^{i\mathcal{W}} \quad \langle 0, \mathrm{out}|0, \mathrm{in}\rangle$ between the out-vacuum and the in-vacuum. Further, the scattering matrix can be expressed by the Bogoliubov coe cients between two vacua in the electromagnetic field background. As quantum field theory, some renormalization scheme should be taken to regularize the UV divergences by renormalizing the vacuum energy and the charge. With the renormalization taken for granted, the task for the one-loop e ective action is to find the Bogoliubov transformation between the in-vacuum and the out-vacuum. This implies that the one-loop e ective action becomes complex when an electric field present regardless of the presence of a magnetic field. In other words, the complex e ective action consists of the vacuum polarization and the vacuum persistence

$$\mathcal{L}^{\mathrm{C}}_{\mathrm{eff}} \quad \mathcal{L}_{\mathrm{vac\,pol}} \quad \frac{i}{2}\mathcal{L}_{\mathrm{vac\,per}}. \tag{1}$$

As will be shown in section 2, the vacuum persistence in the electric field is related to the pair-production rate $\mathcal{N}_{\mathbf{k}}$ for each momentum

$$\mathcal{L}_{\mathrm{vac\,per}} \quad 2\,\mathrm{Im}\mathcal{L}^{\mathrm{C}}_{\mathrm{eff}} \quad \sum_{\mathbf{k}} \ln(1 \quad \mathcal{N}_{\mathbf{k}}). \tag{2}$$

We then apply the quantum invariant theory by Lewis and Riesenfeld [20] to the time-dependent Hamiltonian for a charged spinless scalar in time-dependent electric fields. The Hamiltonian is equivalent to an infinite sum of time-dependent oscillators in homogeneous, time-dependent electric fields and to coupled oscillators in the time-dependent magnetic fields with a fixed direction [21]. The time-dependent annihilation and creation operators, quantum invariant operators, construct not only the time-dependent vacuum but also excited states of multi-pairs and a thermal state. Electron-positron pairs are spontaneously produced from the vacuum and there can be stimulated production of pairs from some pairs or a thermal ensemble of pairs present in the remote past.

2. Review of In-Out Formalism for Scalar QED Action

In this section we review the complex ED action in time-dependent electric fields the in-out formalism. For the sake of simplicity, we consider a charged spinless scalar in a homogeneous, time-dependent electromagnetic field with the vector potential

$$\mathbf{A}(t) \quad -\int^{t} \mathbf{E}(t')dt'. \tag{3}$$

Due to the translational symmetry, the Klein-Gordon equation after the momentum-decomposition becomes a second order equation (in units of $\hbar \quad c \quad 1$)

$$\phi_{\mathbf{k}}(t) \quad \omega^2_{\mathbf{k}}(t)\phi_{\mathbf{k}}(t) \quad 0 \tag{4}$$

with a time-dependent frequency squared

$$\omega^2_{\mathbf{k}}(t) \quad (\mathbf{k} - q\mathbf{A}(t))^2 \quad m^2. \tag{5}$$

By connecting the positive and the negative frequency solutions to eq. (4) satisfying appropriate boundary conditions in the remote future and the remote past, we find the Bogoliubov transformation between the particle and antiparticle annihilation operators $a_{\mathbf{k},\text{out}}, b_{\mathbf{k},\text{out}}$ in the out-vacuum and those $a_{\mathbf{k},\text{in}}, b_{\mathbf{k},\text{in}}$ in the in-vacuum:

$$\begin{aligned} a_{\mathbf{k},\text{out}} &\quad \alpha_{\mathbf{k}} a_{\mathbf{k},\text{in}} \quad \beta^*_{\mathbf{k}} b^\dagger_{\mathbf{k},\text{in}} \quad U_{\mathbf{k}} a_{\mathbf{k},\text{in}} U^\dagger_{\mathbf{k}}, \\ b_{\mathbf{k},\text{out}} &\quad \alpha_{\mathbf{k}} b_{\mathbf{k},\text{in}} \quad \beta^*_{\mathbf{k}} a^\dagger_{\mathbf{k},\text{in}} \quad U_{\mathbf{k}} B_{\mathbf{k},\text{in}} U^\dagger_{\mathbf{k}}. \end{aligned} \tag{6}$$

Here, the evolution operator can be expressed in terms of a two-mode squeezing operator and a phase operator [18]

$$U_{\mathbf{k}} \quad \exp\Big[r_{\mathbf{k}}(a_{\mathbf{k},\text{in}} b_{\mathbf{k},\text{in}} e^{-2i\vartheta_{\mathbf{k}}} - \text{H.C.})\Big] \exp\Big[i\theta_{\mathbf{k}}(a^\dagger_{\mathbf{k},\text{in}} a_{\mathbf{k},\text{in}} \quad b_{\mathbf{k},\text{in}} b^\dagger_{\mathbf{k},\text{in}})\Big], \tag{7}$$

where the squeeze parameter $r_{\mathbf{k}}$ and angles $\theta_{\mathbf{k}}, \vartheta_{\mathbf{k}}$ are given by

$$\alpha_{\mathbf{k}} \quad e^{-i\theta_{\mathbf{k}}} \cosh r_{\mathbf{k}}, \quad \beta_{\mathbf{k}} \quad -e^{i\theta_{\mathbf{k}}}(e^{-2i\vartheta_{\mathbf{k}}} \sinh r_{\mathbf{k}}). \tag{8}$$

These operators satisfy the commutation relations

$$[a_{\mathbf{k},\text{in}}, a^\dagger_{\mathbf{p},\text{in}}] \quad [b_{\mathbf{k},\text{in}}, b^\dagger_{\mathbf{p},\text{in}}] \quad \delta(\mathbf{k} - \mathbf{p}), \tag{9}$$

and a similar expression holds for $a_{\mathbf{k},\text{out}}, b_{\mathbf{k},\text{out}}$ but all the other commutators vanish. The out-vacuum as the squeeze vacuum

$$|0, \text{out}\rangle \quad \prod_{\mathbf{k}} e^{i\theta_{\mathbf{k}}} \exp\Big[r_{\mathbf{k}}(a_{\mathbf{k},\text{in}} b_{\mathbf{k},\text{in}} e^{-2i\vartheta_{\mathbf{k}}} - \text{H.C.})\Big] |0, \text{in}\rangle, \tag{10}$$

is equivalent to the summation of all one-loop diagrams with even number of external legs. Then, the out-vacuum is the multi-pair states of the in-vacuum

$$|0, \text{out}\rangle \quad \prod_{\mathbf{k}} e^{i\theta_{\mathbf{k}}} \text{sech} r_{\mathbf{k}} \Big[\sum_{n_{\mathbf{k}}=0}^{\infty} e^{i2n_{\mathbf{k}}\vartheta_{\mathbf{k}}} \tanh^{n_{\mathbf{k}}}(r_{\mathbf{k}}) |n_{\mathbf{k}}, \bar{n}_{\mathbf{k}}, \text{in}\rangle\Big]. \tag{11}$$

The complex one-loop e ective action at the zero temperature ($T \quad 0$) integrated over three-volume $\mathcal{V}$ and time is given by [18, 19]

$$\mathcal{W} \quad -i \ln\langle 0, \text{out}|0, \text{in}\rangle \quad i\mathcal{V} \sum_{\mathbf{k}} \alpha^*_{\mathbf{k}}, \tag{12}$$

with $\sum_{\mathbf{k}} \quad \int d^3\mathbf{k}/(2\pi)^3$, and the action at finite temperature T is given by [22]

$$e^{\mathcal{W}_{\text{T}}} \quad \frac{\text{Tr}\Big(\prod_{\mathbf{k}} U^\dagger_{\mathbf{k}} \rho_{\mathbf{k}}\Big)}{\text{Tr}\Big(\prod_{\mathbf{k}} \rho_{\mathbf{k}}\Big)}. \tag{13}$$

Here, $\rho_{\mathbf{k}}$ denotes the density operator for an initial thermal ensemble of pairs with the energy $\omega_{\mathbf{k}}$ present in the background field. In a similar way, we may find from eq. (13) the e ective action stimulated by the initial pairs in the state $\rho_{\mathbf{k}} \quad |n_{\mathbf{k}}, \bar{n}_{\mathbf{k}}, \text{in}\rangle\langle n_{\mathbf{k}}, \bar{n}_{\mathbf{k}}, \text{in}|$, which is the ED analog of stimulated emission of photons. Then, the e ective action at T per unit volume has the expression [22]

$$\mathcal{L}^{\text{C}}_{\text{eff}}(T) \quad -i \sum_{\mathbf{k}} \Big[\ln(1 \quad e^{-(\omega_{\mathbf{k}} - z_{\mathbf{k}})/kT}) - \frac{z_{\mathbf{k}}}{kT} - \ln(1 \quad e^{-\omega_{\mathbf{k}}/kT})\Big], \tag{14}$$

where $z_{\mathbf{k}}$ is the chemical potential

$$z_{\mathbf{k}} \quad kT(i\theta_{\mathbf{k}} - \ln\cosh r_{\mathbf{k}}). \tag{15}$$

The action (14) has the following interpretation: the first term is the e ective action with the complex chemical potential $z_{\mathbf{k}}$, the second term is the zero-temperature action and the last term is the free energy for the initial ensemble. In the zero-temperature limit, eq. (14) reduces to the vacuum e ective action. Remarkably, the vacuum persistence is expressed by the initial Bose-Einstein distribution n_{BE} and the complex chemical potential as [22]

$$2\mathrm{Im}\mathcal{L}^{\mathrm{C}}_{\mathrm{eff}}(T) \quad -\sum_{\mathbf{k}}\sum_{l=0}^{\infty}\frac{(n_{BE})^l}{l}[(e^{z_{\mathbf{k}}/kT}-1)^l \quad (e^{z^*_{\mathbf{k}}/kT}-1)^l]. \tag{16}$$

3. Complex QED Action in a Time-dependent Electric Field

We now illustrate the in-out formalism by working out the scalar ED in a time-dependent electric field along the z-direction

$$E_z(t) \quad E_0 e^{\frac{t}{\tau}}, \quad A_z(t) \quad -E_0\tau(e^{\frac{t}{\tau}}-1). \tag{17}$$

The positive frequency solution to eq. (4) in the remote future is given by the Whittaker function

$$\phi^{(+)}_{\mathbf{k}}(t) \quad e^{-i\pi\lambda/2}\sqrt{\frac{\tau}{\xi}}W_{\lambda,-\mu}(\xi), \tag{18}$$

where

$$\xi \quad 2iqE_0\tau^2 e^{\frac{t}{\tau}}, \quad \mu \quad i\tau\bar{\omega}_{\mathbf{k}}, \quad \lambda \quad -i\tau\bar{k}_z, \tag{19}$$

with

$$\bar{k}_z \quad k_z - qE_0\tau, \quad \bar{\omega}_{\mathbf{k}} \quad \sqrt{m^2 \quad \mathbf{k}^2_{\perp} \quad \bar{k}^2_z}. \tag{20}$$

One may be tempted to expect a catastrophic pair production since the model electric field (17) increases beyond the critical field when $t \geq \tau\ln(m^2/qE_0)$. In quantum field theory, however, the out-vacuum is defined as an asymptotic state with respect to the solution (18) in the remote future. And the solution (18) can be linearly superposed as the positive and the negative frequency solutions in the remote past, corresponding to the Minkowski vacuum. Then, the Bogoliubov coe cients are

$$\alpha_{\mathbf{k}} \quad \sqrt{\frac{2\bar{\omega}_{\mathbf{k}}\tau e^{\pi\tau\bar{\omega}_{\mathbf{k}}}}{e^{\pi\bar{k}_z}}}\frac{(2\mu)}{(\frac{1}{2} \quad \mu-\lambda)},$$
$$\beta_{\mathbf{k}} \quad \sqrt{\frac{2\bar{\omega}_{\mathbf{k}}\tau}{e^{\pi\tau\bar{\omega}_{\mathbf{k}}}e^{\pi\bar{k}_z}}}\frac{(-2\mu)}{(\frac{1}{2}-\mu-\lambda)}. \tag{21}$$

The pair-production rate is given by

$$\mathcal{N}_{\mathbf{k}} \quad |\beta_{\mathbf{k}}|^2 \quad \frac{e^{-2\pi\tau\bar{\omega}_{\mathbf{k}}} \quad e^{-2\pi\tau\bar{k}_z}}{e^{2\pi\tau\bar{\omega}_{\mathbf{k}}} - e^{-2\pi\tau\bar{\omega}_{\mathbf{k}}}}. \tag{22}$$

We find the complex e ective action (12) from the in-out formalism

$$\mathcal{L}^{\mathrm{C}}_{\mathrm{eff}} \quad i\sum_{\mathbf{k}}[\ (2\mu^*) - \ (\frac{1}{2} \quad \mu^* - \lambda^*) - \cdots], \tag{23}$$

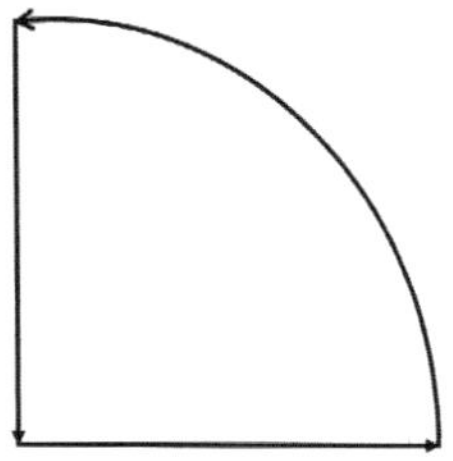

Figure 1. One of the contour integrals in the complex plane of proper time for the gamma function regularization [23].

where dots denote terms to be regulated away through renormalization. Following refs. [18, 19], the complex action can be written as the proper-time integral

$$\mathcal{L}_{\rm eff}^{\rm C} \quad i\sum_{\mathbf{k}} \int_0^\infty \frac{ds}{s}\Big[\frac{e^{2i\tau\bar{\omega}_{\mathbf{k}}s}}{1-e^{-s}} - \frac{e^{i\tau(\bar{\omega}_{\mathbf{k}}+\bar{k}_z)s}}{e^{\frac{s}{2}}-e^{-\frac{s}{2}}} - 1 \quad \frac{2}{s}\Big], \tag{24}$$

where the Schwinger subtraction scheme is employed to renormalize the vacuum energy and the charge. The Cauchy theorem for the contour in figure 1 leads to the complex e ective action

$$\mathcal{L}_{\rm eff}^{\rm C} \quad \frac{1}{2}\sum_{\mathbf{k}} \int_0^\infty \frac{ds}{s}\Big[\frac{e^{-2\tau\bar{\omega}_{\mathbf{k}}s}}{\cos(\frac{s}{2})} - \frac{e^{-\tau(\bar{\omega}_{\mathbf{k}}+\bar{k}_z)s}}{\sin(\frac{s}{2})} - 1 \quad \frac{2}{s}\Big] \quad \frac{i}{2}\ln(1 \quad \mathcal{N}_{\mathbf{k}}). \tag{25}$$

The imaginary part comes from the residues at simple poles $(2i\pi n)$ along the imaginary axis and thus satisfies the consistency condition from the in-out formalism. It is interesting that the first proper-time integral has the form for spinor ED while the second integral has the form for scalar ED [2]. Using the Bogoliubov coe cients (21), we may find the complex e ective action (14) at finite temperature, which will not be pursued here.

4. Quantum Invariant Theory for Schrödinger Equation

In the time-dependent electric field (3), the charged field has the time-dependent Hamiltonian [21]

$$H(t) : \sum_{\mathbf{k}} H_{\mathbf{k}}(t) \quad \sum_{\mathbf{k}} [\pi_{\mathbf{k}}^* \pi_{\mathbf{k}} \quad \omega_{\mathbf{k}}^2(t)\phi_{\mathbf{k}}^*\phi_{\mathbf{k}}], \tag{26}$$

where $\pi_{\mathbf{k}} \quad \phi_{\mathbf{k}}^*$, $\pi_{\mathbf{k}}^* \quad \phi_{\mathbf{k}}$. In the Schrödinger picture, quantum states may be found from the time-dependent functional Schrödinger equation

$$i\frac{\partial}{\partial t} \quad (t,\phi,\phi^*) \quad H(t) \quad (t,\phi,\phi^*). \tag{27}$$

The main advantage of the Schrödinger picture is the diversity of quantum states as in quantum mechanics and quantum optics. We further employ the quantum invariant theory for time-dependent oscillators by Lewis and Riesenfeld, which solves the Liouville-von Neumann equation for the density operator [20]

$$i\frac{\partial \rho_{\mathbf{k}}(t)}{\partial t} \quad [\rho_{\mathbf{k}}(t), H_{\mathbf{k}}(t)] \quad 0. \tag{28}$$

The eigenstate of the invariant (28) is an exact solution to eq. (27) up to a time-dependent phase factor [20]. We adopt a pair of invariant operators that play the role of time-dependent annihilation and creation operators [21, 24, 25]

$$A_{\mathbf{k}}(t) \quad \frac{i}{\sqrt{2}}\Big(\varphi^*_{\mathbf{k}}(\pi^\dagger_{\mathbf{k}} \quad \pi_{\mathbf{k}}) - \varphi^*{}_{\mathbf{k}}(\phi^\dagger_{\mathbf{k}} \quad \phi_{\mathbf{k}})\Big) \quad \frac{1}{\sqrt{2}}(a_{\mathbf{k}}(t) \quad b_{\mathbf{k}}(t)),$$
$$A^\dagger_{\mathbf{k}}(t) \quad -\frac{i}{\sqrt{2}}\Big(\varphi_{\mathbf{k}}(\pi^\dagger_{\mathbf{k}} \quad \pi_{\mathbf{k}}) - \varphi_{\mathbf{k}}(\phi^\dagger_{\mathbf{k}} \quad \phi_{\mathbf{k}})\Big) \quad \frac{1}{\sqrt{2}}(a^\dagger_{\mathbf{k}}(t) \quad b^\dagger_{\mathbf{k}}(t)). \tag{29}$$

Here, $a_{\mathbf{k}}(t)$ and $b_{\mathbf{k}}(t)$ are the annihilation operators for particle and antiparticle, respectively, and φ_α is a complex solution to the mode equation (4) and satisfies the Wronskian condition

$$\varphi_{\mathbf{k}}(t)\dot{\varphi}^*_{\mathbf{k}}(t) - \varphi^*_{\mathbf{k}}(t)\dot{\varphi}_{\mathbf{k}}(t) \quad i. \tag{30}$$

Then, the time-dependent annihilation and creation operators satisfy the equal-time commutator

$$[A_{\mathbf{k}}(t), A^\dagger_{\mathbf{p}}(t)] \quad \delta(\mathbf{k}-\mathbf{p}). \tag{31}$$

Note that $a_{\mathbf{k}}(t)$ and $b_{\mathbf{k}}(t)$ are not independently invariants and become an invariant as the pair. The fields are given by

$$\phi(t,\mathbf{x}) \quad \sum_{\mathbf{k}}[\varphi_{\mathbf{k}}(t)a_{\mathbf{k}}(t)e^{i\mathbf{k}\cdot\mathbf{x}} \quad \varphi^*_{\mathbf{k}}(t)b^\dagger_{\mathbf{k}}(t)e^{-i\mathbf{k}\cdot\mathbf{x}}],$$
$$\pi(t,\mathbf{x}) \quad \sum_{\mathbf{k}}[\dot{\varphi}_{\mathbf{k}}(t)a_{\mathbf{k}}(t)e^{i\mathbf{k}\cdot\mathbf{x}} \quad \dot{\varphi}^*_{\mathbf{k}}(t)b^\dagger_{\mathbf{k}}(t)e^{-i\mathbf{k}\cdot\mathbf{x}}]. \tag{32}$$

Note that the antiparticle carries the opposite momentum of the particle.

Using eq. (29), we can construct coherent states, squeezed states as well as multi-pair states [26, 27, 28]. The quantum invariant theory has been used to derive the nonadiabatic quantum Vlasov equation [8, 29]. The time-dependent vacuum is the product of all zero-particle states

$$|0,t\rangle \quad \prod_{\mathbf{k}}|0_{\mathbf{k}},t\rangle, \quad A_{\mathbf{k}}(t)|0_{\mathbf{k}},t\rangle \quad 0. \tag{33}$$

The fields (32) have the equal-time correlation functions

$$\langle 0,t|\phi^\dagger(t,\mathbf{y})\phi(t,\mathbf{x})|0,t\rangle \quad \sum_{\mathbf{k}}|\varphi_{\mathbf{k}}(t)|^2 e^{i\mathbf{k}\cdot(\mathbf{x}-\mathbf{y})},$$
$$\langle 0,t|\pi^\dagger(t,\mathbf{y})\pi(t,\mathbf{x})|0,t\rangle \quad \sum_{\mathbf{k}}|\dot{\varphi}_{\mathbf{k}}(t)|^2 e^{i\mathbf{k}\cdot(\mathbf{x}-\mathbf{y})}. \tag{34}$$

The correlation functions (34) may be used to find the energy-momentum tensor and even the e ective action, which requires independent works. For the fixed $\mathbf{k}$, acting $A^\dagger_{\mathbf{k}}(t)$ n-times on the vacuum excites the n-pair state [21]

$$|n_{\mathbf{k}},t\rangle \quad \frac{(A^\dagger_{\mathbf{k}}(t))^{n_{\mathbf{k}}}}{\sqrt{n_{\mathbf{k}}!}}|0,t\rangle \quad \frac{1}{\sqrt{2^{n_{\mathbf{k}}}n_{\mathbf{k}}!}}\sum_{l_{\mathbf{k}}=0}^{n_{\mathbf{k}}}\binom{n_{\mathbf{k}}}{l_{\mathbf{k}}}|n_{\mathbf{k}}-l_{\mathbf{k}},\bar{l}_{\mathbf{k}},t\rangle, \tag{35}$$

where the time-dependent particle-antiparticle states are defined as

$$|k_{\mathbf{k}},\bar{l}_{\mathbf{p}},t\rangle \quad \frac{(a^\dagger_{\mathbf{k}}(t))^{k_{\mathbf{k}}}}{\sqrt{k_{\mathbf{k}}!}}\frac{(b^\dagger_{\mathbf{p}}(t))^{l_{\mathbf{p}}}}{\sqrt{l_{\mathbf{p}}!}}|0_{\mathbf{k}},\bar{0}_{\mathbf{p}},t\rangle. \tag{36}$$

The multiparticle states are orthonormal to each other. The n-pair state is completely symmetric with respect to the particle and antiparticle numbers and is an entangled state. The number of particles or antiparticles is

$$\langle n_{\mathbf{k}}, t|a^{\dagger}_{\mathbf{k}}(t)a_{\mathbf{k}}(t)|n_{\mathbf{k}}, t\rangle \quad \langle n_{\mathbf{k}}, t|b^{\dagger}_{\mathbf{k}}(t)b_{\mathbf{k}}(t)|n_{\mathbf{k}}, t\rangle \quad \frac{n_{\mathbf{k}}}{2}. \tag{37}$$

In fact, the n-pair state (35) contains the total number $n_{\mathbf{k}}$ of particles and antiparticles. The thermal state may be given by

$$\rho_{\mathbf{k}}(t) \quad \frac{1}{Z_{\mathbf{k}}} e^{-2\frac{\omega_{\mathbf{k}i}}{kT}\hat{A}^{\dagger}_{\mathbf{k}}(t)\hat{A}_{\mathbf{k}}(t)}. \tag{38}$$

Here, $Z_{\mathbf{k}} \quad \mathrm{Tr}e^{-2(\omega_{\mathbf{k}i}/kT)\hat{A}^{\dagger}_{\mathbf{k}}(t)\hat{A}_{\mathbf{k}}(t)}$ is the partition function and $\omega_{\mathbf{k}i}$ is the initial energy for particle and antiparticle. The factor of two makes the density operator practically equivalent to a product of $e^{-(\omega_{\mathbf{k}i}/kT)\hat{a}^{\dagger}_{\mathbf{k}}(t)\hat{a}_{\mathbf{k}}(t)}$ and $e^{-(\omega_{\mathbf{k}i}/kT)\hat{b}^{\dagger}_{\mathbf{k}}(t)\hat{b}_{\mathbf{k}}(t)}$. The thermal density (38) is also a quantum invariant and may be used to calculate the e ective action at T in section 2.

5. Spontaneous and Induced Pair Production

In the quantum invariant theory for time-dependent systems, using the time-dependent annihilation and creation operators (29), we can construct various exact quantum states from time-dependent number states to the squeezed number state [27] and even to the thermal squeezed-coherent state [28]. The time-dependent vacuum (33) is an exact state for eq. (27). In the limit of $\mathbf{A} \quad 0$ in the remote past ($t_0 \quad -\infty$), the vacuum (33) becomes the Minkowski vacuum. Further, the invariant operators (29) define the number operator with the given momentum $\mathbf{k}$

$$N_{\mathbf{k}}(t) \quad A^{\dagger}_{\mathbf{k}}(t)A_{\mathbf{k}}(t). \tag{39}$$

Now, we may ask a question how many pairs of Minkowski particle and antiparticle are contained in the vacuum (33)

$$\mathcal{N}_{\mathbf{k}}(t) \quad \langle 0, t|A^{\dagger}_{\mathbf{k}}(t_0)A_{\mathbf{k}}(t_0)|0, t\rangle, \tag{40}$$

or how many pairs defined by the number operator (39) are contained in the Minwkowski vacuum

$$\mathcal{N}_{\mathbf{k}}(t) \quad \langle 0, t_0|A^{\dagger}_{\mathbf{k}}(t)A_{\mathbf{k}}(t)|0, t_0\rangle. \tag{41}$$

The rate of spontaneous production of pairs (40) and (41) by the electric field is given by the same formula [21]

$$\mathcal{N}_{\mathbf{k}}(t) \quad |\varphi_{\mathbf{k}}(t)|^2|\varphi_{\mathbf{k}}(t_0)|^2 \quad |\varphi_{\mathbf{k}}(t)|^2|\varphi_{\mathbf{k}}(t_0)|^2 - \frac{1}{2}. \tag{42}$$

It is symmetric with respect to the exchange of times, t and t_0. The production of pairs induced from pairs already present in the remote past is now given by

$$\mathcal{N}_{\mathbf{k}}(t) \quad \langle n_{\mathbf{k}}, t_0|A^{\dagger}_{\mathbf{k}}(t)A_{\mathbf{k}}(t)|n_{\mathbf{k}}, t_0\rangle \quad (|\varphi_{\mathbf{k}}(t)|^2|\varphi_{\mathbf{k}}(t_0)|^2 \quad |\varphi_{\mathbf{k}}(t)|^2|\varphi_{\mathbf{k}}(t_0)|^2)(n_{\mathbf{k}} \quad 1) - \frac{1}{2}. \tag{43}$$

We may use the Hamiltonian expectation value to count the number of quanta with the given energy. The Hamiltonian expressed in terms of particle and antiparticle operators

$$\begin{aligned} H_{\mathbf{k}}(t) \quad & (|\varphi_{\mathbf{k}}(t)|^2 \quad \omega^2_{\mathbf{k}}(t)|\varphi_{\mathbf{k}}(t)|^2)(a^{\dagger}_{\mathbf{k}}(t)a_{\mathbf{k}}(t) \quad b_{\mathbf{k}}(t)b^{\dagger}_{\mathbf{k}}(t)) \\ & (\varphi^2_{\mathbf{k}}(t) \quad \omega^2_{\mathbf{k}}(t)\varphi^2_{\mathbf{k}}(t))a_{\mathbf{k}}(t)b_{\mathbf{k}}(t) \quad (\varphi^{*2}_{\mathbf{k}}(t) \quad \omega^2_{\mathbf{k}}(t)\varphi^{*2}_{\mathbf{k}}(t))a^{\dagger}_{\mathbf{k}}(t)b^{\dagger}_{\mathbf{k}}(t). \end{aligned} \tag{44}$$

has the expectation value at the equal time

$$\langle n_{\mathbf{k}}, t|H_{\mathbf{k}}(t)|n_{\mathbf{k}}, t\rangle \quad (|\varphi_{\mathbf{k}}(t)|^2 \quad \omega_{\mathbf{k}}^2(t)|\varphi_{\mathbf{k}}(t)|^2)(2n_{\mathbf{k}} \quad 1). \tag{45}$$

In the zero-field limit $\mathbf{E}(\pm\infty) \quad 0$, the asymptotic solutions

$$\varphi_{\mathbf{k}}^{\text{in}}(t) \quad \frac{e^{-i\omega_{\mathbf{k}\text{i}}t}}{\sqrt{2\omega_{\mathbf{k}\text{i}}}}, \quad \varphi_{\mathbf{k}}^{\text{out}}(t) \quad \frac{e^{-i\mathbf{k}\text{o}t}}{\sqrt{2\omega_{\mathbf{k}\text{o}}}}, \tag{46}$$

lead to the in-vacuum and the out-vacuum

$$\begin{aligned} H_{\mathbf{k}\text{i}}|n_{\mathbf{k}}, \text{in}\rangle &\quad \omega_{\mathbf{k}\text{i}}\Big(n_{\mathbf{k}} \quad \frac{1}{2}\Big)|n_{\mathbf{k}}, \text{in}\rangle, \\ H_{\mathbf{k}\text{o}}|n_{\mathbf{k}}, \text{out}\rangle &\quad \omega_{\mathbf{k}\text{o}}\Big(n_{\mathbf{k}} \quad \frac{1}{2}\Big)|n_{\mathbf{k}}, \text{out}\rangle. \end{aligned} \tag{47}$$

Now, we use eq. (42) to compare with the pair-production rate in the in-in formalism and the in-out formalism [29]. Substituting eq. (46) into eq. (42), we find the pair-production rates

$$\mathcal{N}_{\mathbf{k}}^{\text{in}}(t) \quad \frac{|\varphi_{\mathbf{k}}(t)|^2 \quad \omega_{\mathbf{k}\text{i}}^2|\varphi_{\mathbf{k}}(t)|^2}{2\omega_{\mathbf{k}\text{i}}} - \frac{1}{2}, \tag{48}$$

and

$$\mathcal{N}_{\mathbf{k}}^{\text{out}}(t) \quad \frac{|\varphi_{\mathbf{k}}(t)|^2 \quad \omega_{\mathbf{k}\text{o}}^2|\varphi_{\mathbf{k}}(t)|^2}{2\omega_{\mathbf{k}\text{o}}} - \frac{1}{2}. \tag{49}$$

On the other hand, the adiabatic solution

$$\varphi_{\mathbf{k}}^{\text{ad}}(t) \quad \frac{e^{-i\int^t \omega_{\mathbf{k}}(t')dt'}}{\sqrt{2\omega_{\mathbf{k}}(t)}} \tag{50}$$

leads to another pair-production rate

$$\mathcal{N}_{\mathbf{k}}^{\text{ad}}(t) \quad \frac{|\varphi_{\mathbf{k}}(t)|^2 \quad \omega_{\mathbf{k}}^2(t)|\varphi_{\mathbf{k}}(t)|^2}{2\omega_{\mathbf{k}}(t)} - \frac{1}{2}. \tag{51}$$

The pair-production rate (48) is the same as eq. (22) in the in-in Vlasov equation and eq. (51) is the same as eq. (12) in the in-out Vlasov equation of ref. [29]. Note that eq. (51) may be obtained by

$$\mathcal{N}_{\mathbf{k}}^{\text{ad}} \quad \frac{1}{2\omega_{\mathbf{k}}(t)}\langle 0_{\mathbf{k}}, t|H_{\mathbf{k}}(t)|0_{\mathbf{k}}, t\rangle - \frac{1}{2}. \tag{52}$$

The pair production (52) is the mean energy divided by the instantaneous energy, which counts the number of quanta with the energy $\omega_{\mathbf{k}}$.

6. Pair Production Via Polons

Various approximation methods have been developed to calculate the pair-production rate. The contour integral method evolves the Hamiltonian in the complex plane of time or space when the gauge potential (3) has an analytical continuation. Then, the magnitude of the in-in scattering matrix approximately gives the pair-production rate [14]

$$\mathcal{N}_{\mathbf{k}} \quad \Big|\sum_J \exp\Big[-i\oint_{C_J^{(1)}} \omega_{\mathbf{k}}(\zeta)d\zeta\Big]\Big|, \tag{53}$$

where $\omega_{\mathbf{k}}(\zeta)$ is the frequency in the complex plane and $C_J^{(1)}$ runs over all the possible contours of winding number one around simple poles, finite and infinite. Hence the pair production is entirely determined by simple poles, which may be dubbed polons.

We consider a constant electric field with the vector potential $A_{\|}(z) \quad -qE_0 z$ in the complex plane of time. Then, the frequency for the spinless charge

$$\omega_{\mathbf{k}}(\zeta) \quad \sqrt{m^2 \quad \mathbf{k}_{\perp}^2 \quad (k_{\|} \quad qE_0 z)^2} \tag{54}$$

does not have any finite simple pole but has a simple pole at the infinity. Under a conformal transformation $z \quad 1/\zeta$, the contour integral gives the Schwinger formula [13]

$$e^{iqE_0 \oint \frac{d\zeta}{\zeta^3} \sqrt{(1-z_0\zeta)(1-z_0^*\zeta)}} \quad e^{-\frac{\pi(m^2+\mathbf{k}_{\perp}^2)}{qE_0}}, \tag{55}$$

where $z_0 \quad (k_{\|} - i\sqrt{m^2 \quad \mathbf{k}_{\perp}^2})/(qE_0)$ and the Cauchy theorem is applied to the contour integral after the square root is expanded. Next, we consider the electric field and vector potential (17), which has the complex frequency in the complex plane

$$\omega_{\mathbf{k}}(z) \quad \sqrt{m^2 \quad \mathbf{k}_{\perp}^2 \quad (\bar{k}_{\|} \quad qE_0 e^{\frac{z}{t}})^2}. \tag{56}$$

Under another conformal transformation $\zeta \quad e^{z/\tau}$, the pair-production rate is given by

$$e^{-iqE_0 \oint \frac{d\zeta}{\zeta} \sqrt{(\zeta-z_0)(\zeta-z_0^*)}} \quad e^{-2\pi\tau(\bar{\omega}_0+\bar{k}_{\|})}, \tag{57}$$

where the contour integral of winding number one is taken around the simple poles at $z \quad 0$ and $z \quad \infty$. The contour integral of winding number two around the simple pole at $z \quad 0$ yields $e^{-4\pi\tau\bar{\omega}_0}$. These two contour integrals are exclusive and thus sum to the numerator of eq. (22). It is remarkable that the exact pair-production rate

$$\mathcal{N}_{\mathbf{k}} \quad (e^{-(2\pi\tau\bar{\omega}_0)(2)} \quad e^{-2\pi\tau(\bar{k}_{\|}+\bar{\omega}_0)}) \sum_{l=0}^{\infty} e^{-(2\pi\tau\bar{\omega}_0)(2l)} \tag{58}$$

may be obtained by summing the possible contour integrals of all winding numbers.

7. Conclusion

We have reviewed and elaborated the ED e ective action in the in-out formalism and the quantum invariant theory for the Schrödinger equation for a charged spinless scalar in time-dependent electric fields. We have also applied the contour integral method to the pair-production rate. In the in-out formalism by Schwinger and DeWitt [4, 5], the one-loop e ective action is expressed by the Bogoliubov coe cients, which relate the out-vacuum to the in-vacuum. Many configurations of electric fields have the Bogoliubov coe cients in terms of the gamma functions and the gamma-function regularization leads the ED action in the proper-time integral in the constant and the Sauter-type electric fields [18, 19].

In this paper we have extended the ED action to an exponentially increasing electric field. The ED action is complex, whose real part is the vacuum polarization and whose imaginary part is related to the pair-production rate. We have also applied the quantum invariant theory to the time-dependent Hamiltonian for the charged spinless field in time-dependent electric fields. The time-dependent annihilation and creation operators construct various quantum states such as number states of pairs and a thermal state. The connection of two representations of the Vlasov equation with the formulae for pair production is also clarified. The quantum invariant theory can be applied to the Hamiltonian in the momentum space for a localized pulse of lasers. Finally, we have explained the contour integral method to calculate the pair production in the constant electric field and further elaborated the method to the exponentially increasing electric field. These recent methods seem to be not only convenient but also powerful in explicitly obtaining the QED actions and pair-production rates in intense lasers, which wait for extensive investigation.

Acknowledgments

The author would like to thank Professor Nikolay Narozhny for the invitation to LPHYS'14 and Antonino Di Piazza and Christian Schubert for useful discussions. This work was supported in part by Basic Science Research Program through the National Research Foundation of Korea (NRF) funded by the Ministry of Education (NRF-2012R1A1B3002852).

References

[1] Heisenberg W and Euler H 1936 *Z. Phys.* **98** 714
[2] Schwinger J 1951 *Phys. Rev.* **82** 664
[3] Dunne G V 2004 *From Fields to Strings : Circumnavigating Theoretical Physics* ed M Shifman, A Vainshtein and J Wheater (Singapore: Worldscientific) (Prepint arXiv:hep-th/0406216)
[4] DeWitt B S 1975 *Phys. Rept.* **19** 295
[5] DeWitt B S 2003 *The Global Approach to Quantum Field Theory* vol 1 and vol 2 (New York: Oxford University Press)
[6] Kluger Y, Mottola E and Eisenberg J M 1998 *Phys. Rev.* D **58** 288 125015
[7] Hebenstreit F, Alkofer R and Gies H 2008 *Phys. Rev.* D **78** 304 061701
[8] Kim S P and Schubert C 2011 *Phys. Rev.* D **84** 125028
[9] Dunne G V and Schubert C 2005 *Phys. Rev.* D **72** 105004
[10] Dunne G V, Wang Q-h, Gies H and Schubert C 2006 *Phys. Rev.* D **73** 065028
[11] Kim S P and Page D N 2002 *Phys. Rev.* D **65** 105002
[12] Dumlu C K and Dunne G V 2010 *Phys. Rev. Lett.* **104** 250402
[13] Kim S P and Page D N 2007 *Phys. Rev.* D **75** 045013
[14] Kim S P 2013 *Phys. Lett.* **725** 500
[15] Dunne G and Hall T 1998 *Phys. Rev.* D **58** 105022
[16] Schubert C 2001 *Phys. Rep.* **355** 73
[17] Hebenstreit F, Alkofer R and Gies H 2010 *Phys. Rev.* D **82** 105026
[18] Kim S P, Lee H K and Yoon Y 2008 *Phys. Rev.* D **78** 105013
[19] Kim S P, Lee H K and Yoon Y 2010 *Phys. Rev.* D **82** 025015
[20] Lewis H R Jr and Riesenfeld W B 1969 *J. Math. Phys.* **10** 1458
[21] Kim S P 2014 *Ann. Phy.* **351** 54
[22] Kim S P, Lee H K and Yoon Y 2010 *Phys. Rev.* **82** 025016
[23] Kim S P 2012 *Int. J. Mod. Phys. Conf. Ser.* **12** 310
[24] Malkin I A, Man'ko V I and Trifonov D A 1970 *Phys. Rev.* D **2** 1371
[25] Kim J K and Kim S P 1999 *J. Phys.* A **32** 2711
[26] Kim S P and Lee C H 2000 *Phys. Rev.* D **62** 125020
[27] Kim S P and Page D N 2001 *Phys. Rev.* A **64** 012104
[28] Kim S P and Page D N 2013 *Phys. Lett.* B **723** 393
[29] Huet A, Kim S P and Schubert C 2014 *Phys. Rev.* D **90** in press (Preprint arXiv:1411.3074)

Journal of Physics: Conference Series **594** (2015) 012051 doi:10.1088/1742-6596/594/1/012051

Pair Creation of Scalar Particles in Intense Bichromatic Laser Fields

Martin J A Jansen and Carsten Müller

Institut für Theoretische Physik I, Heinrich-Heine-Universität, 40225 Düsseldorf, Germany

E-mail: martin.jansen@tp1.uni-duesseldorf.de

Abstract. The creation of scalar particle pairs due to the decay of a highly energetic gamma quantum traveling in a laser field comprising two independent modes is considered. Employing the framework of strong-field scalar QED, detailed calculations aiming at two different aspects of the process are presented. We first investigate interference effects between reaction channels with absorption of different numbers of laser photons with commensurate frequency and demonstrate the possibility of coherent phase control. Then we study numerically a scheme of dynamical assistance, which offers prospects for an experimental realization of a Schwinger-like pair production process.

1. Introduction

The collision of two energetic photons can lead to the creation of electron-positron pairs and is known as the Breit-Wheeler process [1]. Employing instead a laser of frequency ω, a certain number n of laser photons may contribute to the pair creation process [2, 3, 4], which can be written symbolically as

$$\omega_\gamma + n\omega \to e^+e^-, \tag{1}$$

where ω_γ is the frequency of the gamma quantum. This multiphoton Breit-Wheeler process has already been verified experimentally in the SLAC-E144 experiment [5].

Proceeding to a high laser field-strength E while keeping its photon energy small, such that the field-strength parameter $\xi = eE/m\omega$ clearly exceeds unity while $\chi = 2\xi\,\omega_\gamma\omega/m^2 \ll 1$, the particle creation rate scales as [2, 3]

$$\mathcal{R} \sim \frac{\alpha m^2}{\omega_\gamma}\chi^{3/2}\exp\left(-\frac{8}{3\chi}\right) \tag{2}$$

revealing a non-analytic dependence on the laser field-strength. Here, α denotes the fine-structure constant, m the electron mass, and we use relativistic units with $\hbar = c = 1$. A similar behavior is seen for the famous Schwinger effect [6, 7], which refers to the creation of electron-positron pairs from the vacuum resulting from the application of a constant electric field. This fundamental effect has not yet been seen experimentally due to the high field strength $E \sim E_{\text{cr}} = m^2/e \approx 10^{16}$ V/cm required.

In this contribution, we will investigate a setup comprising two co-propagating laser beams of different frequencies and intensities which both give contributions to the pair creation process induced by the collision with an additional counter-propagating gamma quantum. Symbolically, we have

$$\omega_\gamma + n_1\omega_1 + n_2\omega_2 \to e^+e^-, \tag{3}$$

where the indices $1, 2$ refer to the respective laser mode. The addition of the second reaction channel facilitates quantum mechanical interferences [8], which are most prominent for low-intensity lasers and may affect the angular distribution of the particles as well as the total rate; see [9, 10, 11] for related studies on the Bethe-Heitler process. We will demonstrate these effects by investigating the influence of a relative phase shift between the laser modes.

Furthermore, following the proposal of dynamical assistance [12], we demonstrate how a suitable combination of a high-intensity yet low-frequency laser with an additional low-intensity yet high-frequency laser can drastically increase the Schwinger-like pair creation process and finally render the effect visible in today's labs.

We will present detailed calculations following an S matrix approach within strong-field scalar QED, allowing us to treat the laser field non-perturbatively. For the sake of analytical simplicity, we neglect the particle spin. Nevertheless, we shall denote the produced particles as electrons and positrons. For an investigation of spin effects in intense fields, we refer to [13, 14]. Recent studies on Breit-Wheeler pair production in intense laser pulses can be found in [15, 16, 17].

2. Theoretical Framework

The pair creation process is formally described as the transition of a laser-dressed (spinless) electron from the negative to the positive continuum induced by the absorption of the gamma quantum, which is described as a single mode of the quantized radiation field. The depleted particle state with negative energy finally corresponds to the anti-particle.

The transition amplitude is given by the S matrix element

$$S = -i \int dt \, \langle \Psi_{p_-} | \hat{H}_{\text{int}} | \Psi_{p_+,\gamma} \rangle \tag{4}$$

where the initial state $|\Psi_{p_+,\gamma}\rangle = \Phi_{p_+}|\mathbf{k}_\gamma\lambda_\gamma\rangle$ is a product state of the laser-dressed electron with negative energy given by the state Φ_{p_+} and a singly occupied radiation mode with wave four vector $k_\gamma^\mu = (\omega_\gamma, \mathbf{k}_\gamma)$ and mode index λ_γ. The final state $|\Psi_{p_-}\rangle = \Phi_{p_-}|0\rangle$ contains the laser-dressed electron in the positive continuum and the depopulated radiation mode. The absorption of the gamma quantum is described by the field operator

$$\hat{\mathbf{A}}_\gamma = \sqrt{\frac{2\pi}{V\omega_\gamma}} \, \mathrm{e}^{-i(k_\gamma x)} \, \boldsymbol{\epsilon}_\gamma \, \hat{c}_{\mathbf{k}_\gamma \lambda_\gamma} \tag{5}$$

with polarization vector $\boldsymbol{\epsilon}_\gamma$, annihilation operator $\hat{c}_{\mathbf{k}_\gamma\lambda_\gamma}$ and a normalizing volume V. The interaction Hamiltonian thus reads

$$\hat{H}_{\text{int}} = -ie\left(\hat{\mathbf{A}}_\gamma \cdot \overrightarrow{\boldsymbol{\nabla}} - \overleftarrow{\boldsymbol{\nabla}} \cdot \hat{\mathbf{A}}_\gamma\right) + 2e^2 \mathbf{A}_L \cdot \hat{\mathbf{A}}_\gamma \tag{6}$$

where $\mathbf{A}_L$ is the vector potential of the laser field.

The laser field consists of two independent modes of frequencies ω_j, $j = 1, 2$ which are linearly polarized in orthogonal directions $\mathbf{e}_j$ and propagate in the same direction $\mathbf{n}$. Its vector potential in radiation gauge accordingly reads

$$\mathbf{A}_L = \mathbf{A}_1 + \mathbf{A}_2 \; , \; \mathbf{A}_j(\eta_j) = a_j \cos(\eta_j - \varphi_j)\, \mathbf{e}_j \tag{7}$$

where we have introduced phase variables $\eta_j = (k_j x) = \omega_j(t - \mathbf{n}\cdot\mathbf{r})$, which are related to the wave four vectors $k_j^\mu = \omega_j(1, \mathbf{n})$, and phase shifts φ_j. We measure the amplitudes in a Lorentz-invariant manner with $\xi_j = \frac{ea_j}{m}$.

We obtain the laser-dressed states for a particle with charge $\pm e$ and free four-momentum $p_\pm^\mu = (E_{p_\pm}, \mathbf{p}_\pm)$ from the Gordon-Volkov states

$$\Phi_{p_\pm}(x) = \frac{1}{\sqrt{2E_{p_\pm}V}} \mathrm{e}^{i[\pm(p_\pm x) - \Lambda_1 - \Lambda_2]} \tag{8}$$

which, due to the perpendicular polarization directions, separate into contributions from each laser mode alone with

$$\Lambda_j = \frac{1}{(k_j p_\pm)} \int^{\eta_j} [e\mathbf{p}_\pm \cdot \mathbf{A}_j(\tilde{\eta}) \mp e^2 \mathbf{A}_j^2(\tilde{\eta})] d\tilde{\eta}\,. \tag{9}$$

The interaction between the particles and the laser-field is hereby taken into account non-perturbatively.

The time-averaged motion of the particles in the laser field is reflected in the laser-dressed momenta $q_\pm^\mu = p_\pm^\mu + 2k_1^\mu y_1^\pm + 2k_2^\mu y_2^\pm$, which occur naturally in this approach and are expressed using the abbreviation $y_j^\pm = \frac{e^2 a_j^2}{8(k_j p_\pm)}$. The particles experience a mass increase by virtue of $(q_\pm q_\pm) = m_*^2$, leading to the laser-dressed mass $m_* = m[1 + \frac{1}{2}(\xi_1^2 + \xi_2^2)]^{1/2}$.

In order to carry out the space-time integration in Eq. (4), the integrand can be Fourier decomposed using generalized Bessel functions $\tilde{J}_n(u,v)$ which fulfill $\mathrm{e}^{i(u\sin(\phi) + v\sin(2\phi))} = \sum_{n=-\infty}^{\infty} \tilde{J}_n(u,v)\mathrm{e}^{in\phi}$. The generalized Bessel functions can be written in terms of ordinary Bessel functions as $\tilde{J}_n(u,v) = \sum_{l=-\infty}^{\infty} J_{n-2l}(u) J_l(v)$ [18]. Applying this method to each of the laser modes separately, we finally arrive at

$$S = S_0 \sum_{n_1,n_2=-\infty}^{\infty} M_{n_1,n_2}\, \mathrm{e}^{i(n_1\varphi_1 + n_2\varphi_2)} \mathrm{e}^{i(h_1+h_2)} (2\pi)^4 \delta^4\left(q_- + q_+ - k_\gamma - n_1 k_1 - n_2 k_2\right) \tag{10}$$

with the prefactor $S_0 = -iem\sqrt{\pi/(2V^3 E_{p_+} E_{p_-} \omega_\gamma)}$ and abbreviations $h_j = h_j^+ + h_j^-$, $h_j^\pm = x_j^\pm \sin(\varphi_j) + y_j^\pm \sin(2\varphi_j)$ and $x_j^\pm = \mp\frac{ea_j \mathbf{p}_\pm \cdot \mathbf{e}_j}{(k_j p_\pm)}$ for $j = 1, 2$. We introduce the reduced matrix element

$$\begin{aligned} M_{n_1,n_2} \;=\; & \frac{(\mathbf{p}_- - \mathbf{p}_+)\cdot\boldsymbol{\epsilon}_\gamma}{m} \tilde{J}_{n_1}\tilde{J}_{n_2} \\ & + \xi_1(\mathbf{e}_1\cdot\boldsymbol{\epsilon}_\gamma)\left(\tilde{J}_{n_1+1} + \tilde{J}_{n_1-1}\right)\tilde{J}_{n_2} \\ & + \xi_2(\mathbf{e}_2\cdot\boldsymbol{\epsilon}_\gamma)\left(\tilde{J}_{n_2+1} + \tilde{J}_{n_2-1}\right)\tilde{J}_{n_1} \end{aligned} \tag{11}$$

where the arguments of the generalized Bessel functions $\tilde{J}_{n_j}$ and $\tilde{J}_{n_j\pm1}$ are $u_j = -(x_j^+ + x_j^-)$ and $v_j = -(y_j^+ + y_j^-)$. The δ-function in Eq. (10)

$$d_n = (2\pi)^4 \delta^4\left(q_- + q_+ - k_\gamma - n_1 k_1 - n_2 k_2\right) \tag{12}$$

provides energy-momentum conservation for the process and motivates us to interpret the indices n_j as numbers of photons being absorbed in order to create the pair. Note that the conservation conditions include the dressing of the particles. From now on we abbreviate $n = (n_1, n_2)$.

The total particle creation rate $\mathcal{R}$ is obtained by integrating the amplitude squared over all possible particle states, averaging over the polarization states of the gamma quantum and finally dividing by the interaction time T, leading to

$$\mathcal{R} = \frac{1}{2}\sum_{\lambda_\gamma}\int\frac{Vd^3p_+}{(2\pi)^3}\int\frac{Vd^3p_-}{(2\pi)^3}\frac{|S|^2}{T}\ . \tag{13}$$

In order to obtain the square of the amplitude, we introduce primed indices n_1', n_2' and note that the product $d_n d_{n'}$ only allows contributions if

$$(n_1 - n_1')\omega_1 = (n_2' - n_2)\omega_2 \tag{14}$$

such that we are left with

$$|S|^2 = |S_0|^2 TV \sum_{\substack{n_1,n_2,n_1',n_2': \\ (14)}} M_n M_{n'}^* d_n \mathrm{e}^{i[(n_1-n_1')\varphi_1+(n_2-n_2')\varphi_2]} \tag{15}$$

where the factors TV stem from the square of the δ-functions.

We see that for incommensurate frequencies, condition (14) can never be fulfilled, such that no interference terms appear and only ordinary terms with $n_j = n_j'$ contribute. On the other hand, for frequencies that are in a commensurate ratio, interference terms can be expected. Their contribution to the total process is modulated by the phase factor $\mathrm{e}^{i[(n_1-n_1')\varphi_1+(n_2-n_2')\varphi_2]}$. The sum over all possible photon number combinations includes the term (n_1, n_2, n_1', n_2') as well as the related term (n_1', n_2', n_1, n_2), where n_j and n_j' have been interchanged. Since the product of the reduced matrix elements is insensitive to this exchange, it is sufficient to regard in the following the sum of these two terms, which is modulated by $2\cos[(n_1 - n_1')\varphi_1 + (n_2 - n_2')\varphi_2]$. The distinction between primed and unprimed photon numbers is a purely mathematical effect, which cannot be seen in the experiment. Additionally we remark that for a term with $n_j \neq n_j'$, the numbers n_j are lacking a straightforward interpretation.

3. Interference Effects

In this section, we investigate interference effects which occur in a setup of two co-propagating lasers with a frequency ratio of 2. The intensities are chosen small ($\xi_j \ll 1$), such that the process can be regarded perturbatively. This means, by Taylor expanding the generalized Bessel function, we see that the term with photon numbers (n_1', n_2', n_1, n_2) scales as

$$\mathcal{R}_{(n_1',n_2',n_1,n_2)} \sim \xi_1^{|n_1|+|n_1'|}\xi_2^{|n_2|+|n_2'|}\ . \tag{16}$$

Therefore only terms featuring the smallest possible non-negative photon numbers and fulfilling the condition (14) contribute significantly. Additionally, we choose the intensities of the two modes such that the total particle count rates obtained for each mode alone are of the same order. This way, we keep the two modes in balance in order to maximize the visibility of interference effects.

Due to the small intensities applied here, the dressing of the particles may be neglected. Assuming the second mode to have the larger frequency, we fix the ratio of the frequencies $\nu = \omega_2/\omega_1$. Calculating the total particle count rate for each mode alone, we see that the condition of balanced contributions leads to

$$\xi_2 \sim \xi_1^\nu\ , \tag{17}$$

such that Eq. (16) reads

$$\mathcal{R}_{(n_1',n_2',n_1,n_2)} \sim \xi_1^{2N} \tag{18}$$

for non-negative photon numbers. We have introduced the number $N = n_1 + \nu n_2$ and we see that $n_1' + \nu n_2' = N$ in accordance with condition (14).

We carry out the calculations in the center of mass system, which implies

$$n_1\omega_1 + n_2\omega_2 = N\omega_1 = \omega_\gamma \tag{19}$$

and

$$n_1\omega_1 + n_2\omega_2 + \omega_\gamma = 2N\omega_1 = E_{p_+} + E_{p_-} \tag{20}$$

We see that processes with constant ν and N are of the same order and share a common center of mass system. The number N is related to the minimal photon numbers required: We would need either $n_1^{min} = N$ photons from the first or $n_2^{min} = N/\nu$ photons from the second mode in order to create the pair if the lasers were operated separately.

In order to measure the frequencies, we use the relativistic β parameter of the particles in their c.m. frame, and from Eq. (20) we obtain

$$\gamma = \frac{1}{\sqrt{1-\beta^2}} = \frac{N\omega_1}{m} \tag{21}$$

In the following, we investigate the angular distributions of the positrons, i.e. $\frac{d\mathcal{R}}{d\theta d\phi}$, with usual spherical coordinates θ and ϕ which describe the emission direction of the positron. Since we are working in the center of mass system, the electron is emitted in the opposite direction. The polar angle θ is measured with respect to the propagation direction of the lasers, while the azimuthal angle ϕ is measured in the polarization plane with $\phi = 0$ $(\pi/2)$ being the polarization axis of the first (second) mode. We carry out our calculations analytically and present our results without a common prefactor $\frac{\alpha m}{16\pi}\xi_1^{2N}$. The amplitude of the second laser is always chosen as $\xi_2 = 0.78\xi_1^2$, ensuring balanced contributions for both of the following cases.

3.1. Case I

In the first case, we choose the frequencies of our lasers such that either two photons from the first or one photon from the second mode are sufficient to create the pair, such that $\nu = 2 = N$. In addition to the ordinary processes with photon numbers $(2,2;0,0)$ and $(0,0;1,1)$, we have an interference term with $(2,0;0,1)$. Figure 1 shows the angular distributions for different values of the frequencies (left to right) and for different values of the phase shift δ_1 (top to bottom), while $\delta_2 = 0$. The vertical axis shows the angle θ, while ϕ is depicted on the horizontal axis. The frequencies of the lasers are subject to upper limits at which even less photons would be sufficient to create the pair. For a fixed number N, this threshold is related to a maximum beta parameter

$$\beta_N = \frac{1}{\sqrt{N}}. \tag{22}$$

As the figures reveal, the emission happens mostly in the plane of polarization. The distribution is generally asymmetric about ϕ, but for certain values of the phase shift $\delta_1 = \pi/4$, it becomes symmetric. The asymmetry is caused by the interference contribution, which vanishes identically at this point. We see that the angular distribution strongly depends on the frequencies of the lasers; for higher frequencies, the regions of highest emission probability become more pronounced.

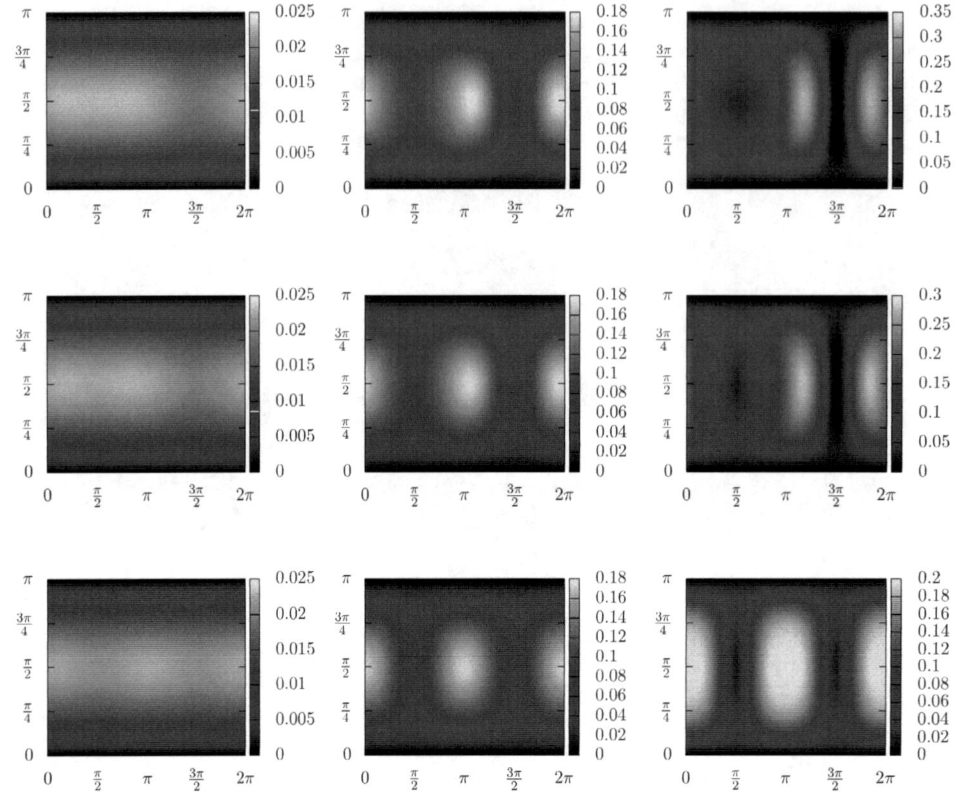

Figure 1. Differential rates $\frac{d\mathcal{R}}{d\phi d\theta}$ for $\frac{\beta}{\beta_2} = 0.1, 0.5$ and 0.9, corresponding to a center-of-mass energy of $E_{cm} = 2.01m, 2.14m$ and $2.59m$ (left to right), and $\delta_1 = 0, \frac{\pi}{8}$ and $\frac{\pi}{4}$ (top to bottom).

3.2. Case II

In the second case, we strongly reduce the frequencies of the lasers with respect to the first case, such that four photons from the first mode or two photons from the second mode are required, leading to $N = 4$ while keeping $\nu = 2$. The increase in N allows for more terms to contribute. In addition to the ordinary processes $(4, 4; 0, 0)$ and $(0, 0; 2, 2)$, we have a mixed term $(2, 2; 1, 1)$, which does not depend on the phase shifts. Furthermore, we have three interference terms with photon numbers $(4, 2; 0, 1)$, $(2, 0; 1, 2)$ and $(4, 0; 0, 2)$. As Fig. 2 shows, the angular distributions of the total process generally exhibit the same properties as in the first case, but they possess a richer structure. The interference contributions have a strong impact, allowing for example for $\beta = 0.5\beta_4$ to force the positrons to be emitted predominantly either in parallel or anti-parallel to the polarization vector of the second laser.

3.3. Total Rates

Now we investigate the total particle count rates. For case I (which is not presented here), we have found a dependence on the frequency, but not on the phase shift. This is caused by the fact that the interference term vanishes after integration over the azimuthal angle.

The second case is examined in Fig. 3. The first panel shows that the ordinary processes

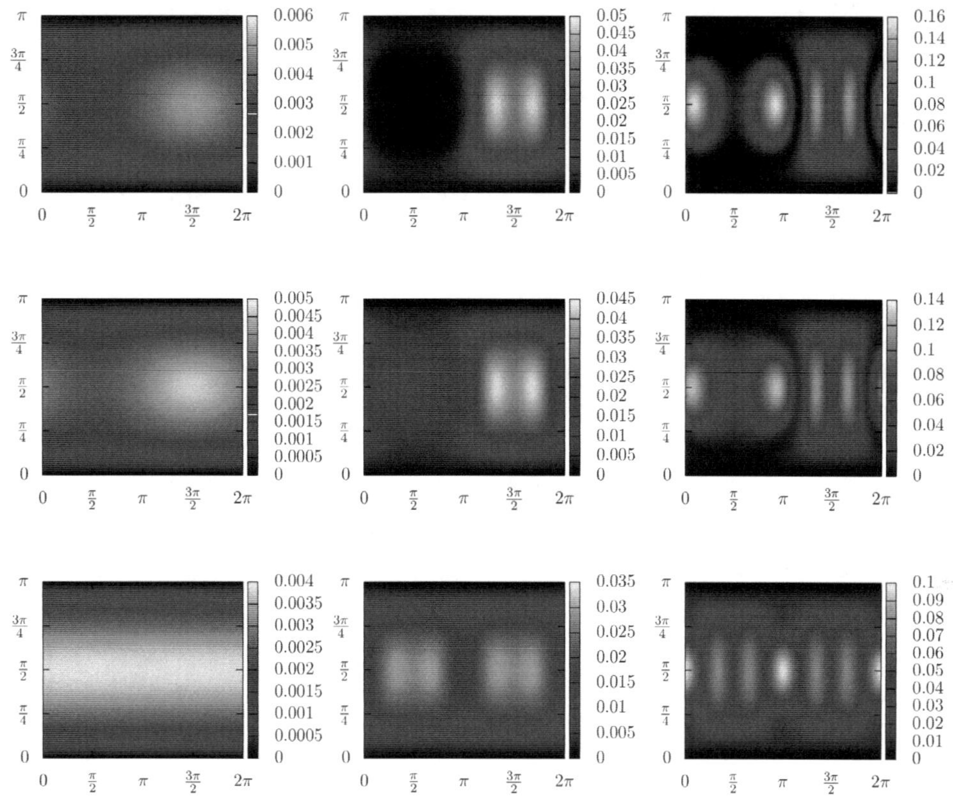

Figure 2. Differential rates $\frac{dR}{d\phi d\theta}$ for $\frac{\beta}{\beta_4} = 0.1, 0.5$ and 0.9, corresponding to $E_{cm} = 2.003m, 2.07m$ and $2.24m$ (left to right), and $\delta_1 = 0, \frac{\pi}{8}$ and $\frac{\pi}{4}$ (top to bottom).

are dominated by the mixed contribution, which does not depend on the phase shift. When we consider the interference contributions, only the term with photon numbers $(4, 0; 0, 2)$ does not vanish after ϕ-integration, and finally contributes to the total rate. The central panel shows its dependence on the frequencies and on the phase shift, where the latter is given by the phase factor $\cos(4\delta_1)$. Finally the third panel shows the total rate as a function of the frequencies and for various phase shifts, which can be seen to have a significant influence.

To conclude this section, we note that by inspection of the Taylor series of the generalized Bessel functions, a necessary condition for non-vanishing interference contributions to the total rate can be derived. By noting that

$$\int_0^{2\pi} \cos^k(\phi) \sin^l(\phi) d\phi \neq 0 \tag{23}$$

if and only if both integers k and l are even, we arrive at the conclusion that only terms where both sums $n_j + n_j'$ are even can contribute to the total rate. This criterion was not fulfilled for the interference term in case I, whereas in case II, the term with photon numbers $(4, 0; 0, 2)$ introduced the possibility of manipulating the total particle count rate by changing the phase shift.

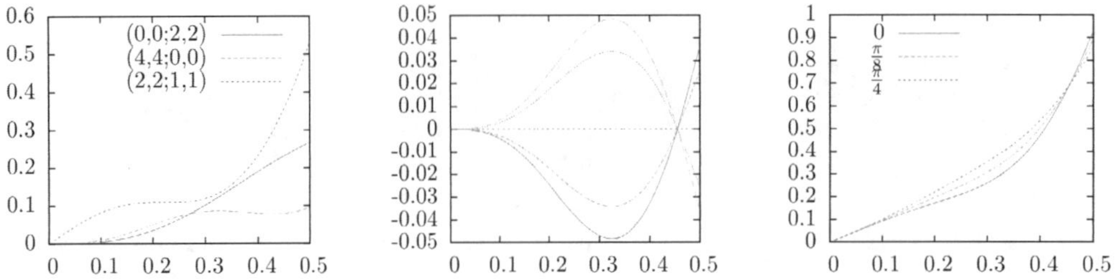

Figure 3. Total rates in units of $\frac{\alpha m}{16\pi}\xi_1^8$ for case II in dependence of β.
Left: Partial contributions. **Center**: Phase-dependence of the interference-contribution $(4,0;0,2)$. The phase shift is increased in steps of $\pi/16$ starting at $\delta_1 = 0$ (red solid curve) up to $\delta_1 = \pi/4$ (cyan dash-dotted curve). **Right**: Total rate for different phase shifts.

4. Dynamical Assistance

The classical Schwinger effect can be regarded as a tunneling transition of an electron from the negative to the positive continuum, transversing a tunneling barrier of height $\sim 2m$ and width $\sim 2m/eE$, where E is the field strength of the (static) electric field. Since even with today's laser technology, the tunneling distance is much bigger than the Compton wavelength, the Schwinger process is exponentially suppressed and escapes experimental observation. In order to lower the tunneling barrier, the scheme of dynamical assistance proposes the addition of high-frequency radiation to the setup [12]. Absorbing these highly energetic photons, the electron is pushed into the energy gap and sees an effectively reduced tunneling barrier.

In the following, we will investigate this scheme by numerically evaluating our expressions for the Breit-Wheeler pair creation process. To begin with, we look at a setup in which one high-intensity optical laser collides with gamma radiation. To avoid numerical problems, we evaluate the expressions in a frame of reference in which the energy of the gamma radiation is heavily reduced to $\omega_\gamma = 0.93869m$, while the frequency of the optical laser is up-shifted to $\omega_2 = 0.07m$. These parameters correspond, for instance, to $\omega_2 = 2.4$ eV and $\omega_\gamma = 7.2$ GeV in the laboratory frame. The red circles in Fig. 4 show the total particle count rate as a function of the inverse of the amplitude ξ_2 of the optical laser. An exponential fit as indicated by the solid line gives a dependence of the form $\mathcal{R} \sim \exp(-16.87/\xi_2)$. The analytic result (2) obtained for Dirac particles was derived under the assumption of $\xi \gg 1$, yet it is known to be approximately valid even for $\xi \approx 1$, and it gives $\mathcal{R} \sim \exp(-20.3/\xi_2)$, which is in rather good agreement with our numerically obtained results. In order to apply the enhancement scheme, we introduce a second laser mode with frequency $\omega_1 = \omega_\gamma$ (which is chosen such that no interferences occur) and amplitude $\xi_1 = 0.004$. In this combined laser field, particles can be created due to the absorption of one photon of frequency ω_1, several photons from the optical laser and the gamma quantum. The corresponding rate is depicted by the blue crosses in Fig. 4. It is enhanced by several orders of magnitude as compared to the case in which only optical photons (and the gamma quantum) have been absorbed[1]. An exponential fit to the data yields $\mathcal{R} \sim \exp(-1.54/\xi_2)$. The total particle rate, as depicted by the open squares in Fig. 4, is mainly given by the channel involving the absorption of one additional high-energy photon.

In order to gain further insight, we investigate the combined process in more detail. The following considerations are still carried out in the frame of reference in which $\omega_\gamma = \omega_1$. Having absorbed the assisting photon and the gamma quantum, the remaining tunneling barrier to be

[1] The data in Fig. 4 is calculated for a setup in which both laser modes are present. The rate labeled by $n_1 = 0$ is essentially the same as for the case when the assisting mode is turned off, $\xi_1 = 0$.

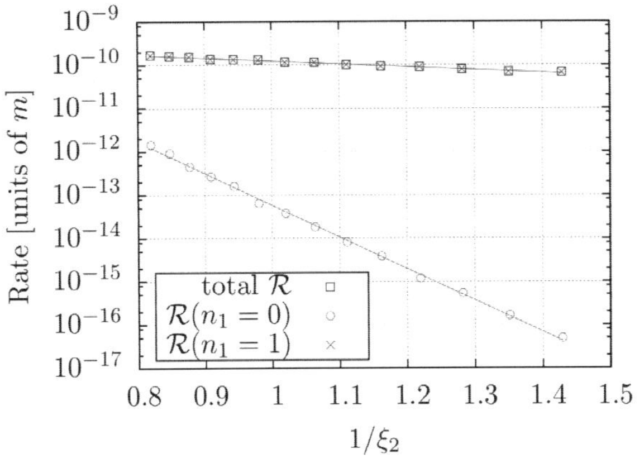

Figure 4. Particle creation rates for $\omega_2 = 0.07m$, $\omega_1 = \omega_\gamma = 0.93869m$, $\xi_1 = 0.004$. The red circles show the contribution from the optical laser (mode 2) alone. The blue crosses depict the rate obtained in the process where one additional photon with frequency ω_1 has been absorbed. The total rate (black squares) is dominated by the channel which is facilitated by the additional radiation. The solid lines are exponential fits to the data.

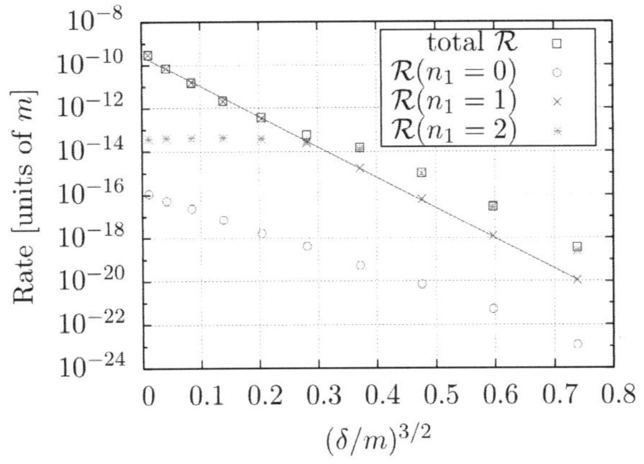

Figure 5. Particle creation rates (black squares) for $\omega_2 = 0.07m$, $\xi_2 = 0.7$, $\xi_1 = 0.004$ and $\omega_1 = \omega_\gamma = \frac{2m-\delta}{2}$ as a function of the gap parameter δ. The red circles, blue crosses and green asteriks show the respective contribution to the rate from the absorption of zero, one or two assisting photons. The solid blue line is an exponential fit.

bridged by the optical laser field is $\delta = 2(m - \omega_1)$. Replacing its slowly oscillating electric field by a constant field $\mathbf{E} = E_0\mathbf{e_2}$, the rate of this process can be obtained by a WKB approximation and yields $\mathcal{R} \sim \exp(-\mathcal{G})$ with the Gamow factor (see also [19])

$$\mathcal{G} = 2\int_0^\ell \sqrt{2m(\delta - eE_0 y)}\,dy = \frac{4\sqrt{2}}{3}\frac{E_{cr}}{E_0}\left(\frac{\delta}{m}\right)^{3/2} \tag{24}$$

where we have used the remaining tunneling distance $\ell = \delta/eE_0$. For our setup, this calculation yields $\mathcal{G} = -1.16/\xi_2$, which is in reasonable agreement with our numerical results. Finally we remark that the pair creation rate induced by the gamma quantum and the assisting laser alone is $4.06 \times 10^{-14} m$.

In order to validate our picture, we have numerically investigated the effect of a change in the tunneling barrier. In Fig. 5 we vary the frequencies of the gamma quantum and the assisting radiation according to $\omega_1 = \omega_\gamma = \frac{2m-\delta}{2}$. The partial rates for the channel comprising one assisting photon can be fitted to a form given by Eq. (24) and we obtain numerically $\mathcal{R} \sim \exp(-32.3\,(\delta/m)^{3/2})$, while the model calculation yields a Gamow factor of $38.5\,(\delta/m)^{3/2}$. The plot clearly shows the enhancement effect, and additionally we see that for bigger values of the gap parameter the channel involving two assisting photons provides the dominating contribution to the total rate.

Journal of Physics: Conference Series **594** (2015) 012051 doi:10.1088/1742-6596/594/1/012051

5. Conclusion

Two different aspects of the Breit-Wheeler pair production process in bichromatic laser fields were investigated. For low-intensity fields with commensurate frequencies, interference effects between production channels involving different numbers of photons have been shown to affect the angular distributions of the emitted particles and in some cases also the total particle count rate. A necessary condition for the latter effect has been derived.

Furthermore, Schwinger-like pair production induced by the decay of a gamma quantum traveling in a high-intensity optical laser wave was studied. Following the scheme of dynamical assistance, the addition of high-frequency yet low-intensity laser radiation was shown to strongly enhance the particle yield, offering good prospects for an experimental realization of Schwinger-like pair production. For more details, the interested reader is referred to [20].

References

[1] Breit G and Wheeler J A 1934 *Phys. Rev.* **46**(12) 1087–1091
[2] Reiss H R 1962 *J. Math. Phys.* **3** 59–67
[3] Nikishov A I and Ritus V I 1964 *Sov. Phys. JETP* **19** 529–541
[4] Narozhny N B, Nikishov A I and Ritus V I 1965 *Sov. Phys. JETP* **20** 622–628
[5] Burke D L, Field R C, Horton-Smith G, Spencer J E, Walz D, Berridge S C, Bugg W M, Shmakov K, Weidemann A W, Bula C, McDonald K T, Prebys E J, Bamber C, Boege S J, Koffas T, Kotseroglou T, Melissinos A C, Meyerhofer D D, Reis D A and Ragg W 1997 *Phys. Rev. Lett.* **79**(9) 1626–1629
[6] Sauter F 1931 *Z. Phys.* **69** 742–764
[7] Schwinger J 1951 *Phys. Rev.* **82**(5) 664–679
[8] Narozhny N B and Fofanov M S 2000 *J. Exp. Theor. Phys.* **90** 415–427
[9] Krajewska K and Kamiński J Z 2012 *Phys. Rev.* A **85**(4) 043404
[10] Augustin S and Müller C 2013 *Phys. Rev.* A **88**(2) 022109
[11] Augustin S and Müller C 2014 *J. Phys. Conf. Ser.* **497** 012020
[12] Schützhold R, Gies H and Dunne G 2008 *Phys. Rev. Lett.* **101**(13) 130404
[13] Villalba-Chávez S and Müller C 2013 *Phys. Lett.* B **718** 992 – 997
[14] Ivanov D Y, Kotkin G L and Serbo V G 2005 *Eur. Phys. J.* C **40** 27–40
[15] Krajewska K and Kamiński J Z 2012 *Phys. Rev. A* **86**(5) 052104
[16] Titov A I, Kämpfer B, Takabe H and Hosaka A 2013 *Phys. Rev. A* **87**(4) 042106
[17] Meuren S, Hatsagortsyan K Z, Keitel C H and Di Piazza A 2014 (*Preprint* `arXiv:1406.7235`)
[18] Reiss H R 1980 *Phys. Rev.* A **22** 1786–1813
[19] Di Piazza A, Lötstedt E, Milstein A I and Keitel C H 2009 *Phys. Rev. Lett.* **103**(17) 170403
[20] Jansen M J A and Müller C 2013 *Phys. Rev.* A **88**(5) 052125

Numerical modelling of Compton scattering in ultra-intense laser pulses

C N Harvey[1] and D G Green[2]

[1] Department of Applied Physics, Chalmers University of Technology, SE-41296 Gothenburg, Sweden

[2] Joint Quantum Centre (JQC) Durham/Newcastle, Department of Chemistry, Durham University, DH1 3LE, United Kingdom

E-mail: christopher.harvey@chalmers.se, dermot.green@balliol.oxon.org

Abstract. We examine current methods of numerically implementing Compton scattering in the context of intense laser-matter interactions. In a recent publication [1] it has been shown that a commonly used approach generates the correct spectra in nearly all cases, except those when the harmonic structure is important. Here we provide an explanation for this using an alternative, classical argument.

1. Introduction

Recent technological advances have led to the development of lasers with unprecedented powers and intensities. The current record was set by the HERCULES laser in 2008, when it achieved a peak focal intensity of $2 \times 10^{22}\mathrm{W/cm^2}$ [2]. This upward trend is expected to continue into the future, with numerous facilities being developed to provide ultra-intense laser fields. These include the Vulcan 10 PW upgrade [3] which is expected to provide peak intensities of $10^{23}\,\mathrm{W/cm^2}$. In the longer term we look forward to the European "Extreme Light Infrastructure" (ELI) Facility [4] and the XCELS project [5], both of which may surpass this intensity by a further order of magnitude.

These developments have led to a renewed interest in probing strong-field quantum electrodynamics (QED) using high-intensity laser fields [6, 7, 8] as well as in a number of applications [9, 10]. However, the new generation of facilities will achieve such ultra-intense conditions, in part, by a strong focusing of the beam. So far, the majority of analytical QED calculations are limited to plane-wave models which do not capture the structural features of strongly focussed fields. Additionally, multiple QED processes can take place while the particle is in the field, even leading to runaway cascading. Describing such situations analytically is almost impossible and so one is forced to resort to numerical methods (However, see the recent paper [11]). The standard numerical techniques for modelling QED processes in a laser field involves propagating a particle classically thorough a field and then using statistical event generators to determine various QED transition processes [12, 13, 14].

If we are to rely on these numerical methods, we must first verify their accuracy and gain a full understanding of their limitations. This is best done by considering problems that can also be solved analytically to which explicit comparisons can be made. Such problems are, however, few in number. In this paper we will study the accuracy of these numerical techniques, restricting ourselves to one such process, the Compton scattering of an electron, for which an analytic solution can be obtained. We outline the differences between the exact analytical calculation of the emission spectra and those obtained via numerics based on discrete time approximations. We also present a classical argument explaining why discrepancies occur.

2. Outline of numerical techniques

While all of the codes used by the high-intensity laser community have their differences, they are all designed with the same general principles at their heart (see, for example, [15]). For the purposes of this study we consider the single-particle QED program SIMLA [16, 17], recently developed by the authors, whose operation we now briefly describe.

Working in natural units where $\hbar = c = 1$, we describe the laser field using the dimensionless intensity parameter $a_0 = eE/\omega m$, where e is the electron charge, m the rest mass, E the peak amplitude of the electric field and ω the laser frequency. The importance of quantum effects can be measured by considering the (invariant) *quantum efficiency parameter* $\chi_e \equiv \sqrt{(F^{\mu}{}_{\nu}p^{\nu})^2}/m^2 \sim \gamma E/E_{\text{cr}}$, where $E_{\text{cr}} = 1.3 \times 10^{16}\,\text{Vcm}^{-1}$ is the QED 'critical' field ('Sauter-Schwinger' field) [18, 19, 20]. This parameter is equal to the work done by the laser field on the particle over the distance of a Compton wavelength. Thus, when $\chi_e \gtrsim 1$ quantum effects dominate and processes such as vacuum pair production occur. For the purposes of this work we will restrict our attention to the regime where $a_0 \gg 1$, $\chi_e \leq 1$, such that quantum effects will play a role, but at the same time we can neglect processes such as pair production. In the high-intensity limit $a_0 \gg 1$ the size of the radiation formation region is of the order $\lambda/a_0 \ll \lambda$, where $\lambda = 2\pi/\omega$ is the laser wavelength [21]. Thus the laser varies on a scale much larger than the formation region and so can be approximated as locally constant and crossed, allowing us to determine the probability of photon emission using the expression for the constant field rate Γ [21]

$$d\Gamma = \frac{\alpha m}{\sqrt{3}\pi\gamma\chi_e}\left[\left(1-\eta+\frac{1}{1-\eta}\right)K_{2/3}(\tilde{\chi}) - \int_{\tilde{\chi}}^{\infty} dx K_{1/3}(x)\right] d\chi_\gamma, \tag{1}$$

where α is the fine structure constant, K_ν is the modified Bessel function of order ν, $\eta \equiv \chi_\gamma/\chi_e$, $\tilde{\chi} \equiv 2\eta/\left[3\chi_e\left(1-\eta\right)\right]$, and $\chi_\gamma \equiv \sqrt{(F^{\mu}{}_{\nu}\kappa^{\nu})^2}/m^2$ is the analogous quantum efficiency parameter for a photon emitted photon with momentum κ^ν.

Note that although $d\Gamma$ diverges at small χ_γ, the total differential probability of photon emission (i.e., of any χ_γ), $dW = \Gamma dt$, where $\Gamma \equiv \int_0^{\chi_e} d\Gamma$, is finite (see also [22]). It is common to model the trajectory of the electron through the field *classically* and then to correct its momentum after every time step using statistical routines to calculate the photon emission rates using the constant field rate. It is important to note that this is not equivalent to what happens in a full, analytical calculation. In normal QED calculations scattering amplitudes are determined using asymptotic 'in' and 'out' states (see e.g., [23]). No assumptions are made about the electron's trajectory in the laser. One simply begins with an in-state momentum at $t = -\infty$ and calculates the out-state momentum at $t = \infty$ without determining how the electron moves in between. In contrast, in the numerical method we are continuously tracking the electron orbit.

The statistical routines that determine the photon emissions are implemented in the following manner. At each time step a uniform random number $r \in [0,1]$ is generated, and emission deemed to occur if the condition $r \leq \Gamma dt$ is satisfied, under the requirement $\Gamma dt \ll 1$. Note that during the simulation $d\Gamma$ (and thus Γ) is a time-dependent quantity owing to the effect of the temporally varying laser pulse and electron motion. Given an emission event, the photon χ_γ is determined as the root of the sampling equation $\zeta = \Gamma(t)^{-1}\int_0^{\chi_\gamma} d\Gamma(t)$, where ζ is a uniform random number $\zeta \in [0,1]$ (In practice we implement an infra-red cut-off so that the code does not include the emission of large numbers of low energy photons; the integral is performed from a lower limit $\varepsilon \sim 10^{-5}$, rather than zero. The emission of soft photons of energy below this cut off does not appreciably affect the electron dynamics (see e.g., Ref. [24]).). Next, we calculate the photon momentum from χ_γ assuming that the emission is in the direction of motion of the electron. This is valid for $\gamma \gg 1$, since in reality the emissions will be in a cone of width γ^{-1} [25, 26]. Finally, the photon momentum is subtracted from the electron momentum and the simulation proceeds by propagating the particle via the Lorentz equation to the next time step.

3. Comparisons

We now present a comparison of the numerical photon spectra produced by the SIMLA program with that obtained via the full analytic calculation. The emission spectra can be calculated exactly for an electron in an infinite plane-wave field, details of the calculation can be found in Refs. [27, 28, 26]. Although we can't run a numerical simulation for an infinite plane-wave field, we can closely approximate such a field using a plane wave with a super-Gaussian temporal profile. The emission spectra in such a field closely resembles that of an infinite plane wave, retaining most of the structural features [29]. In figure 1 we show the emission spectrum calculated analytically (for an infinite plane wave) together with that calculated numerically for an electron in a degree-8 super-Gaussian field. The laser intensity $a_0 = 20$ with wavelength

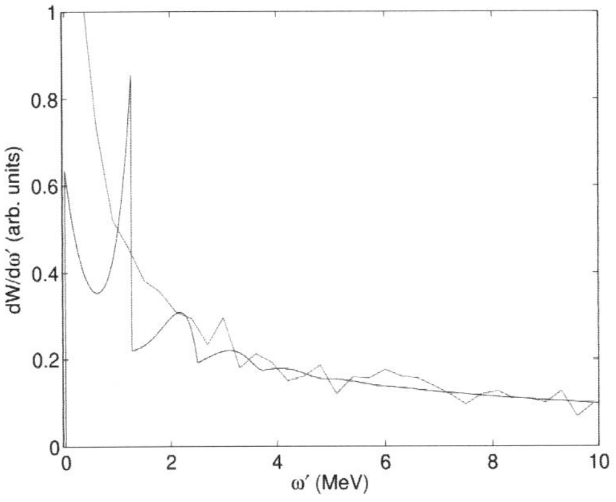

Figure 1. Comparison of the analytical and numerical frequency spectra. Parameters are $a_0 = 20$, $\gamma_0 = 9000$. Black line: analytical spectrum, blue line numerical spectrum.

$\lambda = 0.8\,\mu\text{m}$ and the initial electron $\gamma_0 = 9000$. The parameters have been deliberately chosen to emphasis certain structural features in the spectrum. It can be seen that while in general the numerical spectrum matches the analytical spectrum very well, it fails to reproduce certain structural features (caused here by the first and second harmonics). It turns out to be true in general that the numerical method reproduces the correct spectra except that it averages over the structural features. A detailed analysis for a wide range of parameter regimes is presented in Ref [1].

4. Classical analysis

In these proceedings we explain the discrepancy between the full analytically calculated spectra and that obtained using the numerical method by employing a classical argument.

The expression describing the classical radiation emission of a particle can be found in most textbooks on electrodynamics (see, e.g., [25]). Here we adopt the less common covariant formulation which can be found, for example, in [30]. The four-momentum of the emitted radiation can be written

$$P^\mu = -\int \frac{d^4k'}{(2\pi)^3}\text{sgn}(k'^0)\delta(k'^2)k'^\mu j(k') \cdot j^*(k'), \tag{2}$$

where sgn is the sign function, $k'^{\mu} = \omega'(1, \sin\theta\cos\phi, \sin\theta\sin\phi, \cos\theta)$ is the vector of the emitted radiation, where θ and ϕ are the polar angles that determine the direction of the photon, and

$$j^{\mu}(k') = e \int d\tau \, u^{\mu}(\tau) e^{-ik'\cdot x(\tau)}, \tag{3}$$

is the Fourier transform of the electron current, where u and x are the four-velocity and four-position of the electron, respectively. However, the boundary conditions can result in an infra-red divergence which means it is advisable to regulate the integral: see [31] for a fuller discussion. The resulting expression is

$$j^{\mu}(k') = e \int d\tau \, e^{-ik'\cdot x(\tau)} \frac{d}{d\tau}\left(\frac{u^{\mu}}{ik'\cdot u}\right). \tag{4}$$

Performing an integration by parts we find

$$j^{\mu}(k') \;=\; e\frac{u^{\mu}}{ik'\cdot u} e^{-ik'\cdot x(\tau)}\Bigg|_{\tau=-\infty}^{\tau=+\infty} + e\int d\tau \, u^{\mu}(\tau) e^{-ik'\cdot x(\tau)}, \tag{5}$$

which is (4) with the addition of some boundary terms. Now, returning to our calculation, the radiated energy is given by the zero component of (2)

$$P^0 \equiv \int d\omega' d\Omega \, \omega' \frac{d^2 N_{\gamma}}{d\omega' d\Omega}, \tag{6}$$

where

$$\rho \equiv \frac{d^2 N_{\gamma}}{d\omega' d\Omega} = -\frac{\omega'}{16\pi^3} j(k')\cdot j^*(k'), \tag{7}$$

is the classical spectral density, which gives the 'number of photons' N_{γ} radiated per unit frequency per unit solid angle.

It is possible to formulate a classical equivalent of the QED numerical scheme by dividing the particle orbit into discrete segments, evaluating (6) for each segment, and then adding the resulting spectra together to obtain the total spectrum. Proceeding this way, the electron four-current (4) becomes

$$j^{\mu} = \int_{\tau_1}^{\tau_2} (\cdots) + \int_{\tau_2}^{\tau_3} (\cdots) + \ldots + \int_{\tau_{n-1}}^{\tau_n} (\cdots). \tag{8}$$

Clearly, dividing the integral up into segments and summing the result from each segment as such will not alter the final result of the current. However, when it comes to calculating the spectral density (7) this is not the case, as, of course,

$$\rho \sim j\cdot j \;=\; \left[\int_{\tau_1}^{\tau_2} (\cdots) + \ldots\right] \cdot \left[\int_{\tau_1}^{\tau_2} (\cdots) + \ldots\right]^* \tag{9}$$

$$\neq \left[\int_{\tau_1}^{\tau_2} (\cdots)\right]^2 + \left[\int_{\tau_2}^{\tau_3} (\cdots)\right]^2 + \ldots + \left[\int_{\tau_{n-1}}^{\tau_n} (\cdots)\right]^2, \tag{10}$$

since, by calculating the spectral density in stages using each discrete segment of the four-current at a time, we are in effect throwing away the cross terms.

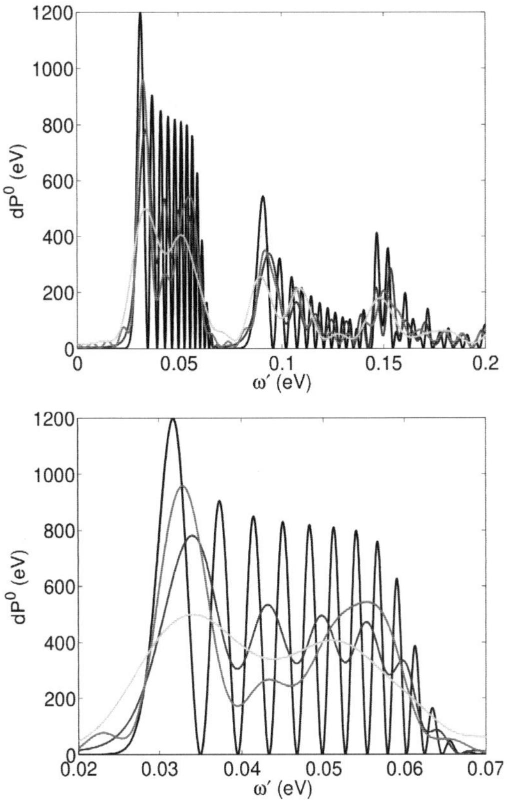

Figure 2. Top panel: The classical emission spectra calculated for an electron of initial $\gamma_0 = 100$ colliding with a linearly polarised laser of intensity $a_0 = 1.5$, wavelength $\lambda = 0.8\,\mu$m and with a Gaussian time envelope of duration 30 fs FDHM. Black line: spectrum calculated using full orbit. Blue line: spectrum calculated by splitting to orbit into two segments, calculating the spectrum using each half-orbit and then summing. Red line: spectrum calculated by splitting orbit into four. Green line: spectrum calculated by splitting orbit into eight. Bottom panel: Zoom into the first harmonic of the top panel.

The evaluation of (6) is a very numerically expensive procedure, especially for large a_0. We are aided in this respect by the fact that the program SIMLA uses a nonlinear time grid. (Other techniques to improve efficiency can be found in refs. [32, 33].) However, we do not have the same constraints in the classical modelling as we do in the QED case. This is because we have the full orbit (and, by implication, the full properties of the field) over each segment, so it is not necessary for us to make any *constant* field assumptions. We just require that the classical 'formation length' (see, e.g., [34, 32]) is shorter than the size of the discretisation. Thus we can perform our analysis with a much lower a_0 than is valid for the QED method.

In figure 2 we show the classical emission spectra, calculated by dividing the orbit into various numbers of segments. It can be seen very clearly that, as we divide the orbit, the neglect of the cross terms in (10) causes a smoothing of the structural features of the spectrum, just as we saw in the QED case. It appears that the discretisation causes an *averaging* of the emitted energy. To prove this we consider the total radiated energy as given by Larmor's formula [25]

$$P = -\frac{2}{3}\frac{e^2}{m^2}\int d\tau\, \frac{dp_\mu}{d\tau}\frac{dp^\mu}{d\tau}. \tag{11}$$

Since (11) is a function of the acceleration only, it is possible to split the integral up, evaluate each segment separately, and add the parts together without changing the result. This means that the total radiated power is *conserved* by the discretisation, even though the spectral density is changed. Hence the discretisation *averages* over the structural features of the spectrum.

5. Summary

We have analysed the commonly used numerical technique based upon the idea of splitting a classical electron orbit into discrete segments, and then correcting the momentum over each of these segments using a QED-derived photon emission rate. In a previous paper [1] it has been shown that this method reproduces the correct spectra (calculated analytically) reasonably well, except that it fails to reproduce certain structural features of the spectra. By adopting an analogous classical approach we have shown that, in dividing the orbit up into discrete segments, we neglect cross terms which provide information about the field. The shorter these segments become the more information is thrown away and the flatter the spectrum becomes. However, the total radiated energy remains constant, meaning that the result of discretisation is an averaging over the structural peaks and troughs of the spectrum.

Acknowledgments

The authors thank Anton Ilderton, Ben King and Adam Noble for useful discussions. The authors were supported by the EPSRC, grant EP/I029206/1–YOTTA (DG and CH) and by the Swedish Research Council, Contract No. 2012-5644 (CH).

References

[1] Harvey C, Ilderton A and King B 2014 (*Preprint* arXiv:1409.6187)
[2] Yanovsky V, Chvykov V, Kalinchenko G, Rousseau P, Planchon T, Matsuoka T, Maksimchuk A, Nees J, Cheriaux G, Mourou G and Krushelnick K 2008 *Opt. Express* **16** 2109–2114
[3] The vulcan 10 petawatt project http://www.clf.stfc.ac.uk/New+Initiatives/The+Vulcan+10+Petawatt+Project/14684.aspx
[4] The extreme light infrastructure (ELI) project http://www.eli-laser.eu/index.html
[5] The exawatt center for extreme light studies (XCELS) http://www.xcels.iapras.ru/
[6] Di Piazza A, Müller C, Hatsagortsyan K Z and Keitel C H 2012 *Rev. Mod. Phys.* **84** 1177–1228
[7] Heinzl T and Ilderton A 2009 *Eur. Phys. J. D* **55** 359–364
[8] Gonoskov A, Gonoskov I, Harvey C, Ilderton A, Kim A *et al.* 2013 *Phys. Rev. Lett.* **111** 060404 (*Preprint* arXiv:1302.4653)
[9] Gonoskov A, Bashinov A, Gonoskov I, Harvey C, Ilderton A *et al.* 2014 *Phys. Rev. Lett.* **113** 014801 (*Preprint* arXiv:1306.5734)
[10] Sarri G, Corvan D J, Schumaker W, Cole J M, Di Piazza A, Ahmed H, Harvey C, Keitel C H, Krushelnick K, Mangles S P D, Najmudin Z, Symes D, Thomas A G R, Yeung M, Zhao Z and Zepf M 2014 *Phys. Rev. Lett.* **113** 224801
[11] Di Piazza A, Hatsagortsyan K Z and Keitel C H 2010 *Phys. Rev. Lett.* **105** 220403
[12] Bell A R and Kirk J G 2008 *Phys. Rev. Lett.* **101** 200403
[13] Sokolov I V, Naumova N M, Nees J A and Mourou G A 2010 *Phys. Rev. Lett.* **105** 195005
[14] Elkina N V, Fedotov A M, Kostyukov I Y, Legkov M V, Narozhny N B, Nerush E N and Ruhl H 2011 *Phys. Rev. ST Accel. Beams* **14** 054401
[15] Ridgers C, Kirk J, Duclous R, Blackburn T, Brady C, Bennett K, Arber T and Bell A 2014 *J. Comput. Phys.* **260** 273 – 285
[16] Green D G and Harvey C N 2014 (*Preprint* arXiv:1410.4055)
[17] Green D G and Harvey C N 2014 *Phys. Rev. Lett.* **112** 164801 (*Preprint* arXiv:1307.8317)
[18] Sauter F 1931 *Z. Phys.* **69** 742–764
[19] Heisenberg W and Euler H 1936 *Z. Phys.* **98** 714–732
[20] Schwinger J 1951 *Phys. Rev.* **82** 664–679
[21] Ritus V 1985 *JSLR* **6** 497–617
[22] Ilderton A and Torgrimsson G 2013 *Phys. Rev. D* **87** 085040 (*Preprint* arXiv:1210.6840)
[23] Berestetskii V B, Lifshitz E M and Pitaevskii L P 1982 *Quantum Electrodynamics - 2nd ed. - (Course of Theoretical Physics; V4)* (Pergamon, Oxford)
[24] Duclous R, Kirk J G and Bell A R 2011 Plasma Phys. Contr. F **53** 015009
[25] Jackson J D *Classical electrodynamics* 3rd ed (New York, NY: Wiley)
[26] Harvey C, Heinzl T and Ilderton A 2009 *Phys. Rev. A* **79** 063407
[27] Nikishov A and Ritus V 1964 *Sov. Phys. JETP* **19** 529–541
[28] Nikishov A and Ritus V 1964 *Sov. Phys. JETP* **19** 1191–1199
[29] Harvey C, Heinzl T, Ilderton A and Marklund M 2012 *Phys. Rev. Lett.* **109** 100402

[30] Mitter H 1998 Elektrodynamik unpublished lectures on electrodynamics (in German), available at `http://physik.uni-graz.at/~hem/`
[31] Dinu V, Heinzl T and Ilderton A 2012 *Phys. Rev. D* **86** 085037 (*Preprint* `arXiv:1206.3957`)
[32] Reville B and Kirk J G 2010 *Astrophysical J.* **724** 1283
[33] Thomas A 2010 *Phys. Rev. ST Accel. Beams* **13** 020702 (*Preprint* `arXiv:0906.0758`)
[34] Akhiezer A I and Shul'ga N F 1987 *Soviet Physics Uspekhi* **30** 197

Journal of Physics: Conference Series **594** (2015) 012053 doi:10.1088/1742-6596/594/1/012053

Double Compton scattering in a constant crossed field and approximations used in simulation

B. King

School of Computing and Mathematics, Plymouth University, Plymouth PL4 8AA, United Kingdom

E-mail: b.king@plymouth.ac.uk

Abstract. The double photon emission of an electron in a constant crossed field has been calculated in [1]. By calculating the process of polarised single-photon emission, it is shown that the double-photon process can be unambiguously separated into two- and one- step subprocesses, also known as sequential and simultaneous double-emission. It is found that the assumption used in simulations of neglecting additional electron spin effects is valid to within a few percent, however the assumption that the two-step process is dominant becomes questionable for shorter background fields.

1. Introduction

The emission of radiation by an accelerated electron or positron is described classically by the process of Thomson scattering [2]. When the energy of the emitted photon in the rest frame of the electron is of the order of the electron rest energy or higher, a quantised theory is necessary to describe the change in emission. When such emission occurs in an intense electromagnetic background field, it is often referred to as "[non-linear] Compton scattering". This was first calculated in monochromatic (hence infinite) plane waves [3, 4] and more recently in finite plane waves [5, 6, 7, 8, 9, 10] with electron spin [11], photon polarisation [12] and external-field polarisation [13, 14] all having been investigated. Experimentally, Compton scattering has been measured in the weakly non-linear regime and recent advances in laser technology have fuelled interest in calculating experimental signals of the two-photon emission process in a pulsed plane-wave background [15, 16] (a review of strong-field QED effects can be found in [17, 18, 19]).

In the current proceedings, we report on results in [1] for the calculation of the two-photon emission process in a constant crossed field. The constant crossed background was chosen at is the one used overwhelmingly in numerical simulations in laser-particle physics, which propagate charged particles using the particle-in-cell method and generate "quantum" events using Monte Carlo techniques [20, 21, 22, 23, 24, 25, 26, 27]. By calculating the two-photon emission rate analytically, the accuracy of various numerical approximations can be assessed. Although the total rate for this process has been found in a calculation for the two-loop electron mass operator [28], using our direct method of calculation, also differential rates can be found that could be used, for example, to calculate the transverse momenta of the emitted particles. Although our analysis is for electrons, the results also apply to positrons. Unless otherwise stated, $c = \hbar = 1$ in our system of units.

2. Polarised single Compton scattering

We begin by presenting the highlights of the calculation of single photon emission, when the polarisation of all particles are taken into account. This is the process:

$$e^- \to e^- + \gamma \tag{1}$$

where e^- refers to an electron and γ to a photon, in a constant crossed field. Let p, s_p, q, s_q, k and ε_k denote the incoming electron momentum and spin, the emitted electron momentum and spin and the photon momentum and polarisation respectively, where $s_p^2 = s_q^2 = \varepsilon_k^2 = -1$. The corresponding Feynman diagram is given in Fig. 1. The electron in a classical plane-wave

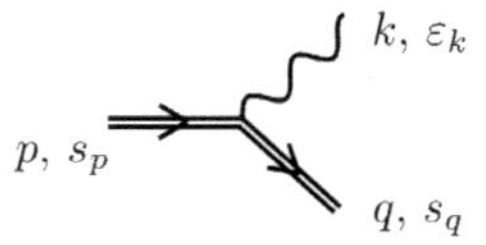

Figure 1. The Feynman diagram for single non-linear Compton scattering.

background is described using the Volkov solution [29] for the wavefunction of the Dirac equation ψ_p and an on-shell photon by the wavefunction A_k:

$$\psi_p(x) = \left[1 + \frac{e\not{\varkappa}\not{A}[\varphi(x)]}{2(\varkappa\cdot p)}\right]\frac{u_p}{\sqrt{2p^0V}}\mathrm{e}^{iS(p,\varphi(x))} \qquad A_k = \frac{\sqrt{4\pi}\varepsilon_k \mathrm{e}^{i\,k\cdot x}}{\sqrt{2k^0V}}, \tag{2}$$

where $\varkappa$, A and φ are the wavevector, gauge potential and phase of the external field, V is the normalisation volume, $\not{\varkappa} = \gamma^\mu\varkappa_\mu$ and γ^μ are the gamma matrices. The amplitude for the process is then given as usual by:

$$S_{fi}(p \to q + k) = ie\int d^4x\ \overline{\psi}_q(x)\not{A}_k^*\psi_p(x), \tag{3}$$

where the spin of an electron with momentum p is included using the relation [30]:

$$u_p\overline{u}_p = \frac{1}{2}(\not{p} + m)(1 - \gamma^5\not{s}_p). \tag{4}$$

We employed the method developed by Nikishov and Ritus (see [17], and a more detailed application to single Compton scattering in [12]), to calculate Eq. (11). One difference worthy of note was the addition of electron spin. If an arbitrary covariant basis is chosen, divergences seem to appear, associated with violating a symmetry of the total probability in the direction of external electric field. One choice of spin-basis that preserves this symmetry and allows the Nikishov-Ritus method to be straightforwardly applied is:

$$A(\varphi) = a^{(1)}g^{(1)}(\varphi) + a^{(2)}g^{(2)}(\varphi); \quad \varepsilon_k^{(1,2)} = a^{(1,2)} - \frac{k\cdot a^{(1,2)}}{k\cdot\varkappa}\varkappa; \quad \zeta_p = a^{(2)} - \frac{p\cdot a^{(2)}}{p\cdot\varkappa}\varkappa, \tag{5}$$

where $a^{(1)}\cdot a^{(2)} = 0$, $a^{(1)}\cdot a^{(1)} = a^{(2)}\cdot a^{(2)} = -1$ and $g^{(2)}(x) = 0$. The gauge potential of the constant crossed field background can be written $g^{(1)}(\varphi) = (ma_0/e)\varphi$ where $a_0 = e|p\cdot F|/m|\varkappa\cdot p|$ is a gauge- and relativistic- invariant intensity parameter [31], for F the Faraday tensor of the external field, which can be written as $a_0 = mE/\varkappa^0 E_{\text{cr}}$ for electric field amplitude E and critical field $E_{\text{cr}} = m^2/e$. As a_0 is inversely proportional to background field frequency, in a constant field it is formally infinite, but total rates can be written in terms of another gauge- and relativistic invariant for a particle with momentum p, namely the "quantum efficiency" parameter $\chi_p = e|p \cdot F|/m^3$. One can define the rate for single Compton scattering using $R_\gamma = P_\gamma/a^0 \int d\varphi$, which the reader notices is independent of the limit $\varkappa^0 \to 0$.

In order to interpret the results, let us expand the polarisation and spins as:

$$\varepsilon_k = c_1\varepsilon_k^{(1)} + c_2\varepsilon_k^{(2)} \tag{6}$$

$$s_p = \sigma_p\zeta_p \qquad s_q = \sigma_q\zeta_q \tag{7}$$

where $c_{1,2} \in \{0, 1\}$, $\sigma_{p,q} \in \{-1, 0, 1\}$. We then find:

$$R_\gamma = -\frac{\alpha}{\chi_p^2}\int_0^{\chi_p} d\chi_k \left[C^{\cdot}\,\mathrm{Ai}(z) + C'\mathrm{Ai}'(z) + C_1\mathrm{Ai}_1(z)\right], \qquad z = \left[\frac{\chi_k}{\chi_p(\chi_p - \chi_k)}\right]^{2/3} \tag{8}$$

where $\mathrm{Ai}(x) = \frac{1}{\pi}\int_0^\infty dk\, \cos(kx + k^3/3)$ is the Airy function, $\mathrm{Ai}'(x)$ its derivative, $\mathrm{Ai}_1(x) = \int_x^\infty dk\, \mathrm{Ai}(k)$ and

$$\begin{aligned} C^{\cdot} &= z\left[(\sigma_p + \sigma_q)(\chi_k - 2c_k^2\chi_p) - 2(1 - c_k^2)\sigma_q\chi_k\right] \\ C' &= \chi_k z^{1/2}(1 - c_\sigma c_\delta\sigma_p\sigma_q) + \frac{2}{z}(1 + \sigma_p\sigma_q)(1 - c_\sigma c_\delta) \\ C_1 &= 1 + \sigma_p\sigma_q - \frac{c_\sigma c_\delta\sigma_p\sigma_q\chi_k^2}{\chi_p(\chi_p - \chi_k)}, \end{aligned} \tag{9}$$

where $c_k = c_1 = \sqrt{1 - c_2^2}$, $2c_\sigma = c_1 + c_2$ and $c_\delta = c_2 - c_1$ have been used to simplify notation. The analytical form of the rate is of particular interest, as a term containing $\mathrm{Ai}(z)$, which cannot be written simply in terms of the other Airy functions, appears when the polarisation of all particles is taken into account. The affect on the rate of taking into account particle polarisation is illustrated in Fig. 2. From Fig. 2 we note that the "spin-flip" channels of

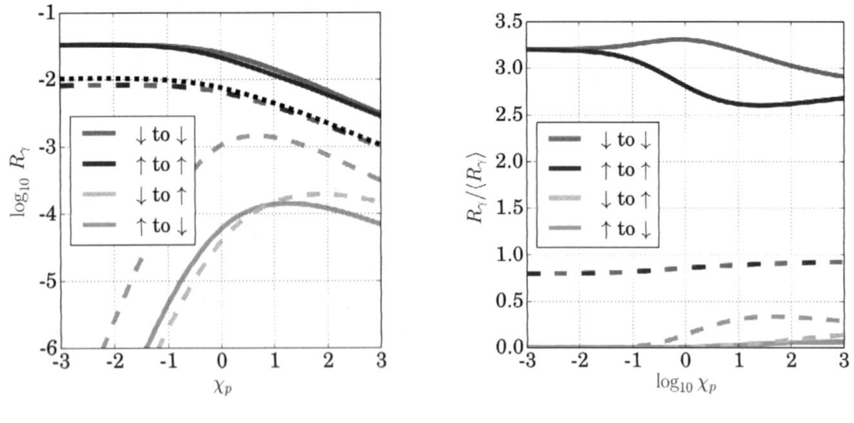

Figure 2. Left: The solid (dashed) lines refer to the rate of polarised single Compton scattering R_γ of an emitted photon in the $\varepsilon_k^{(1)}$ ($\varepsilon_k^{(2)}$) polarisation state and the dotted line refers to the unpolarised rate $\langle R_\gamma\rangle$. Right: The ratio of the rate of each polarisation channel to the unpolarised rate, using the same line scheme as in the left plot.

an electron changing its polarisation state due to the emission of a photon, are significantly suppressed compared to the "no-flip" channels. This also holds for the photon polarisation as those channels in which the photon polarisation is parallel to the external field one are favoured over those where there is an effective "flip". We also note the rate for "no-flip" channels can be more than 300 % the average.

3. Unpolarised double Compton scattering

The calculation for polarised single Compton scattering can be used to better understand the unpolarised rate for double Compton scattering. This is the process:

$$e^- \to e^- + \gamma + \gamma' \tag{10}$$

which can be illustrated using the Feynman diagrams in Fig. 3. The amplitude for this process

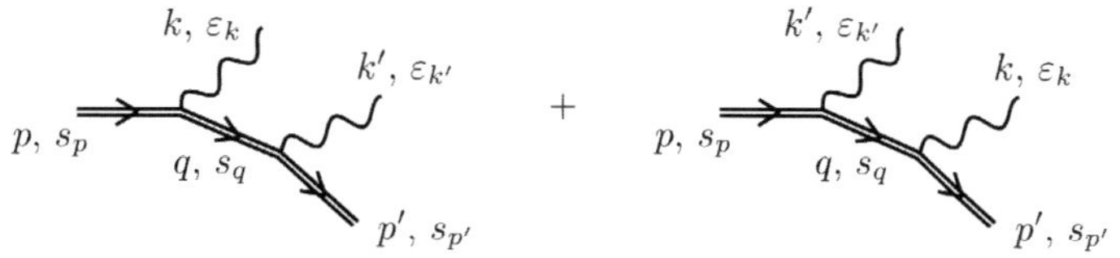

Figure 3. The Feynman diagrams for double non-linear Compton scattering.

$S_{fi} = \overrightarrow{S_{fi}} + \overleftarrow{S_{fi}}$ with:

$$\overrightarrow{S_{fi}} = -e^2 \int d^4x'\, d^4x\, \overline{\psi}_{p'}(x') \not{A}^*_{k'} G_q(x',x) \not{A}^*_k \psi_p(x), \tag{11}$$

where the electron propagator in a plane-wave background is given by [17]:

$$G_q(x',x) = \int \frac{d^4q}{(2\pi)^4} \left[1 + \frac{e\not{\varkappa}\not{A}(\varphi')}{2\,\varkappa\cdot q}\right] \mathrm{e}^{iS(q,\varphi')} \frac{\not{q}+m}{q^2 - m^2 + i0} \mathrm{e}^{-iS(q,\varphi)} \left[1 + \frac{e\not{A}(\varphi)\not{\varkappa}}{2\,\varkappa\cdot q}\right], \tag{12}$$

and $\overleftarrow{S_{fi}}$ is equal to $\overrightarrow{S_{fi}}$ with the photon positions swapped (due to exchange symmetry). The calculation then proceeds as for single Compton scattering, but is slightly complicated by the exchange term in the amplitude and the presence of a propagator (a detailed example of the Nikishov-Ritus approach being applied to a two-vertex process is given in the calculation of electron-seeded pair creation in [32]). It can be shown (e.g. in [1, 32]) that in a constant crossed field, the probability of emission of two photons can be unambiguously written as a probability containing the sequential process plus the probability of the simultaneous process, which we denote the two- and one- step process respectively:

$$R_{\gamma\gamma} = R^{(2)}_{\gamma\gamma} + R^{(1)}_{\gamma\gamma}; \qquad R^{(2)}_{\gamma\gamma} = \mathcal{I}^{(2)}_{\gamma\gamma}\, a_0 \int d\varphi_-; \quad R^{(1)}_{\gamma\gamma} = \mathcal{I}^{(1)}_{\gamma\gamma}, \tag{13}$$

where analogous to the previous section, we have defined the rate $R_{\gamma\gamma} = P_{\gamma\gamma}/a_0 \int d\varphi_x$, φ_x is the phase of the first emission, and $\mathcal{I}^{(1,2)}_{\gamma\gamma} = \mathcal{I}^{(1,2)}_{\gamma\gamma}(\chi_p)$ depend only on the quantum efficiency parameter of the seed electron χ_p. Since the two-step process is sequential, there are two phase points on the electron's trajectory that play a role and in Eq. (13) φ_- is the difference in phase between emissions in the two-step process. In contrast, the one-step process is simultaneous emission and so only a single point of emission, φ_x, is relevant.

The interference between exchange diagrams is an order lower in $a_0 \int d\varphi_- = L/L_*$, where L is the longitudinal spatial extent of the field and L_* is the formation length, and as we are interested in fields of much greater extent than formation length $L_* = \lambda E_{\rm cr}/E$ ($\lambda = 1/m$ is the reduced Compton wavelength), this parameter is formally infinite (later simply assumed large) and the interference terms are neglected. Since we plan to investigate approximations used in simulation, which typically require total rather than differential rates, and since the particle being exchanged is outgoing and hence integrated over, both exchange terms contribute an equal amount and are included in the two- and one- step rates in Eq. (13).

The two-step rate was found to be exactly factorisable as an average over intermediate electron polarisation states, of the twice-iterated rate of single Compton scattering:

$$R_{\gamma\gamma}^{(2)} = \frac{1}{2}\sum_{\sigma_q}\int d\chi_q \frac{\partial R_\gamma(\chi_p \to \chi_k + \chi_q)}{\partial \chi_q} R_\gamma(\chi_q \to \chi_{p'} + \chi_{k'}) \frac{a_0}{2}\int d\varphi_-, \tag{14}$$

where the first factor of a half is an average over the two spin states of the propagating electron, and the second due to time ordering. We note that in choosing the spin states in Eq. (5), which do not precess in the external field, the spin of the electron after the first emission is assumed not to have flipped before the second. If a different spin basis were chosen, which precessed in the external field, it is unclear whether such a straightforward factorisation could be made. (We also mention here, that it is also unclear whether this factorisation is at all possible in a general plane wave.) Photon precession is not included as this is a higher-order affect, first entering at $O(\alpha^2)$ for single-photon emission [33, 34, 35, 36] where $\alpha \approx 1/137$ is the fine-structure constant.

The one-step rate is found to be exclusively negative. This does not contradict the interpretation of the rate as a probability per unit time, as the factor $a_0 \int d\varphi_-$ is formally infinitely large in a constant crossed field and so the positive two-step process being one order higher in this factor, will dominate the one-step process. Later, we will investigate how this changes when, as in simulations, this factor is taken to be finite. An important extra feature in double Compton scattering that is not present in the single version is the appearance of a soft χ_k singularity in the total probability of the one-step process, which diverges as $\sim 1/\chi_k$ as $\chi_k \to 0$. This divergence should be cancelled by self-energy corrections, briefly discussed in [28]. The divergence is particularly interesting as it does not seem to be synonymous with the well-known infra-red divergence of emitted soft photons $k^0 \to 0$ from standard Compton scattering [15, 37, 38], but also appears when the emitted photon is collinear with the external field wavevector. This point is discussed in detail in [39, 40, 41]. A comparison of the one- and two-step processes is given in Fig. 4, where a $\chi_k > 10^{-5}$ cutoff has been chosen to acquire finite results. The comparison of one- and two- step processes was found to be qualitatively the same for cutoffs within $[10^{-7}, 10^{-3}]$. In Fig. 4, the probability for these processes has been compared for an external field of dimension $a_0 \int d\varphi = 1$ ($L = L_*$), for a cutoff of 10^{-5}. It is apparent that for $\chi_p > 0.1$, the magnitude of the one-step process is larger over this length.

4. Approximations used in simulation

Two approximations that are routinely used in the types of numerical simulations of strong-field QED processes mentioned in the introduction are i) neglecting electron spin, ii) neglecting simultaneous processes. Using our results for double Compton scattering in a constant crossed field, both of these could be investigated. Typically, in simulations, the constant crossed field approximation is used, which assumes that a spatio-temporally changing background can be well-approximated as constant during "quantum" processes such as photon emission or pair creation, as long as the scale of variation of the external field Δ fulfills $\Delta \gg L_*$ (we recall

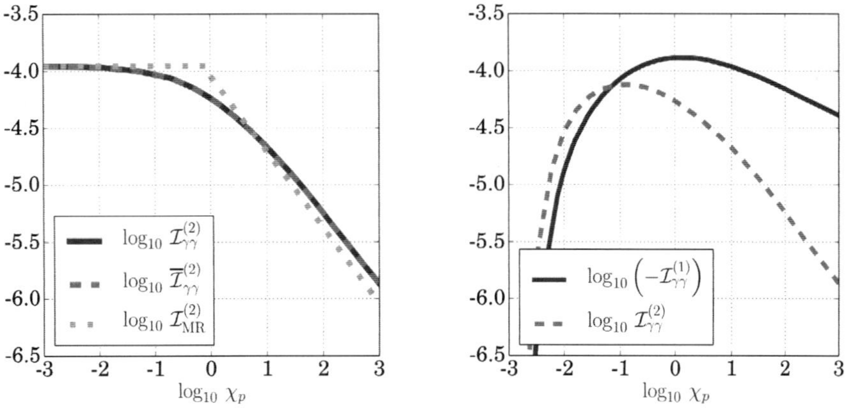

Figure 4. Left: a plot of the two-step part of double Compton scattering (solid blue line), compared to twice-iterated unpolarised single Compton scattering (red dashed line) and the calculation of Morozov and Ritus; Right: the one-step and two-step processes with cutoff $\chi_k > 10^{-5}$.

$L_* = \lambda E_{\mathrm{cr}}/E$ is the formation length). We define the constant crossed field approximation to be when the probability for two-photon emission in such a background is written:

$$P_{\gamma\gamma} \approx P^{(2)}_{\mathrm{ccf}} = \frac{1}{2}\sum_{\sigma_q}\int_{\varphi_0}^{\varphi_\infty} d\varphi_{a_{x'}} \int_{\varphi_0}^{\varphi_{a_{x'}}} d\varphi_{a_x} \int_0^{\chi_p(\varphi_{a_x})} d\chi_q \, P_\gamma[\chi_q(\varphi_{a_{x'}})] \, \frac{\partial P_\gamma[\chi_p(\varphi_{a_x}), \chi_q(\varphi_{a_x})]}{\partial \chi_q}, \quad (15)$$

where $\varphi_{a_x} = \varphi_x \, a_0(\varphi_x)$ and $\varphi_{a'_x} = \varphi_{x'} \, a_0(\varphi_{x'})$ and $\chi(\varphi) = \chi[F(\varphi)]$ now depends on the, in general non-constant, background $F(\varphi)$ and $\varphi_x, \varphi_{x'} \in [\varphi_0, \varphi_\infty]$. If the constant crossed field limit is taken of Eq. (15), then the factorised product in Eq. (14) is recovered.

4.1. Neglecting electron spin

As can be seen from the plot of the two-step process in Fig. 4, when the spin of intermediate electrons is taken into account, the total rate is indistinguishable from when the spin is neglected. To make a fair comparison with numerical simulations, we introduce a cutoff for the minimum χ_k that a scattered photon can have. Then comparing $P^{(2)}_{\mathrm{ccf}}$ from Eq. (15) with a probability that neglects electron spin:

$$\overline{P^{(2)}_{\mathrm{ccf}}} = \int_{\varphi_0}^{\varphi_\infty} d\varphi_{a_{x'}} \int_{\varphi_0}^{\varphi_{a_{x'}}} d\varphi_{a_x} \int_0^{\chi_p(\varphi_{a_x})} d\chi_q \, \langle P_\gamma\rangle[\chi_q(\varphi_{a_{x'}})] \, \frac{\partial \langle P_\gamma\rangle[\chi_p(\varphi_{a_x}), \chi_q(\varphi_{a_x})]}{\partial \chi_q}, \quad (16)$$

where $\langle\rangle$ indicates an average over polarisations has been performed, we find that even for a cutoff as large as $\chi_k = 0.1$, the relative difference in the expected number of double-photon emissions from including the intermediate electron spin is only of the order of a few percent. When looking at polarised single Compton scattering, we saw that the spin-flip channels were greatly suppressed in Fig. 2 and so it should not be a great surprise that they play little role in determining the double Compton scattering rate. We mention here that the precession of electron spin due to interaction with the external field is already included to all orders in the Volkov solution.

4.2. Neglecting simultaneous processes

It was already seen in Fig. 4 that the magnitude of the probability for the one-step process to occur can be larger than for the two-step process when the total length of the external field L is of the order of the formation length L_* so $L \approx L_*$. If we demand the total probability must be positive, this gives a connection between the seed electron's χ-parameter and the length of field that can no longer be described by a constant crossed field approximation. If we demand $L \gg \widetilde{L}$, where $\widetilde{L}$ is the smallest length for which the probability is positive, and we interpret "$\gg$" to be an order of magnitude, we see from Fig. 4, that already for $\chi_p \in [0.1, 1000]$ the minimum acceptable value of $L/L_* \in [10, 300]$. This implies that when L/L_* is smaller than this, current forms of numerical simulation can only reliably reproduce single photon emission for a given background.

We emphasise that single photon emission is dominant in the first stages of an electron's propagation in an external field, but if $L/L_* \gtrsim 2/\alpha$, two-step double photon emission can begin to become as probable as single photon emission. However, since the asymptotic nature of the simultaneous double-photon emission is identical to single-photon emission, there does not seem to arise a situation where simultaneous double-photon emission can be dominant. Therefore the present calculation implies that if the extent of the external field is large enough, the two-step process is still dominant.

If a measurement were sensitive to the generation of double-emitted photons (perhaps through a different transverse distribution as in [16]), an example of when the simultaneous process could influence the number of photons generated is a comparison of the constant crossed field approximation applied to a pulse with electric field of the form $E = E_0 \cos^4(\pi\varphi/2\Phi)\cos\varphi$ for pulse width $\Phi = \varkappa^0\tau$ and amplitude E_0. For a laser of power 500 TW, focussed to a focal width of 10 μm of 910 nm wavelength with peak $a_0 = 10$, and pulse duration 5 fs counterpropagating with 10 GeV seed electrons. The total and sequential probabilities of double photon emission for these parameters are shown in Fig. 5. In Fig. 5 is an example where a qualitative difference

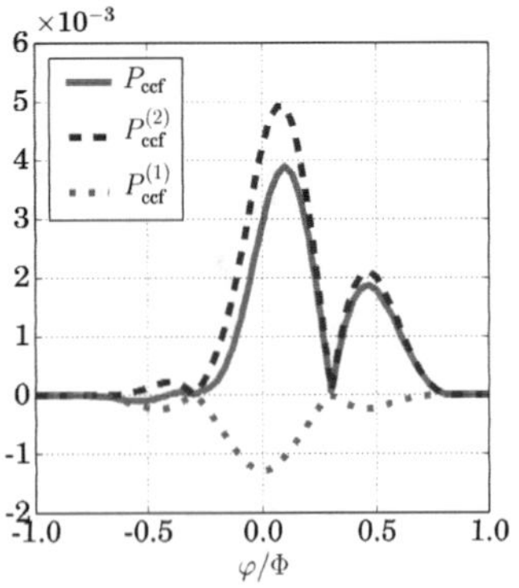

Figure 5. An example of suppression of the sequential process of double photon emission by the simultaneous process.

occurs due to acknowledging the existence of the simultaneous process, namely that double-photon emission is predicted to begin half a cycle later than if only the sequential process is included.

5. Conclusion

We have reported on results from [1] for the calculation of double photon emission in a constant crossed field with an emphasis on the consequences for numerical simulation. First, the results for the rate of electron spin-flipping due to emission of a single photon were presented, where it was found that "flip" channels are suppressed with respect to "no-flip" photon emission in a constant crossed field when a particularly useful non-precessing spin projection is used. Then, results for the calculation of the probability of two-photon emission in a constant crossed field were presented, for which it was shown that the probability can be written in terms of a two- and one- step process. Although the two-step process requires consideration of the intermediate electron's spin, it was found that the assumption in numerical simulations of using unpolarised rates, was accurate to within the few percent level. If numerical simulations include multiple Compton scattering, then the double Compton scattering results imply the constant crossed field can be safely applied when the smallest length scale of the external field is a couple of orders of magnitude larger than the formation length. For many parameter regimes of interest, the sequential process assumed in simulations is dominant.

6. Acknowledgments

B.K. acknowledges the support by the Russian Fund for Basic Research, grant No.13-02-90912.

References

[1] King B 2014 *arXiv:1410.5478*
[2] Jackson J D 1975 *Classical Electrodynamics* (New York: John Wiley & Sons, Inc.)
[3] Nikishov A I and Ritus V I 1964 *Sov. Phys. JETP* **19** 529–541
[4] Brown L S and Kibble T W B 1964 *Phys. Rep.* **133** A705–A719
[5] Narozhny N B and Fofanov M S 1996 *Sov. Phys. JETP* **83**(1) 14–23
[6] Boca M and Florescu V 2009 *Phys. Rev. A* **80**(5) 053403 URL `http://link.aps.org/doi/10.1103/PhysRevA.80.053403`
[7] Harvey C, Heinzl T and Ilderton A 2009 *Phys. Rev. A* **79** 063407
[8] Mackenroth F, Di Piazza A and Keitel C H 2010 *Phys. Rev. Lett.* **105**(6) 063903 URL `http://link.aps.org/doi/10.1103/PhysRevLett.105.063903`
[9] Heinzl T, Ilderton A and Marklund M 2010 *Phys. Lett. B* **692** 250–256
[10] Mackenroth F and Di Piazza A 2011 *Phys. Rev. A* **83**(3) 032106 URL `http://link.aps.org/doi/10.1103/PhysRevA.83.032106`
[11] Krajewska K and Kamiski J Z 2013 *Laser and Particle Beams* **31**(03) 503–513
[12] King B, Elkina N and Ruhl H 2013 *Phys. Rev. A* **87** 042117
[13] Ivanov D, Kotkin G and Serbo V 2004 *Eur. Phys. J. C* **36** 127–145 ISSN 1434-6044 URL `http://dx.doi.org/10.1140/epjc/s2004-01861-x`
[14] Bashmakov V F, Nerush E N, Kostyukov I Y, Fedotov A M and Narozhny N B 2014 *Phys. Plasmas* **21**
[15] Seipt D and Kämpfer B 2012 *Phys. Rev. D* **85**(10) 101701
[16] Mackenroth F and Di Piazza A 2013 *Phys. Rev. Lett.* **110** 070402
[17] Ritus V I 1985 *J. Russ. Laser Res.* **6** 497–617
[18] Marklund M and Shukla P K 2006 *Rev. Mod. Phys.* **78** 591
[19] Di Piazza A *et al.* 2012 *Rev. Mod. Phys.* **84** 1177–1228
[20] Sokolov I V *et al.* 2010 *Phys. Rev. Lett.* **105** 195005
[21] Nerush E N *et al.* 2011 *Phys. Rev. Lett.* **106** 035001
[22] Elkina N V *et al.* 2011 *Phys. Rev. ST Accel. Beams* **14**(5) 054401
[23] Blackburn T G, Ridgers C P, Kirk J G and Bell A R 2014 *Phys. Rev. Lett.* **112**(1) 015001 URL `http://link.aps.org/doi/10.1103/PhysRevLett.112.015001`
[24] Green D G and Harvey C N 2014 *Phys. Rev. Lett.* **112**(16) 164801 URL `http://link.aps.org/doi/10.1103/PhysRevLett.112.164801`
[25] Green D G and Harvey C N 2014 *arXiv:1410.4055*
[26] Harvey C, Ilderton A and King B 2014 *arXiv:1409.6187*
[27] Mironov A A, Narozhny N B and Fedotov A M 2014 *Physics Letters A* **378** 3254–3257
[28] Morozov D A and Ritus V I 1975 *Nucl. Phys. B* **86** 309–332

[29] Volkov D M 1935 *Z. Phys.* **94** 250–260
[30] Berestetskii V B, Lifshitz E M and Pitaevskii L P 1982 *Quantum Electrodynamics (second edition)* (Oxford: Butterworth-Heinemann)
[31] Heinzl T and Ilderton A 2009 *Opt. Commun.* **282** 1879–1883
[32] King B and Ruhl H 2013 *Phys. Rev. D* **88** 013005
[33] Dinu V, Heinzl T, Ilderton A, Marklund M and Torgrimsson G 2014 *Phys. Rev. D* **89**(12) 125003 URL `http://link.aps.org/doi/10.1103/PhysRevD.89.125003`
[34] Dinu V, Heinzl T, Ilderton A, Marklund M and Torgrimsson G 2014 *Phys. Rev. D* **90**(4) 045025 URL `http://link.aps.org/doi/10.1103/PhysRevD.90.045025`
[35] King B, Di Piazza A and Keitel C H 2010 *Phys. Rev. A* **82** 032114
[36] King B, Böhl P and Ruhl H 2014 *Phys. Rev. D* **90**(6) 065018 URL `http://link.aps.org/doi/10.1103/PhysRevD.90.065018`
[37] Lötstedt E and Jentschura U D 2009 *Phys. Rev. Lett.* **103** 110404
[38] Lötstedt E and Jentschura U D 2009 *Phys. Rev. A* **80** 053419
[39] Lavelle M and McMullan D 2006 *JHEP* **03** 026
[40] Dinu V, Heinzl T and Ilderton A 2012 *Phys. Rev. D* **86**(8) 085037 URL `http://link.aps.org/doi/10.1103/PhysRevD.86.085037`
[41] Ilderton A and Torgrimsson G 2013 *Phys. Rev. D* **87**(8) 085040 URL `http://link.aps.org/doi/10.1103/PhysRevD.87.085040`

Perspectives of implementing QED cascade production with the next generation of laser facilities

E G Gelfer[1], A A Mironov[1], A M Fedotov[1], V F Bashmakov[2,3], I Yu Kostyukov[2,3], N B Narozhny[1]

[1] National Research Nuclear University "MEPhI" (Moscow Engineering Physics Institute), 115409 Moscow, Russia
[2] Institute of Applied Physics, Russian Academy of Sciences, 603950 Nizhny Novgorod, Russia
[3] Lobachevsky National Research University of Nizhni Novgorod, 603950 Nizhny Novgorod, Russia

E-mail: egelfer@gmail.com

Abstract. QED cascade in an intense laser field can start developing if the electron proper acceleration expressed in Compton units χ becomes of the order of unity. We derive a general formula expressing the quantum dynamical parameter $\chi(t)$ of an initially slow electron in an arbitrary electromagnetic field, which is valid for a time range $t \ll 1/\omega$, where ω is the field carrier frequency. Using this formula, we identify a special field configuration of colliding focused laser beams, which provides cascade production at intensity of the order of 10^{23}W/cm^2. Such intensity is believed to become attainable with the new generation of laser facilities, particularly the ELI Beamlines, in the coming years.

1. Introduction

Pair creation from vacuum in strong electromagnetic field is being discussed in the literature for alomost 90 years, see e.g. [1–13]. But only nowadays, due to a remarkable progress of laser technology, the possibility to observe such kind of effects has been appeared. There are two conceivable possibilities to probe them experimentally. The first option might be causing direct pair production from vacuum [1–7]. However, calculations in [7] had demonstrated that spontaneous pair creation from vacuum would become observable only for the field intensity above about 10^{26} W/cm^2. Hence, taking into account that the maximum laser intensity attainable nowadays is only $2 \cdot 10^{22}$ W/cm^2 [14], we arrive at conclusion that it would be probably extremely hard to implement such an option in the nearest future.

The second option available would be observation of QED cascades [8–13]. In this paper we consider a cascade seeded by a charged particle that was initially at rest. Laser field accelerates such a particle causing emission of hard photons, that in turn can create electron-positron pairs, and so on. Such an avalanche-like process would then last until the generated particles finally exit the focal area of the laser field. Numerical simulations performed in [8, 10] for some particular models of laser pulses have indicated that the threshold intensity required for cascade development would be still above 10^{24} W/cm^2.

A new new generation of laser facilities, particularly the ELI Beamlines [16] in Czech Republic and VULCAN10 in UK [17] have been recently announced, that would provide intensities of the order 10^{23} W/cm^2. Since it still remains less than the threshold intensity for arising QED cascades as calculated in [8, 10], there arises a question if this threshold could be decreased somehow at least by an order of magnitude.

The probabilities for pair creation and hard photon emission processes [3]

$$W_{e,\gamma}(\chi_{\gamma,e} \gg 1) \sim \frac{\alpha m^2 c^4}{\hbar \varepsilon_{e,\gamma}} \chi_{\gamma,e}^{2/3}, \quad W_e(\chi_\gamma < 1) \sim e^{-\frac{8}{3\chi_\gamma}}, \tag{1}$$

where $\varepsilon_{e,\gamma}$ is the energy of radiating electron or pair creating photon and α is the fine structure constant, are defined by the quantum dynamical parameter χ

$$\chi = \frac{e\hbar}{m^3 c^4}\sqrt{-(F_{\mu\nu}p^\nu)^2}. \tag{2}$$

Here $F_{\mu\nu}$ is the electromagnetic field tensor, m and e are the electron mass and charge, p^ν is the 4-momentum. Since the probability of the pair creation process is exponentially suppressed for $\chi_\gamma < 1$ and since χ_γ for emitted photon is always less than χ_e for an emitting particle, the problem is to find such a field configuration that provides the fastest growth of χ during the period of 'acceleration'[1]. Then by numerical calculation we can find the minimum laser field intensity needed for arising of a QED cascade in such an optimal field configuration.

2. Growth of $\chi(t)$ at the 'acceleration' stage (general case).

Strong electromagnetic field is usually obtained by using focused laser pulses. But a standing wave like the field of two counterpropagating pulses is more efficient for cascade production than a single focused pulse of the same total power [8]. Therefore we consider an arbitrary focused field with the only restriction that the magnetic field vanishes at the center of the focus (the origin $\mathbf{r} = 0$).

For the time range $t \ll 1/\omega$, where ω is the field carrier frequency, the electric and magnetic fields can be both expanded into the Taylor series

$$\mathbf{E}(x) = \mathbf{E}_0 + \mathbf{E}_{,k}(0)x^k, \quad \mathbf{B}(x) = \mathbf{B}_{,k}(0)x^k \tag{3}$$

with respect to time and coordinates $x^k = \{ct, \mathbf{r}\}$. Besides, in what follows we always assume that the laser field is thus strong that the particle becomes ultrarelativistic after acceleration, $eEt/mc \gg 1$, where t is the acceleration time, but at the same time still remains much weaker than the QED critical field, $E \ll E_{cr} = m^2c^3/e\hbar = 1.3 \cdot 10^{16}\mathrm{V/cm}$.

Let the electric field at the focus be initially directed along the z-axis. Then for $t \gg m/eE_0$ [2] the solution of equations of motion accurate to the zeroth order in ωt reads

$$\mathbf{p}^{(0)} = e\mathbf{E_0}t, \quad \varepsilon^{(0)} = eE_0 t, \quad x = y = 0, \quad z = t. \tag{4}$$

where $\mathbf{p}$ and ε are the momentum and the energy of a particle.

Using the first order expressions for the fields in ωt, $\mathbf{E}(x) = \mathbf{E}_0 + \mathbf{E}'t$, $\mathbf{B}(x) = \mathbf{B}'t$, where

$$\mathbf{E}' = \frac{\partial \mathbf{E}}{\partial t}(0) + \frac{\partial \mathbf{E}}{\partial z}(0), \quad \mathbf{B}' = \frac{\partial \mathbf{B}}{\partial t}(0) + \frac{\partial \mathbf{B}}{\partial z}(0) \tag{5}$$

we can now solve the first order equation of motion

$$\frac{d\mathbf{p}^{(1)}}{dt} = e\left(\mathbf{E}' + \frac{[\mathbf{E_0} \times \mathbf{B}']}{E_0}\right)t, \tag{6}$$

[1] Herein and after we drop the induces thus denoting e.g. χ_e as just χ.

[2] We assume below that $\hbar = c = 1$.

Figure 1. Left: two colliding beams; right: $n = 8$ colliding beams.

thus obtaining the formula for $\chi(t)$

$$\chi(t) = \frac{e^2 E_0 t^2}{2m^3}\sqrt{(E'_x - B'_y)^2 + (E'_y + B'_x)^2}. \tag{7}$$

The result (7) is in agreement with the special cases of rotating purely electric field and standing plane wave obtained previously in Refs. [9, 13]. Hence, to optimize the field configuration for cascade production we must increase the coefficient at t^2 in the formula (7) as much as possible.

3. Case of two colliding focused beams

Now consider a realistic model of the focused laser beam suggested in Ref. [6]

$$\mathbf{A} = \frac{\sqrt{P} f(\xi) e^{-i\omega t}}{\omega^3 |\boldsymbol{\epsilon}|} [\hat{\mathbf{n}} \times \nabla_\perp] \left([\boldsymbol{\epsilon} \times \hat{\boldsymbol{n}}] \cdot \nabla_\perp \right) \frac{\sin(\omega \mathcal{R})}{\mathcal{R}}, \tag{8}$$

which is an exact solution of the Maxwell equations. Here, $\mathcal{R} = (r_\parallel - ib)\sqrt{1 + \frac{r_\perp^2}{(r_\parallel - ib)^2}}$, $\hat{\mathbf{n}}$ is a unit vector directed along the focal axis of the beam, $r_\parallel$ and $\mathbf{r}_\perp$ are the components of the radius vector parallel and perpendicular to $\hat{\mathbf{n}}$, respectively, $\boldsymbol{\epsilon}$ is the polarization vector, P is the beam power, b is the Rayleigh length, $\xi = b\omega$ and

$$f(\xi) = \frac{8\sqrt{\pi}\xi^{5/2} e^{\xi}}{\sqrt{e^{4\xi}(4\xi^2 - 6\xi + 3) - 4\xi^2 - 6\xi - 3}}. \tag{9}$$

More details about the model can be found in Appendix A below.

Using (7) for a pair of such colliding beams propagating along and opposite the x-axis, see the left panel of Fig. 1, we obtain

$$\chi(t) = \frac{e^2 P \omega^3}{m^3} \frac{|c_1 c_2|}{c_1^2 + c_2^2} g(\xi) t^2, \tag{10}$$

where the polarization vectors of both beams have been chosen as $\boldsymbol{\epsilon} = c_1\hat{\mathbf{y}} - ic_2\hat{\mathbf{z}}$, P is their total power, and

$$g(\xi) = \frac{128e^{2\xi}(\xi\cosh\xi - \sinh\xi)}{\xi^3\left(e^{4\xi}\left(4\xi^2-6\xi+3\right)-4\xi^2-6\xi-3\right)}\times \\ \left(3\left(\xi^2+4\right)\sinh\xi+\xi\left(\xi^2-12\right)\cosh\xi\right). \tag{11}$$

Note that dependence of χ on both the degree of focusing and polarization have been separated. Therefore from Eq. (10) it follows that the optimal polarization for cascade production, which provides the most rapid growth of $\chi(t)$ is the circular one, $c_1 = c_2$.

4. Case of n colliding focused beams

A method for reducing the total threshold power for spontaneous pair creation from vacuum via the multibeam technology was suggested in Ref. [7], where it was clearly demonstrated how the key parameter of that problem, the magnitude of the electric field strength at the focus, could be increased considerably due to interference by using several pairs of colliding laser beams instead of single one of the same power. In our problem we can not follow this way straightforwardly, because it was expedient in Ref. [7] to use identical linearly polarized beams, but such a field would never produce cascades in the focus.

Nevertheless, let us consider an even number n of beams propagating in the xy plane at angles

$$\varphi_k = 2\pi(k-1)/n, \quad k = 1\dots n \tag{12}$$

to the x axis, see the right panel of Fig. 1, and find their optimal polarization for cascade production. If we choose polarization vectors as $\boldsymbol{\epsilon}_k = e_x^k\hat{\mathbf{x}} + e_y^k\hat{\mathbf{y}} - ie_z^k\hat{\mathbf{z}}$, then using (7) we obtain

$$\chi_n(t) = \frac{e^2\omega^3 Pg(\xi)F}{m^3}t^2, \tag{13}$$

where the factor

$$F = \frac{e_z\sqrt{e_x^2+e_y^2}}{\sum\limits_{k=1}^{n}\left[(e_x^k)^2+(e_y^k)^2+(e_z^k)^2\right]}, \quad \mathbf{e} = \sum_{k=1}^{n}\mathbf{e}^k \tag{14}$$

determines the dependence of χ on polarization and the number of beams.

Due to the symmetry of the problem the polarization vector of the total electric field can lie in any plane that is parallel to z-axis. Let it be the yz-plane. Since for each beam the electric field is transverse to the direction of propagation $\hat{\mathbf{n}}$, we can rewrite the projections of polarization vectors to the xy-plane as $e_x^k = h_k\sin\varphi_k$, $e_y^k = h_k\cos\varphi_k$. Then

$$F = \frac{e_z e_y}{\sum\limits_{k=1}^{n}(h_k^2+(e_z^k)^2)}, \quad e_y = \sum_{k=1}^{n}h_k\cos\varphi_k. \tag{15}$$

Besides, let us define four n-dimensional vectors $\boldsymbol{Z} = \{e_z^k\}$, $\boldsymbol{H} = \{h^k\}$, $\boldsymbol{\alpha} = \{\cos\varphi_k\}$ and $\boldsymbol{\beta}$, such that all the components of $\boldsymbol{\beta}$ equal unity. Then the expression (15) takes the form

$$F = \frac{(\boldsymbol{\beta Z})(\boldsymbol{\alpha H})}{\sqrt{|\boldsymbol{Z}|^2+|\boldsymbol{H}|^2}} \tag{16}$$

Table 1. Power and intensity QED cascade thresholds for configurations with n colliding beams ($\omega = 1eV$, $\xi = 8$).

n	2	4	6	8
P, PW	23	14	10	7.9
$I, 10^{23}\mathrm{W/cm}^2$	17	10	7.3	5.6

The maximization of the expression (15) thus follows from the chain of inequalities

$$F \leq \frac{1}{2}\frac{(\boldsymbol{\beta Z})}{|\boldsymbol{Z}|}\frac{(\boldsymbol{\alpha H})}{|\boldsymbol{H}|} \leq \frac{|\boldsymbol{\alpha}||\boldsymbol{\beta}|}{2} = \frac{n}{2\sqrt{2}}, \tag{17}$$

i.e. the growth rate of the parameter χ is proportional to the number of beams n. Hence F achieves its maximum value when $\mathbf{Z} \parallel \boldsymbol{\beta}$, $\mathbf{H} \parallel \boldsymbol{\alpha}$ and $|\mathbf{Z}| = |\mathbf{H}|$. Consequently,

$$e_z^1 = e_z^2 = \ldots = e_z^n, \quad e_z = ne_z^1, \tag{18}$$

$$h_k = h_1 \cos\varphi_k, \quad e_y = h_1 \sum_{k=1}^{n} \cos^2\varphi_k = \frac{n}{2}h_1 \tag{19}$$

and $e_z^1 = \frac{h_1}{\sqrt{2}}$. Therefore

$$e_z : e_y = \sqrt{2} : 1, \tag{20}$$

i.e. the optimal polarization of the total field is elliptical with the semi-major to semi-minor axis ratio $\sqrt{2} : 1$. The polarization vectors of the beams in this optimal case are

$$\boldsymbol{\epsilon}_k = \{\sqrt{2}\sin\varphi_k\cos\varphi_k, \sqrt{2}\cos^2\varphi_k, -i\}. \tag{21}$$

Note, that Eq. (20) and its derivation are valid if $n > 2$, while for $n = 2$ the optimal polarization is circular, as was already mentioned after Eq. (11).

5. Numerical results

The actual threshold power required for cascade production can be found by numerical simulations. Since such a threshold is never sharp, let us define it by demanding a single pair creation on average by each seeding electron during the passage of laser pulses through the focus. As a model of a focused laser beam we still use the exact solution (8) for the Maxwell equations. The wavelength is taken to be $\lambda = 1.24\mu$m, which corresponds to the laser photon energy $1eV$. Finiteness of pulse duration was introduced artificially by limiting the simulation time. The simulation time is chosen to be 5 laser periods as is typical for the modern laser facilities. The focusing parameter is chosen as $\xi = b\omega = 8$. In this case the FWHM focal spot diameter equals to the wavelength, as at the HERCULES facility [14] (see also Appendix).

For numerical simulation we use a code implementing the Monte-Carlo (MC) algorythm, in which one assumes that e^- and e^+ are moving classically between the acts of the emission events, which are determined in the manner similar to Refs. [10, 12]. The seed particles are represented by a bunch of 10^3 electrons placed initially at rest at the center of the focal region. There exist bounds for a number of focused beams that can be allocated concurrently in a plane avoiding intersections [7, 20]. In our case $\xi = 8$ this number can not exceed 8 (see Appendix

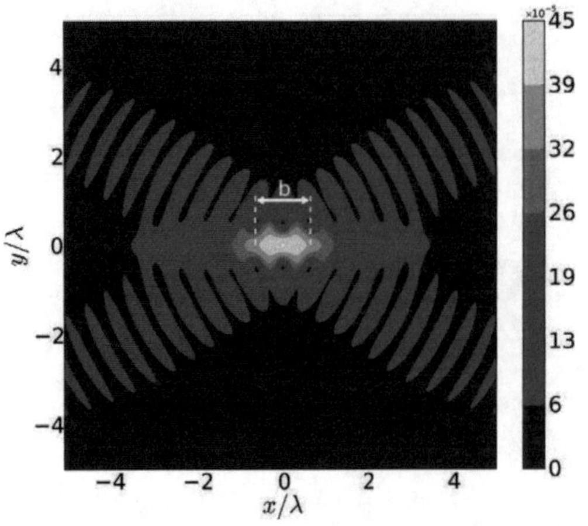

Figure A1. Distribution of the electric field strength (in units of E_{cr}) in the xy-plane for a single beam propagating along the x-axis at $t = 0$, $z = 0$ for $\boldsymbol{\epsilon} = \{0, 1, -i\}$, $\xi = 8$, $\omega = 1eV$ and $P = 1$PW.

A). The resulting thresholds obtained by numerical simulation of QED cascades in the cases of $n = 2, 4, 6$, and 8 beams with optimal polarization (21) colliding in the xy-plane are presented in Table 1. One can observe from the given data that the required laser power and intensity for cascade production decreases significantly as the number of beams increases.

6. Conclusion

We have shown that the QED cascades can arise in laser-target interaction at laser power below 10 PW and intensity below 10^{24}W/cm^2. This is realizable by a multibeam setup with several pairs of colliding laser beams. In order to determine how to arrange different beams, we have derived a general analytical expression for the dynamical quantum parameter $\chi(t)$ on the 'acceleration' stage, which controls the hard photon emission and pair photoproduction. Then, by specifying the case of several colliding beams, we have identified the optimal polarization of the beams with the most rapid growth of $\chi(t)$ in the focus. For such an optimal configuration with the fixed total laser power, χ is proportional to the number of beams. As an illustration, we have computed the QED cascade thresholds for the particular case of 8 colliding beams of optimal polarization. According to our simulations, the threshold total power is $P_{th} = 7.9$PW and the threshold intensity is $I_{th} = 5.6 \cdot 10^{23}$W/cm^2. Such parameters can be very likely achieved with the next generation of laser facilities [16, 17] in the nearest future.

Acknowledgements

This research was supported by the Government of the Russian Federation (Project No. 14.B25.31.0008), the Russian Foundation for Basic Research (grants 13-02-00372 and 13-02-00886) and the Russian Federation President program for support of the leading scientific schools (grant NSh-4829.2014.2).

Appendix A. The focused field model

Following [6], consider a vector potential

$$\mathbf{A}(\mathbf{r}, t) = \int \frac{d\mathbf{k}}{(2\pi)^3} \mathbf{a}(\mathbf{k}) e^{i\mathbf{k}\mathbf{r} - i\omega t}, \quad \mathbf{k} \cdot \mathbf{a}(\mathbf{k}) = 0, \tag{A.1}$$

which provides an exact solution of Maxwell equations $\ddot{\mathbf{A}} - \triangle \mathbf{A} = 0$ with the transverse gauge $\nabla \cdot \mathbf{A} = 0$. If we choose

$$\mathbf{a}(\mathbf{k}) = (\hat{\mathbf{n}} \times \mathbf{k}_\perp)(\mathbf{C} \cdot \mathbf{k}_\perp) e^{b(k_\parallel - \omega)} \delta(k^2 - \omega^2), \tag{A.2}$$

where $\mathbf{k}_\perp$ and $\mathbf{k}_\parallel$ are the components of the wave vector $\mathbf{k}$ parallel and orthogonal to the direction $\hat{\mathbf{n}}$ of the wave propagation, $\mathbf{C} = c[\boldsymbol{\epsilon} \times \hat{\boldsymbol{n}}]$ with some constant c and evaluate the integral (A.1) (see [6] for details), then we arrive at

$$\mathbf{A} = \frac{e^{-\omega(it+b)}}{(2\pi)^2}(\hat{\mathbf{n}} \times \nabla_\perp)(\mathbf{C} \cdot \nabla_\perp)\frac{\sin(\omega\mathcal{R})}{\mathcal{R}}, \tag{A.3}$$

that coincides with the representation (8) up to a constant factor.

In order to calculate the beam power, consider a beam propagating along the x-axis ($\hat{\mathbf{n}} = \hat{\mathbf{x}}$) and assume that $\boldsymbol{\epsilon} = c_1\hat{\mathbf{y}} - ic_2\hat{\mathbf{z}}$. By singling out the outgoing part of the wave, i.e. the part of (A.1) proportional to $e^{i\omega\mathcal{R}}$, in the far region ($r = \sqrt{x^2 + y^2 + z^2} \to \infty$) we obtain

$$\mathbf{E} = c\frac{\omega^3 \sin^2\theta e^{i\omega(ib+\mathcal{R}-t)}(c_1 \sin\phi + ic_2\cos\phi)}{8\pi^2} \times \{0, \sin\phi, -\cos\phi\}, \tag{A.4}$$

$$\mathbf{B} = c\frac{\omega^3 \sin^2\theta e^{i\omega(ib+\mathcal{R}-t)}(c_1 \sin\phi + ic_2\cos\phi)}{8\pi^2} \times \{-\sin\theta, \cos\theta\cos\phi, \cos\theta\sin\phi\}, \tag{A.5}$$

where $\phi = \arctan\frac{z}{y}$ and $\theta = \arctan\frac{x}{\sqrt{z^2+y^2}}$. By noting that $\mathcal{R} \to r\left(1 - \frac{ibx}{r^2}\right)$ for $r \to \infty$, we can compute the beam power $P = \frac{1}{8\pi}\int \mathrm{Re}\,[\mathbf{E} \times \mathbf{B}^*]d\mathbf{S}$

$$P = c^2\frac{\omega\left(c_1^2 + c_2^2\right)e^{-4\xi}\left(e^{4\xi}(2\xi(2\xi-3)+3) - 2\xi(2\xi+3) - 3\right)}{2048\pi^4 b^5}, \tag{A.6}$$

where $\xi = b\omega$. This proves the choice of the normalization factor $g(\xi)$ in (8).

As it was mentioned above, we demand as tight focusing that the circle of the wave length diameter in the focal plane contains a half of the beam power (practical realization of such degree of focusing was reported in Ref. [14]). Calculating the integral for the power over the circle area of the diameter $\lambda = 1.24 \cdot 10^{-4}$ cm in the plane $z = 0$ numerically, we obtain that this condition is satisfied if $\xi = 8$.

Distribution of the electric field magnitude in the xy-plane is displayed in Fig. A1. One can see that it has the distinctive shape of the focused beam. At far distance from the focal region the field is contained inside the cone with the aperture depending on the focusing parameter ξ. It can be shown, that in our case $\xi = 8$, this aperture $\sim \pi/4$, therefore the number of beams that can be allocated in the plane without intersections can not exceed 8.

Finally, to clarify polarization of the beam in our model, consider the electric field at the focus

$$\mathbf{E}(t, 0) = \frac{e^{-i\omega t}\omega f(\xi)\sqrt{P}(\xi\cosh\xi - \sinh\xi)}{\xi^3|\boldsymbol{\epsilon}|}\{0, ic_1, c_2\}. \tag{A.7}$$

This expression proves that the field is polarized elliptically.

References

[1] Sauter F 1931 *Z. Phys.* **69** 742
Sauter F 1932 *Z. Phys.* **73** 547
[2] Schwinger J 1951 *Phys. Rev.* **82** 664
[3] Nikishov A I and Ritus V I 1964 *Sov. Phys. JETP* **19** 529
Nikishov A I and Ritus V I 1967 *Sov. Phys. JETP* **25** 1135
Baier V N and Katkov V M 1967 *Phys. Lett.* **25A** 492
[4] Nikishov A I 1970 *Sov. Phys. JETP* **30** 660
[5] Narozhny N B, Bulanov S S, Mur V D and Popov V S 2004 *Phys. Lett.* A **330** 1
Bulanov S S, Narozhny N B, Mur V D and Popov V S 2006 *JETP* **102** 9
[6] Fedotov A M 2009 *Laser Physics* **19** 214

[7] Bulanov S S, Mur V D, Narozhny N B, Nees J and Popov V S 2010 *Phys. Rev. Lett.* **104** 220404
[8] Bell A R and Kirk J G 2008 *Phys. Rev. Lett.* **101**, 200403
Kirk J G, Bell A R and Arka I 2009 *Plasma Phys. Control. Fusion* **51** 085008
[9] Fedotov A M, Narozhny N B, Mourou G and Korn G 2010 *Phys. Rev. Lett.* **105** 080402
[10] Elkina N V et al 2011 *Phys. Rev. STAB* **14** 054401
[11] Nerush E N et al 2011 *Phys. Rev. Lett.* **106** 035001
[12] Duclous R, Kirk J G and Bell A R 2011 *Plasma Phys. Control. Fusion* **53** 015009
[13] Bashmakov V F et al 2014 *Phys. Plasmas* **21** 013105
[14] Bahk S W et al 2004 *Opt. Letters* **29** 2837
Yanovsky V et al 2008 *Opt. Express* **16** 2109
[15] Strickland D, Mourou G 1985 *Opt. Communications* **55** 447
[16] http://www.extreme-light-infrastructure.eu
[17] http://www.stfc.ac.uk/clf/New+Initiatives/ The+Vulcan+10+Petawatt+Project/14684.aspx
[18] Narozhny N B and Fofanov M S 2000 *JETP* **90** 753
Narozhny N B and Fofanov M S 2000 *Phys. Lett.* A **295** 87
[19] Landau L D and Lifshitz E M 1988 *Theoretical Physics: The Classical Theory of Fields, vol. 2 (Course of Theoretical Physics Series)* (London: Pergamon Press)
[20] Gonoskov I, Aiello A, Heugel S and Leuchs G 2012 *Phys. Rev.* A **86** 053836

Journal of Physics: Conference Series **594** (2015) 012055 doi:10.1088/1742-6596/594/1/012055

Pair production from space- and time-dependent strong fields

Dániel Berényi[1], Sándor Varró[2], Péter Lévai[1] and Vladimir V. Skokov[3]

[1] igner RC , Institute for article and uclear h sics, .O. Box 4 , Budapest 1525, ungar

[2] igner RC , Institute for Solid State h sics and Optics, .O. Box 4 , Budapest 1525, ungar

[3] Department of h sics, estern Michigan ni ersit , 1 3 . Michigan enue, alama oo, MI 4

-mail berenyi.daniel@wigner.mta.hu

Abstract. The recent de elopment of laser technolog and the large number of extreme laser experiments under construction renewed the research related to pair production in strong elds. If the predicted threshold of nonlinear Q D is reached, pair production ma be obser ed and the measurements must be compared with appropriate theoretical predictions. owe er the theoretical side still lac s the understanding of the rele ant uantum processes including the e ect of laser eld parameters on the number and spectrum of created particles. e use the Dirac- eisenberg- igner formalism to in estigate the role of the pure electric inhomogeneit on the spectra of created pairs. This simpli ed model ma also be rele ant in high energ ph sics for the description of string fragmentation in the earl stages of hea ion collisions.

1. Introduction

Pair production from vacuum has been long predicted in ED. The first field theoretical calculation by Schwinger established the threshold for the process in constant and homogeneous electric field. Since then it was shown that time dependent field configurations can significantly lower this threshold thus easing the experimental observation. Since than a few other analytical results appeared, but it seems that realistic field configurations can only be addressed by large scale numerical simulations. Specifically the inhomogeneous field configurations only recently become accessible via world-line instanton techniques and Dirac-Heisenberg-Wigner (DHW) formalism. In this paper we use the latter to investigate the role of electric field inhomogeneity on the number of pair produced and the characteristics that can distinguish such a process from a homogeneous one.

In CD there is an analogous pair production process that happens in heavy ion collisions when the separating quarks create a color ux tube in between each other. As the potential is linearly increasing with the separation distance there is a point when the stored energy exceeds the pair production threshold of a new quark-anti quark pair (or even more composite particles such as diquark-anti diquark pairs) and the separation continues with these new particle. Models based on this idea have proven to be successful in the description of heavy ion spectra but still assume constant and homogeneous fields. As it was shown that the DHW equations have

Journal of Physics: Conference Series **594** (2015) 012055 doi:10.1088/1742-6596/594/1/012055

analogues in CD and contain the ED case as a special case, our choice of the following toy electric field is not completely arbitrary but motivate by the cross section of a color ux tube in heavy ion collisions.

2. Theoretical Framework

The Dirac-Heisenberg-Wigner formalism provides a relativistic quantum description of pair production in phase space and formulates a Boltzmann-like evolution equation set with one time variable starting from the vacuum initial conditions induced by the prescribed external fields. The derivation for ED is introduced in [1] but we prefer to use the notation of [2]. It is worth noting, that the general equations are challenging to solve because they form a partial di erential equation in 7 dimensions, 16 components with non-local di erential operators. In the case when only homogeneous but time dependent electric field is considered reduces to a 3 component ordinary di erential equation investigated widely in the literature as the uantum Kinetic equations. For SU(N) theories like CD the analogous equations are derived in [3] but they have even more complicated structure with more components. However [4] shows that there is a strong Abeilan dominance that ustifies the approximation of the more complicated SU(N) models by U(1) pair production.

As we consider ED electron-positron production we solve the 16 component DHW equations. As a well defined departure from homogeneous models we ust add a transverse inhomogeneity to the electric field still excluding magnetic fields. To compare the yields in the two cases we need the decomposition of the pair density f in the uantum Kinetic models. This can be written in terms of the DHW components as: $f \quad \frac{\epsilon}{4\omega} \quad \frac{1}{2}$ where ϵ is the phase-space energy density and $\omega \quad \sqrt{p^2 \quad m^2}$.

Our choice for the electric field as motivated by the color ux tube cross-section is the following:

$$\vec{E}(x,t) \quad \frac{\vec{e}_z E_0}{\cosh(t/\tau)^2} \frac{1}{2} \left(1 - \tanh\left(\frac{x \quad R}{r}\right) \tanh\left(\frac{x-R}{r}\right)\right) . \tag{1}$$

It describes an electric field linearly polarized in th z direction but having a transverse plateau-like inhomogeneity in the x direction. The temporal shape is a Sauter pulse that is special because in the homogeneous case it has a known analytical solution for the asymptotic phase space distribution [5], so in the case when $R \gg r$ we expect to reproduce the analytical result near $x \quad 0$ and can verify the used computational methods.

We solve the DHW equations numerically with 8th order adaptive Runge-Kutta in the time direction and pseudo-spectral collocation methods on Rational Chebyshev basis for the spatial and momentum derivatives [6].

3. Results

On Figure 1. we show the asymptotic pair density f integrated for the spatial coordinate for two di erent temporal pulse lengths ($\tau \quad 1\lambda_c/c$ *left* and $3\lambda_c/c$ *right*). For short pulses the spectra looks the same as in a homogeneous model but for longer pulses it deviates from the isotropic symmetry and creates particle excess in the backward direction giving the distribution a triangular shape. This deviation may be used as a distinguishing feature in experiments.

The phase-space spectrum can be seen on the on Figure 2. *left.* While there is the expected bulk pair production from the almost homogeneous field in the spatial range $0 - 4\lambda_c$, a higher amplitude more localized source also manifests at the edge of the plateau field that is attributable to the presence of the large field gradients around that position.

To study the scaling of total number of pairs produced with field parameters it is enough to

Journal of Physics: Conference Series **594** (2015) 012055 doi:10.1088/1742-6596/594/1/012055

integrate the spectra in the coordinate and longitudinal momentum direction :

$$n_L \quad \int_{-\infty}^{\infty} \mathrm{d}x \frac{\mathrm{d}p_z}{2\pi} f(x, p_x \quad 0, p_y \quad 0, p_z) \tag{2}$$

because the spectra peaks at zero transverse momentum. Figure 2. *right* shows the di erence of the homogeneous and inhomogeneous models for di erent temporal pulse width. Again, we see, that for shorter pulses they are almost identical but for pulses with larger temporal extent more particles are produced in the final state. The field inhomogeneity acts as another channel for pair production that dominates the total yield over $\tau \quad 1\lambda_c/c$.

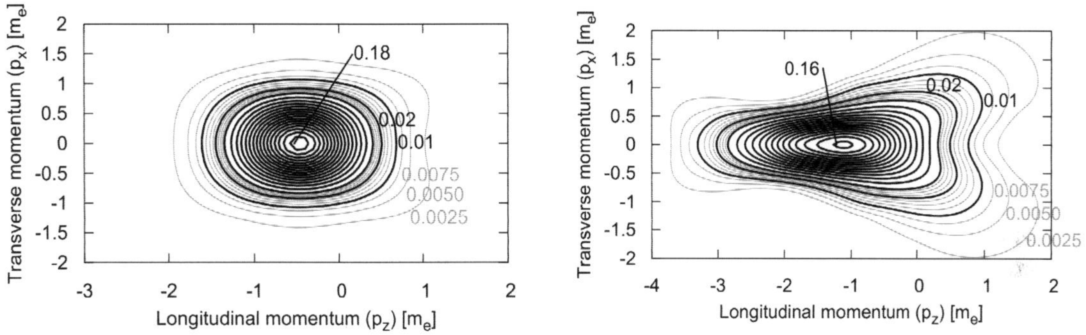

Figure 1. Asymptotic pair density spectra ($f(p_x, p_z)$) integrated for the spatial coordinate x shown for $\tau \quad 1\lambda_c/c$ *left* and $3\lambda_c/c$ *right*. The other field parameters are: $E0 \quad 0.75E_{cr}, R \quad 4\lambda_c, r \quad 0.5\lambda_c$.

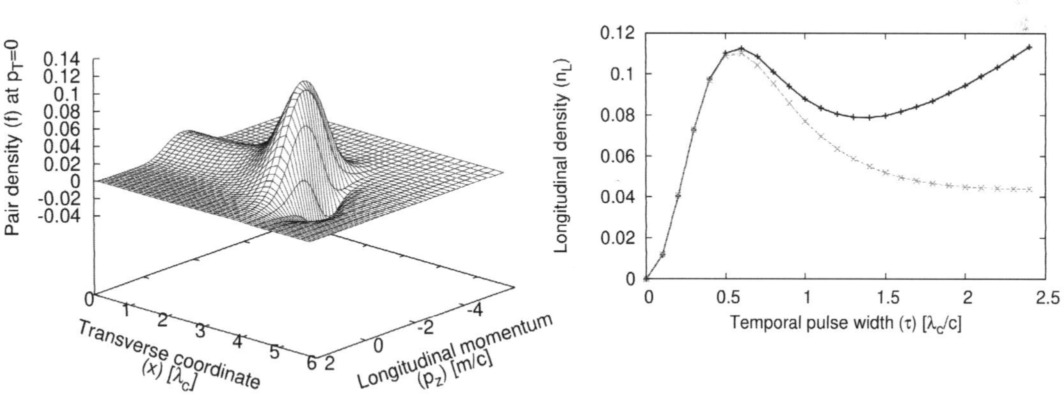

Figure 2. *Left*: phase-space view of the asymptotic pair density with parameters as in Fig. 1. except $\tau \quad 2\lambda_c/c$. Note that the E field has the steepest gradient at $x \quad 4\lambda_c$. *Right*: Comparison of the particle yield in the inhomogeneous model (solid black line) with the homogeneous reference (dashed grey) for varying temporal pulse width (τ).

Journal of Physics: Conference Series **594** (2015) 012055 doi:10.1088/1742-6596/594/1/012055

4. Conclusion

We investigated the effect of a transverse spatial electric field inhomogeneity on pair production yields and spectra in the Dirac-Heisenberg-Wigner formalism. This field configuration is an Abelian analogue of the QCD color string but the results are as well relevant to extreme laser collider experiments. We found that close to the Compton scales significant part of pair production is happening near the edges of the field and results in characteristic deviations from the homogeneous configuration that might be detectable by measuring momentum distribution of pairs around the interaction point. More details can be found in [7].

This work was supported by the OTKA Grants No. 104260, No. 106119.

References

[1] I. Bialynicki-Birula, P. Gornicki, J. Rafelski, Phys. Rev. **D44**, 1825-1835. (1991)
[2] F. Hebenstreit, R. Alkofer, H. Gies, Phys. Rev. **D82** 105026. (2010)
[3] A.V. Prozorkevich, S.A. Smolyansky, S.V. Ilyin, "Progress in Nonequilibrium Green's Functions II", Eds. M. Bonitz and D. Semkat, World Scientific (2003), p. 401.
[4] V. V. Skokov and P. Levai, Phys. Rev. **D78**, 054004. (2008)
[5] S. I. Kruglov, Radiat. Phys. Chem. **75**, 723-728. (2006)
[6] J. P. Boyd, J. Comp. Phys. **69** (1987) 112-142.
[7] D. Berenyi, S. Varró, V. V. Skokov and P. Levai, in preparation arXiv:1401.0039 [hep-ph]

Journal of Physics: Conference Series **594** (2015) 012056 doi:10.1088/1742-6596/594/1/012056

Molecular above-threshold ionization spectra as an evidence of the three-point interference of electron wave packets

Elvedin Hasović[1], Dejan B Milošević[1,2,3], Azra Gazibegović-Busuladžić[1], Aner Čerkić[1] and Mustafa Busuladžić[4]

1 Faculty of Science, University of Sarajevo, Zmaja od Bosne 35, 71000 Sarajevo, Bosnia and Herzegovina

2 Max-Born-Institut, Max-Born-Str. 2a, 12489 Berlin, Germany

3 Academy of Sciences and Arts of Bosnia and Herzegovina, Bistrik 7, 71000 Sarajevo, Bosnia and Herzegovina

4 Faculty of Medicine, University of Sarajevo, Čekaluša 90, 71000 Sarajevo, Bosnia and Herzegovina

E-mail: mustafabusuladzic@bih.net.ba

Abstract. We consider high-order above-threshold ionization (HATI) of polyatomic molecules ionized by a strong linearly polarized laser field. Improved molecular strong-field approximation by which the HATI process on polyatomic molecular species can be described is developed. Using this theory we calculate photoelectron angular-energy spectra for different triatomic molecules. Special attention is devoted to the minima that are observed in the calculated high-energy electron spectra of the ozone and carbon dioxide molecules. A key difference between these minima and minima that are observed in the corresponding spectra of diatomic molecules are presented.

1. Introduction

During laser-atom or laser-molecule interaction a few nonlinear phenomena of higher order can occur (see review articles [1, 2, 3, 4, 5] and references therein). The most important among them are the high-order above-threshold ionization (HATI) and high-order harmonic generation (HHG). These processes and key features of obtained photon and electron spectra can be explained using the so-called three-step model [6, 7]. In the first step of the HATI and HHG, quantum mechanical system (atom or molecule) absorbs more photons from the strong laser field than is necessary for ionization. Some of these electrons can go directly to detector. This process is called direct above-threshold ionization (ATI) and within it any additional interaction between electron and its parent ion is neglected. Due to the influence of the laser field, the ionized electron may also return to the parent molecular ion (second step) and elastically scatter off it (third step), before reaching the detector. In this process, the electron can absorb many more photons from the laser field than in the direct ATI process. This process was named high-order ATI (HATI). In the third step of the HHG process, the electron recombines with the parent ion and one high-energy photon is emitted. The energy spectra of atomic/molecular HATI and atomic/molecular HHG processes are characterized by a plateau which manifests itself as a broad energy interval of the spectrum in which the photoelectron (HATI) or photon (HHG) yield is practically constant. These intervals are followed by abrupt cutoffs.

Above-mentioned molecular laser-induced processes have received much attention in the last few years because of their ability to reveal some features of the internal structure of the molecule under investigation [8, 9]. More precisely, by analyzing their spectra one can extract information about molecular structure. In fact, even direct ionization (direct ATI) without rescattering reveals the initial symmetry of the molecular system. For example, the O_2 molecule shows a suppression in the low-energy electron spectra due to its π_g symmetry, while the N_2 molecule, having σ_g symmetry, does not show such a suppression [10, 11]. A more recent experimental and theoretical study of ATI on different polyatomic molecules was published in [12].

Here we are interested in the HATI process. If rescattering is included, more scenarios are possible than in the ATI process [13]. As molecules are multi-center systems, the ionization as well as rescattering can happen at different centers, causing interference structures in the electron spectrum. In the case of the HATI of arbitrary diatomic molecular species, there are four rescattering contributions to the ionization rate. In other words, our matrix element related to rescattering process in the case of diatomic molecules can be decomposed into four terms, each term describing one of the four possible scenarios [14]. All these contributions interfere in a complicated manner and can produce a complicated spectrum, which depends on the laser parameters, electron emission angle, and the molecular orientation. The most noticeable feature of the calculated HATI spectra is the existence of pronounced minima that are absent in atomic case. We have found an analytical formula for the minima position regardless of symmetry of considered molecules. It is given by the equation [14, 15]

$$R_0 \left| p_{\mathrm{f}} \cos\theta - k_{\mathrm{st}} \cos\theta_{\mathrm{L}} \right| = (2m+1)\pi, \qquad (1)$$

where m is integer (i.e. $m = 0, 1, 2, \ldots$), θ_L is the angle between the laser polarization vector and the internuclear axis R_0, θ is the angle between the final momentum of the detected electron that is denoted by p_f and the internuclear axis R_0, while k_{st} is the stationary intermediate electron momentum between the ionization and the rescattering (it is in the direction of the laser electric field vector). It is important to stress that the previously mentioned two-source two-rescattering-centers interference has been observed in experiments with unaligned molecules [16, 17].

Our next step is to calculate and analyze electronic spectra of polyatomic molecules. We will introduce a theory of strong-field ionization of arbitrary polyatomic molecules. In this sense, we follow the approach presented in [18]. The theory is so formulated that it takes into account a possible rescattering of the emitted electrons on the various atomic centers of polyatomic molecule. Then, we are going to present HATI spectra of some polyatomic molecules, such as O_3 and CO_2, and to give a physical interpretation of the observed minima. Finally, the conditions for the destructive interference of all contributions are provided and the key differences between the HATI spectra of diatomic and polyatomic molecules are discussed. We use atomic system of units.

2. Improved molecular strong-field approximation theory for polyatomic molecules

We have developed a theory describing above-threshold ionization of arbitrary polyatomic molecules. A polyatomic molecule is modeled by an $(N+1)$-particle system, which consists of N heavy atomic (ionic) centers and an electron [19]. After the separation of the center-of-mass coordinate, the dynamics of this system is reduced to the relative electronic and nuclear coordinates. Two forms of the molecular strong-field approximation (MSFA), one with the field-free and the other with the field-dressed initial molecular bound state, are derived. The relevant time-dependent transition matrix element of the modified MSFA in the length gauge, for neutral polyatomic molecules can be written as [19]

$$\mathcal{F}^{q\mathrm{ML}}_{\mathrm{fi,neut}}(t) = \sum_{j=1}^{N} g_q(\boldsymbol{\rho}_j, t) e^{-i[\mathbf{p}_\mathrm{f}+\mathbf{A}(t)]\cdot\boldsymbol{\rho}_j} \sum_a c_{ja} \langle \mathbf{p}_\mathrm{f} + \mathbf{A}(t) | \mathbf{E}(t)\cdot\mathbf{r}|\psi_a\rangle, \tag{2}$$

where $\mathbf{p}_\mathrm{f}$ is the final momentum of the detected electron and the factor $g_q(\boldsymbol{\rho}_k, t)$ enables us to treat both the undressed (q = u) and the dressed (q = d) initial state case. The transition matrix element (2) depends on the coordinates $\boldsymbol{\rho}_k(\{\mathbf{R}\})$ which are given as linear combinations of the relative coordinates $\mathbf{R}_1, \ldots, \mathbf{R}_{N-1}$. The connection of the coordinates $\boldsymbol{\rho}_k(\{\mathbf{R}\})$ and the position vectors $\mathbf{r}_j$ of atoms that are constituent of considered polyatomic molecules can be written as [19]

$$\boldsymbol{\rho}_j = \mathbf{r}_j - \mathbf{R}_\mathrm{c.m.} + \frac{m_e}{M}\mathbf{R}_N, \tag{3}$$

where m_e is the electron mass and M is the mass of all atoms. It can be shown that $\boldsymbol{\rho}_j \approx \mathbf{r}_j$ [19]. The theory presented above was able to describe direct electrons only – electrons that leave the laser field and are collected on the detector right after the ionization.

In order to describe high-energy electrons contribution to the spectrum, we have generalized the above-mentioned theory so that it includes an additional interaction of the ionized electron with the molecular centers [20]. We obtained two terms – the first term describes the direct ATI (2), while the second one corresponds to the rescattered electrons, and, for the dressed case, has the form

$$\begin{aligned}\mathcal{F}^{(1)}_{\mathrm{fi,d}}(t) &= i e^{-iS_{\mathbf{k}_\mathrm{st}}(t)} \int_0^\infty d\tau \left(\frac{2\pi}{i\tau}\right)^{3/2} e^{i[S_{\mathbf{k}_\mathrm{st}}(t') - \Delta E(\{\mathbf{R}_0\})\tau]} \\ &\times \sum_{j=1}^{N} e^{i\mathbf{K}_\mathrm{st}\cdot\boldsymbol{\rho}_j(\{\mathbf{R}_0\})} V^j_{\mathrm{e}\mathbf{K}_\mathrm{st}} \sum_{l=1}^{N} e^{-i\mathbf{k}_\mathrm{st}\cdot\boldsymbol{\rho}_l} \sum_a c_{la}\langle \mathbf{k}_\mathrm{st} + \mathbf{A}(t')|\mathbf{E}(t')\cdot\mathbf{r}|\psi_a\rangle,\end{aligned} \tag{4}$$

where the stationary electron momentum between the ionization and the rescattering is

$$\mathbf{k}_\mathrm{st} = -\frac{1}{\tau}\int_{t'}^{t} dt'' \mathbf{A}(t'') = \frac{1}{\tau}\left[\boldsymbol{\alpha}(t') - \boldsymbol{\alpha}(t)\right], \tag{5}$$

with $\boldsymbol{\alpha}(t) = \int^t dt'' \mathbf{A}(t'')$. The electron-molecular ion scattering potential, according to the independent atom model [21], is presented as the sum of the potentials for scattering off the individual atomic centers of the considered molecule. Also, we have $\mathbf{K}_\mathrm{st} = \mathbf{k}_\mathrm{st} - \mathbf{p}_\mathrm{f}$, while $V^j_{\mathrm{e}\mathbf{K}_\mathrm{st}}$ is the Fourier transform of the rescattering potential at the atomic (ionic) center j. The semiclassical action is denoted by $S_{\mathbf{k}_\mathrm{st}}(t) = \int^t dt''[\mathbf{k}_\mathrm{st} + \mathbf{A}(t'')]^2/2$. Finally, $\sum_a c_{la}|\psi^{(0)}_a\rangle$ is the ground-state electronic wave function, presented as a linear combination of atomic orbitals. The calculation of the highest occupied molecular orbital (HOMO) was done using the GAMESS quantum chemistry package within the cc-pVTZ basis set.

The differential ionization rate with absorption of n photons from the laser field having the angular frequency $\omega = 2\pi/T$ is given by

$$w^{(1)}_{\mathrm{fi},q}(n) = 2\pi p_\mathrm{f} |T^{(1)}_{\mathrm{fi},q}(n)|^2, \tag{6}$$

where $T^{(1)}_{\mathrm{fi},q}(n)$ is the Fourier transform of the matrix element (4) multiplied by the corresponding phase factor [20].

3. Numerical results

Now we are going to demonstrate an application of the presented theory on a simple example. All calculations are performed with the field-dressed initial molecular bound state, i.e. using (4). Let us consider the HATI process for both carbon dioxide and ozone molecule. Carbon dioxide is a linear triatomic molecule with carbon atom placed in the center, and oxygen atoms placed symmetrically on either side of the center. Due to the symmetry, the center of coordinate system is placed at the carbon atom. The distance between the carbon atom and either oxygen atom is 1.1621 Å. The HOMO of the CO_2 molecule is characterized by $1\pi_g$ symmetry, and its vertical ionization energy is $I_p = 13.777$ eV.

Ozone is a planar triatomic molecule consisting of three oxygen atoms with the internuclear distance of 1.278 Å. The O-O-O angle is 116.8°. Therefore, this molecule is more complex than the CO_2 molecule. The structure of the O_3 molecule belongs to the C_{2v} point group and its HOMO has ${}^1A_1$ symmetry. The vertical ionization energy of the ozone molecule is $I_p = 12.73$ eV. For both molecules we assume that the ionization process happens from the HOMO. According to this, the initial electronic state is the corresponding HOMO, written as a linear combination of atomic orbitals. As previously mentioned, the calculation of the HOMO was done using the GAMESS quantum chemistry package within the cc-pVTZ basis set. Such CO_2 and O_3 molecule HOMO wave functions are presented in figures 1 and 2 in the xz plane.

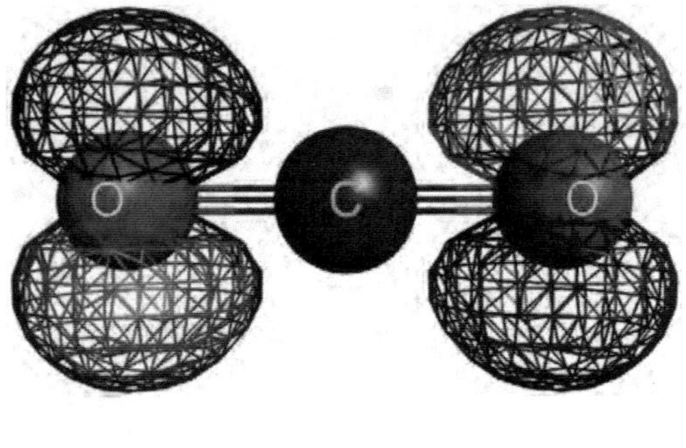

Figure 1. Coordinate-space wave function for the HOMO of the CO_2 molecule, presented in the xz plane.

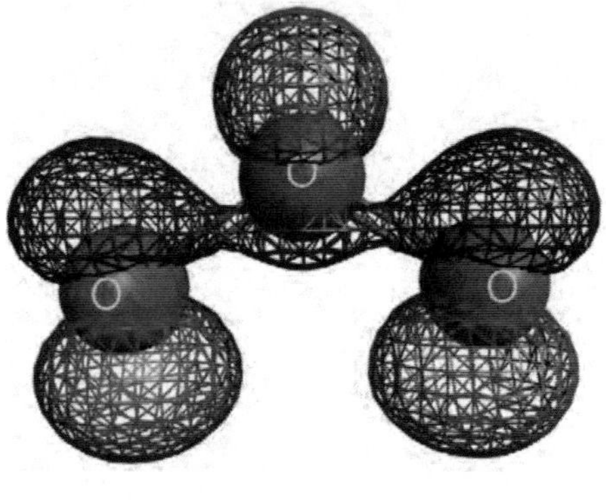

Figure 2. Coordinate-space wave function for the HOMO of the O_3 molecule, presented in the xz plane.

Geometry of the considered process on triatomic molecules is described in the next part of this section. The electron emission angle is defined as the angle between the laser polarization vector and the vector of the final momentum of the emitted electron. For diatomic molecules the internuclear axis is usually chosen as the reference axis (for example, the z axis) and the axial symmetry of the molecule is used to simplify the calculations. For nonlinear polyatomic molecules the situation is more complicated. It is the case for the planar O_3 molecule. Because of the absence of the axial or spherical symmetry we have to consider any possible orientation of the laser polarization vector with respect to the polyatomic molecule. Therefore, for a linearly polarized laser field we define the field vector as

$$\mathbf{E}(t) = E_0 \sin(\omega t)\hat{\mathbf{e}}_\mathrm{L}, \tag{7}$$

where E_0 is the electric field vector amplitude and $\hat{\mathbf{e}}_\mathrm{L} = e_x\hat{\mathbf{i}} + e_y\hat{\mathbf{j}} + e_z\hat{\mathbf{k}}$ is the unit laser-field vector. In practice, we determine the direction of this vector by the azimuthal angle θ_L and the polar angle φ_L, so that, in the Cartesian coordinates, it has the following form

$$\hat{\mathbf{e}}_\mathrm{L} = \sin\theta_\mathrm{L}\cos\varphi_\mathrm{L}\hat{\mathbf{i}} + \sin\theta_\mathrm{L}\sin\varphi_\mathrm{L}\hat{\mathbf{j}} + \cos\theta_\mathrm{L}\hat{\mathbf{k}}. \tag{8}$$

The emitted electrons can be detected in any direction too. The final momentum vector is defined by its intensity p_f, azimuthal angle θ_f and polar angle φ_f. We can write $\mathbf{p}_\mathrm{f} = p_\mathrm{f}\hat{\mathbf{e}}_\mathrm{f}$, where the unit vector of the electron final momentum is defined as

$$\hat{\mathbf{e}}_\mathrm{f} = \sin\theta_\mathrm{f}\cos\varphi_\mathrm{f}\hat{\mathbf{i}} + \sin\theta_\mathrm{f}\sin\varphi_\mathrm{f}\hat{\mathbf{j}} + \cos\theta_\mathrm{f}\hat{\mathbf{k}} \tag{9}$$

For more details about the geometry of the process one can see [19, 20]. Now, we present high-

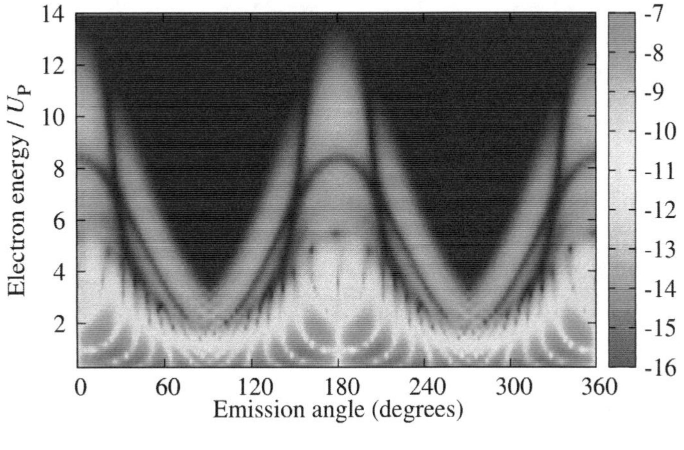

Figure 3. Logarithm of the differential ionization probability of the carbon dioxide molecule as a function of the electron kinetic energy and emission angle. The laser-field intensity is $1.2 \cdot 10^{14}$ W/cm^2, while the wavelength is 800 nm. The field polarization vector is placed in the xz plane with $\theta_\mathrm{L} = 1°$.

energy part of the electron spectra for the both considered molecules. In figure 3, the logarithm of the differential ionization probability of the carbon dioxide molecule as a function of the final electron kinetic energy $E_{\mathbf{p}_\mathrm{f}}$ and azimuthal angle θ_f for linearly polarized laser field of the intensity 1.2×10^{14} W/cm^2 and the wavelength 800 nm is presented. The laser-field polarization vector $\hat{\mathbf{e}}_\mathrm{L}$ is placed in the xz plane with $\varphi_\mathrm{L} = 0°$ and $\theta_\mathrm{L} = 1°$. The most interesting feature of the ATI angular distribution of CO_2 molecule are distinct and sharp minima in the high-energy part of the spectrum. Similar minima are observed in the ATI angular distributions of diatomic molecules and are explained by the destructive interference of the T-matrix partial contributions [15]. The observed suppression in the high-energy part of the photoelectron spectrum of diatomic molecules can be fitted very well by the curve given by equation (1). It can be shown that the observed minima in the case of linear centrosymmetric triatomic molecule, like CO_2 molecule, can be fitted with the curve given by relation

$$x_1 \left| p_\mathrm{f}\cos\theta_\mathrm{f} - k_\mathrm{st}\cos\theta_\mathrm{L} \right| = 2\left(\frac{m}{3} + 1\right)\pi, \tag{10}$$

in analogy to the minima in diatomic molecule case [see equation (1)]. For the laser and molecular parameters we considered in the present paper, only the lowest value $m = 0$ contributes. If we increase the internuclear distance x_1 two times, the minima of the second order will appear

($m = 1$). So, in the case of linear triatomic molecules, minima have the same structure as in the diatomic molecule case. A very similar results were obtained for the HATI of triatomic molecules using the frequency-domain theory [22]. Finally, we conclude that the observed minima have their origin in the three center (three slit) interference, with position given by relation (10).

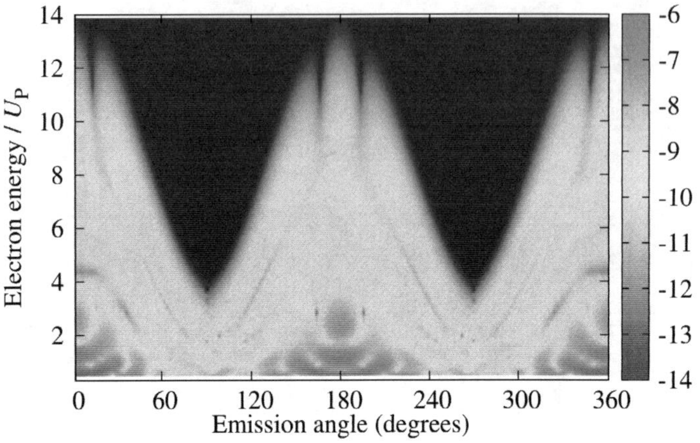

Figure 4. Logarithm of the differential ionization probability of the ozone molecule as a function of the electron kinetic energy and emission angle. The laser-field intensity is $1.2 \cdot 10^{14}$ W/cm^2, while the wavelength is 800 nm. The field polarization vector is directed along the z axis, i.e., $\theta_{\rm L} = 0°$.

The logarithm of the differential ionization probability of the ozone molecule as a function of the final electron kinetic energy $E_{\mathbf{p}_{\rm f}}$ and azimuthal angle $\theta_{\rm f}$ is presented in figure 4. Laser field intensity, wavelength and polarization are the same as in the previous case. The laser-field polarization vector $\mathbf{e}_{\rm L}$ is oriented along the z axis, i.e. $\varphi_{\rm L} = 0°$ and $\theta_{\rm L} = 0°$. The ozone molecule lies in the xz plane. The angular distribution, presented here, has similar features as angular distributions of diatomic molecules or atoms. The high-energy plateau is mostly extended in the laser-field direction, i.e., for angles $\theta_{\rm f} = 0°$ and $\theta_{\rm f} = 180°$, while the high-energy plateau disappears for electrons emitted in direction perpendicular to the laser-field polarization vector. The most interesting feature of the ATI angular distribution of ozone molecule are distinct minima in the high-energy part of the spectrum, detected for azimuthal angles $\theta \approx 10°, 170°, 190°$, and $350°$.

It is interesting to check what happen to the O_3 molecule spectrum and the corresponding minima if the internuclear angle is chosen to be $\alpha = 180°$, i.e. if we assume that the O_3 molecule is linear. Sharp and clear minima appear in the spectrum, minima of the kind that can be fitted with curve given with relation (10), just as in diatomic case. This confirms our previous finding for the CO_2 molecule and the corresponding minima.

Let us go back to the proper ozone molecule, that is more complex, planar molecule. We will consider what mechanism causes the observed minima in the HATI spectrum of the O_3 molecule that are clearly visible on figure 4. Generally, for triatomic molecules there are twelve partial contributions: three partial contributions of the direct amplitude and nine partial contributions of the amplitude which describes rescattered electrons. It can be shown that for triatomic molecules the partial contributions of the electrons rescattered on different atomic centers, which contribute to the high-energy part of the spectrum, interfere destructively. This is also the case for diatomic molecules. Then, the differential ionization probability is proportional to

$$w_{\rm fi}(n) \propto \left| \sum_{j=1}^{N} e^{i\mathbf{K}_{\rm st}\cdot\boldsymbol{\rho}_j(\{\mathbf{R}_0\})} V^j_{\mathrm{e}\mathbf{K}_{\rm st}} \right|^2 . \tag{11}$$

For the ozone molecule the electron scattering amplitudes $V^j_{\mathrm{e}\mathbf{K}_{\rm st}}$ on the three atomic centers are equal, so that the differential ionization probability becomes

$$w_{\rm fi}(n) \propto \left| e^{-iK_x x_1 + iK_z z_1} + e^{iK_z z_2} + e^{iK_x x_1 + iK_z z_1} \right|^2 , \tag{12}$$

where $x_1 = 1.0304148692\,$Å, $z_1 = -0.2012853882\,$Å, and $z_2 = 0.4025707765\,$Å determine the equilibrium position of the oxygen atoms. If we put $z_2 = z_1 + \Delta z$, it can be shown that the differential ionization probability is proportional to

$$w_{\rm fi}(n) \propto \left| e^{-iK_x x_1} + e^{iK_z \Delta z} + e^{iK_x x_1} \right|^2 . \tag{13}$$

This proportion reduces to

$$w_{\rm fi}(n) \propto (3 + 4\cos\alpha\cos\beta + 2\cos 2\alpha), \tag{14}$$

where $\alpha = K_x x_1$ and $\beta = K_z \Delta z$. If the function $f(\alpha, \beta) = 3 + 4\cos\alpha\cos\beta + 2\cos 2\alpha$ is equal to zero, the differential ionization probability will have the absolute minimum. The equation $3 + 4\cos\alpha\cos\beta + 2\cos 2\alpha = 0$ can be transformed to

$$\cos^2\alpha + \cos\beta\cos\alpha + \frac{1}{4} = 0. \tag{15}$$

The solution of this equation can be written as

$$\cos\alpha_{1,2} = \frac{-\cos\beta \pm \sqrt{\cos^2\beta - 1}}{2}. \tag{16}$$

The solutions of the previous quadratic equation will be real only if $\cos^2\beta = 1$, i.e. $\cos\beta = \pm 1$, which leads to $\beta = k\pi$, $k = 0, 1, 2, \ldots$. For even k the values of α become $\alpha = \frac{2m\pi}{3} + 2l\pi$, $m = 1, 2$ and $l = 0, 1, 2, \ldots$, while if k is odd we have $\alpha = \pm\frac{\pi}{3} + 2l\pi$, $l = 0, 1, 2, \ldots$. So, the differential ionization probability becomes zero if the next two conditions are fulfilled simultaneously

$$\begin{aligned} |k_{\rm st}\cos\theta_{\rm L} - p_{\rm f}\cos\theta_{\rm f}|\,\Delta z &= k\pi, \\ |k_{\rm st}\sin\theta_{\rm L} - p_{\rm f}\sin\theta_{\rm f}|\,x_1 &= \begin{cases} \frac{2m\pi}{3} + 2l\pi & \text{for even} \quad k, \\ \pm\frac{\pi}{3} + 2l\pi & \text{for odd} \quad k. \end{cases} \end{aligned} \tag{17}$$

From the previous two conditions we can find the values of electron kinetic energy and emission angles for which the ionization probability becomes zero. In the special case where the angle $\theta_{\rm L} = 0°$ the destructive interference conditions become

$$\begin{aligned} p_{\rm f}\cos\theta_{\rm f} = p_{{\rm f}z} &= k_{\rm st} - \frac{k\pi}{\Delta z}, \\ p_{\rm f}\sin\theta_{\rm f} = p_{{\rm f}x} &= \frac{1}{x_1}\begin{cases} \frac{2m\pi}{3} + 2l\pi & \text{for even} \quad k, \\ \pm\frac{\pi}{3} + 2l\pi & \text{for odd} \quad k, \end{cases} \end{aligned} \tag{18}$$

with the electron kinetic energy and emission angle given by $E_{\mathbf{p}_\mathrm{f}} = \frac{1}{2}(p_{\mathrm{f}x}^2 + p_{\mathrm{f}z}^2)$ and $\theta_\mathrm{f} = \arctan(p_{\mathrm{f}z}/p_{\mathrm{f}x})$, respectively. As the previous two conditions have to be fulfilled at the same time, we cannot find an analytic function $E_{\mathbf{p}_\mathrm{f}} = f(\theta_\mathrm{f})$, which determines the positions of the interference minima in the $(E_{\mathbf{p}_\mathrm{f}}, \theta_\mathrm{f})$ plane, the way we could in the case of diatomic molecules and triatomic linear molecules. The absolute interference minima for the ozone molecule will appear only for certain values of the electron kinetic energy and certain values of the emission angles, and will be represented by a single point in the $(E_{\mathbf{p}_\mathrm{f}}, \theta_\mathrm{f})$ plane. Therefore, the minima in the high-energy part of the angle-resolved electron emission spectrum of the ozone molecule, presented in the figure 4 cannot be explained as the absolute destructive interference minima. However, it does not mean that the interference has nothing to do with the appearance of the minima. If one plots the function $f(\alpha, \beta) = 3 + 4\cos\alpha\cos\beta + 2\cos 2\alpha$ in the (α, β) plane, as it was done in [20], one can conclude that the function $f(\alpha, \beta)$ has an absolute minimum, equal to zero, for $(\alpha, \beta) = (\pm\pi/3, \pm\pi), (\pm 2\pi/3, 0), \ldots$, which was confirmed analytically. However, there are relatively wide intervals of values of α and β, around the zero points, for which the function $f(\alpha, \beta)$ is close to zero. By closer investigation of angular distributions of emitted photoelectrons for fixed electron energies we have found out that there are clear minima for all photoelectron energies. One can extract values of the angles θ_f for which these minima occur. The angles $\alpha = K_x x_1$ and $\beta = K_z \Delta z$ corresponding to those values of θ_f are calculated. We concluded that these values are very close to the zero-point values $(\alpha, \beta) = (\pi/3, -\pi)$, and that the value of the function $f(\alpha, \beta)$ is close to zero, i.e. it is more than ten times smaller than its maximum value. Furthermore, by increasing the electron kinetic energy, the value of the function $f(\alpha, \beta)$ decreases, as well as the differential ionization probability. This is in agreement with the results presented in the energy-angular distribution in figure 4, where the minima in the high-energy part of the spectrum become more pronounced with the increase of the electron kinetic energy. Therefore, the minima in the high-energy part of the HATI spectra of the ozone molecule can be interpreted as the destructive interference of the three rescattered electron wave packets. It should be stressed that these minima do not correspond to the absolute interference minima where the coherent sum of the three wave packets completely vanishes.

4. Conclusions

We have developed an improved-molecular-strong-field-approximation-based theory of high-order above-threshold ionization of polyatomic molecules. The theory is formulated to include the rescattering of the emitted electron off the parent molecular ion. We applied this theory to calculate the energy- and angle-resolved HATI spectra of the ozone and carbon dioxide molecules. Obtained spectra exhibit clear minima for certain emission angles in the high-energy part of the spectrum. These minima originate from the three-center interference of the rescattered electron wave packets. We have concluded that the absolute destructive interference minima, that are characteristic of diatomic as well as triatomic linear molecules, do not cause the appearance of the minima in the HATI spectrum of planar polyatomic molecules.

Acknowledgments

We gratefully acknowledge support by the Federal Ministry of Education and Science, Bosnia and Herzegovina.

References

[1] Becker W, Grasbon F, Kopold R, Milošević D B, Paulus G G, and Walther H 1992 *Adv. At. Mol. Opt. Phys.* **48**, 35
[2] Milošević D B and Ehlotzky F 2003 *Adv. At. Mol. Opt. Phys.* **49**, 373
[3] Becker A and Faisal F H M 2005 *J. Phys.* B **38**, R1
[4] Lein M 2007 *J. Phys.* B **40**, R135

Journal of Physics: Conference Series **594** (2015) 012056 doi:10.1088/1742-6596/594/1/012056

[5] Agostini P and DiMauro L F 2012 *Adv. At. Mol. Opt. Phys.* **61**, 117
[6] Schafer K J, Yang B, DiMauro L F, and Kulander K C 1993 *Phys. Rev. Lett.* **70**, 1599
[7] Corkum P B 1993 *Phys. Rev. Lett.* **71**, 1994
[8] Itatani J, Levesque J, Zeidler D, Niikura H, Pépin H, Kieffer J C, Corkum P B, and Villeneuve D M 2004 *Nature* **432**, 867
[9] Frumker E, Kajumba N, Bertrand J B, Wörner H J, Hebeisen C T, Hockett P, Spanner M, Patchkovskii S, Paulus G G, Villeneuve D M, Naumov A, and Corkum P B 2012 *Phys. Rev. Lett.* **109**, 233904
[10] Muth-Böhm J, Becker A, and Faisal F H M 2000 *Phys. Rev. Lett.* **85**, 2280
[11] Grasbon F, Paulus G G, Chin S L, Walther H, Muth-Böhm J, Becker A, and Faisal F H M 2001 *Phys. Rev.* A **63**, 041402(R)
[12] Durá J, Grün A, Bates P K, Teichmann S M, Ergler T, Senftleben A, Pflüger T, Schröter C D, Moshammer R, Ullrich J, Jaroń-Becker A, Becker A, and Biegert J 2012 *J. Phys. Chem.* A **116**, 2662
[13] Hetzheim H, Figueira de Morisson Faria C, and Becker W 2007 *Phys. Rev.* A **76**, 023418
[14] Busuladžić M, Gazibegović-Busuladžić A, Milošević D B, and Becker W 2008 *Phys. Rev. Lett.* **100**, 203003
[15] Busuladžić M, Gazibegović-Busuladžić A, Milošević D B, and Becker W 2008 *Phys. Rev.* A **78**, 033412
[16] Okunishi M, Itaya R, Shimada K, Prümper G, Ueda K, Busuladžić M, Gazibegović-Busuladžić A, Milošević D B, and Becker W 2009 *Phys. Rev. Lett.* **103**, 043001
[17] Kang H, Quan W, Wang Y, Lin Z, Wu M, Liu H, Liu X, Wang B B, Liu H J, Gu Y Q, Jia X Y, Liu J, Chen J, and Cheng Y 2010 *Phys. Rev. Lett.* *104*, 203001
[18] Milošević D B 2006 *Phys. Rev.* A **74**, 063404
[19] Hasović E and Milošević D B 2012 *Phys. Rev.* A **86**, 043429
[20] Hasović E and Milošević D B 2014 *Phys. Rev.* A **89**, 053401
[21] Xu J, Chen Z, Le A T, and Lin C D 2010 *Phys. Rev* A **82**, 033403
[22] Wang B, Guo Y, Zhang B, Zhao Z, Yan Z C, and Fu P 2010 *Phys. Rev.* A **82**, 043402